D0162721

DIFFUSION IN THE CONDENSED STATE

J. S. Kirkaldy
and
D. J. Young

THE INSTITUTE OF METALS
LONDON

Book 322
published by
The Institute of Metals
1 Carlton House Terrace
London SW1Y 5DB

Distributors:
India
Current Technical Literature Co Pvt Ltd, India House,
opp GP, Box 1374, Bombay 400 001, INDIA
Japan
USACO Corporation, Tsutsumi Building, 13–12
Shimbashi, 1-chome, Minato-ku, Tokyo 105, JAPAN
North America
The Institute of Metals North American Publications Center
Old Post Road, Brookfield, VT 05036, USA

British Library Catologuing in Publication Data

Kirkaldy, J. S.
Diffusion in the condensed state. — (Book; 322).
1. Condensed 2. Diffusion
I. Title II. Young, D. J. III. Institute of
Metals, *1985*. 1985. IV. Series
531′.1137 QC173.4.C65

ISBN 0–904357-87-2

Library of Congress Cataloging in Publication Data

Kirkaldy, J. S. (John Samuel)
Diffusion in the condensed state

Bibliography: p.
Includes index.
1. Diffusion. 2. Diffusion processes.
I. Young, D. J. (David John) II. Title.
QD543.K476 1987 530.4′1 87–26296
ISBN 0-904357-87-2

Typeset by Tech-Set, Gateshead, Tyne & Wear.
Printed by The Universities Press (Belfast) Ltd

Contents

Preface

Diffusion as a scientific term has roots in an extremely broad range of disciplines and is therefore a highly unifying concept within general science. The concept subsumes the transport of such entities as culture, language, populations, genes and technologies as well as heat, charge and atoms because all of these processes involve a strong element of randomicity (or diffuseness). In thermodynamic terms, diffusion is a paradigm for positive entropy change and thus for the second law of thermodynamics. Consequently a general theory of diffusion must be firmly based within the universal principle of entropy increase. In the most elementary diffusion processes, entropy is the only extensive variable which changes during spontaneous isothermal mixing, the volume, energy and total mole numbers remaining constant. This is the case for mixtures of dilute gases and approximately the case for isothermal mixing of high atomic mass isotopes in chemically pure systems (self-diffusion). It should not, therefore, be surprising that the Gaussian and error integral functions from probability theory play an important role in elementary diffusion theory.

The cooperative or non-entropic nature of the diffusion process becomes increasingly apparent as the chemical and physical distinction between diffusing species is delineated. The effect of this distinction is enhanced by an increased state of condensation of the matter and therefore a lowering of the temperature. While it is the case for dilute gases that both the purely entropic diffusion process and its interaction with thermal diffusion can be accurately predicted within the statistical physics of Chapman and Enskog, diffusion in liquid and solid mixtures condensed from such gases submits at best to semiempiricism or molecular dynamics computer experiments. Although we emphasize the condensed state in this volume, gaseous behaviour will often serve as a useful and rigorous theoretical basis for comparison.

As the state of condensation is increased, more and more detailed information about the species and their interactions is required for predictive purposes. This is exemplified on the one hand by specific physical parameters such as relative mass, crystal structure, and lattice defects and their dynamics, and on the other hand by the generalities of chemical kinetics and thermodynamics. Although the thermokinetic approach is the most rigorous, it also requires the most system-specific set of experimental data. Thus, approximate physical models and arguments must often be combined with the general approach for achievement of economy in the establishment of a predictive database.

The generalization from the purely entropic isotope or self-diffusion process to isothermal chemical diffusion is of such a complicating character that with even the simplest binary chemistry, a plethora of new concepts and structural entities is encountered. Interstitial mechanisms, size effects, relative viscosities, clustering effects, gradient energy effects, complex formation, trapping, vacancy correlation effects, ordering, and preprecipitation are essential elements in the description of binary diffusion. Consequently an exhaustive examination of binary diffusion must precede a study of ternary and higher order systems. For the latter the new features pertain to Onsager cross effects due to microscopic reversibility, to uphill diffusion due to thermodynamic interactions in chemically stable solutions (which occur only in ternary and higher order systems), and to phase transformations in ternary and higher order multiphase systems as exemplified by the oxidation of binary alloys. The latter processes exhibit a class of interfacial instabilities which are non-existent in binary systems. Such unique multi-component effects will carry a strong emphasis in the later chapters of the present treatment.

Ternary and higher order kinetic effects in conductive materials such as ionic solutions, salts or crystals demonstrate some significant simplifications as a consequence of the fact that charge is conserved, local neutrality is maintained, and the electrical conductivity is an independently measurable parameter. There is thus a rather complete closure between some aspects of the theory and experiment for aqueous solutions, in part because the correlations which define the Onsager cross-effects are due primarily to prompt correlated exchanges of charge and mass.

Besides the vectorial process of diffusion, which is most commonly considered, there are also non-vectorial diffusion processes as epitomized by order–disorder reactions in solids. This is a special case of homogeneous chemical reaction. Systems in which vectorial diffusion and chemical reaction occur simultaneously are of particular current interest as they can exhibit periodic spatiotemporal patterns (Belousov–Zhabotinskii reactions). The Liesegang instability in ternary precipitating solutions has similar characteristics and will be examined as an example of a unique ternary diffusion phenomenon. Preprecipitation and spinodal decomposition in higher order systems offer an elegant theory for a process with considerable practical application, so will be dealt with in some detail. These latter topics are collected in Chapters 11 and 12 under the general concepts of pattern-forming or self-organizing systems. Current general developments in the latter areas are greatly enhancing our understanding of life processes.

An important practical problem has to do with predicting or estimating ternary diffusion coefficients and mass transfer coefficients from the system thermodynamics and measured or predicted coefficients for tracers and constituent binary pairs. The methods and their limitations are therefore thoroughly examined in Chapters 6–9. The understanding of the binary coefficients is clearly central to the ternary and higher order predictive problem, which further justifies the devotion of a large introductory section of the book to binary diffusion. The latter summarizes material which may in part be found in the monographs on diffusion in solids by Seith, by Shewmon, by Manning, and by Stark and the elegant and comprehensive two-volume work 'Diffusion dans les Solides' by Adda and Philibert. Barrer's book, 'Diffusion in and Through Solids', Jost's 'Diffusion in Solids, Liquids, Gases', Cussler's 'Diffusion: Mass Transfer in Fluid Systems', Tyrrell and Harris's 'Diffusion in Liquids' together with Dunlop, Steel and Lane's comprehensive review article on 'Experimental Methods' in diffusion broadly cover the remaining binary material. Cussler's volume

'Multicomponent Diffusion' deals with the historical record and the chemical engineering approach to the multicomponent field. A bibliography of a large number of specialized reviews and monographs is appended.

One of our departures from these primarily research-oriented offerings is to place a particular emphasis on applications and useful initial and boundary value problems from a wide range of disciplines. The mathematical treatises 'Heat Conduction' by Carslaw and Jaeger and 'The Mathematics of Diffusion' by Crank are essential reference works in developing and extending such practical applications. The extension of classical boundary value problems and solutions to multicomponent systems of relevance to the science of materials in metallurgy, chemistry, solid state physics, geology, biology and physiology is one of the main thrusts of the volume.

The seminal work on multicomponent diffusion is to be found in Onsager's papers of 1931 and 1946 and the foundations within irreversible thermodynamics are treated in monographs by Denbigh, de Groot, Prigogine, de Groot and Mazur, and Fitts.

The present volume is designed as a graduate and undergraduate textbook or reference work and as a source book on multiphase, multicomponent research and applications for metallurgists, materials scientists, physical chemists, geologists, physicists, biologists, physiologists and engineers. The lengthy Chapters 1 and 2 together with Chapter 5 and Appendix 1 are appropriate for an undergraduate course on diffusion and phase transformations and as such have been provided with exercises which emphasize machine computation. The material of the full text is a suitable textbook for a graduate course under the same title and has been used as such in the Department of Materials Science and Engineering at McMaster University for over two decades. SI Units are used in the text. A list of symbols precedes the Table of Contents and an Appendix on data sources and a bibliography appear at the end of the book.

This volume is a testimony to the research of a generation of graduate students and post-doctoral fellows, all of whose names appear as coauthors of publications referenced. We have benefited greatly over the years from collaboration and correspondence with many national and international researchers. Among others we owe a great debt of gratitude to John Lane, Sieb Radelaar, Geoff Kidson, Gary Purdy, Lawrence Brown, Doug Chambers, Denton Coates, Mike Dignam, Joe Goldstein, Peter Dunlop, Don Miller, Paul Shewmon, Mats Hillert, Jean Philibert, K S Goto, Al Cooper, Alan Lidiard, Walt Smeltzer, Al Guy, Dave Whittle, Herman Schmalzreid, Tom Ziebold, John Cahn, Norm Ridley, Bob Ogilvie, Didier de Fontaine, John Manning, John Morral, L-O Sundelöf, and D Dayananda.

A special thanks goes to Kathy Goodram, Janis Hudak, and Vi Weatherill for helping us through many typescripts, to Larry Murphy for the literature search, and to our wives and families for their forbearance.

Finally, this is dedicated in memoriam to

Lars Onsager
Carl Wagner
Larry Darken
and
Lou Gosting

our mentors in the pursuit of excellence in the science of materials.

LIST OF OFTEN-USED SYMBOLS

A total or molar Helmholtz free energy
A area
A chemical affinity
A intensive parameter
A_0 intensive parameter evaluated at equilibrium
a thermodynamic activity
c mass concentration
C total atoms per unit volume, $C = \Sigma C_i$
C_i atoms, molecules per unit volume of species i
D diffusion coefficient
D_{ij}^{k} diffusion coefficient relating flux of component i to concentration gradient of component j in solvent k
$(D)_V$ diffusion coefficient in volume-fixed frame
$(D)_S$ diffusion coefficient in solvent-fixed frame
$(D)_R$ diffusion coefficient in reference frame R
D_{A*} tracer or self-diffusion coefficient for species A
D_A^i intrinsic diffusion coefficient
$\widetilde{D}$ chemical diffusion coefficient
D_V, D_B volume, boundary diffusion coefficients
d membrane thickness, atom spacing
E an energy
$\boldsymbol{F}$ an external force (vector)
F the Faraday (96 500 coulomb)
f correlation factor; fraction of defective material
G total or molar Gibbs' free energy
$[H]$ hessian matrix
h Planck's constant
$J, \boldsymbol{J}$ a flux (bold indicates vector)
$J_i, \boldsymbol{J}_i$ matter flux (bold indicates vector)
$\boldsymbol{J}_q$ heat flux (vector)
$(J_i)_R$ component flux relative to reference frame R
K chemical equilibrium constant
k Boltzmann constant
k partition coefficient
$[L]$ matrix of phenomenological coefficients in flux-force equations
L general mobility coefficient
l_{ij} phenomenological coefficient for diffusion of a single ion in electrolyte solution
m molar concentration
m_i molar concentration of component i
M_i molar mass of component i
N_A Avogadro's number ($R = N_A k$)
N_i atom numbers
n_i mole numbers
p pressure
p_i partial pressure of component i
P permeability, membrane permeability

Q	heat or activation energy
q	charge
R	general gas constant
S	a structural spacing parameter
S	total or molar entropy
s	entropy per unit mole
s_V	entropy per unit volume
$\mathcal{S}$	path entropy
$[S]$	concentration of empty active sites
$[i\|S]$	concentration of active sites with bound ions
S_M^x	species S located on crystal lattice site M with effective charge x
T	temperature
t	time
t_i	transference number for charged component i
U	internal energy
u	internal energy per unit mole
V	volume
v	velocity
v	volume per unit mole
v	rate of reaction
$\boldsymbol{v}$	vector velocity
$\bar{v}$	mean velocity of gases
$\bar{V}_i$	partial molar volume of component i
v_{ic}	velocity of reference frame i with respect to frame c
w	work
$\boldsymbol{X}$	A thermodynamic force (vector)
X	mole fraction
x, y, z	position coordinates
z	valence
α	deviation of state parameter from equilibrium
α	linear thermal expansion coefficient
β	diffusion cell constant
β	wave number
β	liquid phase slip coefficient
$\beta_i^{(j)}$	distribution coefficient between liquid and membrane for component i at interface j
Γ_{ij}	kinetic rate of transition through activated complex
Γ	viscosity
γ	activity coefficient
δ	a distance or thickness
ε_{ij}	Wagner's solution interaction parameter
ε	electrostatic permittivity
ζ	viscous resistance
η	electrochemical potential
Θ_D	Debye temperature
λ	eigenvalue of diffusion matrix
λ	mean free path, interplanar distance, or jump distance
λ	specific electrical conductance

λ $x/t^{1/2}$, for parametric solutions to Fick's equation
λ wavelength
μ chemical potential
μ_{ij} thermodynamic function (see equation (4.32))
ν stoichiometric coefficient in chemical reaction
ν collision frequency
ν_{ij} frequency factor
ξ extent of reaction
π osmotic pressure
π Poisson's ratio
ρ density
ρ a radius
σ time rate of entropy change per unit volume
ϕ phase angle
ψ electrostatic potential

CHAPTER 1

Binary diffusion theory and its applications

1.1 INTRODUCTION

Although it is tempting to imagine a process of diffusion in a single component system at equilibrium, the conception has no operational meaning. That is to say, from the quantum mechanical point of view the interchange of site occupants does not represent a change in thermodynamic state. It is thus necessary to mark a proportion of the constituents so that relative transport can be observed. The substitution of a different isotope or isotopically marked component, i.e. a 'tracer', affords the observation of relative motion in the absence of chemical change and thus access to a kinetic characteristic of a chemically homogeneous system. In atomically pure systems this is best applied to higher atomic masses where isotopic mass effects may be deemed negligible. In nuclear magnetic resonance methods the atoms are marked by their nuclear spin states, which are ideally benign. Diffusion measurements of this kind are always carried out within the context of Fick's law for interdiffusion[1] which states that the flux of tracer is strictly proportional to the concentration gradient, i.e.

$$J = -D\nabla C \tag{1.1}$$

where J is the flux in atoms per unit area per second, ∇C is the gradient of the number of atoms per unit volume, and in the dilute tracer case D is a constant characteristic of the pure material. The diffusion coefficient D is ultimately relatable in the case of solids to the frequency of discrete jumps and the lattice parameter in the crystal, and in gases to the mean molecular velocity and mean free path. The linear Fick relation, which is analogous to Fourier's law for heat conduction and Ohm's Law for electrical conduction, is accurately derivable within the kinetic theory of dilute gases and with considerable rigour within various thermokinetic models for the condensed states of matter. In the sections which follow somewhat condensed versions of the classical arguments are presented and order-of-magnitude evaluations of the derived coefficients and their applications to some typical diffusion-controlled reactions are introduced. These anticipate ternary counterparts to be discussed in later chapters. In Chapter 2 the more detailed classical theories and the modern approaches to binary diffusion which pretend to one level or another of intellectual rigour are reviewed. These include the geometric, random walk procedures which in crystalline solids may, on the assumption of discrete defects, jump distances and times, be developed into a substantial theory of correlation and isotope effects. They also include the purely phenomenological approach based on the principles of irreversible thermodynamics. Bridging these extremes is the theory of activated rate processes with its modern couterparts in lattice, defect and molecular dynamics. Interestingly, the transition state theory approach, which was for many years maligned by theoreticians, now lies at the heart of the more rigorous and sophisticated approaches to this very complex problem.

1.2 DIFFUSION IN GASES

The kinetic theory of gases in the Chapman–Enskog formulation[2,3] is very successful in predicting transport properties such as diffusion coefficients, thermal conductivities and viscosity for dilute and not so dilute gases. The model for dilute systems consists of molecules of about $d = 0{\cdot}3$ nm in diameter in random motion as indicated in Fig. 1.1. The important parameters are the diameters, d, which define the collision cross-section and thus the mean free path (about $300d$ at NTP) the density (number of atoms per mm^3) and the mean velocity, $\bar{v}$, which is close to the speed of sound. The mean free path, λ, divided by the velocity gives the mean time between collisions and the inverse of this, the collision frequency, i.e.

$$\nu = \bar{v}/\lambda \tag{1.2}$$

Consider a constant volume enclosure and within it planes 1 and 2 located a distance λ apart. Let the mean number of marked or tracer molecules A^* at planes 1 and 2 per unit area be n_1 and n_2. The flux of marked atoms per unit area in the x direction can then be estimated as

$$J = \tfrac{1}{3}\nu(n_1 - n_2) \tag{1.3}$$

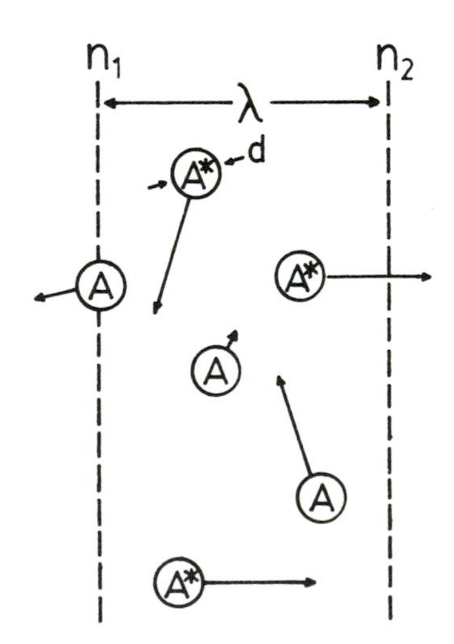

Fig. 1.1 Schematization of the interdiffusion of gas molecules A and A*

the factor 1/3 indicating that those molecules moving in the y and z directions do not contribute to the diffusive flow in the x direction. Multiplying this by $\lambda^2/\lambda^2 = 1$ it can be rearranged to read

$$J = -\frac{1}{3}\nu\lambda^2\left(\frac{\frac{n_2}{\lambda} - \frac{n_1}{\lambda}}{\lambda}\right) \tag{1.4}$$

But $n/\lambda = C$ (the number per unit volume) so the bracketed quantity is the atom or molecular concentration gradient. Hence with the help of equation (1.2) this becomes

$$J = -D\frac{\mathrm{d}C}{\mathrm{d}x} \tag{1.5}$$

with

$$D = \tfrac{1}{3}\nu\lambda^2 = \tfrac{1}{3}\bar{v}\lambda \tag{1.6}$$

identified earlier as Fick's Law. This equation divides the flux into a part determined by external conditions (the gradient) and a part which is a property of the material (D). The temperature and pressure dependence of D are implicit within $\bar{v}$ and λ.

It is to be noted that the foregoing derivation could have been carried out with an arbitrary distance, λ', provided the mean time between collisions, τ', or collision frequency, $\nu' = l/\tau'$, corresponding to it were to be evaluated. Thus

$$D = \frac{1}{3}\frac{\lambda^2}{\tau} = \frac{1}{3}\frac{\lambda'^2}{\tau'} = \frac{1}{3}\frac{\lambda''^2}{\tau''} = \text{etc.} \tag{1.7}$$

This informs us that in a diffusion experiment, the average diffusion distance, l, of a molecule in a time, t, will be given approximately by

$$l = 1{\cdot}7\sqrt{Dt} \tag{1.8}$$

Table 1.1 Comparison of predicted and observed binary gaseous diffusion coefficients at atmospheric pressure according to the Chapman–Enskog theory

Mixture	Temp. K	D_{CALC} mm^2 sec^{-1}	D_{OBS} mm^2 sec^{-1}
Ar–SO_2	263	9	7·7
Ar–NH_3	295	23	23·2
He–C_2H_5OH	298	47	49·4
H_2–C_2H_6	298	54	53·7
H_2–O_2	273	67	69·7
NH_3–He	297	81	84·2
He–CH_3OH	423	109	103·2
Ar–He	448	160	175·0

Note the law of diminishing transport within the $\sqrt{t}$ dependency of the distance. As an example mass differences may be neglected and the diffusion coefficient be estimated for interdiffusion of O_2 and N_2 in air at standard temperature and pressure (STP) via $\bar{v}$ = the speed of sound $\cong 300$ m s^{-1} and $\lambda \cong 300$ d $\cong 90$ nm, as $D = 10$ mm^2 s^{-1} (actual value, 18 mm^2 s^{-1}). Thus, according to equation (1.8), an average molecule in air will diffuse about 6 mm in the first second. It would take several months to diffuse across a room of 10 metres. The fact that odours in a room move much faster than this shows clearly that the process of transport must be convection not diffusion. True diffusion in gases can be measured accurately in fine isothermal capillaries where convection is damped out by wall friction. Table 1.1 shows comparisons between predicted and measured D, for a selection of dilute gaseous mixtures according to the complete Chapman–Enskog theory.[3,4,5]

1.3 DIFFUSION IN LIQUIDS

Figure 1.2 schematically illustrates that liquids are almost as compact as solids. One can think of the diffusion process as a gradual insinuation of the marked molecules through the matrix as a result of thermal vibrations and expansion. Using the same model as for gases we adopt formula (1.6) and attempt to estimate $\bar{v}$ and λ. For many atomic liquids there is equipartition between potential energy and kinetic energy modes yielding a Dulong and Petit value of the molar heat capacity equal to $3RT$ where R is the gas constant. Thus we can assign $3kT/2$ per atom to the kinetic energy, where k is the Boltzmann constant and estimate[6] with m_0 equal to the molecular mass

$$\bar{v} = \sqrt{8kT/\pi m_0} \tag{1.9}$$

a value similar to that for a monatomic gas. The mean free path depends on the free volume in the somewhat loosely packed structure and its change due to thermal expansion. In an estimate developed later (§2.16) we have obtained

$$\lambda \simeq 3\alpha d T \tag{1.10}$$

where α is the linear thermal expansion coefficient, and thus

$$D \simeq \alpha d \sqrt{\frac{8k}{\pi m_0}}\, T^{3/2} \tag{1.11}$$

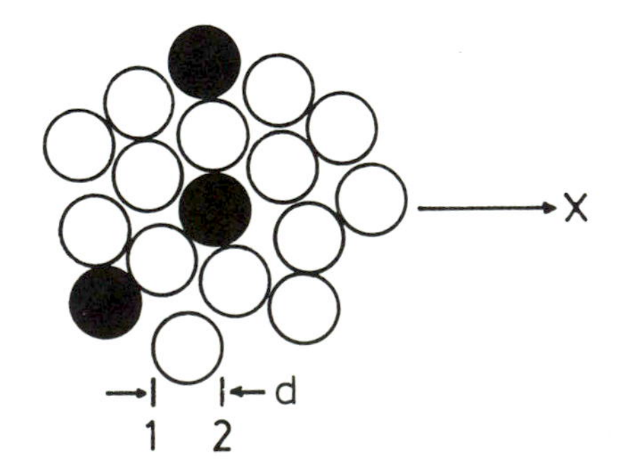

Fig. 1.2 Schematization of a liquid mixture

This formula yields typical melting point values for D of 10^{-3} mm^2 s^{-1} in accord with observations. The weak $T^{3/2}$ temperature dependence in this formula is about right for most liquids. This is in contrast to solids, where it is exponential. A typical liquid diffusion distance in 3600 s (1 h) will be approximately 3·5 mm, which is to be compared to the value in a gas of $l \cong 350$ mm. While there are no accurate fundamental theories of liquid diffusion it is possible to computer-model the molecular dynamics of a system of thousands of particles with empirical forces of interaction and to plot the trajectories (see Fig. 1.3)[7] and from these obtain the self (tracer) diffusion coefficient. For simple liquids, like argon at low temperatures, the predictions are exceptionally accurate.

The tracer coefficient for diffusion in dilute gases is exactly related to the viscosity Γ by the formula[6]

$$D = \Gamma/Mm = \Gamma/\rho \tag{1.12}$$

where M is the molecular mass and m is the molar concentration, and one might expect something similar to hold for liquid self-diffusion. Instead, the hydrodynamic theory in the elementary form of Stokes–Einstein[8, 9] yields[6]

$$D \simeq kT/3\pi d\Gamma \tag{1.13}$$

where d is the molecular diameter so the functional relationship is quite different. Formulas of the latter type are very important in extending the data base and will be treated in detail later (§9.2).

The foregoing estimates refer to roughly spherical molecules. Since the vibration frequency of a quasi-crystalline oscillator varies as (mass)$^{-1/2}$ one might conjecture that diffusion coefficients vary in the same way. Thus a sugar molecule ($C_{12}H_{22}O_{11}$) with a molecular weight of 342 may be expected to have a coefficient $\sqrt{17/342} \simeq 0{\cdot}2$ times the value for an OH^- ion and a protein molecule of weight 170 000 to have a coefficient of 0·01 times the OH^- value. Of course, if, as is likely, the protein molecules are not spherical, then the coefficient could be much smaller again. Actually, the $\sqrt{M}$ law fails in specific cases such as hydrogen isotopes.[10]

The reader should recognize that many useful processes involving liquid diffusion simultaneously involve convection and circulation so diffusion coefficients will appear in the theory of mass transfer in a composite disposition. For example, in stagnant film theories of convection mass transport, the transfer occurs across a surface film according to pure diffusion principles, but the film thickness is determined by fluid dynamical principles (see, for example, § 1.29).

1.4 DIFFUSION IN SOLIDS

The diffusion coefficient model of the previous sections is here broadly applicable. As a first guess, however, one might suppose that solid crystals are so compact that no diffusion is possible. In fact, many solid solutions are interstitial (see Fig. 1.4) and interstitial atoms are able to move through the narrow interatomic channels. Furthermore all crystals contain vacancies (the number increasing rapidly with temperature up to about 0·01% of sites near the melting point) so even substitutional atoms like X in Fig. 1.4 can move by interchange with the vacancies.

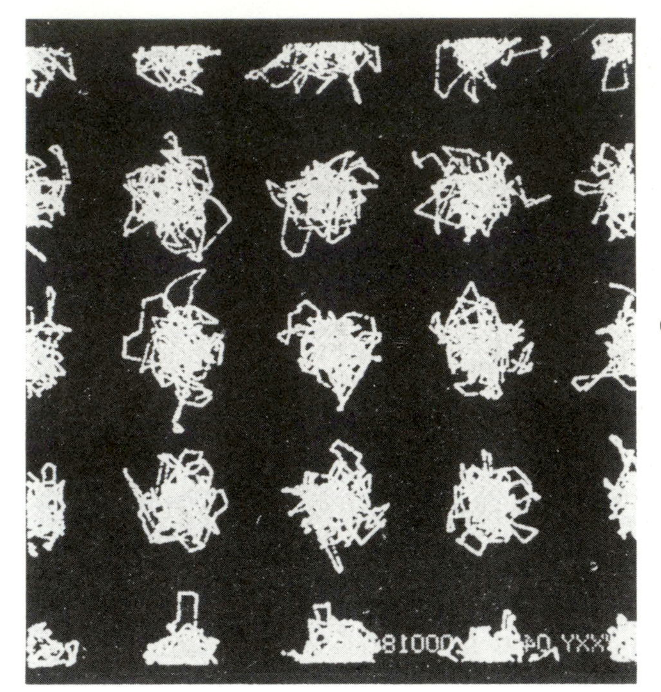

(A)

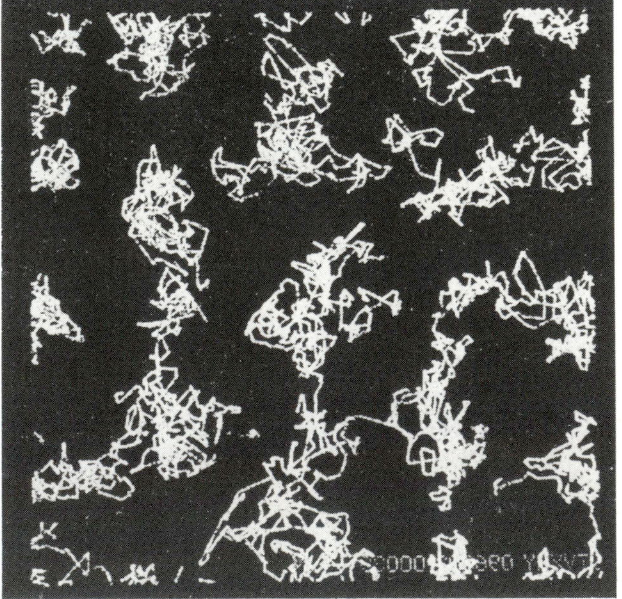

(B)

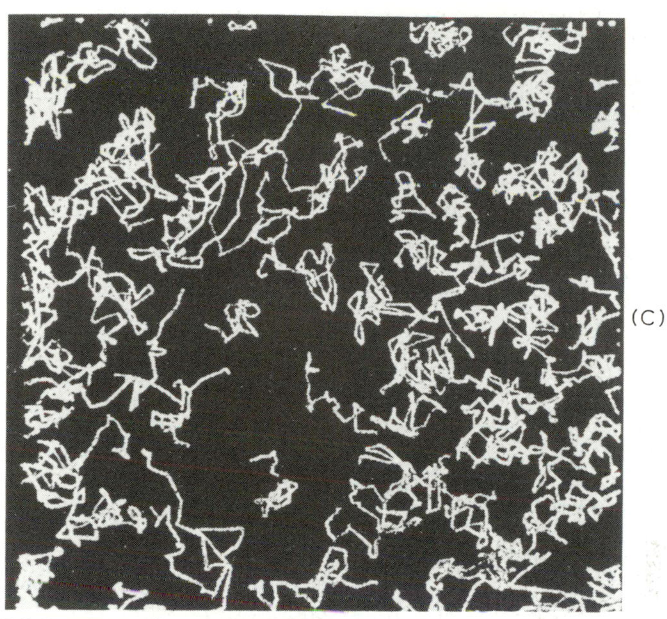

Fig. 1.3 Molecular trajectories calculated by the molecular dynamics method: (*a*) 32 hard spheres in the solid phase for about 3 000 collisions; (*b*) same as (*a*) after it has transformed to the fluid phase; (*c*) 108 particles after about 3 000 collisions in the liquid–vapour region. Reproduced with permission from Alder and Wainwright[7]

The diffusion coefficients for interstitials approach liquid values ($\sim 10^{-4}$ mm^2 s^{-1}) near the melting point and decrease rapidly with temperature due to the decrease in thermal kinetic energy available for mounting the potential barriers in the interatomic channels. For vacancy diffusion the conversion rate of lattice vibrations, which have frequencies of 10^{13}–10^{14} s^{-1}, into diffusion jumps must be discounted in proportion to the number of vibrations needed per exchange with a vacancy (say 100–1 000) and the fraction of adjacent vacancies. Thus, near the melting point the net discount factor will be around 10^7 and the diffusion coefficient will therefore be about 10^{-5} mm^2 s^{-1}. This number also decreases very rapidly with temperature, due to the decreasing vibrational kinetic energy.

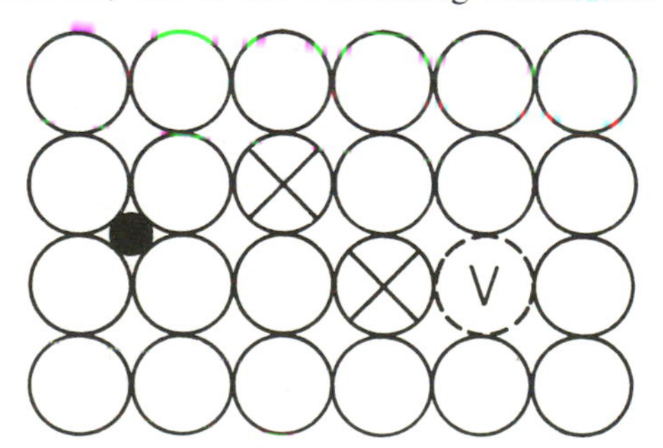

Fig. 1.4 Schematization of a solid crystal with impurity atoms and a vacant site

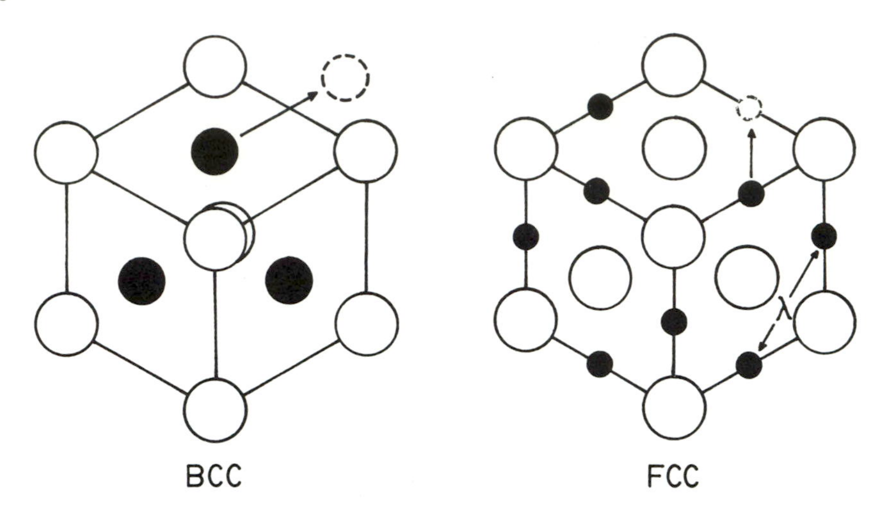

Fig. 1.5 Illustration of interstitial diffusion steps in fcc and bcc crystals

Carbon (and probably nitrogen and oxygen as well) is an interstitial solute in both the body-centred cubic (BCC) and face-centred cubic (FCC) forms of iron. It resides in the octahedral holes (O) of both structures, as indicated in Fig. 1.5. Because the BCC structure is looser than the FCC (which is close-packed) the trajectories are freer and the diffusion coefficient is correspondingly higher ($\sim 100X$). Figures 1.6 and 1.7 summarize some solid state data as a function of inverse temperature. Note that for both solid structures D approaches the liquid values near the melting point and that the temperature dependence is exponential to a high degree of accuracy over substantial temperature ranges. The exceptional range of linearity in the logarithmic plots exhibited here remains a matter of puzzlement to many theorists.

Consider now the diffusion steps indicated by the arrows in Fig. 1.5 which schematically represent carbon jumps in body-centred and face-centred cubic iron, respectively. Redrawing this in two dimensions the configurational and energy relationships of Fig. 1.8 are obtained. Elementary statistical mechanics of solids tell us that for particles in a random vibrational array the kinetic energy or enthalpy (Q) distribution function is proportional to

$$\phi = \exp(-Q/kT) \tag{1.14}$$

(the Boltzmann factor) and that the fraction of particles with energy in excess of Q is the same quantity. Consequently the probability that an interstitial particle at potential energy A will gain enough from the lattice kinetic energy distribution to surmount the barrier at C will be proportional to

$$P = \exp(-Q/RT) \tag{1.15}$$

The frequency with which particles cross the barrier will accordingly be proportional to P and we can write

$$D = D_0 \exp(-Q/RT) \tag{1.16}$$

where D_0 is an empirical quantity measuring the fundamental vibrational

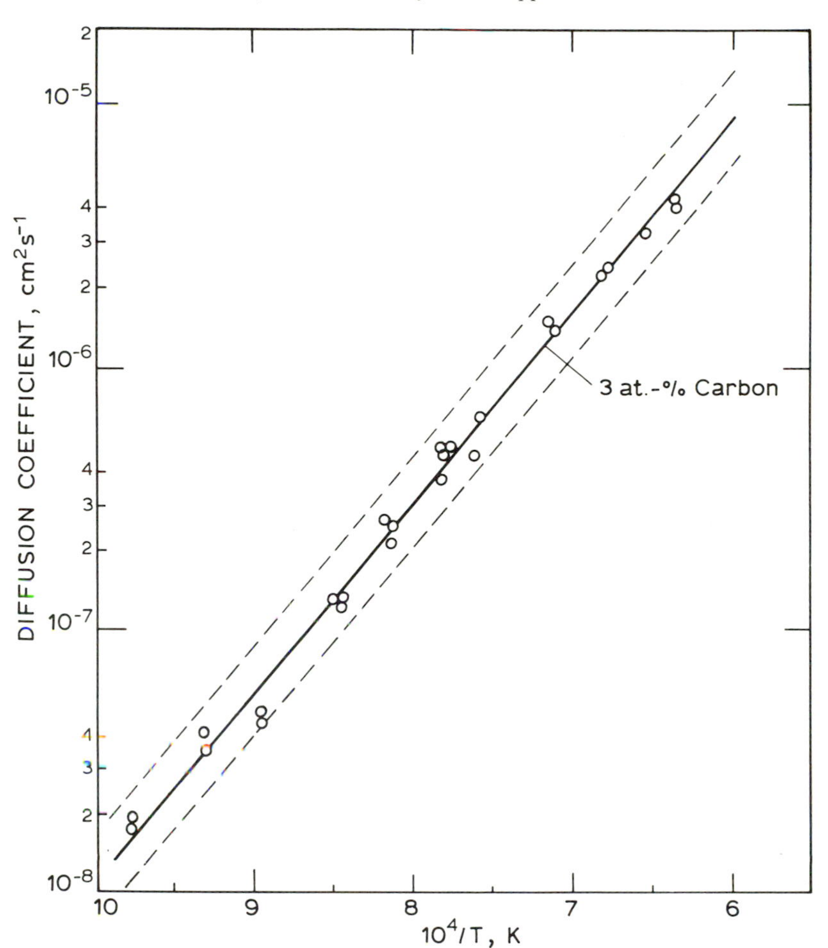

Reprinted with permission from TRANSACTIONS OF THE METALLURGICAL SOCIETY, Vol. 188, p. 553, (1950), a publication of The Metallurgical Society, Warrendale, Pennsylvania

Fig. 1.6 Arrhenius plot for the diffusion of carbon in austenite. After Wells *et al.*[11]

frequency of the lattice and Q is the so-called heat of 'activation' or barrier energy. The values of D_0 and Q corresponding to D^{γ} and D^{α} in Figs. 1.6 and 1.7 are $D_0^{\gamma} = 17{\cdot}5\ \text{mm}^2\ \text{s}^{-1}$ and $Q^{\gamma} = 138{\cdot}3\ \text{kJ mol}^{-1}$ and $D_0^{\alpha} = 3{\cdot}2\ \text{mm}^2\ \text{s}^{-1}$ and $Q^{\alpha} = 83{\cdot}1\ \text{kJ mol}^{-1}$, respectively. Note particularly the accuracy of the fit to equation (1.16) with Q constant as applied to α-iron, where the relationship extends over 18 decades in the diffusion coefficients. Plots of $\ln D$ vs $1/T$, whose slopes are Q/R, are called Arrhenius plots after the Swedish chemist who demonstrated the generality of such correlations.[13] The intercepts at $1/T = 0$ are $\ln D_0$. The theoretical justification for Arrhenius-type behaviour in solids will be made more rigorous via absolute reaction rate theory in Chapter 2.

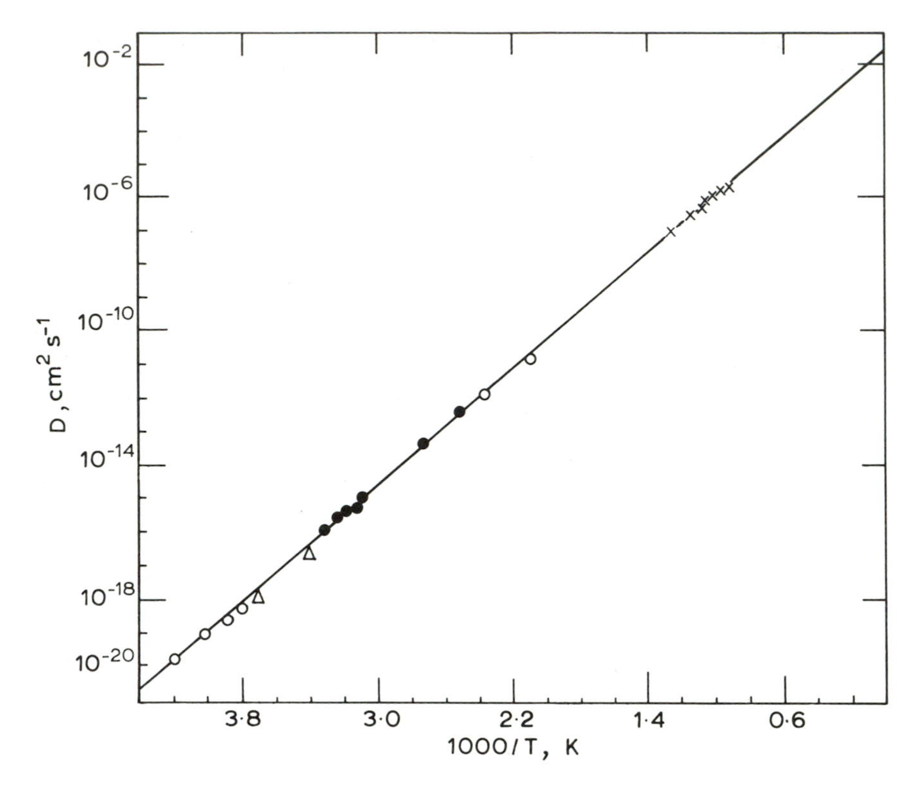

Fig. 1.7 Arrhenius plot for the diffusion of carbon in ferrite. Reproduced with permission from Wert[12]

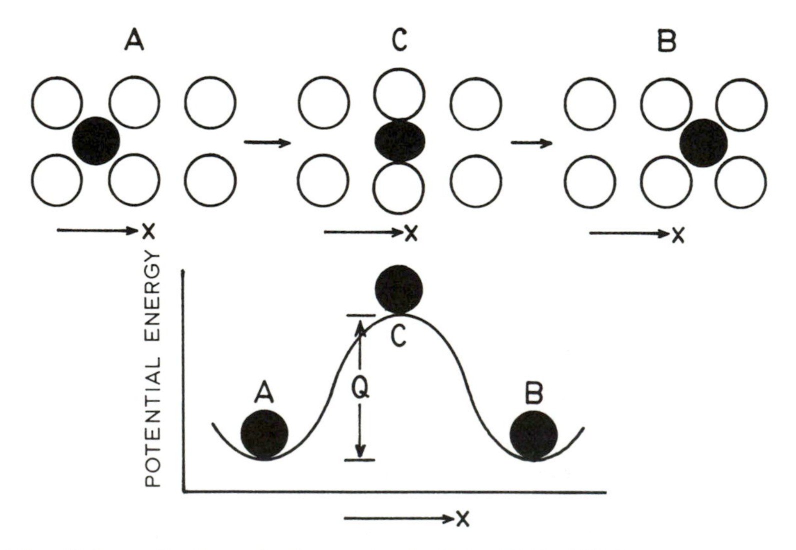

Fig. 1.8 Schematization of a jump step in interstitial diffusion

Most solid metallic solutions are substitutional in character (Fig. 1.4). It can be shown that, since exchange and ring mechanisms[14] are energetically unlikely, diffusion is almost certain to occur by interchange with the intrinsic vacancies of the lattice.[15] In contradistinction to line defects (i.e. dislocations) in crystalline solids, vacancies are an equilibrium defect, and for each temperature there is a unique and predictable concentration which is sufficient to rationalize diffusion rates in such materials. The heat of activation Q now includes an activation energy and a heat of formation of vacancies, and the total tends to be about twice the magnitude as for interstitial diffusion. Consequently the diffusion coefficients are generally much lower, ranging from 10^{-5} mm^2 s^{-1} at the melting point to 10^{-23} mm^2 s^{-1} at ambient temperatures for high melting point (refractory) materials. The latter correspond to the very low individual jump frequencies of 10^{-9} s^{-1} or 10^{15} mol^{-1} s^{-1}, significant only over geological time spans.

These lower limit numbers apply to perfect crystals, which do not in fact exist. All crystals contain dislocations and (or) grain boundaries and these defective regions define volumes which are more liquidlike than solidlike. The effect of such defects can be represented by a mean tracer diffusion coefficient of the form[16] (cf §2.3)

$$D = (1-f)D_V + fD_B \tag{1.17}$$

where D_V is the perfect crystal or lattice value, D_B is the mean value in defective material and f is the fraction of the total volume which is defective. A typical metal crystal can contain 10^{12} dislocations per square metre. This corresponds to a defective volume fraction of about $f = 10^{-7}$. Thus if a diffusion coefficient of 10^{-10} mm^2 s^{-1} is assigned to a grain boundary at ambient temperatures (appreciably less than a liquid value) where D_V is vanishingly small, D may nonetheless approach 10^{-17} mm^2 s^{-1}, which is quite significant in relation to fine precipitate reactions. Low-temperature reactions can also be accelerated by high vacancy densities which have been retained on cooling crystals from high temperatures. Of course, all these considerations are relative, depending absolutely on the melting point of the material. The effect of dislocations on bulk diffusion has recently been reviewed by LeClaire and Rabinovich[17] and LeClaire[18] (cf §1.15).

Anomolously high diffusion rates in normally substitutional alloys are also observed, particularly at temperatures approaching the melting point.[19] This is thought to be due to the activation of a parallel interstitial component in the diffusion process. Anomolously fast diffusion in metals has been examined in detail by Warburton and Turnbull.[20] Hydrogen diffusion in certain metals such as iron and palladium is exceptionally fast. Here the quantum theory of diffusion is most relevant.[21]

1.5 THE SIGNIFICANCE OF DIFFUSION COEFFICIENT MAGNITUDES

As has been seen, the quantity $\sqrt{Dt}$, characterized as the 'diffusion length' describes a mean diffusion distance in terms of the diffusion time. With typical gas, liquid, and solid coefficients of 10, 10^{-3}, and 10^{-7} mm^2 s^{-1}, respectively, the diffusion distances in 10 s are 10, 0·1 and 0·001 mm, respectively. In the absence of circulation these can be taken as the upper limits of precipitate sizes which might evolve in 10 s in a supersaturated solution of the given state (e.g. snow crystals in air, sugar crystals in water, and iron carbide crystals in steel). These upper limits will be discounted in a proper theory of crystal growth by a relative supersaturation factor which is typically 0·1 (see later sections of this chapter). As has been indicated, gas and liquid coefficients do not vary strongly with species and temperature. By contrast, solid coefficients always have a strong exponential dependence, as indicated in Fig. 1.6, and values of 10^{-18} mm^2 s^{-1} in normal defective material are not uncommon for room-temperature diffusion. Despite the minuteness of this quantity it is still significant, for over a period of three years ($\sim 10^8$ s) the diffusion length is $\sim 0{\cdot}01\ \mu$m. Precipitates even smaller than this play a role in dispersion hardening of materials. If the same process were to take place over geological times ($\sim 10^{15}$ s), precipitates up to 1 mm are to be expected through solid state diffusion at ambient temperatures.

Currently, much attention is being given to measurement of the very small diffusion coefficients in ambient or near-ambient solids. The next generation of microcircuitry will possess component dimensions of the order of 100 nm so it is imperative in assessing lifetimes and the required degree of temperature control that values be available for the pertinent diffusion parameters. Some attention will be devoted to the problem and related research in Chapter 5. Such practical problems are a preoccupation of this book.

1.6 THERMOKINETIC FRAMEWORK FOR BINARY DIFFUSION

To most students, scientists, and engineers, the study of diffusion is a means to an end, the end being an initial or boundary value problem associated with a phase transformation or mass transfer process. To solve such a problem it is necessary to obtain a reasonably accurate diffusion data set. Since physical theories of diffusion in the condensed state are still inadequate to the quantitative need, one must rely on a semiempirical approach which employs correlations, irreversible thermodynamics, and absolute reaction rate theory to interpolate and extrapolate the limited accurate experimental data which is available.

The thermodynamics of irreversible processes, which is developed more generally in Chapter 3, is a key element in this practical activity. Since this discipline is based on classical chemical thermodynamics we recommend a brief review of the terminology and the elements of the latter discipline as presented in Appendix 1. In the most elementary form of irreversible thermodynamics the Gibbs' energy equation for a two-component isothermal, constant volume system is written as

$$\mathrm{d}E = T\mathrm{d}S + \mu_1\mathrm{d}n_1 + \mu_2\mathrm{d}n_2 \qquad (1.18)$$

where n_1 signifies mole numbers and the μ_1 are chemical potentials.

This is converted to a rate equation via division by $\mathrm{d}t$ yielding

$$\frac{\mathrm{d}E}{\mathrm{d}t} = T\frac{\mathrm{d}S}{\mathrm{d}t} + \mu_\mathrm{A}\frac{\mathrm{d}n_\mathrm{A}}{\mathrm{d}t} + \mu_\mathrm{B}\frac{\mathrm{d}n_\mathrm{B}}{\mathrm{d}t} \qquad (1.19)$$

The basic assumption of the linear theory of irreversible thermodynamics is associated with the validity of this relation in kinetic situations. Since equation (1.18) applies only to quasi-equilibrium or reversible processes the kinetic equation will only be valid when local equilibrium or near-equilibrium can be established for the process of interest. In diffusion this implies that the gradients should not be too great.

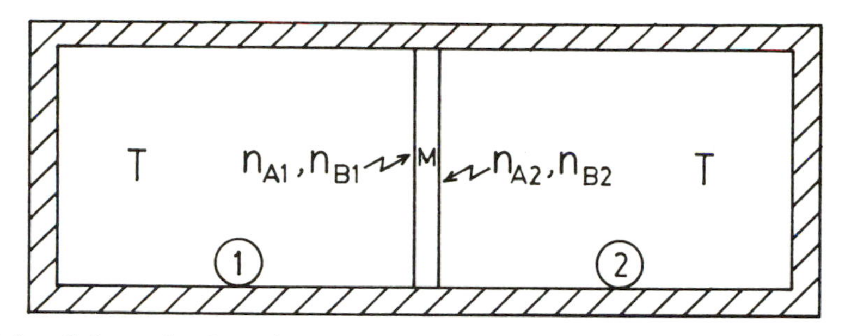

Fig. 1.9 Schematization of a membrane diffusion process in isolation. The thermal reservoirs are large enough that the temperature change is negligible during the diffusion time

Now the composite isolated system of Fig. 1.9 is referred to, where the two mass-heat baths are very much larger than the separating binary diffusion membrane, and countercurrent diffusion of A and B is assumed. Next equation (1.19) can be applied to subsystems 1 and 2 and the two expressions added to yield

$$\frac{\mathrm{d}(E_1 + E_2)}{\mathrm{d}t} = T\frac{\mathrm{d}(S_1 + S_2)}{\mathrm{d}t} + (\mu_{A2} - \mu_{A1})\frac{\mathrm{d}n_{A2}}{\mathrm{d}t} + (\mu_{B2} - \mu_{B1})\frac{\mathrm{d}n_{B2}}{\mathrm{d}t} \tag{1.20}$$

recognizing that $\mathrm{d}n_{A1} = -\mathrm{d}n_{A2}$ and $\mathrm{d}n_{B1} = -\mathrm{d}n_{B2}$. Since the system is isolated, the left-hand side of Equation 1.20 is zero, so that with $S = S_1 + S_2$

$$T\frac{\mathrm{d}S}{\mathrm{d}t} = -\Delta\mu_A\frac{\mathrm{d}n_{A2}}{\mathrm{d}t} - \Delta\mu_B\frac{\mathrm{d}n_{B2}}{\mathrm{d}t} \tag{1.21}$$

Now letting the volume of the membrane be $V = A\,(x_2 - x_1)$ where A is its area and $x_2 - x_1$ is its thickness and recognizing that the countercurrent fluxes are defined by

$$\boldsymbol{J}_A = \frac{\mathrm{d}n_{A2}}{\mathrm{d}t} \quad \text{and} \quad \boldsymbol{J}_B = \frac{\mathrm{d}n_{B2}}{\mathrm{d}t} \tag{1.22}$$

multiplying from the left by $1/V$ and assuming a reference frame where $J_A = -J_B$ (cf §4.3) results in

$$T\sigma = \frac{\mathrm{d}s_v}{\mathrm{d}t} = -J_B\nabla(\mu_B - \mu_A) \tag{1.23}$$

where s_v is the entropy per unit volume, ∇ is the gradient (cf Appendix 2), and $\sigma = \dot{s}_v$ is the entropy production rate per unit volume. Because this is an isolated system σ is positive definite for a non-equilibrium system, i.e.

$$\sigma > 0 \tag{1.24}$$

as an expression of the second law of thermodynamics. Conventionally the thermodynamic force is defined as

$$\boldsymbol{X}_B = -\nabla(\mu_B - \mu_A) \tag{1.25}$$

and $T\sigma$ takes the standard bilinear form

$$T\sigma = \boldsymbol{J}_B\boldsymbol{X}_B > 0 \tag{1.26}$$

In this book there is no distinction made symbolically between one-dimensional and n-dimensional fluxes and forces. This distinction is always clear from the context.

In this strictly non-mechanistic (i.e. phenomenological) theory one must postulate a relation between $\boldsymbol{J}$ and $\boldsymbol{X}$. It is consistent with all that has gone before to adopt the linear relation

$$\boldsymbol{J}_{\mathrm{B}} = L_{\mathrm{B}}\boldsymbol{X}_{\mathrm{B}} \tag{1.27}$$

with the mobility coefficient

$$L_{\mathrm{B}} > 0 \tag{1.28}$$

for through the quadratic relation which results, i.e.

$$\sigma = L_{\mathrm{B}}\boldsymbol{X}_{\mathrm{B}}^2 = \boldsymbol{J}_{\mathrm{B}}^2/L_{\mathrm{B}} \tag{1.29}$$

the positive definite character of σ is assured. It is important to note here that mechanistic theories for processes which sustain local equilibrium invariably confirm the linear relation (1.27).

Next, referring to local equilibrium thermodynamics (Appendix 1) the binary isothermal relation for the chemical potential differences is found

$$\mu_{\mathrm{B}} - \mu_{\mathrm{A}} = \frac{\mathrm{d}G}{\mathrm{d}X_{\mathrm{B}}} \tag{1.30}$$

where G is the molar Gibbs free energy and X_{B} is the mole fraction of component B. In condensed systems the same relation holds to a sufficiently good approximation when G is replaced by A, the Helmholz free energy (see Appendix 1). If equations (1.25) and (1.30) are substituted into (1.27) we recover Fick's first equation, transforming from the atomic to the molar form,

$$J_{\mathrm{B}} = -\frac{L_{\mathrm{B}}}{m}\frac{\mathrm{d}^2G}{\mathrm{d}X_{\mathrm{B}}^2}\frac{\partial m_{\mathrm{B}}}{\partial x} = -\frac{RTL_{\mathrm{B}}}{m_{\mathrm{B}}}\left(1 + \frac{\mathrm{d}\ln\gamma_{\mathrm{B}}}{\mathrm{d}\ln X_{\mathrm{B}}}\right)\frac{\partial m_{\mathrm{B}}}{\partial x} \tag{1.31}$$

where m is the mean molar density or concentration, i.e. $m_{\mathrm{B}} = mX_{\mathrm{B}}$, $m_{\mathrm{A}} = mX_{\mathrm{A}}$ and $m = m_{\mathrm{A}} + m_{\mathrm{B}}$. These units are the most convenient from the point of view of thermodynamics. Note that m is approximately constant for many condensed systems of interest on account of molecular or atomic site conservation. It was for this reason that it was possible to insert it as an approximation within the derivatives of equation (1.31) without generating extra terms. The mutual or chemical diffusion coefficient relations so-defined,

$$\tilde{D}_{\mathrm{B}} = \frac{L_{\mathrm{B}}\,\mathrm{d}^2G}{m\;\mathrm{d}X_{\mathrm{B}}^2} \tag{1.32}$$

or

$$\tilde{D}_{\mathrm{B}} = \frac{RTL_{\mathrm{B}}}{m_{\mathrm{B}}}\left(1 + \frac{\mathrm{d}\ln\gamma_{\mathrm{B}}}{\mathrm{d}\ln X_{\mathrm{B}}}\right) \tag{1.33}$$

make adequately clear that the chemical interdiffusion coefficient is strongly dependent on the thermodynamic solution properties. For one thing it is known that $\mathrm{d}^2G/\mathrm{d}X_{\mathrm{B}}^2$ in equation (1.32) goes negative as a solution is quenched through a critical point to within the miscibility gap, and since $L_B > 0$, $\tilde{D}_b$ must go from positive to negative. This implies that diffusion can occur against a concentration gradient. This is called 'uphill diffusion', although it is impossible in a binary system for diffusion to occur against the true thermodynamic force defined by equation (1.25). Such processes of 'spinodal decomposition' will be dealt with in detail in this and later chapters.

Equation (1.33) gives us further insight into the concentration dependence of the chemical diffusion coefficient. If we think of $\nabla(\mu_B - \mu_A)$ as a potential energy gradient and thus a force in the sense of Newtonian mechanics, and also recognize that the flux can be written in terms of a mean species velocity, i.e.

$$J_B = m_B v_B \tag{1.34}$$

then we may reasonably conjecture that the velocity is proportional to the force. This implies that L_B/m_B will tend to be constant, a result borne out by elementary kinetic theories. Thus the main concentration dependence of $\tilde{D}$ should reside in the thermodynamic factor $(1 + \mathrm{d}\ln\gamma_B/\mathrm{d}\ln X_B)$. While the general trend has been validated the simplification is not sufficiently quantitative to serve in any general context (see Fig. 1.10). Fortunately, the mobilities can be related rather rigorously to experiments carried out with tracers. This calculation, known as the Darken analysis, will be presented in §2.14.[23]

As already noted, the temperature dependence of binary coefficients in solids is usually of the Arrhenius type, as in the case of interstitial Fe-base solutions (Figs. 1.6 and 1.7). A partial theoretical rationalization is to be found in Eyring's absolute reaction rate theory[24, 25, 26] as applied to diffusion. The hypothesis is that there is always a very flat saddle point potential energy configuration assignable to the unit step in diffusion transfer or interchange so that the dilute atom

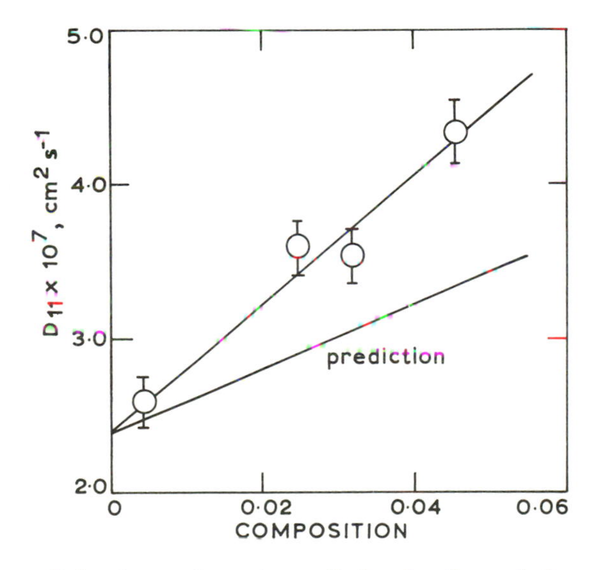

Fig. 1.10 Test of the thermodynamic prediction for the variation of D for carbon in fcc iron. Reproduced with permission from Kirkaldy[22]

population of saddle point states can be assumed to be in equilibrium with the populations in the potential wells (Fig. 1.8) (cf Bennett[27]). Thus one can write the equilibrium constant for the reaction (see Appendix 1)

$$\text{Component (in trough)} \rightleftarrows \text{Component (at saddle)} \quad (1.35)$$

as

$$K = a_B^\dagger / a_B = \exp(-\Delta G_B^\dagger / kT) \quad (1.36)$$

where a_B is the solution activity, † represents the dilute saddle point population and $\Delta G_B^\dagger$ is the average standard free energy of formation of the saddle point or activated state. Now the number of diffusion jumps, and therefore the diffusion coefficient, will be proportional to the concentration of activated states, which according to equation (1.36) is

$$X_B^\dagger = a_B e^{\Delta S^\dagger/k} e^{-\Delta H^\dagger/kT} / \gamma^\dagger \quad (1.37)$$

where $\gamma^\dagger$ is the activity coefficient of the activated population and the approximately constant parameters $\Delta S^\dagger$ and $\Delta H^\dagger$ are the entropy and heat of formation of the activated state. $\Delta H^\dagger$ is equivalent to the heat or energy of activation Q previously introduced. This calculation may be taken as a more rigorous derivation of equation (1.16) and a justification for Arrhenius behaviour over wide ranges of temperature. A more detailed discussion and specific models will be referred to in §§2.10, 7.2 and 7.3.

In closing this section on methods for correlating diffusion data over concentration and temperature, note that there are a number of empirical rules for estimating coefficients over species in liquids and solids[6, 28, 5, 29, 30] as well as numerous general and specialized compendia of constant and variable coefficients (see Appendix 6). The easily accessible ones pertinent to the non-ionic materials dealt with in this chapter are as follows: For metallic solids and liquids we refer to Seith's 'Diffusion in Metallen',[31] Smithell's 'Metals Reference Book',[32] and to Adda and Philibert, 'La Diffusion dans Les Solides'.[33] For non-ionic liquids we refer to Bird *et al.*[6] and to Hildebrand's 'Viscosity and Diffusivity'.[34] It is to be noted that for solids the reported coefficients often vary between researchers or laboratories by a factor of 2 or more, so a certain creative selectivity is required in using the data. Appendix 6 goes into data sources in more detail. The data in the aforementioned compendia are, however, a quite adequate base for the elementary diffusion solutions and applications to be presented in the following section.

1.7 SOLUTIONS OF FICK'S FIRST EQUATION: THE STEADY STATE

In the development of Fick's first equation (1.1) we emphasized a one-dimensional flux. For the steady state this can be generalized to three dimensions by the simple device of defining the flow direction along the radius vector $\boldsymbol{r}$ and expressing the total flux across an arbitrary area perpendicular to the radius (see Fig. 1.11). Thus we write for the total flux $\boldsymbol{J}_T$ with concentration C in atoms per unit volume

$$\boldsymbol{J}_T = -DA \frac{dC}{dr} \quad (1.38)$$

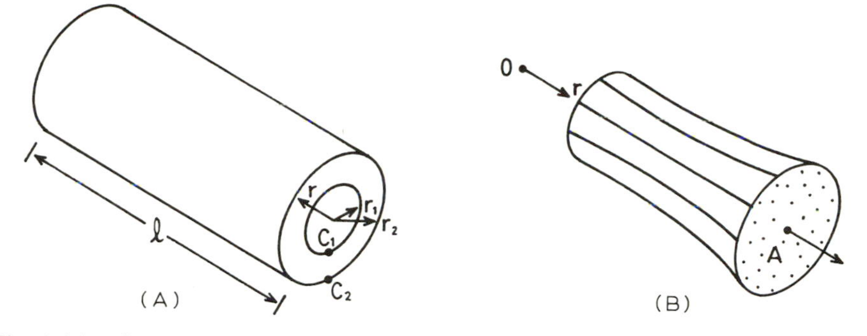

Fig. 1.11 Schematic representation of a cylindrical and a general diffusion path

The inclusion of the directional area $\boldsymbol{A}$ with this disposition assures that the dilution of the unit area flux due to the divergence of the radius vector is exactly compensated for by the expansion of the area of access to the total flux at the steady state. This is in fact a mass conservation relation. For linear, cylindrical, and spherical symmetry we substitute $\boldsymbol{A} = A$; $\boldsymbol{A} = 2\pi rl$ and $\boldsymbol{A} = 4\pi r^2$, respectively. Integrating with constant D for a membrane defined by the coordinates r_1 and $r > r_1$ where $C = C_1$ at $r = r_1$ the solutions for the concentration in these three cases are

$$C = C_1 - J_T \frac{(r - r_1)}{DA} \tag{1.39}$$

$$C = C_1 - \frac{J_T}{2\pi l D} \ln \frac{r}{r_1} \tag{1.40}$$

and

$$C = C_1 - \frac{J_T}{4\pi D}\left(\frac{1}{r_1} - \frac{1}{r}\right) \tag{1.41}$$

Thus in an experiment with $r = r_2$ and $C = C_2$ the measurement of the steady state flux yields a value for the diffusion coefficient D. On the other hand, given D and the boundary values the diffusive rate of escape of matter J_T can be calculated. There are many practical containment problems (e.g. of hydrogen gas and liquid containers) which can be analysed in this fashion. If $D = D(C)$ an average must be used or else numerical integration of equation (1.38) is necessary. In certain containment problems $C_2 \simeq 0$ while C_1 is fixed by the solubility of the diffuser in the containment material. The corresponding process of escape is designated as 'permeation'. A new rate parameter characteristic of the material and diffuser known as the 'permeability' is thus definable. This is

$$P = C_1 D \tag{1.42}$$

which allows evaluation of escape rates via

$$J_T = \frac{PA}{r_2 - r_1} \tag{1.43}$$

and similarly for the other relations. Figure 1.12 shows the permeability of hydrogen in various metals as a function of temperature. Such data is important

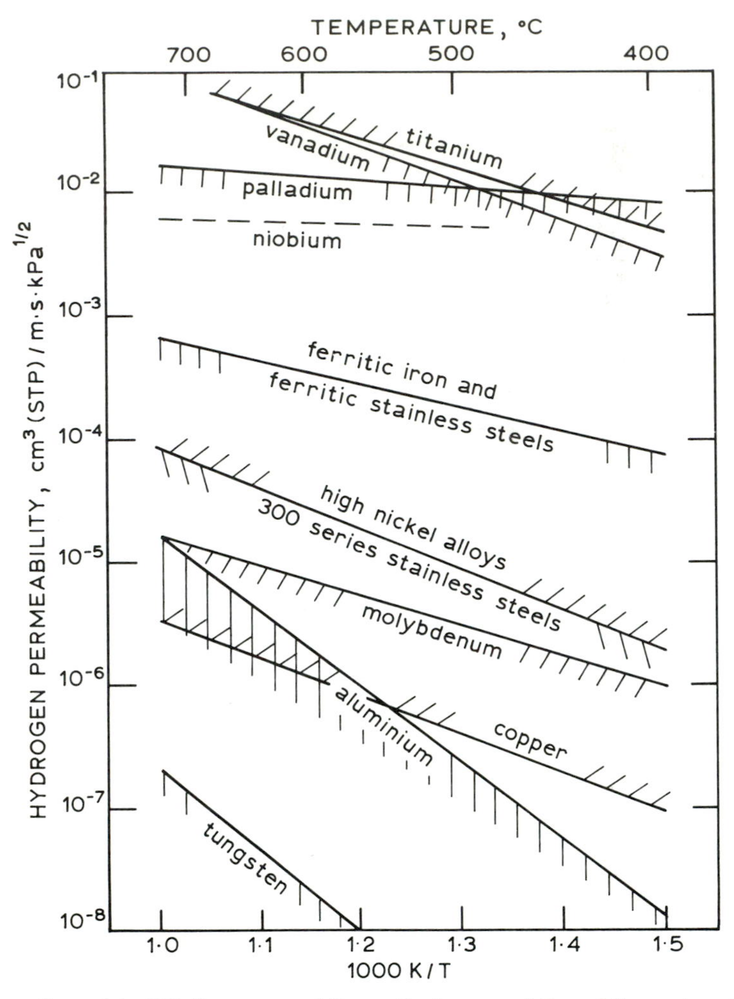

From *Proceedings of the U.S. Department of Energy Environmental Control Symposium*, Nov. 28–30, 1978, Vol. 2, V. A. Maroni, 'Control of tritium permeation through fusion reactor structural materials'. Used with permission

Fig. 1.12 Hydrogen permeability curves for selected metals and alloys. Solubility parameters in terms of Sievert's Law. Abridged with permission from Maroni[35]

for the selection of container materials for hydrogen and its isotopes, and particularly for fusion reactor vessels and peripherals.[35]

1.8 FICK'S SECOND EQUATION: THE NON-STEADY STATE

For the non-steady, or time-dependent state of a diffusion process the mass balance must be entered in an explicit fashion. Referring to Fig. 1.13 it can be seen that the mass balance requires that the increase in mass per unit volume/unit time be the negative of the excess of outflow over inflow per unit length, i.e.

$$\frac{\partial C}{\partial t} = -\frac{\partial J_x}{\partial x} \tag{1.44}$$

This clearly generalizes to three dimensions in the form

$$\frac{\partial C}{\partial t} = -\nabla \cdot \boldsymbol{J} \tag{1.45}$$

where ∇ is the divergence symbol of the vector calculus (Appendix 2). Substituting Fick's first equation

$$\boldsymbol{J} = -D\nabla C \tag{1.46}$$

we obtain Fick's second equation

$$\frac{\partial C}{\partial t} = \nabla \cdot D\nabla C \tag{1.47}$$

where D appears inside the divergence symbol if D is a function of C. For relative small concentration differences ($\Delta X \sim 0{\cdot}1$) it is often adequate to employ the average.[36]

$$D = \frac{1}{\Delta C}\int D(C)\,\mathrm{d}C \tag{1.48}$$

For constant or average D in one dimension

$$\frac{\partial C}{\partial t} = D\frac{\partial^2 C}{\partial x^2} \tag{1.49}$$

Note that dimensionally $\partial C \sim \partial^2 C$ so that

$$(\partial x)^2 \sim D\,(\partial t) \tag{1.50}$$

in agreement with equation (1.8). Steady state solutions are obtainable from equation (1.47) with $\partial C/\partial t = 0$ and constant D, i.e. via

$$\nabla^2 C = 0 \tag{1.51}$$

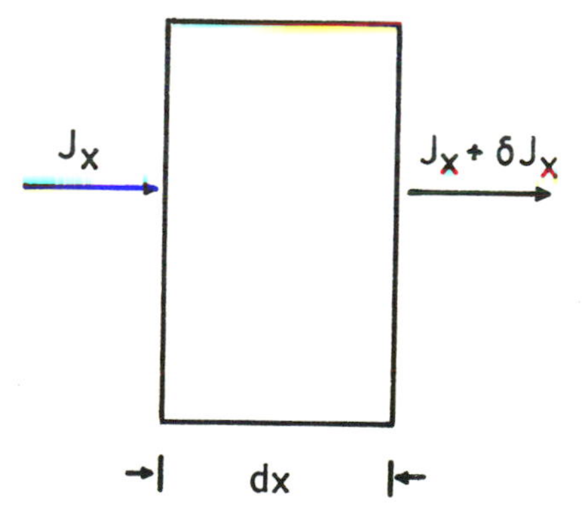

Fig. 1.13 Illustration of relations in differential mass balance

which is Laplace's equation (see Appendices 2 and 3). The solutions of this are identical with equations (1.39)–(1.41) for the special geometries considered. Equation (1.51) can, of course, be integrated numerically for arbitrary boundary conditions.

1.9 AN INTEGRAL OF FICK'S SECOND EQUATION FOR VARIABLE D

Solutions of equation (1.47) in one dimension are now sought, i.e.

$$\frac{\partial C}{\partial t} = \frac{\partial}{\partial x}\left[D(C)\frac{\partial C}{\partial x}\right] \tag{1.52}$$

such that

$$C = C(\lambda) \quad \text{where} \quad \lambda = x/\sqrt{t} \tag{1.53}$$

Upon substitution this converts the partial differential equation (1.52) to the ordinary differential equation

$$-\frac{\lambda}{2}\frac{\mathrm{d}C}{\mathrm{d}\lambda} = \frac{\mathrm{d}}{\mathrm{d}\lambda}D\frac{\mathrm{d}C}{\mathrm{d}\lambda} \tag{1.54}$$

which is known as the Boltzmann equation.[37,38]

Provided the boundary and initial conditions on x and t, respectively, can be expressed in terms of λ, the solutions of equation (1.54) will be unique solutions of equation (1.52).[31,39] Differentiating equation (1.54) to

$$-\frac{\lambda \mathrm{d}\lambda}{2D} = \frac{\mathrm{d}D}{D} + \frac{\mathrm{d}^2C}{\mathrm{d}\lambda^2}\bigg/\frac{\mathrm{d}C}{\mathrm{d}\lambda} \tag{1.55}$$

and integrating once with respect to λ we obtain

$$-\int_0^\lambda \frac{\lambda}{2D}\,\mathrm{d}\lambda = \ln D + \ln\frac{\mathrm{d}C}{\mathrm{d}\lambda} - \ln k_2 \tag{1.56}$$

Integrating once more gives

$$C = k_1 + k_2\int_{-\infty}^{\lambda}\frac{\mathrm{d}\lambda}{D}\exp\int_0^\lambda\left[-\frac{\lambda}{2D}\,\mathrm{d}\lambda\right] \tag{1.57}$$

Note that this is an integral equation, not strictly a solution, since C (through D) appears inside and outside the integral sign. To obtain a solution we must start with a trial solution (usually for constant D), evaluate $D(C)$ numerically, integrate numerically to obtain a new C and repeat until C converges. This is the so-called process of iteration. Before the easy accessibility of fast computers, iterative integral methods of solution were widely investigated and used. Since partial differential equations such as equation (1.47) can now be routinely integrated on a computer (even the handheld variety) the latter procedure is now to be preferred (cf Appendix 5).

The initial boundary conditions to which these solutions apply are for the infinite or semi-infinite diffusion couples specified in Fig. 1.14

$$C(x > 0, 0) = C_2 \quad \text{and} \quad C(x < 0, 0) = C_1 \tag{1.58}$$

and

$$C(\infty, t > 0) = C_2 \quad \text{and} \quad C(-\infty, t > 0) = C_1$$

for the infinite couple and

$$C(x > 0, t = 0) = C_2 \qquad C(x = 0, t = 0) = C_1 \tag{1.59}$$

$$C(\infty, t > 0) = C_2 \quad \text{and} \quad C(x = 0, t > 0) = C_1$$

for the semi-infinite couple. These are the most widely used conditions in the diffusion literature so will receive intensive examination in this and other chapters. Other methods of solution where D is a function of t or C are given by Crank.[40]

It is to be noted that if D in equation (1.57) can be approximated by an average constant then after a change of variables and application of the boundary conditions the solution can be written as

$$C = C_2 + \frac{(C_1 - C_2)}{2}\left(1 - \operatorname{erf}\frac{x}{2\sqrt{Dt}}\right) \tag{1.60}$$

where

$$\operatorname{erf}\frac{x}{2\sqrt{Dt}} = \frac{2}{\sqrt{\pi}}\int_0^{x/2\sqrt{Dt}} e^{-\xi^2}\, d\xi$$

where ξ is a dummy, unitless variable.

The error function $y = \operatorname{erf}(z)$ is a tabulated function having the general form of Fig. 1.14*a*. A brief table of values, its series decomposition, its nomographic representation, and its differential and integral properties are given or discussed in Appendix 3. For most calculations the following analytic representation is sufficiently accurate:[41, 42]

$$y = [1 + \pi \exp(-2z^2/3)/4z^2]^{-\frac{1}{2}}; \; z > 0 \tag{1.61}$$

with $y(-z) = -y(z)$ for values of $z < 0$. Its maximum error for all values $-\infty < z < \infty$ is 0·4%. The quantity $1 - \operatorname{erf}(z)$ is often designated as the error function complement, or simply as $\operatorname{erfc}(x)$.

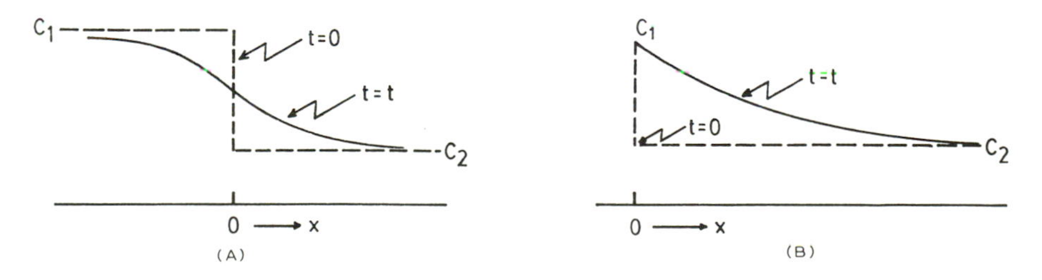

Fig. 1.14 Boundary and initial conditions for (*a*) infinite and (*b*) semi-infinite diffusion couples

That equation (1.60) is a unique solution of equation (1.52) can be verified by checking that conditions (1.58) and (1.59) can be expressed in terms of λ. Thus for the infinite couple

$$C(\lambda = \infty) = C_2 \quad \text{and} \quad C(\lambda = -\infty) = C_1 \tag{1.62}$$

are completely equivalent to relations (1.58).

1.10 MATHEMATICAL AND COMPUTATIONAL EXERCISES

1. Starting with equation (1.52) explicitly carry out all mathematical stages in arriving at the integral solution (1.57).
2. Assuming D = constant explicitly carry out the transformation of integral (1.57) to solution (1.60) through a change of variables and application of boundary and initial conditions (1.58).
3. Letting $z = x/2\sqrt{Dt}$ in relation (1.60) evaluate $y = \text{erfc}(z)$ by numerical integration and compare the results with the table in Appendix 3. Also compare these results with approximation (1.61).
4. Evaluate $y = \text{erfc}(z)$ using the series method and compare (see Appendix 3).
5. Given that $D = 10^{-3}(1 + 100X^2)\text{mm}^2\,\text{s}^{-1}$ for a mole fraction $0 < X < 0{\cdot}1$ determine the diffusion profile as a function of $\lambda = x/\sqrt{t}$ for a couple as in Fig. 1.15 with $X_1 = 0{\cdot}1$ and $X_2 = 0$ at $x = \pm\infty$, respectively. Give a qualitative explanation of the shape obtained. Note that k_2 must be evaluated by normalization on each iteration.
6. A solid state diffusion coefficient is typically $10^{-18}\,\text{mm}^2\,\text{s}^{-1}$ at ambient temperatures. What is the maximum possible size of crystal which can be precipitated in 10^8 years? What geological reasons can you give that many natural crystalline structures have become larger than this in shorter times?
7. Solve the profile in exercise 5 by direct numerical integration of equation (1.52) (cf Appendix 5).

Nb The exercises in this volume have been designed primarily to give practice in machine computation. For a wider range of practice questions, some with solutions, refer to Adda and Philibert[33] and Shewmon.[43]

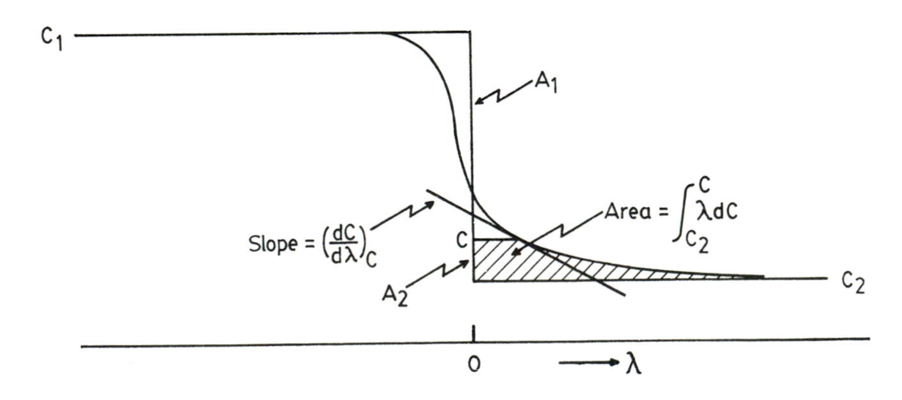

Fig. 1.15 Illustration of the Boltzmann–Matano analysis

1.11 EVALUATION OF TRACER AND CHEMICAL DIFFUSION COEFFICIENTS FROM DIFFUSION PENETRATION DATA

The experimental infinite diffusion couple is widely used in both liquid and solid studies for evaluation of diffusion coefficients (see Chapter 5 for experimental methods). If concentration measurements are very accurate then small differences between the couple extremes can be utilized and the coefficients will be closely constant over the range observed in each couple. This is always the case for tracer or isotope diffusion measurements and occasionally the case for the binary chemical experiments. In such cases the constant diffusion coefficients can be extracted from penetration curves by fitting to solutions of type (1.60). There are two equivalent shorthand methods for achieving the same end. One involves plotting the data on probability paper (see Appendix 3) which converts the error function into a straight line, and the other involves the determination of the intercept of the tangent at the origin on a regular plot of concentration versus penetration distance (see Fig. 1.16). The slope at the origin can be evaluated as the derivative of equation (1.60) or alternatively as $-(C_1 - C_2)/2z$. The equality yields the exact relation

$$D = z^2/\pi t \tag{1.63}$$

The accuracy of such an evaluation is limited by the accuracy to which the intercept z is established by the experimental points near the origin. A least-squares fit via probability paper (cf Appendix 3) or a trial and error fit via the complete penetration curves is obviously to be preferred for other than estimation purposes. Note that the straight line in Fig. 1.16 is a fair approximation to the erfc curve.[44] This line is in fact the first order term in the series expansion (cf Appendix 3).

If D cannot be assumed constant a second integral of equation (1.54) due to Boltzmann[37] can be put to use. Multiplying equation (1.54) by $\mathrm{d}\lambda$ and integrating from C_2 to C as indicated in Fig. 1.15 obtains

$$-\tfrac{1}{2}\int_{C_2}^{C} \lambda\, \mathrm{d}C = D\frac{\mathrm{d}C}{\mathrm{d}\lambda}\bigg|_{C_2}^{C} = D\frac{\mathrm{d}C}{\mathrm{d}\lambda}\bigg|_{C} \tag{1.64}$$

since $\mathrm{d}C/\mathrm{d}\lambda|_{C_2} = 0$. Hence

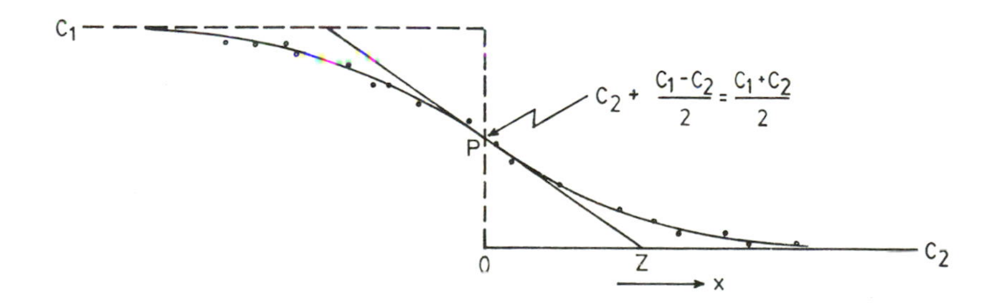

Fig. 1.16 Construction for evaluation of the diffusion distance z

$$D(C) = -\frac{1}{2\mathrm{d}C/\mathrm{d}\lambda|_C}\int_{C_2}^{C}\lambda\,\mathrm{d}C \tag{1.65}$$

which can be obtained from an empirical penetration curve by evaluating the hatched area and slope at each point C. Matano[38] was the first to use this technique for evaluating variable Ds from experiment. Note the potential large source of error as $C \to C_2$ or C_1. Note also, that due to lattice drift associated with the Kirkendall effect (see §2.13), weld markers are not a reliable location of the origin of λ for evaluating $D(C)$. The reliable origin is that which satisfies

$$\int_{C_2}^{C_1}\lambda\,\mathrm{d}C = 0 \tag{1.66}$$

i.e. equal included areas A_1 and A_2 (the mass balance). This is designated as the Matano Condition.

1.12 MATHEMATICAL AND COMPUTATIONAL EXERCISES

1. Prove relation (1.63) in a detailed calculation. The tracer coefficient for Cu^{64} in Cu is summarized by $D_0 = 20\ \mathrm{mm^2\ s^{-1}}$ and $Q = 197{\cdot}2\ \mathrm{kJ\ mol^{-1}}$. Plot a representative diffusion profile from this using an infinite couple at $T = 800°\mathrm{C}$ and time = 1 000 s. Plot this curve on probability paper and comment. From the regular linear plot use equation (1.63) to recover the original value of D.
2. Table 1.2 gives the penetration curve for the diffusion of alloy B into A in an infinite diffusion couple for a time $t = 10^5$ s. Write a program using computer library routines to process this curve as a best fit to a ten power polynomial (or other suitable function) satisfying the Matano condition (1.66). Then using a quadratic form for the unknown $D(X)$, iterate by computer the three coefficients in $D(X)$ to generate a best fit to the derived polynomial via integral expression (1.57) or via the numerical integration of equation (1.54).
3. Write a program for calculating $D(X)$ based on exercise 2 according to the Boltzmann integral (1.65) and record the results for the two cases. Compare these with average results obtained via the intercept method, equation (1.63).

Table 1.2 Penetration curve for Alloy *B* diffusing into *A* for time = 10^5 s

Distance μm	Atom %B	Distance μm	Atom %B	Distance μm	Atom %B
0	28·0	10	6·8	20	0·8
1	28·0	11	5·4	21	0·7
2	28·0	12	4·4	22	0·6
3	27·9	13	3·4	23	0·5
4	27·6	14	2·7	24	0·4
5	26·7	15	2·3	25	0·3
6	25·0	16	1·9	26	0·3
7	21·8	17	1·5	27	0·2
8	16·0	18	1·2	28	0·1
9	9·8	19	1·0	29	0·0

1.13 SOLUTION OF FICK'S EQUATION WITH CONSTANT D BY SEPARATION OF VARIABLES

Solutions of equation (1.49) are sought in one dimension of the form

$$C = X(x)\,T(t) \tag{1.67}$$

Substituting in equation (1.49) yields

$$\frac{1}{T}\frac{\partial T}{\partial (Dt)} = \frac{1}{X}\frac{\partial^2 X}{\partial x^2} = -\lambda^2 \tag{1.68}$$

Since x and t are independent the left- and right-hand sides must be equal to the same constant. This was chosen to be negative ($-\lambda^2$) to assure that the solutions for C converge with time, i.e. each term in the sum of product solutions (equation (1.67))

$$C = C_0 + \sum_{n=0}^{\infty} (A_n \sin \lambda_n x + B_n \cos \lambda_n x) \exp(-\lambda_n^2 Dt) \tag{1.69}$$

converges. The fact that a sum of solutions of a linear equation is also a solution has been used here. This particular form is convenient since many initial conditions of interest ($t = 0$) have analytic Fourier series transforms and A_n and B_n can thus be explicitly evaluated (for a discussion of Fourier series and integral transforms refer to Appendix 2). For example, for diffusion out of a finite plate of thickness h with the surface concentration held at zero (e.g. decarburization of a steel sheet), the solution is

$$C = \frac{4C_0}{\pi} \sum_{j=0}^{\infty} \frac{1}{2j+1} \sin \frac{(2j+1)\pi x}{h} \exp\left[-\left(\frac{(2j+1)\pi}{h}\right)^2 Dt\right] \tag{1.70}$$

The fraction of matter lost at the surfaces $x = 0, h$ can be evaluated by forming the integral

$$I = \frac{1}{C_0 h} \int_0^h (C_0 - C)\,\mathrm{d}x \tag{1.71}$$

or by substituting equation (1.70) in Fick's first equation (1.1) and integrating twice the surface flux up to the time of interest.

For other boundary conditions, solutions related to the Fourier Integral are more appropriate, i.e.

$$C(x, t) = \int_{-\infty}^{\infty} A(\beta) \exp(-\beta^2 Dt) \exp(\mathrm{i}\beta x)\,\mathrm{d}\beta \tag{1.72}$$

where $A(\beta)$ is the Fourier integral transform of $C(x, 0)$. If the initial condition is proportional to a Dirac δ-function (i.e. an infinitely narrow source of unit solute content, see Appendix 3) then equation (1.72) yields

$$C = \frac{\alpha}{2\sqrt{\pi Dt}} \exp\left(-\frac{x^2}{4Dt}\right) \tag{1.73}$$

where α is the number of atoms per unit area in the initial spike. This representation is generally preferred to the error function solution (1.60) for the determination of tracer coefficients since in practice even the initial spike is sufficiently dilute for D to be constant. A simple logarithmic plot of the penetration data allows the extraction of the diffusion constant (see Chapter 5).

This solution can also be used in the construction of the solution for the infinite diffusion couple (cf equation (1.60)). Consider the initial step, as shown in Fig. 1.17 to be made of juxtaposed δ-functions each of which decays independently according to equation (1.73). All that is needed to obtain the overall solution is the addition (integration) of the contributions from each of the spikes.

Another interesting solution based on equation (1.72) has to do with the relaxation of an initial rectangular pulse of the form $C = C_0$ for $-a < x < a$ and zero elsewhere. The Fourier integral transform for this function at $t = 0$ is $\sin \beta a/\beta$ so the solution can be written

$$C(x, t) = C_0 \int_{-\infty}^{\infty} \frac{\sin \beta a}{\beta} \exp(-\beta^2 Dt) \exp(i\beta x)\, d\beta \qquad (1.74)$$

For numerical purposes it is convenient to note that for even functions such as here, we can substitute $\cos \beta x$ for $\exp(i\beta x)$.

1.14 MATHEMATICAL AND COMPUTATIONAL EXERCISES

1. Prove the foregoing statement that for even functions $\cos \beta x$ can be substituted for $\exp(i\beta x)$ in equation (1.74).
2. With respect to solution (1.74) define a typical rectangular concentration spike and compute a set of curves describing its decay with time. Format the solution so a graphical output is recorded.
3. A more realistic version of exercise 2 concerns the decay of a three-dimensional array of localized rectangles, simulating a precipitate dissolution process. This problem can be phrased exactly in terms of Fourier series, which are appropriate to periodic initial conditions. Write down the appropriate Fourier series for one- and three-dimensional cubic arrays.[45] By computational analysis determine whether the one-dimensional model gives an adequate quantitative representation of the three-dimensional dissolution.
4. Evaluate integral I in equation 1.71 both analytically and numerically and present a graph of $I(t)$ *vs* t on a logarithmic plot. Explain the shape.

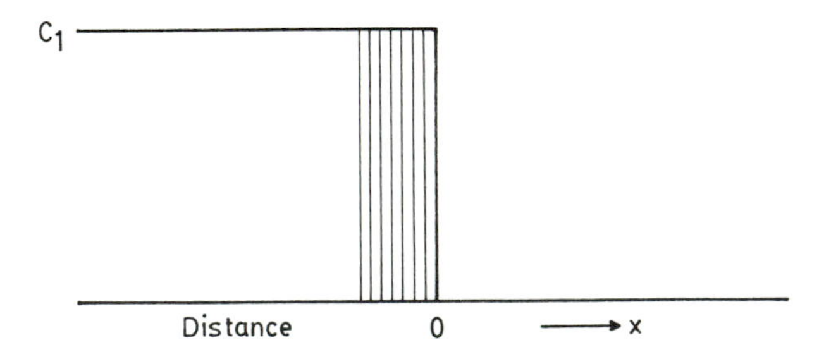

Fig. 1.17 Dirac δ-function decomposition of a step function

1.15 DIFFUSION IN SURFACES, GRAIN BOUNDARIES AND DISLOCATIONS

Surface or line defects on or in condensed solids define sites of weaker bonding for solute or tracer atoms and thus accommodate accelerated diffusion rates. Such defective regions can usually be assigned an approximate share of the total volume of the solid to which an augmented diffusion coefficient can be assigned, the net effect being an enhanced diffusion penetration. Thus diffusion in a defective medium can in principle be described as a complex internal and external boundary value problem. This approach is a more rigorous (though not necessarily more useful) alternative to a statistical formula for evaluating an effective diffusion coefficient such as equation (1.17) due to Hart.[16] Figure 1.18 shows the effect of diffusion enhancement due to grain boundaries in Ag. Since the activation energy for defect diffusion is generally smaller than for bulk diffusion its effect tends to dominate at lower temperatures. Recent comprehensive analyses of defect-enhanced diffusion have been given by Huang *et al.*,[47,48] LeClaire and Rabinovich,[17] and Balluffi.[49]

The simplest case of diffusion of adsorbed insoluble species on a uniform external surface can be treated as a boundary value problem using the Fick equations and concentrations expressed per unit area. In practice even this simplest of problems is complicated by the fact that solid surfaces are themselves defective, containing atomic ledges and kinks in ledges which serve as traps for

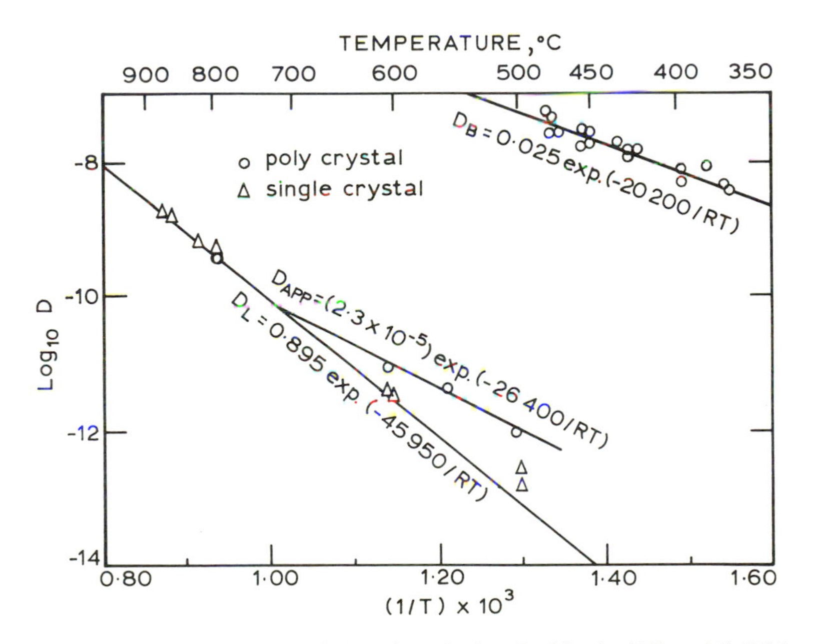

From D. Turnbull, 'Atom Movements', American Society for Metals, 1951, p. 141. With permission

Fig. 1.18 Values of the self-diffusion coefficient for Ag using single crystal and polycrystal samples. After Turnbull[46]

diffusing species (cf § 1.16). An interesting and representative boundary value problem of this nature has to do with the advance of a growth spiral anchored at a screw dislocation as first described by Frank[50] (see also Cabrera and Burton[51]).

When surface adsorbed species are easily soluble in the substrate, such as with isotope tracer diffusion then the process consists of simultaneous surface and volume diffusion, and a more complex boundary value problem must be formulated. This problem is very similar to that for tracer diffusion in a planar grain boundary, which is dealt with in the following paragraphs.

The purpose of this analysis is to describe an experimentally feasible boundary value problem via which a boundary diffusion coefficient can be extracted from the diffusion profiles, given the bulk diffusion coefficient. Ideally, it is considered that an isotopic tracer is deposited from a large volume, well-mixed isotopic vapour mixture with total pressure equal to the vapour pressure of the pure crystal at the temperature of interest. The boundary concentration in the immediate solid surface is thus held at X_0, the gas phase mole fraction. Any chemical isotope effect is ignored so the X_0, or corresponding C_0 boundary value also applies to the grain boundary width, δ. The capillarity effect at the grain boundary groove[52] is ignored. A schematic of the expected evolving concentration profile is given in Fig. 1.19. This boundary value problem has also been deemed applicable to the case where a thin layer of diffusant is deposited on the free surface, but the justification for this inference is by no means clear.[43, 53] The formulation of the two-region boundary value problem of Fig. 1.19 is due to Fisher.[54] It assumes that the concentration in the boundary slab is uniform on the cross-section and that there is an excess divergence of material from a boundary volume element which is proportional to the flux flowing laterally into the bulk at the slab boundary. Thus the bulk concentration distribution on either side of the boundary must be solved via the two-dimensional form of Fick's second equation

$$D\left(\frac{\partial^2 C}{\partial x^2} + \frac{\partial^2 C}{\partial y^2}\right) = \frac{\partial C}{\partial t} \tag{1.75}$$

subject to $C(x, O, t) = C_0$ and $C(x, y > O, O) = O$, where $D\ (= D_V)$ is the lattice or bulk diffusion coefficient evaluated from a profile far from the boundary, and at the slab wall C equals that value in the boundary given by

$$\frac{2D}{\delta}\frac{\partial C}{\partial x} + D_B\frac{\partial^2 C}{\partial y^2} = \frac{\partial C}{\partial t} \tag{1.76}$$

where D_B is the boundary diffusion coefficient. The leading term here expresses the lateral divergence of boundary flux per unit volume into the solid.

A closed computable solution of this problem can be developed using the Fourier sine (F_s) and Laplace transformations (after Whipple;[55] cf Appendix 2) of the relative concentration $C' = C/C_0 \rightarrow \psi$, i.e.

$$\psi(x, \mu, \lambda) = L[F_s\{C'(x, y, t)\}_{y\rightarrow\mu}]_{t\rightarrow\lambda} \tag{1.77}$$

and the inverse transformation

$$C'(x, y, t) = F_s^{-1}\{L^{-1}[\psi(x, \mu, \lambda)]_{\lambda\rightarrow t}\}_{\mu\rightarrow y} \tag{1.78}$$

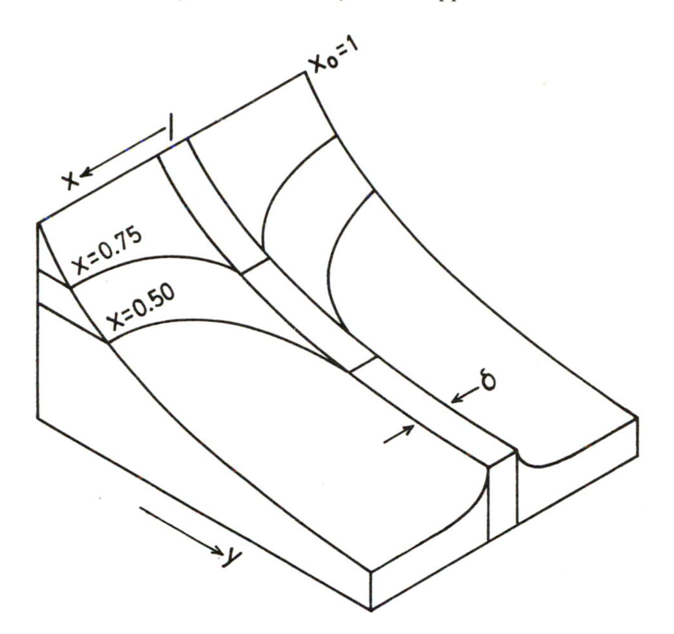

Fig. 1.19(*a*) Schematic of diffusion profile representing enhanced diffusion along a grain boundary

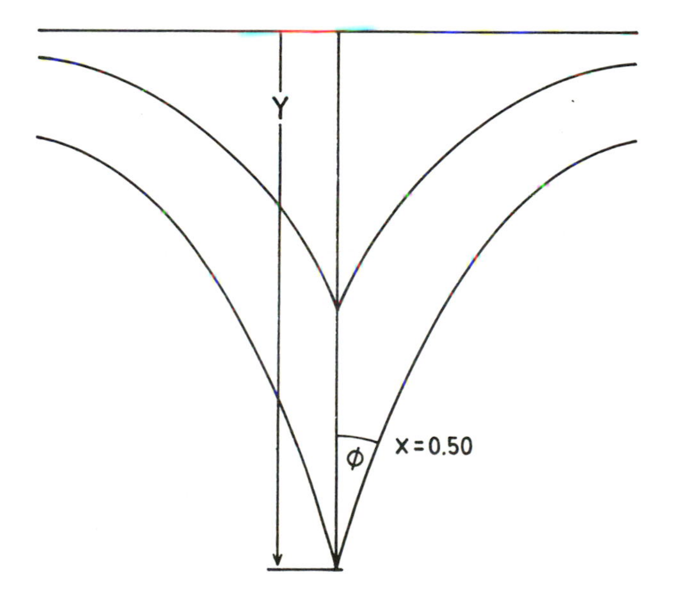

Fig. 1.19(*b*) Isoconcentration line for enhanced grain boundary diffusion defining the observable parameters Y and ϕ

The first application of the operators in relation (1.77) transforms equation (1.75) to

$$\frac{\partial^2\psi}{\partial x^2} - \alpha^2\psi = -\frac{\mu}{\lambda} \tag{1.79}$$

where

$$\alpha^2 = \mu^2 + \lambda/D \tag{1.80}$$

and defining $\Delta = D_B/D$ equation (1.76) transforms to

$$D_B\frac{\partial^2\psi}{\partial x^2} - \frac{2D}{\delta}\frac{\partial\psi}{\partial x} - (\Delta - 1)\lambda\psi = 0 \tag{1.81}$$

with

$$x = +\delta/2 \tag{1.82}$$

The solution of equation (1.79) is

$$\psi = \frac{\mu}{\lambda\alpha^2} + h\mathrm{e}^{-\alpha x} \tag{1.83}$$

To obtain h, ψ is substituted into equation (1.81) whence

$$h = \frac{\mu(\Delta - 1)\mathrm{e}^{\alpha\delta/2}}{\alpha^2\left(D_B\mu^2 + \dfrac{2D\alpha}{\delta} + \lambda\right)} \tag{1.84}$$

Thus the complete solution in transformed coordinates is

$$\psi(x, \mu, \lambda) = \frac{\mu}{\lambda\alpha^2} + \frac{\mu(\Delta - 1)\exp[-\alpha(x - \delta/2)]}{\alpha^2\left(D_B\mu^2 + \dfrac{2D}{\delta}\alpha + \lambda\right)} \tag{1.85}$$

Since most mainframe computer libraries give access to L^{-1} and F_s^{-1} routines the solution in concentration space C' can be easily evaluated numerically. Note, however, that application of the operator $F_s^{-1}L^{-1}$ to the leading term leads via tabulated transforms to erfc $y/2\sqrt{Dt}$ (cf Appendix 2), which represents the solution in the absence of a boundary. A generalization of the Whipple equations, which allows for alloy partition to the boundary, and the corresponding solutions has been given by Young and Funderlic.[56]

A number of transparent approximate solutions to this boundary value problem have been given, that of Whipple[55] (see also Levine and MacCallum[57]) being the most rigorous. Setting

$$\eta = y/\sqrt{Dt} \qquad \xi = \frac{x - \delta/2}{\sqrt{Dt}} \qquad \beta = (\Delta - 1)\frac{\delta}{2\sqrt{Dt}} \tag{1.86}$$

he obtains via the inverse transformation $\psi \rightarrow C/C_0$ the integral solution

$$\frac{C}{C_0} = \operatorname{erfc}\frac{\eta}{2} + \frac{\eta}{2\sqrt{\pi}}\int_1^{\infty}\frac{\mathrm{d}\sigma}{\sigma^{3/2}}\exp\left(-\frac{\eta^2}{4\sigma}\right)\operatorname{erfc}\frac{1}{2}\left(\frac{\sigma - 1}{\beta} + \xi\right) \tag{1.87}$$

which is applicable provided the normally attainable condition $\eta \ll D_B/D$ for all interesting values of y is met.[53] As already stated, the first term represents the volume diffusion solution in the absence of a boundary so the second term containing a definite integral represents the internal volume accumulation of solute due to the boundary short circuit. Note that at typical high temperatures the integral accumulation within the boundary is negligible due to its infinitesimal lateral extent and the efficient draining of the short-circuited material into the adjacent solid through its large surface area. Fig. 1.20 shows a schematic contour penetration diagram where the distance Y to a particular contour and the corresponding angle of intersection with the boundary ϕ are identified. Given D and C_0, measurements of either of these quantities can be analysed to yield the unknown $D_B\delta$. While LeClaire[53] and Borisov *et al.*[58, 59] discuss various approximate analytical and graphical procedures, an effective and accurate procedure in this age of fast computers would be simply to iterate a set of profiles corresponding to C_0 and D_D over trial values of $D_B\delta$ aiming for a best fit to the observed contours (see § 1.16, exercise 2). Alternatively, when $\beta > 1$ (LeClaire states that experiments with $\beta > 10$ show the best discrimination) Whipple's asymptotic solution for the second term in equation (1.87) is

$$\frac{C'}{C_0} = \frac{1{\cdot}159}{(\eta\beta^{-1/2})^{2/3}} \exp\left\{-0{\cdot}473(\eta\beta^{-1/2})^{4/3} + 0{\cdot}396\,\frac{(\eta\beta^{-1/2})^{2/3}}{\beta}(1-\beta\xi)\right\} \tag{1.88}$$

and this can be solved iteratively for $D_B\delta$.

The analogous boundary value problem for short-circuit diffusion along a single edge dislocation represents the latter as a 'pipe' of radius a and uniform high diffusivity $D_p \gg D$ with the coordinate x replaced by the lateral radius vector r. The transformed solution analogous to equation (1.85) is[60, 17]

$$\psi(r,\mu,\lambda) = \frac{\mu}{\lambda\alpha^2} + \frac{\mu(\Delta-1)}{\alpha^2\left[D_p\mu^2 + \lambda + \dfrac{2D\alpha}{a}\dfrac{K_1(\alpha a)}{K_0(\alpha a)}\right]} \cdot \frac{K_0(\alpha r)}{K_0(\alpha a)} \tag{1.89}$$

where K_0 and K_1 are modified Bessel functions of the second kind and order 0 and 1, respectively (cf McLachlan[61]). As before, this is machine computable using standard routines. However, single dislocation experiments are not really feasible since the amount of tracer short-circuited is less than 10^{-7} that in the grain boundary case. The solutions for arrays of dislocations given by LeClaire and Rabinovich[17] are accordingly more relevant to experimental investigations (see also LeClaire[18]). These analyses allow the extraction of key parameters such as dislocation density from diffusion profiles as in Fig. 1.20. Here we note the excess penetration known as dislocation tails.

While early reviews found rather poor closure between theory and experiment[43] current assessments[18] based on increasingly accurate experiments are more salutory.

Because of current interest in low-temperature diffusion behaviour in relation to the integrity of multilayered thin film devices, Balluffi and co-workers[47, 48] have developed an analysis for extracting grain boundary diffusion data from very sensitive measurements based upon surface accumulation. At sufficiently low temperatures (say much less than half the melting temperature) the leakage into the bulk is minimal so the Whipple-type analyses are inappropriate. These authors

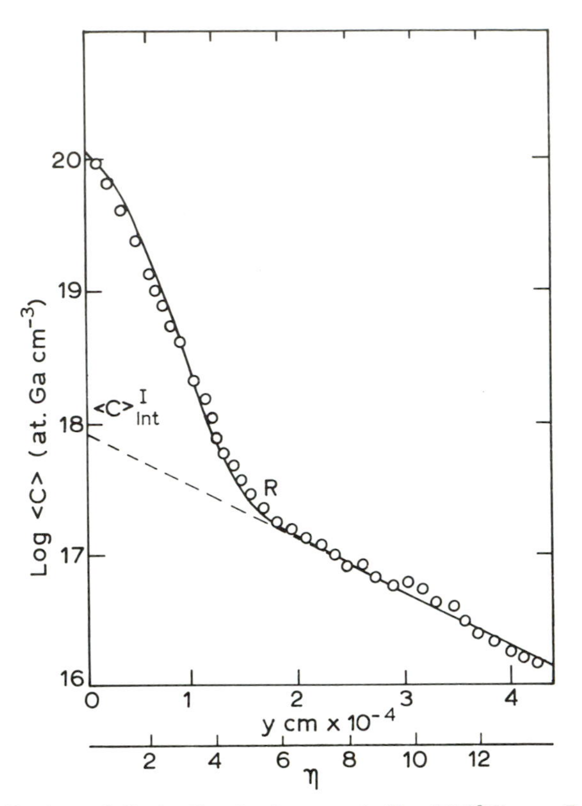

Fig. 1.20 Diffusion of Ga in Ge single crystal. T = 713°C, t = 5·784 × 10^5 s. Circles: experimental data; full curve; theoretical. After LeClaire and Rabinovich[17]

have presented mathematical analyses for 'constant source' and 'instantaneous source' boundary conditions and have reported experiments in the alloy gold–silver. In general, grain boundary diffusion experiments are becoming increasingly accurate.[62]

The Hart formula (1.17) which is based on a segmented random walk argument (cf §2.3), can be used to obtain experimental estimates of short-circuit tracer diffusion coefficients through analysis of data such as Fig. 1.18. If one has an estimate of the relative defect volume f (grain boundaries and (or) dislocations) and D (from single crystal or high temperature experiments, say) then a fit of equation (1.17) will yield a value for the mean defect diffusion constant (cf § 1.20).

All of the foregoing treatments are complicated by a chemical process, for the model must account for adsorption at as well as diffusion along defects. However, as a first approximation one can assign a segregation coefficient, k, to the boundary, and replace D_B or D_P by kD_B or kD_P in the foregoing analyses. If, for example, $k > 1$ then the short circuit driving force and corresponding flux is increased relatively. As regards the Fisher–Whipple-type solutions the substitution of kD_B for D_B represents a rigorous correction to the boundary condition (1.76) for the bulk leakage provided the process is slow enough that local equilibrium can be sustained. The same modification to the Hart equation can also be justified, for boundary segregation is equivalent to an increased residence time in the defects and thus to a proportionately larger boundary diffusion coefficient relative to the bulk (see also Young and Funderlic[56]).

The possibility of a more rigorous computation of bulk effects, given D, D_P, D_B and the distribution of defects has received considerable recent attention from applied mathematicians. In particular, methods applicable to the analogous problem of heat transfer in inhomogeneous materials have been explored by Aifantas, Hill and co-workers.[63, 64, 65, 66, 67] In their methodology a concentration C_1 and flux J_1 is assigned to non-defective regions and a concentration C_2 and flux J_2 to defective regions with the mean total of $C = C_1 + C_2$. The mass balances are written as

$$\frac{\partial C_1}{\partial t} + \nabla \cdot \boldsymbol{J}_1 = \dot{m}_1 \qquad (1.90)$$

and

$$\frac{\partial C_2}{\partial t} + \nabla \cdot \boldsymbol{J}_2 = \dot{m}_2 \qquad (1.91)$$

where $\dot{m}_1$ and $\dot{m}_2$ are source or sink terms corresponding to mass interchange between regions subject to

$$\dot{m}_1 + \dot{m}_2 = 0 \qquad (1.92)$$

It is then postulated that the fluxes in the two regions interact in general according to

$$\boldsymbol{J}_1 = -D_{11}\nabla C_1 - D_{12}\nabla C_2 \qquad (1.93)$$

and

$$\boldsymbol{J}_2 = -D_{21}\nabla C_2 - D_{22}\nabla C_2 \qquad (1.94)$$

where the Ds are functions of concentration, and that the source term $\dot{m}_1 = -\dot{m}_2$ is linear in the concentrations (i.e. a first order chemical reaction)

$$\dot{m}_1 = -\dot{m}_2 = -\kappa_1 C_1 - \kappa_2 C_2 \qquad (1.95)$$

The κs necessarily contain information on volume fractions and equilibrium segregration to the defects. With constant coefficients (e.g. $D_{11} = D$ and $D_{22} = D_B$) the differential equations are

$$\frac{\partial C_1}{\partial t} = D_{11}\nabla^2 C_1 + D_{12}\nabla^2 C_2 - \kappa_1 C_1 - \kappa_2 C_2 \qquad (1.96)$$

and

$$\frac{\partial C_2}{\partial t} = D_{21}\nabla^2 C_1 + D_{22}\nabla^2 C_2 + \kappa_1 C_1 + \kappa_2 C_2 \qquad (1.97)$$

These are very similar in form to those for dual diffusion-reaction in ternary systems as described by Toor.[68] Hill,[65] generalizing on Hart[16] has developed a two-path random walk model with probabilistic path interchanges which leads to the general form of these coupled differential equations and appropriate signs for the coefficients.

Exploration of the solutions of such equations for specific boundary and initial conditions[63, 64, 65, 66, 67] demonstrates that they are not in general decomposable into a total concentration profile indexed by a single, constant, volume-weighted diffusion coefficient. It would appear therefore that such an ideal limit as described by Hart can be realized accurately only in the tracer diffusion case where equilibrium segregation to the defects and boundary or pipe wall activation barriers can be ignored.

There are other complications which appear in the case of alloy short circuit diffusion. The possibility of a grain boundary Kirkendall effect (cf §2.12) can lead to convective intrusions and extrusions at grain boundaries in films.[69] There is also the currently researched phenomenon of diffusion induced grain boundary migration (DIGM).[70, 71] Here, an externally supplied alloying element preferentially diffusing into a substrate along grain boundaries (e.g. Zn into Al) builds up a lateral chemical or elastic driving force which causes the boundary to migrate. The process is closely related to the phenomenon of discontinuous precipitation[72, 73] (see also §12.8). Back-calculated boundary diffusion coefficients are found to be exceptionally large, suggesting that a moving boundary has a looser structure than a stationary one (see, however, Balluffi[49]).

1.16 MATHEMATICAL AND COMPUTATIONAL EXERCISES

1. Based on Appendix 2 write down the explicit operations which lead from equations (1.75) and (1.76) to equation (1.85) (cf Whipple[55]).
2. Write a program which generates C/C_0 as a function of the unitless quantities η, ξ and β and format the output so as to print out a set of isometric surfaces at representative values of β ($0 < \beta < 3$).
3. Reformat exercise 2 to yield contour diagrams of C/C_0 at representative values of β.
4. Repeat exercise 3 using library programs for the inversion of equation (1.85).
5. Discuss the accuracy and range of applicability of solution (1.88) by comparing it with the results of exercise 4.
6. Refer to the Hart formula (1.17) and its derivation in §2.3. Using the following estimated data for Cu^{64} diffusion in Cu

$$D_v = 20 \exp(-197\,000/RT) \qquad D_B = 10^{-6}(T/1\,000)^2 \text{ mm}^2 \text{ s}^{-1}$$

and assuming that the metal is severely cold worked so that $f \simeq 10^{-4}$, graph the average D between 25 and 1083°C on an Arrhenius plot ($\ln D$ *vs* $1/T$). Is it realistic that f would remain constant in the experiments to evaluate D? At approximately what temperature does defect diffusion become dominant?

1.17 DIFFUSION DEPLETION, DOPING, TRAPPING, AND SURFACE CONDITIONING

Some of the most important and common applications of diffusion theory to industrial processes in the first approximation require only elementary solutions of the binary diffusion equations such as those already presented. Important processes include diffusive escape from containers, membrane mass transfer, doping of semiconductors, surface conditioning such as carburizing, nitriding, boriding, aluminizing, chromizing, oxidation and sulphidation, passivation, liquid–solid mass transfer and liquid–liquid mass transfer. Some of these, however, involve phase changes, and the latter two are further complicated by convective mixing. Accordingly representative examples of these more complicated processes will be dealt with in subsequent sections of this chapter.

The depletion or escape processes include permeation from containers, rigid membrane mass transfer, diffusion-controlled surface evaporation and carburizing. All of these can be generally understood in terms of one-dimensional, single-phase diffusion with a constant (but temperature-dependent) diffusion coefficient. Steady state formulas like equations (1.39) and (1.43) are appropriate to permeation and membrane mass transfer while parabolic error function solutions within a semi-infinite boundary condition are appropriate to evaporation and carburizing (cf Fig. 1.14(*b*)). We have constructed predictive exercises which for each demonstrate the elementary principles and have appended them to this section. As stated, these elementary considerations are appropriate to a first approximation, but in practice matters are often more complicated.

Consider as an example the problems of permeation of hydrogen and its isotopes. Since the fuel in first generation fusion reactors will be a 50:50 mixture of the hydrogen isotopes, deuterium and tritium, the development of techniques for storing, handling and purifying the isotopes becomes of paramount importance. Tritium presents special problems because it is very rare and expensive, and dangerously radioactive. It will thus be very important to precisely monitor the inventory of tritium, particularly that part which diffuses into vacuum wall and blanket materials and to recover this wherever possible by out-diffusion or degassing treatments. Finally, hydrogen permeated into many constructional materials causes embrittlement, a fact which militates against the use of the highly favoured austenitic stainless steel for the walls of power reactor vacuum vessels.

Hydrogen diffuses exceptionally fast in many materials; bcc iron and palladium are well-known examples (see Fig. 1.12[35]). Furthermore, it is highly subject to trapping at lattice defects such as vacancies, dislocations and trace impurity atoms. This has a marked effect on permeation rates and retention. Fig. 1.21 represents schematically the potential energy configuration which leads to trapping. If X_H, X_T and X_P are the mole fractions of hydrogen atoms, trap and hydrogen-trap pairs, all assumed to be dilute, the mass law for formation of pairs can be written as

$$\frac{X_P}{(X_H - X_P)(X_T - X_P)} = K = \exp(-\Delta G/RT) \tag{1.98}$$

where $X_H - X_P$ and $X_T - X_P$ represent those fractions available for reaction and ΔG is the free energy of formation based on Henrian activities (see Appendix 1). X_H and X_T are the gross quantities which would be measured in a chemical analysis. The solution of equation (1.98) is

$$X_P = \tfrac{1}{2}\{(X_T + X_H + 1/K) - \sqrt{(X_T + X_H + 1/K)^2 - 4X_TX_H}\} \tag{1.99}$$

where the negative root is chosen to ensure that $X_P < X_T, X_H$. For shallow traps

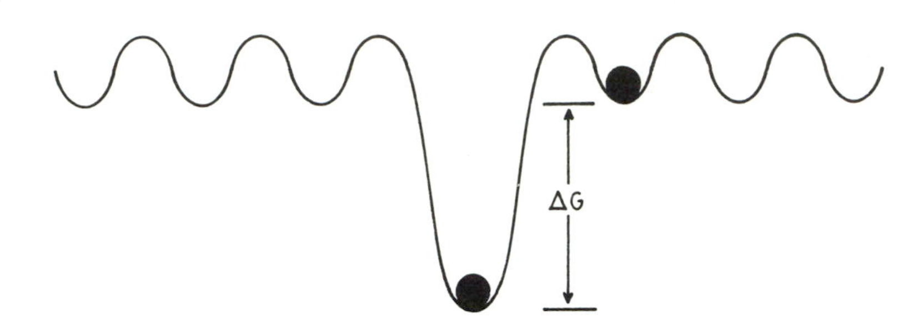

Fig. 1.21 Potential representation of a solute trap in a crystal

ΔG is negative and small, so $1/K \rightarrow 1$, and it can be concluded for the dilute case that $X_T, X_H \ll 1/K$. Then by binomial expansion of the radical in equation (1.99) we obtain

$$X_P \cong X_T X_H K \tag{1.100}$$

Thus

$$\frac{dX_P}{dt} \cong X_T K \frac{dX_H}{dt} \tag{1.101}$$

Now the one-dimensional mass balance for a non-conserved quantity like C_H can be written as

$$\frac{\partial J_x}{\partial x} + \frac{\partial C_H}{\partial t} = \alpha \tag{1.102}$$

where α is the source of H production in the diffusion field and J_x is the flux defined in the absence of a source. Clearly α is represented by $-dX_P/dt$ (actually a sink) so from equation (1.101) and the first Fick equation (1.1) is obtained

$$\frac{\partial C_H}{\partial t} \cong \frac{D_H}{1 + X_T K} \frac{\partial^2 C_H}{\partial x^2} \tag{1.103}$$

Accordingly, the effective coefficient and the penetration distance are reduced by trapping. To calculate the relative amounts of hydrogen absorbed care must be taken in defining the boundary conditions. Here it is assumed that a semi-infinite diffusion couple, as in Fig. 1.14*b*, is being considered supplied by a fixed partial pressure of H_2 gas at the origin. Thus if the boundary value of C_H is C_H^0 without trapping it will, according to equation (1.100), be $C_H^0(1 + X_T K)$ with trapping. The amount absorbed will be proportional to the product of the boundary value and the diffusion distance so that the ratio of amounts absorbed, with and without trapping, will be

$$R = \frac{(1 + X_T K)}{\sqrt{1 + X_T K}} = \sqrt{1 + X_T K} \tag{1.104}$$

Thus there is a net accumulation due to trapping. In a dehydrogenation treatment at higher kT (cf equation (1.98)) K will be reduced, so the diffusion coefficient will more closely approach the no-trapping value. Since the hydrogen is nearer the surface in the trapped case it should therefore exhibit an accelerated depletion rate. Experiments on trapping of vacancies and other point defects are discussed by Wichert *et al.*[74] and Howe and Swanson[75].

As a second example consider the carburization process for selectively surface hardening a steel. This is applied to many machine parts, including gear teeth. In the process of pack carburizing, the part, which contains a core concentration ~0·2% C by weight, is placed in a sealed container in contact with a mixture of cast iron chips, which provide the excess carbon, and $BaCO_3$ which reacts to produce the carburizing gas mixture of CO and CO_2. This arrangement provides a surprisingly constant surface potential and source of carbon which quickly stabilizes at temperature (~ 800°C). The diffusion problem is again very close to that identified for the semi-infinite diffusion couple (Fig. 1.14*b*). After diffusion for periods of hours to a day at temperatures between 700 and 900°C the part is removed and quenched in water or oil, which causes the carbon-rich surface region, which reaches about 0·8% C, to harden significantly. The zone of hardening or case depth is determined by the diffusion length and can be represented to reasonable accuracy by a formula like equation (1.63). Indeed the formula due to Harris[76]

$$\text{Case Depth (mm)} = 13{\cdot}4\sqrt{t}\exp(-8\,572/T) \tag{1.105}$$

where t is in seconds and T is in K is widely used in industry. If more precise predictions are desired, one must, for a given gas composition, calculate the surface boundary value for carbon as a function of alloy content (the carbon activities are affected) and integrate the diffusion equation with a variable diffusion coefficient (it is particularly sensitive to carbon and chromium levels). Further complications are introduced when the large-scale gas carburization process is employed. Here the furnace and loads are immense so the surface gas mixture takes many hours to reach the stable, set value and produce, say, 0·8% C. A calculation must accordingly utilize a diffusion solution which allows for the time variation of surface carbon content. A useful integral solution, replacing the error function profile is[45]

$$C - C_c = \frac{x}{2\sqrt{\pi D}}\int_0^t \frac{e^{-x^2/4D(t-t')}}{(t-t')^{\frac{3}{2}}}[C_s(t') - C_c]\,dt' \tag{1.106}$$

where C_c is the core carbon concentration and $C_s(t')$ is the surface concentration as a function of time. This latter must be determined empirically for a given installation.

To increase productivity, engineers press for higher diffusion rates. One way is to increase the temperature, but this has disadvantages in terms of furnace costs, poor surface finish and heat treatment effectiveness. Another way is to increase the carbon potential via a richer gas mixture — the so-called 'boost-diffuse' method. With the higher potential more carbon reaches a given depth in a given time. Now, however, the carbon level, of 1·5% say, is much too high for an effective, non-brittle case after heat treatment. Thus the concentration profile is subsequently levelled for a short time at lower potential where some carbon diffuses back out, or in vacuum where the surface flux goes to zero. All of these variations can be easily accommodated in a computer program which numerically integrates the non-steady diffusion equation with variable boundary conditions.[77] The nitriding process and the doping processes for semiconductor device production are very similar to carburizing.

Decarburizing processes are also important, most often in a negative sense. Hot iron and steel must often be handled in oxidizing atmospheres such as ambient air where the carbon tends to react to CO and CO_2 and be depleted from the surface of the metal. The mathematical treatment of the out-diffusion process is of course similar to that for inflow. There are very similar processes of diffusion depletion of liquid materials, but these are complicated by circulation, so mass transfer rates, although involving diffusion, tend to be accelerated. In the next few sections we turn to diffusion processes which simultaneously involve phase change and/or circulation.

1.18 MATHEMATICAL AND COMPUTATIONAL EXERCISES

1. Justify relation (1.105) using the data of Fig. 1.6 and diffusion theory. Hint: refer to equation (1.63) and its derivation.
2. Verify the validity of equation (1.106) by comparison with a direct integration of Fick's second equation with varying $\mathrm{d}C_s/\mathrm{d}t \rightarrow C_s$.
3. From the data of Fig. 1.12 estimate the maximum temperature allowable for tritium stored at 100 atm in 1 cm thick walled, 1 m diameter vessels of ferritic and austenitic stainless steels to avoid a leakage of more than 1 millicurie in 1 hour at temperature. This is an important question *vis-a-vis* the shipping of tritium by air or truck where accident may lead to overheating of the vessel (1 g of tritium ≡ 20 000 curies).
4. Using diffusion solution (1.70) and the data of Fig. 1.7 calculate how long it will take for the carbon to be 90% depleted from a 1 mm thick plate of bcc iron in an oxidizing atmosphere at 700°C. An oxidizing atmosphere may be assumed to reduce the carbon concentration at the surface to zero.
5. It has been suggested that hydrogen isotopes can be mass separated using a series of metallic membranes via the differential diffusion rates of protium, deuterium and tritium. Assuming that a $1/\sqrt{\mathrm{m}}$ rule is valid for the relative permeation data for hydrogen of Fig. 1.12 assess this proposition in terms of number of stages required for the effective separation of 50:50 deuterium and tritium to the 90:10 separation level.
6. Refer to equation (1.106) which deals with the diffusion profile in doping when the surface concentration is a function of time. Let a large carburizing furnace deliver a surface carbon percentage according to the measured relation

   $$C_s(t) - C_c = 0{\cdot}8(1 - \mathrm{e}^{-10^{-4}t})$$

   Taking data from Fig. 1.6 assume that the surface temperature rapidly reaches a constant value of 850°C. Plot a set of carbon penetration curves between zero and saturation time ($\sim 3{\cdot}10^4$ s). Compare these on a graph with curves obtained assuming that $C_s - C_c$ reaches 0·8 instantly. Comment on the relative shapes from an applications point of view.
7. D_c in fcc iron (austenite) increases rather rapidly with carbon concentration. How would this affect the shape of the penetration curves in exercise 6? Can you describe a computation scheme where this effect might be taken into account?

8. A special case of the 'boost-diffuse' method described above involves a diffuse cycle in vacuum. This has the effect of adjusting the surface *flux* to zero. Why? Write a program for evaluating the profiles in this cycle as a numerical integration of Fick's second equation with the initial condition provided by the profile at the end of the boost cycle.
9. The diffusion coefficient for B in Si is approximately independent of dopant concentration in the range [B] = 10^{17} to 10^{19} atom cm^{-3} at 1200°C where $D_B \simeq 10^{-12}$ cm^2 s^{-1}.[78] One method of doping the Si is to expose its surface to a constant activity of (B ≡ boron) (*a*). Calculate the depth at which the dopant concentration reaches 10% of the surface concentration after diffusion times of 1 h and 2 h at 1200°C, and plot the concentration profiles achieved over these distances.

   An alternative method of doping is to implant the B in a subsurface region of the Si, using ion bombardment. The initial distribution of implanted B can be approximated as being uniform within the bombardment affected zone and zero elsewhere. Subsequent diffusion of the B out of this 'extended source' is described by

$$C = \tfrac{1}{2}C_0\left\{\operatorname{erf}\frac{d-x}{2\sqrt{Dt}} + \operatorname{erf}\frac{d+x}{2\sqrt{Dt}}\right\}$$

   where C_0 is the initial concentration of B within a zone of depth d (*b*). Putting $d = 0{\cdot}75\ \mu$m, calculate the value of C_0 required to yield a surface concentration of 7×10^{18} atom cm^{-3} after diffusion times of 1 h and 2 h at 1200°C. For these conditions, calculate the resultant concentration profiles.
10. This describes a simple experiment to estimate a liquid state diffusion coefficient, that of OH^- ion in H_2O. Reference is made to Fig. 1.22. Prepare a rich gel containing a small amount of OH^- indicator (phenolphthalein), half fill an ordinary test tube and chill. At time zero add a *very* dilute solution of NaOH to the other half of the container. Measure the advance of the reaction front l (the pink region) as a function of time for about four hours taking care to correct for the contraction meniscus. Plot a curve of l *vs* $\sqrt{t}$ and from the

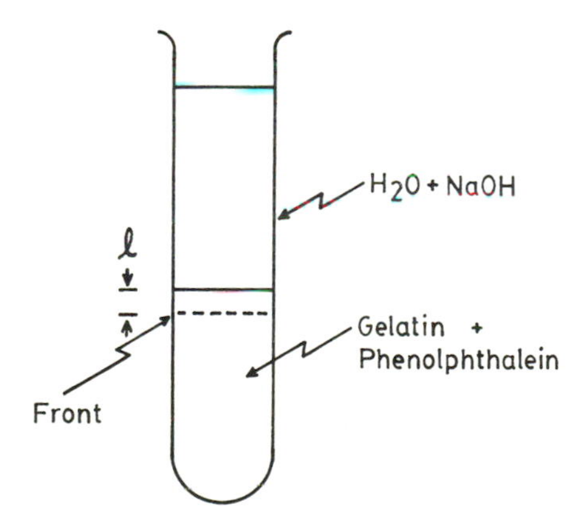

Fig. 1.22 Illustration of an experiment to measure the diffusion rate of OH^- ion in water

slope of the curve obtain an estimate of the diffusion coefficient. Why is a gel used? Why do we use half-and-half proportions? What kind of an average diffusion distance does l represent? Why must the NaOH solution be very dilute? How would you carry out an accurate experiment?

1.19 STEADY STATE, FORCED VELOCITY SOLIDIFICATION

The controlled solidification of a binary solution represents an excellent example of applied diffusion theory and relates to two very important industrial processes, zone refining and zone levelling. Zone refining in the first instance provided the very pure semiconductor materials required for device manufacture. A bar of impure material is placed in a ceramic boat and a narrow radiant-heating coil is cycled unidirectionally from end to end so as to produce and force from end to end a narrow molten zone. In the early stages of the process, after an initial transient, the trailing edge of the molten zone undergoes steady state solidification as described by the phase diagram parameters and profiles in Figs. 1.23*a* and *b*. Here it is assumed that the impurity lowers the melting point of the solvent and is separated from the liquid according to a constant partition ratio $k < 1$ (Fig. 1.23*a*).[79] The temperature profile and concentration profile are indicated in Fig. 1.23*b* and these can be mapped onto the phase diagram by eliminating the distance x as shown. In this, as in the subsequent examples, we will assume that local equilibrium exists at the interface, which is equivalent to saying that the interface chemical reaction offers no resistance to the process (see however §12.10).

Anticipating that the interface may not be flat, we refer to the three-dimensional diffusion equation

$$\frac{\partial C}{\partial t} = D\left(\frac{\partial^2 C}{\partial x^2} + \frac{\partial^2 C}{\partial y^2} + \frac{\partial^2 C}{\partial z^2}\right) \tag{1.107}$$

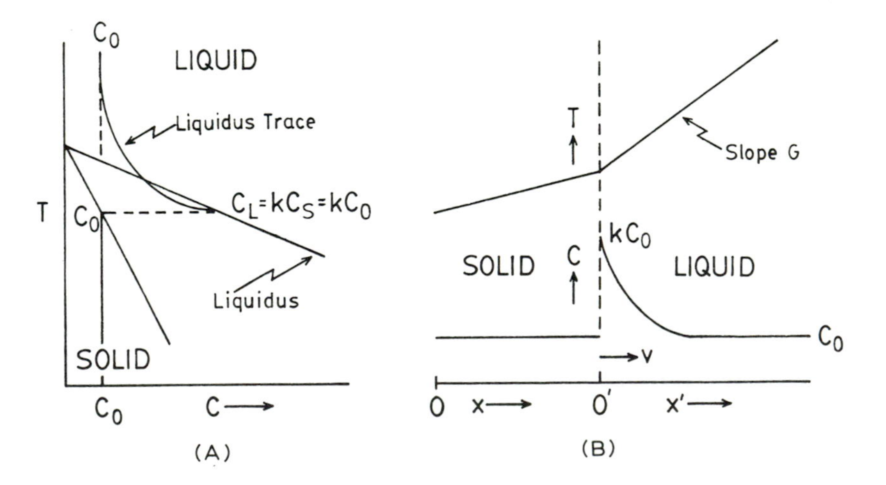

Fig. 1.23 Parameters relevant to forced velocity binary solidification: (*a*) phase diagram; (*b*) thermal and concentration distributions

which applies in the laboratory frame of reference and transform this to a steadily moving frame at velocity v in the x-direction, specified by the coordinates (x', y', z')

$$y = y'; \qquad z = z'; \qquad x = x' + vt \tag{1.108}$$

this leads to the differential transforms

$$\partial x' = -v\partial t; \qquad \partial x^2 = \partial x'^2 \tag{1.109}$$

so equation (1.84) becomes

$$-v\frac{\partial C}{\partial x'} = D\left(\frac{\partial^2 C}{\partial x'^2} + \frac{\partial^2 C}{\partial y'^2} + \frac{\partial^2 C}{\partial z'^2}\right) \tag{1.110}$$

This general steady state equation will be later employed in some two- or three-dimensional problems, but for the moment we restrict attention to a planar interface where the y' and z' variations vanish. Equation (1.110) becomes the ordinary differential equation (dropping the prime)

$$\frac{\mathrm{d}^2 C}{\mathrm{d}x^2} + \frac{v}{D}\frac{\mathrm{d}C}{\mathrm{d}x} = 0 \tag{1.111}$$

which according to the boundary conditions of Fig. 123 has the solution

$$C = C_0 + C_0\frac{(1-k)}{k}\exp\left(-\frac{v}{D}x\right) \tag{1.112}$$

In Fig. 1.23*b* the area lying above the line $C = C_0$ represents the amount of solute that in the early stages of the process is carried end to end in a single pass. For the later stages a more elaborate time-dependent theory is required. This will clearly depend on C_0, v and k which will determine how many passes must be taken to achieve the desired degree of removal of the impurity.[80]

In Fig. 1.23*a* we have sketched a possible liquidus trace which dips inside the two-phase region. Since this implies the possibility of precipitation in advance of

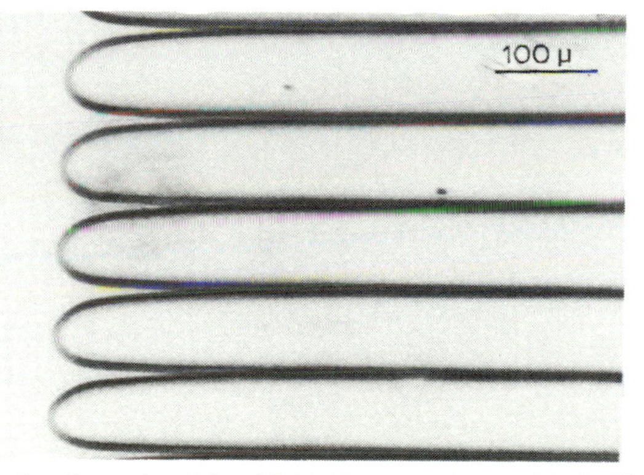

Reprinted with permission from *Acta Met.*, **32**, D. Venugopalan and J. S. Kirkaldy, p. 893, copyright 1984, Pergamon Press Ltd

Fig. 1.24 Cellular instability of binary solid–liquid interface when *v/G* is large. After Venugopalan and Kirkaldy[81]

the solid–liquid interface we may conclude that the interface enters a regime of instability, i.e. non-planar interfaces as in Fig. 1.24 occur. The marginal state in accord with Fig. 1.23 is when the liquidus trace and the liquidus coincide, which is given by the equality in[79]

$$\frac{v}{G} \geqslant \frac{Dk}{|m|C_0(1-k)} \tag{1.113}$$

where G is the temperature gradient in the liquid and m is the liquidus slope (see exercise 2, § 1.28). The inequality defines the unstable region, which clearly must be avoided in a zone purification process for with non-uniformity of interface a non-uniform solid accrues. This relation accordingly restricts tne zone purification process to low velocities and/or high thermal gradients. The unstable process, which produces some remarkable interface shape patterns (Figs. 1.24 and 12.31), is of great interest to the general theory of pattern formation. We will return to this in § 12.9.

1.20 ONE-DIMENSIONAL PARABOLIC PRECIPITATION

Wagner constructed a large number of useful diffusion solutions which can be expressed in terms of the 'parabolic' parameter $\lambda = x/\sqrt{t}$ (compare the infinite and semi-infinite diffusion couple, Fig. 1.14). Grain boundary precipitation represents an interesting process in heat treatment technology, and can be approximated by a one-dimensional parabolic diffusion model (cf Jost[82]). Consider alloy composition C_0 quenched to temperature T as indicated in Fig. 1.25, which defines boundary value concentrations for the parent (γ) and product (α) phases.

The solution of the time-dependent diffusion equation is

$$C = C_0 + (C_0^{\gamma\alpha} - C_0)\left(1 - \operatorname{erf}\frac{x}{2\sqrt{Dt}}\right) \tag{1.114}$$

Since the intercept $C_0^{\gamma\alpha}$, as defined in Fig. 1.25*b*, is not known, another condition is needed. This is the requirement that at the precipitate front ($x = \eta$)

$$C(\eta) = C^{\gamma\alpha} \tag{1.115}$$

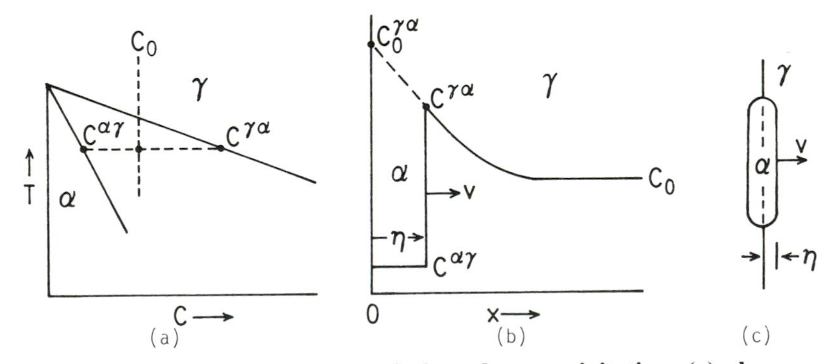

Fig. 1.25 Parameters relevant to grain boundary precipitation: (*a*) phase diagram; (*b*) concentration distribution; (*c*) grain boundary configuration

i.e.

$$C^{\gamma\alpha} = C_0 + (C_0^{\gamma\alpha} - C_0)\left(1 - \operatorname{erf}\frac{\eta}{2\sqrt{Dt}}\right) \tag{1.116}$$

from which

$$C_0^{\gamma\alpha} - C_0 = \frac{C^{\gamma\alpha} - C_0}{\left(1 - \operatorname{erf}\frac{\eta}{2\sqrt{Dt}}\right)} \tag{1.117}$$

and thus

$$C = C_0 + (C^{\gamma\alpha} - C_0)\frac{\left(1 - \operatorname{erf}\frac{x}{2\sqrt{Dt}}\right)}{\left(1 - \operatorname{erf}\frac{\eta}{2\sqrt{Dt}}\right)} \tag{1.118}$$

The velocity of transformation $v(t)$ can be calculated from the mass balance

$$v(C^{\gamma\alpha} - C^{\alpha\gamma}) = J(\eta) = -D\left.\frac{\partial C}{\partial x}\right|_{\eta} \tag{1.119}$$

or

$$v = \frac{D(C^{\gamma\alpha} - C_0)}{\sqrt{\pi Dt}((C^{\gamma\alpha} - C^{\alpha\gamma})} \cdot \frac{\exp\left[-\frac{\eta^2}{4Dt}\right]}{\left(1 - \operatorname{erf}\frac{\eta}{2\sqrt{Dt}}\right)} \tag{1.120}$$

Now define

$$\eta = 2\alpha\sqrt{Dt} \tag{1.121}$$

so

$$v = \frac{d\eta}{dt} = \alpha\sqrt{D/t} \tag{1.122}$$

Thus equation (1.120), can be written in the form

$$\alpha = \frac{1}{\sqrt{\pi}}\frac{(C^{\gamma\alpha} - C_0)}{(C^{\gamma\alpha} - C^{\alpha\gamma})}\frac{\exp[-\alpha^2]}{(1 - \operatorname{erf}\alpha)} \tag{1.123}$$

which can be solved numerically for $\alpha (\gtrsim 1)$ and therefore for $\eta(t)$ and $v(t)$. Thus all parameters are determined in terms of the diffusion coefficient and boundary conditions. Similar solutions can be used for the analysis of oxide formation during oxidation of pure material. Fig. 1.26 shows an experimental test of relation (1.121) for transformation of carbon-alloyed bcc to fcc iron. A more general solution for an interface velocity corresponds to a prepared two-phase diffusion couple and therefore involves diffusion in both phases (cf Fig. 1.27). This is

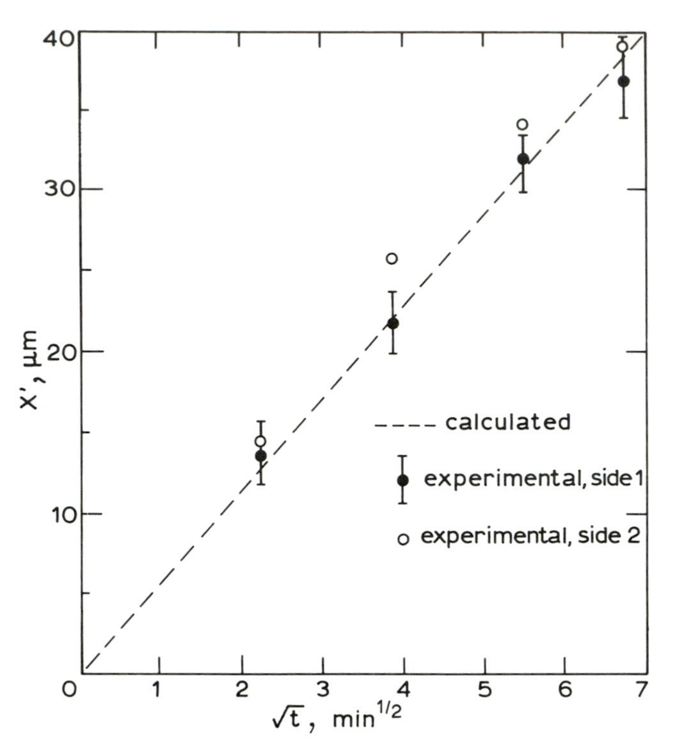

Reprinted with permission from TRANSACTIONS OF THE METALLURGICAL SOCIETY, Vol. 227, p. 1255, (1963), a publication of The Metallurgical Society, Warrendale, Pennsylvania

Fig. 1.26 Predicted *vs* experimental interface motion for a bcc → fcc phase transformation in carbon steel at T = 792°C. After Purdy and Kirkaldy[83]

$$\beta = \frac{\eta}{\sqrt{t}} = 2\left[\frac{(C_1^{\mathrm{I}} - C_{\mathrm{e}}^{\mathrm{I}})}{C_{\mathrm{e}}^{\mathrm{I}} - C_{\mathrm{e}}^{\mathrm{II}}}\sqrt{\frac{D_{\mathrm{I}}}{\pi}}\frac{\exp\{-\beta^2/4D_{\mathrm{I}}\}}{1 + \operatorname{erf}\beta/2\sqrt{D_{\mathrm{I}}}} - \frac{(C_{\mathrm{e}}^{\mathrm{II}} - C_2^{\mathrm{II}})}{C_{\mathrm{e}}^{\mathrm{I}} - C_{\mathrm{e}}^{\mathrm{II}}}\sqrt{\frac{D_{\mathrm{II}}}{\pi}}\frac{\exp\{-\beta^2/4D_{\mathrm{II}}\}}{1 - \operatorname{erf}\beta/2\sqrt{D_{\mathrm{II}}}}\right] \quad (1.124)$$

1.21 AN ELEMENTARY THEORY OF OXIDATION

Consider the growth of an oxide layer by diffusion control either by metal (cation) diffusion outwards or oxygen (anion) diffusion inwards (Fig. 1.28). It is assumed that the oxygen solubility in the metal (M) substrate is essentially zero, so the diffusion problem does not have to be solved there. Let V_{MO} be the volume of oxide formed per oxygen or metal atom combined. We can therefore write:

$$\frac{\mathrm{d}x}{\mathrm{d}t} = V_{\mathrm{MO}} D \frac{\Delta C}{x} \quad (1.125)$$

The solution is therefore

$$x^2 = 2V_{MO}\Delta Ct = k_p t \qquad (1.126)$$

where

$$k_p = 2V_{MO}D\Delta C \qquad (1.127)$$

is called the parabolic rate constant. Note the similarity of structure between this and the permeability (equation (1.42)). Although pure parabolic growth of oxides is sometimes observed, complex growth patterns are more often the case, as the following example and analysis indicates.

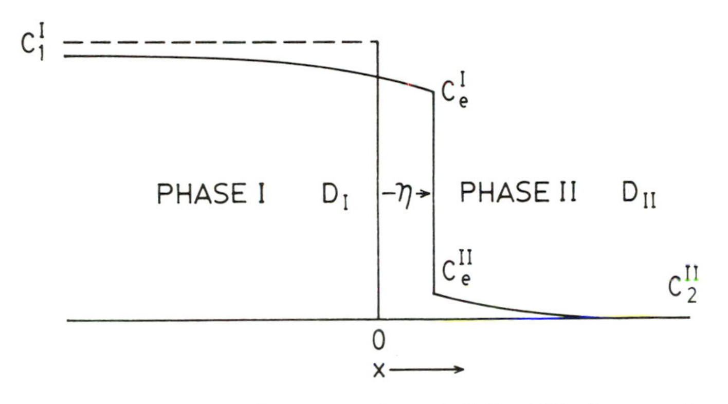

Fig. 1.27 Penetration curve for a two-phase infinite diffusion couple

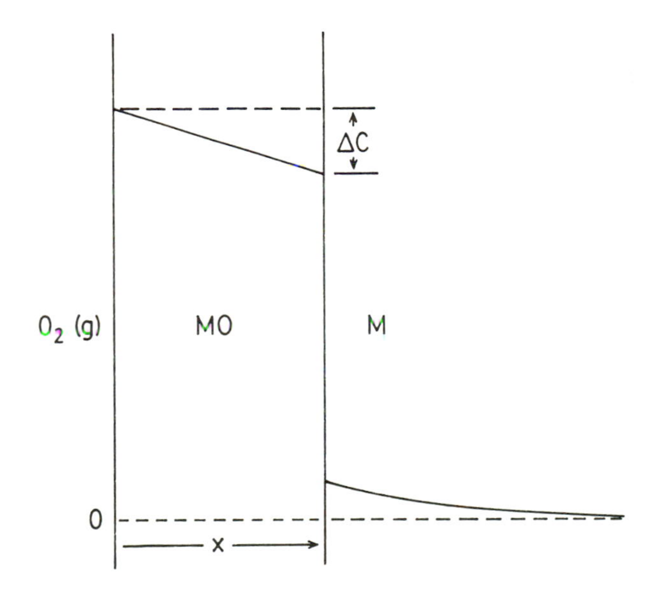

Fig. 1.28 Schematization of an oxide layer on metal defining oxidation model parameters

1.22 THEORY OF OXIDATION WITH SHORT-CIRCUIT DIFFUSION

We refer again to Fig. 1.28 and assume two parallel transport processes for the inward migration of oxygen through the film to the metal–oxide interface: lattice diffusion and diffusion within a random array of low resistance paths.[84] The driving force for oxidation is again the concentration gradient of oxygen between the inner and outer surfaces maintained by the free energy change for oxide formation. A uniform gradient which is inversely proportional to the film thickness is assumed according to the usual linear approximation to the error function. Interfacial inhibitions due to chemical reaction are assumed to be negligible. It is further hypothesized that the number of effective leakage points for oxygen in an oxide film decreases by grain growth or dislocation annihilation as oxidation proceeds. Accordingly we assume a time dependence of the fraction of available oxygen sites within the array of short circuit paths according to the first order rate expression,

$$-\frac{\mathrm{d}f}{\mathrm{d}t} = kf \tag{1.128}$$

or

$$f(t) = f^0 \mathrm{e}^{-kt} \tag{1.129}$$

Here, f^0 is the initial fraction of total oxygen sites within the low-resistance paths. The precise decay law chosen is of secondary consequence since k is an adjustable phenomenological coefficient. However, the first order rate expression does appear plausible and it leads to a tractable mathematical analysis (cf Perrow *et al.*[85]).

We next assume for the effective diffusion constant, Hart's expression (1.17)

$$D = D_\mathrm{V}(1-f) + D_\mathrm{B} f \tag{1.130}$$

and for the rate of film growth, equation (1.125).

Here D_V and D_B are constants for lattice and short circuit diffusion, and f is the fraction of the total available oxygen sites lying within low-resistance paths in the oxide film.

Upon substitution of equations (1.129) and (1.130) into (1.125) we obtain

$$\frac{\mathrm{d}x}{\mathrm{d}t} = V_\mathrm{MO}\{D_\mathrm{V}(1-f^0\mathrm{e}^{-kt}) + D_\mathrm{B} f^0 \mathrm{e}^{-kt}\}\frac{\Delta C}{x} \tag{1.131}$$

In terms of k_p (equation (1.127)) and in the approximation that $D_\mathrm{B} \gg D_\mathrm{V}$ equation (1.131) becomes

$$2x\frac{\mathrm{d}x}{\mathrm{d}t} = k_\mathrm{p}\{1 + f^0\delta\mathrm{e}^{-kt}\} \tag{1.132}$$

where

$$\delta = D_\mathrm{B}/D_\mathrm{V} \tag{1.133}$$

The growth law for oxidation is obtained by integrating equation (1.132) to

$$x^2 = k_\mathrm{p}\left\{t + \frac{f^0\delta}{k}(1-\mathrm{e}^{-kt})\right\} \tag{1.134}$$

This equation may now be investigated for various limiting cases which are useful in the analysis of the experimental data:

For thin films ($t \rightarrow 0$):

$$x^2 = (1 + f^0\delta)\, k_p t \tag{1.135}$$

For thick films ($e^{-kt} \ll 1$):

$$x^2 - x_0^2 = k_p t \tag{1.136}$$

where the intercept on a parabolic plot is

$$x_0^2 = k_p \frac{f^0\delta}{k} \tag{1.137}$$

We next refer to the data set of Kofstad *et al.*[86] (cf Smeltzer *et al.*[84]) for oxidation of Ti between 338° and 536°C (Fig. 1.29). The fit of equation (1.136) to the data for the late time process yields x_0 and k_p as a function of temperature. The fit of equation (1.135) to the short time data with the measured k_p then yields $f^0\delta(T)$.

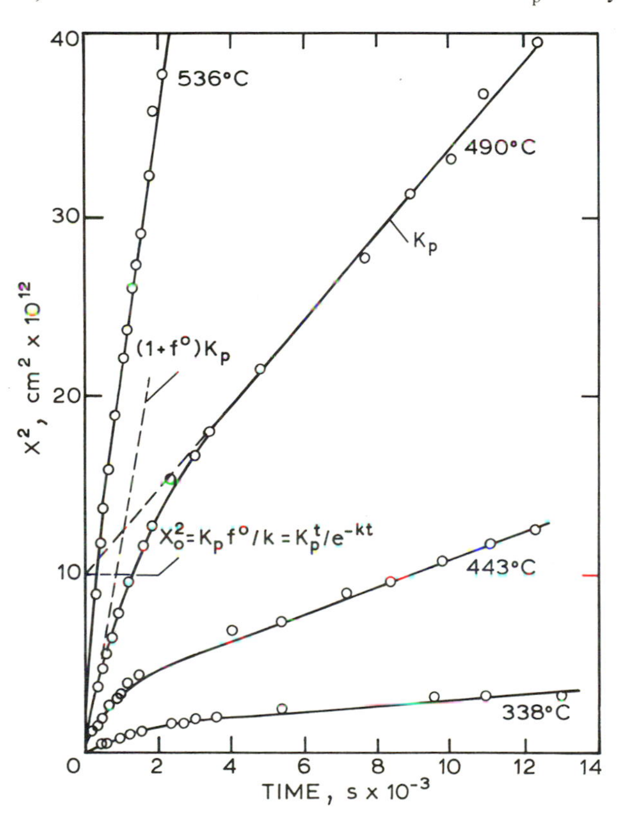

Fig. 1.29 The oxidation of titanium over the temperature range 338–536°C. After Smeltzer *et al.*[84]

Finally, $k(T)$ can be extracted from equation (1.137). The experiments yield values of $f^0\delta$ lying between 5 and 10 for the range of temperatures covered. Since the ratio of coefficients δ is typically of the order 10^5, the initial fraction of short circuit sites is $f^0 \sim 10^{-4}$ which is comparable to that in a highly deformed (heavily cold worked) metal. Dislocation arrays and grain boundaries representing comparable values of f^0 have been observed in NiO, which exhibits similar kinetic curves.[85]

1.23 THE GROWTH OF A SPHERICAL PRECIPITATE

Zener[44] has demonstrated that there are analogues to the one-dimensional Boltzmann diffusion equation (equation (1.54)) for cylindrical and spherical symmetry. The general parabolic equation is

$$-\frac{\lambda}{2}\frac{dC}{d\lambda} = \frac{D}{\lambda^{p-1}}\frac{d}{d\lambda}\left[\lambda^{p-1}\frac{dC}{d\lambda}\right]; \qquad \lambda = r/\sqrt{t} \tag{1.138}$$

where $p = 1, 2$ and 3 for linear, cylindrical and spherical symmetries respectively. Next to the linear precipitate analysis, which led us to equation (1.123), the spherical case is most important. If we assume in relation to Fig. 1.30 the boundary conditions at an interface radius $\rho = \beta\sqrt{t}$, are

$$C(\beta-) = C^{\alpha\gamma}; \qquad C(\beta+) = C^{\gamma\alpha}; \qquad C(\infty) = C_0 \tag{1.139}$$

the general solution is

$$C = C_0 + (C^{\gamma\alpha} - C_0)\phi(\lambda)/\phi(\beta) \tag{1.140}$$

where

$$\phi(\lambda) = \int_{\lambda}^{\infty} \lambda^{1-p} e^{-\lambda^2/4D}\, d\lambda \tag{1.141}$$

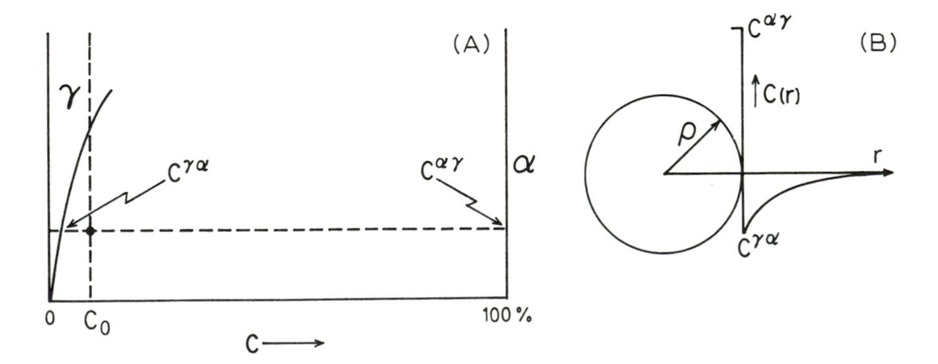

Fig. 1.30 Parameters relevant to precipitation of a spherical particle: (*a*) phase diagram; (*b*) concentration distribution relative to sphere dimensions

To determine the location of the interface in λ-space, β, the transcendental flux continuity relation

$$D\frac{dC}{d\lambda}\bigg|_{\beta} = \frac{\beta}{2}(C^{\alpha\gamma} - C^{\gamma\alpha}) \qquad (1.141)$$

is invoked. Such problems must in general be solved by numerical methods. The following, however, offers an approximate analytic solution for the spherical case.

If as defined by the phase diagram and alloy composition of Fig. 1.30 the relative supersaturation

$$\delta = \frac{C_0 - C^{\gamma\alpha}}{C^{\alpha\gamma} - C^{\gamma\alpha}} \ll 1 \qquad (1.143)$$

then by the mass balance the radius of the sphere will advance at a rate very much slower than the expansion of the surrounding diffusion field. We can therefore proceed to seek an analytic quasi-steady solution to the diffusion equation in spherical coordinates, i.e.

$$\frac{\partial C}{\partial t} = D\frac{1}{r^2}\frac{\partial}{\partial r}\left(r^2\frac{\partial C}{\partial r}\right) \simeq 0 \qquad (1.144)$$

This has the quasi-stationary solution

$$C = -\frac{\rho}{r}(C_0 - C^{\gamma\alpha}) + C_0 \qquad (1.145)$$

where the sphere radius ρ is a very slowly varying function of the time.

Now we can apply the mass balance to the entire sphere of volume V and area A in the form

$$(C^{\alpha\gamma} - C^{\gamma\alpha})\frac{dV}{dt} = A\,D\frac{\partial C}{\partial r}\bigg|_{\rho} \qquad (1.146)$$

i.e.

$$(C^{\alpha\gamma} - C^{\gamma\alpha})\frac{d}{dt}\left(\frac{4}{3}\pi\rho^3\right) = D\frac{\rho}{\rho^2}(C_0 - C^{\gamma\alpha}) \cdot 4\pi\rho^2 \qquad (1.147)$$

The velocity of interface advance will therefore be

$$v = \frac{d\rho}{dt} = \frac{D}{\rho}\frac{(C_0 - C_0^{\gamma\alpha})}{(C^{\alpha\gamma} - C^{\gamma\alpha})} = \frac{D}{\rho}\delta \qquad (1.148)$$

and upon integration

$$\rho = \sqrt{2\delta\,Dt} \qquad (1.149)$$

Not surprisingly, a form has been recovered which is close to the original estimation of a diffusion length (equation (1.8)). Note also that equation (1.148) is made up of a characteristic velocity of the system (D/ρ) discounted by the relative supersaturation $\delta < 1$.

If account had been taken of surface tension or capillarity, which retards growth, it would have been necessary to discount v by a further factor of the magnitude $1 - \rho_c/\rho$, to give

$$v \simeq \frac{D}{\rho}\delta\left(1 - \frac{\rho_c}{\rho}\right) \tag{1.150}$$

ρ_c being the radius where the reaction free energy released by the volume increase is exactly equal to the free energy gained by the increasing surface area. This is known as the critical radius for nucleation, i.e. if $\rho > \rho_c$, $v > 0$. The effects of capillarity in phase transformations will be dealt with fully in Chapter 12.

1.24 GROWTH OF INTERFERING SPHERICAL PRECIPITATE PARTICLES

Wert and Zener[25] have presented a calculation for the growth of a collection of small spherical particles from a dilute uniform solution. They assume that the solute density very near each growing particle has the same form as the steady state solution of the diffusion equation near a particle of fixed dimensions equal to the instantaneous dimensions of the growing particle. This steady state solution is normalized at infinite distance to the average instantaneous value of the solute concentration in the solution. Such an approximation is accurate for spherical particles at low volume fraction.

With Wert and Zener we denote $C(r, t)$ as the concentration of solute atoms in units of numbers per unit volume. Let C in the precipitate be C_0 and the precipitate boundary value in the matrix be C_1. The rate of growth is then

$$(C_0 - C_1)\,\mathrm{d}V/\mathrm{d}t = 4\pi\rho^2 D(\partial C/\partial r)|_{r=\rho} \tag{1.151}$$

where V and ρ are the volume and radius of the particle and D is the diffusion coefficient for solute atoms. For a dilute solution the particle grows so slowly that the growth rate has a negligible effect on the gradient $(\partial C/\partial r)|_{r=\rho}$, so we can give it the steady state value it would have if ρ were constant. The steady state solution of the diffusion equation (Laplace's equation in this case; Appendices 2 and 3) is

$$C(r) = C_\infty - (C_\infty - C_1)(\rho/r) \tag{1.152}$$

and

$$(\partial C/\partial r)_{r=\rho} = (C_\infty - C_1)/\rho \tag{1.153}$$

This differs from the previous single particle solution in that $C_0 = C_\infty(t)$.

For dilute solutions the precipitate particles are far apart so that each particle is surrounded by a concentration field which approaches the average, $C_\infty(t)$, at long distances. This average concentration is related to the fraction W of the precipitation which has already occurred.

$$\{C_\infty(t) - C_1\}/\{C_\infty(0) - C_1\} = 1 - W \tag{1.154}$$

If we let ρ_0 denote the final radius of the precipitate particles and observe that

$$W = (\rho/\rho_0)^3 \tag{1.155}$$

then equations (1.151), (1.153) and (1.154) lead to

$$\tau(\mathrm{d}W/\mathrm{d}t) = \tfrac{3}{2}(1 - W)\,W^{1/3} \tag{1.156}$$

where the time of relaxation τ is defined as

$$\tau = [(C_0 - C_1)/(C_\infty(0) - C_1)]\, \rho_0^2/2D \tag{1.157}$$

For the limit $t \rightarrow 0$ and therefore $W \rightarrow 0$, this has the solution

$$W = 1 - \exp\{t/\tau\}^{3/2} \cong (t/\tau)^{3/2} \tag{1.158}$$

The quantity $W(t)$ is accessible to determination by internal friction methods (cf Chapter 5). However, in the iron-Fe_3C system studied by Wert[12] ρ_0 was not accessible to direct observation. Accordingly, Wert and Zener could only check the general form of a $(1 - W)$ versus t/τ plot using an experimentally determined scale factor for t/τ. Their results, as shown in Fig. 1.31, were regarded by the authors as a fair test of the theory, and by inference, to give a reasonable estimate of particle density and final radius via equation (1.157).

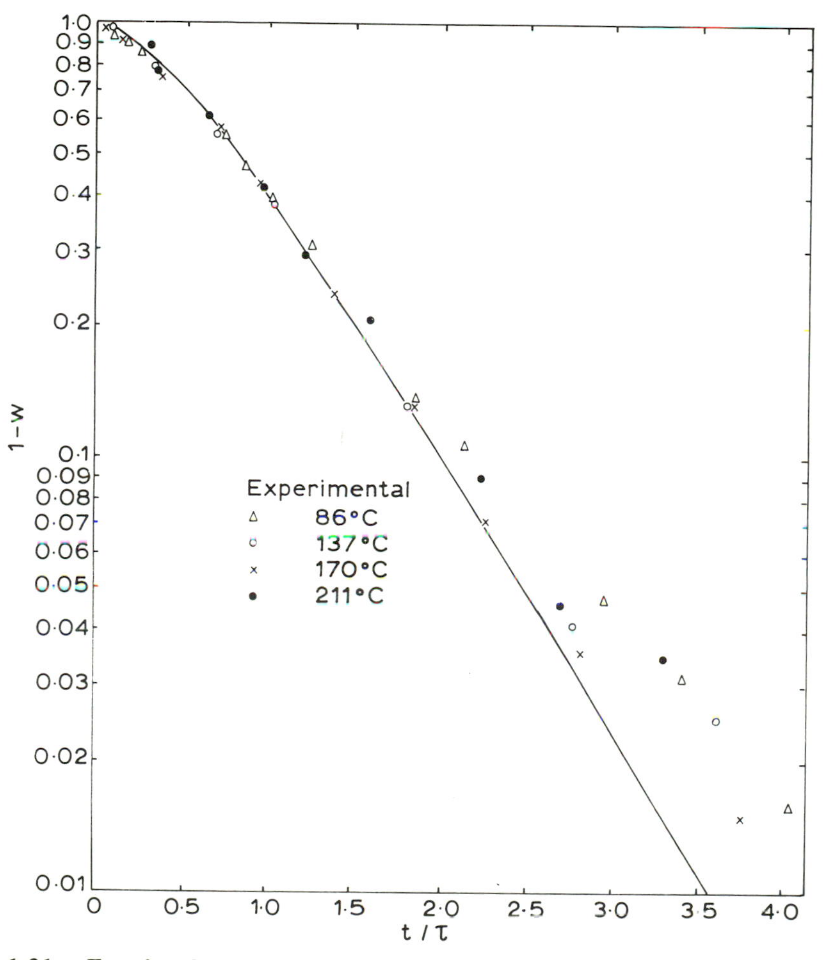

Fig. 1.31 Fractional carbon precipitation in α-iron as a function of time. After Wert and Zener[25]

1.25 MIXED DIFFUSION AND REACTION CONTROL IN TRANSFORMATIONS

A very common situation in solid–solid transformations occurs when an essentially diffusion-limited process is influenced by a sluggish interfacial reaction. In many diffusion situations the inclusion of reaction influence creates a rather complex mathematical problem. However, in certain cases satisfactory elementary solutions exist.[87] Assume that there exists a single particle of concentration C_0 in an infinite medium, the concentration on the parent side of the interface takes the value C_1 which is different from the equilibrium value C_0, and that the flux transfer across the interface is governed by the velocity rate relation

$$J = (C_0 - C_1)\,v\rho = (C_0 - C_1)\,\mathrm{d}\rho/\mathrm{d}t \tag{1.159}$$

or by Fick's law for mass delivery to or from the interface

$$J = -D\partial C/\partial x\,|_{r=\rho} \tag{1.160}$$

Using the steady state approximation for dilute solutions,

$$\partial C/\partial x\,|_{r=\rho} = (C_\infty - C_1)/\rho \tag{1.161}$$

or

$$J = D(C_\infty - C_1)/\rho \tag{1.162}$$

where C_∞ is the average concentration far from the surface.

Now J must also satisfy

$$J = G(C_1 - C_e) \tag{1.163}$$

where G is the interface transfer coefficient, assumed independent of C_1 and C_e is the equilibrium interface concentration. Solving for C_1 between these and assuming $C_\infty \ll C_0$, one can write via equation (1.159)

$$C_0\frac{\mathrm{d}\rho}{\mathrm{d}t} = \frac{DG(C_\infty - C_e)}{\rho G + D} \tag{1.164}$$

with the solution

$$\frac{\rho^2}{2D} + \frac{\rho}{G} = \frac{C_\infty - C_e}{C_0}t \tag{1.165}$$

This clearly demonstrates for a single non-interacting particle that reaction will tend to be controlling for early times (linear growth) while diffusion will be controlling for long times (parabolic growth).

If we now consider interacting particles, C_∞ is not constant, and the above integration is invalid. We can, however, follow Wert and Zener's model of the previous section and express the transformation kinetics for a pure reaction mechanism ($C_1 = C_\infty$) in terms of the precipitated fraction W. The interface mass balance is now written in the form

$$C_0\frac{\mathrm{d}V}{\mathrm{d}t} = 4\pi\rho^3(C_\infty - C_e)\,G \tag{1.166}$$

For a very dilute solution, we can write on the average

$$[C_\infty(t) - C_e]/[C_\infty(0) - C_e] = 1 - W \tag{1.167}$$

and noting that $V = 4\pi\rho^3/3$ and that

$$W = (\rho/\rho_0)^3 \tag{1.168}$$

where ρ_0 is the final equilibrium radius, we obtain

$$\tau(\mathrm{d}W/\mathrm{d}t) = 3(1 - W)\, W^{2/3} \tag{1.169}$$

where

$$\tau = \frac{\rho_0}{G}\left[\frac{C_0}{C_\infty(0) - C_e}\right] \tag{1.170}$$

For $t \rightarrow 0$ this integrates to

$$W = 1 - \exp\{-(t/\tau)^3\} \cong (t/\tau)^3 \tag{1.171}$$

This result for pure reaction control is to be compared with Wert and Zener's result with its $(t/\tau)^{3/2}$ exponent for pure volume diffusion control. Turnbull[87] has observed this t^3 dependence for $W < 0{\cdot}04$ in the precipitation of $BaSO_4$ from aqueous solution. A rigorous solution of the general diffusion-reaction problem is presented in §12.4.

1.26. PRECIPITATION OF A LAMELLAR BICRYSTAL: THE PEARLITE REACTION

A binary system which possesses the Gibbs' constitution diagram of Fig. 1.32 often exhibits a lamellar diffusion-controlled decomposition (a eutectoid or eutectic reaction) when a uniform solution at temperature T_0 is quenched to the supersaturated solution T (Figs. 1.33 and 1.34). The uniform growth model considered is shown in Fig. 1.34.[89] The precise interface shape is not at first

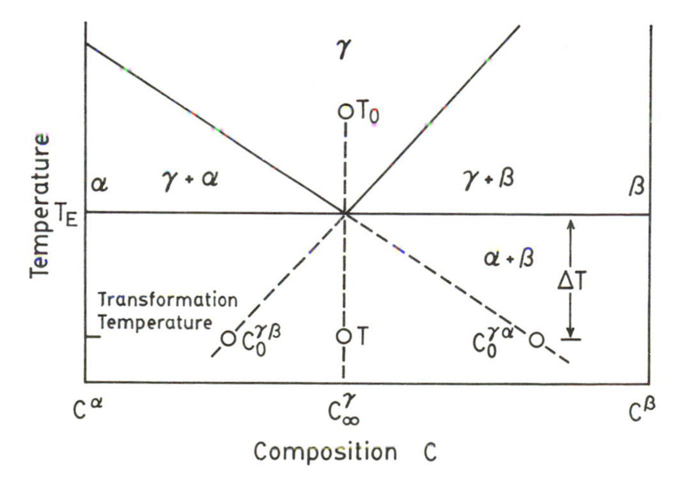

Fig. 1.32 Phase diagram for a binary system which exhibits eutectic or eutectoid (pearlite) reactions

specified although it is assumed that it does not deviate very far from the plane. It is also assumed that negligible diffusion occurs behind the interface, that the interface is sufficiently mobile for the chemical reaction not to inhibit the kinetics, and that division of spacing between the layers of width S_α and S_β can be determined *a priori* to sufficient precision on the basis of an assumption of equilibrium segregation in the transformed material. This latter is strictly an assumption of convenience since it could be replaced by the mass balance in the transformed material[90]

$$\int_{S_\alpha} C^\alpha \, dy + \int_{S_\beta} C^\beta \, dy = C^\gamma S \tag{1.172}$$

S_α (or $S_\beta = S - S_\alpha$) now being determined simultaneously with the diffusion solution for the γ-phase.

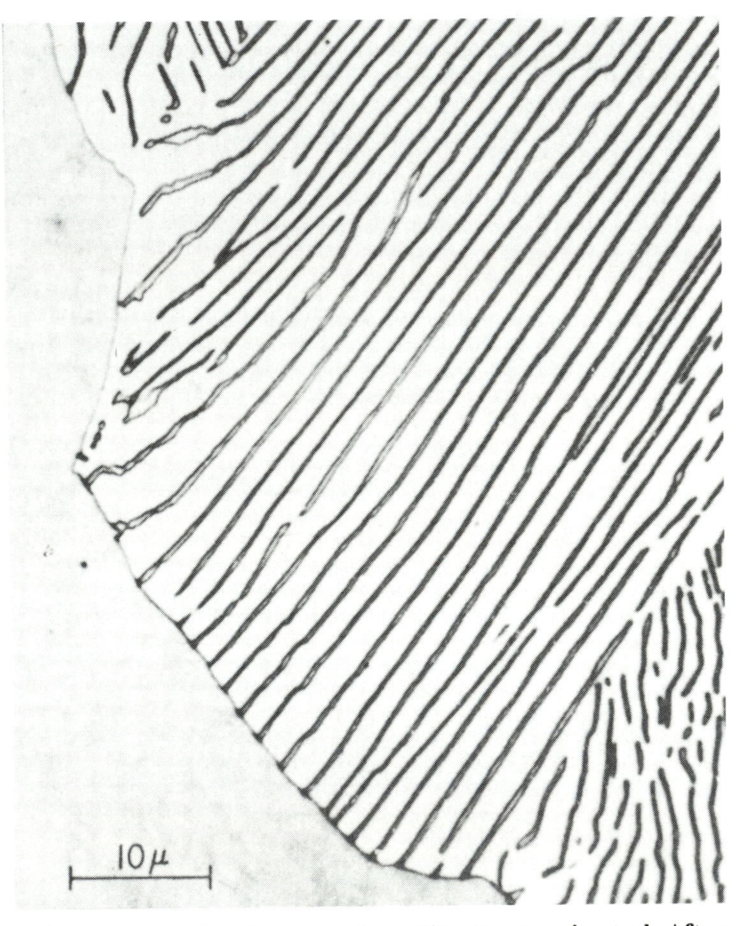

Fig. 1.33 Micrograph of an isothermal pearlite structure in steel. After Vilella[88]

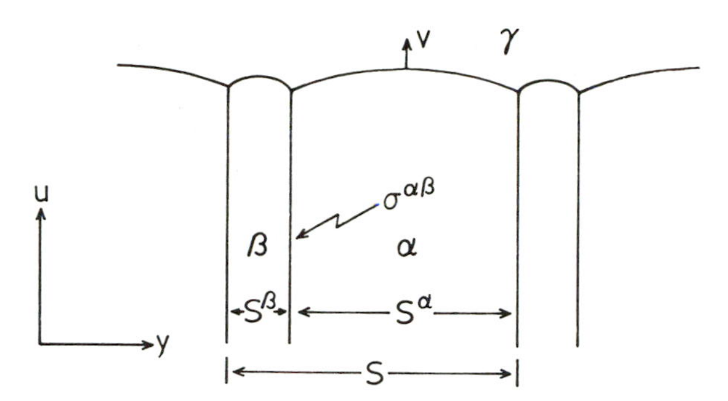

Fig. 1.34 Schematic of lamellar pearlite structure which together with Fig. 1.32 defines the growth parameters

The general two-dimensional diffusion equation applicable to the γ-phase is (cf equation (1.75))

$$\frac{1}{D}\frac{\partial C^{\gamma}}{\partial t} = \frac{\partial^2 C^{\gamma}}{\partial x^2} + \frac{\partial^2 C^{\gamma}}{\partial y^2} \tag{1.173}$$

or transforming to the uniformly moving system, $u = x - vt$, (cf equation (1.110))

$$-\frac{v}{D}\frac{\partial C^{\gamma}}{\partial u} = \frac{\partial^2 C^{\gamma}}{\partial u^2} + \frac{\partial^2 C^{\gamma}}{\partial y^2} \tag{1.174}$$

Separating variables by the assumption

$$C^{\gamma}(u, y) = g(u)\, h(y) \tag{1.175}$$

yields the separated ordinary equations

$$\frac{1}{g}\left(\frac{\mathrm{d}^2 g}{\mathrm{d}u^2} + \frac{v}{D}\frac{\mathrm{d}g}{\mathrm{d}u}\right) = -\frac{1}{h}\frac{\mathrm{d}^2 h}{\mathrm{d}y^2} = b^2 \tag{1.176}$$

where b^2 is the separation constant. These have the particular solutions

$$g = \mathrm{e}^{\lambda u}, \lambda = -\frac{v}{2D}(1 + \sqrt{1 + 4b^2D^2/v^2}) \tag{1.177}$$

and

$$h = \cos by \tag{1.178}$$

where b can take any constant value. Since the above are linear equations and we require a solution that is periodic in the y direction, the appropriate general solution is

$$C^{\gamma} - C^{\gamma}_{\infty} = \sum_{n=0}^{\infty} A_n \mathrm{e}^{\lambda_n u} \cos b_n y \tag{1.179}$$

The Fourier coefficients A_n and the interface shape must be simultaneously determined in such a way that the steady state lateral flux balance,

$$\int_{-\infty}^{\infty} (\operatorname{div} \boldsymbol{J})_{y}\, du = v(C_{\infty}^{\gamma} - C^{\alpha \text{ or } \beta}) \tag{1.180}$$

the curvature dependence of $C^{\gamma\alpha}$ and $C^{\gamma\beta}$, and the triple point surface tension balance are all satisfied. Because the γ-concentration must be continuous along the interface, the model must automatically provide a triple (eutectoid) point in the pressure–temperature–composition diagram located at the γ–α–β junction. Hillert has carried out this ambitious program in approximation obtaining very plausible shapes for the interface and an equation relating growth velocity v to the assumed spacing S (cf Zener[91])

$$v = \frac{1}{a}\frac{C_0^{\gamma\alpha} - C_0^{\gamma\beta}}{C^{\beta} - C^{\alpha}}\frac{D}{S}\left(1 - \frac{S_c}{S}\right) \tag{1.181}$$

where

$$a = \frac{1}{\pi^3}\left(\frac{S^2}{S_\alpha S_\beta}\right)^2 \sum_{1}^{\infty} \frac{1}{n^3} \sin^2 n\pi S_\alpha / S \sim 1 \tag{1.182}$$

Here S_c is a critical spacing which is a complex function of the interfacial tensions, the spacings, and the thermodynamic properties of the solutions. Note the formal similarity between this relation and that for a precipitating sphere (equation (1.150)). However, the problem differs in that for the spherical case there is a second relation which determines ρ, so the solution is unique. Because equation (1.181) represents an infinity of steady state solutions there exists an indeterminancy or degeneracy which must be removed if a unique solution of the problem is to be found. This and other related problems of stationary pattern formation and capillarity will be discussed in Chapter 12.

1.27 UPHILL DIFFUSION: SPINODAL DECOMPOSITION

In § 1.6 we obtained an expression for the binary chemical diffusion coefficient which contains the second derivative of the Gibbs free energy, i.e.

$$\tilde{D} = \frac{L}{m}\frac{d^2G}{dX^2} \tag{1.183}$$

where X is the mole fraction of the specified independent component. This relation anticipates two kinds of diffusion instability as the temperature is lowered. If there is a strong increase in the attraction between unlike atoms then a disorder → order chemical reaction can be initiated.[92, 93, 94] If, on the other hand, strong repulsion between unlike atoms accrues then a miscibility gap appears (see § A1.4). For a solution with a miscibility gap and critical point, T_c (e.g. Fig. 1.35), a typical set of free energy curves appears as in Fig. 1.36. Thus between the inflection points or 'spinodes' within the miscibility gap d^2G/dX^2 becomes negative and 'uphill' diffusion or continuous phase separation results. In solids and liquids there are gradient energy or incipient surface tension effects which tend to damp this decomposition, and in solids there are strain energy restrictions as well.[92, 93, 95] These will be dealt with later. The three-dimensional solution of Fick's second equation which describes the initial linear growth of an arbitrary fluctuation in a uniform solution of concentration C_0 is

$$C - C_0 = \int A(\boldsymbol{\beta})\, e^{-\beta^2 Dt} \cos \boldsymbol{\beta} \cdot \boldsymbol{r}\, d\beta \tag{1.184}$$

where $\boldsymbol{\beta}$ is a variable wave number.

A convenient exploratory one-dimensional initial condition is represented by

$$C - C_0 = \frac{\sin ax}{ax} \tag{1.185}$$

since its Fourier transform is $A(\beta) = 1/2a$ for $-a < \beta < a$ and zero elsewhere. Although equation (1.185) does not conserve mass we can always suppose that an infinitesimal inhomogeneity makes up the deficit. The low-amplitude solution is thus represented by

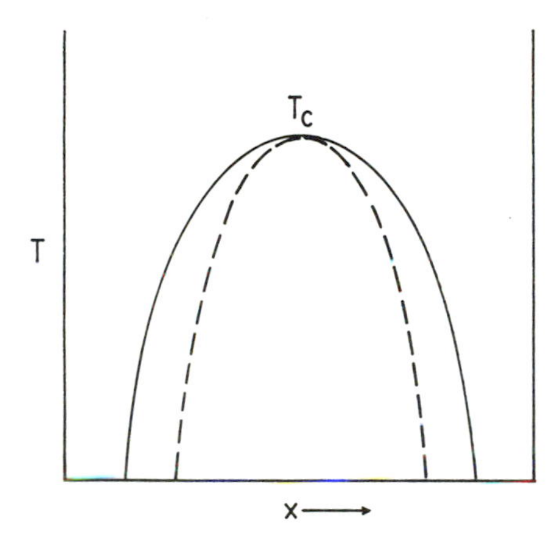

Fig. 1.35 Schematic miscibility gap illustrating trace of spinodes (----)

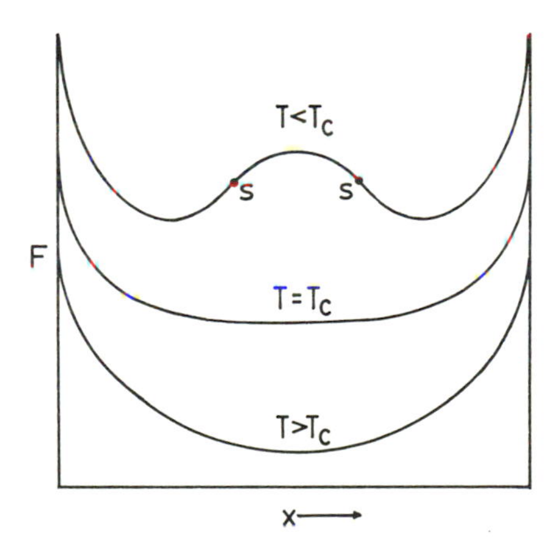

Fig. 1.36 Schematic free energy curves corresponding to Fig. 1.35 identifying the spinodes

$$C - C_0 = \frac{1}{2a}\int_{-a}^{a} e^{-D\beta^2 t} \cos \beta x \, d\beta \qquad (1.186)$$

where D according to equation (1.183) is negative so the solution amplitude grows and spreads as indicated in Fig. 1.37. Because the damping and non-linear terms have been omitted the initial dominant wave number $\beta = a$ persists.

When strain energy and gradient energy terms are included the Fick equation is generalized in the low-amplitude approximation, using equation (1.32) to evaluate D, to obtain the fourth-order linear differential equation

$$\frac{\partial C}{\partial t} = \frac{L}{m}\left[\frac{d^2G}{dX^2} + \frac{2\eta^2 Y}{1-\pi}\right] \nabla^2 C - \frac{2L\chi}{m}\nabla^4 C \qquad (1.187)$$

where η is the strain energy coefficient due to the change in lattice parameter with composition, Y is Young's modulus, π is the Poisson ratio from elasticity theory and $\chi > 0$ is the gradient energy coefficient associated with the onset of surface tension[92, 95] (cf § 12.5). Note that the effective diffusion coefficient

$$D = \frac{L}{m}\left[\frac{d^2G}{dX^2} + \frac{2\eta^2 Y}{1-\pi}\right] \qquad (1.188)$$

contains a positive constant term which acts against negative changes of the second derivative. Thus spinodal decomposition can only be initiated by a continuous process at and below a definite undercooling as measured from the critical point and spinodal line. Furthermore, since $\chi > 0$, the fourth order term is always negative indicating a strong damping of the amplitude of short wavelength (steep gradient) Fourier components of the initial condition. This is seen most clearly by noting that solution (1.186) is modified to

$$C - C_0 = \frac{1}{2a}\int_{-a}^{a} A(\beta)\, e^{-(D\beta^2 + (2L\chi/m)\beta^4)t} \cos \beta x \, d\beta \qquad (1.189)$$

The damping term in the exponent is to the fourth order in β which strongly discriminates against large wavenumbers (short wavelengths). Indeed, if D is negative the exponent $R(\beta)$ possesses a sign change as a function of β at

$$\beta_c = \sqrt{-\frac{Dm}{2L\chi}} \qquad (1.190)$$

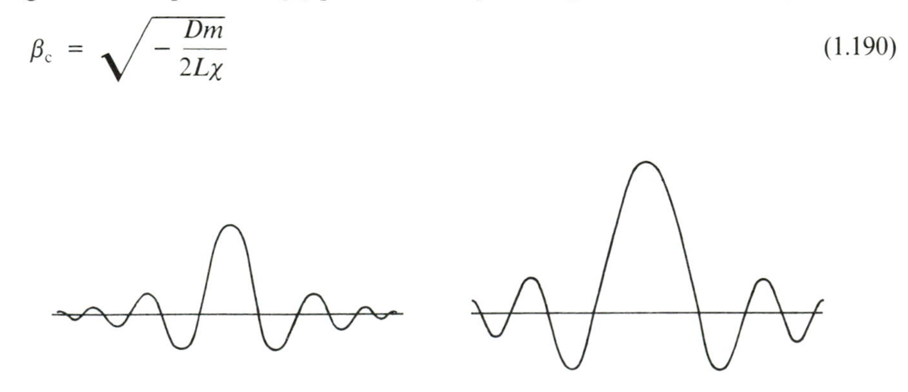

Fig. 1.37 Schematic representation of fluctuation growth

and a positive maximum at

$$\beta_m = \beta_c/\sqrt{2} \tag{1.191}$$

(see Fig. 1.38). One may therefore conclude that if β_m has a sufficiently strong Fourier component within the initial condition it will be amplified to a dominant position in the Fourier spectrum in a time of the order of $[(D\beta_m^2 + (2L\chi/m)\beta_m^4)]^{-1}$. More precise statements about the decomposition require the introduction of non-linear theory[96, 97] (refer also to § 12.5).

1.28 MATHEMATICAL AND COMPUTATIONAL EXERCISES

1. Verify the coefficients in equation (1.112) for the boundary conditions of Fig. 1.23*b*.
2. Derive the marginal relation (1.113).
3. Graph the solutions of equation (1.123) via a computer program.
4. Derive expression (1.124).
5. Derive expression (1.138).
6. Demonstrate that equation (1.141) is the solution of equation (1.138) for boundary conditions (1.139).
7. Write a program and graph the numerical solution of (1.156).
8. Modify the program of exercise 7 and graph the numerical solution of equation (1.171).
9. Introduce dimensionless parameters into equation (1.189) via equation (1.191) and by numerical integration graph the solution $C - C_0$ for initial condition (1.185) as a function of unitless time (cf Fig. 1.37).
10. In many one-dimensional phase transformations the error function part of the solution can be represented approximately by a triangle[44] (cf Figs. 1.15 and 1.25*b*). Use this approximation to solve the problem described by Fig. 1.25*b* and compare this solution numerically with that obtained in exercise 3. This straight line solution is used in the oxidation problem of § 1.21.

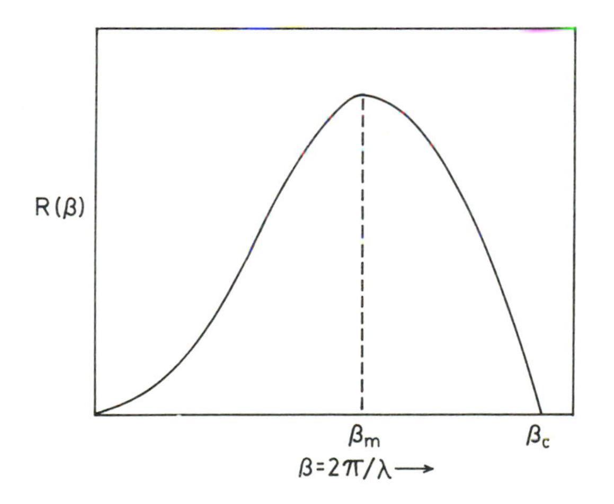

Fig. 1.38 Amplification factor *R vs* wave number *β*

11. The industrially important process of diffusion layer formation from the vapour phase can be described quantitatively by a variant of equation (1.124). In this problem one is given the surface value C_0^{I} rather than C_1^{I}. Using a solution of form (1.114) find an expression for C_1^{I} in terms of C_0^{I} and η thereby making equation (1.124) applicable. If D_{II} is $9D_{\mathrm{I}}$ solve (1.124) for the dimensionless growth rate parameter $\beta/\sqrt{D_{\mathrm{I}}}$. If D_{I} is $10^{-14}\ \mathrm{m^2\,s^{-1}}$ and the diffusion time is 10^5 s evaluate the diffusion layer thickness η.
12. Equation (1.112) identifies a characteristic diffusion length D/v. Supposing that this is an estimate for solidification cell spacing as in Fig. 1.24 compare its predictions with the data of Fig. 12.24. Use $D = 2 \times 10^{-3}\ \mathrm{mm^2\,s^{-1}}$.
13. Refer to equation (1.181) for the frontal velocity of the eutectoid (pearlite) reaction. Here the critical spacing S_c, which sets the velocity to zero, is known to vary inversely with the undercooling ΔT. If the optimum spacing is that for which the velocity is a maximum, find a relationship for v as a function of temperature T and graph as a function of ΔT. Take D according to equation (1.16).

1.29. SOLID–LIQUID MASS TRANSFER WITH CONVECTION

When phase change involves a fluid phase, the mathematical description becomes exceptionally complicated due to the presence of convection as well as diffusion and temperature gradients associated with the transfer of the latent heat. Furthermore, in ternary and higher order liquid–liquid systems the interfacial instabilities can become vigorous due to diffusion interactions and variable surface tensions (the Marangoni effects). Such surface tension related phenomena are discussed in Chapter 12.

Despite the complication and mathematical difficulties, understanding of general mass transfer is so important to physiology and the chemical, materials and metallurgical processing industries that a fully fledged, semiempirical discipline has evolved, with many books devoted entirely to the subject. As an elegant primer we recommend the volume of Bird *et al.*[6] wherein is considered the prediction of mass transfer coefficients based on the film, penetration and boundary layer theories. For many useful applications we refer the reader to Cussler.[5]

In this section we consider alloy solidification in the presence of natural convection, a problem of theoretical and practical interest to materials scientists, chemical engineers and metallurgists. This treatment of transient solidification due to Wagner[98] is a generalization of the steady state solidification theory presented in §1.18.

Referring to the one-dimensional mass transfer problem depicted by the binary concentration profile of Fig. 1.39 we imagine a planar interface 0 which has originated at 0′ and which is advancing at a non-steady velocity $u(t)$. The equilibrium partition ratio as before is

$$\frac{X_{\mathrm{S}}}{X_{\mathrm{L}}} = k \tag{1.192}$$

where k is usually less than unity. The effective diffusion distance δ defined by the intercept of the slope at $x' = 0$ (Fig. 1.39) is determined by diffusion and (or) convection and is estimated approximately via the mass balance. If, for example,

$u(t)$ is given from a heat transfer calculation then δ is estimated via Fick's equation (1.1) and the instantaneous mass balance at the interface where convection vanishes, i.e.

$$u(X_L - X_S) = D(X_L - X_L^0)/\delta \tag{1.193}$$

or

$$\delta = \frac{D}{u}\frac{1 - k(X_L^0/X_S)}{1 - k} \tag{1.194}$$

Since X_S cannot exceed X_L^0, δ takes its maximum value D/u when segregation disappears ($X_S = X_L^0$). An example is the case of steady state solidification (cf §1.119). However, the interest here is in the general transient problem. Expressing the fractional segregation X_S/X_L^0 as a measure of the mass transfer in terms of δ/δ_{max} obtains

$$\frac{X_S}{X_L^0} = \frac{k}{1 - (\delta/\delta_{max})(1 - k)} \tag{1.195}$$

First the case is considered where the interface is horizontal and the density relationships in the diffusion field are such as to oppose convection. Furthermore, the growth conditions are such that the interface is stable (planar) and the heat transfer as determined by the planar mold wall is such that the interface position follows the commonly observed parabolic law

$$x_0 = (\alpha t)^{1/2} \tag{1.196}$$

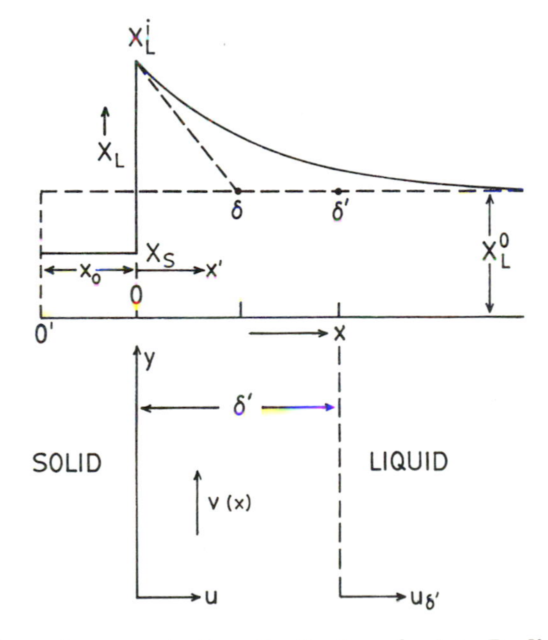

Fig. 1.39 Schematic concentration distribution of solute *B* adjacent to the liquid–solid interface. After Wagner[98]

where the coefficient α has the units of a diffusion coefficient. The diffusion field in the liquid satisfies Fick's second equation (1.49) subject to the initial condition

$$X_L = X_L^0 \quad \text{for} \quad x > 0' \qquad t = 0 \tag{1.197}$$

The surface boundary condition is

$$X_S = kX_L \quad \text{at} \quad x = (\alpha t)^{1/2} \tag{1.198}$$

and the mass balance corresponding to equation (1.193) is

$$\frac{1}{2}\left(\frac{\alpha}{t}\right)^{1/2} (1 - k) X_L \bigg|_{x = (\alpha t)^{1/2}} = -D \frac{\partial X_L}{\partial t}\bigg|_{x = (\alpha t)^{1/2}} \tag{1.199}$$

Except for the end of the solidification process where the liquid is depleted, equations (1.49) and (1.199) are satisfied by the relation

$$\frac{X_L}{X_L^0} = 1 + \frac{(1 - k)\,\mathrm{erfc}\,(x/2\sqrt{Dt})}{(2\sqrt{D/\pi\alpha})\exp(-\alpha/4D) - (1 - k)\,\mathrm{erfc}\,(x/2\sqrt{Dt})} \tag{1.200}$$

or the segregation

$$\frac{X_S}{X_L^0} = \frac{k}{1 - \dfrac{1}{2}\sqrt{\dfrac{\pi\alpha}{D}}\dfrac{\mathrm{erfc}\sqrt{\alpha/4D}}{\exp(-\alpha/4D)}(1 - k)} \tag{1.201}$$

From equations (1.195) and (1.201) is identified δ/δ_{max} which is a constant determined by the values of k, α and D.

From Appendix 3 comes the approximation

$$\mathrm{erfc}\, z = \pi^{-1/2}(\mathrm{e}^{-z}/z)(1 - \tfrac{1}{2}z^{-2} + \ldots) \tag{1.202}$$

for large arguments z. Accordingly, since in general $\alpha \gg 2D$, we obtain the segregation factor

$$\frac{X_S}{X_L^0} = \frac{k}{1 - (1 - k)(1 - 2D/\alpha)} = \frac{1}{1 + 2(1 - k)\,D/k\alpha} \tag{1.203}$$

and

$$\frac{\delta}{\delta_{max}} = 1 - \frac{2D}{\alpha} \simeq 1 \tag{1.204}$$

It may therefore be concluded from equation (1.203) that the segregation factor can take a wide range of values depending on the partition, diffusion and heat transfer conditions.

In the case of a vertical solid–liquid interface natural convection is unavoidable. If, for example, the solute decreases the density of the liquid mixture then there will be an upward convective flow near the interface (cf Fig. 1.40). The general effect of convection is to flatten the diffusion profiles and therefore to increase the segregation. Wagner,[98] following von Kármán,[99] has undertaken a quasi-steady state analysis of natural convection effects and assumes that the shearing stress in the liquid at the phase interface, which is the product of

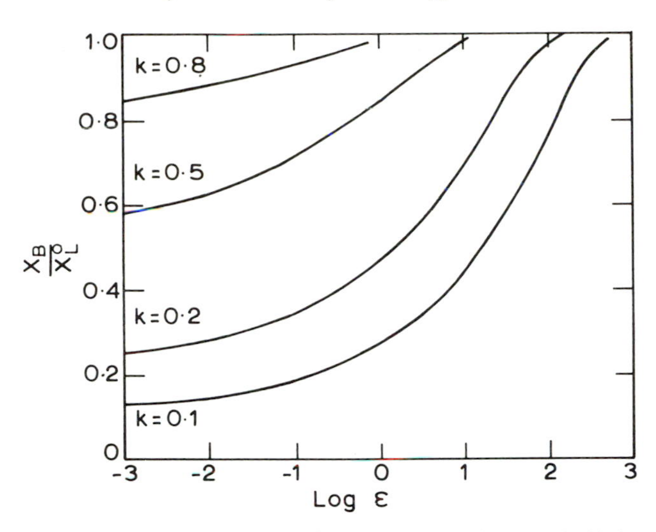

Reprinted with permission from JOURNAL OF METALS, Vol. 6, No. 2, 1954, a publication of The Metallurgical Society, Warrendale, Pennsylvania

Fig. 1.40 Concentration ratio X_B/X_L^0 as a function of the dimensionless group ξ. After Wagner[98]

dynamic viscosity μ of the liquid and the velocity gradient $(\partial v/\partial x)_{x=x_0}$, is the sum of the buoyancy of the boundary layer per unit interface area and the change in the flow of momentum per unit width per unit length in the flow direction. The momentum term can be disregarded since when $\mu/\rho_L D \gg 1$ the effect of inertia is significant only at a substantial distance from the phase interface. Thus

$$\mu(\partial v/\partial x)_{x=x_0} = g\int_0^{\delta} \Delta\rho \, dx' \tag{1.205}$$

where g is the gravitational acceleration, $\Delta\rho$ is the absolute value of the difference between the local and the bulk density of the liquid, and δ' is the distance where the bulk concentration is effectively reached (Fig. 1.39).

This density difference can be expressed as

$$\Delta\rho = \beta\rho_L(X_L - X_L^0) \tag{1.206}$$

where ρ_L is the density (mass/unit volume) of the bulk liquid and

$$\beta = \frac{1}{\rho_L}\left|\frac{\partial\rho_L}{\partial X_L}\right| \tag{1.207}$$

is constant.

On substituting equation (1.206) into equation (1.205) and introducing the kinematic viscosity $\nu = \mu/\rho_L$, it follows that

$$\nu(\partial v/\partial x)_{x=x_0} = g\int_0^{\delta'} \beta(X_L - X_L^0)\, dx' \tag{1.208}$$

Both the flow velocity v and the thickness of the boundary layer increase in the direction y parallel to the solid–liquid interface. Hence the mass rate also increases in the flow direction. Simultaneously, the boundary layer loses the mass $\rho_S u$ per unit of area and time owing to solidification, where ρ_S is the density of the solid. Consequently there is a flow of liquid into the boundary layer with a velocity v_n at $x' = \delta'$ normal to the solid–liquid interface. The equation for conservation of mass reads

$$\frac{d}{dy}\int_0^{\delta'} \rho_L v \, dx' + \rho_S u = \rho_L v_n \tag{1.209}$$

In equation (1.209), local density differences and the density difference between the liquid and the solid can be disregarded. Equation (1.209) becomes

$$v_n \simeq \frac{d}{dy}\int_0^{\delta'} v \, dx' + u \tag{1.210}$$

The balance of solute in a volume element of the boundary layer of width b and thickness δ' between y and $y + dy$ is calculated according to

1. The rate of gain of solute due to inflow of liquid normal to the phase interface is $\rho_L^0 X_L^0 u_n b \, dx$.
2. The rate of gain of solute due to flow parallel to the phase interface at distance y is

$$b\left(\int_0^{\delta'} \rho_L X_L v \, dx'\right)_y$$

3. The rate of loss of solute due to flow parallel to the solid–liquid interface at distance $y + dy$ is

$$b\left(\int_0^{\delta'} \rho_L X_L v \, dx'\right)_y + b\left[\frac{d}{dy}\int_0^{\delta'} \rho_L X_L v \, dx'\right] dy.$$

4. The rate of loss of solute due to its transfer from the liquid to the solid is $\rho_S X_S u b \, dy$. Thus the steady state solute balance reads

$$v_n \rho_L^0 b \, dy = \left[\frac{d}{dy}\int_0^{\delta'} \rho_L X_L v \, dx'\right] b \, dy + u \rho_S X_S b \, dy \tag{1.211}$$

On substituting equation (1.210) into equation (1.211), dividing equation (1.211) by $b \, dx$ and applying the equal density approximation as in equation (1.210) it follows that

$$X_L^0 \frac{d}{dy}\int_0^{\delta'} v \, dx' + X_L^0 u = \frac{d}{dy}\int_0^{\delta'} X_L v \, dx' + u X_S \tag{1.212}$$

To obtain an approximate solution power series expressions for the velocity and the concentration in the boundary layer as functions of distance x' from the phase interface are introduced, i.e.

$$v = v_{\delta'}[2(x'/\delta') - (x'/\delta')^2] \quad \text{for} \quad 0 \leqslant x' \leqslant \delta' \tag{1.213}$$

$$X_L = X_L^0 + [X_S/k - X_L^0](1 - x'/\delta')^2 \quad \text{for} \quad 0 \leqslant x' \leqslant \delta' \tag{1.214}$$

where $v_{\delta'}$, is the velocity parallel to the phase interface at the outer side of the diffusion boundary layer ($x' = \delta'$) and X_S/k is the concentration of solute in the liquid phase at $x = 0$ according to equation (1.192). From equation (1.214) it follows that

$$\begin{aligned}(\partial X_L/\partial x')_{x'=0} &= -2[X_S/k - X_L^0]/\delta' \\ &= -2[X_L - X_L^0]/\delta'\end{aligned} \tag{1.215}$$

The value of δ' based on equation (1.214) is therefore twice the value of δ defined by equation (1.194), i.e.

$$\delta' = 2\delta \tag{1.216}$$

Substituting equation (1.216) into equation (1.195) we obtain

$$X_L = \frac{X_L^0 k}{1 - (\frac{1}{2}u\delta'/D)(1 - k)} \tag{1.217}$$

Substituting equations (1.213), (1.214) and (1.217) into equation (1.208) and solving for $v_{\delta'}$, yields

$$v_{\delta'} = \frac{g\beta X_L^0 u(1 - k)\,\delta'^3}{12\nu D[1 - (\frac{1}{2}u\delta'/D)(1 - k)]} \tag{1.218}$$

Substitution of equations (1.213) and (1.214) into equation (1.212) gives

$$(X_L^0 - X_S)\,u = \frac{2}{15}\frac{\mathrm{d}}{\mathrm{d}y}\{[X_S/k - X_L^0]\,v_{\delta'}\delta'\} \tag{1.219}$$

Substituting equations (1.217) and (1.218) into equation (1.219), dividing by uX_L^0 and rearranging yields

$$1 - \frac{k}{1 - (\frac{1}{2}u\delta'/D)(1 - k)} = \frac{\mathrm{d}}{\mathrm{d}y}\left\{\frac{1}{180}\;\frac{g\beta X_L^0 u(1 - k)^2\,\delta'^5}{\upsilon D^2[1 - (\frac{1}{2}u\delta'/D)(1 - k)]^2}\right\} \tag{1.220}$$

On introduction of the dimensionless groups

$$\sigma = \tfrac{1}{2}u\delta'/D = \delta/\delta_{max} \tag{1.221}$$

$$\xi = \frac{45\upsilon u^4 x}{8g\beta X_L^0 D^3} \tag{1.222}$$

equation (1.220) becomes

$$1 - \frac{k}{1 - \sigma(1 - k)} = \frac{\mathrm{d}}{\mathrm{d}\xi}\left\{\frac{(1 - k)^2\sigma^5}{[1 - \sigma(1 - k)]^2}\right\} \tag{1.223}$$

Transforming and integrating equation (1.223) yields the solution

$$\xi = (1 - k)\int_0^{\sigma} \frac{[5\sigma^4 - 3(1 - k)\,\sigma^5]\,\mathrm{d}\sigma}{(1 - \sigma)[1 - \sigma(1 - k)]^2} \tag{1.224}$$

Substitution of equation (1.221) into equation (1.217) finally yields the segregation fraction

$$\frac{X_B}{X_L^0} = \frac{k}{1 - \sigma(1 - k)} \tag{1.225}$$

Figure 1.40 is a plot of X_B/X_L^0 versus $\log \xi$ for various values of k.

Since $\xi \propto u^4$ it is seen that the segregation effect of natural convection will be greatest for u small, i.e., $\xi < 1$. In a transient heat transfer situation as described above, natural convection will tend to increase as the velocity decreases in time so the segregation effect superposed upon the purely diffusive effect will build up in time. With forced convection the effect is greatly enhanced.[98]

The foregoing should serve to illustrate how complicated mass transfer calculations become when convection is superposed upon the diffusion field. Crystal growers and chemical engineers are tackling such difficult problems by increasingly powerful methods.[100, 101]

REFERENCES

1 A. E. FICK: *Pogg. Ann.*, 1855, **94**, 59
2 D. ENSKOG: *Arkiv. Met. Astronomi Fyz.*, 1922, **16**, 16
3 S. CHAPMAN AND T. G. COWLING: 'The mathematical theory of non-uniform gases', 1939, Cambridge University Press
4 J. HIRSCHFELDER, C. F. CURTISS, AND R. B. BIRD: 'Molecular theory of gases and liquids', 1954, New York, John Wiley
5 E. L. CUSSLER: 'Diffusion: mass transfer in fluid systems', 1984, Cambridge University Press
6 R. B. BIRD, W. E. STEWART, AND E. N. LIGHTFOOT: 'Transport phenomena', 20, 253 *et seq.*, 1960, New York, John Wiley
7 B. J. ALDER AND T. E. WAINWRIGHT: *J. Chem. Phys.*, 1959, **31**, 459
8 A. EINSTEIN: *Ann. Phys. (Leipzig)*, 1905, **17**, 549
9 A. EINSTEIN: *ibid.*, 1906, **19**, 371
10 J. VOLKL AND G. ALEFIELD: in 'Diffusion in solids: recent developments' (ed. A. S. Nowick and J. B. Burton), 231 *et seq.*, 1975, New York, Academic Press
11 C. WELLS, W. BATZ, AND R. F. MEHL: *Trans. AIME*, 1950, **188**, 553
12 C. A. WERT: *Phys. Rev.*, 1950, **79**, 601
13 S. ARRHENIUS: *Z. Phys. Chem. (Leipzig)*, 1892, **10**, 51
14 C. ZENER: *Acta Cryst.*, 1950, **3**, 346
15 F. SEITZ: *Ibid.*, 355
16 E. HART: *Acta Met.*, 1957, **5**, 597
17 A. D. LECLAIRE AND A. RABINOVICH: 'Diffusion in crystalline solids', (ed. G. E. Murch and A. S. Nowick), 259, 1984, Orlando FA, Academic Press
18 A. D. LECLAIRE: in 'Solute-defect interaction: theory and experiment', (ed. S. Saimoto, G. R. Purdy, and G. V. Kidson), 251, 1986, Toronto, Pergamon
19 G. V. KIDSON AND J. S. KIRKALDY: *Phil. Mag.*, 1969, **20**, 1057
20 W. K. WARBURTON AND D. TURNBULL: in 'Diffusion in solids: recent developments', (ed. A. S. Nowick and J. B. Burton), 171, 1975, New York, Academic Press
21 A. M. STONEHAM: *Ber. Bunsenges.*, 1972, **76**, 816
22 J. S. KIRKALDY: in 'Energetics in metallurgical phenomena, Vol. 4', (ed. W. M. Mueller), 197, 1968, New York, Gordon and Breach
23 L. S. DARKEN: *Trans. AIME*, 1948, **174**, 184
24 S. GLASSTONE, K. J. LAIDLER, AND H. EYRING: 'The theory of rate processes', Ch. 9, 1941, New York, McGraw-Hill
25 C. WERT AND C. ZENER: *J. Appl. Phys.*, 1950, **21**, 5

26 J. E. LANE AND J. S. KIRKALDY: *Can. J. Phys.*, 1964, **42**, 1643
27 C. H. BENNETT: in 'Diffusion in solids: recent developments', (ed. A. S. Nowick and J. B. Burton), 91, 1975, New York, Academic Press
28 A. M. BROWN AND M. F. ASHBY: *Acta Met.*, 1980, **28**, 1085
29 H. BAKKER: in 'Diffusion in crystalline solids', (ed. G. E. Murch and A. S. Nowick), 189, 1984, Orlando FA, Academic Press
30 T. YAMAMOTO: D. Eng. Thesis, 1984, Hokkaido Institute of Technology, Sapporo, Japan
31 W. SEITH: 'Diffusion in Metallen', 1955, Berlin, Springer
32 C. J. SMITHELLS: 'Metals reference book', 6th edn, (ed. E. A. Brandes), 13-1-97, 1983, London, Butterworths
33 Y. ADDA AND J. PHILIBERT: 'La diffusion dans les solides, Vol. II', 1966, Paris, Presse Universitaires de France
34 J. H. HILDEBRAND: 'Viscosity and diffusivity, a predictive treatment', 1977, New York, John Wiley
35 V. MARONI: 'US Department of Energy environmental control symposium, Vol. 2', Nov. 28–30, 1978
36 R. TRIVEDI AND G. M. POUND: *J. Appl. Phys.*, 1967, **38**, 3569
37 L. BOLTZMANN: *Ann. Phys.*, 1894, **53**, 960
38 C. MATANO: *Japan Physics*, 1933, **8**, 109
39 J. S. KIRKALDY: *Acta Met.*, 1956, **4**, 92
40 J. CRANK: 'Mathematics of diffusion', 1956, Oxford University Press
41 T. S. BUYS AND K. DE CLERK: *Anal. Chem.*, 1976, **48**, 585
42 S. H. ALGIE: *ibid.*, 1977, **49**, 186
43 P. G. SHEWMON: 'Diffusion in solids', 164 *et seq.*, 1963, New York, McGraw-Hill
44 C. ZENER: *J. Appl. Phys.*, 1949, **20**, 950
45 H. S. CARSLAW AND J. C. JAEGER: 'Conduction of heat in solids', 2nd edn, 1959, Oxford, Clarendon Press
46 D. TURNBULL: in 'Atom movements', (ed. J. H. Hollomon), 129, 1951, Cleveland, ASM
47 J. C. M. HWANG AND R. W. BALLUFFI: *J. Appl. Phys.*, 1979, **50**, 1339
48 J. C. M. HWANG, J. D. PAN, AND R. W. BALLUFFI: *ibid.*, 1349
49 R. W. BALLUFFI: in 'Diffusion in crystalline solids', (ed. G. E. Murch and A. S. Nowick), 320, 1984, Orlando FA, Academic Press
50 F. C. FRANK: *Disc. Faraday Soc.*, 1949, **5**, 48
51 N. CABRERA AND W. K. BURTON: *ibid.*, 33, 40
52 W. W. MULLINS: *J. Appl. Phys.*, 1959, **30**, 77
53 A. D. LECLAIRE: *Brit. J. Appl. Phys.*, 1963, **14**, 351
54 J. C. FISHER: *J. Appl. Phys.*, 1951, **22**, 74
55 R. T. WHIPPLE: *Phil. Mag.*, 1954, **45**, 1225
56 G. YOUNG AND R. E. FUNDERLIC: *J. Appl. Phys.*, 1973, **44**, 5151
57 H. S. LEVINE AND C. J. MACCALLUM: *ibid.*, 1960, **31**, 595
58 V. G. BORISOV, V. M. GOLIKOV, AND B. Y. LYMBOV: 'Radioisotopes in scientific research', (ed. R. C. Extermann), 212, 1957, New York, United Nations
59 V. G. BORISOV AND B. Y. LYMBOV: *Fiz. Met. Mettallov.*, 1958, **1**, 298
60 A. D. LECLAIRE AND A. RABINOVICH: *J. Phys. C. (Solid State Phys.)*, 1981, **14**, 3863
61 N. W. MCLACHLAN: 'Bessel functions for engineers', 1955, London, Oxford University Press
62 C. HERTZIG, P. NEUHAUS, AND J. GEISE: 'Solute–defect interaction: theory and experiment', (ed. S. Saimoto, G. R. Purdy, and G. V. Kidson), 271, 1986, Toronto, Pergamon
63 E. C. AIFANTIS: *Acta Met.*, 1979, **27**, 783
64 E. C. AIFANTIS: *J. Appl. Phys.*, 1979, **50**, 1334
65 J. H. HILL: *Scripta Met.*, 1979, **13**, 1027
66 A. I. LEE AND J. H. HILL: *J. Math. Anal. Appl.*, 1982, **89**, 530
67 A. I. LEE AND J. H. HILL: *Acta Mechanica*, 1983, **46**, 23
68 H. L. TOOR: *Chem. Eng. Sci.*, 1965, **20**, 941

69 R. J. BRIGHAM: M.Sc. Thesis, 1963, McMaster University
70 G. R. PURDY AND M. HILLERT: *Acta Met.*, 1978, **26**, 833
71 K. SMIDODA, W. GOTTSCHALK, AND H. GLEITER: *ibid.*, 1833
72 I. G. SOLORZANO AND G. R. PURDY: *Metall. Trans. A*, 1984, **15**, (6), 1055
73 I. G. SOLORZANO, G. R. PURDY, AND G. C. WEATHERLY: *Acta Met.*, 1984, **32**, (10), 1709
74 T. WICHERT, G. GRÜBEL, AND M. DEICHER: in 'Solute–defect interaction: theory and experiment', (ed. S. Saimoto, G. R. Purdy, and G. V. Kidson), 75, 1986, Toronto, Pergamon
75 L. M. HOWE AND M. L. SWANSON: 'Solute-defect interaction: theory and experiment', (ed. S. Saimoto, G. R. Purdy, and G. V. Kidson), 43, 1986, Toronto, Pergamon
76 F. E. HARRIS: *Met. Progr.*, Aug. 1943
77 J. S. KIRKALDY: *Trans. SAE*, 1977, **86**, 770415
78 C. KIM: *J. Electrochem. Soc.*, 1984, **131**, 885
79 W. A. TILLER, K. A. JACKSON, J. W. RUTTER, AND B. CHALMERS: *Acta Met.*, 1953, **1**, 428
80 B. CHALMERS: 'Principles of solidification', 1964, New York, John Wiley
81 D. VENUGOPALAN AND J. S. KIRKALDY: *Acta Met.*, 1984, **32**, 893
82 W. JOST: 'Diffusion in solids, liquids, gases', 1952, New York, Academic Press
83 G. R. PURDY AND J. S. KIRKALDY: *Trans. AIME*, 1963, **227**, 1255
84 W. W. SMELTZER, R. R. HAERING, AND J. S. KIRKALDY: *Acta Met.*, 1961, **9**, 880
85 J. M. PERROW, W. W. SMELTZER, AND J. D. EMBURY: *ibid.*, 1968, **16**, 1209
86 P. KOFSTAD, K. HAUFFE, AND H. KJOLLESDAL: *Acta Chim. Scand.*, 1958, **12**, 239
87 D. TURNBULL: *Acta Met.*, 1953, **1**, 764
88 R. VILELLA: (reproduced by L. S. Darken and W. C. Leslie) in 'Decomposition of austenite by diffusional processes', (ed. V. F. Zackay and H. I. Aaronson), 253, 1962, New York, Interscience
89 M. HILLERT: *Jernkont. Ann.*, 1957, **141**, 757
90 G. BOLZE, M. P. PULS, AND J. S. KIRKALDY: *Acta Met.*, 1972, **20**, 73
91 C. ZENER: *Trans. AIME*, 1946, **167**, 550
92 M. HILLERT: D.Sc. Thesis, 1956, MIT, Cambridge, Mass.
93 M. HILLERT: *Acta Met.*, 1961, **9**, 525
94 R. KIKUCHI: *Ann. Phys.*, 1960, **10**, 127
95 J. W. CAHN: *Acta Met.*, 1961, **9**, 795
96 J. W. CAHN: *ibid.*, 1966, **14**, 1685
97 J. S. LANGER: *ibid.*, 1973, **21**, 1649
98 C. WAGNER: *J. Met.*, Feb. 1954, 154
99 T. VON KÁRMÁN: *Letech. angew. Mathematik Mechanik*, 1921, **1**, 233
100 S. R. CORIELL, M. R. CORDES, W. J. BOETLINGER, AND R. F. SEKERKA: *J. Cryst. Growth*, 1980, **49**, 13
101 P. A. CLARK AND W. R. WILCOX: *ibid.*, **50**, 461

CHAPTER 2

Diffusion mechanisms in solids and liquids

2.1 INTRODUCTION

In principle, diffusion in the condensed state can be classified as a multibody problem in quantum mechanics. Through sheer complexity such a multibody viewpoint is not attractive from a theoretical standpoint, although single particle quantum effects cannot be ignored in the diffusion of helium and hydrogen isotopes.[1] Thus from the start an averaging structure is usually introduced which is in one way or another derivative of equilibrium statistical mechanics. The most non-committal approach as regards mechanisms (Onsager's Theory, cf Chapter 3) invokes the statistical theory of equilibrium fluctuations and the average laws of their decay as applicable to irreversible processes such as diffusion.

For low density gases the laws of large numbers are most successful since both free paths and momenta (or velocities) can be represented as statistical variables with precisely defined distributions and averages. The correspondingly short collision times obviate the need for precise knowledge of interparticle forces, so good predictions are possible (Chapman–Enskog Theory[2,3,4]). As the state of liquid condensation is approached, a clear conception of free trajectories disappears and collisions in course occupy most of the time. Indeed a theory of diffusion becomes intimately entwined with a theory of condensation with its attendant physical singularities (discontinuities in state variables or their derivatives). Clearly, the nature of the phase transition singularities and the force laws which define them now become of prime importance. The approach to diffusion through the methodology of Molecular Dynamics[5,6,7,8,9] which attempts to characterize actual trajectories through specific force laws, has enjoyed some success where such force laws can be independently established. For example, a condensed rare gas such as argon with its Van der Waals-like (fluctuating dipole) forces can be represented by hard sphere or Lennard-Jones potentials[10] (see § §2.8 and 2.15) to good effect and thereby accurate tracer diffusion coefficients can be predicted for the liquid state. Such simple potentials, while often applied to metallic systems, are however difficult to conceptualize in the context of the metallic (delocalized electron) bond.

Linear Response Theory, which is considered briefly in § §2.9 and 9.5, is an attempt to bridge the generality of the Kinetic Theory of Gases and the specificity of Molecular Dynamics. It is, in fact, a sophisticated and detailed version of Onsager's Theory.[11]

The ordered configurational nature of the solid state, which is a consequence of the free energy trend towards saturation of the bonds at low temperatures, exhibits a distinct simplification through the appearance of site singularities including point defects. These define discrete diffusion 'free' paths and 'collision' or jump times, thus abetting a quantitative random walk approach to the problem. The other side of the coin lies in the pre-eminence of the detailed, and at best semi-empirical, force laws required to determine the jump or 'activation' barriers.

2.2 POINT DEFECT FORMATION AND ITS ROLE IN DIFFUSION

Interestingly enough, the idea that point defects might play a part in diffusion was first emphasized in detail for liquids. Eyring and co-workers[12] hypothesized that diffusion occurred via vacant sites which existed in equilibrium with the liquid in accordance with the mass law and the free energy of formation of such sites. While the importance of such defects in liquids has now been discounted,[13] they are known to play an essential role in diffusion in solids. Following Howard and Lidiard,[14] the Gibbs Free Energy for a pure monatomic solid consisting of N atoms and N_V vacancies can be written as

$$G(T,p) = G_0(T,p) + N_V g_V - kT \ln \frac{(N+N_V)!}{N!\,N_V!} \tag{2.1}$$

where G_V is the free energy of formation of a vacancy. One may think of this as arising from the dominant enthalpy change of breaking the bonds and carrying an atom to the surface, corrected via a thermal entropy term associated with the perturbed lattice vibrations. In practice, the vacancies are more likely to emerge from sources and sinks such as dislocations and grain boundaries. The trailing term in equation (2.1) represents the excess configurational entropy introduced by N_V vacancies, assuming that there are no interactions between the vacancies. The free energy minimum which represents the stable state of the array is given by the chemical potential of the vacancies

$$\left(\frac{\partial G}{\partial N_V}\right)_{T,p,N} = \mu_V = 0 \tag{2.2}$$

Evaluating this via equation (2.1) using Stirling's formula to approximate the factorials (Appendix 3) one obtains

$$X_V = \frac{N_V}{N+N_V} = \exp(-g_V/kT) \tag{2.3}$$

Since g_V for metals or inert gas solids is typically large (of order $9kT$ near the melting point) X_V is generally $<0{\cdot}001$, decreasing rapidly with temperature. Notwithstanding this small concentration, it provides the vehicle for diffusion in most substitutional solids.

In binary alloys or mixtures the vacancies may bond preferentially to one or the other constituent. Here, following Howard and Lidiard, the free energy for a dilute solution of atoms b can be written as

$$G = G_0(p,T) + N_b g_b + N_V g_V + N_b \pi \Delta g_V - kT \ln \Omega \tag{2.4}$$

where $k \ln \Omega$ is the configurational entropy, π is the number of vacancy solute pairs, g_V is the free energy of formation of the vacancy in the pure solvent, $g_V + \Delta g_V$ is the free energy of formation of a vacancy next to an impurity atom and g_b is the free energy of addition of a solute atom to the crystal. The minimization of the free energy with respect to the number of pairs yields the formula

$$\frac{X_\pi}{(X_V - X_\pi)(X_b - X_\pi)} = z \exp(-\Delta g_V/kT) \tag{2.5}$$

where z is the coordination number. Comparing this with equation (1.98) one recognizes equation (2.5) as The Law of Mass Action for pair association. Minimization with respect to N_V yields

$$\left(\frac{\partial G}{\partial N_V}\right)_{p,T,N_a,N_b,N_p} = \mu_V = 0 \tag{2.6}$$

Together with equation (2.5), assuming weak association as in metals, this yields

$$X_\pi \cong X_b z \exp[-(g_V + \Delta g_V)/kT] \tag{2.7}$$

and

$$X_V \cong \exp[-g_V/kT)\{1 - zX_b[1 - \exp(-\Delta g_V/kT)]\} \tag{2.8}$$

When the vacancy is attracted to the solute atom $(\Delta g_V < 0)\, X_V$ is enhanced, which can lead to an enhancement of the solvent self-diffusion coefficient. This effect has been reported by Hehenkamp.[15] The most important conclusions of the alloy calculation are that the formula for X_V is unchanged from the pure material case if association is weak and/or the solution is very dilute ($X_b \ll 1$) and the chemical potential of the vacancies vanishes. This latter fact is of considerable consequence to the theory of local equilibrium diffusion. The existence of vacancies is assumed in the random walk considerations of the following section.

Because of charge neutrality requirements in ionic solids, the important point defects are somewhat more complex. These are discussed in later sections on diffusion in such materials (cf §2.18 and §8.2).

2.3 RANDOM WALK TREATMENT OF DIFFUSION

An actual diffusion process during a finite time τ involves a discrete number of atom trajectories, as indicated in Fig. 1.3. The Fick equation (1.1) may be used to define rigorously the diffusion coefficient, i.e.

$$D = -\lim_{J \to 0} J/\nabla C \tag{2.9}$$

and is recognized thereby to deal with an average over a sufficient number of trajectories. Within the same ideal zero gradient limit we can take Fick's second equation (1.49) with D constant as exact and obtain from its δ-function solution (1.73) the single particle distribution function

$$P = \frac{1}{2(\pi Dt)^{1/2}} \exp\left(-\frac{x^2}{4Dt}\right) \tag{2.10}$$

which for fixed time has the form of a Gaussian. One can now evaluate the mean square distance of one-dimensional diffusion as[16]

$$\langle x^2 \rangle = \int_{-\infty}^{\infty} x^2 P(x)\,\mathrm{d}x = 2Dt \tag{2.11}$$

Random walk theory, due originally to Smoluchowski[17] and Einstein,[18, 19] deals with the direct estimation of $\langle x^2 \rangle$ from the statistical analysis of atom trajectories for models corresponding to various structures and mechanisms of diffusion. In general, where such calculations are feasible, rather sophisticated mathematical methods are required.[16] However, for crystals of high symmetry where distance and mean time increments are constant, the appropriate trajectory sums can be easily evaluated.

For an individual square displacement of n increments

$$x^2 = (x_1 + x_2 + x_3 + \ldots x_n)^2 \tag{2.12}$$

the diagonal product terms can be collected so that

$$x^2 = \sum_{i=1}^{n} x_i^2 + 2\sum_{j=1}^{n-1} \sum_{k=j+1}^{n} x_j x_k \tag{2.13}$$

Vacancies and dilute interstitials in close packed crystals follow close to random walk trajectories. In such processes the terms in the second summation are equally likely to be negative or positive, so when averaged over many trajectories the contribution to the mean approaches zero. Accordingly the mean displacement can be written

$$\langle x^2 \rangle = \left\langle \sum_{i=1}^{n} x_i^2 \right\rangle = \sum_{i=1}^{n} \langle x_i \rangle^2 \tag{2.14}$$

since the mean of the sum equals the sum of the means. Thus, for a single jump distance λ equation (2.14) yields the root mean square displacement $\sqrt{\langle x^2 \rangle} = \sqrt{n}\lambda$ indicating the 'law of diminishing displacement' for diffusion mentioned previously. If $\langle x^2 \rangle$ is divided by $2t$ to define the diffusion coefficient via equation (2.11) and it is recognized that the forward jump frequency $n/t = \nu/3$ where ν is the isotropic jump frequency as defined in cubic crystals, the well-known expression

$$D = \tfrac{1}{6}\nu\lambda^2 \tag{2.15}$$

is obtained. This is in fact a slightly modified version of expression (1.6) obtained earlier via the kinetic theory of gases. The two expressions differ only in the interpretation of the frequency ν. While these random walk theoretical structures are very elegant and have been extended to anisotropic structures by Manning and others they give no clue as to the concentration and temperature dependencies of the frequency ν, so are of limited value in quantifying the applications of diffusion theory.

In § 1.22 formula (1.17) due to Hart[20] was applied to a problem of tracer diffusion through a defective oxide film containing a fraction of high diffusivity sites (in grain boundaries or dislocations). This formula, which has a certain intuitive compulsion, can be justified on the basis of random walk arguments.[20, 21] One supposes that an average tracer atom makes a total of n jumps, n_L in the lattice and n_B in defective material, and assumes a detailed balance between the defective and non-defective regions such that the partitioned time of occupancy of an average atom in a given region is equal to the number of jumps in that region times the mean jump time (1/frequency). Assuming that tracer atoms do not segregate to defects and that the site density is the same in defective and non-defective material, the site fraction f is equal to the ratio of the time t_B which an average atom spends in defective regions to the total time ($t = t_B + t_L$). To obtain the Hart equation (1.5) one then simply partitions the jumps within the sum of squares in equation (2.13) according to n_L and n_B, dropping the cross terms, divides the entire expression by $2t$, multiplies the partitioned terms by $t_B/t_B = 1$ and $t_L/t_L = 1$, respectively, and completes the averaging. As already noted in §1.15, Aifantis[22] and Hill[23] have examined more rigorous models of this process.

A synthetic approach to the random walk method has been developed by Lidiard[24] among others. Koiwa and Ishioka[25] have developed advanced applications for defective and alloyed crystals.

2.4 CORRELATION FACTORS DEFINED

The two examples of random walk systems which were cited above (vacancy and dilute interstitial diffusion) were assured randomicity by universal site accessibility and the condition of solution diluteness. When the diluteness condition is not met or when multiple species interactions are involved, the trailing terms in equation (2.13) do not average to zero. The interactions are prone to destroy randomicity and so are identified as correlation effects. An auto- or self-correlation factor f (not to be confused with the site fraction f in the previous section) can be defined, and evaluated for fixed jump distance λ as in cubic crystals, as

$$f = \langle x^2 \rangle_c / \langle x^2 \rangle_r = \langle x^2 \rangle_c / n\lambda^2 \qquad n \to \infty \tag{2.16}$$

where the numerator is defined by the average of relation (2.13) and the denominator by equation (2.14). Since correlations imply restrictions on accessible trajectories we may conclude that $f \leqslant 1$. The precise analytic evaluation of f is generally a very involved procedure.[16, 26, 27, 28] However, Monte Carlo simulations of diffusion via the computer allow the accurate evaluation of the correlation factors of crystals of arbitrary complexity.[29]

2.5 TRACER AUTOCORRELATIONS

Tracer or self-diffusion coefficients are usually determined by measuring the evolving concentration profile of a dilute radioactive isotope A^* in pure isotope A. Because of the high sensitivity of concentration measurements by radioactive emissions the solution of A^* can be exceptionally dilute. Now if the diffusion process is by exchange with conserved vacancies in a metal say, which specie is also very dilute, then the conditions for evaluating (2.16) with rigour are optimal. Indeed, the exact calculations are based on the successive encounters of one vacancy with one tracer atom. If for simple systems f is interpreted as the ratio of effective vacancy exchanges to actual vacancy exchanges then, following Shewmon,[21] a useful estimate can be obtained.

Consider, for example, the set of encounters between individual A^* atoms and vacancies in a crystal for which the coordination number is z. Of the z random interchanges which the vacancy can make after a designated random exchange with a chance A^* in its coordination shell, two exchanges, the random and a correlated one, will on the average be annulled because of prompt reinterchange of these species. Thus the correlation factor can be estimated as

$$f = 1 - \frac{2}{z} \tag{2.17}$$

Table 2.1 gives precise values of f[30] in comparison with this simple formula for cubic crystals.

Precise values of f for other structures are summarized by Shewmon,[21] Manning,[16] and Murch.[29] The significance of the correlation factor in binary non-ionic systems lies in the fact that from the measured tracer coefficient D_{A^*} we can exactly predict a vacancy diffusion coefficient D_V which is applicable when this species is initially non-uniform in a pure material and conserved. Indeed, we recognize from the foregoing random walk considerations that for a given time τ

$$\frac{\langle x^2 \rangle_{A^*}}{\langle x^2 \rangle_V} = \frac{D_{A^*}}{D_V} = fX_V \tag{2.18}$$

This formula will be needed in later discussions of ternary diffusion and of the Darken Equation.

There is a complex theory which attempts to relate f to the difference in the diffusion coefficients of the two isotopes in a mixture, but generally speaking one or more of the key assumptions has not been validated (cf Manning[16]). In ionic solid diffusion, where drift can be initiated by an electric field, there is an analysis which accords a clear empirical significance to f. It will be introduced in §2.18.

2.6 VACANCY ASSOCIATED CORRELATIONS IN PURE MATERIALS

There are vacancy associated correlation or wind effects in pure, or nearly pure, materials which are very small in magnitude relative to the tracer effects, but which are nonetheless important because of the proportionately high mobility.

Let us consider the diffusion of pure A countercurrent to conserved vacancies in the limit of vanishing concentration differences so that X_V is roughly constant. As usual, it is assumed that each and every vacancy jump can be designated as random. Furthermore, the random walk of a single vacancy leaves the A-manifold

Table 2.1 Comparison of approximate and exact values of *f*

Structure	$1 - \frac{2}{z}$	f
Simple cubic	0·67	0·655 5
BCC	0·75	0·721 5
FCC	0·83	0·781 4

essentially unchanged so the A jumps in such a process must be 100% correlated.

To evaluate the solvent correlation factor, one proceeds as with the tracer argument. We consider a given independent, conserved vacancy and designate a random interchange between the vacancy and an A. But the latter, by the very act of interchange, is marked as a tracer, say as $A^{\dagger}$. Thus the previous argument applies to this individual and we may assign to it the tracer autocorrelation factor f. Since the mole fraction of such individuals is X_V at any instant, the fraction of uncorrelated A jumps or the average auto-correlation factor for the manifold of A atoms is

$$\phi = X_V f \tag{2.19}$$

The fraction of correlated jumps is $1 - X_V f$ and, as expected, this approaches unity as $X_V \rightarrow 0$.

In simple countercurrent diffusion between A and vacancies, this abstract effect has no consequence since the A diffusion coefficient must be exactly equal to the random vacancy coefficient D_V. However, if with essentially uniform A ($\partial C_A/\partial X \simeq 0$) a vacancy wind is created by countercurrent diffusion of very dilute tracer A* and vacancies ($X_{A^*} \ll X_V$) then the fraction of the vacancy wind intercepted by the uniformly disposed A atoms is $1 - X_V f$, which is the correlated fraction of A. The remainder of the vacancies are exchanged with the tracers under the condition that

$$\frac{\partial C_A}{\partial X} = -\frac{\partial C_V}{\partial X} \quad \text{or} \quad -\frac{J_{A^*}}{J_V} = \frac{D_A^*}{D_V} = X_V f \tag{2.20}$$

This accords with the tracer correlation results of §2.5. The extension and application of this solvent correlation effect appears in §6.4.2.

2.7 CORRELATION FACTORS IN BINARY ALLOYS

In alloys it is necessary to assign correlation factors to the tracer components A* and B* as well as the vacancies, and since the same jump frequencies depend on specific configurations of the three species in the first, if not the second coordination shell their evaluation is extremely difficult for other than dilute alloys.[16,31] To simplify the calculations for rich alloys Manning has defined a 'random' alloy which is normalized to the properties of an isotopic mixture and therein he strongly averages over the spectrum of frequencies which must exist in a real alloy. It is first assumed that the frequencies ω_A and ω_B for exchange with a vacancy depend only on the identity of the jumping atom, ignoring the identity of adjacent atoms. Equivalently, there can be no tendency for vacancy binding to a particular species. Thus an average equilibrium vacancy jump frequency can be defined as

$$\omega = \omega_A X_A + \omega_B X_B \tag{2.21}$$

Next one estimates with Manning[16] the alloy tracer autocorrelation factors by proportioning the annulled fraction of alloy jumps from the isotopic alloy fraction, $1 - f$, in accord with the relative tracer coefficients of the two tracer species, i.e.

$$f_A = 1 - \frac{D_A^*}{\overline{D}}(1-f) \qquad f_B = 1 - \frac{D_B^*}{\overline{D}}(1-f) \tag{2.22}$$

where $\overline{D}$ is the weighted tracer average analogous to equation (2.21).

These imply that the average elemental autocorrelation factor is

$$\overline{f} = \omega_A f_A + \omega_B f_B = f \tag{2.23}$$

In §6.4.4 these relations are given a different interpretation.

Since, in an alloy, the vacancies do not possess homogeneous trajectories, successive jumps can be correlated. Suppose that $\omega_A > \omega_B$. The vacancy which exchanges with A has a greater than average probability of making a reverse jump while that which exchanges with a B has a lower than average probability and there will be a net decrement in the effective jumps of vacancies. To evaluate the average decrement the effective jump frequency of vacancies as determined by the effective vacancy–alloy exchange frequencies is formulated as

$$\omega_V = X_A \omega_A f_A + X_B \omega_B f_B \tag{2.24}$$

and it is recognized that within the isotope-normed model with conserved vacancies the uncorrelated effective vacancy jump frequency is $\omega\overline{f}$. Thus the vacancy autocorrelation factor, defined as the ratio of actual to effective jump frequency is[16]

$$f_V = \omega_V/\omega f = X_A \frac{\omega_A}{\omega} \cdot \frac{f_A}{f} + X_B \frac{\omega_B f_B}{\omega\ f} = X_A \frac{\omega_A}{\omega} f_V^A + X_B \frac{\omega_B}{\omega} f_V^B \tag{2.25}$$

where

$$f_V^A = f_A/f \qquad \text{and} \qquad f_V^B = f_B/f \tag{2.26}$$

are the partial vacancy correlation factors for exchange with A and B atoms, respectively. As expected f_V is always less than unity. However, it can reach negative values at low concentrations and highly differing frequencies. This apparently anomolous behaviour has been discussed by Manning,[16] Heumann,[32] and Kirkaldy.[33] The random alloy model has received considerable support from Monte Carlo calculations[29] and experiments in fcc alloys.[34]

Correlation factors have been the subject of an exceptionally large volume of sophisticated fundamental research. Yet this research has not made a strong impact on applications of diffusion theory to practical problems. Potentially, one of the most important applications has pertained to the Darken equation which relates mutual to tracer diffusion coefficients. However, as will be noted in Chapter 3, such correlation effects have not made a substantial quantitative impact on practical diffusion predictions at the current level of experimental precision. Currently argued vacancy correlation corrections to the chemical diffusion coefficient are of the order of 10 or 20% which will be reflected in diffusion penetration distances at the 5 and 10% level, which is generally well within the reproducibility between laboratories.

2.8 DIFFUSION STUDIES VIA MOLECULAR DYNAMICS

The advent of the fast, high capacity computer has made feasible a rather complete modelling of collision processes in gaseous and condensed matter (cf Fig. 1.3). This is a semiempirical rather than a theoretical approach since for precision it is necessary to include realistic potential functions for the interaction of diffusing entities. Reliable numerical potentials of this kind can only be obtained by independent experiments on compressibility and like functions.

The calculation of multibody trajectories by high speed computer in gas, liquid and solid model systems has been extensively explored by Alder, Rahman, and their associates.[5, 6, 7, 8, 35, 36] Since molecular diffusion is generally a high-temperature phenomenon, classical particle mechanics, together with realistic average intermolecular potentials may be deemed an adequate representation of the molecular dynamics. The exceptions are for He and H isotopes. The evaluation of realistic potential functions which must ultimately accord with thermodynamic, kinetic phase transformation and spectroscopic data, and thermomechanical properties of the condensed state is a complex problem in itself[37] so most of the early investigations have assumed model pair potentials such as the Lennard-Jones 6–12[10] (cf § 1.3) or square well potentials. For example, the potential

$$\phi(r) = 4\varepsilon\left[\left(\frac{\sigma}{r}\right)^{12} - \left(\frac{\sigma}{r}\right)^{6}\right] \tag{2.27}$$

with $\varepsilon/k = 119{\cdot}8$ K and $\sigma = 0{\cdot}340\,5$ nm[38, 39] is a widely used model of the argon pair potential. Systems of 1 000 or more atoms can be easily dealt with using modern computers and periodic boundary conditions. In this procedure the array of particles is confined to a parallelepiped in such a way that an individual exiting through one face instantaneously re-enters through the opposite face at the same velocity, the potential propagating across the boundary in such a way as to simulate an infinite system. Energy, linear momentum, volume and atom numbers are conserved via such periodic boundary conditions. The input data are the initial position and velocities of all the N particles together with the classical Hamiltonian governing the motion. The N equations of motion are solved simultaneously by extrapolating the trajectories for periods $\sim 10^{-14}$ s, then re-evaluating the forces and accelerations at the new positions, and so on. The calculations are simpler and faster if a discontinuous potential such as a square well is used, for then the extrapolation can extend between one binary collision and the next. This was the procedure used by Alder and Wainwright[5] in arriving at the results of Fig. 1.3.

Alder and Wainwright[5] emphasize that one of the aims of the methodology is to compare the results with analytical theories, the computer 'experiment' being more clear-cut, informative and detailed than an actual experiment. The method is ideally suited to the study of phase transitions and nucleation behaviour. Since all the trajectories are recorded (Fig. 1.3) both kinetic and equilibrium averages can be extracted by appropriate averaging over trajectories which have randomized (more than ten collisions).

In a more recent study Dymond and Alder[9] have optimized a numerical pair potential for argon which generates consistent and experimentally valid self-diffusion and thermodynamic data. This potential is not, however, successful in predicting other observable properties.[39] Nonetheless, it lends some credence to the main tenet of Onsager's Fluctuation Theory (cf Chapter 3) and the related Linear Response Theory to the effect that non-equilibrium kinetics is uniquely determined by the laws of fluctuations at equilibrium. Rahman[35, 36] has demonstrated the importance of correlated motions in liquid argon by this methodology. We indicate below how Molecular Dynamics studies have influenced the theories of liquid diffusion (§2.16).

The application of the Molecular Dynamics methodology and the related Monte Carlo procedures[29] to the study of solids, with their pervasive point and line defects, is developing rapidly.[29, 38] The increasing capacity and speed of computers will undoubtedly help research in this area.

2.9 LINEAR RESPONSE THEORY

Physicists and chemists have long sought an accurate analytic theory of transport processes in the condensed state based on non-equilibrium statistical mechanics. Linear response theory is such a theory and consists of a first order perturbation expansion of the equilibrium distribution function. It is in fact an elegant synthesis of Einstein's idealized random walk theory and Onsager's general theory based on the decay rate of fluctuations at equilibrium[11] (cf §3.8). The foundations are to be found in the works of Kubo[40] who coined the phrase 'linear response', and of Kirkwood and associates.[41, 42] Allnatt[43, 44, 45, 46, 47] has undertaken the application to solids with point defects. Hertz[48] has focused on diffusion in ionic liquids. The treatment of Kirkwood and Fitts[49] for multicomponent diffusion processes with applications by McCall and Douglass[50] and Douglass and Frisch[51] are particularly transparent. The key conclusion is that the rate coefficients in irreversible processes can be expressed in terms of momentum or velocity time correlation functions. For example, a tracer coefficient can be expressed as an integral over the velocity–time autocorrelation function, i.e.

$$D^* = \frac{1}{3}\int_0^{\tau} \langle v_\alpha(s)\, v_\alpha(0)\rangle \, \mathrm{d}s \tag{2.28}$$

where the average $\langle\,\rangle$ is taken over all atoms α following a trajectory from time $t = 0$ to $t = s$. Such trajectories are determined ultimately by the intermolecular forces and can be evaluated on the average through cluster expansion and Green's function methods or by molecular dynamics calculations.[42, 46] How such a formula comes about through a rigorous statistical analysis can be comprehended by comparing it with the elementary random walk formula (2.15). Indeed, the correlation corrected formula (2.18) for the tracer coefficient can be obtained through operations such as described by equation (2.28).[46] We will return to the applications of this methodology in Chapter 9.

Allnatt and Lidiard[52] have presented a comprehensive review on statistical theories of atomic transport in crystalline solids.

2.10 ABSOLUTE REACTION RATE THEORY APPLIED TO BINARY NON-IONIC DIFFUSION

In §1.6 the methodology of Eyring's Absolute Reaction Theory[12] was outlined. An application to solid state diffusion has been given by Wert and Zener.[53] As typical of binary diffusion an alloy single crystal of isotropic structure is postulated in which the average movement of species is restricted to direct interchange of nearest neighbours in parallel planes. This requires distortion of the lattice, which is confined mainly to a region within a few atomic diameters of the interchanging atoms. During the distortion, an activated complex, consisting of the two interchanging species, occupies a metastable position midway between the two lattice sites originally occupied by the moving species. Since many liquid structures are quite close-packed, we expect that this model has some bearing on ternary liquid diffusion as well.[54]

As usual, in the transition state theory, the complex consisting of species i and species j in transition, $ij^\dagger$, is considered to be in thermodynamic equilibrium with the adjacent region of the alloy. The equilibrium constant $K^\dagger$ is given by (cf Appendix 1)

$$K_{ij}^\dagger = a_{ij}^\dagger/(a_i a_j) = \exp[-\Delta G_{ij}^\dagger/RT] \tag{2.28}$$

where a are the activities of each species referred to the pure components as standard states, and $\Delta G_{ij}^\dagger = \Delta G_{ji}^\dagger$ is the standard free energy of formation of a single complex from pure i and j. The frequency of transition per atom through the transition state is then given by

$$\Gamma_{ij} = \nu_{ij} X_{ij}^\dagger \tag{2.29}$$

where ν_{ij} is the frequency per atom of forward transitions through the activated state and $X_{ij}^\dagger$ is the mole fraction of the complex, obtained from equation (2.28) as

$$X_{ij}^\dagger = \exp[-\Delta G_{ij}^\dagger/RT]\, a_i a_j/\gamma_{ij}^\dagger \tag{2.30}$$

where $\gamma_{ij}^\dagger$ is the activity coefficient of the complex. The symmetry of Γ_{ij} is an expression of detailed balance which is related to Onsager Reciprocity (cf § 3.8).

We first consider the jump frequency $\Gamma_{i(a\to b)}^{i-j}$ of i atoms from a plane (a) by interchange with j atoms from the adjacent plane (b). By equations (2.29) and (2.30) the flux is

$$J_{i(a\to b)}^{i-j} = C\lambda\nu_{ij} \exp[-\Delta G_{ij}^\dagger/RT]\, a_i^{(a)} a_j^{(b)}/\gamma_{ij}^\dagger \tag{2.31}$$

C being the total number of atoms per unit volume of (111) plane, and λ the (111) lattice spacing. Similarly, the rate of the reverse process will be

$$J_{i(b\to a)}^{i-j} = C\lambda\nu_{ij} \exp[-\Delta G_{ij}^\dagger/RT]\, a_i^{(b)} a_j^{(a)}/\gamma_{ij}^\dagger \tag{2.32}$$

The overall flux J_i^{i-j} due to this process, in the positive direction a→b, is given by subtracting equation (2.32) from equation (2.31)

$$J_i^{i-j} = +C\lambda\nu_{ij} \exp[-\Delta G_{ij}^\dagger/RT](a_i^{(a)} a_j^{(b)} - a_i^{(b)} a_j^{(a)}/\gamma_{ij}^\dagger \tag{2.33}$$

If the activities with superscript (b) are written as a Taylor expansion in the lattice spacing λ with reference to plane (a), this becomes

$$J_i^{i-j} = -\frac{C\lambda^2}{RT}\nu_{ij} \exp[-\Delta G_{ij}^\dagger/RT]\frac{a_i a_j}{\gamma_{ij}^\dagger}\left(\frac{\partial\mu_i}{\partial x} - \frac{\partial\mu_j}{\partial x}\right) \tag{2.34}$$

the superscripts to the activities now being discarded as inconsequential. This detailed Fick-type formula can be applied to pairs of interacting species, including vacancies in binary and higher order alloys.

2.11 BINARY DIFFUSION BY AN INTERSTITIAL MECHANISM

For illustrative purposes carbon diffusion is considered in fcc iron (cf Figs. 1.5 and 1.8) with λ defined by adjacent (111) planes in the interstitial sublattice (Fig. 1.5). In this system the dilute carbon atoms designated j interchange with the interstitial sites which will be assigned the index i. Let us describe the solution as a ternary mixture of iron, interstitial carbon and interstitial sites and recognize that since the latter cannot enter the Gibbs–Duhem equation, $\partial\mu_i = 0$. Furthermore, the standard chemical potential $\mu_i^0 = 0$ so the activity $a_i = 1$. The frequency within equation (2.34) is estimated within statistical mechanics as

$$\nu_{ij} = \kappa\frac{kT}{h} \quad \text{or} \quad \kappa\frac{k\theta_D}{h} \tag{2.35}$$

where h is Planck's constant, θ_D is the Debye temperature and κ is a transmission coefficient which can be estimated through random walk considerations as 1/6 (1/3 from geometric considerations and 1/2 from the fact that the transition state relaxes with equal probability in the forward and backward direction). Equation (2.34) can now be expressed as

$$\begin{aligned} J_i &= -\left[\frac{C}{6}\frac{k\theta_D}{h}\lambda^2 \exp\left(\frac{\Delta S^\dagger}{R}\right)\frac{\gamma_j}{\gamma_{ij}^\dagger}\right]\left(1 + X_j\frac{d\ln\gamma_j}{dX_j}\right)\exp\left(-\frac{\Delta H^\dagger}{RT}\right)\nabla X_j \\ &= -D_0 \exp\left(-\frac{\Delta H^\dagger}{RT}\right)\nabla C_j \end{aligned} \tag{2.36}$$

where for iron–carbon D_0 and $\Delta H^\dagger$ take the empirical values corresponding to Fig. 1.7. The base frequency $k\theta_D/6h \sim 10^{13}\ s^{-1}$ and $\lambda^2 \cong 2 \times 10^{-14}\ mm^2$. Typical values of D_0 in metals are $\sim 1\ mm^2\ s^{-1}$. Hence the trailing term in D_0 involving the entropy of activation is of the order of 5. If $\gamma_j/\gamma_{ij}^\dagger$ is assumed to be unity and $X_j \rightarrow 0$ then $\Delta S^\dagger/R \sim 1{\cdot}6$.

One of course expects from the Second Law that there will be excess vibrational entropy associated with random activated complexes ($\Delta S^\dagger > 0$) and that the magnitude of this of will depend on the elastic modulus μ of the lattice. Furthermore, from the model of Fig. 1.8 it is apparent that $Q \simeq \Delta H^\dagger$ will also depend on the elastic modulus so there will exist a relation between $\Delta S^\dagger$, μ and $\Delta H^\dagger$. Zener[55] has estimated this relation to be

$$\Delta S^\dagger \cong -\frac{d(\mu/\mu_m)}{d(T/T_m)}\frac{\Delta H^\dagger}{T_m} = \beta\frac{\Delta H^\dagger}{T_m} \tag{2.37}$$

where T_m is the melting temperature of the solvent and β is typically 0·4. This suggests that $\Delta S^\dagger$ is always a positive fraction of $\Delta H^\dagger$. Formula (2.37) predicts $\Delta S^\dagger$ on the basis of empirical values of $\Delta H^\dagger$ to within a factor of about 3, which error, in view, of the assumption $\gamma_j/\gamma_j^\dagger = 1$ and the uncertainty in the base frequency (equation (2.35)), is not unexpected. The limited success of this argument suggests that reasonable predictions of $\Delta S^\dagger$, $\Delta H^\dagger$ and D might ultimately be achieved via the molecular dynamics methodology, appropriate pair potentials taking the place of the elastic modulus.

2.12 BINARY DIFFUSION BY A VACANCY MECHANISM (CF § 1.4)

Referring to equations (2.21) and (2.34) and identifying the vacancy species as j we recognize that relations of form (2.34) must apply to both species A and B and that in general $J_A \neq -J_B$. The theory is accordingly being applied in a frame of reference which is moving at a velocity v relative to the laboratory frame, the latter being defined by

$$J_A = -J_B \tag{2.38}$$

Since the theory refers the flows to definite paired lattice planes, this is designated as the lattice-fixed frame of reference and is formally defined by

$$J'_A + J'_B = -J_V \tag{2.39}$$

Now the laboratory frame fluxes can be written as[56]

$$J_A = J'_A + C_A v \tag{2.40}$$

and

$$J_B = J'_B + C_B v \tag{2.41}$$

where v is the velocity of the lattice-fixed frame with respect to the laboratory frame (see § 4.3). Movement of the lattice relative to the laboratory can be ascribed to a hydrostatic pressure gradient which would become non-zero in the event of unequal diffusive flows. Noting equation (2.38) we solve for

$$v = -\frac{J'_A + J'_B}{C_A + C_B} \tag{2.42}$$

or upon resubstitution

$$J_A = -J_B = X_B J'_A - X_A J'_B \tag{2.43}$$

Notwithstanding the local or prompt conservation of vacancies as implied by equation 2.39, the local equilibrium condition $\partial\mu_j = \partial\mu_V = 0$ is applied, assuming that non-interacting vacancies form a Henrian solution and that therefore the vacancy activity is equal to its unique standard value

$$a_V = X_V^0 \tag{2.44}$$

Also, in the lattice-fixed frame

$$J'_A = -D_A \nabla C_A \tag{2.45}$$

and

$$J'_B = -D_B \nabla C_B \tag{2.46}$$

Thus we finally obtain

$$J_A = -(X_B D_A + X_A D_B)\nabla C_A = -J_B \tag{2.47}$$

where from equation (2.34)

$$D_A = \frac{X_V^0}{6}\frac{k\theta_D}{h}\lambda^2 \exp\left(\frac{\Delta S_{AV}^{\dagger}}{R}\right)\exp\left(-\frac{\Delta H_{AV}^{\dagger}}{RT}\right)\left(1 + X_A\frac{\mathrm{d}\ln\gamma_A}{\mathrm{d}X_A}\right)(\gamma_A/\gamma_{AV}^{\dagger}) \tag{2.48}$$

and

$$D_B = \frac{X_V^0 k\theta_D}{6 \quad h}\lambda^2 \exp\left(\frac{\Delta S_{BV}^{\dagger}}{R}\right)\exp\left(-\frac{\Delta H_{BV}^{\dagger}}{RT}\right)\left(1 + X_B\frac{d\ln\gamma_B}{dX_B}\right)(\gamma_B/\gamma_{BV}^{\dagger}) \quad (2.49)$$

The quantities D_A and D_B are designated as the intrinsic coefficients[57] while

$$\tilde{D} = X_B D_A + X_A D_B \quad (2.50)$$

is designated as the chemical, mutual or interdiffusion coefficient.[56, 58] The intrinsic coefficients are both strongly depressed by the very small concentration X_V^0, which states simply that no jumps can occur unless a vacancy lies adjacent to a diffusing species. This strong restriction against diffusion is somewhat compensated for by the fact that the activation enthalpy for exchange between a vacancy and an atom need not be large (see below).

It will now be realized what a very complex function of concentration and temperature is the mutual diffusion coefficient, even if the entropy and enthalpy of vacancy formation, which appear in the X_V^0 term (equation (2.3)) and the entropy and enthalpy of migration are independent of temperature. Of course, many of the experiments have been carried out in relatively dilute terminal or tracer solutions so one or the other terms in equation (2.50) dominates, and an accurate Arrhenius behaviour is observed (equation (2.47) *et seq.*) over quite wide ranges of temperature. Such data are analysed via

$$\tilde{D} = D_0 \exp\left[-\frac{(\Delta H_f + \Delta H^{\dagger})}{RT}\right] \quad (2.51)$$

where the enthalpy of vacancy formation appears through X_V^0 in equations (2.48) and (2.49) (cf equation (2.3)) and where both enthalpies pertain to the solvent lattice. Clearly, diffusion measurements alone cannot yield separate values for ΔH_f and $\Delta H^{\dagger}$.

Simmons and Balluffi[59] conceived of a very delicate experiment for determining ΔH_f and ΔS_f independently of the migration values. In principle, according to equation (2.3), one needs to measure the concentration of vacancies as a function of temperature. This cannot be done by simple volume or linear expansion measurements since thermal expansion is also involved. However, one can substract this latter effect out by simultaneously measuring the lattice parameter d. Thus the atom fraction of vacancies expressed in terms of the fractional elongation $\Delta l/l$ and the fractional increase in the lattice parameter $\Delta d/d$ is

$$X_V = 3\left(\frac{\Delta l}{l} - \frac{\Delta d}{d}\right) \quad (2.52)$$

Figure 2.1 shows the results of such an experiment for aluminium. From this data Simmons and Balluffi obtain via equation (2.52)

$$\Delta H_V = 73{\cdot}2 \text{ kJ mol}^{-1} \qquad \text{and} \qquad \Delta S_V/R = 2{\cdot}4 \quad (2.53)$$

From the measured total ΔH for self-diffusion in aluminium this yields a value of $\Delta H^{\dagger} \simeq 45{\cdot}0$ kJ mol^{-1} or about 60% of the formation value. This direct observation of vacancy concentrations, together with the Kirkendall effect described below, can be taken as experimental proof of the prime role of vacancies in diffusion in

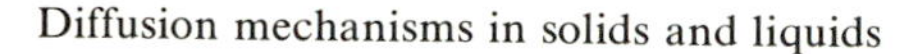

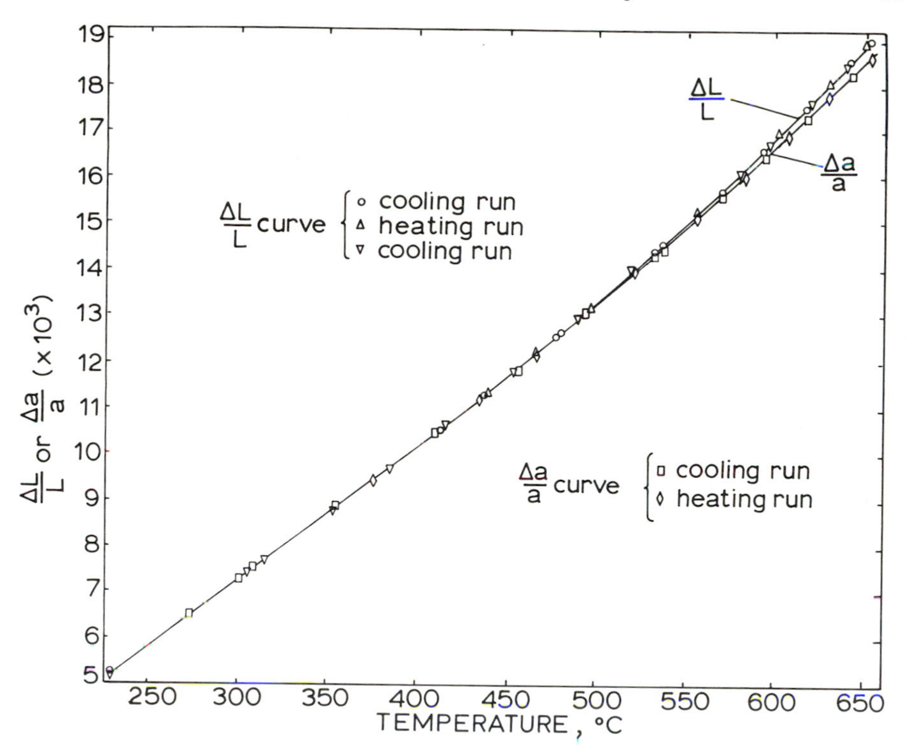

Fig. 2.1 Measured length and lattice parameter expansions of aluminium *vs* temperature. After Simmons and Balluffi[59]

substitutional alloys. Notwithstanding this knowledge, fundamental theory reaches an impasse in dealing quantitatively with this complex three-species problem. Darken[56] found a partial solution to the impasse by relating the intrinsic coefficients to tracer coefficients and the activity coefficient of the solution.

2.13 SIMULTANEOUS DIFFUSION AND DRIFT: THE KIRKENDALL EFFECT

The concept of a drift velocity v arising from differing intrinsic diffusion coefficients was known to be operative in gases and liquids long before it was comprehended for solids. However, because the effect is most unexpected and dramatic in solids it bears the name of the person to first observe it. We first noted its importance for rigorous diffusion analysis in §1.11.

Consider a thin film for which B atoms are diffusing countercurrent to A atoms with jump frequencies different from those of the A atoms ($D_A > D_B$, for example) as indicated in Fig. 2.2. A little reflection will convince one that for $D_A > D_B$ and a film that was initially A-rich on the left there will be an excess flux of A atoms to the right over B atoms to the left. If the concentration of vacancies is kept reasonably constant through the action of equilibration with sources and sinks, the sample as a whole must drift to the right (interfaces 1 moving to 2, etc.).

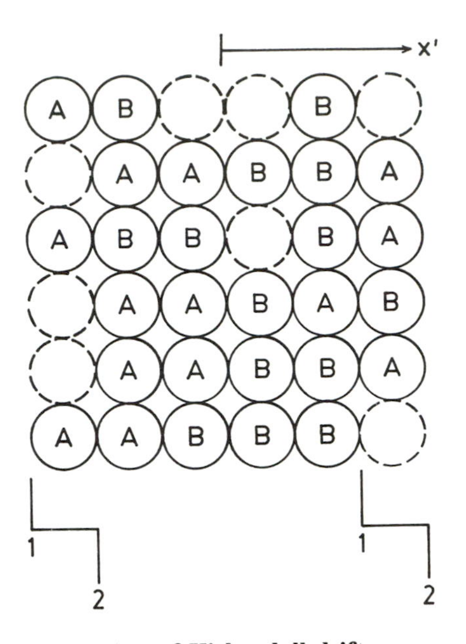

Fig. 2.2 Discrete representation of Kirkendall drift

In a typical diffusion experiment between alloy samples the diffusion zone is restricted to a narrow region adjacent to the original weld, as indicated in Fig. 2.3. Consider a small volume element ΔV, comparable to that in Fig. 2.3, and attach a coordinate system x' to some lattice plane, as indicated in Figs. 2.2 and 2.3. Clearly, for $D_A > D_B$ the coordinate frame x' (the lattice-fixed or Kirkendall frame) will try to drift to the left relative to the instantaneous left-hand edge of the model crystal with a velocity v, which is equivalent to saying that planes 1 and 2 are forced to remain coincident in Fig. 1.2. This represents a mechanistic description of the drift process introduced formally in the previous paragraph.

We proceed here according to equations (2.39), (2.40), (2.46) and (2.49) and generalize that discussion by noting that the partial molar volumes of the two species may not be equal. The laboratory frame flux balance must therefore be written as

$$\overline{V}_A J_A + \overline{V}_B J_B = 0 \tag{2.54}$$

With the assumption that the molar volumes $\overline{V}_A$ and $\overline{V}_B$ are independent of composition (or that the concentration extremes are small) we can write

$$\overline{V}_A \Delta C_A = -\overline{V}_B \nabla C_B \tag{2.55}$$

and from equations (2.40) and (2.41)

$$v = -\frac{\overline{V}_A J_A + \overline{V}_B J_B}{\overline{V}_A C_A + \overline{V}_B C_B} = \overline{V}_A (D_A - D_B)\frac{\partial C_A}{\partial x} = \overline{V}_B (D_B - D_A)\frac{\partial C_B}{\partial x} \tag{2.56}$$

where for an isobaric system

$$\overline{V}_A C_A + \overline{V}_B C_B = 1 \tag{2.57}$$

Now for the infinite diffusion couple boundary conditions of Fig. 1.14,

$$C_A = C_A(\lambda) \qquad \text{where} \qquad \lambda = x/\sqrt{t} \tag{2.58}$$

so we may write

$$v = \frac{\overline{V}_A(D_A - D_B)}{\sqrt{t}} \frac{dC_A}{d\lambda} \tag{2.59}$$

A row of inert markers placed initially at the join (0 in Fig. 2.3) migrates at some point of constant C_A and therefore of $dC_A/d\lambda$. We may therefore integrate (2.59) to obtain the distance of marker migration

$$x_0 = \int_0^t v\, dt = 2\overline{V}_A \frac{(D_A - D_B)\, dC_A}{d\lambda} \cdot \sqrt{t} \tag{2.60}$$

It may seem surprising that according to equation (2.59) the velocity vanishes when $D_A = D_B$, even though $\overline{V}_A \neq \overline{V}_B$. The explanation lies in relation (2.55) which assures the conservation of volume in any volume element. In the rare event where the molar volumes are variable and the concentration differences are large, bulk volume changes of mixing occur and the diffusion sample changes dimensions. Balluffi[60] and Prager[61] have described how to deal rigorously with this unlikely occurrence. In substitutional metals it is usually sufficient to assume that molar volumes are equal and constant.

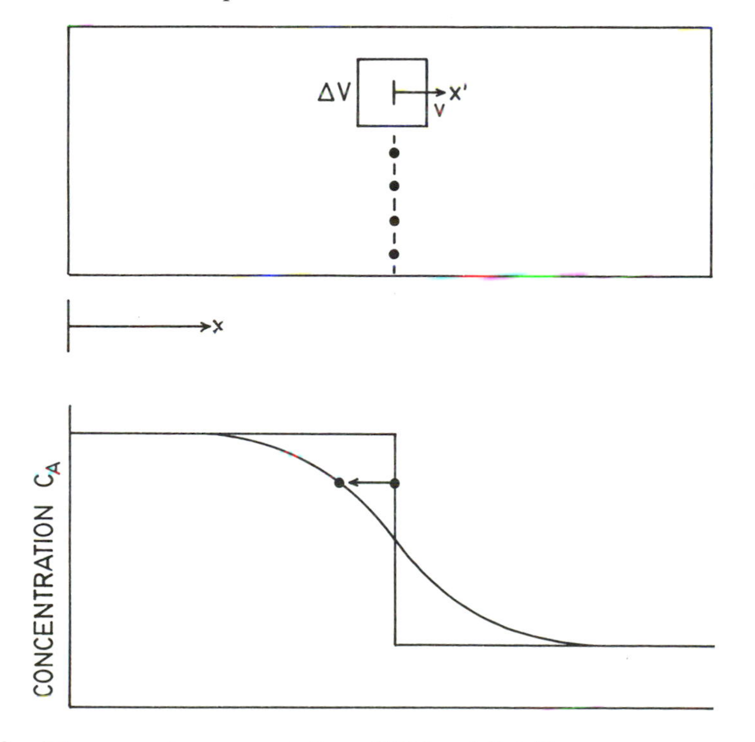

Fig. 2.3 Macroscopic representation of Kirkendall drift

Smigelskas and Kirkendall[62] carried out an experiment which anticipated the functional dependence of equation (2.60). They wrapped Mo wires around a block of Cu–30Zn and plated a thick layer of Cu over the whole to produce a double couple, as shown in Fig. 2.4. They then measured the change in distance d, Δd, which is $2|x_0|$, as a function of t and plotted $|x_0|$ versus $\sqrt{t}$ obtaining the result in Fig. 2.5, thus verifying relation (2.60). As a matter of sign and magnitude the markers moved together ($x_0 < 0$) indicating that component A (zinc) has a higher intrinsic diffusion coefficient D_{Zn} than component B (D_{Cu}) by a factor of about 3. Correa da Silva and Mehl[63] have recorded a large number of similar measurements in fcc and bcc alloys.

The foregoing structure is rigorous as applied to liquids (provided there is no convection) since the free volume (equivalent to vacancies) relaxes to its local equilibrium value promptly and continuously. This is not always the case for solids since the sources and sinks for vacancies may be of insufficient density to sustain a continuous close approach to the equilibrium distribution specified by equation (2.3). For example, when zinc is evaporated from the surface of a copper–zinc alloy single crystal there is a rapid countercurrent flux of vacancies towards the interior (see the Kirkendall experiment) which under some circumstances condenses into bubbles through supersaturation.[64] Since the diffusion rate of vacancies is $\sim 1/X_V$ times that of the atoms in a solution and their concentration is $\sim X_V$ times that of the atoms they will approach equilibrium (their super- or under-saturations will become negligible) for all diffusion times for which the diffusion length $2\sqrt{\tilde{D}t} > d/2$, the mean half-distance between sources, provided concentration differences are not too large. At a moderate dislocation density of 10^{12} m^{-2} the mean separation of sources is 1 μm. Thus for $\tilde{D} = 10^{-8}$ mm^2 s^{-1}, close to local equilibrium concentrations of vacancies will be sustained for times in excess of 10 s. The early time dependence of the diffusion coefficient due to finite vacancy density relaxation times was recognized by Balluffi.[64] Henceforth it will be assumed that for all long term diffusion processes in normal crystals local equilibrium concentrations of vacancies are sustained.

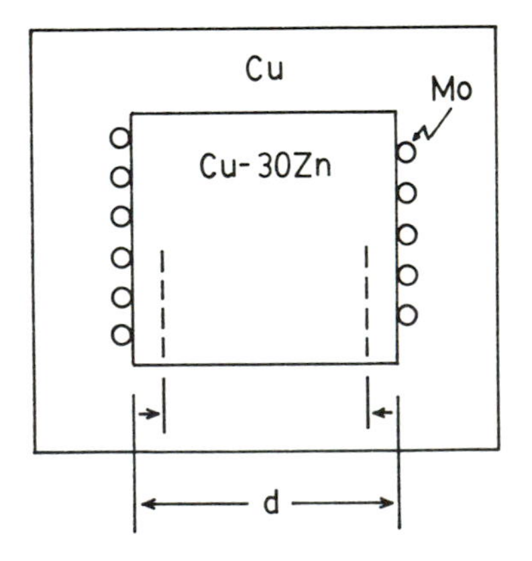

Fig. 2.4 Experiment of Smigelskas and Kirkendall

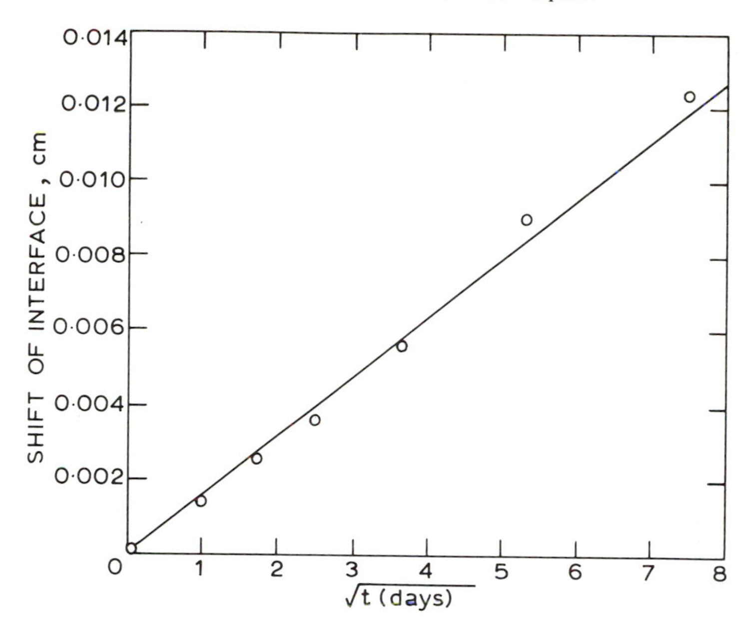

Reprinted with permission from TRANSACTIONS OF THE METALLURGICAL SOCIETY, Vol. 171, p. 130, (1947), a publication of The Metallurgical Society, Warrendale, Pennsylvania

Fig. 2.5 Shows linear relation between shift in position of inert markers and square root of time of diffusion. Specimens consist of electroplated copper on 70-30 brass. Diffusion temperature is 785°C. After Smigelskas and Kirkendall[62]

2.14 RELATIONSHIP BETWEEN THE INTRINSIC AND TRACER COEFFICIENTS: THE DARKEN–HARTLEY–CRANK EQUATION

Let us refer back to equations (2.48) and (2.49) and their derivation via equation (2.34). Now consider an experiment involving the countercurrent flow of A and a dilute isotope A* at a constant and non-interacting concentration X_B. A predicted value of D_A^* in this experiment can be obtained by letting B → A* in equation (2.49), noting that the thermodynamic terms on the right tend to unity since $X_{A*} \rightarrow 0$. Now the activation entropies and enthalpies will be the same for the two isotopes, $\gamma^{\dagger}_{A^*V} = \gamma^{\dagger}_{AV}$ on account of diluteness of the activated states and $\gamma_{A^*} = \gamma_A$ on account of the Gibbs–Duhem equation (cf Howard and Lidiard[14]), so

$$\frac{\gamma_{A^*}}{\gamma^{\dagger}_{A^*V}} = \frac{\gamma_A}{\gamma^{\dagger}_{AV}} \tag{2.60}$$

and it follows that

$$D_A = D_{A^*}\left(1 + \frac{d \ln \gamma_A}{d \ln X_A}\right) \tag{2.61}$$

and symmetrically,

$$D_B = D_{B*}\left(1 + \frac{d \ln \gamma_B}{d \ln X_B}\right) \tag{2.62}$$

Since further, by the Gibbs–Duhem equation

$$1 + \frac{d \ln \gamma_A}{d \ln X_A} = 1 + \frac{d \ln \gamma_B}{d \ln X_B} = 1 + \frac{d \ln \gamma}{d \ln X} = \phi \tag{2.63}$$

the thermodynamic terms in equations (2.62) and (2.63) are identical. Thus equation (2.50) can be factored to

$$\tilde{D} = (X_B D_{A*} + X_A D_{B*})\phi \tag{2.64}$$

which is known as the Darken–Hartley–Crank (D–H–C) equation.[56, 58] This relates the mutual or chemical coefficient directly to the tracer coefficients and

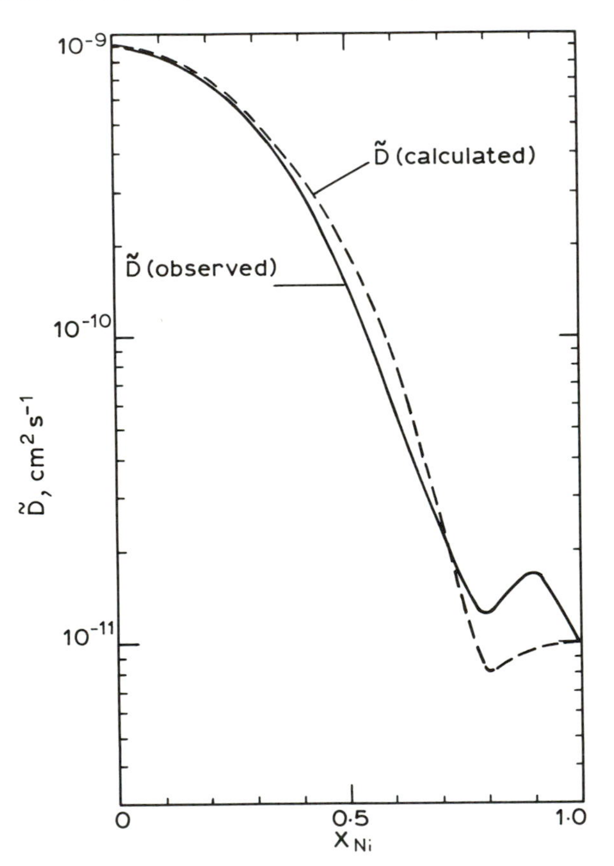

Reprinted with permission from *Acta Met.*, **5**, J. E. Reynolds *et al.*, p. 29, copyright 1957, Pergamon Press Ltd

Fig. 2.6 Calculated and observed interdiffusion coefficients in gold–nickel alloys at 900°C. After Reynolds *et al.*[66]

the thermodynamics of the solution. This is in precise agreement with experiment in the dilute limits. The derivation, however, lacks rigour for rich alloys because of some delicate correlation effects which generally imply an asymmetric availability of vacancies to A and B atoms, and the X_V terms in equations (2.48) and (2.49) must reflect this asymmetry.[16, 65] Since these effects can be expressed most clearly as cross-interactions in a quaternary A–B–A*–B* complex they will be dealt with in the later discussions of multicomponent diffusion (cf §3.11).

The D–H–C Equation has been tested in a number of instances, as for example in Fig. 2.6.[66] The closure between theory and experiment can be deemed adequate for many other binary solids.[67] Equation (2.64) may thus be regarded as a useful interpolation formula. Some recent data, however, suggests that for certain systems the Kirkendall velocities are higher than predicted (see references in Dayananda[68]), in order of magnitude agreement with corrections suggested by Manning.[16]

The $\tilde{D}$ data for equimolar liquids (Table 2.2[69]) indicates positive discrepancies from the D–H–C equation.

2.15 EFFECT OF PRESSURE AND STRESS ON SOLID STATE SELF-DIFFUSION

The focus here is upon a vacancy model of self-diffusion in a solid metal, say, which is described by equations (2.48) and (2.49) with B as the tracer A* and thus $D_A = D_{A^*}$ in equation (2.47). The vacancy concentration X_V is given by equation (2.3). If it is assumed to the first approximation that vibration frequencies are

Table 2.2 Deviations between observed and predicted values of $\tilde{D}$ based on the D–H–C equation. After Tyrrell and Harris.[69] $\Delta = 100[1 - (X_1 D_2^* + X_2 D_1^*)\phi/\tilde{D}]$

System	Δ
Benzene–chlorobenzene	0·7
Cyclopentane–cyclooctane	3·9
n-Dodecane–n-octane	3·2
n-Dodecane–n-hexane	2·9
Benzene–n hexane	12·1
Benzene–n heptane	11·7
Benzene–cyclohexane	9·5
Benzene–OMCTS[a]	9·0
Tetrachloromethane–OMCTS	4·3

[a]The intradiffusion coefficient of octamethyl cyclotetrasiloxane (OMCTS) used was obtained for a mixture with perdeuteriobenzene.

independent of pressure, which is equivalent to a persistent equipartitioning between kinetic and potential modes, then the main influence of a hydrostatic pressure p is on the free energy of formation of vacancies g and the standard free energy of formation of activated complexes $\Delta G^{\dagger}$. Thus

$$-RT\left(\frac{\partial \ln D_{A*}}{\partial p}\right)_T = \Delta V_V + \Delta V^{\dagger} = \Delta V_a \tag{2.65}$$

can be evaluated where the thermodynamic identity

$$\left(\frac{\partial \Delta G}{\partial p}\right)_T = -\Delta V \tag{2.66}$$

has been applied. ΔV_a is called the activation volume. By the definition of g it follows that ΔV_V is the partial molar volume of vacancies, which is a positive quantity, and through necessary relaxation, less than the atomic molar volume $\bar{V}_{A*}$. It accordingly contributes to a decrease in D_{A*} with p. $\Delta V^{\dagger}$ is thought to be positive as well, but negligible compared to ΔV_V. Experiments on solid Na and Pb yield, in accordance with equation (2.65), linear decrements in $\ln D_{A*}$ at all temperatures and ratios of $\Delta V_a/\bar{V}_{A*} \sim \Delta V_V/\bar{V}_{A*}$ of 0·54 and 0·64, respectively.[70] For the liquid metals Hg and Ga, however, the corresponding experimental ratios are 0·04 and 0·05, respectively.[70] The latter result is an indication that a vacancy model for diffusion in liquid metals is probably untenable (see §2.16).

Larché and Cahn[71] have drawn attention to the possible role of self-stress due to concentration gradients in modifying the effective diffusion coefficient. This phenomenon is related to the gradient energy correction which appears in treatments of spinodal decomposition (cf §§1.27 and 12.5). It is possible that such effects will be moderated by dislocation climb or grain boundary migration in most practical systems.

2.16 MODELS FOR MUTUAL AND SELF-DIFFUSION IN NON-IONIC LIQUIDS

During recent years the methods of molecular dynamics and computer simulation have greatly clarified the nature of liquid diffusion in systems with simple pair potentials.[7, 8, 72, 73] This work has discounted the earlier approach using the concept of 'holes' generated by fluctuations together with activated rate theory[12] concluding in general that the mean free path for diffusion is appreciably less than a molecular diameter. The computer experiments have confirmed the views of Hildebrand[13] which spring from the fact that diffusion coefficients in non-electrolytes are apparently linearly dependent on temperature. He has shown that the coefficients in this linear expansion are consistent with changes in relative free volume associated with thermal expansion.

Swalin, who has focused on liquid metals[74, 75] has attempted to combine elements of rate theory with the free volume concept and has, depending on specific assumptions, arrived at dependencies which are linear or quadratic in T. He appears, however, to have underrated the fact that for small displacements (which dominate the transport process) within a pair approximation there is a potential well rather than an activation barrier at intermediate positions of the average trajectory. The transport is thus more gaslike than solidlike. It seems therefore that an elementary theory for self-diffusion in metallic and Van der

Waals liquids should be based upon the classical gas equation[76] (§ 1.2)

$$D = \tfrac{1}{3}\bar{v}\lambda \tag{2.67}$$

where $\bar{v}$ is an appropriate mean molecular velocity and λ is a mean free path which is intimately associated with the thermal expansion coefficient.

Since the heat capacity per atom of many liquids is well represented by the Dulong and Petit value ($3k$) it is reasonable to associate $\tfrac{3}{2}kT$ of kinetic energy with the linear motion of a molecule and thereby estimate the average velocity as the perfect gas value[76]

$$\bar{v} = (8kT/\pi m)^{1/2} \tag{2.68}$$

where m is the mass of the molecule. The estimation of λ requires a more elaborate scheme.

It is informative firstly to obtain an expression for the thermal expansion in terms of representative pair potentials, for this establishes the expectation of generality for a law of linear expansion with absolute temperature and the effect of binding on the diffusion coefficient. Here potentials of the Lennard-Jones[10] form (Fig. 2.7) are explored in terms of separation of centres, d,

$$P = \frac{k_1}{d^{2n}} - \frac{k_2}{d^n} \tag{2.69}$$

where k_1 and k_2 are assumed to be independent of temperature, and specifically examine the case $n = 6$, which is a deep well potential, and the case $n = 4$, which is much shallower. For $n = 6$ the minimum is evaluated to yield

$$d_0 = (2k_1/k_2)^{1/6} \tag{2.70}$$

at a potential

$$P_0 = -k_2^2/4k_1 \tag{2.71}$$

Then assigning a potential energy partition of ξkT to the pair, where $\xi \simeq \tfrac{1}{2}$ for a coordination number of 6 (cubic; $\xi \simeq \tfrac{3}{8}$ for bcc and $\tfrac{1}{4}$ for fcc) the potential at temperature T is

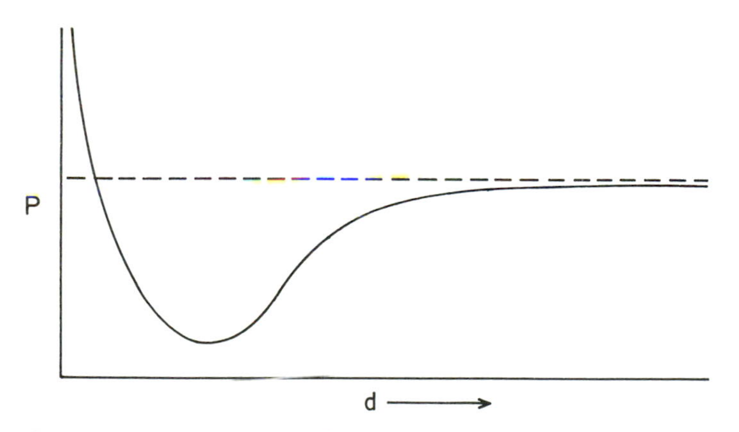

Fig. 2.7 Lennard-Jones-type potential function

$$\xi kT - \frac{k_2^2}{4k_1} = \frac{k_1}{(d^6)^2} - \frac{k_2}{(d^6)} \tag{2.72}$$

This has two solutions and yields a positive displacement Δd from the mean d_0 which to a second order binomial expansion yields the linear expansion ratio

$$\frac{\Delta d}{d_0} = \frac{19}{36}\frac{\xi k_1}{k_2^2} \cdot kT = \delta T \tag{2.73}$$

For $n = 4$ this becomes

$$\frac{\Delta d}{d_0} = \frac{13}{16}\frac{\xi k_1}{k_2^2} \cdot kT = \delta T \tag{2.74}$$

Thus the form of the expansion law is independent of well curvature for this common type of potential. For simple cubic coordination the coefficient δ is equal to the linear expansion coefficient α. For bcc the linear expansion coefficient is $0{\cdot}84\delta$ and for fcc, $0{\cdot}707\delta$.

Since the coefficients k_1 and k_2 in the potentials cannot as yet be evaluated from first principles α must henceforth be regarded as empirical. From equations (2.73) and (2.74) we can write down the distribution function for Δd since factoring ξkT on the right and assigning this as a fluctuating potential energy identifies the unnormalized probability function

$$P \sim \exp\left(-\frac{\alpha}{\xi}\frac{\Delta d}{d_0}/kT\right) \tag{2.75}$$

Normalizing this, then evaluating the mean value of Δd we recover the equilibrium value $\alpha T d_0$ as given by equations (2.73) or (2.74). It is this mean extension which determines the free path. Unfortunately, in the spatially fluctuating milieu the relationship is not a simple one.

For a numerical estimate of λ the case is considered of packing equivalent to simple cubic coordination of spheres where we can say that $\Delta d = \alpha T d_0$ is the lower limit of the mean free path since the average molecule steered by potential focusing through the centre of the coordination sphere can reach to the radius $d_0/2 + \Delta d$ (Fig. 2.8). One must then consider that there is a purely geometrical extension of the free paths determined by the average interstitial spaces provided by the molecules in the first coordination sphere. For cubic coordination this provides a maximum excess free path of $\sim\Delta d$ and the average contribution to the

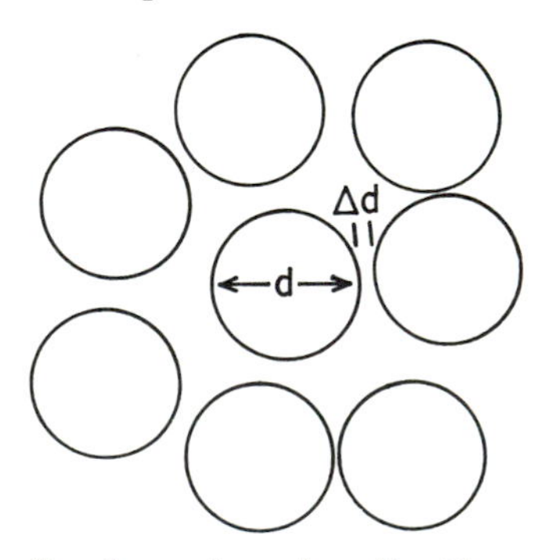

Fig. 2.8 Schematic coordination sphere in a liquid

mean free path will be no more than half of this. Finally, one must take account of the deeper interstitial penetrations of the coordination shell, to a maximum distance of less than d_0 (the hole, or vacant site case) which might be accommodated by spacing fluctuations in the coordination sphere. Molecular dynamics and delicate physical experiments[77] indicate, however, that hole events are of exceptionally low probability. Indeed, the maximum jump distance is at least one order of magnitude smaller than d. Hence we conclude that the average contribution of this term will be of the same magnitude as the first two, say $1{\cdot}5\Delta d$ for a total of $3\Delta d$. For closer packed structures the first term is larger since the linear expansion is projected onto diagonals, the second term is about the same (shallower penetrations but more sites) while the third term is smaller, due to the tighter packing in the coordination spheres. We thus conclude that a diffusion coefficient expressed in terms of the empirical expansion coefficient (which is inclusive of packing effects) requires no further corrections for packing effects.

The mean free path for tracer diffusion is accordingly estimated to be

$$\lambda \simeq 3\alpha d_0 T \tag{2.76}$$

and our final expression for the self-diffusion coefficient from equation (2.67) is

$$D = \alpha d_0 \left(\frac{3k}{m}\right)^{1/2} T^{3/2} \tag{2.77}$$

We put this formula to the test for liquid mercury where the Dulong and Petit value of the heat capacity obtains and the diffusion coefficient is known between 273 and 512K and is adequately, if not accurately, represented over this range by (Fig. 2.9[78])

$$D = 3{\cdot}5 \times 10^{-7}\, T^{3/2}\ \text{mm}^2\ \text{s}^{-1} \tag{2.78}$$

Meyer[78] finds that a $T^{1{\cdot}85}$ law gives the best fit. The linear expansion coefficient for Hg is $6{\cdot}06 \times 10^{-5}$ K^{-1} and the molecular weight is $200{\cdot}59$ g mol^{-1}. d_0 can be estimated from the density of Hg or the Van der Waals volume correction[79] to be about 3×10^{-7} mm. These values yield the coefficient in equation (2.77) equal to $2{\cdot}0 \times 10^{-7}$.

Apart from the success of formula (2.77) as regards magnitude and approximate temperature dependence, it has further merits in providing a widely recognized weak mass dependence and a rationalization of the fact that near the melting point values of D do not vary widely among approximately spherical species. The reason for the latter observation is to be found in equations (2.71) and (2.73) or (2.74), where it is recognized that the linear expansion coefficient is inversely proportional to the depth of the potential well. Since deeper wells imply higher melting points, lower α are compensated for by higher T in formula (2.77), thus abetting the invariance of magnitude.

Finally, the model allows one to estimate the isothermal effect of a pressure increase ΔP, for it suggests that one should simply substitute $\alpha T - \beta\Delta P/3$ for αT in equation (2.76) for the mean free path, where $\beta = -\partial V/V\,\partial P$ is the isothermal compressibility. For Hg this is reasonably independent of temperature and pressure and equal to $0{\cdot}42 \times 10^{-10}$ m^2 N^{-1} at 80°C.[79] Nachtrieb and Petit[80] observed a maximum 18% linear decrease in the diffusion coefficient at 303K

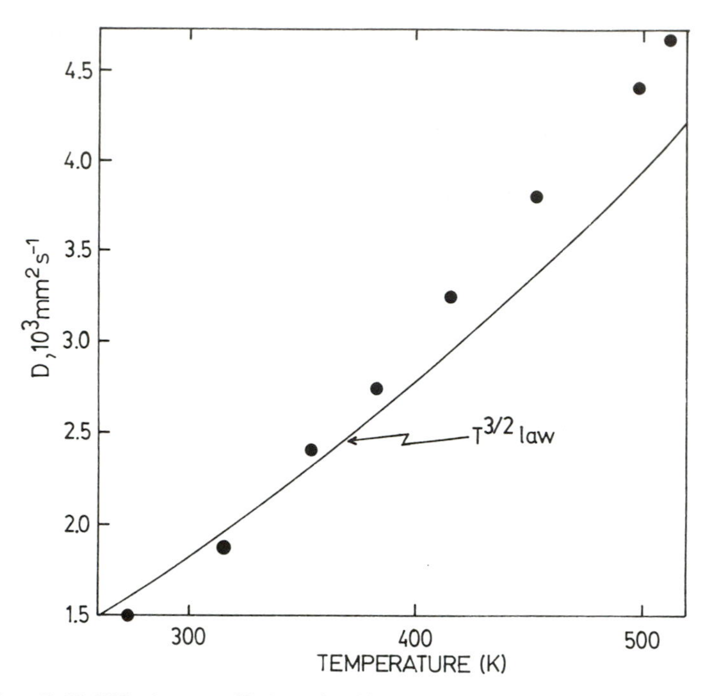

Fig. 2.9 Self-diffusion coefficient for liquid mercury. Experimental points after Meyer[78]

under pressures up to $8{\cdot}4 \times 10^5$ kg m^{-2}. The predicted decrement at this limit is

$$100\left(\frac{\alpha T - \beta \Delta P/3}{\alpha T}\right) = 37\% \tag{2.79}$$

which is of the right magnitude and almost within the range of measurement error.[78, 80] The relative success of this estimate as compared to that obtained via the vacancy model (cf §2.15) is further evidence of the latter's inapplicability to liquid diffusion.

To quantitatively improve on the foregoing it is necessary either to invoke further empiricism, or to develop kinetic theories which more accurately represent molecular trajectories and their correlations. The various hydrodynamic theories tend towards empiricism by trading upon explicit relationships between the self- or mutual diffusion coefficients and the solvent viscosity. The original version of these theories combined the theory of Brownian motion with Stoke's Law for the laminar free gravitational drift of a sphere in a continuous viscous fluid. The resulting relation for the chemical diffusion coefficient of a dilute solution of spherical particles is

$$\widetilde{D} = kT/(3\pi\eta\sigma) \tag{2.80}$$

where σ is the molecular diameter and η is the viscosity. This is known as the Stokes–Einstein relation. It should be strictly applicable for relatively large diameter solutes such as macromolecules. If the solute and solvent molecules have similar diameters then one can anticipate 'slip' between the solute and the solvent, effectively increasing the diffusion coefficient. The diffusion coefficient is then estimated to be

$$D = kT/(2\pi\eta\sigma) \tag{2.81}$$

Corrections have been developed for non-spherical molecules and non-dilute solutions.[69] For oxygen in water at 25°C equation (2.81) yields $D = 1{\cdot}9 \times 10^{-3}$ mm^2 s^{-1}.[81] The experimental value is $1{\cdot}8 \times 10^{-3}$ mm^2 s^{-1}. Many other examples of empirical correlations have been provided by Hildebrand[13] and Cussler.[81]

The more advanced kinetic theories take as their point of departure the hard sphere ideal gas relation 2·67 or

$$D^{\infty} = \frac{3}{8C_{\infty}\sigma^2}\left(\frac{kT}{\pi m}\right)^{1/2} \tag{2.82}$$

where C_{∞} is the number density at a typical dilute condition, and its Chapman–Enskog generalization for non-dilute density C[69]

$$D_{\mathrm{CE}} = \frac{C_{\infty}}{C} D^{\infty}/g(\sigma) \tag{2.83}$$

where $g(\sigma)$ is the radial distribution function when the hard spheres are in contact. This assumes instantaneous binary collisions and an absence of correlated motions and in fact has features in common with our introductory model. Enskog obtained good estimates of diffusion coefficient magnitudes in 1922. Since in these theories the diffusion coefficient is normalized to a hypothetical state of close packing, the temperature dependence is bound to be different from ours. For fcc or hexagonal close packing the contact radial distribution function is given approximately by

$$g \simeq \left(1 - 0{\cdot}370\,2\,\frac{V_0}{V}\right)\bigg/\left(1 - 0{\cdot}740\,5\,\frac{V_0}{V}\right)^3 \tag{2.84}$$

where V_0/V is the ratio of the volume occupied by close packed spheres, V_0, to the actual (expanded) volume of the fluid, V. The remaining theoretical problem is to correct for correlation effects and to establish V_0/V as a function of temperature. The theoretical program has been only marginally successful from the point of view of analytic physics, requiring extensive investigations via molecular dynamics, dimensional analysis and experiment to identify a universal melting point (m) scaling factor

$$\xi_{\mathrm{m}} = V_{0\mathrm{m}}/V_{\mathrm{m}} \simeq 0{\cdot}472 \tag{2.85}$$

The program has generated a formula for melting point tracer coefficients

$$D^{*}_{\mathrm{m}} = \frac{3}{8\pi C_{\mathrm{m}}}\left(\frac{C_{\mathrm{m}}}{6\xi_{\mathrm{m}}}\right)^{2/3}\left(\frac{\pi R T_{\mathrm{m}}}{M}\right)^{1/2}\left(\frac{D_{\mathrm{HS}}}{D_{\mathrm{CE}}}\right) \tag{2.86}$$

where D_{HS} is obtained via the molecular dynamics of hard sphere assemblies. Table 2.3 demonstrates the efficacy of such a correlation for a group of metallic liquids and argon.

2.17 DIFFUSION IN DISPERSE STRUCTURES

There are a large number of condensed materials of industrial importance which consist of dispersions of molecular regions of different packing. In the broadest

Table 2.3 Comparison of experimental self-diffusion coefficients with predictions for some elements at their melting points. After Protopappas *et al.*[82]

Element	D^*_m (10^3 mm^2 s^{-1})	
Li	7·00	7·01
Na	4·22	4·24
K	3·82	3·85
Cu	3·96	3·40
Rb	2·62	2·68
Ag	2·55	2·77
Zn	2·05	2·55
Cd	1·78	2·00
Hg	1·17	1·07
Ga	1·72	1·73
In	1·74	1·77
Sn	2·05	1·96
Pb	1·68	1·67
Ar	1·53	1·57

sense of the term these can be classified as gels. Jello, or gelatin, the most familiar form of gel, consists of linked filaments of polymeric protein with occluded water. The three subgroups which capture our attention here are polymer solutions, membranes and zeolites (or molecular sieves). The dispersed regions or 'phases' are usually distinct and to a greater or lesser extent ordered. The zeolites are in fact perfectly crystalline, containing channels when dry with diameters well in excess of typical atomic diameters. Intercalation compounds are two dimensional analogues of zeolites in which parallel lattice planes are weakly bonded and widely separated.[83] In applications, membranes and molecular sieves operate while submerged in some fluid medium so their diffusive properties are partly determined by constituents of the medium which can be absorbed up to some stable concentration. This fact of stability allows us to identify the state of these mixed 'phases' as a kind of metastable equilibrium consisting of a balance of osmotic (pressure) or electrostatic forces in the absence of concentration equalization. For example, many hygroscopic membranes of natural or synethetic origin swell when placed in an aqueous solution until the point where network back forces balance the osmotic pressure (cf Appendix 1). In all of the three structures considered, chemical reaction or complexing often occurs simultaneously with diffusion, so process evolution may be quite complex.

2.17.1 Polymeric diffusion

Because of the topographical complexity of large compound molecules (polymers) one must from the start recognize at least four distinct kinds of diffusion in polymeric solutions. Firstly, if the solution is crystalline, typical solid state diffusion via defects, including amorphous intercrystalline regions, must be recognized (cf §2.18). Secondly, if the polymeric component of a binary fluid solution is rich or possesses strong cross-linkages then there will exist a more or less rigid network with continuous but tortuous fluid channels (a gel- or membranelike structure) to which significant diffusion of non-polymeric or light polymeric fractions is confined. This process is the subject of the volume *Diffusion in Polymers*.[84] Thirdly, when the polymeric component is relatively dilute then the classical Stokes–Einstein, viscous drop model for large particles in a small particle solvent is applicable (see below). Fourthly, when there are two polymeric species or a tracer mixture of one species, the very slow interdiffusion process which may occur requires very detailed, indeed, creative modelling such as through the 'reptation' process. Of course, since fluid polymer solutions always consist of a size distribution and since complexing or cross-linking between molecules is very common (e.g. vulcanization of latex to rubber) elementary Fickian type processes of diffusion will be the exception rather than the rule.

The mathematical problem of diffusion in disperse polymer aggregates has been recognized by Barrer[85] as a special case of diffusion and permeation in heterogeneous media. The structure or tortuosity factors which have been derived check well with measurements on electrical analogues.

As noted, dilute solutions of polymers in monomer solvents can fulfil the conditions for the validity of the Stokes–Einstein Law as the molecules are obviously large and the solvent can, by comparison, be regarded as a continuous medium. Indeed, diffusion measurements are used in conjunction with the Stokes relation for a sphere to estimate polymer size.[86] Because the average number N of monomer units in a polymer of a particular type is almost infinitely variable, there is a strong interest in relating polymer properties to this number. The relationships which exist between the various properties and N are known as scaling laws[87] and much of the literature on polymer diffusion is devoted to the theoretical development and experimental testing of these laws.

If the solute polymer molecule is thought of as having a spherical shape, the dependence on the polymer multiplicity of the radius of that sphere, R, may be calculated from polymer statistics[88] as

$$R \propto N^{1/2} \tag{2.87}$$

when the polymer behaves as an ideal chain. Non-ideal chains in well-behaved solvents are in fact somewhat larger, due to the inclusion of some solvent within the sphere and consequent stretching of the chain. In this case

$$R \propto N^{3/5} \tag{2.88}$$

and this is typical of dilute solutions. A large body of experimental evidence[87] is consistent with the relationship

$$D \propto N^{-x} \tag{2.89}$$

where $0{\cdot}55 \leqslant x \leqslant 0{\cdot}57$, and the scaling law 2·88 is thereby confirmed via the Stokes–Einstein equation (2.80) because $R = \sigma/2$.

When polymers form concentrated solutions or melts, the polymer chains obviously cannot be regarded as non-interacting. The polymer chains are said to be entangled and a schematic illustration of the structure is shown in Fig. 2.10. Clearly the points of chain overlap represent entanglement and prevent free

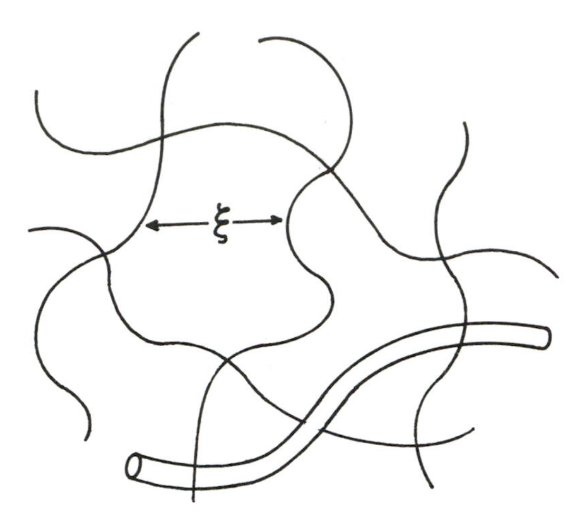

Fig. 2.10 Entangled polymer chains designating an average mesh size and identifying a portion of a reptation tube

molecular motion. Self-diffusion of a polymer molecule therefore involves the sinuous motion of a chain through the entangled mesh in a process known as reptation.[87, 89, 90] The process is analysed in terms of the diffusion motion of a particle along a curvilinear pipe representing the tortuous pathway through the mesh.

The mobility of a polymer chain along a pipe is expected to be inversely proportional to the chain length, or number of monomer units in the chain, N. The time taken for a chain to diffuse along a path a distance equal to its own length L can then be estimated via equation (2.15) as

$$\tau \propto L^2/D_{\text{chain}} \tag{2.90}$$

$$\propto L^2 N$$

Since L must be linear in N for an unbranched polymer,

$$\tau = \tau_1 N^3 \tag{2.91}$$

where τ_1 reflects the diffusion characteristics of monomer in the same tubular path. The actual displacement R_0 achieved when a chain moves along its contorted tube a distance L is calculated from polymer statistics[91] using the fact that a chain in a melt is ideal and therefore equation (2.87) applies. Substitution of equations (2.91) and (2.87) into equation (2.15) then yields an estimate for the reptating diffusion coefficient

$$D \propto N^{-2} \tag{2.92}$$

The inverse dependence on molecular weight arises through (*a*) the larger frictional coefficient of a longer molecule, and (*b*) the more circuitous path of motion. The entire argument, it must be stressed, depends on lack of branching in the polymer chain.

Experimental determinations of polystyrene and polyethylene self-diffusion in the melt have been reviewed by Tirrell[92] and the data for polystyrene is reproduced here in Fig. 2.11. It is seen that the data reflects an inverse dependence of D on the square of molecular weight, which confirms equation (2.92). Similar studies have been carried out on non-dilute solutions of polystyrene in various solvents.[92] This body of work demonstrates that $D \propto M^{-2.0 \pm 0.1}$, further confirming the success of equation (2.92).

When a polymer solution is formed, an additional variable, concentration, can affect diffusion behaviour. Below a certain critical concentration C^* the solution is dilute and Stoke's Law (2.80) applies. Above this concentration, entanglement commences. Since longer chains commence to entangle at lower concentration, C^* is molecular weight dependent. Once entanglement occurs, the mesh size, ξ (see Fig. 2.10) clearly depends on concentration. Mesh size determines the diameter of the tube through which reptation occurs and thereby affects the rate of the process. Dimensional arguments[92, 93, 94] lead to the relationship

$$D \propto N^{-2}C^{-7/4} \tag{2.93}$$

for good solvents, i.e., solvents for which equation (2.88) applies, and

$$D \propto N^{-2}C^{-3} \tag{2.94}$$

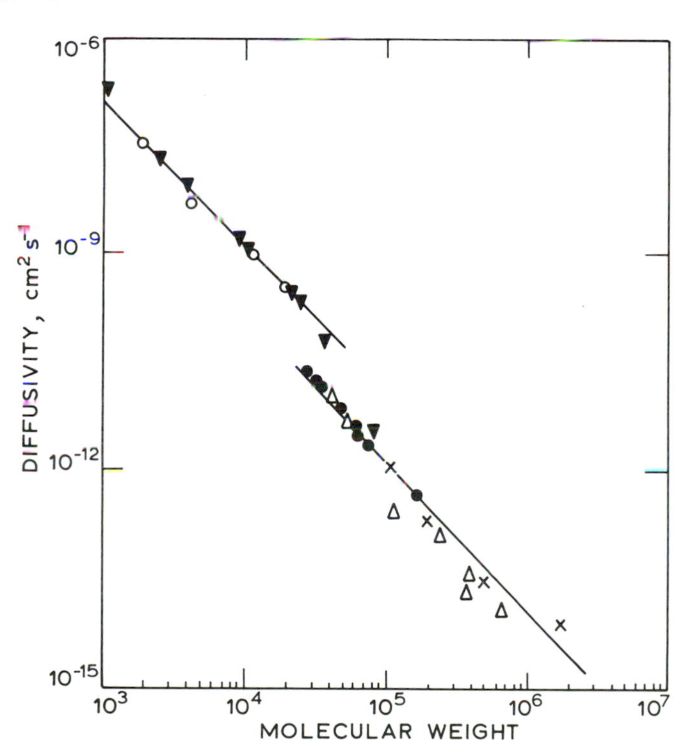

Reproduced with permission of Rubber Chemistry and Technology

Fig. 2.11 Polystyrene self-diffusion coefficient in $cm^2\ s^{-1}$ in the melt as a function of molecular weight at or near 175°C. After Tirrell[92]

for theta solvents. A theta solvent is one in which, at a particular temperature, polymer solutes behave in a quasi-ideal fashion and the scaling law (2.87) is obeyed. The utility of these relationships is illustrated in Figs. 2.12 and 2.13, taken from Tirrell.[92] The failure to include concentration effects on monomer friction within the reptation tube is thought[95] to account for some of the deviation observed at high concentrations. Nonetheless, the predictions of equations (2.93) and (2.94) are seen to be broadly confirmed.

When a tangled polymer network exists with a binary concentration gradient within a non-dilute solution, then diffusion occurs *en masse*. That is to say, the diffusion process involves the network rather than individual molecules. Alternatively, one may think of the motion of each chain as augmenting the motion of the others.[96] This description can be developed[87] to yield an analogue of the Stokes–Einstein equation:

$$\tilde{D} \propto \frac{T}{\eta_s \xi} \propto \frac{\phi^{3/4}}{\eta_s} \tag{2.95}$$

where η_s is solvent viscosity and ϕ is the polymer volume fraction in solution. Data on polystyrene in benzene[97] yield

$$\tilde{D} \propto \phi^{0.67 \pm 0.02}$$

which is a somewhat less sensitive dependence on ϕ. A similar expression is found for gel diffusion and has been verified approximately by Munch.[98]

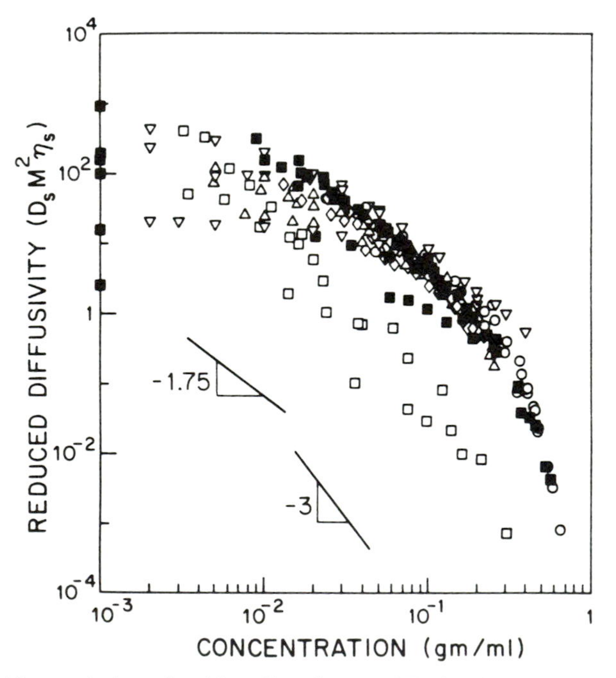

Reproduced with permission of Rubber Chemistry and Technology

Fig. 2.12 Polystyrene self-diffusion coefficient in solution in good solvents presented as log $D_s M^2 \eta_s$ *vs* log concentration. After Tirrell[92]

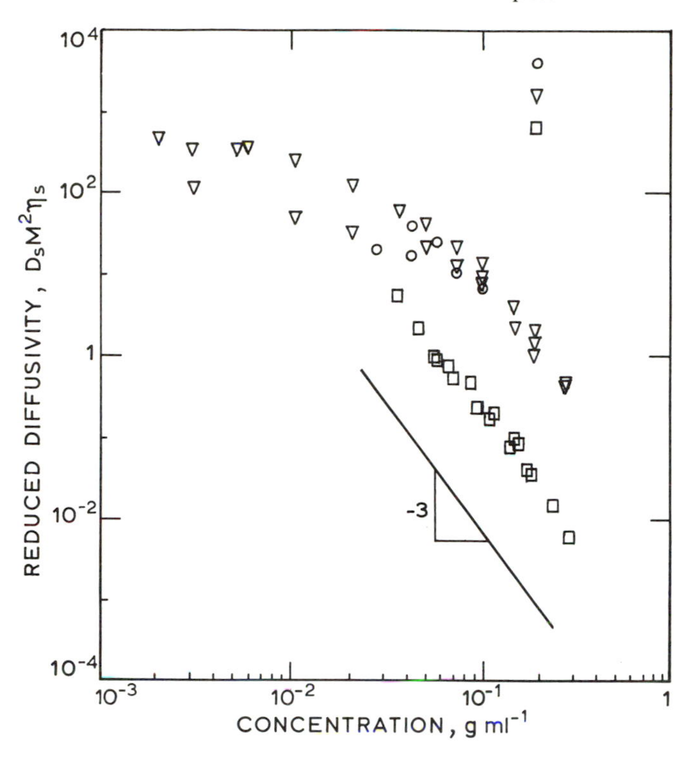

Reproduced with permission of Rubber Chemistry and Technology

Fig. 2.13 Polystyrene self-diffusion coefficient in solution in theta solvents presented as log $D_sM^2\eta_s$ *vs* log concentration. After Tirrell[92]

The theory of liquid diffusion is fundamental to the understanding of the diffusion of light elements within polymer networks as it is to the understanding of membrane permeation. The process is complicated, however, by tortuosity of diffusion paths, by heat of sorption and desorption effects and swelling of the mixture. A substantial quantification of the various processes has, however, been achieved.[84] This is particularly the case for the diffusion of gases, which can be regarded as analogous to interstitial diffusion in solids (cf § §1.4 and 2.11). Michaels and Bixler[99] have demonstrated a very close correlation between the diffusion coefficient and molecular diameter for diffusion of molecular gases in rubber (Fig. 2.14). Various free volume theories have provided a deeper understanding of the process (e.g. Cohen and Turnbull[100]).

The temperature dependence of observed diffusion coefficients often follows Arrhenius behaviour over significant ranges of temperature. However, slow crystallization at low temperatures and the glass transition at intermediate temperatures leads to anomolous observations and hysteresis. Figure 2.15 shows the relative permeability of He and H_2 in crystallizing terylene as a function of annealing time, indicating the formation of spherulitic crystals which block direct paths and confine the diffusion to the tortuous paths in the remnant amorphous intercrystalline material.[101] Figure 2.16 shows the Arrhenius plot for

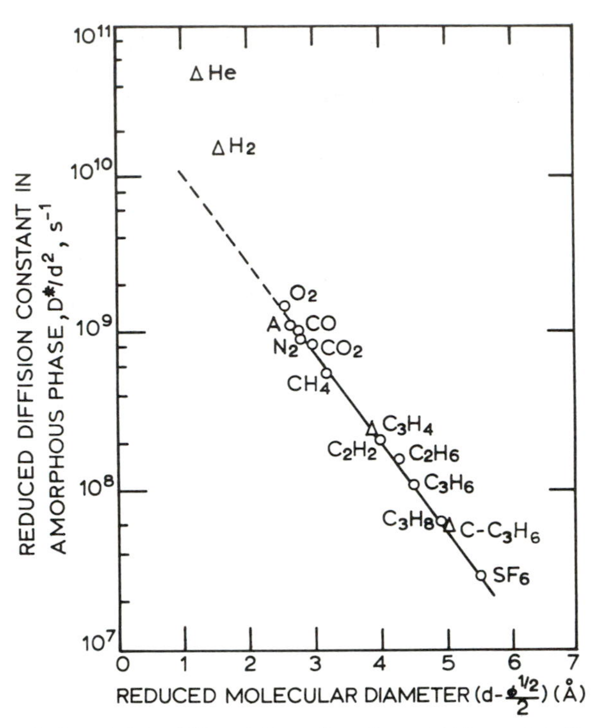

Fig. 2.14 Correlation of the diffusion constants with the reduced molecular diameter for gases in natural rubber. Reproduced with permission from Michaels and Bixler[99]

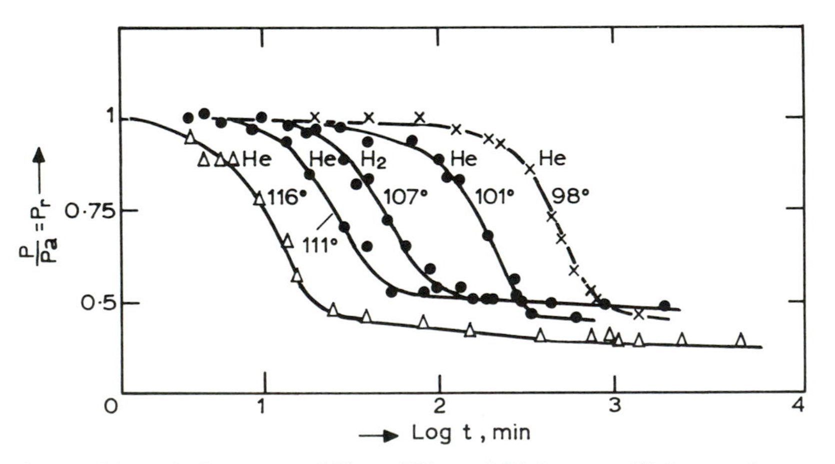

Fig. 2.15 The relative permeability of He and H_2 in crystallizing terylene as a function of annealing time. Reproduced with permission from Jeschke and Stuart[101]

the coefficients of argon and oxygen in polyvinyl acetate.[102] This behaviour, which shows an ultimate decrease in the activation energy for the glassy state, is explained by the fact that there is a smaller jump distance in the glassy state. The effect, however, is not universal.

The diffusion of organic vapours in polymers has been extensively studied.[103] Since organics are commonly solvents or swelling agents for polymers a non-destructive experiment must be carried out within the gas phase. Nonetheless, some swelling will occur. Above the glass transition temperature organic diffusants behave somewhat like the simple gases (H_2, Ar, etc.) modified by diffusion rate enhancement with concentration due to the swelling reaction, but below they exhibit very complex behaviour. A related very practical problem has to do with the dyeing of polymeric fibres and sheets.[104]

Dissolution of polymers in organic solvents represents an unusual and interesting phase transition and makes clear by microscopic observations that there exists a discrete gel temperature.[105] In the initial stages at higher

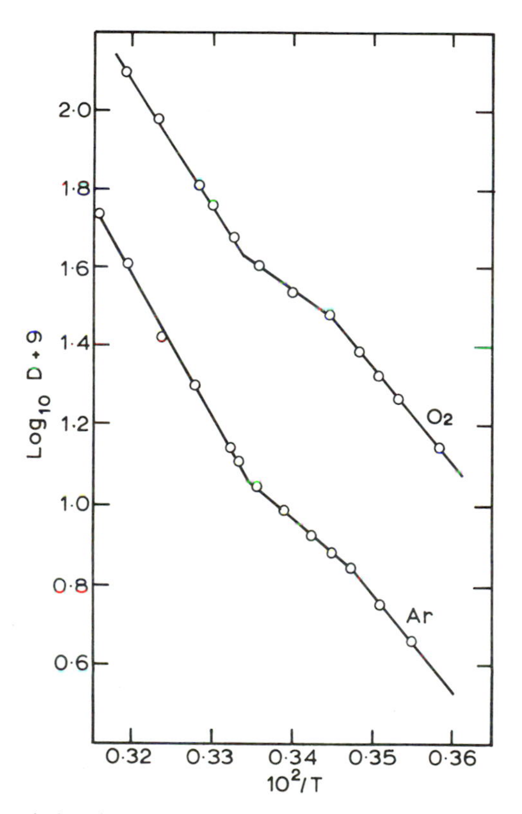

Reprinted with permission from P. Meares, *J. Am. Chem. Soc.*, 1954, **76**, 3415. Copyright 1954 American Chemical Society

Fig. 2.16 Temperature dependence of the diffusion constants for argon and oxygen in polyvinyl acetate. Reproduced with permission from Meares[102]

temperatures, but below the glass transition, the mutual sorption of the two regions produces first a swollen layer, then a gel layer consisting of increasingly tenuous polymer streamers and occluded solvent. If the temperature is lowered sufficiently, the gel layer does not appear and dissolution occurs by stress cracking which releases discrete glassy blocks from the surface.

Water is distinguished from the majority of organic penetrants in that the molecule is relatively small and self-interacts through the hydrogen bonds.[106] In contrast to organic penetrants, for which the diffusion coefficient usually increases with concentration, the water diffusion coefficient often shows marked decreases with concentration. Isothermal diffusion is by and large Fickian, sorption showing a normal $\sqrt{t}$ dependence. The temperature dependence of the permeability is variable from polymer to polymer and often anomolous. Figure 2.17 shows the Arrhenius plot for water in natural rubber. Contrast this with Fig. 2.16 for argon and oxygen in polyvinyl acetate. Again, results such as Fig. 2.17 are not general.

Barker and Thomas[108] have reported the ionic conductivity of cellular acetate doped in 0·1 M aqueous chloride solutions of Li, Na, K, Rb, and Cs above and below the glass transition temperature, T_g, which varied slightly with the ionic volume of the dopant. Figure 2.18 shows the dependence of the activation energies above and below T_g as a function of alkali ion volumes. Kumins and Kwei[109] have rationalized this behaviour on the basis of free volume theory.

The class of polymers known as ion-exchange resins offer a suitable bridge to our discussions of zeolites and membranes which follow. These are usually insoluble synthetics which consist of a three-dimensional polymer chain network with dangling, ionizable groups. The ion exchangers may be chain monomers with an ionizable group or be introduced by chemical sorption. These are usually derivatives of phenol + formaldehyde and are produced in bead or granular form. Extensive swelling in aqueous solution is a common property and essential to high exchange rates. The swelling rate is indeed so large as to prevent the

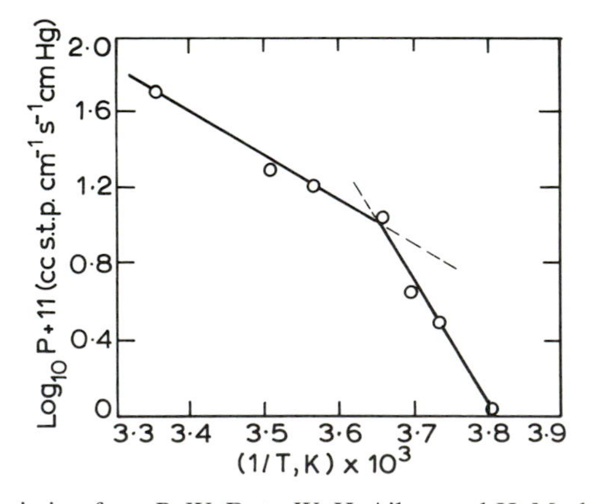

Reprinted with permission from P. W. Doty, W. H. Aiken and H. Mark, *Ind. Eng. Chem. ind. (int.) Edn*, 1946, **38**, 788. Copyright 1946 American Chemical Society

Fig. 2.17 Temperature dependence of the permeability coefficient for water in natural rubber. Reproduced with permission from Doty *et al.*[107]

estimation of the initial diffusion coefficient; but it is clear that this increases rapidly with concentration of solution in the resin.

The performance of a resin is determined by the counter-ion exchange equilibrium between a solution which contains two kinds of ions whose concentrations at equilibrium are different in the resin and in the solution. Thus in principle the diffusion-controlled processes forcing the system towards equilibrium can be used to separate appropriate ions. This type of equilibrium superposed upon the osmotic equilibrium, which ultimately saturates along with the swelling, is known as 'Donnan equilibrium'[110] and is surveyed in Appendix 1. It is accompanied by a 'cell' potential which expresses the balance of electrical forces in the system. Figure 2.19[111] shows the very effective separation of Sr into resin from Sr in solution in strontium–sodium exchange on the commercial resin Zeo-Kart. Further aspects of the theory are developed immediately below and in Chapter 10.

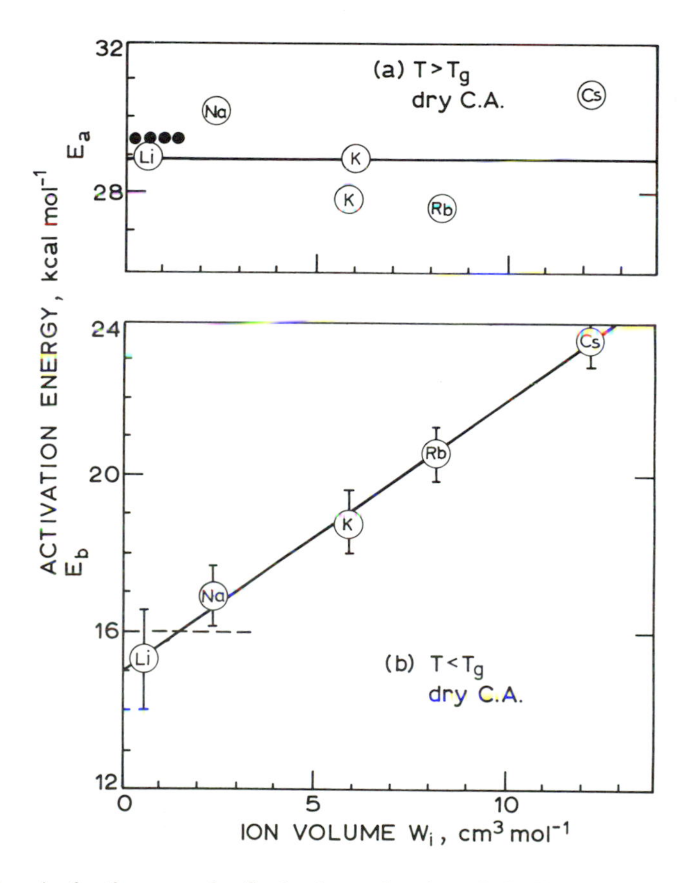

Fig. 2.18 Activation energies for ionic conduction of alkali chloride impregnated cellulose acetate, (*a*) above T_g and (*b*) below T_g *vs* alkali ion volumes. For the undoped *CA*, $E_a \approx 29{\cdot}3$ kcal mol^{-1} and $E_b \approx 11$–16 kcal mol^{-1}. After Barker and Thomas[108]

2.17.2 Diffusion in molecular sieves

Zeolites or molecular sieves were discovered over 200 years ago as natural alumino-silicate minerals.[83] The structural units in natural zeolites consist of SiO_4 and AlO_4 tetrahedra. The crystal structures are formed by the linking of these units into three-dimensional anion networks in which all oxygen atoms are shared between units. Hence for every Si^{4+} which is substituted for by Al^{3+} a negative charge is created which must be neutralized by a cation, typically Ca^{2+}, K^+, Ba^{2+}, Na^+, Li^+, or NH_4^+. These building blocks with shared oxygens generally force the structure to be rather open (Fig. 2.20) so the channels can accommodate water molecules as well as the charge neutralizing cations, both of which can readily diffuse. The cations can also be readily exchanged with other cations and water can be continuously infiltrated or removed.

Practically, zeolites are used as sorbents for gases, vapours and liquids, i.e., as dessicants, as catalysts, as cation exchangers such as water softeners (cf §2.17.1) or as ion or molecular sieves. Exceptional isotope separation ratios have been obtained in particular cases.

Zeolites are most widely available and used in practice as fine powders, loose or compacted. The overall diffusion processes are accordingly very complex consisting of interdiffusion, surface reaction and intradiffusion compounded by the effects of sorption heating or desorption cooling. Overall indices of transport are determined by volumetric or gravimetric means. Intradiffusion parameters for powders can be obtained by NMR, dielectric relaxation, infrared absorption

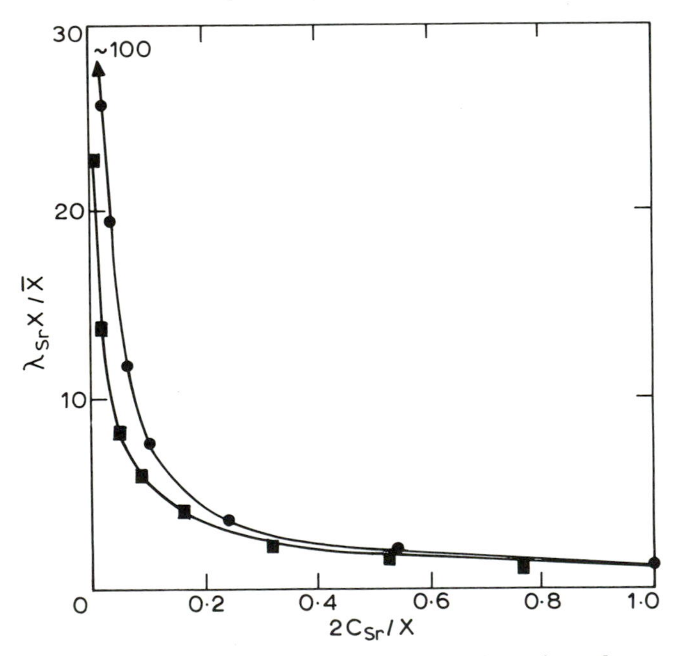

Fig. 2.19 Ratio of equivalent fractions of strontium in resin and strontium in solution ($\lambda_{Sr} X/\overline{X}$) against equivalent fraction in solution $2C_{Sr}/X$ in strontium ⇌ sodium exchange on Zeo-Kart 315 at 25°. ● $X = 0{\cdot}02$; $X/\overline{X} = 0{\cdot}035$; ■ $X = 0{\cdot}10$; $X/\overline{X} = 0{\cdot}164$. After Meares[111]

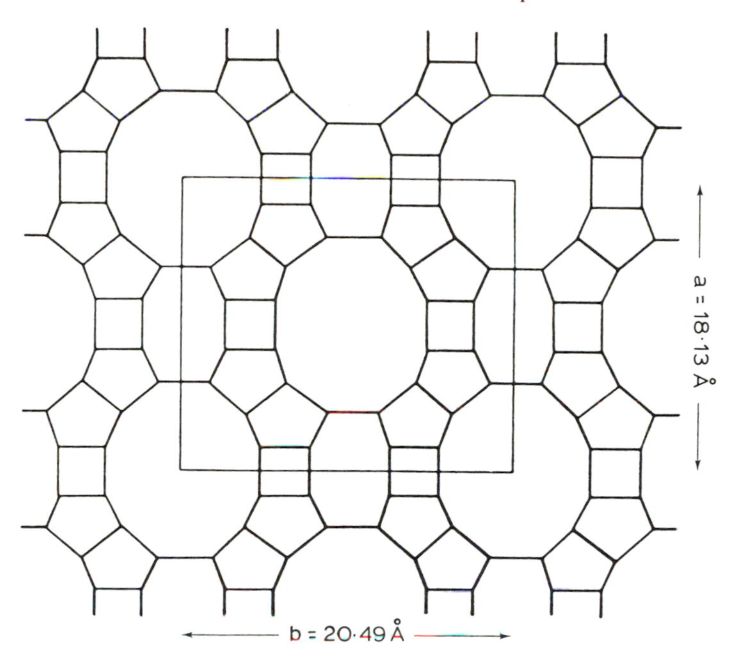

Fig. 2.20 The framework structure of mordenite in cross-section normal to the wide channels. Al or Si atoms are at each corner and oxygens near the mid point of each edge. The cations are not shown. After Barrer[112]

and neutron scattering spectroscopy. Optical birefringent methods can be used to study intradiffusion in transparent single crystals (cf §5.6 and Fig. 5.9).

Intradiffusion of a single species into a zeolite is similar to interstitial diffusion in other types of crystalline solids; the rate law for initial sorption generally follows the Fickian $\sqrt{t}$ dependencies and the Arrhenius plots are usually linear. Boundary value problems are accordingly simplified.[83]

In the first instance the diffusion or sorption rates are determined by the mesh window size, the tortuosity of pathways (one-, two-, or three-dimensional), the degree of distortion caused by the cations, and the molecular size of the diffusant. As a sieve, crystals with small windows can be used to distinguish between groups of small molecules and exclude big ones altogether (see the examples given by Barrer[113]).

The Darken equation for binary interdiffusion is also greatly simplified since the component B (the zeolite) can be assigned a zero mobility (cf equation (2.64)). However, the corrections to the Darken equation associated with correlations represented by Onsager cross-effects in a ternary solution of zeolite, water and D_2O (cf §3.11) have been observed to be substantial[114] (Table 2.4).

Accommodating the Donnan membrane equilibrium for zeolites immersed in mixed salts, varying amounts of a given cation can be charged to the zeolite. The sorption rates and sieving properties can be drastically inhibited by doping with a polar molecule like NH_3[115] (cf Fig. 2.21).

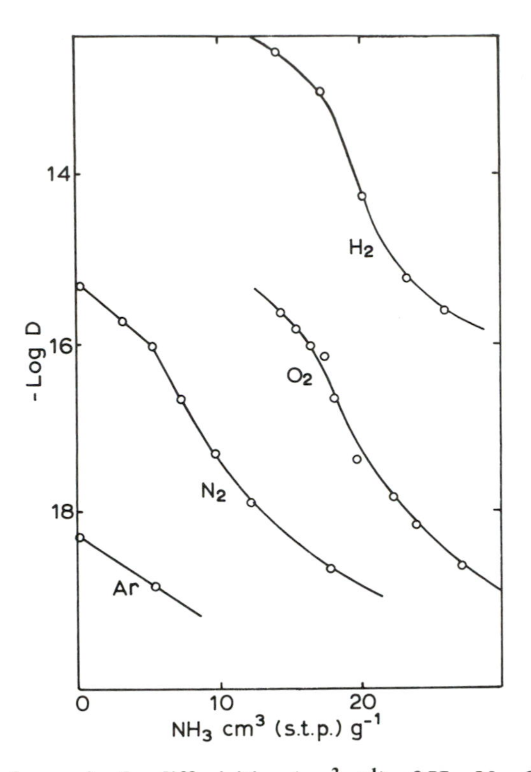

Fig. 2.21 The change in the diffusivities ($cm^2\ s^{-1}$) of H_2, N_2, O_2, and Ar in mordenite at −183°C as a function of the amount of ammonia presorbed. After Barrer and Rees[115]

Ruthven[116] has presented a review of recent measurements on diffusion in zeolites. In his own current research he has demonstrated Fickian type (i.e. intracrystalline) diffusion behaviour for uptake of N_2 in synthetic zeolite 4A crystals of different size fractions (Figs. 2.22 and 2.23). He notes, however, that there remains a discrepancy between coefficients deduced from sorption uptake and from NMR experiments.

There exists a second major class of minerals which accommodate high rates of sorption. These are the layer silicates or clay minerals which technically belong to the class of intercalation, or two-dimensional crystal compounds. Unlike the zeolites, which are rigid in three directions, these crystals possess weak Van der Waals bonds between structural layers so are subject to swelling under osmotic forces. This class of absorbers and molecular sieves is dealt with in detail by Barrer.[83]

Although amenable in principle, zeolites have not been widely used in ion exchange applications. However, it has been recently discovered that Zeolite A in its sodium form is appropriate as a substitute for triphosphates as a builder for detergents.[117] There is a significant multicomponent diffusion problem here which will be discussed in Chapter 10. A review of current and future zeolite applications is given by Breck.[118]

Table 2.4 Relation between D^* and $\tilde{D}$ for water in several zeolites near saturation of the zeolite. After Barrer and Fender[114]

Zeolite	T, °C	$D^* \times 10^8$, $cm^2\,s^{-1}$	$\tilde{D} \times 10^6$, $cm^2\,s^{-1}$	$\frac{\partial \ln a}{\partial \ln C}$	$\frac{CL_{AA^*}}{C^*L_{AA}}$
Chabazite B	75	46	10·7	23·0	$0{\cdot}09_1$
	65	32	7·6	24·0	0·15
	55	21	5·5	25·5	0·17
	45	14	3·8	27·0	0·21
	35	9·0	2·5	28·0	0·27
Heulandite	75	9·8	3·0	30·0	0·39
	65	6·1	2·0	32·0	0·44
	55	3·7	1·26	34·0	0·48
	45	2·2	0·78	35·5	0·51
	35	1·24	0·47	37·0	0·56
Gmelinite	55	7·3	2·0	26·5	0·15
	45	5·0	1·40	28·0	$0{\cdot}06_5$
	35	3·3	0·97	29·5	0·030

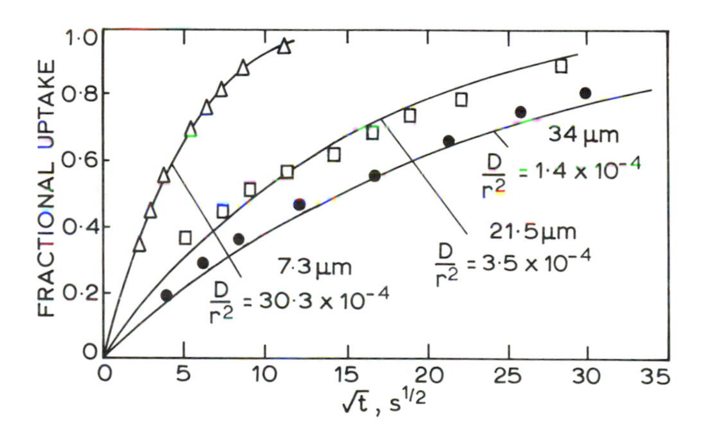

Fig. 2.22 Representative uptake curves for N_2 at 273K in three different size fractions of 4A zeolite crystals: points are experimental; curves are calculated for isothermal diffusion, with due allowance for crystal size distribution ($D = 4 \times 10^{-10}$ $cm^2\,s^{-1}$ for all curves). After Ruthven[116]

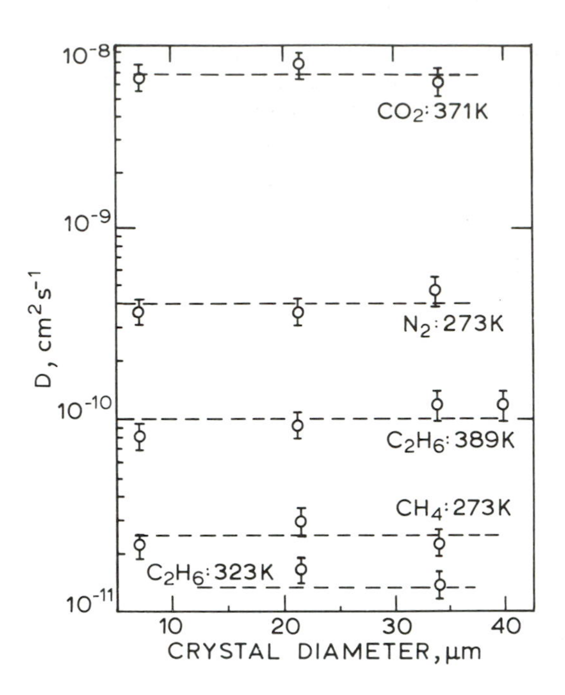

Fig. 2.23 Diffusivities calculated from uptake curves for different size fractions of 4A zeolite crystals. Error bars denote ± 15%. Data are for low concentration region in which *D* is not concentration dependent. After Ruthven[116]

2.17.3 Diffusion in membranes

As was indicated earlier, a membrane can be regarded as a localized, metastable gel, i.e. as a colloidal structure consisting of a dispersed solid impregnated with a liquid phase. Strictly speaking, membrane processes exclude hydrodynamic flow.[119] Thus for an immersed membrane, interdiffusion is fundamentally no different than liquid diffusion. For synthetic membranes at least, the diffusion process can be completely understood in terms of the processes of polymer interdiffusion and polymer dissolution described in §2.17.1. In particular, membranes usually undergo osmotic swelling, and are capable of sustaining an osmotic pressure between solutions of different composition (cf Appendix 1). Furthermore, membrane channels are usually ion specific, so under ionic concentration gradients exchange will occur. As with polymeric solutions the Donnan equilibrium (Appendix 1) is essential to a base analysis. The membrane function is thus predicated upon a relaxation towards some synthesis of an osmotic (pressure) and a Donnan (electrostatic) equilibrium. As already indicated, there are both molecular and ionic screening actions which are subsumed by the term dialysis.[119] Relatively minor changes in structural conditions, such as cross-linking and isomeric complement of the polymer chains, have a profound effect upon the tenuous balance between the forces of cohesion, solvation and electrical neutralization.

Natural, or biological (e.g. cell) membranes have served as the norm for scientific analysis in the science of 'membranology' over most of its history.[120] They possess the properties of swelling and screening which we have attributed to synthetic membranes and dissolving polymers, but they are generally much more specific with respect to separation or screening processes, a characteristic which can be attributed to their more highly organized structure.

This enhanced organization can in turn be attributed to a process of secondary crystallization. The primary crystals consist of single or bilayers of the most highly insoluble components of the aqueous vital fluids, the lipids. These are essentially impermeable to the same vital fluids and, by definition, they dissolve freely in organic solvents other than water. The secondary crystals consist of gel-like, but ordered protein networks which are dispersed in a loose pattern within the lipid regions (cf Fig. 2.24) and provide screening pathways between extracellular and intercellular fluids (the cytoplasm) or between the intercellular and nuclear fluids. The schematic model of Fig. 2.25 was developed for the nuclear envelope of the organism Tetrahymena in accordance with the high power electron microscopy of Figs. 2.26. Such membranes are generally 6–9 nm or about 30 atoms thick with penetrable protein structures or 'gap junctions' (GJ) spaced by about 100 nm. A computer-modelled sodium channel in the membrane of a nerve cell is offered by Weinberg.[124] The phenomenological theory of membrane transport in many of its manifestations is deferred until Chapter 10, with its emphasis on the thermodynamics of irreversible processes. However, as an introduction to diffusion in disperse systems, we offer in the following the analysis of a particular serial membrane process of relevance to mammalian biological function.

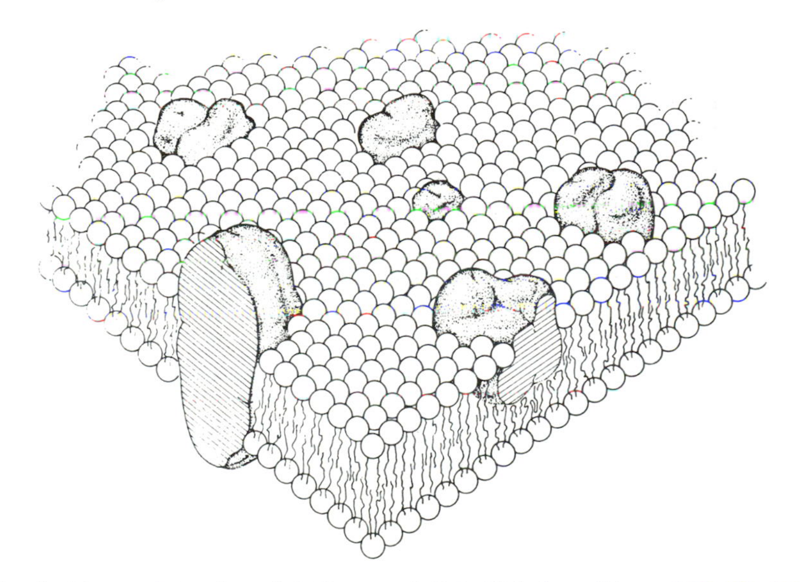

Reprinted with permission from S. J. Singer and G. L. Nicholson, *Science*, 1972, **175**, 720. Copyright 1972 by the AAAS

Fig. 2.24 The fluid mosaic model of membrane structure. Reproduced with permission from Singer and Nicolson[121]

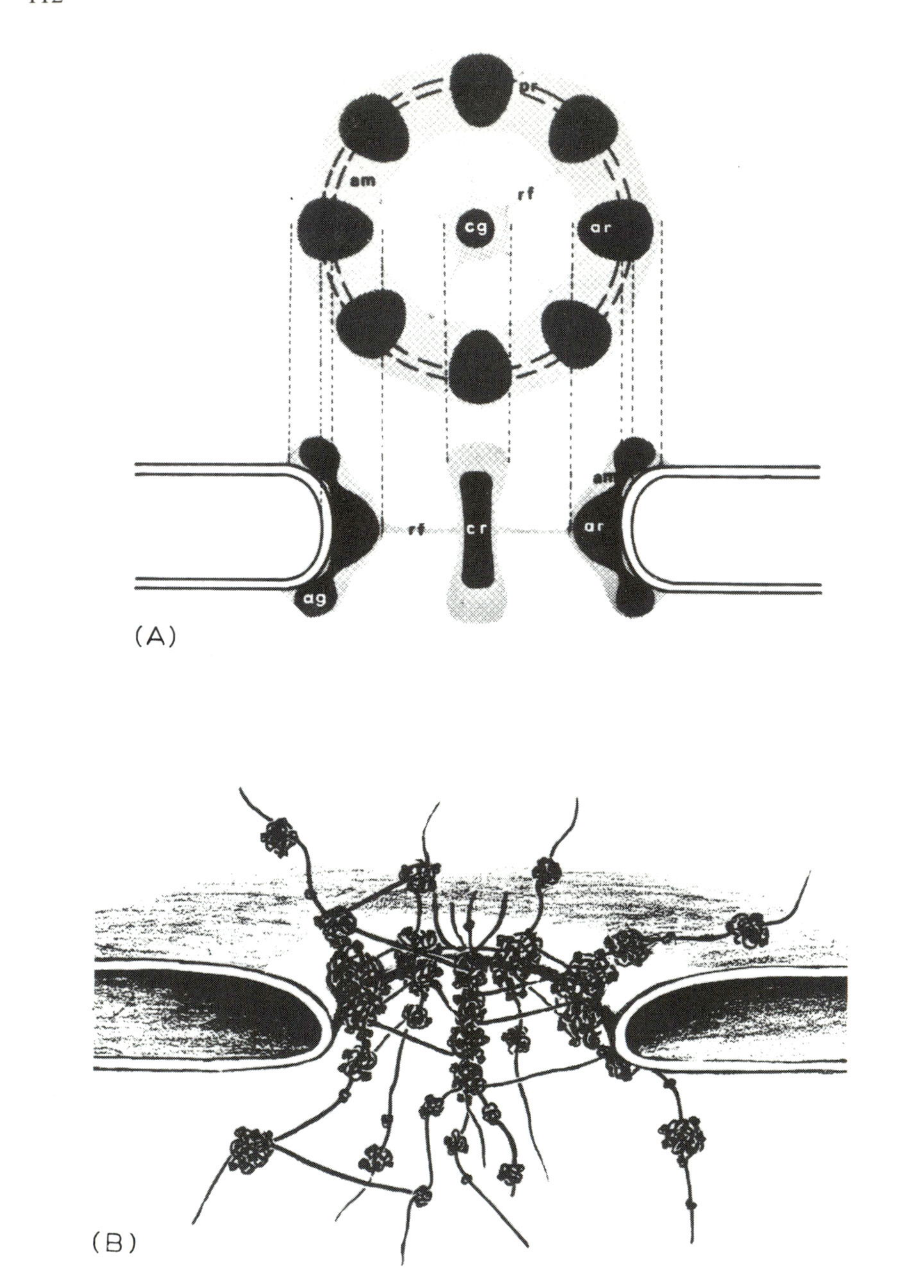

Fig. 2.25 **Nuclear pore complex model based on thin sectioning, negative staining and freeze etching observations of *Tetrahymena* pore complexes: am, annular material; pr, pore complex margin; rf, radial filaments; cr, central rod; cg, central granule; ar, annular rodlet; ag, annular globules. After Wunderlick *et al.*[122]**

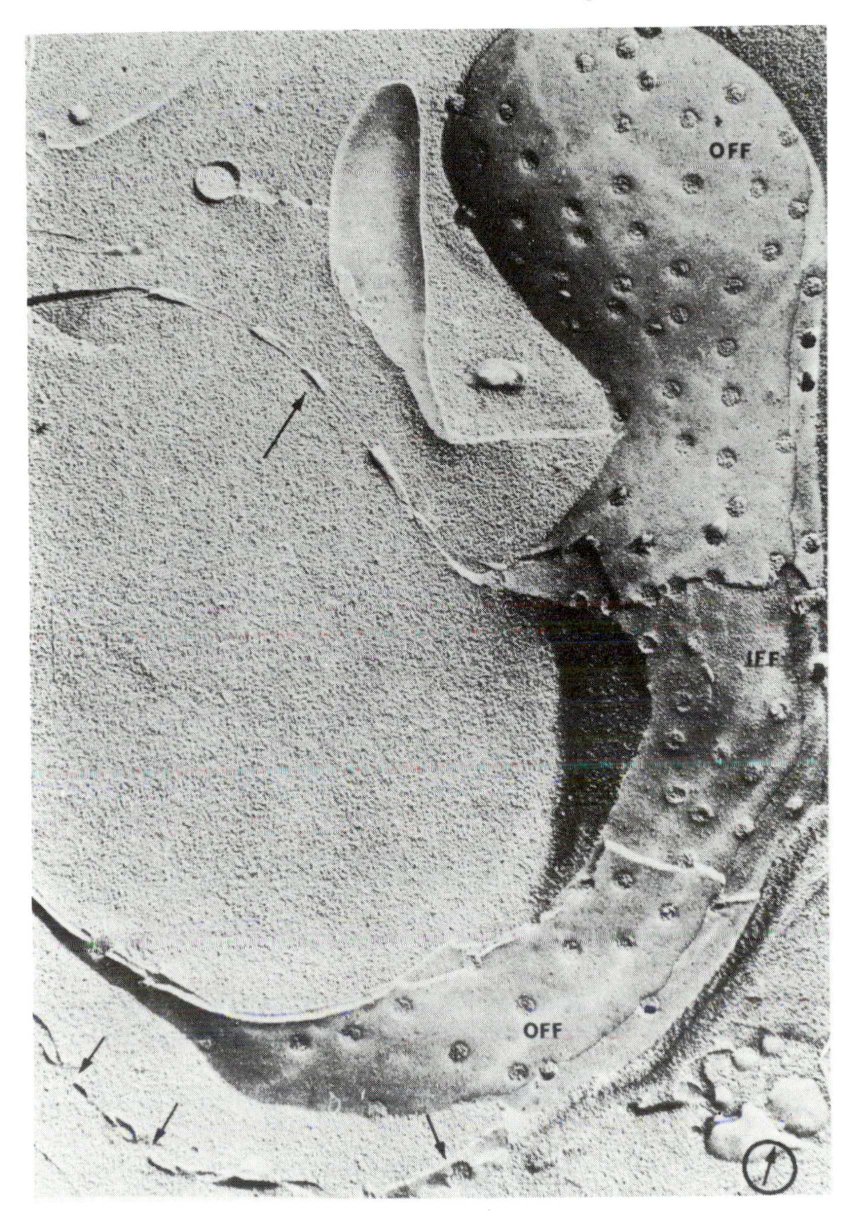

Fig. 2.26 Nuclear envelope of one nucleus of an onion root tip cell is cross- and tangential-fractured: IFF, inner fracture face of the inner nuclear membrane; OFF, outer fracture face of the outer nuclear membrane. Long arrow indicates the membranous margin of a cross-fractured nuclear pore complex. Small arrows show cisternal pore complexes in the endoplasmic reticulum. The shadowing direction is marked by the encircled arrow (×30 000). After Wunderlick and Speth[123]

2.17.4 A quantitative model for intercellular communication via membrane diffusion between muscle cells

Following the hypothesis that the onset of labour in rats, i.e. muscular contractions, is connected with an increase in the diffusive communication along muscle cell chains, and that this in turn is to be attributed to the multiplication of the number of gap junctions in intercell membranes, the tracer diffusion coefficient of the organic tracer 2-DG has been determined in semi-infinite diffusion couple experiments (cf § 1.8) along a set of strips of muscle prepared at times spanning the parturition event.[125] Figure 2.27 presents a set of typical penetration curves which, when plotted on probability paper (top frame),

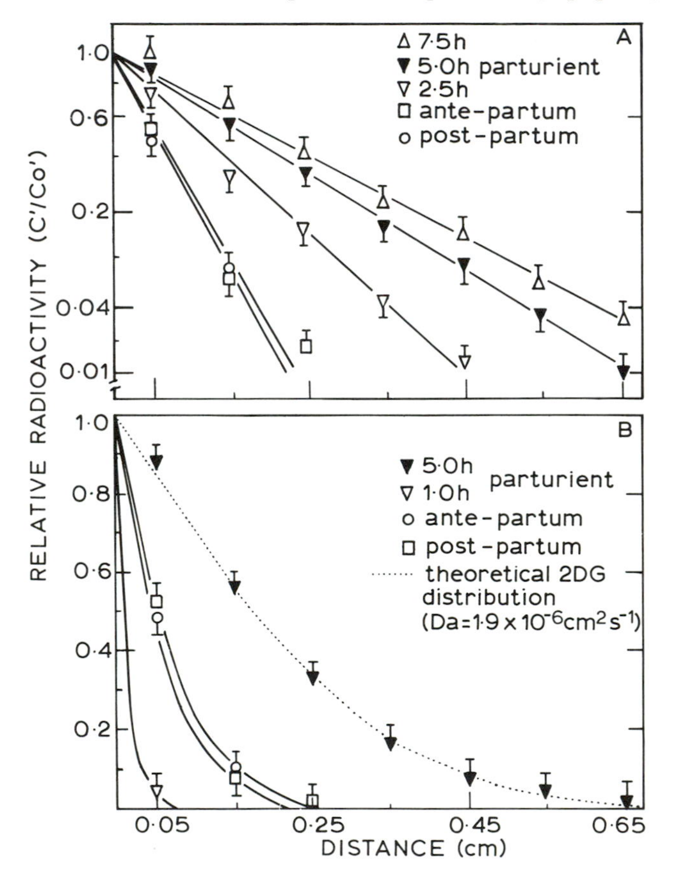

Fig. 2.27 **Longitudinal distribution of 2-[^{3}H]deoxy-D-glucose (2-DG) in myometrial strips from 17 to 20 days pregnant, (n = 41 strips), parturient (n = 46 strips), and 2 to 3 days postpartum (n = 14 strips) rats after 5 h diffusion time plotted on arithmetic probability graph paper (a) and normal graph paper (b). Data are included in (a) for parturient tissues after 2·5 and 7·5 h diffusion periods, and in (b) a theoretical plot of 2-DG distribution is shown and a value for the apparent diffusion coefficient (D_a) of $1{\cdot}9 \times 10^{-6}$ cm^2 s^{-1}**

illustrates the conventional error function profile for this boundary value problem (§ 1.8). Figure 2.28 summarizes the apparent diffusion coefficient D_a as a function of the day of gestation and indicates a dramatic increase at parturition.

To relate such observations to morphology as determined by electron microscopy it is necessary to disaggregate the average coefficient D_a. Our model (Fig. 2.29) assumes that GJs represent the cell-to-cell pathway for the exchange of small molecules. The GJs are treated as aggregates of proteins, each protein possessing a central hydrophilic channel of uniform dimensions. In calculating the diffusivity of tracers in the cell-to-cell channel, we have assumed that transfer occurs by a restricted diffusion mechanism and have incorporated correction factors into the model to account for the steric and frictional interactions between the tracer and channel.

Longitudinal diffusion of 2-DG in a strip of the myometrium comprising chains of smooth muscle cells lying in parallel is equivalent to 2-DG diffusion in a single chain of cells (i.e. the bundle of fibres is treated as though it were a single cable with a uniform core) (Fig. 2.29*a*). The longitudinal movement of 2-DG is subject to the serial barriers to diffusion imposed by the aqueous fluid

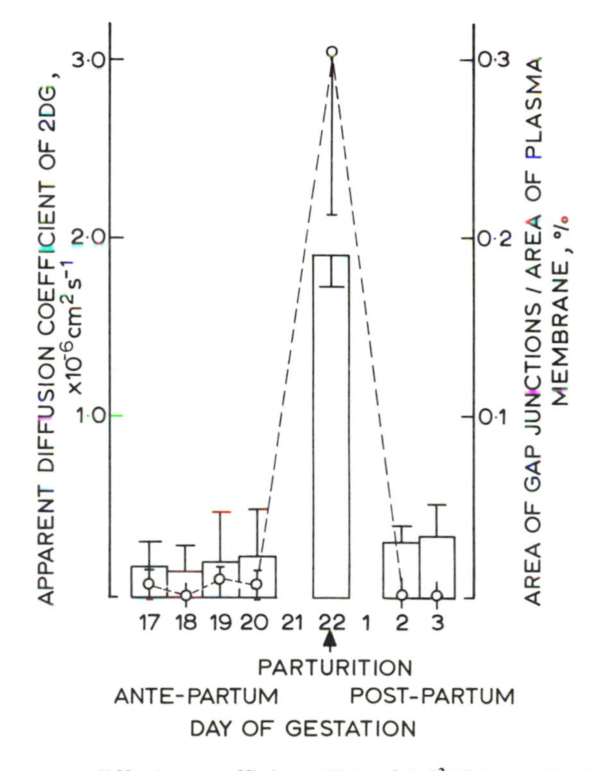

Fig. 2.28 Apparent diffusion coefficient (D_a) of 2-[^{3}H]deoxy-D-glucose (2-DG) (*histogram*) and area of gap junctions (GJs) as a percentage of plasma membrane as functions of day of gestation. D_a and area of GJs are expressed as means ± SE for several tissues

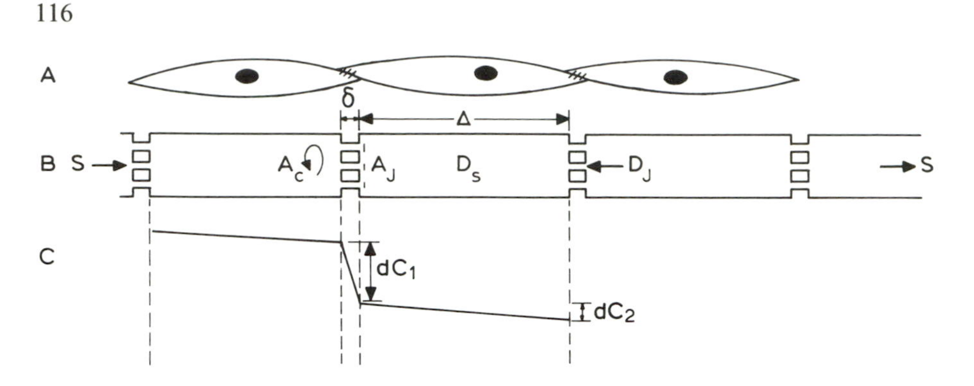

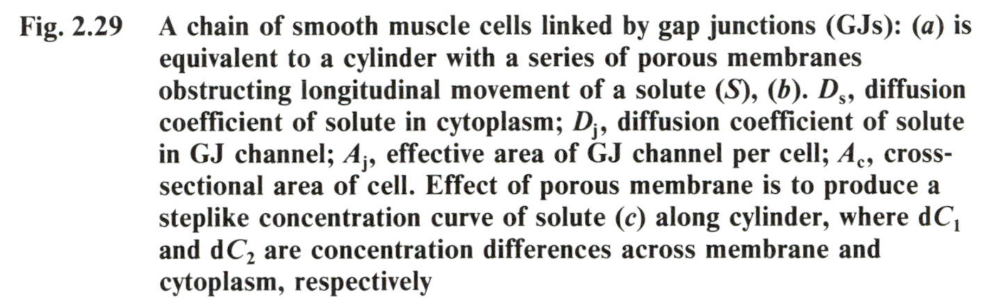

Fig. 2.29 A chain of smooth muscle cells linked by gap junctions (GJs): (*a*) is equivalent to a cylinder with a series of porous membranes obstructing longitudinal movement of a solute (*S*), (*b*). D_s, diffusion coefficient of solute in cytoplasm; D_j, diffusion coefficient of solute in GJ channel; A_j, effective area of GJ channel per cell; A_c, cross-sectional area of cell. Effect of porous membrane is to produce a steplike concentration curve of solute (*c*) along cylinder, where dC_1 and dC_2 are concentration differences across membrane and cytoplasm, respectively

(cytoplasm) and GJ channels between the cells. Although the GJs are shown to lie at the ends of the smooth muscle cells in Fig. 2.29*b*, there is no evidence for this in either the myometrium or intestine. The presence of GJs over the entire surface of the cells, rather than just at the ends, should not significantly affect the mathematical description of average solute diffusion. D_a is dependent upon the coefficients of diffusion in the cytoplasm (D_s) and GJ channels (D_j) (Fig. 2.29*b*), and these combine via a steplike concentration curve of solute in the cells.

From Fick's Law, the flux (J) through one cell is

$$-J = D_a[(dC_1 + dC_2)/(\delta + \Delta)] \tag{2.96}$$

where dC_1 and dC_2 are the differences in 2-DG concentration across the GJ channel and cytoplasm, respectively (Fig. 2.29*c*), and Δ and δ are the lengths of the smooth muscle cell and channel, respectively (Fig. 2.29*b*). If Δ is much less than the diffusion penetration then we can assume a local quasi-steady state with the flux continuous and

$$-J = D_a[(dC_1 + dC_2)/(\Delta + \delta)] = D_s(dC_2/\Delta) = D_j(dC_1/\delta) \tag{2.97}$$

where D_j is the diffusion coefficient of solute in the GJ channel. Hence, we have three expressions for the flux. If we solve first for dC_1 and dC_2 and subsequently invert the equation, an expression for D_a is obtained, i.e.

$$1/D_a = [1/(\Delta + \delta)][(\Delta/D_s) + (\delta/D_j)] \tag{2.98}$$

In the smooth muscle cell, δ is very much less than Δ; therefore we may eliminate δ from $1/(\delta + \Delta)$ and achieve a new expression

$$1/D_a = 1/D_s + \delta/D_j\Delta \tag{2.99}$$

If it is further assumed that the junctional channels are filled with cytoplasm, then

$$D_j = D_s A_j / A_c \tag{2.100}$$

Thus D_a is determined by the variables D_s, Δ, δ, A_c, and A_j as

$$D_a = D_s/(1 + \delta A_c/\Delta A_j) \tag{2.101}$$

For delivering myometrium we assume the following values from microscopic observation and independent diffusion experiments:[126] $\Delta = 450\ \mu m$; $\delta = 15$ nm; $A_c = 78{\cdot}5\ \mu m^2$; $D_s = 3{\cdot}3 \times 10^{-4}$ mm^2 s^{-1} or about one half that of glucose in water. A value for the effective area of cell-to-cell channel per cell, A_j, was determined by calculating the fractional area of GJ membrane occupied by cell-to-cell channels and accounting for steric and fractional hindrance to the movement of permeant molecules through the channel, which results because the molecular dimensions of 2-DG approach those of the channel. The area of GJ membrane per cell is ~0·2% of 7 200 μm^2 of plasma membrane surface area or 14·5 μm^2. The packing density of intramembrane protein particles in myometrial GJs is ~6 500 μm^{-2};[127] thus in 13·2 μm^2 of GJ membrane there are 94 000 particles. The exact dimensions of the cell-to-cell channel are unknown, but the effective diameter is thought to lie between 1·5 and 2·0 nm. We may assume these values to calculate upper and lower limits for the effective area of channel per cell and D_a for 2-DG.

If there are 94 000 particles per cell and their diameter is 1·5 or 2·0 nm then the pore area per cell is either 0·2 or 0·38 μm^2. Correction coefficients for steric and frictional interactions introduced into the calculation of A_j include variables for the molecular radius of the permeant molecule (a) and that of the channel (r). The coefficient for steric hindrance at the channel opening (S) is equal to $(1 - a/r)^2$, and frictional drag between the molecule and channel wall (f) is equal to $1 - 2{\cdot}019\,(a/r) + 2{\cdot}09(a/r)^3 - 0{\cdot}95(a/r)^5$.[128] A value of 0·4 nm was used for the molecular radius of 2-DG, and the two limiting values of 0·75 and 1·00 nm were used for the channel radius. If the channel is 1·5 nm in diameter, then S and f are 0·218 and 0·199, respectively, and for a 2·0 nm channel they are 0·360 and 0·316, respectively. A_j is the product of the area of junctions per cell, S, and f. For 1·5 channels, A_j is 0·009 μm^2, and for 2·0 nm channels A_j is 0·040 μm^2. Thus in theory, D_a should lie between $2{\cdot}6 \times 10^{-4}$ and $3{\cdot}0 \times 10^{-4}$ mm^2 s^{-1} for the limiting values of 1·5 and 2·0 nm for the channel diameter, respectively. Our empirical value of $1{\cdot}85 \pm 0{\cdot}19 \times 10^{-4}$ mm^2 s^{-1} is compatible with that predicted by the model. The fact that the observed value is slightly less than that predicted would follow if our assumption that all the channels are open is not entirely correct. It may be that there is always a distribution of channels in open and closed conformations, which would account for the slightly lower value for D_a.

2.18 DIFFUSION IN ORDERED AND IONIC SOLIDS

There is extensive literature on diffusion in ordered alloys and intermetallic crystals. This material can be accessed through the recent work of Bakker and co-workers.[34, 129, 130, 131] Some, but not all of the features of such materials are shared by ionic solids, which will be emphasized in this volume (cf Chapter 8).

As in metals, the existence of crystalline point defects as equilibrium entities is essential to the diffusion process in ionic solids. Indeed, they are of such a variety that special notations must be introduced (cf Chapter 8). Besides the possibility of mobile or free carriers such as electrons (e^-) and positive holes ($h^{\cdot}$) in a pure material, there can be matched anion and cation vacancies (Schottky defects[132]) and anion or cation interstitials with matching vacancies (Frenkel defects[33]). For the alkali halides and related species there are negligible free carriers so we can focus on the formation of the other defects, which on account of charge neutrality, must always appear in pairs. These eventualities are illustrated in Fig. 2.30. Note that the cations in many such structures have much smaller ionic radii than the anions, because the latter possess an electron outside a closed shell while the

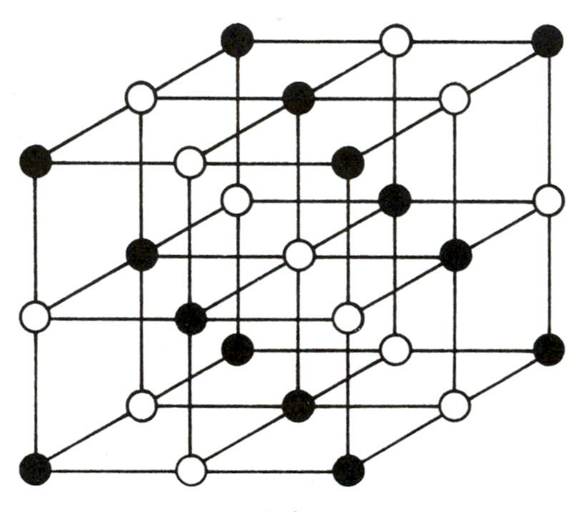

(a)

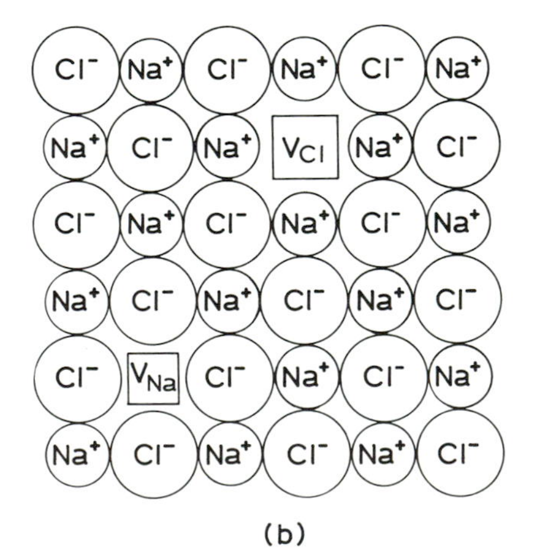

(b)

Fig. 2.30 **(*a*) Arrangement of the anions (●) and cations (○) in the NaCl unit cell; (*b*) schematic drawing of the NaCl lattice containing anion (Cl) and cation (Na) vacancies. After Shewmon[21]**

former possess an extra core proton relative to the number of electrons so the closed shell electron cloud contracts. Accordingly a close-packed structure of fcc sub-lattices is common (e.g. NaCl). On the other hand, with ionic radii more nearly equal, average ionic bonds can be more effectively saturated with bcc sublattices (e.g. CsCl).

Following the statistical procedures of Howard and Lidiard[14] used in the derivation of the equilibrium vacancy concentration in metals (equations (2.3) and (2.8)) one obtains for completely dissociated Schottky defects the concentration relations

$$X_{Va}X_{Vc} = \exp(-g_{Va} - g_{Vc})/kT = \exp(-g_V/KT) \quad (2.102)$$

and for completely dissociated Frenkel defects

$$X_{Va}X_{ia} = \exp(-g_{Va} - g_{ia})/kT = \exp(-g_i/KT) \quad (2.103)$$

and

$$X_{Vc}X_{ic} = \exp(-g_{Vc} - g_{ic})/kT = \exp(-g_i/kT) \quad (2.104)$$

If the activation free energies are such that only one type of defect appears then at electrical neutrality the defects are paired and

$$X_{Va} = X_{ia} = \exp(-g_V/2kT) \quad (2.105)$$

for Schottky defects and

$$X_{Vc} = X_{ic} = \exp(-g_i/2kT) \quad (2.106)$$

for the Frenkel defect of equation (2.104). More generally, all three configurations can occur and the charge balance must be applied to a sum over all species. Actually, on account of the generally large size of the anions and the high activation energy for interstitials, there is usually a clear bias towards case (2.102) which includes the alkali halides, and towards case (2.104) which includes the silver halides. In either case, the equilibrium vacancies for a pure material are designated as *intrinsic* vacancies to distinguish them from *extrinsic* vacancies injected with impurities of different valence (e.g Ca^{++} in LiCl). The latter, which are often in the form of unintentional impurities, have a rather drastic effect on diffusion rates (see Figs. 2.31, 2.32, and Chapter 8).

Identification of the concentration of defect species in a crystal is only the starting point for a discussion of tracer or chemical diffusion. For a crystal with Schottky defects only, the anions and their vacancies, due to their large sizes, are relatively immobile so, by analogy to the situation in metals, diffusion will occur via cation–vacancy interchange. In the case of impurity diffusion (e.g. LiCl in NaCl) there is a significant difference, however, for the cations will possess different jump frequencies. Yet because the anion sublattice is rigid there can be no corresponding drift or Kirkendall effect. One can only conclude that the faster of the species must continually undergo decelerations along its jump trajectories and the slower must undergo accelerations to sustain electroneutrality.[135] This implies the generation of a distribution of dipole charge ρ' (as opposed to real charge, which is zero), and in accord with Poisson's equation of electrostatistics, an electrostatic potential defined by

$$\nabla^2\psi = \rho'/\varepsilon_0 \quad (2.107)$$

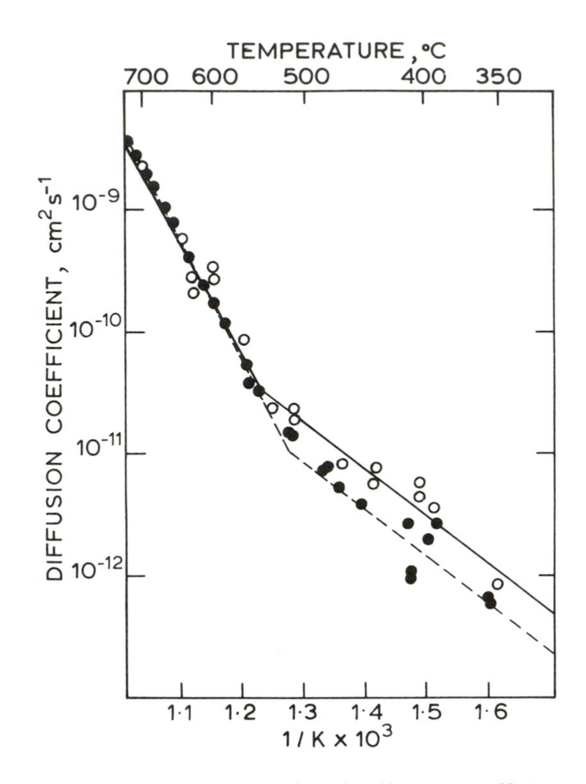

Fig. 2.31 Temperature dependence of self-diffusion coefficient of sodium in sodium chloride: ○, directly measured; ●, calculated from measured conductivity. After Mapather *et al.*[134]

This potential ψ is known as the *diffusion potential* and must be explicitly entered into the diffusion flux–force relations. Irreversible thermodynamics (Chapters 3 and 8) informs us that the correct way to do this is to substitute the electrochemical potential of the ith species

$$\eta_i = \mu_i + z_i F \psi \tag{2.108}$$

for the chemical potential μ_i wherever it appears. Here z_i is the valence of the species i and F is the Faraday. The η_i like the μ_i satisfy the Gibbs–Duhem equation. This structure implies in general that diffusion can be influenced by its self-generated dipole field and/or by an externally applied field. Analysis of experiments of the latter kind in the absence of chemical gradients leads to various relationships between the diffusion coefficient and the electrical conductivity σ. Consider, for example, a crystal like NaCl which we suppose to have mobile cations and cation vacancies only. The application of an external field E to the pure crystal yields a current expressible in terms of the vacancy flux J_V of the form

$$J_e = -z_V J_V = \sigma E = -\sigma \nabla \psi \tag{2.109}$$

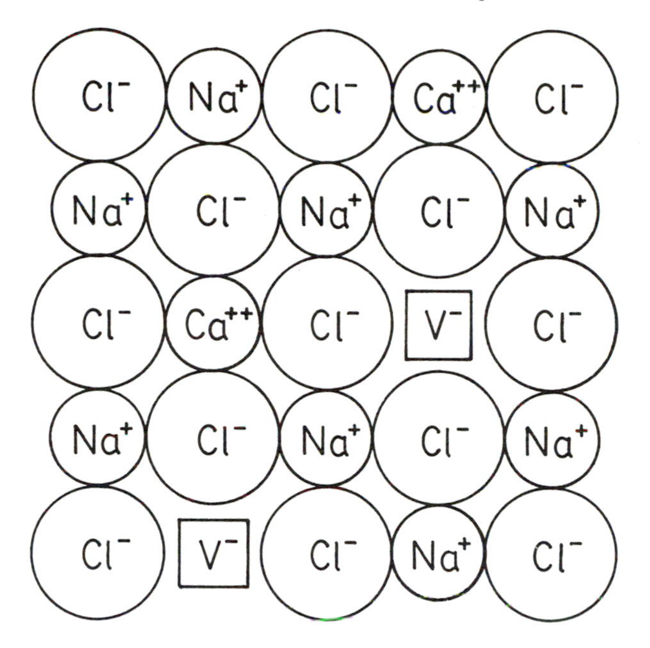

Fig. 2.32 Effect of aliovalent impurity Ca^{++} producing extrinsic vacancies

Since the cations and vacancies are countercurrent we can focus on the latter through the appropriate version of equation (1.27), i.e.

$$J_V = -L_V \nabla(\eta_V - \eta_c) \cong -L_V \nabla \eta_V \tag{2.110}$$

the dilute solution approximation being validated through the Gibbs–Duhem equation. From equation (1.31), L_V can be expressed in terms of the diffusion coefficient as

$$L_V = m_V D_V / RT \tag{2.111}$$

Now in the electrical experiment $\Delta\mu_V = 0$, so we deduce from equations (2.109) and (2.110) the conductivity

$$\sigma = m_V D_V F^2 z_V^2 / RT \tag{2.112}$$

Now D_V can be expressed in terms of the cation tracer coefficient D_c^* via equation (2.19), so if m_c is the molar concentration of cations ($m_V/m_c = X_V$)

$$\sigma = \frac{m_c F^2 z_V^2}{RTf} D_c^* \tag{2.113}$$

where f is the correlation factor for the particular crystal structure (Table 2.1). Since σ and D_c^* are independently measurable and f can be calculated exactly, such relations if validated serve to verify the assumed diffusion mechanism.[16, 29]

For a crystal with Frenkel defects there are at least three modes of diffusion possible, vacancy interchange, interstitial and interstitialcy mechanisms. The latter involves the displacement of a neighbouring ion to an interstitial site and exchange with the resulting vacancy (Fig. 2.33). With a pure interstitial mechanism for cation diffusion the analogue of equation (2.113) is

$$\sigma = \frac{m_c F^2}{RT} z_V^2 D_c^* \tag{2.114}$$

while for a pure interstitialcy mechanism it is

$$\sigma = \frac{2m_c F^2}{RTf} z_V^2 D_c^* \tag{2.115}$$

The experimental relation can accordingly be used for diagnostics on the mix of mechanisms.

The complications introduced by association (or pairing) of ions and vacancies have been discussed by Howard and Lidiard.[14] They also review the experimental record. Chapter 8 deals with other aspects of diffusion in ionic solids and semiconductors including ternary interactions.

Liquid electrolytic solutions have much in common with ionic solids, and conductivity experiments play much the same role in their characterization. We note here that the mass–electrical interactions discussed above could have been analyzed with completeness and greater rigour in terms of the Onsager theory of irreversible processes (cf Chapter 3), invoking independent mass and electrical fluxes and forces of the form[135]

$$J_V = -\frac{D_V m_V}{RT} \nabla \mu_V - \frac{D_V m_V z_V F}{RT} \nabla \psi \tag{2.116}$$

$$J_e = -\frac{D_V m_V z_V F}{RT} \nabla \mu_V - \frac{D_V m_V z_V^2 F^2}{RT} \nabla \psi \tag{2.117}$$

and subject to the Onsager reciprocal relations

$$L_{eV} = L_{Ve} \tag{2.118}$$

In Chapter 9 diffusion in binary and ternary electrolytes is examined from the irreversible thermodynamic point of view.

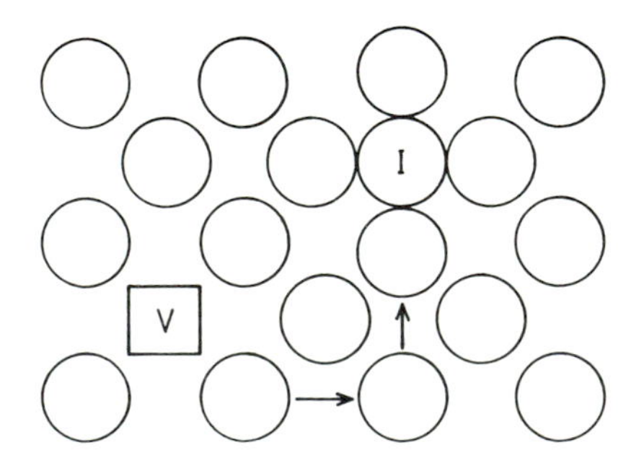

Fig. 2.33 Illustrating Frenkel defects and interstitialcy mechanism

2.19 MATHEMATICAL AND COMPUTATIONAL EXERCISES

1. Derive the Hart equation (1.5) using a detailed argument aided by diagrams. Base this upon the sketched argument on p. 73.

2. Construct a rigorous derivation and interpretation of the right-hand equality in equation (2.16) based on the arguments of p. 72.

3. The generalized three-dimensional form of equation (2.16) is

$$f = \langle R^2 \rangle / \sum_{\alpha=1}^{N} n_\alpha \lambda_\alpha^2 \qquad n \rightarrow \infty$$

Modify the base equation, then derive and interpret this formula.

4. Murch[136] has made available a program for Monte Carlo simulations. Access this and present the results of a typical evaluation of f.

5. Write down the full derivation of equation (2.25) and demonstrate that this quantity is always less than unity.

6. Carry out the manipulation of equation (2.33) as instructed to obtain the detailed Fick equation (2.34).

7. Justify in detail the relationships between the linear expansion coefficient α and the parameter δ in equation (2.74).

8. Rederive equation (2.99) using the series resistance concept from conduction heat transfer. A metallic two-layer structure consisting of metals A and B of serial thickness 0·1 and 1 μm respectively is bonded to the wall of a highly permeable ceramic container. If the coefficients of diffusion for the metals are D_A and D_B, what is the effective diffusion coefficient of the entire wall to be used in permeation calculations?

9. Refer to §2.13 and Fig. 2.5. Given that the mutual diffusion coefficient is $5{\cdot}23 \times 10^{-9}$ at 785°C, would you say that the Kirkendall effect in α-brass is a strong or weak effect?

REFERENCES

1 M. J. PUSKA AND R. M. NIEMINEN: *Phys. Rev. B, Conden. Matt.*, 1984, **29**, 5382
2 D. ENSKOG: Dissertation, 1917, Uppsala University
3 D. ENSKOG: *Arkiv. Met. Astron. Fys.*, 1922, **16**, 16
4 S. CHAPMAN AND T. G. COWLING: 'The mathematical theory of non-uniform gases', 1939, Cambridge University Press
5 B. J. ALDER AND T. E. WAINWRIGHT: *J. Chem. Phys.*, 1959, **31**, 459
6 B. J. ALDER AND T. E. WAINWRIGHT: *ibid.*, 1960, **33**, 1439
7 B. J. ALDER, D. M. GLASS, AND T. E. WAINWRIGHT: *ibid.*, 1970, **53**, 3813
8 B. J. ALDER AND T. E. WAINWRIGHT: *Phys. Rev.*, 1970, **A1**, 18
9 J. H. DYMOND AND B. J. ALDER: *J. Chem. Phys.*, 1969, **51**, 308
10 J. E. LENNARD-JONES: *Trans. Faraday Soc.*, 1932, **28**, 334
11 L. ONSAGER: *Phys. Rev.*, 1931, **37**, 405; **38**, 2265
12 S. GLASSTONE, K. J. LAIDLER, AND H. EYRING: 'The theory of rate processes', Ch. 9, 1941, New York, McGraw-Hill
13 J. H. HILDEBRAND: 'Viscosity and diffusivity, a predictive treatment', 1977, New York, John Wiley

14 R. E. HOWARD AND A. B. LIDIARD: *Rep. Prog. Phys.*, 1964, **27**, 161
15 T. HEHENKAMP: in 'Solute-defect interaction: theory and experiment', (ed. S. Saimoto, G. R. Purdy, and G. V. Kidson), 241, 1986, Toronto, Pergamon
16 J. R. MANNING: 'Diffusion kinetics for atoms in crystals', 36 *et seq.*, 83 *et seq.*, 1968, Princeton, D. van Nostrand & Co.
17 M. VON SMOLUCHOWSKI: *Ann. Phys. (Leipzig)*, 1908, **25**, 205
18 A. EINSTEIN: *ibid.*, 1905, **17**, 549
19 A. EINSTEIN: *ibid.*, 1906, **19**, 371
20 E. HART: *Acta Met.*, 1957, **5**, 597
21 P. G. SHEWMON: 'Diffusion in solids', 175 *et seq.*, 1963, New York, McGraw-Hill
22 E. C. AIFANTIS: *Acta Met.*, 1979, **27**, 683
23 J. M. HILL: *Scripta Met.*, 1979, **13**, 1027
24 A. B. LIDIARD: in 'Solute-defect interaction: theory and experiment', (ed. S. Saimoto, G. R. Purdy, and G. V. Kidson), 186, 1986, Toronto, Pergamon
25 M. KOIWA AND S. ISHIOKA: *ibid.*, 232
26 G. V. KIDSON: *Can. J. Phys.*, 1975, **53**, 1054
27 G. V. KIDSON: *Phil. Mag.*, 1978, **37**, 305
28 G. V. KIDSON: in 'Solute-defect interaction: theory and experiment', (ed. S. Saimoto, G. R. Purdy, and G. V. Kidson), 35, 1986, Toronto, Pergamon
29 G. E. MURCH: in 'Diffusion in crystalline solids', (ed. G. E. Murch and A. S. Nowick), 379, 1984, Orlando FA, Academic Press
30 K. COMPAAN AND Y. HAVEN: *Trans. Faraday Soc.*, 1956, **52**, 786
31 A. B. LIDIARD: *Phil. Mag.*, 1955, **46**, 1218
32 T. HEUMANN: *J. Phys F (Met. Phys.)*, 1979, **9**, 1997
33 J. S. KIRKALDY: in 'Solute-defect interaction: theory and experiment', (ed. S. Saimoto, G. R. Purdy, and G. V. Kidson), 214, 1986, Toronto, Pergamon
34 H. BAKKER: in 'Diffusion in crystalline solids', (ed. G. E. Murch and A. S. Nowick), 189, 1984, Orlando FA, Academic Press
35 A. RAHMAN AND F. H. STILLINGER: *J. Chem. Phys.*, 1971, **55**, 3336
36 A. RAHMAN, R. H. FOWLER, AND A. H. NARTAN: *ibid.*, 1972, **57**, 3010
37 'Interatomic potentials and simulation of lattice defects', (ed. P. C. Gehlen, J. R. Beeler Jr, and R. I. Jaffee), 1972, New York, Plenum
38 C. H. BENNETT: in 'Diffusion in solids: recent developments', (ed. A. S. Nowick and J. B. Burton), 73 *et seq.*, 1975, New York, Academic Press
39 R. O. WATTS AND L. J. MCGEE: 'Liquid state chemical physics', 9 *et seq.*, 90 *et seq.*, 182 *et seq.*, 1976, New York, John Wiley
40 R. KUBO: *J. Phys. Soc. Jpn.*, 1957, **12**, 570
41 J. G. KIRKWOOD: *J. Chem. Phys.*, 1946, **14**, 180
42 J. G. KIRKWOOD: 'Selected topics in statistical mechanics', 1967, New York, Gordon and Breach
43 A. R. ALLNATT: *J. Chem. Phys.*, 1965, **43**, 1855
44 A. R. ALLNATT: *J. Phys. C (Solid State Phys.)*, 1981, **14**, 5453
45 A. R. ALLNATT: *ibid.*, 5467
46 A. R. ALLNATT: *ibid.*, 1982, **15**, 5605
47 A. R. ALLNATT AND Y. OKAMURA: 'International seminar on solute-defect interaction', Aug. 1985, Kingston, Canada
48 H. G. HERTZ: 'Diffusion and conductance in ionic liquids, a linear response treatment', 1982, Wiesbaden, Akademische Verlagsgesellschaft
49 J. G. KIRKWOOD AND D. D. FITTS: *J. Chem. Phys.*, 1960, **33**, 1317
50 D. W. MCCALL AND D. C. DOUGLASS: *ibid.*, 1967, **71**, 987
51 D. C. DOUGLASS AND H. L. FRISCH: *ibid.*, 1969, **73**, 3039
52 A. R. ALLNATT AND A. B. LIDIARD: *Rep. Prog. Phys.*, 1987, **50**, 373
53 C. WERT AND C. ZENER: *J. Appl. Phys.*, 1950, **21**, 5

54 J. E. LANE AND J. S. KIRKALDY: *Can. J. Phys.*, 1964, **42**, 1643
55 C. ZENER: in 'Imperfections in nearly perfect crystals', (ed. W. Shockley), 289, 1952, New York, John Wiley
56 L. S. DARKEN: *Trans. AIME*, 1948, **175**, 184
57 O. E. MEYER: 'The kinetic theory of gases', (transl. R. E. Baynes), 248, 255, 1899, London, Longmans Green
58 G. S. HARTLEY AND J. CRANK: *Trans. Faraday Soc.*, 1949, **45**, 801
59 R. SIMMONS AND R. BALLUFFI: *Phys. Rev.*, 1960, **117**, 52; **119**, 600
60 R. W. BALLUFFI: *Acta Met.*, 1961, **8**, 871
61 S. PRAGER: *J. Chem. Phys.*, 1953, **21**, 1344
62 A. SMIGELSKAS AND E. KIRKENDALL: *Trans. AIME*, 1947, **171**, 130
63 L. C. CORREA DA SILVA AND R. MEHL: *ibid.*, 1951, **191**, 155
64 R. W. BALLUFFI: *Acta Met.*, 1954, **2**, 195
65 J. BARDEEN AND C. HERRING: 'Atom movements', 87, 1951, Cleveland, ASM
66 J. E. REYNOLDS, B. L. AVERBACH, AND M. COHEN: *Acta Met.*, 1957, **5**, 29
67 A. V. VIGNES AND M. BADIA: 'Diffusion processes, proceedings', 1971, **1**, 275
68 M. A. DAYANANDA: *Acta Met.*, 1981, **29**, 1151
69 H. T. A. TYRELL AND K. H. HARRIS: 'Diffusion in liquids', 1984, London, Butterworths
70 N. NACHTRIEB: in 'Liquid metals and solidification', 49, 1958, Cleveland, ASM
71 F. C. LARCHÉ AND J. W. CAHN: *Acta Met.*, 1985, **33**, 331
72 B. J. ALDER, W. E. ALLEY, AND J. H. DYMOND: *J. Chem. Phys.*, 1974, **61**, 14
73 D. CHANDLER: *Acc. Chem. Res.*, 1974, **7**, 246
74 R. A. SWALIN: *Acta Met.*, 1959, **7**, 736
75 R. A. SWALIN: *Z. Naturforsch.*, 1968, **23a**, 805
76 R. B. BIRD, W. E. STEWART, AND E. N. LIGHTFOOT: 'Transport phenomena', 20, 253 *et seq.*, 1960, New York, John Wiley
77 S. L. RUBY, J. C. LOVE, P. A. FLINN, AND B. J. ZABRANSKY: *Appl. Phys. Lett.*, 1975, **27**, 320
78 R. E. MEYER: *J. Phys. Chem.*, 1961, **65**, 567
79 'Handbook of chemistry and physics', 8th edn, 1977–78, West Palm Beach, Florida, CRC Press
80 N. H. NACHTRIEB AND J. PETIT: *J. Chem. Phys.*, 1956, **24**, 746
81 E. L. CUSSLER: 'Diffusion: mass transfer in liquid systems', 1984, Cambridge University Press
82 P. PROTOPAPPAS, H. C. ANDERSON, AND N. A. D. PARLEE: *J. Chem. Phys.*, 1973, **59**, 15
83 R. M. BARRER: 'Zeolites and clay minerals as sorbents and molecular sieves', 1978, London, Academic Press
84 G. S. PARK: in 'Diffusion in polymers', (ed. J. S. Crank and C. S. Park), 141, 1968, London, Academic Press
85 R. M. BARRER: *ibid.*, 165
86 H. YAMAKAWA: 'Modern theory of polymer solutions', 1971, New York, Harper and Row
87 P-G. DE GENNES: 'Scaling concepts in polymer physics', 1979, Ithica, Cornell University Press
88 P. FLORY: 'Statistics of chain molecules', 1969, New York, Interscience
89 S. F. EDWARDS: *Proc. Phys. Soc.*, 1967, **92**, 9
90 P-G. DE GENNES: *J. Chem. Phys.*, 1971, **55**, 572
91 P. FLORY: *ibid.*, 1949, **17**, 303
92 M. TIRRELL: *Rubber Chem. Tech.*, 1984, **57**, 523
93 P-G. DE GENNES AND L. LEGER: *Ann. Rev. Chem.*, 1982, **33**, 49
94 F. BROCHARD AND P-G. DE GENNES: *Macromolecules*, 1977, **10**, 1157
95 J. D. FERRY: 'Viscoelastic properties of polymers', 3rd edn, 1980, New York, John Wiley
96 J. G. KIRKWOOD AND S. RISEMANN: *J. Chem. Phys.*, 1948, **16**, 565
97 M. ADAMS AND M. DELSANTI: *Macromolecules*, 1977, **10**, 1229
98 J. P. MUNCH: *J. Polym. Sci.*, 1976, **14**, 1097

99 A. S. MICHAELS AND J. J. BIXLER: *ibid.*, 1961, **50**, 413
100 M. H. COHEN AND T. TURNBULL: *J. Chem. Phys.*, 1959, **31**, 1164
101 D. JESCHKE AND H. STUART: *Z. Naturforsch.*, 1961, **16**, 37
102 P. MEARES: *J. Am. Chem. Soc.*, 1954, **76**, 3415
103 H. FUJITA: in 'Diffusion in polymers', (ed. J. Crank and G. S. Park), 75, 1968, London, Academic Press
104 R. H. PETERS: *ibid.*, 315
105 K. UEBERREITER: *ibid.*, 220
106 J. A. BARRIE: *ibid.*, 259
107 P. M. DOTY, W. H. AIKEN, AND H. MARK: *Ind. Eng. Chem. ind. (int.) edn*, 1946, **38**, 788
108 R. E. BARKER AND C. R. THOMAS: *J. Appl. Phys.*, 1964, **35**, 87
109 C. A. KUMINS AND T. K. KWEI: in 'Diffusion in Polymers', (ed. J. Crank and G. S. Park), 107, 1968, London, Academic Press
110 F. G. DONNAN: *'Z. Electrochemie'*, 1911, **17**, 572. See also translation in 'Cell membrane permeability and transport', (ed. G. R. Kepner), 211, 1979, Stroudsberg PA, Dowden, Hutchinson and Ross
111 P. MEARES: in 'Diffusion in polymers', (ed. J. Crank and G. S. Park), 373, 1968, London, Academic Press
112 R. M. BARRER: *Ber. Bunsenges.*, 1965, **69**, 786
113 R. M. BARRER: in 'The properties and applications of zeolites', (ed. R. P. Townsend), 3, 1980, London, The Chemical Society
114 R. M. BARRER AND B. E. F. FENDER: *J. Phys. Chem. Solids*, 1961, **21**, 12
115 R. M. BARRER AND L. V. C. REES: *Trans. Faraday Soc.*, 1954, **50**, 852, 989
116 D. M. RUTHVEN: in 'The properties and applications of zeolites', (ed. R. P. Townsend), 43, 1980, London, The Chemical Society
117 L. V. C. REES: *ibid.*, 218
118 D. W. BRECK: *ibid.*, 423
119 S. B. TUWINER, L. P. MILLER, AND W. E. BROWN: 'Diffusion and membrane technology', ACS Monograph Series, 1962, New York, Reinhold
120 'Cell membrane permeability and transport', (ed. G. R. Kepner), 1979, Stroudsberg PA, Dowden, Hutchinson and Ross
121 S. J. SINGER AND G. L. NICHOLSON: *Science*, 1972, **175**, 720
122 F. WUNDERLICK, R. BEREZNEY, AND H. KLEINIG: in 'Biological membranes, Vol. III' (ed. D. Chapman and F. H. Wallach), 241, 1976, London, Academic Press
123 F. WUNDERLICK AND V. SPETH: *J. Microscopie*, 1972, **13**, 361
124 R. A. WEINBERG: *Sci. Am.*, 1985, **253**, 48
125 W. C. COLE, R. E. GARFIELD, AND J. S. KIRKALDY: *Am. J. Physiol.*, 1985, **249**, *Cell Physiol.*, **18**, C20
126 W. R. LOEWENSTEIN: *Phys. Rev.*, 1981, **61**, 829
127 W. C. COLE AND R. E. GARFIELD: unpublished observations, 1984
128 J. R. PAPPENHEIMER, E. M. RENKIN, AND L. M. BORRERO: *Am. J. Physiol.*, 1951, **167**, 13
129 H. BAKKER: *Phil. Mag.*, 1979, **40**, 525
130 H. BAKKER: in 'Diffusion in metals and alloys', (ed. F. J. Kedres and D. L. Beke), Diffusion and defect monograph series No. 7, 1985, Trans. Tech. Publications
131 N. A. STOLWIJK, M. VAN GERD, AND H. BAKKER: *Phil. Mag.*, 1980, **42**, 783
132 C. WAGNER AND W. SCHOTTKY: *Z. Phys. Chem. (B)*, 1930, **11**, 163
133 J. FRENKEL: *Z. Phys.*, 1926, **35**, 652
134 D. MAPATHER, H. N. CROOKS, AND R. MAURER: *J. Chem. Phys.*, 1950, **18**, 1231
135 J. S. KIRKALDY: *Can. J. Phys.*, 1979, **57**, 717
136 G. E. MURCH: *Am. J. Phys.*, 1979, **47**, 78

CHAPTER 3

Thermokinetic foundations of a multicomponent diffusion formalism

3.1 INTRODUCTION

In § 1.6 a thermokinetic formulation was presented for a binary system based on the entropy evolution within a composite isolated system (Fig. 1.9). The generalization of that structure to multicomponent, multiprocess systems in isolation was first given by Onsager,[1,2] and the procedures whereby the structure can be generalized to continuously varying systems was also suggested by him.[1,2,3] More complete accounts are to be found in de Groot and Mazur,[4] de Groot,[5] Prigogine,[6] and Fitts.[7] In the following we present the Onsager structure for isolated systems, since it is needed in his proof of the reciprocity theorem (cf § 3.8). The most elementary example of homogeneous chemical reaction is then discussed since it bears on later practical problems, followed by the rigorous continuum structure specialized to combined multicomponent reaction and diffusion. A proof of the Onsager reciprocity theorem is then offered with a detailed justification of its application to multicomponent diffusion. Finally, the rigorous phenomenology is developed for ternary diffusion in the alloy A, B plus vacancies (V). This will be applied to kinetic models in Chapter 6.

3.2 THE ENTROPY PRODUCTION AND IRREVERSIBLE PROCESSES

Because of the intimate connection between diffusion and entropy evolution a rational approach to the subject must begin with the second law of thermodynamics. Consider then a small subsystem embedded within an isothermal heat bath B which in turn has rigid insulating and impermeable walls (an isolated enclosure). Thus, for the system as a whole the entropy evolution is described by

$$dS_i = dS + dS_B \geqslant 0 \tag{3.1}$$

Now the irreversible processes are conjectured to occur in the subsystem only, producing heat dQ which is transferred isothermally and reversibly to the heat bath according to

$$dS_B = -\frac{dQ}{T} \tag{3.2}$$

where the positive flow direction has been defined as into the subsystem. Thus the entropy change of the subsystem is

$$dS = \frac{dQ}{T} + dS_i = dS_e + dS_i \tag{3.3}$$

Now this can be interpreted for any subsystem in the universe (deemed to be an isolated system) as a balance equation for the non-conservative extensive parameter S consisting of an entropy flow term dS_e and a positive definite source term dS_i. The latter represents the entropy change of the universe which may by the foregoing construction be attributable entirely to the small subsystem. In § 1.6, we demonstrated how this term could be evaluated for a binary diffusion process. In the following, that structure is generalized to multicomponent diffusion-reaction systems.

3.3 THE ENTROPY PRODUCTION (RATE) FOR A GENERAL ISOLATED SYSTEM

Consider an isolated system whose state is described by intensive variables (pressure, temperature, concentrations, etc.), denoted by $A_1, A_2 \ldots A_n$. If the equilibrium values of these quantities are $A_1^0, A_2^0 \ldots A_n^0$, then the deviations of the state parameters from equilibrium can be defined as $\alpha_i = A_i - A_i^0$. At equilibrium the entropy is maximal and the α_i are zero by definition. The value of the entropy in non-equilibrium states may be expressed as the Taylor expansion of the deviation ΔS from the optimal equilibrium value,[1,2]

$$\Delta S = -\frac{1}{2}\sum_{i,k} \frac{\partial^2 \Delta S}{\partial \alpha_i \, \partial \alpha_k} \alpha_i \alpha_k = -\frac{1}{2}\sum_{i,k} m_{ik} \alpha_i \alpha_k \tag{3.4}$$

where m_{ik} is a positive definite matrix. The internal entropy source term, $\dot{S}_i$, is evaluated as

$$\dot{S}_i = \frac{d(\Delta S)}{dt} = -\sum_{i,k} m_{ik} \alpha_i \frac{d\alpha_k}{dt} \tag{3.5}$$

which is of the form

$$\dot{S}_i = \sum_k J_k X_k \tag{3.6}$$

with

$$J_k = \frac{d\alpha_k}{dt} = \dot{\alpha}_k \tag{3.7}$$

and

$$X_k = -\sum_i m_{ik} \alpha_i = \frac{\partial(\Delta S)}{\partial \alpha_k} \tag{3.8}$$

Thus a set of conjugate fluxes (or reaction rates) and forces (or derivatives of a potential function) are identified which enter into a bilinear expression for the entropy production (rate), $\dot{S}_i$. Note that α-variables can be related to variations of intensive variables over space, as in § 1.6, in which case each α_i is to be associated with an invariant such as mass or energy, or they can be related to internal order parameters, such as extents of reaction, as discussed in the following section.

3.4 ENTROPY PRODUCTION IN HOMOGENEOUS, REACTING SYSTEMS

From the definition of entropy given by equation (3.1), and the first law of equilibrium thermodynamics,

$$dU = dQ - p\,dV \tag{3.9}$$

the entropy for a system in which the mole numbers of all constituents are constant may be written

$$dS = dU/T + p\,dV/T \tag{3.10}$$

where U is the internal energy, p the pressure and V the volume of the system. This relation was generalized by Gibbs to account for changes in the mole numbers n_1, $n_2 \ldots n_c$. The procedure is outlined in Appendix 1 and leads to the Gibbs equation

$$T\,dS = dU + p\,dV - \sum_i \mu_i\,dn_i \tag{3.11}$$

where the μ_i are the chemical potentials (i.e. the partial molar free energies) of the components. This relation is of fundamental importance to the derivation of the entropy source term in multicomponent systems. Its validity in non-equilibrium situations must be carefully examined in each case. Such considerations must utilize an advanced kinetic theory, but no such theory of irreversible processes is yet at hand. Hirschfelder[8] has demonstrated for a reacting mixture that the Gibbs equation can be used provided that the chemical changes do not seriously perturb the Maxwell distribution of particle velocities. In other words, the reaction velocities must be small and the system must not deviate too far from equilibrium.

Assuming the restrictions are met, we apply the Gibbs equation to a homogeneous, closed system undergoing chemical reaction. Using the notation of de Donder,[9] each chemical reaction is written as

$$\nu_1 R_1 + \nu_2 R_2 + \ldots \nu_r R_r = \nu_{r+1} R_{r+1} + \ldots \nu_j R_j \tag{3.12}$$

where the R_i, $i \leqslant r$ are the reactant species and the R_i, $i \geqslant r+1$ are the product species. The stoichiometric coefficients ν_i are positive for the reaction products ($i \geqslant r+1$) and negative for the reactants. This notation facilitates the use of the stoichiometric equation in the form

$$\sum_i \nu_i M_i = 0 \tag{3.13}$$

where the M_i are the masses of the participating species. We now define the 'extent of reaction', ξ, through the equation

$$dn_i = \nu_i\,d\xi \tag{3.14}$$

which in turn leads to the definition of reaction rate

$$v = d\xi/dt \tag{3.15}$$

These equations have the benefit of conciseness regardless of how complex the reaction might be. An example is provided by the oxidation of propanol:

$$C_3H_7OH + \tfrac{9}{2}O_2 = 3\,CO_2 + 4\,H_2O$$

$$\nu_{C_3H_7OH} = -1 \qquad \nu_{O_2} = -\tfrac{9}{2} \qquad \nu_{CO_2} = 3 \qquad \nu_{H_2O} = 4$$

$$d\xi = -\,dn_{C_3H_2OH} = -\tfrac{9}{2}\,dn_{O_2} = 3\,dn_{CO_2} = 4\,dn_{H_2O}$$

This description is readily extended to the case of z simultaneous reactions

$$dn_i = \sum_{j=l}^{z} \nu_{ij}\,d\xi_j \tag{3.16}$$

The entropy change can now be evaluated via the Gibbs equation together with equation (3.16) for the case of a single reaction

$$TdS = dU + p\,dV - (\sum_i \nu_i\,\mu_i)\,d\xi \tag{3.17}$$

At equilibrium this leads to the familiar result that $\Sigma\nu_i\,\mu_i = 0$. In an irreversible reaction the right hand term is non-zero and we proceed via the first law of thermodynamics to rewrite equation (3.17) as

$$dS = \frac{dQ}{T} + \frac{A\,d\xi}{T} \tag{3.18}$$

where A is the chemical affinity as defined by de Donder, namely

$$A = -\sum_i \nu_i\,\mu_i \tag{3.19}$$

Consistent with equation (3.3), the entropy has two parts

$$dS_e = dQ/T \tag{3.20}$$

and

$$dS_i = A\,d\xi/T \tag{3.21}$$

and hence the entropy source term is given by

$$\dot{S}_i = Av/T \tag{3.22}$$

Again this term has the expected form of a product of a flux, v, and a force, A. That the chemical affinity as defined may properly be regarded as a force is seen by writing the chemical potential as a partial molar derivative of free energy (Appendix 1), whence

$$A = -\sum_i \nu_i \left(\frac{\partial G}{\partial n_i}\right)_{p,T,n_j} \tag{3.23}$$

Combination of equations (3.14) and (3.23) yields

$$A = -\left(\frac{\partial G}{\partial \xi}\right)_{p,T} \tag{3.24}$$

i.e., the affinity is the free energy gradient with respect to the extent of reaction. The positive definite (Appendix 5) character of $\dot{S}_i$ implies through equation (3.22) that the reaction proceeds in the direction which reduces the system free energy.

Extension of the description to a system undergoing simultaneous reactions leads to conclusions of a less trivial nature. For example, the entropy source term becomes

$$\dot{S}_i = \frac{1}{T}\sum_{j=l}^{z} A_j v_j > 0 \tag{3.25}$$

where

$$A_j = -\sum_i \nu_{ij}\mu_i \tag{3.26}$$

Equation (3.25) states that the overall entropy production rate is the sum of the entropy productions attributable to each of the reactions. The thermodynamic theory requires that this sum be positive, but places no such constraint on the individual terms. In the case of two simultaneous reactions it is possible that

$$A_1 v_1 < 0 \qquad A_2 v_2 > 0$$

provided that the sum

$$A_1 v_1 + A_2 v_2 > 0$$

Thus one reaction may proceed in a direction contrary to that indicated by its own (conjugate) affinity. It is this kind of 'coupling' of irreversible processes which is the principal subject of this book (cf Prigogine[6]). For reactions proceeding in dilute solution near equilibrium the reaction velocities or rates v_i can be taken as linear in all of the affinities or forces A_j (e.g. equations (1.95) and (12.222)).

3.5 ENTROPY PRODUCTION AND MASS TRANSFER IN CONTINUOUS SYSTEMS

Thus far systems have been considered for which it appears intuitively reasonable to apply the equations of equilibrium thermodynamics even though irreversible processes are occurring. In the binary diffusion case, the subsystems were in equilibrium everywhere except at the intervening boundary. In the chemical reaction case, the system was homogeneous and underwent relatively slow changes in composition. In both cases the irreversible state of the system depended upon time but not upon spatial position. To deal with mass transfer, we must take explicit account of compositional variation with position in the system. The theoretical technique is to consider a microscopic volume element δV which is nonetheless sufficiently large to contain a statistically meaningful number of particles. The simultaneous fulfilment of these two requirements is readily structured for a condensed phase provided that gradients are not too high.

As diffusive mass transfer is often studied in the absence of forces other than chemical composition gradients, we first consider an isothermal, isobaric, field-free, reaction-free materially closed system which is initially nonhomogeneous. Such a system undergoes spontaneous irreversible change as it approaches the eqilibrium state of chemical homogeneity. The description of the rate at which the various constituents move in space during this process is the aim of multi-component diffusion theory. The description commences, as usual, with the Gibbs equation (3.11) for entropy change. However, as seen above we must apply this equation to a small volume element within which it is supposed that a 'local' state of equilibrium exists. Since we will ultimately have to accommodate the fact that different elements contain different amounts of substance, it is convenient to write the Gibbs equation in the centre of mass (barycentric) form

$$T\,\mathrm{d}s = \mathrm{d}u + p\,\mathrm{d}v - \sum_{i}^{r} \mu_i\,\mathrm{d}X_i \tag{3.27}$$

where s, u and v are respectively the entropy, internal energy and volume per unit mole and X_i is the mole fraction defined in terms of the mole numbers, n_i, as

$$X_i = n_i / \sum_{j}^{r} n_j \tag{3.28}$$

If as before the derivatives are taken with respect to time we have for a system of r components.

$$T\frac{\mathrm{d}s}{\mathrm{d}t} = \frac{\mathrm{d}u}{\mathrm{d}t} + \frac{p\,\mathrm{d}v}{\mathrm{d}t} - \sum_{i}^{r} \mu_i \frac{dX_i}{\mathrm{d}t} \tag{3.29}$$

For an isothermal system at mechanical equilibrium

$$\mathrm{d}u = -p\,\mathrm{d}v$$

and equation (3.29) becomes

$$T\frac{\mathrm{d}s}{\mathrm{d}t} = -\sum_{i}^{r} \mu_i \frac{dX_i}{\mathrm{d}t} \tag{3.30}$$

We now relate the rate of concentration changes to the fluxes of matter which in the absence of chemical reaction are the sole causes of such change. The required relationship is found from the law of conservation of mass.

In non-reacting systems, the law of conservation of mass applies to each component and may be expressed in terms of the mole numbers. Defining the number of moles of a component per unit volume as

$$m_i = n_i/V \tag{3.31}$$

and the velocity of each component with respect to some fixed point outside the system by the vector $\mathbf{v}_i$, then conservation of mass is expressible as the continuity equation

$$\frac{\partial m_i}{\partial t} = -\,\mathrm{div}\,(m_i \mathbf{v}_i) \tag{3.32}$$

Summing over all components

$$\frac{\partial m}{\partial t} = -\operatorname{div}(m\boldsymbol{v}) \tag{3.33}$$

where the overall density $m = \Sigma m_i$ and the centre of mass velocity is defined by

$$\boldsymbol{v} = \sum_i (m_i \boldsymbol{v}_i)/m \tag{3.34}$$

Defining a component flux relative to the centre of mass velocity by

$$\boldsymbol{J}_i = m_i(\boldsymbol{v}_i - \boldsymbol{v}) \tag{3.35}$$

it is seen by comparison with equation (3.34) that

$$\sum_i \boldsymbol{J}_i = 0 \tag{3.36}$$

Addition of div $(m_i\boldsymbol{v})$ to both sides of equation (3.32) yields

$$\frac{\partial m_i}{\partial t} + \operatorname{div}(m_i\boldsymbol{v}) = -\operatorname{div}\boldsymbol{J}_i \tag{3.37}$$

This result is simplified further by the introduction of the mole fractions

$$m_i = m X_i \tag{3.38}$$

to yield

$$m\left\{\frac{\partial X_i}{\partial t} + v \operatorname{grad} X_i\right\} + X_i\left\{\frac{\partial m}{\partial t} + \operatorname{div}(m\boldsymbol{v})\right\} = -\operatorname{div}\boldsymbol{J}_i \tag{3.39}$$

Substitution of equation (3.33) and the vector calculus expression

$$\mathrm{d}/\mathrm{d}t = \partial/\partial t + \boldsymbol{v} \operatorname{grad} \tag{3.40}$$

into equation (3.39) yields

$$m\frac{\mathrm{d}X_i}{\mathrm{d}t} = -\operatorname{div}\boldsymbol{J}_i \tag{3.41}$$

Combination of this result with equation (3.30) generates

$$\mathrm{m}T\,\mathrm{d}s/\mathrm{d}t = \sum_i \mu_i \operatorname{div}\boldsymbol{J}_i \tag{3.42}$$

which reduces with the aid of the vector identity

$$\operatorname{div}\left(\frac{\mu\boldsymbol{J}}{T}\right) = \frac{\mu}{T}\operatorname{div}\boldsymbol{J} + \boldsymbol{J}\operatorname{grad}\left(\frac{\mu}{T}\right) \tag{3.43}$$

to

$$m\frac{\mathrm{d}s}{\mathrm{d}t} = \operatorname{div}\sum_i \frac{\mu_i\boldsymbol{J}_i}{T} - \sum \boldsymbol{J}_i \operatorname{grad}\left(\frac{\mu_i}{T}\right) \tag{3.44}$$

The specific entropy production consists therefore of a divergence of entropy flow resulting from the flows of material into and out of the volume element under discussion and an entropy source term σ defined by

$$T\sigma = \sum_i \boldsymbol{J}_i\boldsymbol{X}_i \tag{3.45}$$

where

$$X_i = -T \text{ grad } (\mu_i/T) \tag{3.46}$$

Clearly the source term has the expected form: a sum of products of fluxes J_i and forces X_i.

In the more general case where a thermal gradient exists and where chemical reactions are possible, but in the absence of viscous dissipation, equation (3.44) becomes[4]

$$m\frac{\mathrm{d}s}{\mathrm{d}t} = -\operatorname{div}\left\{\frac{J_q - \sum_i \mu_i J_i}{T}\right\} + \frac{J_q X_u + \sum_i J_i X_i + \sum_j A_j v_j}{T} \tag{3.47}$$

where J_q is the heat flux and

$$X_u = -(\text{grad } T)/T \tag{3.48}$$

$$X_i = F_i - T \text{ grad } (\mu_i/T) \tag{3.49}$$

and F_i is the external force per unit mole of substance i (e.g. electrical).

In the case of isothermal diffusion in the absence of reaction and external forces the last term of equation 3.47 reduces to a simple result

$$T\sigma = \sum_i J_i X_i \qquad X_i = -\text{grad } (\mu_i)_T \tag{3.50}$$

where σ is the entropy production rate per unit volume. The total entropy production rate is

$$\dot{S}_i = \int_v \sigma \, \mathrm{d}V > 0 \tag{3.51}$$

It is to be noted at this point that the entropy production terms for diffusion and homogeneous chemical reaction can be combined without cross terms on account of Curie's principle (cf §3.10). The combined process has been investigated with rigour by Toor.[10]

To proceed to applications within all of the foregoing structures it is necessary to establish a relationship between the fluxes and forces through theory or experimentation. First, however, we consider the important situation in which the diffusing species are electrically charged, e.g. an electrolyte solute or the constituent particles of an ionic solid.

3.6 ENTROPY PRODUCTION AND CHARGE TRANSFER

To extend the thermodynamic treatment to cover ionic species, we consider the work required to change the amount of charge within a volume element under consideration. If the region in question has an electrostatic potential ψ and the change in charge is $\mathrm{d}q$ then the work done is

$$\mathrm{d}w = -\psi \, \mathrm{d}q \tag{3.52}$$

Then the total internal energy change within the region is given by

$$\mathrm{d}U = T\,\mathrm{d}S - p\,\mathrm{d}V + \sum_i^r \mu_i \, \mathrm{d}n_i + \psi \, \mathrm{d}q \tag{3.53}$$

which is the Gibbs equation (3.11) for a system of r species generalized to include electrical work. Assuming that the change in charge of the system is due to changes in the mole numbers of the ionic species, then

$$\mathrm{d}q = \sum_{i}^{r} z_i F \,\mathrm{d}n_i \qquad (3.54)$$

where z_i is the valence (which may be positive, negative or zero) of species i and F is the Faraday, the amount of charge carried by one mole of singly charged ions, numerically equal to 96 500 c. Insertion of equation (3.54) into equation (3.53) yields

$$\mathrm{d}U = T\,\mathrm{d}S - p\,\mathrm{d}V + \sum_i (\mu_i + z_i F\psi)\,\mathrm{d}n_i \qquad (3.55)$$

It is thus seen that the work performed in changing the amount of a substance is composed of two parts, a chemical term $\mu_i\,\mathrm{d}n_i$ and an electrical term $z_i F\psi\,\mathrm{d}n_i$. Since it is impossible to do other than simultaneously transport charge and mass, the terms are so inextricably bound that they cannot be distinguished experimentally.[11] This rigid association has led to the definition of a new quantity called the electrochemical potential

$$\eta_i = \mu_i + z_i F\psi \qquad (3.56)$$

The calculation of the entropy production proceeds precisely as before with the electrochemical potential η_i substituted for the chemical potential μ. In the case of isothermal diffusion equation (3.46) is written as

$$\boldsymbol{X}_i = -\,\mathrm{grad}\,\eta_i \qquad (3.57)$$

and the dissipation function is as before

$$T\sigma = \sum_i \boldsymbol{J}_i \boldsymbol{X}_i \qquad (3.58)$$

3.7 RELATIONSHIPS BETWEEN FLUXES AND FORCES

The irreversible thermodynamic structure discussed above is of interest in so far as it permits the identification of conjugate fluxes and forces. While this information affords some insight into the physics of a system an additional statement is required as to the functional relationship between fluxes and forces, namely

$$\boldsymbol{J}_i = \mathrm{f}(\boldsymbol{X}_i) \qquad (3.59)$$

In general this relationship might be represented by a polynomial expansion

$$\boldsymbol{J}_i = a_1 \boldsymbol{X}_i + a_3 \boldsymbol{X}_i^3 + \ldots \qquad (3.60)$$

Since in the absence of forces there are no fluxes, the zeroth order term in the polynomial has been omitted. There are strong indications from kinetic theory and experiment (cf Chapter 1) that the linear form of equation (3.60) provides an accurate description of most transport processes, i.e.

$$\boldsymbol{J}_i = a_i \boldsymbol{X}_i \qquad (3.61)$$

Indeed, the experimental relations for heat, electricity, and mass transport (due respectively to Fourier, Ohm, and Fick) are all linear: heat flow is proportional to the negative of the temperature gradient, electric current to the negative of the electric potential gradient, and material flow to the negative of the chemical potential gradient. Whilst these laws might appear inconsistent with Newtonian mechanics in that no acceleration occurs, this is due to the fact that the media are not frictionless. In a material medium there exists a resistance to any flow, such that in the steady state the velocity, and not the acceleration, is proportional to the applied force. This is equally true in mechanics, as is revealed by a consideration of the motion of objects falling through air at their terminal velocity.

In the development of the entropy source term it was assumed that the system under discussion was not too far removed from equilibrium and hence that the processes were occurring slowly. The linear form of equation (3.61) is appropriate to this restriction. A useful consequence of the linear relationship and the fact that the dissipation function can be expressed as a sum of products of forces and fluxes is that the dissipation function can also be expressed as a quadratic function of the forces or the fluxes. Provided that the forces are defined in terms of state variables, this result is consistent with Onsager's entropy function equation (3.4). It is of course to be understood that the insistence on linearity imposes a limitation on the breadth of application of the theory. This is particularly evident in the case of chemical reactions where the equations of reaction kinetics, which are well founded in statistical mechanics, are in most cases non-linear. Thus the linearity assumption is only consistent with reaction kinetic theory in the often uninteresting case when the system is very close to equilibrium. For transport processes the limitation is of much less consequence.

Although equation (3.61) adequately represents the correlation beetween a flux and its conjugate force it does not include the effects on that flux of the other system forces which are to be expected in real systems. There are in fact many examples known of forces producing non-conjugated flows. These so-called 'cross-effects' always occur in reciprocal pairs. An example is the coupling between diffusion and heat conduction. There is the Soret effect in which the imposition of a temperature gradient on an initially homogeneous system causes concentration gradients to form and its inverse, the Dufour effect, where a temperature gradient develops as a result of a concentration gradient. Charge and heat conduction are also coupled. In the Seebeck effect a temperature difference between two bimetallic junctions leads to the generation through charge transport of an electromotive force between the junctions. In its inverse, the Peltier effect, the passage of charge through such a system causes an isothermal transfer of heat from one junction to the other. The character of these and other phenomena may be summarized in the statement that coupling exists between flows of one type and forces of other types.[5]

Onsager[1,2] postulated that an equation of form (3.61) can be generalized so that each thermodynamic flux is linearly related to every thermodynamic force. This postulate generates the phenomenological equations

$$\boldsymbol{J}_i = \sum_{j=1}^{n} L_{ij}\boldsymbol{X}_j \qquad (i = 1, 2, \dots n) \tag{3.62}$$

They are of course in accord with the equations of Fourier, Ohm and Fick in that each flux is linearly proportional to its conjugate force, the coefficients representing these relationships being the L_{ii} which appear on the diagonal of the matrix of coefficients. The off-diagonal coefficients L_{ij} $(i \neq j)$ are known as the 'cross-coefficients' or 'coupling coefficients' because they represent the relationships between fluxes and non-conjugated forces.

If the forces are independent equation (3.62) leads to an inverse representation,[12,13,14]

$$\boldsymbol{X}_i = \sum_{j=1}^{n} R_{ij}\boldsymbol{J}_j \qquad (i = 1, 2 \ldots n) \tag{3.63}$$

where the matrix $[R]$ is the inverse of the matrix $[L]$ (cf Appendix 4), i.e.,

$$[R] = [L]^{-1} \tag{3.64}$$

and specific terms in R may be found from

$$R_{ij} = |L_{ij}|/|L| \tag{3.65}$$

where $|L|$ is the determinant of the matrix of coefficients $[L_{ij}]$ and $|L_{ij}|$ is the determinant of the minor corresponding to the element L_{ij}. The coefficients L_{ik} have the dimension of flux per unit force and hence, the characteristics of a generalized mobility. The coefficients R_{ij} have the dimension force per unit flux and represent generalized resistances or frictions.[12,13] In this treatment, the form of equation (3.62) is preferred.

In the particular case of diffusion in an isothermal, isobaric, field-free system we have

$$\boldsymbol{J}_i = \sum_{j=1}^{n} L_{ij} \operatorname{grad}(-\mu_j) \qquad (i = 1, 2 \ldots n) \tag{3.66}$$

and it is seen that the flux of any particular component is related to the chemical potential gradients of all components. Note, however, that neither the forces nor the fluxes in equation (3.66) are in general independent.

There are important cross effects with thermal gradients (the Soret effect or thermal diffusion) but these will not be explored in this volume (cf Shewmon[15]).

3.8 ONSAGER'S LAW OF RECIPROCITY

The description provided by equation (3.66) generally requires some reduction via choice of the frame of reference and solution restrictions on the chemical potentials. In the simplest case the system of n equations with n^2 coefficients is reduced to a set of independent equations containing $(n-1)^2$ coefficients. The structure is further reduced by the application of Onsager's reciprocity theorem[1,2] which states that if a proper choice is made for the independent fluxes J_i and the forces X_i, the reduced matrix of phenomenological coefficients is symmetric, i.e.

$$L_{ij} = L_{ji} \qquad (i,j = 1, 2 \ldots n) \tag{3.67}$$

Thus for a simple ternary system only three of an original nine dependent coefficients are required for a complete description.

Onsager recognized that this symmetry law for the interaction of irreversible processes such as the flow of mass, energy, and electricity derives from the time reversal invariance of the mechanical laws of particle motion. He thus sought a general proof of the theorem: given n linear flux-force relations

$$J_i = \sum_{k=1}^{n} L_{ik} X_k \tag{3.68}$$

where the n processes are of the same tensorial character and independent, prove the symmetry of the L-matrix. Provided the independent fluxes and forces are expressible within the self-consistent formalism of §3.3, thus forming an expansion for the irreversible rate of entropy production of the bilinear form

$$\dot{S}_{\mathrm{i}} = \sum_{i=1}^{n} J_i X_i \tag{3.69}$$

then precise symmetry obtains and is experimentally observed in all cases[16] (cf §7.8).

Onsager developed his proof from the conventional description of the chemical monomolecular triangle reaction designated by

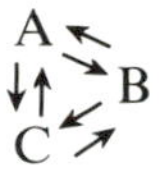

$$\begin{array}{lcl} \mathrm{A} & \nwarrow\searrow & \\ \downarrow\uparrow & & \mathrm{B} \\ \mathrm{C} & \swarrow\nearrow & \end{array} \tag{3.70}$$

in which net equilibrium transitions of the form

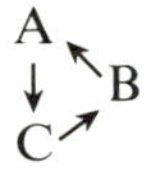

$$\begin{array}{lcl} \mathrm{A} & \nwarrow & \\ \downarrow & & \mathrm{B} \\ \mathrm{C} & \nearrow & \end{array} \tag{3.71}$$

are denied by the 'principle of detailed balance'. That is, each sub-reaction in equation (3.71) is averred to equilibrate independently. He noted that 'the idea of equilibrium maintained by a mechanism like equation (3.71) ... is not in harmony with our notion that molecular mechanics has much in common with the mechanics of ordinary conservative dynamical systems ... the dynamical laws of familiar conservative systems are always reversible ... if the velocities of all the particles present are reversed simultaneously the particles will retrace their former paths, reversing the entire succession of configurations. This implies that if we wait a long time so as to make sure of thermodynamic equilibrium, in the end every type of motion is just as likely to occur as the inverse'. One consequence of the 'principle of dynamical reversibility' is 'detailed balance' as invoked by the chemists for the example above (cf §2.10).

Onsager's induction concerns the proposition that the entire physical content of the linear theory of irreversible processes is contained within a description of fluctuation phenomena at the equilibrium state. In particular his reciprocity theorem assumes that the laws of decay of fluctuations are identical with those for artificially imposed initial conditions, namely, equation (3.68). On this basis he sought a proof of relation (3.67) which is independent of mechanism. He was very careful to point out that the symmetry principle is rather trivially derivable when the species or molecular trajectories are known. It is only for the cases where nothing is known about the mechanism that the elaborate fluctuation formalism is required for a proof.

In the first step, it is recognized from the Boltzmann relation

$$\Delta S = k \ln W < 0 \tag{3.72}$$

(where k is the Boltzmann constant and $0 < W < 1$ is the relative state probability) that the most probable fluctuations lie within

$$-k < \Delta S < 0 \tag{3.73}$$

For a random process it is then easy to demonstrate via equations (3.4) and (3.72) that

$$\overline{\alpha_i X_j} = -k\delta_{ij} \tag{3.74}$$

is the temporal mean, where δ_{ij} is the Krönecker δ-matrix (Appendix 3).

Microscopic reversibility is expressed by the requirements that the two-particle collision transitions expressed in terms of before and after velocies v' and v''

$$(v_1', v_2') \rightarrow (v_1'', v_2'') \quad \text{and} \quad (-v_1'', -v_2'') \rightarrow (-v_1', -v_2') \tag{3.75}$$

occur equally often at equilibrium, or as the temporal relation

$$\overline{\alpha_i(t)\, \alpha_j(t+\tau)} = \frac{1}{t}\int_0^{t\rightarrow\infty} \alpha_i(t)\, \alpha_j(t+\tau)\, \mathrm{d}t$$

$$= \frac{1}{t}\int_0^{t\rightarrow\infty} \alpha_j(t)\, \alpha_i(t+\tau)\, \mathrm{d}t = \overline{\alpha_j(t)\, \alpha_i(t+\tau)} \tag{3.76}$$

where t is any instant of time and τ is a time interval much greater than a molecular collision time.

The mean value on the left differs from that on the right by the temporal order of the fluctuations, so by microscopic reversibility ($t \rightarrow -t$) they must be equal.

Subtracting the term $\overline{\alpha_i(t)\, \alpha_j(t)}$ from both sides of equation (3.76) and dividing by τ one obtains in the limit of τ small (but greater than the molecular collision time τ_0)

$$\overline{\alpha_i(t)\, \dot{\alpha}_j(t)} = \overline{\alpha_j(t)\, \dot{\alpha}_i(t)} \tag{3.77}$$

Substituting $\dot{\alpha}_{i,j} = J_{i,j}$ from equation (3.68), which is assumed applicable to the decay of fluctuations, yields

$$\sum_k L_{jk}\, \overline{\alpha_i X_k} = \sum_k L_{ik}\, \overline{\alpha_j X_k} \tag{3.78}$$

whence from equations (3.74) and (3.78) follows the symmetry relation (3.67).

Since the foregoing structure was formulated for isolated systems and appears to refer to scalar parameters α_i, debate has often arisen as to whether vectorial and tensorial continuum processes are subsumed by the Onsager theorem.[17] For flows of conserved quantities such as mass, energy and electricity it is easy to demonstrate that this is rigorously the case. A detailed proof is given in the following section for multicomponent diffusion.

3.9 APPLICATION OF THE RECIPROCITY THEOREM TO MULTICOMPONENT DIFFUSION

It will be recalled that Onsager's theorem is based on the hypothesis that at and near local equilibrium the laws governing the average rate of regression of fluctuations are identical with the macroscopic laws of regression and that the proof involves a set of scalar variables α_i which represent fluctuation amplitudes of meaningful state variables for the test volume as a whole. This can be the total volume of an isolated enclosure, as in the original theorem, or an incremental unit volume of an isothermal, isobaric system as in the present case. These α-variables are necessarily integrals over the test volume and are specified in the first instance by the invariants of the system, namely, energy, mass, charge, etc. In our case, the n independent mass components, are conserved according to equation (3.41). Let us accordingly form the integrals[18]

$$\int \boldsymbol{J}_i \cdot \boldsymbol{n}\,\mathrm{d}A = \int \operatorname{div} \boldsymbol{J}_i\,\mathrm{d}V = -\int m\frac{\mathrm{d}X_i}{\mathrm{d}t}\,\mathrm{d}V \tag{3.79}$$

and in one dimension evaluate the left-hand side for the bounds of the test volume with one end impermeable so that the average flux

$$\overline{J_i} = -\frac{\mathrm{d}}{\mathrm{d}t}\int \frac{m}{A} X_i\,\mathrm{d}V \tag{3.80}$$

with m taking the average molar density value for the test volume.

This average flux has a simple meaning. If we recall that the rms fluctuation of the number of particles N_i in any test sub-volume is $\sim\sqrt{N_i}$, there must be a corresponding boundary flux which designates the influx or efflux of particles. For the purposes of this calculation the flux is to be associated with $\overline{J}_i$ only when an influx follows a decrement of particles or an efflux follows an increment of particles for it is the nature of fluctuations that $\overline{J}_i$ is discontinuous.[1,2] We can now define our regressing α_i variables as

$$\alpha_i = -\int \frac{m}{A}(X_i - X_i^0)\,\mathrm{d}V = -\frac{mV}{A}\overline{(X_i - X_i^0)} \tag{3.81}$$

which assures the form (3.7) for the flux, with the α_i as integrals. Here X_i^0 is the average composition at equilibrium. The next step is to form the linear expression of the macroscopic forces in terms of α_i via

$$X_i = -\frac{1}{T}\nabla(\mu_i - \mu_n) = \sum_{j=1}^{n}\frac{1}{T}\frac{\partial(\mu_i - \mu_n)}{\partial X_j}\overline{\nabla X_j} = -\frac{1}{T}\sum_{j=1}^{n}\mu_{ij}\,\overline{\nabla X_j} \tag{3.82}$$

where, by analogy with our operations on the fluxes, we have introduced the averages of the gradients of X_i over the fluctuations to make the connection with the average macroscopic X_i. Now since the entropy production rate must be invariant under our transformation of variables with $J_i = \overline{J}_i$, i.e.

$$\sigma = \sum_i J_i X_i = \sum_i \dot{\alpha} \sum_k m_{ik}\alpha_k \tag{3.83}$$

it follows that the full association can be made with

$$\overline{\nabla X_i} = \beta \alpha_i = -\frac{\beta m V}{A}\overline{(X_i - X_i^0)} \tag{3.84}$$

where β is a constant. The proportionality between $\overline{\nabla X_i}$ and $\overline{(X_i - X_i^0)}$, which is a corollary of the present argument, is identified by Onsager as a quantifiable theorem within general fluctuation theory due to Einstein.[19] Furthermore, the matrix m_{ik} is uniquely defined as the positive definite matrix

$$m_{ik} = \beta \mu_{ik}/T \tag{3.85}$$

where the μ_{ik} (equation (3.82)) is closely related to the Hessian of the Gibbs free energy (cf §4.4).

3.10 THERMODYNAMIC AND SYMMETRY RESTRICTIONS ON THE ONSAGER COEFFICIENTS

The second law requirement that $\sigma > 0$ imposes the restriction that the matrix $[L]$ be positive definite (cf Appendix 4). That is to say, the coefficients are real and the determinant of coefficients and all principal minors must be positive,

$$L_{ii} > 0$$

$$L_{ii}L_{jj} - L_{ij}L_{ji} > 0$$

$$|L_{ij}| > 0 \tag{3.86}$$

De Groot and Mazur[4] have expanded on a general symmetry principle due to Curie[20] which suggests that processes of different tensorial rank cannot interact through non-zero off-diagonal terms in the L-matrix. For example, the off-diagonal L-coefficients relating to homogeneous chemical reaction (a scalar process) and diffusion or heat conduction (vector processes) must be zero. Notwithstanding, reaction and diffusion interact strongly through the sharing of concentration variables and the consequent alterations in boundary values (e.g., 'active transport', cf Chapter 10).

Additional restrictions will emerge from the discussion in the next chapter which is aimed at the formulation of generalized diffusion equations in their most compact form.

3.11 CORRECTIONS TO THE DARKEN EQUATION WITHIN A QUATERNARY FORMALISM

For at least three decades, it has been accepted that there is a lack of precision in the Darken equation relating the binary chemical diffusion coefficient to the component tracer coefficients[21,22] (cf §2.14) and this has been associated with correlation corrections in random walk treatments of chemical diffusion, otherwise characterized as 'vacancy wind' effects, and expressible as Onsager cross terms in a particular form of the diffusion mobility matrix. The currently accepted treatment, attributable to Manning[23] and Heumann[24] among others, identifies the correlations with local non-equilibrium vacancy concentrations caused by the Kirkendall vacancy flux and arrives at a value which represents an increase in the predicted Kirkendall velocity in terms of the tracer coefficients which is 28% for fcc and greater for less close-packed structures. In this section, we follow Kirkaldy[25] in a strict local equilibrium phenomenological treatment of the problem, modifying calculations due to LeClaire[26] and Howard and Lidiard.[27] We identify L_{AA^*} and L_{BB^*} as the two non-zero independent mobility coefficients which are to be associated with the corrections. Our result, based on a manipulation of the L-matrix for the lattice-fixed or Kirkendall frame of reference, is analogous to that presented by Tyrrell and Harris[28] for a liquid mass-fixed frame with molar concentration units.

Referring to Darken's treatment of the Kirkendall Effect (cf §2.13) which assumes the validity of marker experiments in establishing the lattice-fixed frame of reference,[21] we recall that the intrinsic coefficients are defined at local equilibrium by the independent particle flux equations

$$J_a = -D_a N \nabla X_a = -L_a \nabla \mu_a \tag{3.87}$$

and

$$J_b = -D_b N \nabla X_b = -L_b \nabla \mu_b \tag{3.87}$$

where N is the number of lattice sites per unit volume ($NX_i = C_i$), $X_{a,b}$ are mole fractions and the $\mu_{a,b}$ are the chemical potentials subject to

$$X_a \nabla \mu_a + X_b \nabla \mu_b = 0 \tag{3.89}$$

Here a and b refer to chemical species and therefore can apply either to species A and B or to A + A* and B + B* where * designates a tracer. The Onsager intrinsic mobility coefficients are uniquely connected to the Ds by

$$D_a = \frac{kT}{N} \frac{L_a}{X_a} \phi \tag{3.90}$$

and

$$D_b = \frac{kT}{N} \frac{L_b}{X_b} \phi \tag{3.91}$$

where by the Gibbs–Duhem equation (3.89)

$$\phi = 1 + \frac{d \ln \gamma_a}{d \ln X_a} = 1 + \frac{d \ln \gamma_b}{d \ln X_b} = 1 + \frac{d \ln \gamma}{d \ln X} \tag{3.92}$$

Thus, subject to the applicability of equation (3.89), L_a and L_b are hereby defined exactly. The Darken equations for the mutual diffusion coefficient $\widetilde{D}$ and the Kirkendall velocity v are

$$\widetilde{D} = \frac{kT}{N}\left(\frac{X_b}{X_a} L_a + \frac{X_a}{X_b} L_b\right)\phi = D_a X_b + D_b X_a \tag{3.93}$$

and

$$v = \frac{kT}{N}\left(\frac{L_a}{X_a} - \frac{L_b}{X_b}\right)\phi \nabla C_a = (D_a - D_b)\nabla X_a \tag{3.94}$$

The choice of the diagonal form of the L-matrix in equations (3.87) and (3.88) is in part predicated upon the fact that the matrix elements are uniquely defined via equations (3.93) and (3.94) and the experimental values of $\widetilde{D}$, v and the thermodynamic parameter ϕ. We thus insist that all operational phenomenological coefficients are experimentally accessible in such a manner (cf Tyrrell and Harris[28]). Notwithstanding the clarity of the logic in the foregoing, there remains an ambiguity in the literature (Howard and Lidiard,[27] Manning,[23] Heumann,[24] Lidiard,[29] Manning[30]) whereby it is implied that the L-matrix in equations (3.87) and (3.88) contains unique non-zero off-diagonal terms. Here we undertake a rigorous exploration of the kinetic equations for the A–B–V solution and its local equilibrium limit. It turns out that the ambiguity arises from the absence of analytic continuation between the binary local equilibrium A–B matrix and the ternary non-local equilibrium A–B–V matrix.

Consider an A–B–V solution with a non-zero chemical potential gradient of vacancies, $\nabla\mu_V$. In the first instance this does not imply that the vacancies are necessarily conserved. We start with Anthony's rigorous presentation pertaining to the 3×3 L-matrix based on the dependent flux–force equations (a = A and b = B)[31]

$$J_A = -L_{AA}\nabla\mu_A - L_{AB}\nabla\mu_B - L_{AV}\nabla\mu_V \tag{3.95}$$

$$J_B = -L_{BA}\nabla\mu_A - L_{BB}\nabla\mu_B - L_{BV}\nabla\mu_V \tag{3.96}$$

and

$$J_V = -L_{VA}\nabla\mu_A - L_{VB}\nabla\mu_B - L_{VV}\nabla\mu_V \tag{3.97}$$

These can be reduced to the independent forms

$$J_A = -L_{AA}\nabla(\mu_A - \mu_V) - L_{AB}\nabla(\mu_B - \mu_V) \tag{3.98}$$

and

$$J_B = -L_{BA}\nabla(\mu_A - \mu_V) - L_{BB}\nabla(\mu_B - \mu_V) \tag{3.99}$$

via the total flux balance and arbitrary symmetry constraints on the L-matrix yielding the non-trivial reciprocal relations[5]

$$L_{AB} = L_{BA} \tag{3.100}$$

and dependently

$$J_V = -L_{VA}\nabla(\mu_A - \mu_V) - L_{VB}\nabla(\mu_B - \mu_V) \tag{3.101}$$

with

$$L_{VV} = L_{AA} + L_{BB} + 2L_{AB} \tag{3.102}$$

and

$$L_{AV} = -(L_{AA} + L_{AB}) \quad \text{and} \quad L_{BV} = -(L_{BB} + L_{AB}) \tag{3.103}$$

These are applicable to any frame of reference where the sum of atom and vacancy fluxes is zero. The chemical potentials satisfy the Gibbs–Duhem equation

$$X_A\nabla\mu_A + X_B\nabla\mu_B + X_V\nabla\mu_V = 0 \tag{3.104}$$

Again we emphasize that the vacancies need not be conserved. There can be simultaneous scalar chemical reactions involving vacancy sources and sinks which do not inject cross-terms into the vectorial flux equations (Curie's theorem, cf § 3.10). These effects enter explicitly into the divergence of the vacancy flux but not into the flux itself.[33] First we use equation (3.104) to transform equations (3.98) and (3.99) to

$$J_A = -\left[L_{AA}\left(1+\frac{X_A}{X_V}\right)+L_{AB}\frac{X_A}{X_V}\right]\nabla\mu_A - \left[L_{AB}\left(1+\frac{X_B}{X_V}\right)+L_{AA}\frac{X_B}{X_V}\right]\nabla\mu_B \tag{3.105}$$

and

$$J_B = -\left[L_{AB}\left(1+\frac{X_A}{X_V}\right)+L_{BB}\frac{X_A}{X_V}\right]\nabla\mu_A - \left[L_{BB}\left(1+\frac{X_B}{X_V}\right)+L_{AB}\frac{X_B}{X_V}\right]\nabla\mu_B \tag{3.106}$$

Next let us evaluate the vacancy flux

$$J_V = -(J_A + J_B) \tag{3.107}$$

as

$$J_V = \left[L_{AA}\left(1+\frac{X_A}{X_V}\right)+L_{AB}\left(1+\frac{2X_A}{X_V}\right)+L_{BB}\frac{X_A}{X_V}\right]\nabla\mu_A + \left[L_{AA}\frac{X_B}{X_V}+L_{AB}\left(1+\frac{2X_B}{X_V}\right)+L_{BB}\left(1+\frac{X_B}{X_V}\right)\right]\nabla\mu_B \tag{3.108}$$

At this juncture we make a decision whether to consider the vacancies as strictly conserved, in which case we must retain all the terms divided by X_V, or whether to consider the case where the vacancies approach local equilibrium, whence

$$\nabla\mu_V \to 0 \tag{3.109}$$

and thus all terms divided by X_V disappear (cf equations (3.95) and (3.96)). Note, however, that $\nabla\mu_V = 0$ must not be allowed exactly for then there is an ambiguity as to whether or not L_{VV} is defined. This restriction is in accord with the actual situation which obtains within an alloy diffusion couple. Further discussion of the first case will be deferred to Chapter 6 and we will now focus on the local equilibrium case. Thus from the definitional relations (3.87) and (3.88) with $a \to A$ and $b \to B$ (no tracer components) we can write

$$J_V \simeq L_A\nabla\mu_A + L_B\nabla\mu_B \tag{3.110}$$

where L_A and L_B according to equations (3.90) and (3.91) are unique, and according to equation (3.89) $\nabla\mu_A$ and $\nabla\mu_B$ are of opposite sign. Next assuming that the non-diagonal matrix is also unique we equate coefficients on the right-hand sides of equations (3.108) and (3.110) (with terms in $1/X_V$ removed) and substituting the result in equation (3.102) obtain

$$L_{VV} \simeq L_A + L_B \tag{3.111}$$

Next, rewriting the general expressions for J_A, J_B, and J_V in terms of the independent forces $\nabla\mu_A$ and $\nabla\mu_V$, or $\nabla\mu_B$ and $\Delta\mu_V$, we obtain

$$J_A = -\left(L_{AA}-\frac{X_A}{X_B}L_{AB}\right)\nabla\mu_A + \left[L_{AA}+L_{AB}\left(1+\frac{X_V}{X_B}\right)\right]\nabla\mu_V \tag{3.112}$$

and symmetrically for J_B, together with

$$J_V = \left[\frac{(L_{AA} + L_{AB})}{X_A} - \frac{(L_{BB} + L_{AB})}{X_B}\right] X_A \nabla \mu_A$$

$$- \left[L_{AA} + 2L_{AB} + L_{BB} + \frac{X_V}{X_B}(L_{BB} + L_{AB})\right] \nabla \mu_V \quad (3.113)$$

These equations will be further referred to in §6.4. Now when $\nabla\mu_V \to 0$ we can compare equation (3.112) and its B counterpart to J_A and J_B according to equations (3.87) and (3.88) to obtain by matching coefficients

$$L_A \simeq L_{AA} - \frac{X_A}{X_B} L_{AB} \quad (3.114)$$

and

$$L_B \simeq L_{BB} - \frac{X_B}{X_A} L_{AB} \quad (3.115)$$

Finally substituting these in equation (3.111) and equating to the expression given by equation (3.102) we obtain

$$L_{AA} + 2L_{AB} + L_{BB} \simeq L_{AA} - \frac{X_A^2 + X_B^2}{X_A X_B} L_{AB} + L_{BB} \quad (3.116)$$

If L_{AA} and L_{BB} are the same on both sides this can only be satisfied if

$$L_{AB} \simeq 0 \quad (3.117)$$

and according to equations (3.114) and (3.115)

$$L_{AA} \simeq L_A \quad \text{and} \quad L_{BB} \simeq L_B \quad (3.118)$$

What has been explicitly demonstrated via equation (3.116) and its derivation in relation to equations (3.87) and (3.88) is that if the forces are subject to the Gibbs–Duhem constraint (3.89), and we nonetheless insist that the L-matrix be non-diagonal, then that matrix is non-unique. If we insist that the L-matrix be unique then it is diagonal. This was already obvious from our discussion of equations (3.87)–(3.94). Note, however, that if the vacancies are conserved the system becomes a ternary rather than a binary, and the L-matrix takes a different meaning in accord with equations (3.105)–(3.108) with off-diagonal coefficients $L_{AB} \neq 0$. Clearly the L-matrix does not analytically continue in transforming from the ternary to the binary representation.

Let us now introduce isotopic mixtures A + A* and B + B* into the system such that

$$X_A + X_{A^*} = X_a \quad (3.119)$$

and

$$X_B + X_{B^*} = X_b \quad (3.120)$$

Consider first the conventional 2-species flux equations for the ternary systems A–B–B* and A*–B–B* at identical chemical compositions,

$$J_A = -L_{AA}\nabla\mu_A - L_{AB}\nabla\mu_B - L_{AB^*}\nabla\mu_{B^*} \tag{3.121}$$

and

$$J_{A^*} = -L_{A^*A^*}\nabla\mu_{A^*} - L_{A^*B}\nabla\mu_B - L_{A^*B^*}\nabla\mu_{B^*} \tag{3.122}$$

In view of the Gibbs–Duhem equation, the L-matrices are non-unique and therefore not necessarily symmetric. Noting in the two cases that a ≡ A or a ≡ A* and comparing with equations (3.87) and (3.88), a unique correspondence for $J_a \equiv J_{A^*}$ or $J_a \equiv J_A$ can be expressed in the ternary systems by

$$L_{AA} = L_a \qquad \text{or} \qquad L_{A^*A^*} = L_a \tag{3.123}$$

with

$$L_{AB}\nabla\mu_B + L_{AB^*}\nabla\mu_{B^*} = 0 \tag{3.124}$$

and

$$L_{A^*B}\nabla\mu_B + L_{A^*B^*}\nabla\mu_{B^*} = 0 \tag{3.125}$$

Since these are ternaries, $\nabla\mu_B$ and $\nabla\mu_{B^*}$ are independent so

$$L_{AB} = L_{AB^*} = L_{A^*B} = L_{A^*B^*} = 0 \tag{3.126}$$

A similar argument pertains to the ternaries B–A–A* and B*–A–A* so

$$L_{BA} = L_{B^*A} = L_{BA^*} = L_{B^*A^*} = 0 \tag{3.127}$$

As will be seen later, this correspondence will assure that the number of unknown phenomenological coefficients is exactly equal to the number of relations among them which are established by the experimental constraints (cf the earlier remark concerning equations (3.87) and (3.89)). These unknown coefficients appear in the full quaternary matrix[27] and can be introduced by disaggregating the total fluxes, namely,

$$J_a = J_A + J_{A^*} \tag{3.128}$$

to

$$J_A = -L_{AA}\nabla\mu_A - L_{AA^*}\nabla\mu_{A^*} \tag{3.129}$$

and

$$J_{A^*} = -L_{A^*A}\nabla\mu_A - L_{A^*A^*}\nabla\mu_{A^*} \tag{3.130}$$

and

$$J_b = J_B + J_{B^*} \tag{3.131}$$

to

$$J_B = -L_{BB}\nabla\mu_B - L_{BB^*}\nabla\mu_{B^*} \tag{3.132}$$

and

$$J_{B^*} = -L_{B^*B}\nabla\mu_B - L_{B^*B^*}\nabla\mu_{B^*} \tag{3.133}$$

Since as before $\nabla\mu_B$, $\nabla\mu_{B^*}$, and $\nabla\mu_A$, $\nabla\mu_{A^*}$, and their associated fluxes are independent through free variations in $\nabla\mu_a$ and $\nabla\mu_b$, Onsager reciprocity applies to both matrices.

Here the chemical potentials are related and take the forms

$$\mu_a = \mu_{0a}(T,p) + kT\ln(X_a\gamma_a) \qquad \mu_b = \mu_{0b}(T,p) + kT\ln(X_b\gamma_b) \tag{3.134}$$

$$\mu_A = \mu_{0A}(T,p) + kT\ln(X_A\gamma_a) \qquad \mu_{A^*} = \mu_{0A^*}(T,p) + kT\ln(X_{A^*}\gamma_a) \tag{3.135}$$

$$\mu_B = \mu_{0B}(T,p) + kT\ln(X_B\gamma_b) \qquad \mu_{B^*} = \mu_{0B^*}(T,p) + kT\ln(X_{B^*}\gamma_b) \tag{3.136}$$

Following Howard and Lidiard,[27] we seek the relations between the various L-coefficients attributable to the chemical indistinguishability of A–A* and B–B* and proceed by analysing a number of symmetric thought experiments. First, however, we consider the tracer diffusion measurements wherein A* (or B*) are surface injected at the limit of infinite dilution.

In these conventional experiments, the flow of A* will be countercurrent to a minute flow of vacancies and we can write (cf Howard and Lidiard[27])

$$D_A^* = \frac{kT}{NX_A^*}L_{A^*A^*} \tag{3.137}$$

and similarly,

$$D_B^* = \frac{kT}{NX_B^*}L_{B^*B^*} \tag{3.138}$$

which contain the appropriate correlation factors (§6.4). Further, in an A–A* experiment

$$J_B = 0 = J_{B^*} \qquad J_A + J_{A^*} = 0 \tag{3.139}$$

and similarly for the B–B* experiment. Applying these conditions to equations (3.129)–(3.133), we prove that

$$(L_{AA} + L_{A^*A})/X_A = (L_{AA^*} + L_{A^*A^*})/X_{A^*} \tag{3.140}$$

and

$$(L_{BB} + L_{B^*B})/X_B = (L_{BB^*} + L_{B^*B^*})/X_{B^*} \tag{3.141}$$

Finally, we consider the interdiffusion of A*–B–A in the absence of B* and arrange the conditions of a diffusion couple such that A* is initially at uniform chemical potential, that is, the step functions ∇ in A and B are adjusted so that

$$X_B\nabla\mu_B + X_A\nabla\mu_A = -X_{A^*}\nabla\mu_{A^*} = 0 \tag{3.142}$$

In other words, choose an X_B and X_A and a small $\nabla X_A \ll X_A$ and from these calculate ∇X_B and the isopotential values of X_{A^*} which accord with the site balance. During diffusion, μ_{A^*} will possess a very shallow maximum and matching shallow minimum on respective sides of the origin due to the cross interactions (cf §6.4). At such points, $\nabla\mu_{A^*} = 0$ and from equations (3.87), (3.128), (3.129), and (3.130) we obtain in the dilute tracer limit ($X_{A^*} \to 0$; $\nabla\mu_A = \nabla\mu_a$)

$$L_a = L_{AA} + L_{A^*A} \tag{3.143}$$

Similarly,

$$L_b = L_{BB} + L_{B^*B} \tag{3.144}$$

Eliminating $L_{A^*A^*}$ and $L_{B^*B^*}$ in equations (3.140) and (3.141) via equations (3.137) and (3.138), and substituting equations (3.143) and (3.144) to eliminate L_{AA} and L_{BB} which yields L_a and L_b, we obtain the corrected Darken equations via equations (3.93) and (3.94) in the approximations $X_a = X_A$, $X_b = X_B$ i.e.

$$\tilde{D}/\phi = (X_B D_A^* + X_A D_B^*) + \frac{kT}{N}\left(X_B \frac{L_{AA^*}}{X_{A^*}} + X_A \frac{L_{BB^*}}{X_{B^*}}\right) \tag{3.145}$$

and

$$v/(\phi \nabla C_A) = (D_A^* - D_B^*) + \frac{kT}{N}\left(\frac{L_{AA^*}}{X_{A^*}} - \frac{L_{BB^*}}{X_{B^*}}\right) \tag{3.146}$$

Note on account of equations (3.143) and (3.144) the remarkable parallel which arises between the form of the trailing correction terms and the form of the exact equations (3.93) and (3.94). Expanded relations retaining a non-zero diagonal element L_{AB} have been given by Howard and Lidiard.[27] Our expressions are obtainable from theirs by adopting the present binary, local equilibrium condition $L_{AB} = 0$. The minute perturbations supplied by simultaneous diffusion of A* or B* should of course have no effect on the interdiffusion of A and B.

Until recently there had been no published attempt to evaluate L_{AA^*} and L_{BB^*} directly in accord with equations (3.129), (3.130), (3.132), and (3.133) and in relation to a local equilibrium vacancy flux created by countercurrent A + A* and B + B*. Manning's treatment of the Kirkendall effect[23] is, however, closely related to such a development, and, after further preliminaries, will be discussed in §6.4. In §6.4 we will offer a direct derivation of L_{AA^*} and L_{BB^*}.[34]

REFERENCES

1 L. ONSAGER: *Phys. Rev.*, 1931, **37**, 405
2 L. ONSAGER: *ibid.*, **38**, 2265
3 L. ONSAGER: *Ann. N.Y. Acad. Sci.*, 1945–6, **46**, 241
4 S. R. DE GROOT AND P. MAZUR: 'Non-equilibrium thermodynamics', 1962, Amsterdam, North-Holland
5 S. R. DE GROOT: 'Thermodynamics of irreversible processes', 1952, Amsterdam, North-Holland
6 I. PRIGOGINE: 'Introduction to thermodynamics of irreversible processes', 1955, Springfield, Ill., C. C. Thomas
7 D. D. FITTS: 'Nonequilibrium thermodynamics', 1962, New York, McGraw-Hill
8 J. A. HIRSCHFELDER: *J. Chem. Phys.*, 1957, **26**, 274
9 I. PRIGOGINE AND R. DEFAY: 'Chemical thermodynamics', 39 *et seq.*, 1954, New York, Longmans
10 H. L. TOOR: *Chem. Eng. Sci.*, 1965, **20**, 941
11 E. A. GUGGENHEIM: 'Mixtures', 1952, Oxford University press
12 A. KLEMM: *Z. Naturforsch.*, 1953, **8a**, 397
13 O. LAMM: *J. Phys. Chem.*, 1947, **51**, 1064
14 O. LAMM: *ibid.*, 1957, **61**, 948
15 P. G. SHEWMON: 'Diffusion in solids', 1963, New York, McGraw-Hill
16 D. G. MILLER: *Chem. Rev.*, 1960, **60**, 15
17 B. C. COLEMAN AND C. TRUESDELL: *J. Chem. Phys.*, 1960, **33**, 28
18 J. S. KIRKALDY: *Scripta Met.*, 1985, **19**, 1307
19 A. EINSTEIN: *Ann. Phys. (Leipzig)*, 1910, **33**, 1275
20 P. CURIE: *J. Phys. (Paris)*, 1894, **3**, 393
21 L. S. DARKEN: *Trans. AIME*, 1948, **174**, 184
22 J. BARDEEN AND C. HERRING: 'Atom movements', 87, 1951, Cleveland, ASM

23 J. R. MANNING: 'Diffusion kinetics of atoms in crystals', 121 *et seq.*, 139 *et seq.*, 182 *et seq.*, 218 *et seq.*, 1968, Princeton, NJ, D. Van Nostrand
24 T. HEUMANN: *J. Phys. F (Met. Phys.)*, 1979, **9**, 1997
25 J. S. KIRKALDY: in 'Solute-defect interaction: theory and experiment', (ed. S. Saimoto, G. R. Purdy, and G. V. Kidson), 214, 1986, Toronto, Pergamon
26 A. D. LECLAIRE: *Prog. Met. Phys.*, 1953, **4**, 265
27 R. E. HOWARD AND A. B. LIDIARD: *Rep. Prog. Phys.*, 1964, **27**, 161
28 H. J. V. TYRRELL AND K. H. HARRIS: 'Diffusion in Liquids', 78 *et seq.*, 1984, London, Butterworths
29 A. B. LIDIARD: Discussion to reference 25
30 J. R. MANNING: Discussion to reference 25
31 T. R. ANTHONY: in 'Diffusion in solids, recent developments', (ed. A. S. Nowick and J. J. Burton), 353, 1975, New York, Academic Press
32 As ref. 27, footnote p. 209
33 R. W. BALLUFFI: *Acta Met.*, 1954, **2**, 195
34 J. S. KIRKALDY: *Scripta Met.*, 1987, **21**, 33

CHAPTER 4

The multicomponent Fick equations and their solutions

4.1 INTRODUCTION

In principle it is possible to develop a quantitative diffusion theory in terms of chemical potentials or activities, and indeed there is a certain attraction in operating with a variable that is continuous in multiphase systems.[1] However, the phenomenological coefficients are discontinuous in such systems producing discontinuous derivatives, and more important, it is very difficult to introduce explicit mass conservation which is essential to time dependent formulations. This chapter is accordingly concerned with the rigorous transformation to a concentration formulation.

4.2 REDUCTION OF THE DEPENDENT *L*-MATRIX

It was found from the preceding thermodynamic argument that isothermal diffusion in an n-component system can be described by the flux equations

$$\boldsymbol{J}_i = \sum_{j=1}^{n} L_{ij}\boldsymbol{X}_j \qquad (i = 1, 2, \ldots n) \tag{4.1}$$

where the forces $\boldsymbol{X}_j$ are the chemical potential gradients

$$\boldsymbol{X}_j = -\operatorname{grad}\mu_j \tag{4.2}$$

However the thermodynamic forces are related by the Gibbs–Duhem equation (Appendix 1), i.e.

$$\sum_{i=1}^{n} X_i\boldsymbol{X}_i = 0 \tag{4.3}$$

and the fluxes have a dependency determined by the frame of reference in which they are measured. The presence of dependencies among the fluxes and forces means that the matrix $[L]$ is non-unique and that the validity of the Onsager reciprocity theorem is therefore not guaranteed. It has been shown by Hooyman and de Groot[2] that nonetheless it is always possible to choose the coefficients in such a way that the Onsager relations hold. We seek to use this fact and simultaneously to eliminate dependent variables, and hence coefficients, so as to arrive at an irreducible number of coefficients in the description.

The Gibbs–Duhem equation is first used to eliminate component n from the description yielding

$$\boldsymbol{X}_n = -\sum_{i}^{n-1} X_i \boldsymbol{X}_i / X_n \tag{4.4}$$

and the entropy source term becomes

$$T\sigma = \sum_{i}^{n-1} \boldsymbol{J}_i \boldsymbol{X}_i - \frac{\boldsymbol{J}_n}{X_n} \sum_{i}^{n-1} X_i \boldsymbol{X}_i \tag{4.5}$$

or

$$T\sigma = \sum_{i}^{n-1} \boldsymbol{J}'_i \boldsymbol{X}_i \tag{4.6}$$

where

$$\boldsymbol{J}'_i = \boldsymbol{J}_i - \frac{m_i}{m_n} \boldsymbol{J}_n \tag{4.7}$$

and use has been made of the fact that the mole concentration ratio equals the mole fraction ratio. The fluxes defined in this way may be interpreted by rewriting equation (4.7) as

$$\boldsymbol{J}'_i = m_i \left(\frac{\boldsymbol{J}_i}{m_i} - \frac{\boldsymbol{J}_n}{m_n} \right) \tag{4.8}$$

Equation (4.8) then becomes

$$\boldsymbol{J}'_i = m_i(\boldsymbol{v}_i - \boldsymbol{v}_n) \tag{4.9}$$

where $\boldsymbol{v}$ are average species velocities. Thus the fluxes $\boldsymbol{J}'_i$ are measures of the motion of the species i *relative* to species n (the convenient choice for n is as solvent). In a closed system which is mass conservative, it is always possible to specify the absolute flux of the components in terms of all other component fluxes. Thus the set of absolute fluxes contains a linear dependency of the form

$$\sum_{i}^{n} a_i \boldsymbol{J}_i = 0 \tag{4.10}$$

where the a_i depend on the reference frame chosen, a specification which will be examined presently. We note at this point that the set $\{\boldsymbol{J}'_1, \ldots \boldsymbol{J}'_{n-1}\}$ has had this dependency removed, and the entropy source term of equation (4.6) is properly specified in so far as the constraints arising out of the conservation of material are concerned.

An alternative procedure is to substitute the constraint (4.10) into the entropy source equation, again eliminating component n from the description and obtaining

$$J_n = -\sum_i^{n-1} a_i J_i / a_n \tag{4.11}$$

$$T\sigma = \sum_i^{n-1} J_i X_i' \tag{4.12}$$

$$X_i' = X_i - (a_i/a_n) X_n \tag{4.13}$$

Since the X_i' are chemical (or electrochemical) potential gradients *relative* to that of species n, the linear dependence arising from the Gibbs–Duhem relationship has been eliminated from the equation set. Expressions (4.6) and (4.12) are of course, equivalent. From equation (4.12) it is seen that the appropriate phenomenological equations are

$$J_i = \sum_i^{n-1} L_{ij} X_j' \tag{4.14}$$

to which the Onsager reciprocity theorem applies, i.e.

$$L_{ij} = L_{ji} \tag{4.15}$$

Thus for an n-component system a matrix of coefficients of order $(n-1)$ is required in the absence of any additional constraints. The relations (4.15) imply that a total of $n(n-1)/2$ independent coefficients are required for a complete description.

Additional constraints can arise in particular systems due to additional conservation laws such as the conservation of sites or charge (see Chapter 8). These will take the general form

$$\sum_i^{n-1} b_i J_i = 0 \tag{4.16}$$

and their application to equation (4.12) results in the descriptions

$$J_{n-1} = -\sum_i^{n-2} b_i J_i / b_{n-1} \tag{4.17}$$

$$T\sigma = \sum_i^{n-2} J_i X_i'' \tag{4.18}$$

and

$$X_i'' = X_i' - (b_i/b_{n-1}) X_n' \tag{4.19}$$

Having completed the reduction of the coefficient matrix it remains to choose a reference frame appropriate to the physical situation, and finally to transform the phenomenological equations (4.18) into diffusion equations in which concentration gradients explicitly appear.

4.3 FRAMES OF REFERENCE

As seen in the preceding section, the choice of frame of reference within which fluxes are to be measured appears explicitly in the phenomenological equations. It is therefore useful to enumerate the more commonly used frames and explore the procedures whereby fluxes can be transformed from one frame to another. This general question has been discussed in detail by Kirkwood *et al.*[3] and by de Groot and Mazur.[4]

Discussion here will be restricted to diffusion in one direction in an isothermal, isobaric, field-free system. The flux $(\boldsymbol{J}_i)_R$ is the flux of component i (in moles per second per unit cross-sectional area) relative to a reference frame R. It is related to $(\boldsymbol{J}_i)_S$, the flux relative to another frame by

$$(\boldsymbol{J}_i)_R = (\boldsymbol{J}_i)_S + m_i \boldsymbol{u}_{SR} \tag{4.20}$$

where u_{SR} is the velocity of frame S relative to frame R at the point in space and time considered. Following Kirkwood *et al.*[3] we denote by the subscripts M, S, V, N and C, respectively, the mass-fixed, solvent-fixed, volume-fixed, number-fixed and diffusion cell (≡ laboratory) fixed frames. Each frame is based on the definition that no net flow of quantity (mass, solvent, etc.) occurs relative to the x-coordinate at the point in question. The first four frames are defined by the relations

$$\Sigma M_i (J_i)_M = 0 \tag{4.21}$$

$$(J_n)_S = 0 \tag{4.22}$$

$$\Sigma \bar{V}_i (J_i)_V = 0 \tag{4.23}$$

$$\Sigma (J_i)_N = 0 \tag{4.24}$$

Recalling that $m = \Sigma m_i$ and noting that

$$\Sigma \bar{V}_i m_i = 1 \tag{4.25}$$

it is found from equations (4.20)–(4.25) that these frames move relative to the cell or laboratory frame at velocities

$$\boldsymbol{u}_{MC} = \Sigma M_i (J_i)_M / \rho \tag{4.26}$$

$$\boldsymbol{u}_{SC} = (J_n)_C / m \tag{4.27}$$

$$\boldsymbol{u}_{VC} = \Sigma \bar{V}_i (J_i)_C \tag{4.28}$$

$$\boldsymbol{u}_{NC} = \Sigma (J_i)_C / m \tag{4.29}$$

These relationships in combination with equation (4.20) may be used to calculate practical diffusion coefficients in one frame from values measured in another (cf §9.6). For dilute solutions they all tend to coincide. Under the circumstances that the molar volumes can be approximated by constants (e.g. with small concentration differences) the volume-fixed and laboratory frames coincide ($\boldsymbol{u}_{VC} = 0$). Indeed, for substitutional solids the $\bar{V}_i$ may be taken as approximately equal constants, so that the volume-fixed, number-fixed and laboratory frames coincide. The mass-fixed (or barycentric) and solvent-fixed frames[5, 6] are usually introduced for reasons of theoretical convenience. For example, the Onsager reciprocal relations are initially defined within the mass-fixed frame while

Hittorf transference (cf §9.4) is defined with respect to the solvent-fixed frame.

When there is an interstitial component I in an otherwise substitutional solution one can set $\overline{V}_I = 0$ and the others equal and constant assuring coincidence of the volume-fixed and laboratory frames and generating the usual substitutional number balance via

$$\sum_{2}^{n} J_i = 0$$

The Kirkendall or lattice-fixed frame of reference is the number-fixed frame in which vacancies in substitutional solids are regarded as a conserved component. There is a linear velocity transformation due to Darken[7] which relates this useful theoretical frame to the laboratory or number-fixed frame for real components (§2.13 and 3.11).

4.4 TRANSFORMATION TO FICK-TYPE DIFFUSION EQUATIONS

We commence with the volume-fixed frame of reference which is usually adequate to the discussion of both solid and liquid diffusion. In this case equations (4.11)–(4.15) hold with $a_i = \overline{V}_i$. To obtain equations in the Fick's law form we must transform the chemical potential gradients to concentration gradients. Here use is made of the local equilibrium relationships for an n-component system, expressing the chemical potentials as a function of molar concentrations, namely

$$\mu_i = \mu_i(m_1, m_2, \ldots m_{n-1}) \qquad i = 1, 2 \ldots n-1 \tag{4.30}$$

whence

$$\operatorname{grad}\left(\mu_i - \frac{\overline{V}_i}{\overline{V}_n}\mu_n\right) = \sum_{j}^{n-1} \mu_{ij} \operatorname{grad} m_j \tag{4.31}$$

where

$$\mu_{ij} = \frac{\partial(\mu_i - (\overline{V}_i/\overline{V}_n)\,\mu_n)}{\partial m_j} \tag{4.32}$$

Substitution of equation (4.31) into (4.13) and (4.14) yields

$$-\boldsymbol{J}_i = \sum_{k}^{n-1} L'_{ik}\left(\sum_{j}^{n-1} \mu_{kj} \operatorname{grad} m_j\right) \tag{4.33}$$

or, upon reversing the order of summation,

$$-\boldsymbol{J}_i = \sum_{j}^{n-1} D_{ij} \operatorname{grad} m_j \tag{4.34}$$

where

$$D_{ij} = \sum_{k}^{n-1} L_{ik}\mu_{kj} \tag{4.35}$$

Equation (4.34) is recognized as the generalized Fick equation for the analysis of experimental data in terms of concentration gradients. The form of equation (4.35) is recognized as a matrix product (Appendix 4) which is written in shorthand notation as

$$D = L\mu \tag{4.36}$$

It is seen that the D matrix is equal to the product of an Onsager matrix of coefficients and a thermodynamic coupling matrix. Furthermore, since the law of multiplication of determinants is the same as the law of multiplication of the matrices from which they derive, equation (4.36) applies to the determinants as well as the D matrix.

Following a procedure suggested by Onsager[8] and later clarified by Hooyman[9] and Kirkaldy[10] it has been found possible to state relationships between the D_{ij} based on the reciprocity relations among the L_{ij}. If the sum

$$\sum_{j}^{n-1} D_{jk}\,\mu_{ji}$$

is expanded with the aid of equation (4.35) and the order of summation reversed, then it can be shown that

$$\sum_{j}^{n-1} D_{jk}\,\mu_{ji} = \sum_{j}^{n-1}\left(\mu_{jk}\sum_{l}^{n-1}\mu_{li}L_{lj}\right) \tag{4.37}$$

From the reciprocal relation (4.15) and equation (4.35) it follows that

$$\sum_{j}^{n-1} D_{jk}\mu_{ji} = \sum_{j}^{n-1}\mu_{jk}D_{ji} \tag{4.38}$$

which represent the reciprocal relations expressed in terms of the D-matrix. Further clarification is sought through an examination of the matrix μ_{kj}.

Recalling its definition in equation (4.32) we note that the matrix μ_{kj} is related to the Hessian matrix for the Gibbs free energy

$$[H_{kj}] = \left[\frac{\partial^2 G}{\partial n_k\,\partial n_j}\right] = \left[\frac{\partial \mu_k}{\partial n_j}\right] \tag{4.39}$$

Since the order of differentiation is irrelevant, this latter matrix is obviously symmetric. Furthermore, it has been shown by Prigogine and Defay[11] that for stable homogeneous solutions, the matrix $[H_{kj}]$ must also be positive definite. It was originally implied by Onsager[8] that the matrix $[D_{ij}]$ would, in the case of a stable solution, have eigenvalues which are real and positive. Noting the apparent relationship between the matrices $[H_{kj}]$ and $[\mu_{kj}]$, Kirkaldy *et al.*[12] proved for the general case with volume change that real, positive eigenvalues for $[D_{ij}]$ are a necessary consequence of the positive definite character of $[H_{ij}]$. In the simpler case where partial molar volumes are equal, Kirkaldy and Purdy[13] showed that the matrix

$$[\mu_{jk}] = \frac{1}{m}\frac{\partial(\mu_j - \mu_n)}{\partial X_k} \tag{4.40}$$

is positive definite. The matrix $[D_{ij}]$ is then formed as a product of two positive definite matrices in equation (4.36). It thus follows from matrix theory that since the eigenvalues of positive definite matrices are positive and the eigenvalues of the product matrix satisfy

$$\lambda = \lambda_L \lambda_\mu \tag{4.41}$$

the eigenvalues λ of the D matrix must be real and positive (Appendix 4).

The implications of this can be seen from the definition of the eigenvalues of a matrix as the roots λ of the characteristic (or determinantal) equation for the matrix[1, 14, 15, 16, 17, 18, 19, 20]

$$|D - \lambda I| = 0 \tag{4.42}$$

where I is the unit matrix. For the case $n = 3$ we have

$$\begin{vmatrix} D_{11} - \lambda & D_{12} \\ D_{21} & D_{22} - \lambda \end{vmatrix} = 0 \tag{4.43}$$

whence

$$\lambda = \tfrac{1}{2}\{(D_{11} + D_{22}) \pm [(D_{11} + D_{22})^{22} - 4(D_{11}D_{22} - D_{21}D_{12})]^{1/2}\} \tag{4.44}$$

and[20]

$$\lambda_+ \lambda_- = D_{11}D_{22} - D_{12}D_{21} \tag{4.45}$$

Thus the requirement that all λ are real and positive is met if, and only if,

$$(D_{11} + D_{22}) > 0 \tag{4.46}$$

$$(D_{11} + D_{22})^2 \geqslant 4(D_{11}D_{22} - D_{21}D_{12}) \tag{4.47}$$

$$(D_{11}D_{22} - D_{21}D_{12}) \geqslant 0 \tag{4.48}$$

This set of kinetic statements is equivalent to the thermodynamic requirements, arising from the character of the Onsager and the Hessian free energy matrices.

It is noted that the above conditions do not imply that the individual on-diagonal coefficients D_{ii} are necessarily positive, only that their sum is positive. It is also apparent that the matrix $[D_{ij}]$ need not, and in general will not, be symmetric. Despite the lack of thermodynamic constraint on the individual D_{ii}, it is an invariable consequence of kinetic theories for terminal solutions, where component 3 is the solvent, that they are positive. Hence to ensure a real value for the quantity

$$[(D_{11} + D_{22})^2 - 4(D_{11}D_{22} - D_{21}D_{12})]^{1/2} = [(D_{11} - D_{22})^2 + 4D_{21}D_{12}]^{1/2} \tag{4.49}$$

it is sufficient that

$$D_{12}D_{21} \geqslant 0 \tag{4.50}$$

which is itself a consequence of elementary dilute solution kinetic models. This together with equation (4.48) and

$$D_{11}, D_{22} > 0 \tag{4.51}$$

can now replace conditions (4.46)–(4.48) as sufficient conditions for thermokinetic stability.

The fact that the eigenvalues of the matrix $[D_{ij}]$ are real and positive has further implications in the common situation where the D_{ij} are assumed constant and where the multicomponent diffusion solutions can be constructed as linear combinations of binary solutions. This eventuality will be explored in §4.6. An up-to-date review of these considerations has been presented by Miller *et al.*[21]

4.5 THE TERNARY DIFFUSION FRAMEWORK: NON-UNIQUENESS OF THE *D*-MATRIX

In a binary system, the chemical diffusion coefficient is a unique quantity, independent of the choice of the solvent species. In a ternary system this is not the case. A different set of four chemical coefficients will be obtained depending on the choice of solvent for the system. Mason[22] has suggested that the solvent should be always designated by attaching a superscript to the coefficient, e.g. D_{ij}^3 where 3 is the solvent, and this convention has been widely adopted (cf Guy and Smith[23]). Mason has also derived the complete relations for transformation between solvents[1, 22] (see also Sanchez[20])

$$D_{ik}^m = D_{mm}^k - D_{kk}^m = D_{ii}^m - D_{ii}^k = -D_{im}^k \tag{4.52}$$

Ziebold and Ogilvie[24] rearranged these into the simple transformations

$$\begin{aligned} D_{11}^2 &= D_{11}^3 - D_{12}^3 \\ D_{13}^2 &= -D_{12}^3 \\ D_{31}^2 &= D_{22}^3 + D_{12}^3 - D_{11}^3 - D_{21}^3 \\ D_{33}^2 &= D_{22}^3 + D_{12}^3 \end{aligned} \tag{4.53}$$

and

$$\begin{aligned} D_{22}^1 &= D_{22}^3 - D_{21}^3 \\ D_{23}^1 &= -D_{21}^3 \\ D_{32}^1 &= D_{11}^3 + D_{21}^3 - D_{22}^3 - D_{12}^3 \\ D_{33}^1 &= D_{11}^3 + D_{21}^3 \end{aligned} \tag{4.54}$$

These apply to solutions which can be approximated by constant molar volumes. Note that the trace of the matrix

$$t = D_{11} + D_{22} \tag{4.55}$$

and the determinant

$$d = D_{11}D_{22} - D_{12}D_{21} \tag{4.56}$$

are invariants of the transformation. This is essential to assure that the eigenvalues, which determine the diffusion profiles (see below), are also invariant. Of course the concentration profiles must not depend on the choice of diffusion formalisms.

The choice of solvent (or superscript) is by no means a trivial matter, for interpretation must often be based on dilute solution kinetics and thermodynamics whence understanding can only be achieved by designating the solvent as the neutral, and usually rich component. Consider the two diffusion matrices recorded by Murakumi *et al.*[25,26] For Ag–Au–Pd at all temperatures and compositions (and their identification: 1 = Au, 2 = Ag, 3 = Pd) they obtain D_{11} negative and almost equal to $-D_{21}$, D_{22} positive and equal to $-D_{12}$, and the determinant is close to zero. Although these satisfy the thermokinetic conditions, it is very difficult to imagine in kinetic terms how nature chose these magnitudes and signs. Let us accordingly select the more natural coordinate system for their average alloys (0·1 Pd and 0·3 Au) which is Ag (2) as solvent. Now via relations (4.53) we obtain

$$D_{11}^2 > 0 \qquad D_{13}^2 > 0 \qquad D_{31}^2 \cong 0 \qquad D_{33}^2 \cong 0 \tag{4.57}$$

Next if we look up the tracer coefficients for dilute Au and Pd in Ag and extrapolate in temperature we find that $D_1^* \gg D_3^*$. This explains the sign of D_{11}^2 and the vanishing values of D_{31}^2 and D_{33}^2, together with the near vanishing of the determinant. Note particularly that in the Murakami framework the on-diagonal coefficient D_{11} is negative, which is thermokinetically acceptable (see §4.4).

As a second example we consider the diffusion coefficient set reported by Murakami *et al.*[26] for near critical point compositions in Ag–Au–Cu at 242°C (Table 4.1).

These coefficients correspond to the identification system 1 = Ag, 2 = Au, 3 = Cu and do not appear to have any rhyme or reason. Following Ziebold and Ogilvie, who studied the same system at 725°C, the more natural choice would have been 1 = Cu, 2 = Ag, 3 = Au, particularly as the mean alloys are towards the Au-rich corner of the Gibbs' triangle. The appropriate transformation of Table 4.1 is Table 4.2. This data representation illustrates the point made in §4.4 that for relatively dilute solutions of 1 and 2 in 3, elementary kinetic theories generate the inequalities

$$D_{11}, D_{22} > 0 \qquad D_{12}D_{21} > 0 \qquad D_{11}D_{22} - D_{12}D_{21} > 0$$

which are sufficient conditions to assure the diffusive stability of the solution. They appear in a very plausible pattern in Table 4.2 and will be examined in more detail in §7.7.

Since ternary diffusion coefficients are functions of two independent concentrations they can conveniently be represented as constant D lines on the Gibbs triangle (cf Fig. 4.1). Relations (4.52), together with dilute solution models to be discussed later, and knowledge of the binary limits place important

Table 4.1 Elements of diffusion matrix ($cm^2\ sec^{-1}$); 1 = Ag, 2 = Au, 3 = Cu; Λ is the diffusion layer thickness

	Microcouple 2	Microcouple 3	
	Λ = 16·1 Å	Λ = 16·4 Å	Λ = 17·2 Å
D_{11}	$4{\cdot}52 \times 10^{-18}$	$2{\cdot}21 \times 10^{-18}$	$1{\cdot}46 \times 10^{-18}$
D_{12}	$8{\cdot}48 \times 10^{-20}$	$-1{\cdot}07 \times 10^{-17}$	$-7{\cdot}18 \times 10^{-18}$
D_{21}	$-2{\cdot}10 \times 10^{-18}$	$-8{\cdot}09 \times 10^{-19}$	$-5{\cdot}37 \times 10^{-19}$
D_{22}	$2{\cdot}36 \times 10^{-19}$	$5{\cdot}73 \times 10^{-18}$	$3{\cdot}96 \times 10^{-18}$

constraints on the D matrix and its composition dependence. For example, let $i = 1$, $k = 2$, and $m = 3$. Then

$$D^3_{12} = D^2_{33} - D^3_{22} = D^3_{11} - D^2_{11} = -D^2_{13} \tag{4.58}$$

Considering the binary limit 3–2 we have that

$$D^3_{12} = D^2_{33} - D^3_{22} = 0 \tag{4.59}$$

Similarly, for the binary limit 3–1

$$D^3_{21} = D^1_{33} - D^3_{11} = 0 \tag{4.60}$$

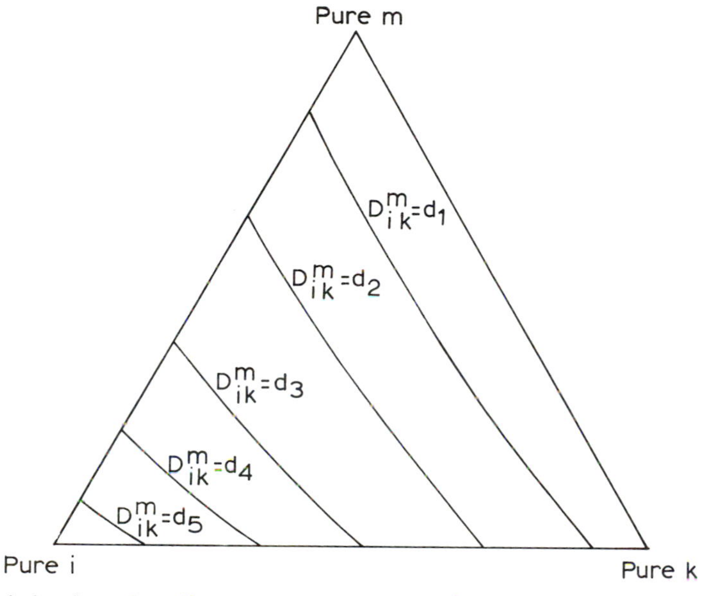

With permission from Canadian Journal of Physics, National Research Council of Canada

Fig. 4.1 Contour map for D^m_{ik} in a ternary system. After Kirkaldy *et al.*[1]

Table 4.2 Elements of diffusion matrix ($cm^2\ sec^{-1}$); 1 = Cu, 2 = Ag, 3 = Au

	Microcouple 2	Microcouple 3	
	$\Lambda = 16{\cdot}1$ Å	$\Lambda = 16{\cdot}4$ Å	$\Lambda = 17{\cdot}2$ Å
D_{11}	$2{\cdot}34 \times 10^{-18}$	$6{\cdot}54 \times 10^{-18}$	$4{\cdot}50 \times 10^{-18}$
D_{12}	$2{\cdot}10 \times 10^{-18}$	$0{\cdot}81 \times 10^{-18}$	$0{\cdot}54 \times 10^{-18}$
D_{21}	$2{\cdot}10 \times 10^{-18}$	$6{\cdot}37 \times 10^{-18}$	$4{\cdot}16 \times 10^{-18}$
D_{22}	$2{\cdot}42 \times 10^{-18}$	$1{\cdot}4 \times 10^{-18}$	$0{\cdot}92 \times 10^{-18}$
Det	$1{\cdot}25 \times 10^{-36}$	$4{\cdot}00 \times 10^{-36}$	$1{\cdot}89 \times 10^{-36}$

From equation (4.52) we see further that

$$D_{11}^{3} - D_{12}^{3} = D_{11}^{2} \tag{4.61}$$

and

$$D_{22}^{3} - D_{21}^{3} = D_{22}^{1} \tag{4.62}$$

where in the limit as $C_3 \to 0$, $D_{11}^{2} = D_{22}^{1}$ is the binary coefficient for 1 in 2 or vice versa. It follows that

$$D_{12}^{3} = D_{11}^{3} - D_{11}^{2} \tag{4.63}$$

where D_{11}^{3} and D_{11}^{2} in the limit as $C_2, C_3 \to 0$, are the tracer coefficients for diffusion of 3 and 2, respectively, in pure 1. Similar relations are obtained from

$$D_{21}^{3} = D_{22}^{3} - D_{22}^{1} \tag{4.64}$$

in the limit as $C_1, C_3 \to 0$. These limits have been discussed further by Kirkaldy,[15] Birchard and Toor,[27] Sabatier and Vignes,[28] and Vignes and Sabatier.[29]

These relations, together with a knowledge of the binary coefficients and dilute solution thermodynamics (cf §6.3) can be used to reduce the required experimentation substantially.

4.6 THE TIME-DEPENDENT TERNARY EQUATIONS AND THEIR SOLUTIONS

In a source free system and in the absence of chemical reactions the atomic or molar mass of each component is conserved so from equation 3.41 with $m \simeq$ constant and $C_i = N_A m_i$

$$\nabla \cdot J_i + \frac{\partial C_i}{\partial t} = 0 \tag{4.65}$$

Hence we obtain the $n - 1$ independent partial differential equations[30]

$$\frac{\partial C_i}{\partial t} = \sum_{k=1}^{n-1} \nabla \cdot D_{ik} \nabla C_k \tag{4.66}$$

For the general case of concentration dependent D the non-linear equations (4.66) must be solved by numerical methods. Castleman and Wo[31] have investigated solutions for the case when D are linear in the concentrations. There exist as well certain important boundary conditions for which these equations may be transformed to ordinary equations for which integral, iterable solutions are available.[1] In the case of dilute solutions, the non-linear equations can in some cases be integrated.[32] Generalizations of equation (4.66) and its solutions to include laminar or turbulent flow have been given by Stewart and Prober[17] and diffusion-reaction coupling by Toor.[16] The latter pertains to coupled equations such as (1.96) and (1.97).

4.6.1 General solutions of the diffusion equations with constant coefficients

If the relative range of concentration difference (that is to say, the maximum change divided by the average value) in a stable diffusing system is not more than about 20% then the diffusion coefficients may be approximated by average constants. Equations (4.66) then simplify to

$$\frac{\partial C_i}{\partial t} = \sum_{k=1}^{n-1} D_{ik} \nabla^2 C_k \tag{4.67}$$

If the boundary conditions on all components are symbolically the same then we may construct the multicomponent solutions as linear combinations of solutions of the binary equations (u_k yet to be defined)

$$\frac{\partial C^k}{\partial t} = u_k \nabla^2 C^k \tag{4.68}$$

of the form

$$C_j = a_{j0} + \sum_{j=1}^{n-1} a_{jk} C^k \tag{4.69}$$

Substitution of equation (4.68) yields the characteristic equations

$$u_k a_{ik} = \sum_{j=1}^{n-1} D_{ik} a_{jk} \tag{4.70}$$

or in matrix notation

$$[D_{ik} - u_k \delta_{kj}][a_k] = 0 \tag{4.71}$$

where the a_k are the column vectors of the coefficient matrix. Recognizing the solution of these as an eigenvalue problem, one obtains the eigenvalues u_k of the matrix D_{ij} as the roots of the polynomial in u[14] (or λ, equation (4.43))

$$\det(D_{ij} - u\delta_{ij}) = 0 \tag{4.72}$$

Onsager,[8] Fujita and Gosting,[30] and Kirkaldy[10,33,14] contributed to the evolution of this solution structure.

One will now appreciate the full significance of the eigenvalues of the diffusion matrix. They are the effective diffusion coefficients in the decomposition of ternary solutions into a sum of solutions of the binary diffusion equations (cf equation (4.69)). For the homogeneous type of boundary conditions selected they are independent of the boundary and initial conditions (equations (4.44)). The disaggregated quantities C^k are designated as the eigenvectors.

The coefficients, a_{jk} ($j, k \neq 0$) are determined by equations (4.70) to within $(n-1)$ conditions. $(n-1)$ of the $2(n-1)$ boundary conditions provide these, the remaining $n-1$ determining the $n-1$ constants a_{j0}. There are quite elegant matrix methods for evaluating and recording these coefficients for boundary conditions which accord with linearity and the superposition principle.[18,34] Since, however, there is a high probability of error in transposing these representations to individual boundary conditions we prefer, with Fujita and Gosting,[30] to record the coefficients in detail for an important and representative set of ternary boundary and initial conditions. Kim[35,36] has investigated the quaternary case.

We now see the relationship between the stability and thermodynamic conditions (4.46)–(4.48) and the form of the solutions of the diffusion equations. Since the $u_k > 0$ for stable mixtures equations (4.68) imply that the solutions of the equations must decay monotonically. This may be seen explicitly by observing the series and integral binary solutions (eigenvectors) adaptable to the most common one-dimensional boundary conditions

$$C^k(x,t) = (4\pi u_k t)^{-1/2} \int_{-\infty}^{\infty} C^k(\xi) \exp\{-(x-\xi)^2/4u_k t\}\, d\xi \tag{4.73}$$

$$C^k(x,t) = \sum_n A_n^k \exp(-n^2 u_k t) \exp(inx) \tag{4.74}$$

and

$$C^k(x,t) = \int_{-\infty}^{\infty} A^k(\beta) \exp(-\beta^2 u_k t) \exp(i\beta x)\, d\beta \tag{4.75}$$

We presume that this will also be the case for the non-linear equations (4.66), but this has not been proven. Thompson and Morral[37] have introduced alternative solutions whose coefficients are expressed in terms of the 'angle' between composition limits and the 'square root' D matrix.

It should be emphasized that the above methods cannot be used when the boundary conditions are mixed, for example, when one component has fixed concentration values at the boundaries while the other has fixed fluxes. It should be further emphasized that multiple eigenvalues lead to singularities in the standard solutions and these cases must be treated separately.[33, 38, 16]

4.6.2 Parametric solutions

There are two types of parametric solutions often encountered in practical problems, the so-called 'parabolic' solutions and the steady state solutions. The first type is obtained by making the substitution in equation (4.67)

$$C = C(\lambda) \qquad \text{where} \qquad \lambda = x/t^{1/2} \tag{4.76}$$

In one dimension this yields the ordinary equations for $n = 3$,

$$\frac{-\lambda}{2}\frac{dC_i}{d\lambda} = \sum_{k=1}^{2} D_{ik} \frac{d^2C_k}{d\lambda^2} \tag{4.77}$$

If the initial and boundary conditions can also be expressed in terms of λ then solutions of equations (4.77) will be unique for these conditions. It is well known that such conditions correspond to the infinite or semi-infinite boundary conditions. That is, the conditions for the infinite couple

$$C_i(x_+, 0) = C_i(+\infty, t) = C_i^0 \tag{4.78}$$

and

$$C_i(x_-, 0) = C_i(-\infty, t) = C_i^1 \tag{4.79}$$

can be summarized as

$$C_i(\lambda = +\infty) = C_i^0 \tag{4.80}$$

and

$$C_i(\lambda=-\infty) = C_i^1 \tag{4.81}$$

and for the semi-infinite couple

$$C_i(0,t) = C_i^1 \tag{4.82}$$

and

$$C_i(x_+,0) = C_i(+\infty,t) = C_i^0 \tag{4.83}$$

can be expressed as

$$C_i(\lambda = 0) = C_i^1 \tag{4.84}$$

and

$$C_i(\lambda = +\infty) = C_i^0 \tag{4.85}$$

The complete solutions (given originally in another form by Fujita and Gosting[30]) are

$$C_1 = a \operatorname{erf}\frac{\lambda}{2u^{1/2}} + b \operatorname{erf}\frac{\lambda}{2w^{1/2}} + c \tag{4.86}$$

$$C_2 = d \operatorname{erf}\frac{\lambda}{2u^{1/2}} + e \operatorname{erf}\frac{\lambda}{2w^{1/2}} + f \tag{4.87}$$

where, for the infinite couple,

$$\begin{aligned}
a &= \{D_{12}(C_2^0 - C_2^1) - [(D_{22} - D_{11}) - D][\tfrac{1}{2}(C_1^0 - C_1^1)]\}/2D \\
b &= \tfrac{1}{2}\{C_1^0 - C_1^1 - 2a\} \\
c &= \tfrac{1}{2}\{C_1^0 + C_1^1\} \\
d &= \{D_{21}(C_1^0 - C_1^1) - [(D_{11} - D_{22}) - D][\tfrac{1}{2}(C_2^0 - C_2^1)]\}/2D \\
e &= \tfrac{1}{2}\{C_2^0 - C_2^1 - 2d\} \\
f &= \tfrac{1}{2}\{C_2^0 + C_2^1\} \\
u &= D_{11} + \tfrac{1}{2}\{(D_{22} - D_{11}) + D\} \\
w &= D_{22} + \tfrac{1}{2}\{(D_{11} - D_{22}) - D\} \\
D &= [(D_{11} - D_{22})^2 + 4D_{12}D_{21}]^{1/2}
\end{aligned}$$

A very useful approximation to this is obtained when $D_{11} \gg D_{22}$, for then $u = D_{11}$ and $w = D_{22}$ and the general solution reduces to[10]

$$C_1 = \frac{1}{2}(C_{10} + C_{11}) + A\left[\operatorname{erf}\frac{\lambda}{2\sqrt{D_{22}}} - \frac{2A - C_{10} + C_{11}}{2A}\operatorname{erf}\frac{\lambda}{2\sqrt{D_{11}}}\right] \tag{4.88}$$

where

$$A = \frac{D_{12}}{D_{11} - D_{22}}\,\frac{C_{21} - D_{20}}{2} \tag{4.89}$$

and

$$\Delta C_2 = C_2 - C_{20} = \frac{1}{2}(C_{21} - C_{20})\left[1 - \text{erf}\frac{\lambda}{2\sqrt{D_{22}}}\right] \tag{4.90}$$

Figure 4.2 contains this approximate solution applied to Darken's[39] data for Fe–C–Si with $D_{11} = 4{\cdot}8 \times 10^{-5}\ \text{mm}^2\ \text{s}^{-1}$ and $D_{22} = 2{\cdot}3 \times 10^{-7}\ \text{mm}^2\ \text{s}^{-1}$.[10] The foregoing boundary conditions are widely used in the determination of ternary diffusion data.

From equations (4.44) we see that a multiple eigenvalue occurs when the discriminant D vanishes. In this case a and b in equation (4.86) become indeterminate. In the special case that $D_{11} = D_{22}$ and $D_{21} = 0$, L'Hospital's rule yields

$$C_1 = \frac{C_1^0 + C_1^1}{2} + \frac{C_1^0 - C_1^1}{2}\text{erf}\frac{\lambda}{2D_{11}^{1/2}} - \frac{C_2^0 - C_2^1}{2}\frac{D_{12}}{2\pi^{1/2}D_{11}^{3/2}}\lambda\exp\left[-\frac{\lambda^2}{4D_{11}}\right] \tag{4.91}$$

and

$$C_2 = \frac{C_2^0 + C_2^2}{2} + \frac{C_2^0 - C_2^1}{2}\text{erf}\frac{\lambda}{2D_{22}^{1/2}} \tag{4.92}$$

Sundelöf and Södervi[38] and Miller *et al.*[21] have discussed the general problem of multiple eigenvalues.

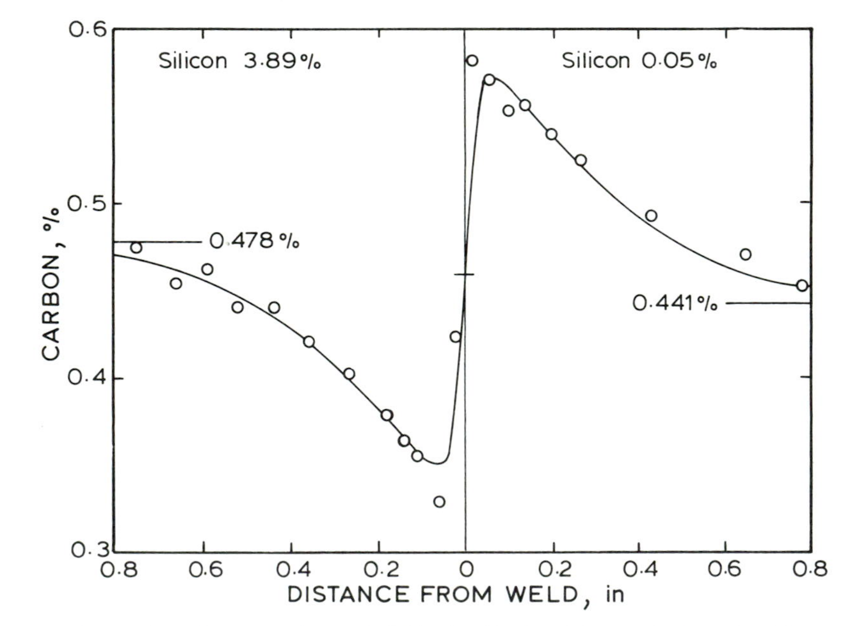

With permission from Canadian Journal of Physics, National Research Council of Canada

Fig. 4.2 Penetration curve for carbon in an iron, carbon, silicon diffusion couple. Experimental points after Darken.[39] After Kirkaldy[10]

Concern has been expressed in the literature that for diffusion paths lying near the binary limits of a ternary isotherm, the off-diagonal coefficients might assume values which yield negative concentrations in the solutions of the equations. However, a careful examination of the concentration dependence of the off-diagonal coefficients shows that these vanish at the limits relative to the on-diagonal coefficients in such a way that negative concentrations cannot occur.[40] The apparent anomaly arises only if one assumes constancy of coefficients outside the limits set in §4.6.1.

The solution for the semi-infinite couple can be obtained from that for the infinite couple by substituting $2C_i^1 - C_2^0$ for C_i^1 throughout.

The second type of parametric solution is obtained by transforming to a system x' moving at a velocity v relative to the x-origin via

$$x' = x - vt \tag{4.93}$$

and requiring that $C_i(x')$ be independent of t. This yields the ordinary equations for steady state diffusion (cf §1.19)

$$-v\frac{dC_i}{dx'} = \sum_{k=1}^{2} D_{ik}\frac{d^2C_k}{dx'^2} \tag{4.94}$$

The boundary conditions are

$$C_i(x' = 0) = C_{iF} \tag{4.95}$$

and

$$C_i(x' = +\infty) = C_{i\infty} \tag{4.96}$$

The ternary solutions, which are constructed from the binary exponential solution are[41,32]

$$C_1 = a_{10} + a_{11}\exp[-vx'/u] + a_{12}\exp[-vx'/w] \tag{4.97}$$

and

$$C_2 = a_{20} + a_{21}\exp[-vx'/u] + a_{22}\exp[-vx'/w] \tag{4.98}$$

where

$$a_{10} = C_{1\infty}$$

$$a_{20} = C_{2\infty}$$

$$a_{11} = \{D_{12}(C_{2F} - C_{2\infty}) + \tfrac{1}{2}(D_{11} - D_{22} + D)(C_{1F} - C_{1\infty})\}/D$$

$$a_{12} = (C_{1F} - C_{1\infty}) - a_{11}$$

$$a_{21} = \{D_{21}(C_{1F} - C_{1\infty}) - \tfrac{1}{2}(D_{11} - D_{22} - D)(C_{2F} - C_{2\infty})\}/D$$

$$a_{22} = (C_{2F} - C_{2\infty}) - a_{21}$$

and where D and the eigenvalues u and w are the same as for solutions (4.86) and (4.87). These solutions are relevant to the steady state growth of ternary alloy crystals from the melt.

4.6.3 A non-parametric solution for a three-dimensional ternary system

Consider the diffusion field about a point source consisting of a mixture of two solute atoms deposited at the r origin in a pure infinite matrix. For the binary case, in which the number of moles deposited is S, the solution of the diffusion equation is

$$C = (S/(4\pi Dt)^{3/2}) \exp[-r^2/4Dt] \tag{4.99}$$

For the ternary case[14]

$$C_1 = a_{10} + \frac{a_{11}}{(ut)^{3/2}} \exp\left[-\frac{r^2}{4ut}\right] + \frac{a_{12}}{(vt)^{3/2}} \exp\left[-\frac{r^2}{4wt}\right] \tag{4.100}$$

and

$$C_2 = a_{20} + \frac{a_{21}}{(ut)^{3/2}} \exp\left[-\frac{r^2}{4ut}\right] + \frac{a_{22}}{(vt)^{3/2}} \exp\left[-\frac{r^2}{4wt}\right] \tag{4.101}$$

If the initial point source consists of S_1 and S_2 atoms of components 1 and 2, respectively, then

$$S_1 = 4\pi \int_0^\infty C_1 r^2 \, \mathrm{d}r \tag{4.102}$$

and

$$S_2 = 4\pi \int_0^\infty C_2 r^2 \, \mathrm{d}r \tag{4.103}$$

Combining these relations with the characteristic equations (4.70) we can readily complete the solutions to obtain

$$a_{10} = a_{20} = 0$$

$$a_{11} = \{S_1[D_{11} - D_{22} + D]/2 + D_{12}S_2\}/8\pi^{3/2}D$$

$$a_{21} = \{D_{21}S_1 - S_2[D_{11} - D_{22} - D]/2\}/8\pi^{3/2}D$$

$$a_{12} = S_1/8\pi^{3/2} - a_{11} \qquad a_{22} = S_2/8\pi^{3/2} - a_{21}$$

where the discriminant D and the eigenvalues u and w are the same as for solutions (4.86) and (4.87).

4.6.4 Solution for a periodic initial condition

For finite systems with an initial step function, trigonometric solutions of the diffusion equations are appropriate,[34] i.e.

$$C_i(x, t) = a_{i0} + \sum_{n=1}^{\infty} \sum_{j=1}^{2} a_{ijn} [\exp(-\xi_n{}^2 u_j t) \cos \xi_n x] \tag{4.104}$$

where the index n runs from 1 to infinity and the u_j are given as before. Referring to the boundary and initial conditions of Fig. 4.3 we find that

$$\xi_n = n\pi/L$$

$$a_{10} = C_{10}$$

$$a_{20} = C_{20}/2$$

and in matrix notation

$$a_{ijn} = \frac{1}{u_1 - u_2}\left[\begin{pmatrix} D_{11} - u_2; & D_{12} \\ D_{21}; & D_{22} - u_2 \end{pmatrix}\begin{pmatrix} \Delta_{1n} \\ \Delta_{2n} \end{pmatrix}; \begin{pmatrix} D_{22} - u_2; & -D_{12} \\ -D_{21}; & D_{11} - u_2 \end{pmatrix}\begin{pmatrix} \Delta_{1n} \\ \Delta_{2n} \end{pmatrix}\right] \quad (4.105)$$

or, completely equivalently,

$$a_{ijn} = \frac{1}{u_1 - u_2}\left[\begin{pmatrix} D_{22} - u_1; & -D_{12} \\ -D_{21}; & D_{11} - u_1 \end{pmatrix}\begin{pmatrix} \Delta_{1n} \\ \Delta_{2n} \end{pmatrix}; \begin{pmatrix} D_{11} - u_1; & D_{12} \\ D_{21}; & D_{22} - u_1 \end{pmatrix}\begin{pmatrix} \Delta_{1n} \\ \Delta_{2n} \end{pmatrix}\right] \quad (4.106)$$

with

$$\Delta_{1n} = 0$$

and

$$\Delta_{2n} = \frac{2(C_{20} - C_{21})}{n\pi} \sin\frac{n\pi}{2}$$

For practical calculations the matrix notation is not very useful, but the remarkable symmetry of the coefficients is most easily observed in this form. The elements appearing are the very matrices, and the matrix of their co-factors, whose determinant defined u_i.

We now restrict consideration to an initial condition where component 1 is initially uniform at a value C_{10} and component 2 is a periodic step-function of wavelength $2L$, varying between C_{20} and C_{21}. This boundary condition may also be used to represent an impervious walled finite system lying between $-L$ and $+L$, since from symmetry all gradients, and therefore fluxes, in the periodic solution must vanish at $\pm L$. Under these conditions component 1 segregates entirely due to the action of the steep gradient of component 2, and therefore its cross-effect on component 2 will be negligible provided the state is sufficiently far from a critical point (i.e. provided $D_{11}D_{22} \gg D_{12}D_{21}$). This can be checked via the coefficients of equations(4.86) and (4.88). The approximation can be obtained simply by setting $D_{21} = 0$ in the differential equation or in the complete solution. In this approximation $u_{1,2} \rightarrow D_{11}, D_{22}$ and the solutions become

$$C_1 = C_{10} + \frac{2}{\pi}\frac{(C_{20} - C_{21})D_{12}}{D_{22} - D_{11}}\sum_{m=1}^{\infty}\frac{(-1)^{m+1}}{(2m-1)} \times \cos\xi_m x$$
$$\times [\exp(-D_{22}\xi_m{}^2 t) - \exp(-D_{11}\xi_m{}^2 t)] \quad (4.107)$$

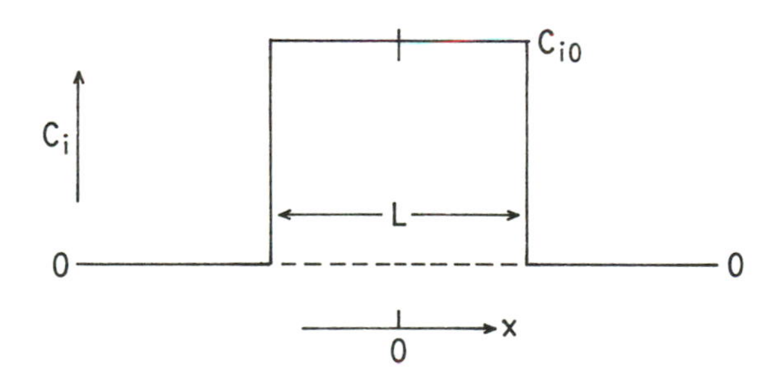

Fig. 4.3 Initial condition for periodic solution of differential equation

and

$$C_2 = \frac{C_{20} + C_{21}}{2} + \frac{2}{\pi}(C_{20} - C_{21}) \sum_{m=1}^{\infty} \frac{(-1)^{m+1}}{(2m-1)} \cos \xi_m x \tag{4.108}$$

where

$$\xi_m = (2m - 1)\pi/L \tag{4.109}$$

Solutions similar to these have been used by Guy *et al.*[42] to discuss segregation and homogenization in multicomponent alloys.

It should be appreciated that the foregoing approximation may be applied to all the solutions thus far presented. If component 1 is initially uniform we may always set $D_{21} = 0$ and let the eigenvalues simplify to D_{11}, D_{22}.

4.6.5 Spinodal solutions in the absence of strain and gradient energy effects

If a critical ternary solution is supersaturated below the critical temperature the determinant of the thermodynamic matrix $[\mu_{kj}]$ goes negative, and it follows from equation (4.36) applied to determinants[33, 43, 44] that

$$D_{11}D_{22} - D_{12}D_{21} \rightarrow \text{negative} \tag{4.110}$$

Thus, according to equation (4.44) one of the eigenvalues $\rightarrow$ negative. Hence if a uniform solution is perturbed by fluctuation or otherwise, the subsequent evolution of the diffusion profile must contain growing (uphill diffusion) components. Eigenvectors of form (4.75) are clearly appropriate to the description of this spinodal decomposition, for when $t = 0$ the $A^k(\beta)$ can be obtained via Fourier transforms of the initial conditions, $C_i(x, 0)$. These solutions will be further explored in §7.6 and with the inclusion of strain and gradient energy effects in §12.5.

4.6.6 Steady and quasi-steady solutions

Referring to equations (4.66) (with constant coefficients) it can be seen that the steady state will be described in part by[14]

$$D_{11}\nabla^2 C_1 + D_{12}\nabla^2 C_2 = 0 \tag{4.111}$$

and

$$D_{21}\nabla^2 C_1 + D_{21}\nabla^2 C_2 = 0 \tag{4.112}$$

By cross-substitution it can be seen that

$$\nabla^2 C_1 = \nabla^2 C_2 = 0 \tag{4.113}$$

That is to say, for fixed boundary values of each component, the distributions will be independent of the cross-effects. Note, however, that the fluxes

$$J_i = -\sum_{k=1}^{n-1} D_{ik} \nabla C_k$$

will reflect the cross-effects. For variable diffusion coefficients, however, this decoupling will not exist.

For a steady state in which one of the components (2) has fixed boundary values and the other (1) is closed within the system, then the steady state is defined by

$$J_1 = -D_{11}\nabla C_1 - D_{12}\nabla C_2 = 0 \tag{4.114}$$

or

$$\nabla C_1 = -\frac{D_{12}}{D_{11}}\Delta C_1 \tag{4.115}$$

This will be true for constant or variable coefficients.

In many systems of interest $D_{11} \gg D_{22}$. If the experimental system is finite and closed, then after times such that the diffusion length of component 1 is appreciably greater than the dimensions of the system, then component 1 will come to a quasi-steady state with respect to the slowly diffusing component 2, and equation (4.114) will hold. Since these conditions on D are also such that the cross-effect of 1 on 2 will be negligible (except through the concentration dependence of the diagonal coefficients), the C_2 distribution can be calculated as a binary and the C_1 distribution in one dimension is given by

$$C_1 = -\int \frac{D_{12}}{D_{11}} \mathrm{d}C_2 + \text{constant} \tag{4.116}$$

where the constant is determined by the conservation of mass. In a finite system as described by Fig. 4.4, assuming constant coefficients, we obtain

$$\frac{D_{12}}{D_{11}} = -\frac{\Delta C_1}{\Delta C_2} \tag{4.117}$$

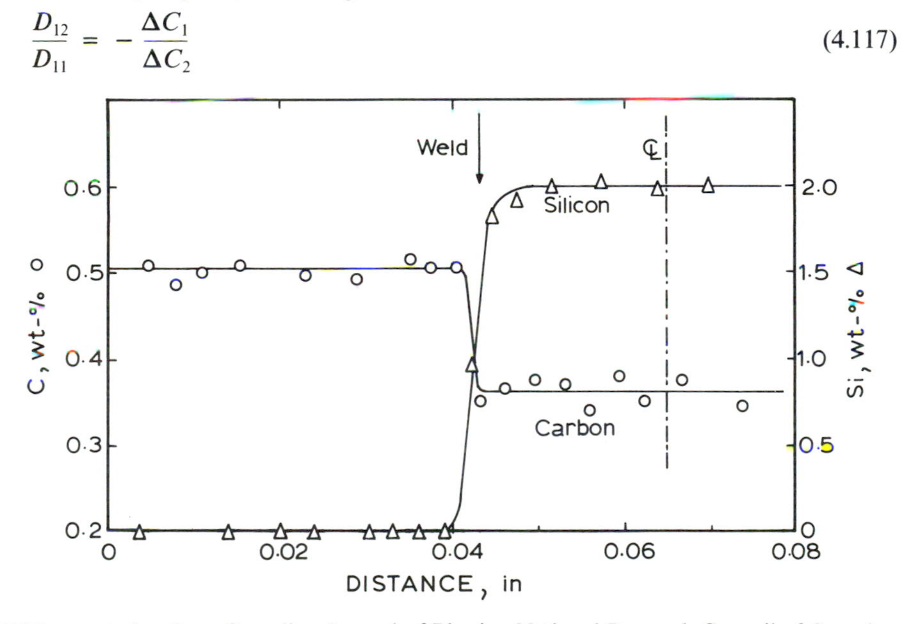

With permission from Canadian Journal of Physics, National Research Council of Canada

Fig. 4.4 Sample penetration curve for Fe–C–Si diffusion layer couple in the 'transient equilibrium' state. The carbons have been determined by microcombustion analysis while the silicons have been determined by wet and X-ray fluorescent analysis. After Kirkaldy and Purdy[45]

where the right-hand side is an easily measurable quantity. This so-called 'transient equilibrium' experiment is discussed more fully in Chapter 5.

It should be emphasized that this is not in general an equilibrium experiment since the activity of component 1 need not be uniform within the couple. Indeed the nearness of approach to partial equilibrium of component 1 will be defined by

$$J_1 = -L_{11}\nabla(\mu_1 - (\bar{V}_1/\bar{V}_3)\mu_3) - L_{12}\nabla(\mu_2 - (\bar{V}_2/\bar{V}_3)\mu_3) = 0 \quad (4.118)$$

As can be seen, both L_{12} and $\bar{V}_1$ must be zero for such a partial equilibrium state. This happens to be the case for the ternary austenites described experimentally above, but this is not generally so. Solutions for steady state liquid diffusion cells are given in §5.7.

4.6.7 Solutions with variable diffusion coefficients

Before fast, cheap computation became available there developed a small literature concerned with analytic or iterable solutions of the ternary diffusion equations.[1,40,32] Nowadays there are finite difference library programs available with most mainframes so there is little incentive for further pursuing this research topic. Indeed, we have recently succeeded in generating numerical ternary solutions of the coupled set of equations (4.66) with variable coefficients using the HP41CV hand-held calculator (see Appendix 5).

REFERENCES

1 J. S. KIRKALDY, J. E. LANE, AND G. R. MASON: *Can. J. Phys.*, 1963, **41**, 2174
2 G. J. HOOYMAN AND S. R. DE GROOT: *Physica*, 1955, **21**, 73
3 J. G. KIRWOOD, R. L. BALDWIN, P. J. DUNLOP, L. J. GOSTING, AND G. KEGELES: *J. Chem. Phys.*, 1960, **33**, 1505
4 S. R. DE GROOT AND P. MAZUR: 'Non-equilibrium thermodynamics', 1962, Amsterdam, North-Holland
5 S. R. DE GROOT: 'Thermodynamics of irreversible process', 1952, Amsterdam, North-Holland
6 D. G. MILLER: *J. Phys. Chem.*, 1966, **70**, 2639
7 L. S. DARKEN: *Trans. AIME*, 1948, **175**, 184
8 L. ONSAGER: *Ann. N. Y. Acad. Sci.*, 1945–46, **46**, 241
9 G. J. HOOYMAN: *Physica*, 1956, **22**, 751
10 J. S. KIRKALDY: *Can. J. Phys.*, 1957, **35**, 435
11 I. PRIGOGINE AND R. DEFAY: 'Chemical thermodynamics', 1954, New York, Longmans
12 J. S. KIRKALDY, D. WEICHERT, AND ZIA-UL-HAQ: *Can. J. Phys.*, 1963, **41**, 2166
13 J. S. KIRKALDY AND G. R. PURDY: *ibid.*, 1969, **47**, 865
14 J. S. KIRKALDY: *ibid.*, 1959, **37**, 30
15 J. S. KIRKALDY: in 'Advances in materials research, Vol. 4', (ed. H. Herman), 1970, New York, John Wiley
16 H. L. TOOR: *Am. Inst. Chem. Eng.*, 1964, **10**, 460
17 W. E. STEWART AND R. PROBER: *Ind. Eng. Chem. Fundamentals*, 1964, **3**, 224
18 A. R. COOPER: in 'Geochemical transport and kinetics', (ed. A. W. Hofmann, B. J. Giletti, H. S. Yoder Jr., and R. A. Yund), 1974, Carnegie Institution of Washington
19 D. DE FONTAINE: *J. Phys. Chem. Solids*, 1973, **34**, 1285
20 J. M. SANCHEZ: M.Sc. Thesis, 1975, UCLA
21 D. G. MILLER, V. VITAGLIANO, AND R. SARTORIO: *J. Phys. Chem.*, 1986, **90**, 1509
22 G. R. MASON: M. Sc. Thesis, 1959, McMaster University, Hamilton, Canada

23 A. G. GUY AND C. B. SMITH: *Trans. ASM*, 1962, **55**, 1
24 T. O. ZIEBOLD AND R. E. OGILVIE: *Trans. AIME*, 1967, **239**, 942
25 M. MURAKAMI, D. DE FONTAINE, J. M. SANCHEZ, AND J. FODOR: *Acta Met.*, 1974, **22**, 709
26 M. MURAKAMI, D. DE FONTAINE, J. M. SANCHEZ, AND J. FODOR: *Thin Solid Films*, 1975, **25**, 465
27 J. K. BIRCHARD AND H. G. TOOR: *J. Phys. Chem.*, 1962, **66**, 2015
28 J. P. SABATIER AND A. VIGNES: *Mém. Sci. Rev. Métall.*, 1967, **64**, 225
29 A. VIGNES AND J. P. SABATIER: *Trans. AIME*, 1969, **245**, 1795
30 H. FUJITA AND L. J. GOSTING: *J. Am. Chem. Soc.*, 1956, **78**, 1099
31 L. S. CASTLEMAN AND G. WO: *Met. Trans.*, 1984, **A15**, 1359
32 G. BOLZE, D. E. COATES, AND J. S. KIRKALDY: *Trans. ASM*, 1969, **62**, 795
33 J. S. KIRKALDY: *Can J. Phys.*, 1958, **36**, 899
34 J. S. KIRKALDY, R. J. BRIGHAM, AND D. H. WEICHERT: *Acta Met.*, 1965, **13**, 907
35 H. KIM: *J. Phys. Chem.*, 1966, **70**, 562
36 H. KIM: *ibid.*, 1969, **73**, 1716
37 M. S. THOMPSON AND J. E. MORRAL: *Acta Met.*, 1986, **34**, 339
38 L. O. SUNDELÖF AND I. SÖDERVI: *Arkiv Kemi*, 1964, **21**, 143
39 L. S. DARKEN: *Trans. AIME*, 1949, **180**, 430
40 J. S. KIRKALDY: *Can. Met. Q.*, 1964, **3**, 279
41 D. E. COATES, S. V. SUBRAMANIAN, AND G. Ř. PURDY: *Trans. AIME*, 1968, **242**, 800
42 A. G. GUY, R. G. BLAKE, AND H. OIKAWA: *ibid.*, 1967, **239**, 771
43 L. O. SUNDELÖF: *Arkiv Kemi*, 1963, **20**, 369
44 D. DE FONTAINE: *Ph. D. Thesis, 1966, Northwestern University*
45 J. S. KIRKALDY AND G. R. PURDY: *Can. J. Phys.*, 1962, **40**, 208

CHAPTER 5

Experimental methods in binary and ternary diffusion

5.1 INTRODUCTION

In reviewing the field of experimental methods in diffusion, it is inevitable that the extremely comprehensive accounts given by Dunlop, Steel and Lane[1] and Tyrrell and Harris[2] are encountered. We accordingly draw heavily upon them in treating liquid diffusion. Our own experience in the field of solid state diffusion enables us to add something in this area. Authoritative general accounts for solids are also to be found in Adda and Philibert,[3] Rothman,[4] and Goldstein *et al.*[5]

Diffusion coefficients are usually measured by observing the movement of a system towards homogeneity and comparing its behaviour with the predictions of the diffusion equations. The differential equations describing diffusion, together with boundary conditions appropriate to the experiment, yield solutions for the molar concentrations of the form

$$m_i = m_i(x, t, D) \tag{5.1}$$

The diffusion coefficient is therefore evaluated by fitting expressions to an experimental curve of the form $m_i = m_i(x, t)$. Two principal methods are in use. In the most widely applied method, two different mixtures are brought into contact at a planar interface and diffusion is observed along a coordinate normal to that plane. The period of observation is such that the concentrations do not change at the ends of the system. For this reason, the experiment is described as an infinite diffusion couple. In the literature on liquid diffusion, the same experiment is usually described as free diffusion. The other method used widely in the study of liquids is associated with steady state diffusion. In the original, still popular quasi-steady version of this method, two large reservoirs of differing but homogeneous composition are separated by a permeable and chemically inert diffusion barrier. The rate at which the compositions of the reservoirs change reflects the rate of diffusion through the barrier and is used to evaluate an average diffusion coefficient. Because the characteristics of the barrier affect the diffusion rate, the method is not absolute and a calibration experiment is required. In the more recent version of this method, a true steady state is achieved by holding concentration differences constant across a permeable diffusion barrier or other defined diffusion path (e.g. a capillary). In this way the flux is held constant and is itself measured, leading directly to the evaluation of a diffusion coefficient.

Of increasing importance are the physical or relaxation methods such as the nuclear magnetic resonance technique. Internal friction measurements in solids fall into this classification.[6] Although these methods suffer some ambiguity as regards interpretation in chemical terms, improved theories are tending to overcome this difficulty. Furthermore, they provide methodologies for determining low temperature diffusion coefficients, quantities which are increasingly significant to the design and lifetimes of the next generation of microelectronic circuitry.

5.2 DIFFUSION IN THE ABSENCE OF A GRADIENT, SELF DIFFUSION

Before discussing diffusion in a chemical gradient it is helpful to consider briefly the techniques used in self-diffusion measurements. The classical techniques for measuring these coefficients all involve observation of the penetration of an isotopically labelled species into an unlabelled species.[4] For solids, a small quantity α of labelled material is deposited, often by electroplating, on the end of a long bar of the same composition and diffusion allowed to occur. Provided that the bar is infinite compared to the depth of diffusion, the concentration profile of labelled material is given in the x direction by

$$m = \frac{\alpha}{2(\pi Dt)^{1/2}} \exp(-x^2/4Dt) \tag{5.2}$$

Measurement of this profile is accomplished by serial sectioning of the bar and isotopic analysis of the samples.[4] For liquids, the diaphragm cell measurement described in §5.7 can be used. Alternatively, the rate at which a labelled solute diffuses out of a narrow capillary tube into a surrounding reservoir of unlabelled solution can be studied.[7] The average concentration, $\overline{m}$ remaining in the capillary is measured as a function of time. If the capillary is of length l and the initial concentration within it is m^0, then

$$\frac{\overline{m}}{m^0} = \frac{8}{\pi^2} \sum_{n=0}^{\infty} \frac{(-1)^n}{(2n+1)^2} \exp\left(-\frac{(2n+1)^2 \pi^2 Dt}{4l^2}\right) \tag{5.3}$$

and for $Dt/l > 0{\cdot}2$[8] only the first term of the series need be considered.

The above methods are necessarily time consuming and can be experimentally rather demanding. Alternative techniques which are much speedier are increasingly being used. The obvious way to decrease measurement times is to decrease the diffusion distance being observed. The quickest diffusion measurements will therefore involve observation of diffusion distances of the order of interparticle distances. An individual event of this sort cannot be observed directly, but there are situations in which the sum of many such events within a macroscopic sample can be detected externally by means of some physical manifestation. Phenomena in this class involve the interaction of individual particles with a transient perturbation, or their subsequent relaxation to a low energy state. Interactions or relaxations which occur via translational particle motion, or are thereby modified, will display a time-dependence which involves particle mobilities. It is the measurement of this time-dependence which leads to a determination of D. All of these techniques can be used to measure self-diffusion in solutions and thus to determine concentration effects. However, because they are not sensitive to position within a sample they cannot reveal gradient effects, either on- or off-diagonal. For this reason they will be described very briefly.

One of the most widely used techniques of this sort is pulsed nuclear magnetic resonance (NMR). The physics of the measurement is well known (see for example Slichter[9]). Its application to liquid phase diffusion has been reviewed by Tyrrell and Harris[2] and to diffusion in solids by Stokes.[10] Briefly, a nuclear spin has a magnetic moment which precesses at the Larmor frequency in an applied magnetic field, $\boldsymbol{H}$. Application of a radio frequency (rf) field oscillating at the Larmor frequency and aligned normal to $\boldsymbol{H}$, produces an additional magnetic vector equal to the sum of two rotating fields. One of these fields rotates in the same direction as the precessing nuclear moment and resonates with it whilst the second rotating field has negligible effect and can be ignored. Interaction between the rotating field and the precessing nuclear moment induces a secondary precessional movement around the axis of the (precessing) primary moment. If the rf excitation is applied as a pulse of appropriate duration, the precession angle is controlled. It is this induced moment which induces an rf voltage in an appropriately aligned receiver coil and gives rise to the NMR signal. The decay of this signal after the pulse ceases can then be observed.

Decay times are determined by the rates of energy exchange between nuclear spins (spin–spin relaxation) and between the nuclear spin and its environment (spin–lattice relaxation). Diffusion causes the nuclear moments to interact with different local fields due to the other nuclei in the sample. If these local fields are random in direction and magnitude, then rapid diffusion, which causes a nucleus to sample many local environments, leads to a low average field and lessened spin–spin relaxation so that the decay time is lengthened. On the other hand, spin–lattice interaction is enhanced, and decay accelerated, by diffusion. A number of pulsed NMR techniques have been devised to measure self-diffusion[2] and the method is now widely use for liquids.

Self-diffusion of point defects and solutes in crystalline and amorphous solids can be studied by observing anelastic relaxation phenomena.[6, 11, 12, 13] The presence of point defects in solids causes partial displacement of adjacent lattice species and hence a change in sample dimensions. Depending on the defect type, these may be anisotropic. In an unstressed solid a random distribution of defect orientations exists. When a stress is applied, the unfavourably oriented defects tend to relax to lower energy states by local migration processes. The resulting change in specimen dimension is an anelastic strain and its rate of development is obviously related to defect mobility which can thereby be measured.

The amount of anelastic strain measured in creep is usually only a small fraction of the elastic strain. Anelastic behaviour is therefore usually detected by observing vibration damping, or internal friction, as a function of temperature. For a single thermally activated relaxation process a single internal friction peak results. Repeating the measurement at a number of temperatures and different vibration frequencies leads to a determination of D, including the activation energy. A technique has been proposed by Cost[14] for analysing anelastic relaxations where a range of relaxation processes (and times) exists.

The related measurement of Gorsky relaxation involves bending an initially flat sheet sample around a form of constant curvature so as to produce a stress gradient across the sheet thickness. Defects then diffuse to an equilibrium distribution under the effect of this gradient. When the stress is removed the sheet relaxes almost, but not quite, to its initial shape. As the defects diffuse back to a uniform distribution the sheet slowly unwinds to its original shape. The time taken for this anelastic relaxation is controlled by the defect diffusion rate. Because the diffusion distance is of the order of the sheet thickness, the technique is limited in application to highly mobile defects like hydrogen in metals (see, for example, Berry and Pritchet[13]). If metal alloy films were formed as diffusion couples with well-characterized concentration gradients before charging with hydrogen, it might be possible to detect diffusional cross-effects by this means.

Quasi-elastic neutron scattering has been used to measure self-diffusion in a number of solids and liquids.[15,16] Unlike inelastic scattering, which arises from neutron interaction with periodic vibrations of particles about their mean positions, quasi-elastic scattering results from interaction with randomly moving, i.e., diffusing particles. It causes broadening of the scattering angle to an extent dependent on the magnitude of D. The result is sensitive to lattice symmetry and can distinguish between different types of interstitial site occupancies. It can also distinguish between continuous (liquidlike) and discrete hopping modes of transport, a distinction of importance for intercalation compounds[17] and superionic conductors[18] where contributions from both mechanisms have been reported. The technique is limited to atoms having significant neutron cross-sections and is best used for an element with significantly different coherent and incoherent scattering efficiencies. For this reason it is particularly well suited to the study of hydrogen diffusion in metals.

Mössbauer spectroscopy can also be used to measure self-diffusion in solids or liquids.[19,20,21] The effect is observed by oscillating a Mössbauer source longitudinally with respect to the radiation direction being observed, so as to induce a Doppler effect. A Mössbauer absorber will then absorb at the resonant frequency and also at frequencies shifted by the Doppler effect. The resonant line intensity is decreased by an amount equal to the sum of the sideband intensities. It is apparent that if the Mössbauer atom moves with a jump time comparable to or less than the nuclear lifetime, then a similar effect on line intensity will result. Measurements of line broadening are used in practice. There are difficulties with the technique[22] and its application is restricted to Mössbauer active nuclei (principally ^{57}Fe). However, the facts that crystal orientation effects are easily studied[23] and that the correlation factor f can be estimated[24] mean that Mössbauer experiments continue to be of interest.

5.3 DIFFUSION COUPLES FOR BINARY AND TERNARY DIFFUSION

When diffusion in the presence of concentration gradients is measured in the laboratory frame, (cf §4.3), the resultant chemical diffusion coefficients are functions of concentration (see equation (2.64)). This suggests that ideally the concentration difference in a diffusion experiment should be vanishingly small. The achievement in other than tracer experiments of such a limit would also have the advantages of eliminating the effects of the volume change of mixing and the development of porosity associated with the Kirkendall effect[25] (cf §2.13). However, differential chemical analysis requires that a finite and often large concentration difference be imposed upon the couple. An appropriate judgement must therefore be made regarding the conflicting demands of optimal analytic sensitivity and minimal corrections and errors due to the artifacts. This question has been discussed by Reynolds *et al.*[26] and Ziebold.[27]

If the diffusion coefficients can be regarded as constant, the measured concentration profiles must satisfy solutions to the diffusion equations obtained under the assumption of constant D or D_{ii} (cf § 1.8). The first analysis of this kind for a ternary solid was provided by Kirkaldy[28] in relation to the data of Darken[29] (Fig. 4.2). For the boundary conditions of a ternary or higher order infinite diffusion couple, there are known analytic solutions (cf § 4.6) and trial values for the D_{ij} can be used to predict the concentration profiles. Comparison of these with the experimental curves leads to better estimates of the D_{ij}. An iterative computation based on minimization of the residuals in concentration, $\Sigma[c_i(\text{measured}) - c_i(\text{calculated})]^2$, is employed. The usual first approximation[30] is to assign the D_{ii} values of the self-diffusion coefficients and to set the D_{ij}, $i \neq j = 0$. Schut and Cooper[31] have proposed a method for determining the full matrix from a single diffusion couple.

Two very elegant methods for determining the D matrix when it is known *a priori* to be constant (concentration differences sufficiently small) have been proposed.[32, 33] These are based on the fact that the solutions (4.86) and (4.87) or, more generally, (4.69), are linear in the boundary concentration differences $C_i^1 - C_i^0$. Consequently, the observables $\partial C_i/\partial x$ or $\int_0^t J_i \, dt$ evaluated at the Matano interface ($\lambda = 0$) can be expressed linearly in the same concentration differences with coefficients as a function of the D values only. Thus two carefully selected diffusion couples at approximately the same average ternary composition yield four equations and can in principle yield the four coefficients of the D matrix at that composition. Clearly, this can be generalized to higher order systems. At the time of writing it is not clear that the sensitivity, reproducibility, and accuracy of concentration measurements is sufficient generally to allow the evaluation of the highly variable off-diagonal coefficients with appropriately small concentration differences. The generalization of the Boltzmann–Matano method for variable coefficients in a ternary system is straightforward and widely employed (see below).

The values for the D_{ij} determined in this way depend on the choice of solvent from among the components. It is therefore essential to specify the solvent when quoting multicomponent diffusion coefficients. A convention is sometimes adopted whereby the solvent component used as reference frame is denoted by a superscript as in D_{ij}^k (cf § 4.5).

As indicated in § 1.9, Fick's second law for a binary system

$$\frac{\partial m}{\partial t} = \frac{\partial}{\partial x}\left(D\frac{\partial m}{\partial x}\right)$$

with the boundary conditions for an infinite diffusion couple

$$\begin{aligned} m(x, 0) &= m(\infty, t) = m^0 \\ m(-x, 0) &= m(-\infty, t) = m^1 \end{aligned} \tag{5.4}$$

may be transformed by making the change of variable

$$\lambda = x/t^{1/2} \tag{5.5}$$

to yield the ordinary differential equation

$$\frac{-\lambda}{2}\frac{\mathrm{d}m}{\mathrm{d}\lambda} = \frac{\mathrm{d}}{\mathrm{d}\lambda}\left(D\frac{dm}{\mathrm{d}\lambda}\right) \tag{5.6}$$

with the conditions

$$m = m^0(\lambda = +\infty) \tag{5.7}$$

$$m = m^1(\lambda = -\infty) \tag{5.8}$$

This is the analysis of Matano[34] and is based on an earlier result of Boltzman.[35] Equation (5.6) integrates to yield

$$\int_{m^0}^{m}\frac{\lambda}{2}\,\mathrm{d}m = -D\left(\frac{\mathrm{d}m}{\mathrm{d}\lambda}\right)_m \tag{5.9}$$

where the fact that $\mathrm{d}m/\mathrm{d}\lambda = 0$ as $m \rightarrow m^0$ has been utilized. Rearrangement of equation (5.9) yields

$$D = -\frac{1}{2}\frac{\mathrm{d}\lambda}{\mathrm{d}m}\int_{m^0}^{m}\lambda\,\mathrm{d}m \tag{5.10}$$

where the origin, known as the Matano interface, is defined by

$$\int_{m^0}^{m^1}\lambda\,\mathrm{d}m = 0 \tag{5.11}$$

This analysis depends for its validity on the assumption that partial molar volumes are independent of concentration.

Kirkaldy[28] demonstrated that the Boltzmann-Matano analysis for binary systems can be extended to multicomponent systems by forming the equation analogous to (5.9)

$$\frac{1}{2}\int_{m_i^0}^{m_i}\lambda\,\mathrm{d}m_i = -\sum_{j}^{n-1}D_{ij}\left(\frac{\mathrm{d}m_k}{\mathrm{d}\lambda}\right)_m \qquad i = 1, 2\ldots(n-1) \tag{5.12}$$

with the origin given by

$$\int_{m_i^0}^{m_i^1}\lambda\,\mathrm{d}m_i = 0 \tag{5.13}$$

where the location given by equation (5.13) must be the same for all components. The boundary conditions for the above solution are again those appropriate to an infinite diffusion couple

$$m_i = m_i^0\,(\lambda = +\infty) \tag{5.14}$$

$$m_i = m_i^1\,(\lambda = -\infty) \tag{5.15}$$

The first reported use of this method was by Ziebold and Ogilvie[36] who applied it to the system Cu–Au–Ag. Other observations analysed in this way have been reported by Vignes and Sabatier[37] for Fe–Co–Ni. Examples of these observations are shown in Figs. 5.1 and 5.2.

For any ternary system the two equations (5.12) contain four D_{ij}. In order to solve for the D_{ij} it is necessary to employ two diffusion couples having different terminal compositions but one common composition. Thus solutions are obtained for the points of intersection of the composition paths. The difficulty of extending this approach to higher order systems is apparent (see however, Dayananda[38]).

If the couples are such that the penetration plots have extrema ($dm_i/d\lambda = 0$) some simplification results. Thus, if component 1 shows extreme points (cf Fig. 5.2), rearrangement of equation (5.12) yields

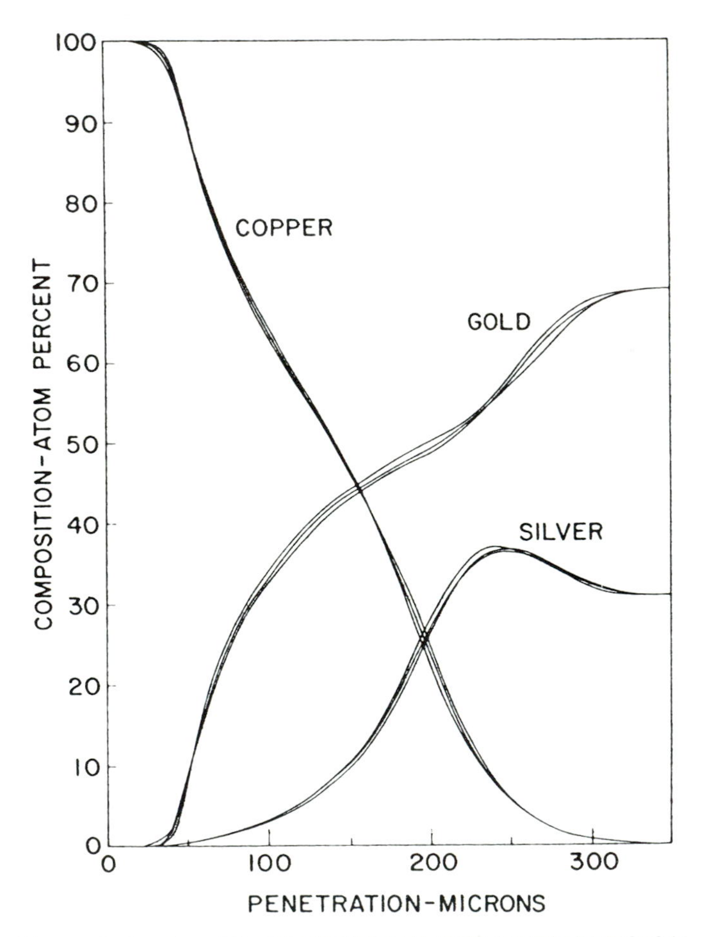

Reprinted with permission from TRANSACTIONS OF THE METALLURGICAL SOCIETY, Vol. 239, p. 9423, (1967), a publication of The Metallurgical Society, Warrendale, Pennsylvania

Fig. 5.1 Penetration curves in Cu–Au–Ag. After Ziebold[27]

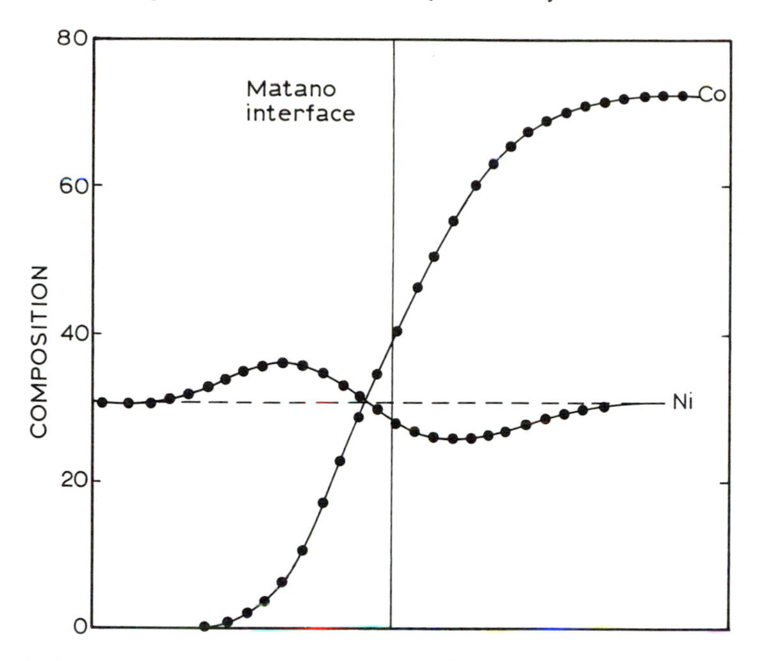

Reprinted with permission from TRANSACTIONS OF THE METALLURGICAL SOCIETY, Vol. 245, p. 1795, (1969), a publication of The Metallurgical Society, Warrendale, Pennsylvania

Fig. 5.2 Penetration curves in Fe–Ni–Co. Points are at 10 μ m intervals. After Vignes and Sabatier[37]

$$D_{12} = -\left(\frac{d\lambda}{dm_2}\right)_e \int_{m_1^0}^{m_1^e} \lambda \, dm_1 \tag{5.16}$$

$$D_{22} = -\left(\frac{d\lambda}{dm_2}\right)_e \int_{m_2^0}^{m_2^e} \lambda \, dm_2 \tag{5.17}$$

where e indicates extremum. In this case, the two concentration profiles from a single diffusion experiment lead to the evaluation of one diagonal and one off-diagonal diffusion coefficient (cf §§7.7 and 7.8). The methodology has been perfected by Ziebold.[27]

5.4 INCREASING THE SENSITIVITY OF DIFFUSION COUPLE MEASUREMENTS

The sensitivity with which off-diagonal coefficients can be measured varies greatly with the experimental design. We consider here a ternary system in which component 3 is the solvent. It has long been recognized[29, 39, 40] as expedient to set the concentration of one component (say m_1) initially constant while providing an initial step change in m_2, as shown in Figs. 4.2 and 5.2. Redistribution of component 1 is then due primarily to the diffusional cross-effect represented by D_{12}. Conversely, the flow of component 2 is not significantly affected by the small gradients which develop in C_1 and, to a reasonable approximation, we may set $D_{21} = 0$. The consequences are explored[41] using equations (4.88) and (4.89) on the assumption that D_{ij} can be regarded as independent of position.

The sensitivity of the measurement for D_{12} is reflected in the size of the maximum excursion in m_1 from its uniform value Δm_1 as shown in Fig. 5.3. The maximum is located by differentiation of equation (4.88) to yield

$$\lambda_m^2 = -2D_{11}\frac{\phi}{\phi - 1}\ln\phi \tag{5.18}$$

$$\Delta m_1 = \frac{D_{12}}{D_{11}}\frac{\Delta m_2}{2}k(\phi) \tag{5.19}$$

where

$$\phi = D_{22}/D_{11} \tag{5.20}$$

and

$$k(\phi) = \frac{\phi}{1-\phi}\left(\operatorname{erf}\frac{\lambda_m}{2\sqrt{D_{11}}} - \operatorname{erf}\frac{\lambda_m}{2\sqrt{D_{22}}}\right) \tag{5.21}$$

The highest sensitivity is therefore realized for high values of Δm_2 and $k(\phi)$.

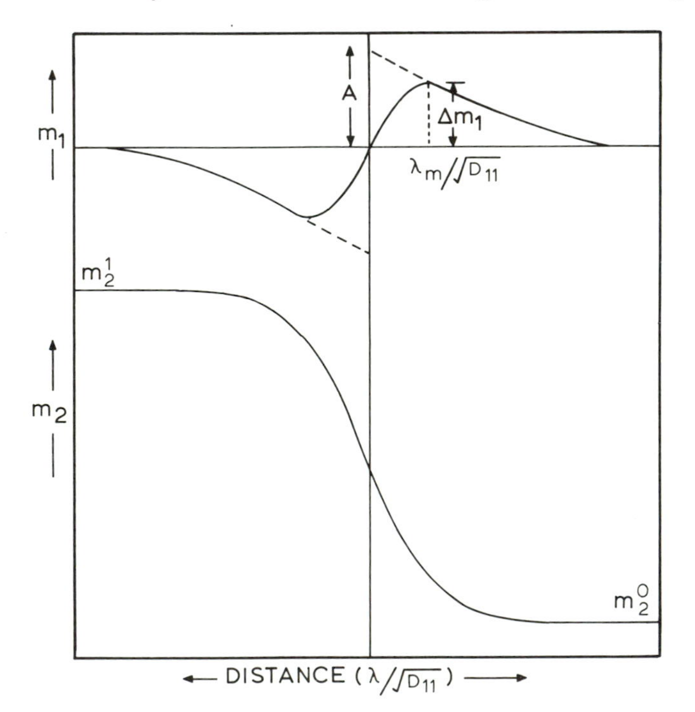

From J. S. Kirkaldy, Zia-Ul-Haq, L. C. Brown, *Trans. ASM*, **56**, American Society for Metals, Metals Park, OH 44073, USA, p. 834, 1963

Fig. 5.3 Schematic representation of segregation of an initially uniform distribution of component 1 in the presence of a concentration gradient of component 2. After Kirkaldy *et al.*[41]

Figure 5.4 shows the variation of k with ϕ and it is seen that k approaches its maximum value of unity as D_{22}/D_{11} becomes very small. If $D_{11} \gg D_{22}$ then D_{12} will be easily measured but the inverse experiment designed to measure D_{21} will fail.

The sensitivity of this type of measurement can be improved under some circumstances by substituting a finite diffusion couple for an infinite couple.[42] The initial and diffused profiles in such an experiment are shown in Fig. 5.5 for alloy layers of thickness $L/2$. A complete solution for the case of concentration-independent D_{ij} is available[42] and where $D_{11} \geqslant D_{22}$ and $t \geqslant L^2/\pi^2 D_{22}$ it may be approximated by

$$\Delta m_1 = \frac{2\Delta m_2 D_{12}}{\pi(D_{22} - D_{11})} [\exp(-D_{22}\lambda_m^2 t) - \exp(-D_{11}\lambda_m^2 t)] \cos \lambda_m x \tag{5.22}$$

This expression may be optimized with respect to time yielding an optimum response

$$\Delta m_1 = \frac{D_{12}}{D_{11}} \frac{\Delta m_2}{2} K(\phi) \tag{5.23}$$

where

$$K(\phi) = \frac{4}{\pi(\phi - 1)} \left[\exp\left(\frac{\phi \ln \phi}{1 - \phi}\right) - \exp\left(\frac{\ln \phi}{1 - \phi}\right) \right] \tag{5.24}$$

after a diffusion time of

$$t = \frac{\ln \phi}{\lambda_m^2 (D_{22} - D_{11})} \tag{5.25}$$

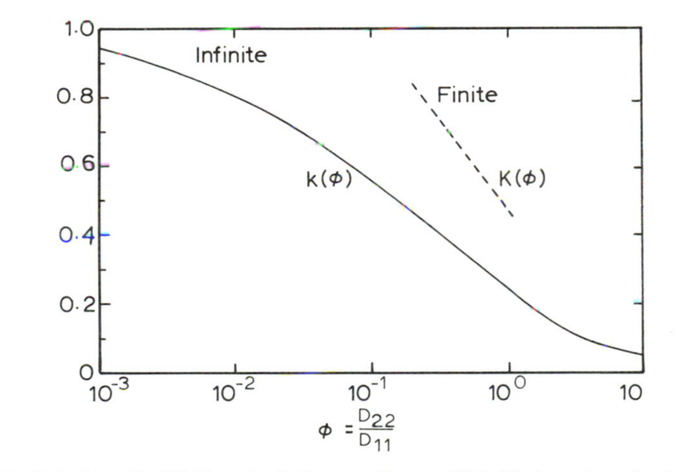

From J. S. Kirkaldy, Zia-Ul-Haq, L. C. Brown, *Trans. ASM*, **56**, American Society for Metals, Metals Park, OH 44073, USA, p. 834, 1963

Fig. 5.4 Theoretical curves for the variation of the sensitivity of diffusion interaction $k(\phi)$ in an infinite couple and $K(\phi)$ in a finite couple as a function of the coefficient ratio $\phi = D_{22}/D_{11}$. After Kirkaldy *et al.*[41]

The function $K(\phi)$ is shown plotted in Fig. 5.4 and it is seen that a substantial increase in sensitivity (~2×) is realized with the finite couple when $0{\cdot}4 < \phi < 1$. Experimental profiles found in this experiment for the Cu–Sn–Zn system are shown in Fig. 5.6.

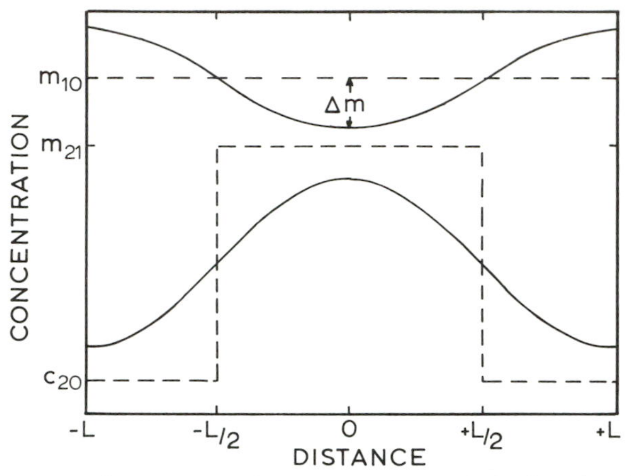

From J. S. Kirkaldy, Zia-Ul-Haq, L. C. Brown, *Trans. ASM*, **56**, American Society for Metals, Metals Park, OH 44073, USA, p. 834, 1963

Fig. 5.5 Schematic representation of the initial (dashed) concentration distributions and the resulting cosine distributions (solid) after diffusion for an optimum time in a finite diffusion couple with reflection from the boundaries at $\pm L$. After Kirkaldy *et al.*[41]

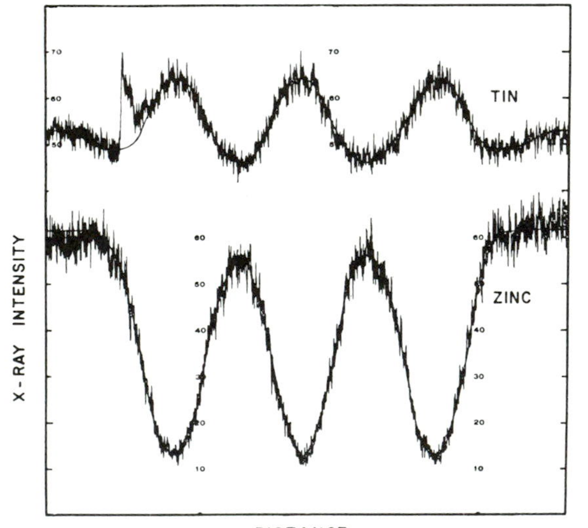

Fig. 5.6 Reproduction of the strip chart showing X-ray intensity as a function of distance from which standardized penetration curves were obtained. Irregularities at the left are the result of traversing a microhardness indentation used as a locating point. After Kirkaldy *et al.*[42]

A quasi-steady variant of the finite couple method, where one of the components diffuses much more rapidly than the other was discussed in detail in §4.6.6. De Fontaine and co-workers[43, 44, 45] have developed an exciting variant of the diffusion layer experiment which is applicable to the determination of exceptionally small (low temperature) diffusion coefficients. Here the layers are evaporated to thicknesses not too far in excess of X-ray wavelengths so X-ray reflections possess sidebands whose amplitudes decay in parallel with diffusion regression of the solute profiles. The approximate reciprocal space analysis[45, 46, 47] is such as to simultaneously yield the full D matrix in a ternary system. Typical results were already discussed in §4.5. A useful discussion on sensitivity has also been given by Cussler.[30]

5.5 ANALYTICAL TECHNIQUES FOR SOLID DIFFUSION COUPLES

Most methods for determining chemical diffusion coefficients involve the analysis of the concentration profile along the diffusion axis. The most accurate methods involve serial sectioning of the couple normal to the diffusion axis and subsequent analysis of the sections. Sectioning methods have been reviewed by Dunlop *et al.*[1] and may be divided into mechanical, chemical and electrochemical techniques. Refinements have since been made in the chemical and electrochemical techniques and the reader is referred in particular to the work of Rothman.[4] Indeed, the general area of sectioning techniques has been reviewed recently by Rothman. A variant on these methods is the sputter sectioning technique in which material sputtered with an ion beam from the solid surface is collected for subsequent analysis.[48] Although subsequent analysis of the sections can be performed to almost any desired degree of precision and/or sensitivity by means of tracer, neutron activation or other chemical techniques, the method is slow and laborious owing to the need for a substantial number of precisely defined sections. For this reason, the bulk of multicomponent diffusional analysis derives from non-sectioning techniques. The most commonly used of these techniques is electron probe microanalysis. An exhaustive review of this and related techniques has been given by Goldstein *et al.*[5] and an extensive bibliography has been given by Romig *et al.*[49]

Electron probe microanalysis was first announced by Castaing[50] and has since been developed into a standard instrumental technique by a number of companies. A brief account of the principles involved is given by Theisen[51] along with tabulated values of the physical constants required in the interpretation of data. More comprehensive descriptions are available from Reed,[52] Goldstein and Yakowitz,[53] and Siegel and Bearman.[54] A schematic diagram of an electron microprobe is shown in Fig. 5.7. In essence, a high energy electron beam is focused to a 0·1–2 μm diameter spot on a flat specimen surface. The electrons excite subsurface atoms which then emit X-rays. The X-rays are characteristic of the atomic number of their parent atoms and X-ray analysis, by wavelength or energy dispersive spectrometry, permits elemental identification. Quantitative measurement of X-ray intensities relative to those excited under the same conditions from standards of known composition leads, although not in a straightforward way, to quantitative elemental analysis.

The conversion of X-ray intensity ratios to elemental mass concentrations requires calculation of the magnitude of a number of physical phenomena occurring within the sample. These are the X-ray excitation efficiency, the absorption within the sample of emerging X-ray beams, and fluorescence due to characteristic radiation or, less importantly, the continuous spectrum. All these effects are functions of the atomic numbers of the constituent atoms within the sample and their concentrations; methods for their calculation are given in the above references. It should be noted that the effects are explicit functions of the 'take-off angle', i.e. the angle between the incident electron beam and the path of the emergent X-rays being detected. In the case of wavelength-dispersive spectrometers this angle is precisely defined and quantitative computation is possible. In the case of energy-dispersive analysis, the detector was originally located adjacent to the sample, subtending a large angle. Consequently, the quantitative analysis was of poor quality. This deficiency has been rectified in

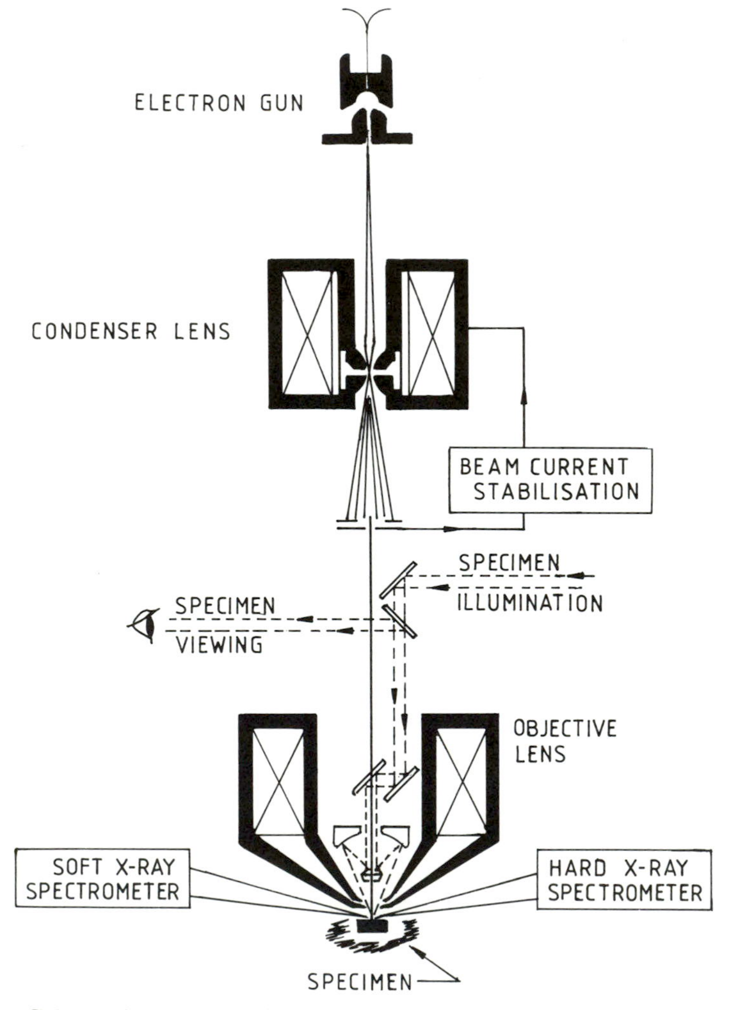

Fig. 5.7 Schematic diagram of electron probe microanalyser

modern instruments by locating the detector at a position remote from the sample.

The analysis of binary systems has at times been carried out with the aid of calibration curves constructed from homogeneous mixtures. It will be apparent that this procedure is inappropriate to multicomponent systems and that computer calculation is mandatory. Nonetheless, it will be found that standards of composition not too far removed from that of the experimental sample are to be preferred. The solid terminals of a diffusion couple are usually suitable.

Absolute errors are usually less than 1 wt-%. As the concentration decreases the absolute error is reduced but the relative error becomes larger. However, since large numbers of replicate analyses are readily obtainable, the shape of the concentration profiles can be determined exactly (cf Figs. 5.1 and 5.2). Reproducibility of chemical diffusion coefficients between independently prepared and heat treated samples, however, remains poor. Deviations of a factor of two between measurements in different laboratories is not uncommon.

Spatial resolution is a function of the electron-sample interactions which lead to electron-scattering within the sample. These are the same processes which determine the X-ray excitation efficiency and are more severe at higher atomic numbers (more scattering) and higher electron beam voltages (longer path length). X-rays will be excited within the volume of solid defined by the electron scattering. A compromise must be reached between the needs for high X-ray intensities and for small excitation volumes. With modern instruments a beam voltage of 20% greater than the critical excitation voltage of the element in question is probably optimum. A spatial resolution of 2–3 μm is usually attainable.

A diffusion couple being prepared for electron probe analysis is sectional normal to the original interface, polished flat to avoid false intensity measurements, and cleaned. If the material is not electrically conductive it must be coated with a thin film of carbon to avoid specimen charging and beam displacement effects. A diffusion sample is by definition non-homogeneous. To minimize the effect on the correction calculation procedures of this artifact, the X-rays should always be collected in a direction normal to the diffusion axis. In this case the emergent X-ray traverses a region of the sample having the same composition as the microvolume from which it was emitted. Point counting must be used for accuracy and multiple scans performed. It is highly desirable that all constituents of the sample be analysed simultaneously.

The disadvantages of the electron microprobe technique are its poor sensitivity to light elements, its inability to detect isotope differences and its limited spatial resolution. The spatial resolution can be greatly improved by analysing thin film samples through which the electron beam is transmitted. Because the electrons traverse only a thin film, the extent of scattering is minimized and, consequently, so is the spread of the X-ray excitation volume. However, obtaining a uniformly thin sample is not easy. The technique is commonly employed with scanning transmission electron microscopes (e.g. Solorzano *et al.*[55]) and the measurement of a diffusion coefficient as low as 10^{-14} mm^2 s^{-1} has been reported.[56] For developments in this field the reader is referred to the Annual conference proceedings of the Electron Microscope Society of America. Satisfactory light element and isotopic analyses have not been achieved with the technique but other methods are available. A further difficulty is encountered with samples which are thermal and electrical insulators. Local electron beam heating leads to enhanced diffusion of mobile species. Analyses for these species show a downward drift and measurement times must be strictly limited.

Greatly enhanced sensitivity to light elements and improved spatial resolution are obtained by combining electron energy loss spectroscopy (EELS) with an electron beam instrument. A particularly promising innovation is the EELS electron microscope.[57]

An alternative method of measuring a concentration profile is provided by the technique of ion microprobe mass analysis, which has now been developed to the stage of commercial instrumentation by Cameca and is proving a very powerful technique. As suggested by the name, a high energy ion beam is focused on the sample surface where it dislodges the constituent atoms of the sample. Thus, unlike the electron probe analysis, ion probe analysis is destructive. The species emerging from the sample are analysed by mass spectrometry. This technique yields very high sensitivities and is much better suited to light elements than the electron probe, although calibration can be a problem. It is obviously appropriate to isotope analysis. A number of reviews exist in the literature and a suitable introduction is given by Colby.[58]

A related and more commonly used technique is that of secondary ion mass spectrometry (SIMS). Here a sample surface, at right-angles to the diffusion direction is continuously eroded, or sputtered, over a relatively large surface area using an ion beam. Ions dislodged from the sample are analysed by mass spectrometry as described above. Slow sputter rates and high analytical sensitivities yield potentially good spatial resolution. Because there is no need for a finely focused ion beam, the instrumentation is relatively cheap. However, there are problems in defining the continuously moving surface as discussed later in connection with Auger electron spectroscopy. The use of SIMS in diffusion measurements has been described in detail by Macht and Naundorf[59] and its application to oxides has been developed by Mitchell *et al.*[60]

An alternative use of ion beams is to implant a profile of diffusant species at concentrations far in excess of thermal solubilities. In this way easily measurable concentrations can be arrived at and, furthermore, any phase boundary kinetic problems associated with a thermal dissolution process are avoided. Observation of the subsequent broadening of implantation profiles leads to estimates of D. However the results are influenced by the degree of ion damage induced during implantation.[61, 62] This disadvantage may be overcome by observing the subsequent diffusion of the implanted species through undamaged regions of the host lattice remote from the implantation zone.[63]

A well known technique of analysis using ion beams is Rutherford back-scattering. High energy ions incident on the sample undergo elastic backscattering when they encounter target nuclei. The scattered ions sustain an energy loss due to target nucleus recoil. The extent of recoil varies with atomic number, thereby leading to element identification. Because the ions lose energy along their path through the solid, depth resolution is possible. Recent applications have been reviewed by Myers.[64]

A technique which is now becoming widely available is Auger electron spectroscopy (AES) in which an electron beam is used to excite Auger electrons within surface atoms (e.g. Hofmann and Erlewein[65]). Because the excitation volume is limited by the escape depth of the electrons (typically of order 1 nm) high resolution is obviously attainable. The technique is one of surface analysis, and carefully cleaned surfaces are essential. Ion sputtering is normally used for this purpose. As a result, depth-profiling can readily be used to analyse diffusion profiles in thin solid regions. In this technique inert gas ions continuously bombard the sample surface and, with time, etch a crater in the material. Auger analysis is performed on the new surface being revealed at the bottom of the crater, yielding a profile of composition versus etching time in what is, in effect, a continuous sectioning experiment. Rendering the technique quantitative is not simple, because of instrument interface broadening, ion beam mixing effects and non-uniform sputtering (see, for example, Cox and Pemsler,[66] Sheasby and Smeltzer[67]). These effects can be reduced in magnitude by reducing sputter ion energies and choosing heavy ions and low sputter angles. Nonetheless, the magnitude of the effects can remain sufficiently high as to render the technique inaccurate when applied to alloy diffusion. These problems are less serious in stable compounds such as base metal oxides.

As noted above, profiling light elements is experimentally difficult. Because of the interest in the diffusion behaviour of non-metallic solute elements in alloys, and in the diffusion of the electronegative species in compounds, considerable effort has gone into devising techniques specifically for these elements. The most widely used involves a gas–solid diffusion couple rather than a solid–solid one and hinges on isotopic exchange, for example between O_2^{18}(g) and a solid oxide which initially contains O^{16} only. Instead of measuring the penetration profile, most workers have measured the integral change in composition of the gas phase. An example of the use of this technique is described by Desmaison and Smeltzer.[68] Alternatively, the integral change in composition of the solid can be determined, for example, by converting the oxygen in a solid to CO_2 which is then subjected to analysis.[69] Serial sectioning analysis of the exchanged solid is also possible, of course. On the other hand, nuclear reactions can be used to determine isotope penetration profiles without the need for serial sectioning.

In the nuclear reaction method, a monoenergetic beam of particles is used to irradiate the isotopically exchanged solid along the direction of the diffusion axis. Nuclear reactions within the solid give rise to the emission of other energetic particles. An example is proton activation of ^{18}O via the reaction $^{18}O(p, \alpha)^{15}N$. The energy of the emitted α-particles is related to the depth at which they were created. The spectrum of α-particle energies corresponding to the range of depths is analysed to yield an ^{18}O concentration profile. Spatial resolution is better than 1 μm and the high sensitivity of the method permits the measurement of shallow profiles, i.e. of low diffusion coefficients. Recent accounts of the use of this technique have been given by Yinnon and Cooper[70] and Reddy *et al.*[71] The use of other nuclear reactions in analysing light element profiles in crystals has been reviewed by Lanford *et al.*[72] A particularly useful technique is that of resonant nuclear reaction analysis. Here a monoenergetic beam incident on the sample surface is attenuated with increasing penetration depth. When the resonant energy is reached, the nuclear reaction occurs and is detected by its characteristic emission. By varying the incident beam energy one varies the depth at which the reaction is excited and hence obtains a concentration profile. In addition to a knowledge of the absorption coefficients for the particles involved and the nuclear cross-section for the reaction, the experimentalist needs access to a relatively high energy particle accelerator. Not surprisingly, the technique has not been widely adopted.

Solid compound–gas diffusion couples can also be used to measure diffusion coefficients of metallic species. Consider a metal oxide in equilibrium with a given value of p_{O_2}. If it is subjected to an abrupt change in p_{O_2} then the process of re-equilibration to a new homogeneous state involves the creation or annihilation of defects at the solid–gas interface (see § 8.2) and their diffusion into or out of the bulk solid. If the surface process is effectively at equilibrium the diffusion of the defects, commonly cation defects, controls the rate of re-equilibration. The process may be monitored thermogravimetrically[73] or by observing the change in electrical properties of the compound[74, 75] as the level of charged defects changes. For long diffusion times the change in weight Δw in a compound re-equilibrating to a new stoichiometry is given by Neuman[76] as

$$(1 - \Delta w/\Delta w_\infty) = \frac{8}{\pi^2}\exp\left(-\frac{\tilde{D}\pi^2 t}{4l^2}\right) \tag{5.26}$$

where the sample is a thin plate of thickness $2l$ and Δw_∞ is the weight change after the new equilibrium state is reached. It has been estimated that $\tilde{D}$ is independent of composition.

A variant on this method has been proposed by Rosenburg.[77] It involves changing the partial pressure of oxidant abruptly during the course of a metal oxidation reaction, thereby perturbing the oxidation rate. A general solution for a sample on which the scale is much thinner than its lateral dimensions has been given by Fryt[78]

$$\frac{\Delta w}{\Delta x_0 \Delta m} = \frac{\tilde{D}t}{x_0^2} + \frac{2}{\pi^2}\sum_{n=1}^{\infty}\frac{1}{n^2}\left\{1 - \exp\left[-\frac{n^2\pi^2\tilde{D}t}{x_0^2}\right]\right\} \tag{5.27}$$

where x_0 is the oxide scale thickness at the time of change in p_{O_2}, Δx_0 is the subsequently observed change and Δm is the molar point defect concentration difference across the scale thickness. Again it has been assumed that $\tilde{D} \neq f(m)$. A number of applications of this technique have been reviewed by Mrowec.[79] Fryt *et al.*[80] have demonstrated that the results obtained by the two methods described by equations (5.26) and (5.27) agree for $Fe_{1-x}S$. The situation where the surface process contributes to rate control has been analysed[81] and the appropriately modified form of equation (5.27) developed. It is clear that these methods are limited in applicability to situations in which only one lattice species is mobile, and hence have value only for binary diffusion studies.

A survey of methods for measuring very low diffusion coefficients ($>10^{-17}$ mm^2 s^{-2}) has been recently presented by Mehrer.[82]

5.6 ANALYTICAL TECHNIQUES FOR FREE DIFFUSION IN LIQUIDS

In performing a diffusion experiment, a sharp boundary is formed between two different solutions, the more dense solution below the less dense, which are then allowed to mix. In transparent materials the different solutions have different refractive indices and their interdiffusion causes the formation of a refractive index gradient. This gradient can be measured by optical interferometry and is related to the concentration gradients within the diffusion couple. Interferometic techniques are probably the most widely used means of measuring absolute diffusion coefficients and have been reviewed comprehensively by Dunlop *et al.*[1] and by Tyrrell and Harris.[2] The principles of these measurements can be schematically illustrated by considering the simple Gouy diffusiometer shown in Fig. 5.8. Monochromatic light is focused by a lens onto a photographic plate. A diffusion cell is placed between the lens and the film. As a light ray passes through the cell it will be bent through an angle proportional to the refractive index gradient. The original liquid–liquid boundary is formed at the optical axis and consequently the refractive index gradient decays symmetrically above and below this axis as diffusion proceeds. Light passing through regions of constant composition above and below the diffusion zone produces an undeviated image. The maximum deflection occurs when a light ray passes through the point of maximum refractive index gradient, i.e. at the optical axis. The size of this deflection decreases as diffusion causes relaxation of the gradient. Consider now two rays symmetrically disposed about the optical axis. These rays sample equal refractive index gradients, are equally deflected and are therefore focused at the plane of the photographic plate. Depending on the relative path lengths of these rays, they interfere constructively or destructively. Consequently a pattern of fringes appears in the focal plane. The appearance of Gouy interference fringes is shown in Fig. 5.9. Various numerical methods for extracting diffusion coefficients from data of this sort have been proposed. The most general and successful is that due to Gosting and Fujita[84] in which both the diffusion coefficient and the refractive index are expressed as Taylor series in concentration. This method and its refinements are reviewed by Dunlop *et al.*[1] and by Tyrrell and Harris.[2] These very thorough reviews include a discussion of the various means of producing an initially sharp boundary and corrections for the inevitable imprecision accompanying the operation. The techniques of recording diffusion all involve time-lapse photography of the interference fringe patterns produced by passing monochromatic light through a cell containing the diffusion couple. Methods for extraction of binary diffusion coefficients from the optical data are given along with a method for calculating ternary diffusion coefficients from Gouy interferometry. A precision and reproducibility of $\pm 0{\cdot}1\%$ is reported, which is two magnitudes better than that achievable in solids. A practical description of the optical equipment involved in Gouy interferometry has been given by Gosting *et al.*[85] Methods have also been given for calculating ternary diffusion coefficients from Rayleigh and Mach–Zender interferometry by Albright and Sherrill[86] and

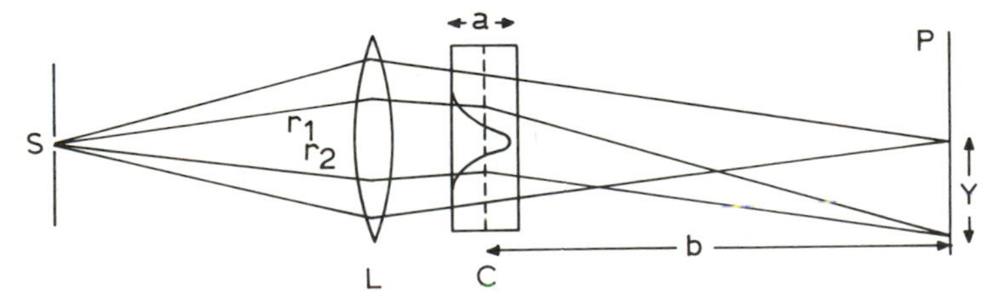

Fig. 5.8 Optical system for the single-lens Gouy diffusiometer. After Gosting[82]

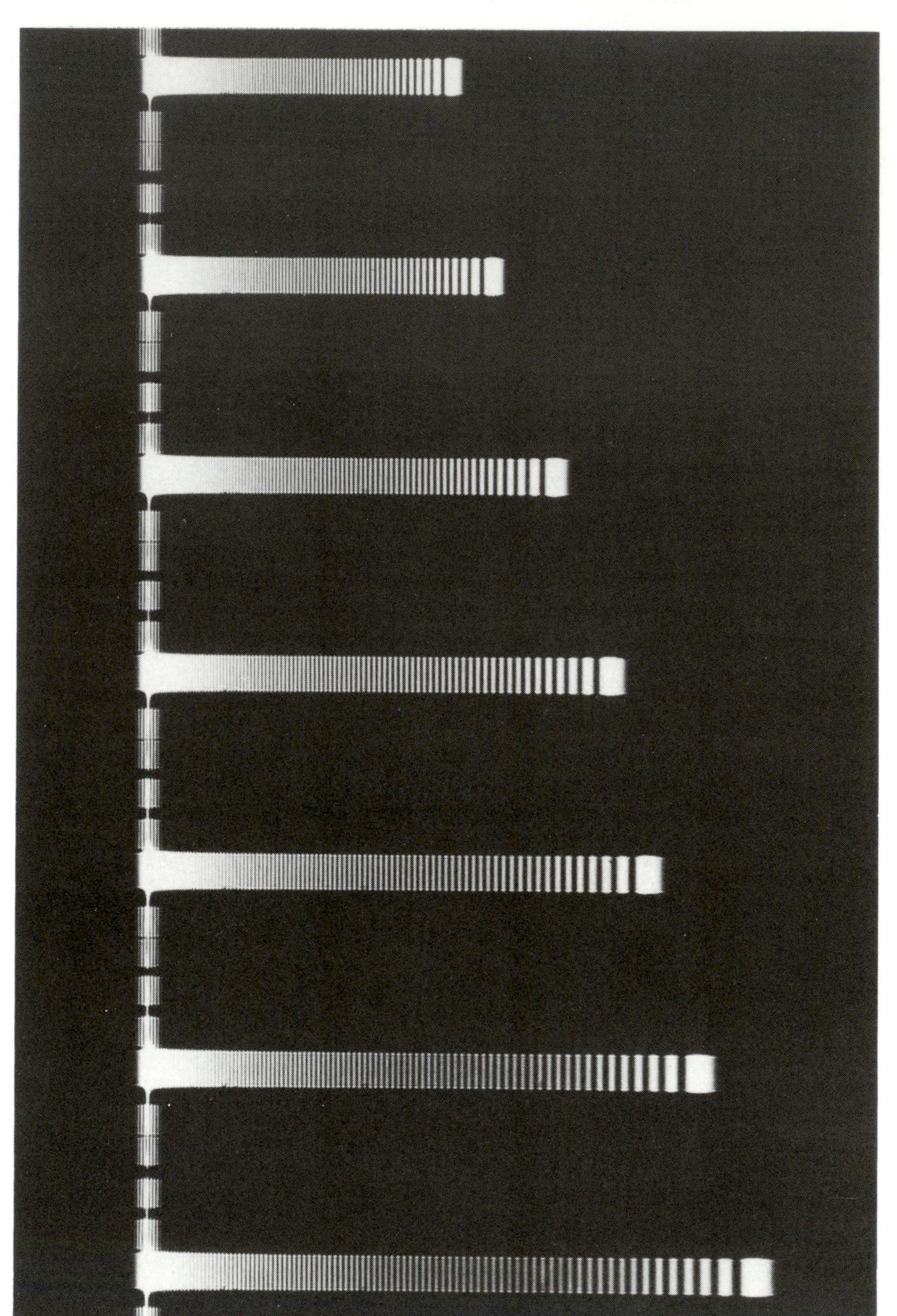

Fig. 5.9 Gouy interference fringes photographed at successively longer times. Rayleigh reference fringes appear at the top of the photograph. Photograph courtesy of P. J. Dunlop, University of Adelaide

by Miller.[87] The latter method has been applied to a series of aqueous electrolyte solutions by Rard and Miller.[88, 89, 90]

All methods used for extracting ternary diffusion coefficients from interferometric data depend ultimately on the form of the solutions to the ternary diffusion equations employed. Solutions for free diffusion boundary conditions in terms of interferometric data have been provided by Fujita and Gosting,[91] assuming that the diffusion coefficients are independent of composition. Since this is not generally the case, interferometric experiments must be carried out using small concentration differences.

Interferometry is appropriate for studies of solutions containing electrolytes, non-electrolytes and polymers. Diffusion in polymer solutions can be difficult to analyse because of the distribution of molecular weights. The particular averaging procedure chosen influences greatly the apparent results and it is therefore preferable where possible for experimental purposes to use monodisperse polymers. The development of laser technology has made possible the measurement of spectral line broadening following light scattering by a solution of diffusing polymers.[92] The method of calculation has been discussed by Chu and Schoenes[93] and Hyde[94] and a recent example of the application of this method has been given by Gulari *et al.*[95]

Systems which are liquid only at very high temperatures or which, like metals, are opaque can be investigated by carrying out an infinite diffusion couple experiment and then rapidly quenching the couple. When the system freezes, diffusion practically ceases and the penetration profiles may be examined at leisure. The older literature on this technique has been reviewed by Yang and Simnad[96] and modern accounts of its application to oxide slags have been given by Sugawara *et al.*[97] and Ukyo and Goto.[98] As might be expected, accuracy and reproducibility, as in solids, are not impressive.

5.7 STEADY STATE DIFFUSION TECHNIQUE FOR LIQUIDS

One of the most widely used methods for obtaining diffusion coefficients for liquids is the diaphragm cell. In this cell, two solutions of differing concentrations are separated by a thin, porous diaphragm of chemically inert material such as glass. The two solutions are stirred so that the only concentration gradient is within the diaphragm. The narrow pores in the diaphragm protect the solution from thermal and mechanical gradients so that true diffusion is the only mechanism of mass transport. Provided that the volume of solution within the diaphragm is very much less than the volume of the cell, the diffusive flux will, at any given time, be independent of position. This pseudosteady state value will change with time as the concentrations in the two compartments change. Measurement of these concentration changes after a period of time allows the evaluation of a diffusion coefficient. The value thus obtained is an average value over the time and concentration ranges covered by the experiment. Experimental procedures are described by Dunlop *et al.*[1] and Tyrrell and Harris[2], as are the methods for calculating diffusion coefficients from experimental data.

For binary diffusion we define[99] a time-dependent concentration averaged diffusion coefficient $\bar{D}$ as

$$\bar{D}(t) = \frac{1}{\Delta m} \int_{m_{\mathrm{T}}}^{m_{\mathrm{B}}} D \, \mathrm{d}m$$

where m_B and m_T are the solute concentrations in the bottom and top compartments of the cell and

$$\Delta m = m_B - m_T$$

It may then be shown that

$$\ln\left(\frac{\Delta m^0}{\Delta m^t}\right) = \beta \int_{t=0}^{t} \bar{D}(t)\, dt \tag{5.28}$$

where the superscripts 0 and t denote initial and final values and β, the cell constant, is given by

$$\beta = \frac{A}{l}\left(\frac{1}{V_B} + \frac{1}{V_T}\right) \tag{5.29}$$

with A the effective diaphragm area and l its thickness, and V_i are the compartment volumes. Defining the time average diffusion coefficient by

$$\bar{\bar{D}} = \frac{1}{t}\int_0^t \bar{D}(t)\, dt$$

we obtain

$$\bar{\bar{D}} = \frac{1}{\beta t} \ln \frac{\Delta m^0}{\Delta m^t} \tag{5.30}$$

Thus the value calculated for $\bar{\bar{D}}$ is a double average over both concentration and time. This treatment contains the approximation of steady state diffusion

$$J = -\bar{D}(t)\, \Delta m/l$$

An exact solution due to Barnes[100] leads to a redefinition of the cell constant as

$$\beta' = \beta(1 - V_D/6V_T) \tag{5.31}$$

where V_D is the volume of the diaphragm and $V_T = V_B$ has been assumed.

For ternary systems, experiments are performed in such a way that steady state conditions have been established in the diaphragm prior to the start of the diffusion period. Combination of the ternary flow equations with the mass conservation equations

$$\frac{d(\Delta m_i)}{dt} = -J_i(t) A \left(\frac{1}{V_B} + \frac{1}{V_T}\right) \tag{5.32}$$

leads to the results

$$\frac{d^2 \Delta m_1}{dt^2} = \beta \left\{ D_{11} \frac{d\Delta m_1}{dt} + D_{12} \frac{d\Delta m_2}{dt} \right\} \tag{5.33}$$

$$\frac{d^2 \Delta m_2}{dt^2} = \beta \left\{ D_{12} \frac{d\Delta m_1}{dt} + D_{22} \frac{d\Delta m_2}{dt} \right\} \tag{5.34}$$

where the Δm_i are concentration differences between the cell compartments and are adopted because of the pseudosteady state description. The cell constant β must be evaluated by calibrating with a binary system for which absolute diffusion data is available. Common choices of calibration solutes are potassium chloride or urea.[101, 102]

Solutions to equations (5.33) and (5.34) are obtained for constant D_{ij} as (cf §4.6.2)

$$\begin{aligned} \Delta m_1 &= P_1 \exp(-\sigma_1 \beta t) + P_2 \exp(-\sigma_2 \beta t) \\ \Delta m_2 &= Q_1 \exp(-\sigma_1 \beta t) + Q_2 \exp(-\sigma_2 \beta t) \end{aligned} \tag{5.35}$$

where

$$\sigma_1 = \tfrac{1}{2}\{D_{22} + D_{11} + [(D_{22} - D_{11})^2 + 4D_{12}D_{21}]^{1/2}\} \tag{5.36}$$

$$\sigma_2 = \tfrac{1}{2}\{D_{22} + D_{11} - [(D_{22} - D_{11})^2 + 4D_{12}D_{21}]^{1/2}\} \tag{5.37}$$

$$P_1 = \frac{(D_{11} - \sigma_2)\,\Delta m_1^0 + D_{12}\Delta m_2^0}{(\sigma_1 - \sigma_2)} \tag{5.38}$$

$$P_2 = \frac{(D_{11} - \sigma_1)\,\Delta m_1^0 + D_{12}\Delta m_2^0}{\sigma_2 - \sigma_1} \tag{5.39}$$

$$Q_1 = \frac{D_{21}\Delta m_1^0 + (D_{22} - \sigma_2)\,\Delta m_2^0}{\sigma_1 - \sigma_2} \tag{5.40}$$

$$Q_2 = \frac{D_{21}\Delta m_1^0 + (D_{22} - \sigma_1)\,\Delta m_2^0}{\sigma_2 - \sigma_1} \tag{5.41}$$

and the Δm_i^0 are the initial concentration differences. To solve the above equations, which contain four unknowns, a minimum of four independent experiments must be undertaken and a non-linear least-squares procedure employed.[103] Because the D_{ij} are, in fact, functions of concentration, Δm_1 must be kept small and considerable precision of chemical analysis is required. The accuracy of the technique is claimed to be better than 0·5%.

As noted above, the diaphragm cell measurement is not a true steady state experiment as the compartment concentrations do in fact change with time. If, however, the concentrations can be held constant by continuously replacing the liquid in the reservoirs then a true steady state will be achieved. Since diffusion occurs a small change in concentration must result. It is this concentration change which is measured and used to calculate the diffusion flux. For small fluxes and large liquid replenishment rates the change in concentration is small and the error in the steady state approximation is not serious. This can be seen from inspection of the mass balance statement

$$-J_i A = \delta m_i \frac{dV}{dt} \tag{5.42}$$

where A is the effective cross-section of the diffusion barrier, dV/dt is the volumetric flow rate of liquid in the reservoir and δm_i is the resultant change in concentration. Provided that sufficiently accurate analytical methods are available, then it can be arranged that

$$|\delta m_i| \ll \Delta m_i$$

If the effective diffusive path length is x then the flux equations may be written with the aid of equation (5.42) as

$$\frac{\delta m_i \, dV}{dt} = \frac{A}{x} \sum_{ij} D_{ij} \Delta m_j \qquad (5.43)$$

and the cell constant $\beta = A/x$ can be evaluated by a calibration procedure. By holding all but one of Δm_j equal to zero it is possible to obtain a diffusion coefficient from each experiment. However, establishing the steady state condition is not easy. Methods have been described by Graff and Drew[104] and Rao and Bennett,[105] and the latter method continues in use.[106]

It should be recognized that many useful processes involving liquid diffusion simultaneously involve convection and circulation, so diffusion coefficients will appear in the theory of mass transfer in a composite disposition. For example, in stagnant film theories of convection mass transport, the transfer occurs across a surface film according to pure diffusion principles, but the film thickness is determined by fluid dynamical principles (see § 1.29 for an example).

The statement made in § 1.3 concerning the tendency for practical processes to involve diffusion and convection in complex juxtaposition is reiterated here. The general subject of mass transfer deals quantitatively and semi-empirically with such composite processes. This chapter is primarily concerned with theories of pure multicomponent diffusion and their tests against accurate experimentation. However, diffusion in molten slags, glasses and salts and in capillaries and fissures, as in membranes, physiological and geological structures is often free of convection so diffusion data is directly relevant to process rates.

5.8 ERRORS IN DIFFUSION COEFFICIENT MEASUREMENTS

Generally speaking, the expected error in currently reported diffusion coefficients is up to two magnitudes greater in solids than in transparent liquids, spanning at least the range of 20–0·1% probable error from solids to transparent liquids. The broad reasons for this are that diffusion in solids is structure sensitive, depending strongly on sample preparation, the coefficients are more temperature sensitive, experimental temperatures tend to be much higher for solids and therefore less accurately controlled, and the analytical techniques for solids are more subject to sectioning error, calibration errors and instrumental instability and drift. Divalent impurities in univalent salts and non-stoichiometry in oxides can cause huge errors.[4]

It is to be recognized that before substructure sensitivity for solids, and particularly metals, was well understood, diffusion coefficients reported by different laboratories varied by as much as an order of magnitude and this is reflected in values given in older compendia such as that of Smithells.[107]

Reproducibility between laboratories for binary mutual coefficients in solids remains relatively poor for the reasons mentioned, a figure of $\pm$ 30% being optimistic. Reproducibility for self-diffusion or tracer coefficients is somewhat better due to the high sensitivity of the tracer method, with a deviation range of $\pm$ 10% between laboratories and 3% within the same laboratory being optimistic estimates.[4] Because structure is not highly controlled in metal processing such errors are of little significance in diffusion calculations which estimate processing performance (e.g., carburization times). On the other hand, the structure of semiconductors is carefully controlled so accurate predictions on doping rates are feasible if the data is reliable. Here it is low temperature data that is particularly required and experimental accuracy is necessarily hard to achieve.

The lack of accuracy and reproducibility of mutual and tracer coefficients in solids becomes evident when one tries to test theory, such as that involved in the corrections to the Darken equations for $\tilde{D}$ (§6.4.4). The composite error and irreproducibility in coefficients of >20% completely masks the corrections, which are expected to be ~ 10% or less. Note that by contrast for transparent liquids, the experimental deviation of $\tilde{D}$ from the Darken equation has been accurately determined (Table 2.2) so the data are ripe for theoretical interpretation.

The high accuracy and reproducibility for transparent liquids stems not only from the absence of artifacts in the materials, processing and analysis, but also from the meticulous care taken in instrumental design, in calibration and interpretation by the small group of chemists who have specialized in this area. The Gouy diffusiometers and like optical instruments, as established within the diffusion industries located in Madison, Adelaide, Livermore and Canberra, were and are paragons of fine instrumental design and stability (cf Fig. 5.9). Data from these laboratories were not collected primarily for engineering application but for testing theories which emerged from the schools of Onsager and Kirkwood, particularly since 1946, and in this respect has served the cause of science well (Chapter 9). There is also a wider literature for diffusion in liquids such as reviewed and analysed by Hildebrand[108] and Ertl *et al.*[109, 110] in which accuracies approaching 1% are the norm. In only a few cases, for example Hg, have accurate values for liquid metal tracer diffusion been determined (see Fig. 2.9 and Protopappas *et al.*[111]) and accurate mutual diffusion results for liquid alloy melts are almost non-existent. Again, this dearth is not of great concern from the point of view of industrial design since estimates are usually sufficient in the planning of processes which will ultimately be optimized empirically.

Measurement of diffusion coefficients in polymers and molten slags and glasses carry large errors which approach those for solids (cf Figs. 2.12 and 2.13) for the reason that polymers have structure which cannot always be completely characterized, and high temperature molten slags must be quenched before composition profiles are evaluated in the solid state.

In the following chapter we investigate the variety of interactions which lead to interesting observations in ternary diffusion experiments.

REFERENCES

1 P. J. DUNLOP, B. J. STEEL, AND J. E. LANE: 'Experimental methods for studying diffusion in liquids, gases and solids', in 'Physical methods of chemistry, Vol. 1', (ed. A. Weissberger and B. W. Rossiter), 1972, New York, John Wiley

2 H. J. V. TYRRELL AND K. R. HARRIS: 'Diffusion in liquids', 1984, London, Butterworths

3 Y. ADDA AND J. PHILIBERT: 'La diffusion dans les solides', 1966, Paris, Presse Universitaires de France

4 S. J. ROTHMAN: in 'Diffusion in crystalline solids' (ed. G. E. Murch and A. S. Nowick), 1–61, 1984, Orlando FA, Academic Press

5 J. I. GOLDSTEIN, M. R. NOTIS, AND A. D. ROMIG JR: 'AIME symposium on atomic transport in concentrated alloys and intermetallic compounds', Sept. 1984, Detroit

6 C. ZENER: 'Elasticity and anelasticity of solids', 1948, Chicago University Press

7 J. ANDERSON AND K. SADDINGTON: *J. Chem. Soc.*, 1949, **S**, 381

8 J. H. WANG: *J. Am. Chem. Soc.*, 1952, **74**, 1182

9 C. P. SLICHTER: 'Principles of magnetic resonance', 2nd edn, 1980, New York, Springer

10 H. T. STOKES: in 'Nontraditional methods in diffusion', (ed. G. E. Murch, H. K. Birnbaum, and J. R. Cost), 39–58, 1984, Warrendale PA, AIME
11 A. S. NOWICK AND B. S. BERRY: 'Anelastic relaxation in crystalline solids', 1972, New York, Academic Press
12 A. S. NOWICK: in 'Physical accoustics', (ed. W. P. Mason and R. N. Thurston), 1–28, 1977, New York, Academic Press
13 B. S. BERRY AND W. C. PRITCHETT: *Phys. Rev. B*, 1981, **24**, 2299
14 J. R. COST: in 'Nontraditional methods in diffusion', (ed. G. E. Murch, H. K. Birnbaum, and J. R. Cost), 111–35, 1984, Warrendale PA, AIME
15 T. SPRINGER: 'Quasielastic neutron scattering for the investigation of diffusive motions in solids and liquids (Springer tracts in modern physics)', 1972, Berlin, Springer
16 H. ZABEL: in 'Nontraditional methods in diffusion', (ed. G. K. Murch, H. K. Birnbaum, and J. R. Cost), 1–37, 1984, Warrendale PA, AIME
17 H. ZABEL, A. MAGERL, A. J. DIANOUX, AND J. J. RUSH: *Phys. Rev. Lett.*, 1983, **50**, 2094
18 K. FUNKE: *Solid State Ionics*, 1981, **85**, 415
19 K. S. SINGWI AND A. SJÖLANDER: *Phys. Rev.*, 1960, **120**, 1093
20 S. L. RUBY, J. C. LOVE, P. A. FLINN, AND B. J. ZABRANSKY: *Appl. Phys. Lett.*, 1975, **27**, 320
21 T. WICHERT, G. GRÜBEL, AND M. DEICHER: in 'Solute-defect interaction: theory and experiment', (ed. S. Saimoto, G. R. Purdy, and G. V. Kidson, 75, 1986, Toronto, Pergamon
22 J. G. MULLIN: in 'Nontraditional methods in diffusion', (ed. G. K. Murch, H. K. Birnbaum, and J. R. Cost), 59–81, 1984, Warrendale PA, AIME
23 S. MANTL, W. PETRY, K. SCHROEDER, AND G. VOGL: *Phys. Rev. B*, 1983, **27**, 5315
24 D. WOLF: *Appl. Phys. Lett.*, 1977, **30**, 617
25 J. S. KIRKALDY: 'Advances in materials research, Vol. 4', (ed. H. Herman), 1970, New York, John Wiley
26 J. E. REYNOLDS, B. AVERBACH, AND M. COHEN: *Acta Met.*, 1957, **5**, 29
27 T. O. ZIEBOLD: Ph.D. Thesis, 1965, Cambridge, Mass., MIT
28 J. S. KIRKALDY: *Can. J. Phys.*, 1957, **35**, 435
29 L. S. DARKEN: *Trans. AIME*, 1949, **180**, 430
30 E. L. CUSSLER: 'Multicomponent diffusion', 72 *et seq.*, 1976, Amsterdam, Elsevier
31 R. J. SCHUT AND A. R. COOPER: *Acta Met.*, 1982, **30**, 1957
32 M. A. KRISHTAL, V. K. AKIMOV, A. P. MOKROV, AND P. N. ZAKHAROV: *Fiz. Met. Metalloved.*, 1974, **35**, 106
33 M. S. THOMPSON AND J. E. MORRAL: *Acta Met.*, 1986, **34**, 339
34 C. MATANO: *Japan Physics*, 1933, **8**, 109
35 L. BOLZMANN: *Ann. Phys. (Liepzig)*, 1894, **53**, 96
36 T. O. ZIEBOLD AND R. E. OGILVIE: *Trans. AIME*, 1967, **239**, 942
37 A. VIGNES AND J. P. SABATIER: *ibid.*, 1969, **245**, 1795
38 M. A. DAYANANDA: in Proc. AIME Symp. on 'Atomic transport in concentrated alloys and intermetallic compounds', Sept. 1984, Detroit
39 V. K. CHANDHOK, J. P. HIRTH, AND E. J. DULIS: *Trans. AIME*, 1962, **224**, 858
40 J. S. KIRKALDY, G. R. MASON, AND W. J. SLATER: *Trans. Can. Inst. Min. Met.*, 1961, **64**, 53
41 J. S. KIRKALDY, ZIA-UL-HAQ, AND L. C. BROWN: *Trans. ASM*, 1963, **56**, 834
42 J. S. KIRKALDY, R. J. BRIGHAM, AND D. H. WEICHERT: *Acta Met.*, 1965, **13**, 907
43 D. DE FONTAINE: *J. Phys. Chem. Solids*, 1972, **33**, 297
44 D. DE FONTAINE: *ibid.*, 1973, **34**, 1285
45 M. MURAKAMI, D. DE FONTAINE, J. M. SANCHEZ, AND J. FODOR: *Acta Met.*, 1974, **22**, 709
46 J. E. MORRAL AND J. W. CAHN: *Acta Met.*, 1971, **19**, 1037
47 M. MURAKAMI, D. DE FONTAINE, J. M. SANCHEZ, AND J. FODOR: *Thin Solid Films*, 1975, **25**, 465
48 K. MAIER AND W. SCHÜLE: *Euroatombericht*, 1974, EUR5234d
49 A. D. ROMIG JR, D. L. HUMPHREYS, J. I. GOLDSTEIN, AND M. R. NOTIS: 'Microbeam analysis', (ed. J. Armstrong), 1985, San Francisco Press

50 R. CASTAING: Thesis, 1951, University of Paris, France
51 R. THEISEN: 'Quantitative electron microbe analysis', 1965, Berlin, Springer-Verlag
52 S. J. B. REED: 'Electron microbe analysis', 1975, Cambridge University Press
53 'Practical scanning microscopy: electron and ion microprobe analysis', (ed. J. I. Goldstein and H. Yakowitz), 1975, New York, Plenum
54 'Physical aspects of electron microscopy and microbeam analysis', (ed. B. M. Siegel and D. R. Bearman), 1975, New York, John Wiley
55 I. G. SOLORZANO, G. R. PURDY, AND G. C. WEATHERLY: *Acta Met.*, 1984, **32**, 1709
56 C. NARAYAN AND J. I. GOLDSTEIN: *Met. Trans. A*, 1983, **14**, 2437
57 D. F. MITCHELL AND M. J. GRAHAM: *J. Vac. Technol.*, 1987, **A5**, 1258
58 J. W. COLBY: in 'Practical scanning microscopy', (ed. J. I. Goldstein and H. Yakowitz), 1975, New York, Plenum
59 M. P. MACHT AND V. NAUNDORF: *J. Appl. Phys.*, 1982, **53**, 7551
60 D. F. MITCHELL, R. J. HUSSEY, AND M. J. GRAHAM: 'Proc. JIMS3: High temperature corrosion', Trans. Jpn Inst. Met. Suppl., 121–5, 1983
61 J. P. BIERSACK AND D. FINK: Proc. 8th Symp. on 'Fusion technology', EUROATOM Report 5182e, 1974
62 W. MÖLLER, M. HUFSCHMIDT, AND T. PFEIFFER: *Nucl. Inst. Methods*, 1978, **149**, 73
63 W. MÖLLER, B. M. V. SCHERZER, AND R. BEHRISCH: *ibid.*, 1980, **168**, 289
64 S. M. MYERS: in 'Nontraditional methods in diffusion', (ed. G. E. Murch, H. K. Birnbaum, and J. R. Cost), 137–54, 1984, Warrendale PA, AIME
65 S. HOFMANN AND J. ERLEWEIN: *Scripta Met.*, 1976, **10**, 857
66 B. COX AND J. P. PEMSLER: *J. Nucl. Mat.*, 1968, **28**, 73
67 J. S. SHEASBY AND W. W. SMELTZER: *Oxid. Met.*, 1981, **15**, 215
68 J. DESMAISON AND W. W. SMELTZER: *J. Electrochem. Soc.*, 1975, **122**, 354
69 Y. OISHI AND W. D. KINGERY: *J. Chem. Phys.*, 1960, **33**, 480
70 H. YINNON AND A. R. COOPER: *Phys. Chem. Glasses*, 1980, **21**, 204
71 K. P. R. REDDY, S. M. OH, L. D. MAJOR, AND A. R. COOPER JR: *J. Geophys. Res.*, 1980, **85**, 322
72 W. A. LANFORD, R. BENENSON, C. BURMAN, AND L. WIELUNSKI: in 'Nontraditional methods in diffusion', (ed. G. E. Murch, H. K. Birnbaum, and J. R. Cost), 155–78, 1984, Warrendale PA, AIME
73 R. L. LEVIN AND J. B. WAGNER: *Trans. AIME*, 1965, **223**, 159
74 J. B. PRICE AND J. B. WAGNER JR: *Z. Phys. Chem. Neue Folge*, 1966, **49**, 257
75 K. WEISS: *Ber. Bunsenges.*, 1969, **73**, 338
76 A. B. NEUMAN: *Trans. Am. Inst. Chem. Eng.*, 1931, **27**, 203
77 A. J. ROSENBURG: *J. Electrochem. Soc.*, 1960, **107**, 795
78 E. M. FRYT: *Oxid. Met.*, 1978, **12**, 139
79 S. MROWEC: in 'Metal–slag–gas reactions and processes', (ed. Z. A. Foroulis and W. W. Smeltzer), 414, 1975, Princeton NJ, Electrochem. Soc.
80 E. M. FRYT, W. W. SMELTZER, AND J. S. KIRKALDY: *J. Electrochem. Soc.*, 1979, **126**, 673
81 F. GESMUNDO, F. VIANI, AND V. DOVI: *Oxid. Met.*, 1985, **23**, 141
82 H. MEHRER: in 'Solute-defect interaction: theory and experiment', (ed. S. Saimoto, G. R. Purdy, and G. V. Kidson, 162, 1986, Toronto, Pergamon
83 L. J. GOSTING: in 'Advances in protein chemistry, Vol. XI', 1956, New York, Academic Press
84 L. J. GOSTING AND H. FUJITA: *J. Am. Chem. Soc.*, 1957, **79**, 1359
85 L. J. GOSTING, H. KIM, M. A. LOEWENSTEIN, G. REINFELDS, AND A. REVSIN: *Rev. Sci. Instrum.*, 1973, **44**, 1602
86 J. G. ALBRIGHT AND B. C. SHERRILL: *J. Solution Chem.*, 1979, **8**, 201
87 D. G. MILLER: *ibid.*, 1981, **10**, 831
88 J. A. RARD AND D. G. MILLER: *ibid.*, 1979, **8**, 701
89 J. A. RARD AND D. G. MILLER: *ibid.*, 755
90 J. A. RARD AND D. G. MILLER: *J. Chem. Eng. Data*, 1980, **25**, 211
91 H. FUJITA AND L. J. GOSTING: *J. Am. Chem. Soc.*, 1956, **78**, 1099

92 I. J. HERBERT, F. D. CARLSON, AND H. Z. CUMMINS: *Biophys. J.*, 1968, **8**, A95
93 B. CHU AND F. J. SCHOENES: *Phys. Rev. Lett.*, 1968, **21**, 6
94 A. J. HYDE: in 'Diffusion processes', (ed. J. N. Sherwood, A. V. Chadwick, W. M. Muir, and F. L. Swinton), 115 *et seq.*, 1971, London, Gordon and Breach
95 E. GULARI, Y. TSUNASHIMA, AND B. CHU: *Polymer*, 1979, **20**, 347
96 L. YANG AND M. T. SIMNAD: in 'General aspects of physicochemical research at high temperatures', (ed. J. O'M. Bockris, J. L. White, and J. D. Mackenzie), 295, 1959, New York, Academic Press
97 H. SUGAWARA, K. NAGATA, AND K. S. GOTO: *Met. Trans. B*, 1977, **8**, 605
98 Y. UKYO AND K. S. GOTO: *ibid.*, 1981, **12**, 449
99 A. R. GORDON: *Ann. N.Y. Acad. Sci.*, 1945, **46**, 285
100 C. BARNES: *Physics*, 1934, **5**, 4
101 R. MILLS: *J. Phys. Chem.*, 1963, **67**, 600
102 K. R. HARRIS, C. PUA, AND P. J. DUNLOP: *ibid.*, 1970, **74**, 3518
103 E. L. CUSSLER AND P. J. DUNLOP: *ibid.*, 1966, **70**, 1880
104 R. A. GRAFF AND T. B. DREW: *Ind. Eng. Chem. Fundam.*, 1968, **7**, 490
105 S. S. RAO AND C. O. BENNETT: *J.A.I.Ch.E.*, 1971, **17**, 75
106 C. ALVEREZ-FUSTER, N. MIDOUX, A. LAURENT, AND J. C. CHARPENTIER: *Chem. Eng. Sci.*, 1981, **36**, 1513
107 C. J. SMITHELLS: 'Metals reference book', 654–6, 1967, London, Butterworths
108 J. H. HILDEBRAND: 'Viscosity and diffusivity, a predictive treatment', 1977, New York, John Wiley
109 H. ERTL AND F. A. L. DULLIEN: *J.A.I.Ch.E.*, 1973, **19**, 1215
110 H. ERTL, R. K. GHAI, AND F. A. L. DULLIEN: *ibid.*, 1974, **20**, 1
111 P. PROTOPAPPAS, H. C. ANDERSON, AND N. A. D. PARLEE: *J. Chem. Phys.*, 1973, **59**, 15

CHAPTER 6

Origins of species cross-effects in multicomponent diffusion

6.1 INTRODUCTION

In this chapter we explore in a general way the origins of correlation or cross-effects in multicomponent diffusion such as presented in Figs. 5.2 and 5.6. As shown in Chapter 4, the diffusion matrix is given by

$$D = L\mu \tag{6.1}$$

where L is the matrix of phenomenological coefficients and μ is the matrix of chemical potential derivatives with respect to concentration (equation (4.32)). It is clear that the existence of non-zero off-diagonal terms in either one of the matrixes L or μ will lead to the appearance of non-zero cross-terms in the diffusion matrix. That is to say, kinetic interactions, as represented by the Onsager coefficients, or thermodynamic interactions, as represented by the variation of chemical potential with composition, lead to diffusional cross-effects. The Onsager cross-effects can be associated with correlations due to electrostatic or point defect interactions. In dilute solutions the relations allow good estimates of the ternary coefficients from the binary coefficients and the ternary interaction parameter (§6.3).

6.2 EFFECT OF REFERENCE FRAME; INTRINSIC DIFFUSION BEHAVIOUR

It will be recalled that many reference frames are defined in terms of conservation equations (§4.3) which express linear dependencies among the fluxes. In particular, for the experimentally convenient volume-fixed frame

$$\sum_i^n \overline{V}_i (J_i)_V = 0 \tag{6.2}$$

An exception is the intrinsic or Kirkendall frame where no such relationship is supposed to exist *a priori* among the $(J_i)_K$ (cf §2.13). Indeed, it is this lack of balance among the intrinsic diffusive flows which, as discussed earlier, leads to a compensating bulk flow of material. We now proceed to demonstrate that this compensatory flow constitutes an interaction or cross-effect as observed in the experimental diffusion frame.

The relationship between the values of a flux expressed in the volume-fixed and intrinsic frames in terms of the relative velocity, v_{KV}, is

$$(J_i)_V = (J_i)_K + m_i v_{KV} \tag{6.3}$$

where the subscript K denotes intrinsic. For the purposes of the demonstration, it is supposed that no interactions exist in the intrinsic frame

$$(J_i)_K = -(L_{ii})_K \nabla\mu_i \tag{6.4}$$

which is consistent with elementary transition state theory (§7.2). The matrix $[L]_K$ is thus supposed to be diagonal and, of course, symmetric. The effect on the flux equations of changing reference frame is shown by equations (4.20)–(4.24) which represent a linear transformation. As such, the transformation must lead to a new matrix of coefficients which is itself symmetric[1]

$$(L'_{ij})_V = (L'_{ji})_V \tag{6.5}$$

Since, however, the coefficients are changed to new values

$$(L'_{ij})_V \neq (L_{ij})_K$$

it follows that the transformation of the diagonal $[L_K]$ matrix yields a new matrix $[L'_V]$ in which $(L'_{ij})_V \neq 0$. As seen from equation (6.1), these non-zero off-diagonal terms will be propagated into the D-matrix. Thus cross-effects are generated by the nature of the transformation and the conservation law expressed by equation (6.2). Examples of the application of this theorem have been given by Kirkaldy[2] and Kirkwood *et al.*[3] An interesting and transparent case is a ternary substitutional solution, either solid or liquid, which has been treated by Lane and Kirkaldy.[4, 5]

Flux equations appropriate to the volume-fixed reference frame with the approximations $\overline{V}_i/\overline{V}_3 = 1$ are

$$(J_1)_V = -(L_{11})_V \Delta(\mu_1 - \mu_3) - (L_{12})_V \nabla(\mu_2 - \mu_3) \tag{6.6}$$

$$(J_2)_V = -(L_{21})_V \Delta(\mu_1 - \mu_3) - (L_{22})_V \nabla(\mu_2 - \mu_3) \tag{6.7}$$

Since for $\overline{V}_i = \overline{V}_3$, equation (6.2) becomes

$$\sum_i^n (J_i)_V = 0 \tag{6.8}$$

which together with equation (6.3) yields

$$v_{KV} = -\Sigma\,(J_i)_K/\Sigma\, m_i \tag{6.9}$$

Expressing the intrinsic fluxes by means of equation (6.4), i.e. with no cross-terms, yields an expression for the reference frame velocity

$$m_i v_{KV} = X_i\{(L_{11})_K \nabla\mu_1 + (L_{22})_K \nabla\mu_2 + (L_{33})_K \nabla\mu_3\} \tag{6.10}$$

Substitution of these quantities in equations (6.3) yields the set

$$(J)_V = -[L'_V]\,\nabla\mu \tag{6.11}$$

where the matrix of coefficients is given by

$$[L'_V] = \begin{bmatrix} -(L_{11})_K(X_1-1) & -(L_{22})_K X_1 & -(L_{33})_K X_1 \\ -(L_{11})_K X_2 & -(L_{22})_K(X_2-1) & -(L_{33})_K X_2 \\ -(L_{11})_K X_3 & -(L_{22})_K X_3 & -(L_{33})_K(X_3-1) \end{bmatrix} \tag{6.12}$$

It is seen that cross-effects in the $[L'_V]$ matrix have arisen through the requirement expressed in equations (6.2) and (6.3) that each flux must balance the sum of all the others. By considering all fluxes we have, however, constructed a set which contains a linear dependence and in consequence the matrix $[L'_V]$ is not unique. It is therefore reduced to a 2×2 matrix by requiring that the fluxes $(J_1)_V$, $(J_2)_V$ be given in terms of the correct gradients as in equations (6.6) and (6.7).

The required reduction is achieved with the aid of the Gibbs–Duhem relationship written with a constant multiplier

$$A_i \sum_j X_j \frac{\partial \mu_j}{\partial x} = 0 \qquad i = 1, 2, 3 \tag{6.13}$$

or, in matrix notation,

$$[A_i X_j] \frac{\partial \mu}{\partial x} = 0 \tag{6.14}$$

Addition of equation (6.14) to equation (6.11) therefore leaves the latter unchanged but leads to the formation of a new matrix

$$[L''_V] = \begin{bmatrix} -(L_{11})_K(X_1-1)+A_1X_1 & -(L_{22})_K X_1 + A_1X_2 & -(L_{33})_K X_1 + A_1X_3 \\ -(L_{11})_K X_2 + A_2X_1 & -(L_{22})_K(X_2-1)+A_2X_2 & -(L_{33})_K X_2 + A_2X_3 \\ -(L_{11})_K X_3 + A_3X_1 & -(L_{22})_K X_3 + A_3X_2 & -(L_{33})_K(X_3-1)+A_3X_3 \end{bmatrix} \tag{6.15}$$

The redundancy is now removed by insisting that the equations be of the form of equations (6.6) and (6.7). This requires that

$$L''_{13} = -L''_{11} - L''_{12} \tag{6.16}$$

$$L''_{23} = -L''_{21} - L''_{22} \tag{6.17}$$

which have the unique solutions

$$A_1 = X_1 \sum_i (L_{ii})_K - (L_{11})_K \tag{6.18}$$

$$A_2 = X_2 \sum_i (L_{ii})_K - (L_{22})_K \tag{6.19}$$

Substitutions for A_1, A_2 in equation (6.15) yields the unique matrix

$$[L_V] = \begin{bmatrix} (L_{11})_K(1-2X_1) + (X_1)^2 \sum_i (L_{ii})_K & -X_2(L_{11})_K - X_1(L_{22})_K + X_1X_2 \sum_i (L_{ii})_K \\ -X_2(L_{11})_K - X_1(L_{22})_K + X_1X_2 \sum_i (L_{ii})_K & (L_{22})_K(1-2X_2) + (X_2)^2 \sum_i (L_{ii})_K \end{bmatrix} \tag{6.20}$$

which is symmetric and contains off-diagonal terms which are, in general, non-zero. Reference to equation (6.1) shows that even if the μ-matrix contains zero off-diagonal terms, the resultant D-matrix will show non-zero cross-effects (cf §7.4).

The discussion above shows that even when intrinsic diffusion is entirely free of correlations, either kinetic or thermodynamic, a diffusional cross-effect is observed in the volume-fixed reference frame. These formal cross-effects reflect unequal species intrinsic diffusion coefficients together with the conservation of atomic volume as expressed by equation (6.2). This kinetic effect is a generalization of that already discussed for binary systems, where it is found that the movement of one species brings about a convective flow of the other, particularly when the latter's intrinsic mobility approaches zero (cf §2.13).

6.3 THERMODYNAMIC INTERACTIONS IN THE ABSENCE OF ONSAGER EFFECTS

Elements of the thermodynamic matrix are given by equation (4.32)

$$\mu_{ij} = \frac{\partial(\mu_i - \overline{V}_i/\overline{V}_n\mu_n)}{\partial m_j} \tag{6.21}$$

indicating that the matrix is, in general, non-symmetric. In order to obtain a symmetric form, we change from mole numbers to mole fractions, X_i. A general treatment including volume change has been given by Kirkaldy *et al.*[6] but for simplicity we present here the derivation for the special case where all molar volumes are equal (see also Miller *et al.*[7]).

The diffusion equations are restated in terms of mole fractions as

$$J_i = -\sum_{j}^{n-1} D_{ij}\, m \operatorname{grad} X_j \tag{6.22}$$

which are to be compared with the phenomenological equations

$$J_i = -\sum_{j}^{n-1} L_{ij} \operatorname{grad}(\mu_j - \mu_n) \tag{6.23}$$

where the subscript on the L-matrix has been dropped. Noting that through the equilibrium relationship, analogous to equation (4.30), we have

$$\mu_i = \mu_i(X_1, X_2 \ldots X_{n-1}) = \mu_i(n_1, n_2 \ldots n_n) \tag{6.24}$$

where the n are mole numbers and applying the argument of equations (4.31)–(4.36) we obtain the relationship between the D-matrix, as determined on a mole fraction scale, and the L matrix

$$m[D] = [L][\overline{\mu}] \tag{6.25}$$

The thermodynamic matrix $[\overline{\mu}]$ is defined in terms of its elements[8]

$$\overline{\mu}_{ij} = \frac{\partial(\mu_i - \mu_n)}{\partial X_j} \tag{6.26}$$

and is next demonstrated to be symmetric.

Each of the differentials appearing in equation (6.26) may be written as

$$\frac{\partial \mu_j}{\partial X_k} = \sum_i^n \frac{\partial \mu_j}{\partial n_i}\frac{\partial n_i}{\partial X_k} \tag{6.27}$$

Since

$$n_i = X_i n \qquad i = 1, \ldots (n-1) \tag{6.28}$$

$$n_n = \left(1 - \sum_i^{n-1} X_i\right) n \tag{6.29}$$

and n is held constant during differentiation, only two of the terms on the right hand side of equation (6.27) are non-zero, namely

$$\frac{\partial \mu_l}{\partial X_k} = \frac{\partial \mu_l}{\partial n_k} n - \frac{\partial \mu_l}{\partial n_n} \tag{6.30}$$

Noting that by the rules of exact differentiation

$$\frac{\partial \mu_l}{\partial n_k} = \frac{\partial \mu_k}{\partial n_l} \tag{6.31}$$

it is readily verified from equation (6.29) that

$$\frac{\partial (\mu_j - \mu_n)}{\partial X_k} = \frac{\partial (\mu_k - \mu_n)}{\partial X_j} \tag{6.32}$$

that is to say, the matrix $[\bar{\mu}]$ is symmetric. We now consider the nature of these derivatives.

Investigation of the thermodynamic matrix requires a description of the solution under discussion. A suitably general description was provided by Wagner[9] for dilute solutions wherein the logarithm of the activity coefficient is represented by a Taylor series expansion in the mole fractions about the state of infinite dilution. In the linear approximation for component i in solvent n (Appendix 1)

$$\ln \gamma_i = \ln \gamma_i^0 + \sum_{j=1}^{n-1} \varepsilon_{ij} X_j \tag{6.33}$$

where γ_i is the activity coefficient and γ_i^0 its value at infinite dilution. Applying equation (6.33) to the appropriate defining equation for chemical potential

$$\mu = \mu^0 + RT \ln \gamma X \tag{6.34}$$

and substituting the result in the expression for $\bar{\mu}_{ik}$ leads to[10, 11]

$$\bar{\mu}_{ij} = RT\varepsilon_{ij} \qquad (i \neq j) \qquad \varepsilon_{ij} = \varepsilon_{ji} \tag{6.35}$$

$$\bar{\mu}_{ii} = RT\left(\frac{1}{X_i} + \varepsilon_{ii}\right) \tag{6.36}$$

The application of these results to a ternary system using equations (6.25), (6.35), and (6.36) yields

$$\begin{aligned} D_{11} &= RT\{L_{11}(\varepsilon_{11} + 1/X_1) + L_{12}\varepsilon_{21}\} \\ D_{12} &= RT\{L_{11}\varepsilon_{12} + L_{12}(\varepsilon_{22} + 1/X_2)\} \\ D_{21} &= RT\{L_{22}\varepsilon_{21} + L_{21}(\varepsilon_{11} + 1/X_1)\} \\ D_{22} &= RT\{L_{22}(\varepsilon_{22} + 1/X_2) + L_{21}\varepsilon_{12}\} \end{aligned} \tag{6.37}$$

In an ideal solution all $\varepsilon_{ij} = 0$ and the D_{ij} reduce to the purely kinetic form

$$D_{ij} = RT\,L_{ij}/X_j \tag{6.38}$$

For real solutions, if no kinetic cross-effects occur, i.e. $L_{ij}\,(i \neq j) = 0$, it is clear that the diffusional cross-terms are non-zero. In this case the dilute solution limit $(X_1, X_2 \rightarrow 0)$ may be described by

$$D_{12}/D_{11} = X_1\varepsilon_{12} \tag{6.39}$$

and

$$D_{21}/D_{22} = X_2\varepsilon_{21} \tag{6.40}$$

Thus the ternary coefficients are determined uniquely by the binary ones in dilute solutions. For interstitial diffusion there are negligible correlations between sublattices so that the approximation $L_{ij}(i \neq j) = 0$ is valid. Furthermore such solutions are always dilute so the thermodynamic interactions described by equations (6.39) and (6.40) are dominant and accurate[10] (refer also to §7.9).

We note finally that where there exists complexing or association between components of a solution the thermodynamic effects are exceptionally strong and such solutions can exhibit diffusion cross-effects even when very dilute (cf §8.12 on semiconductors and §10.4 on carrier-mediated transport in membranes).

6.4 CORRELATION EFFECTS IN INTERDIFFUSION OF CONSERVED VACANCIES AND A BINARY ALLOY

A substantial and difficult literature has built up over the past thirty years concerning the relationship of Onsager cross-effects to correlation effects in diffusion. As an application of the theory, attempts have been made within a multicomponent formalism to calculate corrections to the Darken equation due to these correlation effects (see §3.11). Directly or indirectly, all of the quantitative developments are related to a 'thought' experiment conceived of by Howard and Lidiard.[12] This concerns the non-equilibrium isotopic alloy A–A*–V where V represents the vacancies in a source-sink free crystal. The corresponding A–B–V experiment with conserved vacancies is also of great theoretical interest. We will return to the A–B–V problem following a detailed analysis of A–A*–V interdiffusion.

6.4.1 Phenomenology of ternary diffusion in a non-equilibrium A–A*–V system

We here consider the ideal of one-dimensional diffusion in a single crystal which is free of surface sources or sinks and dislocations.[13] This is accordingly an inherently non-local equilibrium diffusion situation. Species number conservation demands that the fluxes satisfy

$$J_A + J_{A^*} + J_V = 0 \tag{6.41}$$

in the lattice frame of reference, and the mole fractions satisfy

$$X_A + X_{A^*} + X_V = 1 \tag{6.42}$$

The appropriate independent flux equations are

$$J_A = -L_{AA}\left(\frac{\partial \mu_A}{\partial x} - \frac{\partial \mu_V}{\partial x}\right) - L_{AA^*}\left(\frac{\partial \mu_{A^*}}{\partial x} - \frac{\partial \mu_V}{\partial x}\right) \tag{6.43}$$

and

$$J_{A^*} = -L_{A^*A}\left(\frac{\partial \mu_A}{\partial x} - \frac{\partial \mu_V}{\partial x}\right) - L_{A^*A^*}\left(\frac{\partial \mu_{A^*}}{\partial x} - \frac{\partial \mu_V}{\partial x}\right) \tag{6.44}$$

subject to Onsager reciprocity. The chemical potential of the vacancies is adequately represented by

$$\mu_V = g(T) + kT \ln X_V \tag{6.45}$$

Proceeding in strict accord with conservation of vacancies as per relations (6.41) and (6.42) implying that $\nabla \mu_V$ is not everywhere zero, we refer to relations (6.43) and (6.44) and note that since the L-matrix is independent of gradients we can quantify several limiting cases using the weak and spontaneously achievable conditions, ∇X_A or $\nabla X_{A^*} = 0$. First, according to equation (6.41) and the chemical equivalence of A and A* we can write in terms of site density N

$$\begin{aligned} J_V &= L_{AA}\nabla(\mu_A - \mu_V) + L_{AA^*}\nabla(-\mu_V) + L_{A^*A}\nabla(\mu_A - \mu_V) + L_{A^*A^*}\nabla(-\mu_V) \\ &= -D_V N \nabla X_V \end{aligned} \tag{6.46}$$

and

$$\begin{aligned} J_V &= L_{AA}\nabla(-\mu_V) + L_{AA^*}\nabla(\mu_{A^*} - \mu_V) + L_{A^*A}\nabla(-\mu_V) + L_{A^*A^*}\nabla(\mu_{A^*} - \mu_V) \\ &= -D_V N \nabla X_V \end{aligned} \tag{6.47}$$

evaluated in the limit of identical constant composition, i.e. vanishingly small concentration differences

$$\nabla X_A \quad \text{or} \quad \nabla X_{A^*} = -\nabla X_V \tag{6.48}$$

Hence substituting from equation (6.48) rearranging and cancelling ∇X_V ($\neq 0$) one obtains

$$(L_{AA} + L_{A^*A})\left(\frac{1}{X_A} + \frac{1}{X_V}\right) + (L_{A^*A^*} + L_{AA^*})\frac{1}{X_V} = D_V N/kT \tag{6.49}$$

and

$$(L_{AA} + L_{A^*A})\frac{1}{X_V} + (L_{A^*A^*} + L_{AA^*})\left(\frac{1}{X_{A^*}} + \frac{1}{X_V}\right) = D_V N/kT \tag{6.50}$$

or rationalizing via equation (6.42)

$$(L_{AA} + L_{A^*A})\frac{(1 - X_{A^*})}{X_A X_V} + \frac{(L_{A^*A^*} + L_{AA^*})}{X_V} = D_V N/kT \tag{6.51}$$

and

$$\frac{(L_{AA} + L_{A^*A})}{X_V} + (L_{A^*A^*} + L_{AA^*})\frac{(1 - X_A)}{X_{A^*}X_V} = D_V N/kT \tag{6.52}$$

Now eliminating $D_V N/kT$ between equations (6.51) and (6.52) and rearranging yields

$$L_{AA}\frac{(1 - X_A - X_{A^*})}{X_A} + L_{A^*A}\frac{(1 - X_A - X_{A^*})}{X_A}$$

$$= L_{A^*A^*}\frac{(1 - X_A - X_{A^*})}{X_{A^*}} + L_{AA^*}\frac{(1 - X_A - X_{A^*})}{X_{A^*}} \tag{6.53}$$

or

$$\frac{L_{AA} + L_{A^*A}}{X_A} = \frac{L_{A^*A^*} + L_{AA^*}}{X_{A^*}} \tag{6.54}$$

Interestingly, equation (6.54) is identical to the local equilibrium result obtained by setting $\mu_V = 0$, $-J_A = J_{A^*}$, and $X_A \Delta\mu_A = -X_{A^*}\Delta\mu_{A^*}$ in equations (6.43) and (6.44).

Now substituting equation (6.54) into equations (6.51) and (6.52) yields the simplifications

$$L_{AA} + L_{A^*A} = X_A X_V D_V N/kT \quad \text{and} \quad L_{A^*A^*} + L_{AA^*} = X_{A^*} X_V D_V N/kT \tag{6.55}$$

or combining

$$L_{AA} + L_{AA^*} + L_{A^*A} + L_{A^*A^*} \simeq X_V N D_V/kT \tag{6.56}$$

Returning attention now to relations (6.43) and (6.44) we reconsider the two countercurrent conditions and write these equations in the form

$$-L_{AA}\nabla(\mu_A - \mu_V) + L_{AA^*}\nabla\mu_V = g_A D_V N \nabla X_V \qquad \nabla X_A = -\nabla X_V \tag{6.57}$$

and

$$L_{A^*A}\nabla\mu_V - L_{A^*A^*}\nabla(\mu_{A^*} - \mu_V) = g_{A^*} D_V N \nabla X_V \quad \nabla X_{A^*} = -\nabla X_V \tag{6.58}$$

defining $g_A = -J_A/J_V$ and $g_{A^*} = -J_{A^*}/J_V$ ($J_V \neq 0$), respectively. At this point we can directly evaluate the two binary limits

$$g_A(X_{A^*} = 0) = g_{A^*}(X_A = 0) = 1 \tag{6.59}$$

If we take Onsager reciprocity as an *a priori* condition then we possess five conditions on six coefficients together with the weak condition (6.59). Alternatively, if through equations (6.59) and correct kinetic arguments we evaluate g_A and g_{A^*} directly then the four L_{ij} are specified independently and Onsager reciprocity appears within the argument. For shorthand from this point let

$$L_{ik} \equiv \frac{L_{ik} kT}{D_V N} \tag{6.60}$$

so

$$L_{AA} + L_{A^*A} = X_A X_V \tag{6.61}$$

and

$$L_{A^*A^*} + L_{AA^*} = X_{A^*}X_V \tag{6.62}$$

and from equations (6.57) and (6.58) cancelling ∇X_V ($\neq 0$), and normalizing via equation (6.60),

$$L_{AA}\frac{(1 - X_{A^*})}{X_A} + L_{AA^*} = g_A X_V \tag{6.63}$$

and

$$L_{A^*A} + L_{A^*A^*}\frac{(1 - X_{A^*})}{X_{A^*}} = g_{A^*}X_V \tag{6.64}$$

Now eliminating the cross-terms in equations (6.63) and (6.64) via equations (6.61) and (6.62) we can solve simultaneously to obtain

$$L_{AA} = X_A\overline{X}_A \qquad L_{A^*A^*} = X_{A^*}\overline{X}_{A^*} \qquad L_{A^*A} = X_A(X_V - \overline{X}_A)$$
$$L_{AA^*} = X_{A^*}(X_V - \overline{X}_{A^*}) \tag{6.65}$$

where

$$\begin{aligned}\overline{X}_A &= (g_A - X_{A^*})(1 - X_A) + (g_{A^*} - X_A)X_{A^*}\\ \overline{X}_{A^*} &= (g_{A^*} - X_A)(1 - X_{A^*}) + (g_A - X_{A^*})X_A\end{aligned} \tag{6.66}$$

Onsager reciprocity via equations (6.65) specifies

$$X_A(1 - X_A - X_{A^*} - \overline{X}_A) = X_{A^*}(1 - X_A - X_{A^*} - \overline{X}_{A^*}) \tag{6.67}$$

which via equations (6.66) yields

$$X_A(1 - g_A) = X_{A^*}(1 - g_{A^*}) \tag{6.68}$$

Together with equations (6.59), this is highly suggestive of the form which g_A and g_{A^*} must take. Indeed, since X_A and X_{A^*} are independent equation (6.68) can only be satisfied if

$$1 - g_A = h(X_A, X_{A^*})X_{A^*} \qquad \text{and} \qquad 1 - g_{A^*} = h(X_A, X_{A^*})X_A \tag{6.69}$$

or

$$g_A = 1 - hX_{A^*} \qquad \text{and} \qquad g_{A^*} = 1 - hX_A \tag{6.70}$$

where h is invariant under the interchange $X_A \longleftrightarrow X_{A^*}$. Note that equation (6.70) verifies the limiting relations (6.59).

6.4.2 Kinetic evaluation of the Onsager *L*-coefficients

Turning to the kinetics we refer to the discussion of solvent and tracer correlation effects in § § 2.5 and 2.6 and are reminded that rigorous correlation calculations refer to a single tracer atom and a single vacancy (see also Kirkaldy *et al.*[13]).

Let us now consider, as in § 2.6, an experiment with countercurrent pure A–V and identify a fraction X_V of A atoms each in contact with a vacancy at time zero, and in particular, the individual in the vacancy's coordination sphere which jumps first. In principle, we can always retain the identity of these individuals (which we can designate as $A^\dagger$) by noting their subsequent trajectories with respect to their associated vacancies and attribute a correlation factor f to each of them; i.e., the tracer argument applies to each individual and we may identify an uncorrelated fraction of its jumps as f. Since the mole fraction of $A^\dagger$ individuals is X_V the overall uncorrelated fraction of jumps associated with an array of A atoms is $\phi = X_V f$.

Equivalently, the fraction of correlated A motions in very dilute A–V counter-diffusion is $1 - X_V f$ (§2.5). Now for a finite A–A* mixture with A or A* uniform and A* or A countercurrent to V the fraction of the vacancy flux or 'wind' intercepted by A or A* will be $1 - X_V f$ multiplied by the proportion of X_A or X_{A^*} to $X_A + X_{A^*}$, respectively.

Next, consider the evaluation of the ratio of fluxes $g_A = -J_A/J_V$ (which approaches unity for pure A) as uniform A* is added to the mixture. The general relation in accord with the above is

$$g_A = 1 - \frac{X_{A^*}}{X_A + X_{A^*}}(1 - X_V f) \tag{6.71}$$

which evaluates the fractional vacancy availability for the J_A flux by subtracting from unity the fraction of vacancies countercurrent to uniform A* motions. Rewriting this as

$$g_A = \frac{X_A}{X_A + X_{A^*}} + \frac{X_{A^*} X_V f}{X_A + X_{A^*}} \tag{6.72}$$

one recognizes the correction to a pure countercurrent motion ($g_A = X_A/(X_A + X_{A^*})$) as being positive and associated with the frequency of occurrence of A*–V pairs. Note also that in the limit $X_{A^*} \rightarrow 0$, J_A approaches perfect countercurrent motion with the vacancies, and in the limit $X_A \ll X_V \ll X_{A^*}$ (A is here the tracer)

$$g_A = X_V f = D_A^*/D_V \tag{6.73}$$

which is consistent with the conventional random walk interpretation of a tracer diffusion measurement (cf equation (2.18)). We emphasize here the requirement that X_A and X_A^* must both remain non-zero if the symmetry of the structure is to be sustained. Equivalently, non-zero initial increments ΔX_A and ΔX_{A^*} must be attainable at approximately constant composition. This requirement explains why g_A or g_{A^*} can reach constant values in the approach to infinite dilution of A or A* (e.g. equation (6.73)).

Symmetrically we identify

$$g_{A^*} = 1 - \frac{X_A}{X_A + X_{A^*}}(1 - X_V f) \tag{6.74}$$

so we deduce from equation (6.69) that

$$h = \frac{(1 - f X_V)}{X_A + X_{A^*}} \tag{6.75}$$

which possesses the symmetry property required by Onsager reciprocity. Note that the principle of similitude whereby we obtained equation (6.74) from equation (6.71) is equivalent to the statement that the probabilities of an A* atom blocking an A movement and of an A atom blocking an A* movement are equal.

Indeed, we are invoking detailed balance in the form

$$\frac{(1 - X_V f) X_{A^*}}{X_A + X_{A^*}} \cdot X_A = \frac{(1 - X_V f) X_A}{X_A + X_{A^*}} \cdot X_{A^*} \tag{6.76}$$

which is identical with the reciprocity-generated relation (6.68).

Finally, combining equations (6.71) and (6.74) with equations (6.65) and (6.66) we obtain the matrix of L-coefficients

$$\begin{aligned} L_{AA} &= \frac{D_V N}{kT} \frac{X_A X_V}{X_A + X_{A^*}} [1 - X_V - X_{A^*}(1 - f)] \\ L_{A^*A^*} &= \frac{D_V N}{kT} \frac{X_{A^*} X_V}{X_A + X_{A^*}} [1 - X_V - X_A(1 - f)] \end{aligned} \tag{6.77}$$

and

$$L_{AA^*} = L_{A^*A} = \frac{D_V N}{kT} \frac{X_A X_{A^*} X_V}{X_A + X_{A^*}} (1 - f) \tag{6.78}$$

which with trivial rearrangement are identical to a result of Howard and Lidiard[12] obtained via a different argument. These clearly satisfy the second law requirements expressed by relations (3.86).

6.4.3 Solutions of the Fick equations

The ultimate test of the formalism rests on its ability to describe actual ternary diffusion processes with varying initial and boundary conditions. Here we seek solutions of the ternary versions of the Fick equations (4.67) for an infinite diffusion couple. In evaluating the concentration profiles we require the trace of the matrix (cf §4.6)

$$t = D_{AA} + D_{A^*A^*} \tag{6.79}$$

the determinant

$$d = D_{AA} D_{A^*A^*} - D_{AA^*} D_{A^*A} \tag{6.80}$$

the radical

$$r = (t^2 - 4d)^{1/2} = [(D_{AA} - D_{A^*A^*})^2 + 4 D_{AA^*} D_{A^*A}]^{1/2} \tag{6.81}$$

and the two eigenvalues

$$\lambda_1 = \tfrac{1}{2}(t + r) \tag{6.82}$$

$$\lambda_2 = \tfrac{1}{2}(t - r) \tag{6.83}$$

The quantities X_{A^*0}, X_{A^*1}, X_{A0}, and X_{A1} denote the initial conditions defined in Fig. 6.1. Expressing the D-matrix in terms of the L-matrix via equations (6.43) and (6.44)

$$D_{AA} = D_V \left[X_A + X_V - \frac{X_{A^*} X_V (1 - f)}{X_A + X_{A^*}} \right] \quad D_{AA^*} = D_V \left[X_A + \frac{X_V X_A}{X_A + X_{A^*}} (1 - f) \right] \tag{6.84}$$

$$D_{A^*A} = D_V\left[X_{A^*} + \frac{X_V X_{A^*}}{X_A + X_{A^*}}(1-f)\right]$$

$$D_{A^*A^*} = D_V\left[X_{A^*} + X_V - \frac{X_A X_V}{X_A + X_{A^*}}(1-f)\right] \quad (6.85)$$

whence the appropriate Fick second equations become (cf §4.4)

$$\frac{\partial X_A}{\partial t} = D_{AA}\frac{\partial^2 X_A}{\partial x^2} + D_{AA^*}\frac{\partial^2 X_{A^*}}{\partial x^2} \quad (6.86)$$

and

$$\frac{\partial X_{A^*}}{\partial t} = D_{A^*A}\frac{\partial^2 X_A}{\partial x^2} + D_{A^*A^*}\frac{\partial^2 X_{A^*}}{\partial x^2} \quad (6.87)$$

strictly applicable only in the limit of uniform concentrations (D_{ik} = constant). For the general initial conditions of Fig. 6.1, which sustain approximately uniform concentrations, the solutions are (cf §4.6)

$$X_A = a_A \operatorname{erf}\frac{x}{2(\lambda_1 t)^{1/2}} + b_A \operatorname{erf}\frac{x}{2(\lambda_2 t)^{1/2}} + c_A \quad (6.88)$$

and

$$X_{A^*} = a_{A^*} \operatorname{erf}\frac{x}{2(\lambda_1 t)^{1/2}} + b_{A^*} \operatorname{erf}\frac{x}{2(\lambda_2 t)^{1/2}} + c_{A^*} \quad (6.89)$$

where

$$a_A = \frac{1}{2r}\{D_{AA^*}(X_{A^*0} - X_{A^*1}) - [(D_{A^*A^*} - D_{AA}) - r]\,\tfrac{1}{2}(X_{A0} - X_{A1})\} \quad (6.90)$$

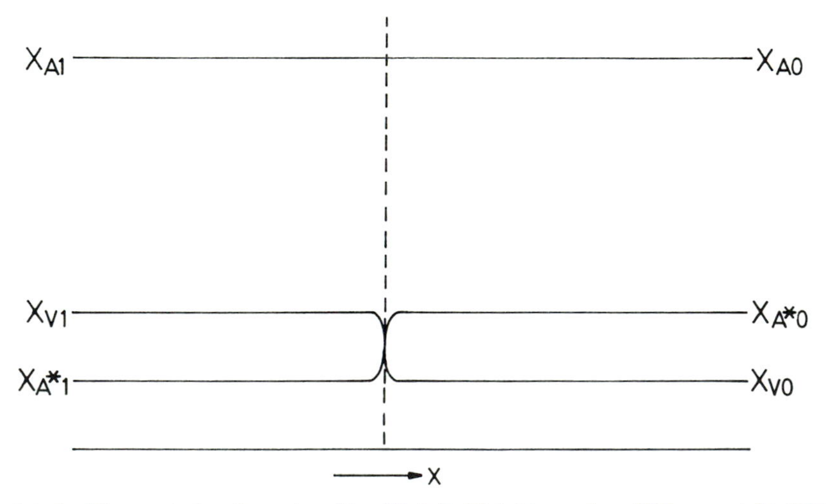

Reprinted with permission from *Acta Met.*, **35**, J. S. Kirkaldy *et al.*, p. 1273, copyright 1987, Pergamon Press Ltd

Fig. 6.1 Schematic initial condition for A–A*–V diffusion couple

$$b_{\mathrm{A}} = \frac{1}{2}\{X_{\mathrm{A0}} - X_{\mathrm{A1}} - 2a_{\mathrm{A}}\} \qquad c_{\mathrm{A}} = \frac{1}{2}\{X_{\mathrm{A0}} + X_{\mathrm{A1}}\} \tag{6.91}$$

$$a_{\mathrm{A^*}} = \frac{1}{2r}\{D_{\mathrm{A^*A}}(X_{\mathrm{A0}} - X_{\mathrm{A1}}) - [(D_{\mathrm{AA}} - D_{\mathrm{A^*A^*}}) - r]\,\tfrac{1}{2}(X_{\mathrm{A^*0}} - X_{\mathrm{A^*1}})\} \tag{6.92}$$

and

$$b_{\mathrm{A^*}} = \frac{1}{2}\{X_{\mathrm{A^*0}} - X_{\mathrm{A^*1}} - 2a_{\mathrm{A^*}}\} \qquad c_{\mathrm{A^*}} = \frac{1}{2}\{X_{\mathrm{A^*0}} + X_{\mathrm{A^*1}}\} \tag{6.93}$$

Note, via the Fickian flux equations for J_{A} and $J_{\mathrm{A^*}}$ with these values inserted, that in an experiment where A* is the tracer and $\nabla X_{\mathrm{A}} \to 0$

$$-J_{\mathrm{A^*}}/J_{\mathrm{V}} \simeq -J_{\mathrm{A^*}}/J_{\mathrm{A}} = D_{\mathrm{A^*}}/D_{\mathrm{V}} = X_{\mathrm{V}}f \tag{6.94}$$

This represents a rigorous proof of the well-known intuitive relation on the right (cf equation (2.18)). The formalism represents the correct and complete A*A description of the 'vacancy wind' including a purely correlation effect.

Since this is a thermodynamically metastable solution we see consistently that to the first order in X_{V}

$$D_{\mathrm{AA}} > 0 \quad D_{\mathrm{A^*A^*}} > 0 \quad t = D_{\mathrm{V}}(1 + X_{\mathrm{V}}f) > 0 \quad d = X_{\mathrm{V}}fD_{\mathrm{V}}^2 > 0 \tag{6.95}$$

with

$$r = D_{\mathrm{V}}(1 - X_{\mathrm{V}}f) \tag{6.96}$$

and equivalently that the eigenvalues are both positive, namely,

$$\lambda_1 = D_{\mathrm{V}} > 0 \tag{6.97}$$

and

$$\lambda_2 = X_{\mathrm{V}}fD_{\mathrm{V}} = D_{\mathrm{A}}^* > 0 \tag{6.98}$$

In an A*-V countercurrent experiment involving the initial condition of Fig. 6.1, equations (6.97) and (6.98) imply via equations (6.90) to (6.93) evolving profiles as in Fig. 6.2, the jog extent being determined by the minor eigenvalue λ_2 and the tails by the major eigenvalue λ_1. Recognizing that there exist parabolic solutions, we can express all derivatives in equations (6.43)–(6.47) with respect to $\lambda = x/\sqrt{t}$. Hence, the appearance of the maximum and minimum in X_{A} verifies *a posteriori* that spontaneous $\mathrm{d}X_{\mathrm{A}} = 0$ (or $\mathrm{d}X_{\mathrm{A^*}} = 0$) states exist which are stationary in λ-space, thus validating the limiting procedures used in evaluating the L-coefficients. Note that because $D_{\mathrm{A}}^* \ll D_{\mathrm{V}}$ one or the other or both of the atom profiles will generally be sharp near the origin. Note that the vacancy profile is much foreshortened in Fig. 6.2.

For the case of Fig. 6.2 as $X_{\mathrm{A^*}}$ approaches a tracer magnitude, the peak-to-peak magnitude of the jog in X_{A} becomes $X_{\mathrm{A^*0}} - X_{\mathrm{A^*1}}$ (equations (6.84), (6.90) and (6.91)). As might be expected, as $X_{\mathrm{A^*}} \to 1$ in accordance with Figs. 6.1 and 6.2 the mass transfer of A must vanish together with $D_{\mathrm{AA^*}}$ and the jog.

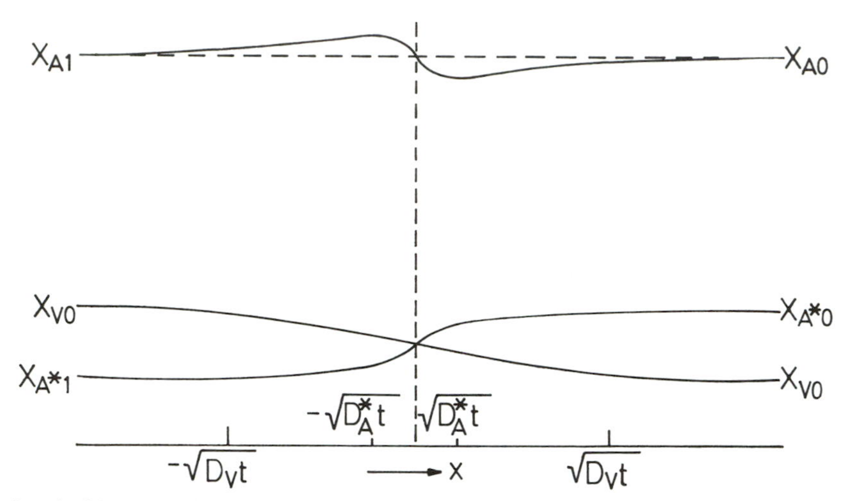

Reprinted with permission from *Acta Met.*, **35**, J. S. Kirkaldy *et al.*, p. 1273, copyright 1987, Pergamon Press Ltd

Fig. 6.2 Schematic profiles according to the A–A*–V diffusion model

6.4.4 Generalization to an A–B–V alloy

If the exchange frequencies of vacancies with A and B are different then the ternary diffusion problem becomes more complex. Manning,[14] however, has found a useful approximation based on an alloy model normalized to the isotope case (see the 'random alloy' model, § §2.7 and 3.11). Here we paraphrase the theoretical structure of Heumann[15] who analytically extrapolates the treatment of § §6.4.1 and 6.4.2 to the A–B–V case. Alternative treatments are given by Kirkaldy *et al.*[13, 16] Heumann in effect recognizes that if one starts with the diagonal elements of the Howard–Lidiard L-matrix (equations (6.77) and (6.78)) in the form

$$L_{AA} = \frac{D_A^* X_A}{kT}\left[1 + X_A \frac{(1-f)}{f}\right] \quad \text{and} \quad L_{A^*A^*} = \frac{D_A^* X_{A^*}}{kT}\left[1 + X_{A^*} \frac{(1-f)}{f}\right] \tag{6.99}$$

these appear as Taylor expansions in the concentrations of dilute A or A*. He then notes that these can be analytically continued to the general case by inducing the new pair of dilute solution series

$$L_{AA} = \frac{D_A^* X_A}{kT}\left[1 + \frac{X_A D_A^* (1-f)}{\overline{D} f}\right] \quad \text{and} \quad L_{BB} = \frac{D_B^* X_B}{kT}\left[1 + X_B \frac{D_B^*}{\overline{D}} \frac{(1-f)}{f}\right] \tag{6.100}$$

where

$$\overline{D} = D_A^* X_A + D_B^* X_B \tag{6.101}$$

with

$$D_A^* = \frac{X_V}{6} \cdot \lambda^2 \omega f_A \quad \text{and} \quad D_B^* = \frac{X_V}{6} \cdot \lambda^2 \omega f_B \tag{6.102}$$

and where $D_A^*X_A$ and $D_B^*X_B$ are the new series expansion variables. These accord with various limiting cases of Manning's random alloy model. Furthermore, in view of the necessary symmetry of the problem, equations (6.100) are expected to interpolate reasonably well across the alloy range.

Heumann's final step is to invoke the general equation (3.102) in the approximation

$$L_{AA} + 2L_{AB} + L_{BB} = X_V D_V/kT \tag{6.103}$$

and solve for

$$L_{AB} = L_{BA} = \frac{X_A X_B D_A^* D_B^* (1-f)}{kT \cdot \overline{D} \cdot f} \tag{6.104}$$

This together with equations (6.100) recovers Manning's L-matrix.[14, 16] Since it is normalized to a structure which is valid for strictly conserved vacancies we may undertake the evaluation of diffusion profiles under boundary conditions which accord with such conservancy. The most interesting case is a countercurrent flow of A and B with X_V initially uniform. This acts as a vacancy pump and produces profiles analogous to those in Figs. 6.1 and 6.2, as shown in Figs. 6.3 and 6.4. Here we consider an ideal or a terminal solution with $\nabla\mu_A$ and $\nabla\mu_V$ as the independent forces and via equations (3.112) and (3.113) obtain the flux equations

$$J_A = -D_A^*\left[1 + \frac{X_A(1-f)}{f} \cdot \frac{(D_A^* - D_B^*)}{\overline{D}}\right]\nabla X_A + \frac{D_A^* X_A}{f X_V}\nabla X_V \tag{6.105}$$

and

$$J_V = \frac{(D_A^* - D_B^*)}{f}\nabla X_A - \frac{(D_A^* X_A + D_B^* X_B)}{f X_V}\nabla X_V \tag{6.106}$$

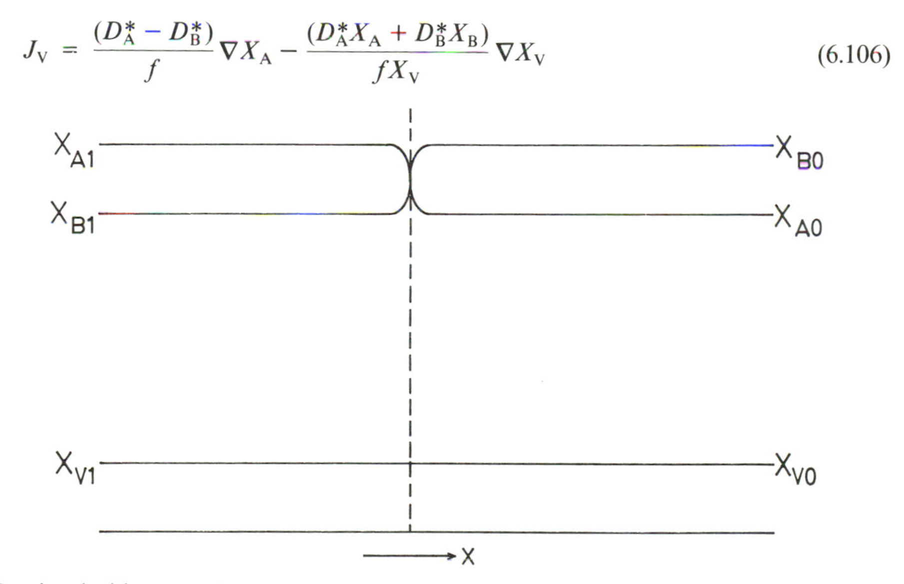

Reprinted with permission from *Acta Met.*, **35**, J. S. Kirkaldy *et al.*, p. 1273, copyright 1987, Pergamon Press Ltd

Fig. 6.3 Schematic initial condition for A–B–V diffusion couple

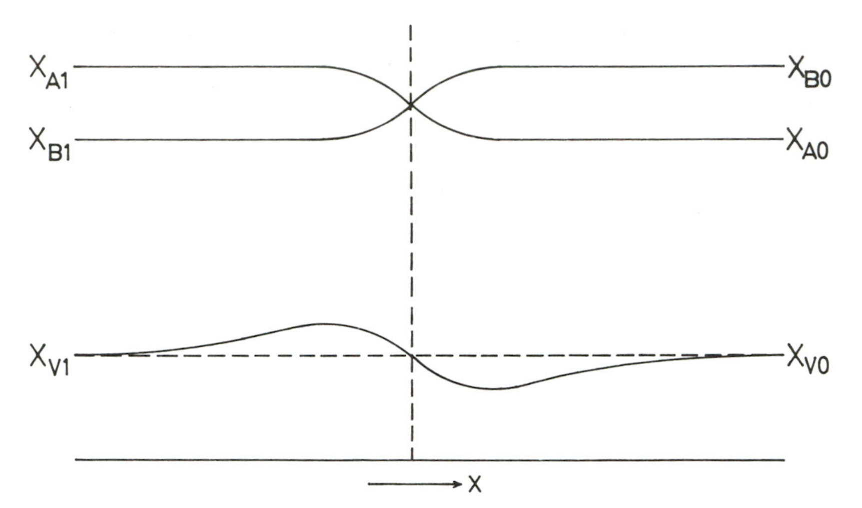

Reprinted with permission from *Acta Met.*, **35**, J. S. Kirkaldy *et al.*, p. 1273, copyright 1987, Pergamon Press Ltd

Fig. 6.4 Schematic profiles according to the A–B–V diffusion model

The eigenvalues of the D-matrix are

$$\lambda_1 = \bar{D}/fX_V \qquad \text{and} \qquad \lambda_2 = D_A^* D_B^*/\bar{D} \tag{6.107}$$

generating an X_V curve according to solutions via equations (6.90)–(6.93) as shown in Fig. 6.4. The peak-to-peak swing of X_V is $X_V(D_A^* - D_B^*)(X_{A0} - X_{A1})/\bar{D}$, which as expected vanishes when the pump term $D_A^* - D_B^*$ vanishes.

Manning[14] has implied that this matrix with the thermodynamic factor ϕ inserted can also be applied to countercurrent diffusion of A–B under local equilibrium conditions, where the latter are attained physically via inclusion of a high density of vacancy sources and sinks (dislocations and grain boundaries) acting to suppress the vacancy profile of Fig. 6.4. This is theoretically effected by setting $\nabla X_V = 0$ in equations (6.105) and (6.106) and leads, via the usual transformation from the Kirkendall to the laboratory frame to the modified Darken equations

$$\tilde{D}/\phi = X_A D_B^* + X_B D_B^* + \frac{X_A X_B (D_A^* - D_B^*)^2 (1 - f)}{\bar{D} f} \tag{6.108}$$

and

$$v/(\phi \nabla C_A) = D_A^* - D_B^* + \frac{(D_A^* - D_B^*)(1 - f)}{f} = (D_A^* - D_B^*)/f \tag{6.109}$$

the correction factor $1/f$ in v being established from Table 2.1 (cf equation (3.145) and (3.146)). Since the derivations of Manning, Heumann and others possess considerable ambiguity we present a new, more general structure based upon the modified phenomenological treatment of Howard and Lidiard presented in §3.11. The primary task is to evaluate the phenomenological coefficients L_{AA^*} and L_{BB^*} which appear in equations (3.145) and (3.146).

The pertinent phenomenological equations in the Kirkendall frame of reference pertaining to A-A*-B interdiffusion are (§3.11)

$$J_B = -L_{BB}\nabla\mu_B$$
$$J_A = -L_{AA}\nabla\mu_A - L_{AA^*}\nabla\mu_{A^*}$$
$$J_{A^*} = -L_{A^*A}\nabla\mu_A - L_{A^*A^*}\nabla\mu_{A^*} \tag{6.110}$$

where J_a is disaggregated to $J_A + J_{A^*}$ and, since B atoms do not distinguish between A and A*, $L_{BB} = L_b$ by definition. There is a similar set for B-B*-A diffusion. Consider a countercurrent A-B experiment with the concentration and chemical potential of dilute tracer A* initially uniform with $D_A^* < D_B^*$ as indicated in Fig. 6.4 with μ_{A^*} replacing X_V. At the optima of the A* curves $\nabla\mu_{A^*} = 0$ so applying equations (3.94) and (6.110)

$$\frac{J_{A^*}}{J_A} = \frac{L_{A^*A}}{L_{AA}} = \frac{X_{A^*}J_V(1-f_A)}{J_A} = \frac{X_{A^*}(D_a - D_b)(1-f_A)\phi}{D_a} \tag{6.111}$$

where f_A is the A* tracer correlation factor and ∇X_A has been cancelled between numerator and denominator on the right. J_{A^*} has been evaluated as the fraction of the vacancy flux (cf equation (3.94)) which is intercepted by uniform A* atoms per unit time $(-X_{A^*}J_V)$ multiplied by the fraction of interceptions which are not random $(1-f_A)$, and therefore contribute to an A* flux in the *uniform* regions. This, in our view, is the correct and very limited context of a so-called 'vacancy wind'. Now in view of dilute A*, $L_{A^*A} \ll L_{AA}$, so from equations (3.90), (3.143) and (3.144), $D_a = kTL_{AA}\phi/NX_A$, and

$$L_{A^*A} = L_{AA^*} = \frac{N}{kT}X_AX_{A^*}(D_a - D_b)(1-f_A)$$

and symmetrically

$$L_{B^*B} = L_{BB^*} = \frac{N}{kT}X_BX_{B^*}(D_b - D_a)(1-f_B) \tag{6.112}$$

Substituting these into equation (3.146) and comparing with equation (3.94), we obtain

$$v/\phi\nabla X_A = D_a - D_b = (D_A^* - D_B^*)/\bar{f} \tag{6.113}$$

where $\bar{f} = X_Af_A + X_Bf_B$. Next, substituting equation (6.113) into (6.112) and these into (3.145), we obtain

$$\tilde{D}/\phi = X_BD_A^* + X_AD_B^* + X_AX_B(D_A^* - D_B^*)(f_B - f_A)/\bar{f} \tag{6.114}$$

Detailed balance can now be invoked for a local equilibrium A*-B*-A-B solution in the form

$$D_A^*/(1-f_A) = D_B^*/(1-f_B) = \bar{D}/(1-\bar{f}\,) \tag{6.115}$$

where the right follows from the left equality. The latter, after cross-multiplying and multiplying by $X_{A^*}X_{B^*}$, states that the probability of B* vacancy exchanges associated with the vacancy wind produced in local A* motions (cf equation (6.111)) is equal to the probability of A* vacancy exchanges associated locally with B* motions since each trajectory set is the other's inverse set. Thus, with simultaneous application of equations (6.115), equation (6.114) can be written as

$$\widetilde{D}/\phi = X_B D_A^* + X_A D_B^* + X_A X_B (D_a - D_b)^2 (1 - \bar{f}\,)/\bar{f}\,\bar{D} \qquad (6.116)$$

with the intrinsic coefficients

$$\begin{aligned} D_a &= D_A^* \phi [1 + X_A (D_A^* - D_B^*)(1 - \bar{f}\,)/\bar{f}\,] \\ D_b &= D_B^* \phi [1 + X_B (D_B^* - D_A^*)(1 - \bar{f}\,)/\bar{f}\,] \end{aligned} \qquad (6.117)$$

With $\bar{f} = f$, the tracer value in a pure material and one of the defining conditions of Manning's random alloy model (cf §2.7), these reduce to his well-known expressions (6.108) and (6.109). Manning's equation for υ predicts values 28% higher than Darken's for fcc and greater for looser crystal structures. The values for $\widetilde{D}$ are typically 10% higher. Experiments for $\widetilde{D}$ in liquids in fact yield values which are up to 10% higher than the Darken prediction (cf Table 2.2), although the significance of this observation is not clear. On the other hand, the experiments in solids are for some cases reported to yield rather high υ values in general accordance with the Manning equation (6.109).[17] Monte Carlo calculations tend to support the random alloy model, especially for fcc crystals (cf §2.7).

From the practical point of view of predicting diffusion penetrations, which vary as $\widetilde{D}^{1/2}$, the Darken prediction is clearly accurate (better than 5%) as compared to current experimental reproducibility. As already noted, it has been shown experimentally to be very accurate at infinite dilution (cf equation (6.108) or (6.116)). It may therefore be concluded that it will also be accurate for terminal solutions. Recent contributions to this problem have been offered by Lidiard[18] and Kirkaldy.[16]

6.5 ELECTROSTATIC INTERACTIONS AND THE DIFFUSION POTENTIAL

The movement of particles having electrical charges z_i is necessarily correlated through the need to preserve electroneutrality where the real or free charge density

$$\rho = \sum_i z_i m_i = 0 \qquad (6.118)$$

This means that in the absence of an externally applied electrostatic field the current of real or free charge

$$I = \sum_i z_i J_i = 0 \qquad (6.119)$$

This is known as the 'zero net current condition'. As seen in §2.18 the forces appearing in the flux equations are the electrochemical potential gradients

$$J_i = -\sum_j L_{ij} \operatorname{grad} \eta_j \qquad (6.120)$$

where the electrochemical potential η is defined through

$$\eta_i = \mu_i + z_i F \psi \qquad (6.121)$$

with ψ being the electrostatic potential and F the Faraday. Application of the constraint (6.119) to the set of flux equations (6.120) implies that

$$\sum_i z_i L_{ij} = 0 \tag{6.122}$$

The consequences of this constraint are now examined for a simple case. More complex examples will be analysed in Chapters 8 and 9.

We consider binary one-dimensional diffusion in a solid of two cations with a common immobile anion and examine for simplicity the case where both cations have the same charge. Flux equations appropriate to the cation sublattice are then

$$J_1 = -L_{11}\nabla(\eta_1 - \eta_V) - L_{12}\nabla(\eta_2 - \eta_V) \tag{6.123}$$

and

$$J_2 = -L_{21}\nabla(\eta_1 - \eta_V) - L_{22}\nabla(\eta_2 - \eta_V) \tag{6.124}$$

where the site conservation condition

$$\sum_i^3 J_i = 0 \tag{6.125}$$

has been used to eliminate the vacancy flux J_V from the description. Application of the condition (6.122) with $z_1 = z_2$ then reduces the set (6.124) to a single expression

$$\begin{aligned} J_1 = -J_2 &= -L_{11}\nabla(\eta_1 - \eta_2) \\ &= -L_{11}\nabla(\mu_1 - \mu_2) \end{aligned} \tag{6.126}$$

The terms involving the electrostatic potential have vanished from the description through simple algebraic subtraction. This is an immediate consequence of the fact that diffusion in the absence of an external electrostatic field must involve no net charge transfer. It further reflects the fact that purely diffusive motion can always be completely described in terms of chemical potential gradients only. This must always be true, as these gradients suffice to define completely the thermodynamic state of the system. Several more complex examples will be given in Chapter 8.

Equation (6.126) shows that only one diffusion coefficient is required to describe this system which is, in its diffusion behaviour, a binary. In an ideal or Henrian solution equation (6.126) may be rewritten as

$$J_1 = -L_{11}RT\left(\frac{\nabla m_1}{m_1} - \frac{\nabla m_2}{m_2}\right) \tag{6.127}$$

It follows from equation (6.118) with zero anion concentration gradient, that $\nabla m_1 = -\nabla m_2$ and hence

$$J_1 = -L_{11}RT\left(\frac{m_1 + m_2}{m_1 m_2}\right)\nabla m_1 \tag{6.128}$$

and the chemical diffusion coefficient is accordingly defined as

$$\tilde{D} = L_{11}RT\left(\frac{m_1 + m_2}{m_1 m_2}\right) \tag{6.129}$$

The question now arises as to what is the relationship between $\tilde{D}$ and the tracer diffusion coefficients of the individual cations, D_1 and D_2.

It will be recalled that the same question arose in connection with binary alloy diffusion. In that case (§2.14) the question was answered from a consideration of the relationship between intrinsic and laboratory reference frames. Differing intrinsic mobilities led to a compensating bulk material flow (the Kirkendall effect) which balanced the otherwise uneven flows. No such effect is possible in the ionic lattice under consideration, because the anion lattice is immobile. It is instead the electrostatic potential gradient or field developed within the solid which brings into balance the fluxes of charged species having different intrinsic mobilities. For this reason the potential developed within a couple is called a 'diffusion potential'. Unfortunately, this potential cannot be measured except in combination with other unknown potentials such as electrode contact potentials. However, it can be calculated in terms of intrinsic mobilities, and the relationship between these quantities and the value of $\tilde{D}$ is thereby established. Since the value of ψ is experimentally inaccessible, it must be possible also to relate $\tilde{D}$ to D_1, D_2 from a consideration of the kinetics of ion movements, as will be demonstrated below.

In the usual approximation, equations (6.120) are written

$$J_i = -\frac{D_i m_i}{RT}\{\nabla\mu_i + z_i F\nabla\psi\} \qquad (6.130)$$

where the neglect of cross-coefficients is appropriate for the binary case. Then, $z_1 = z_2$ and $J_1 = -J_2$ it is found that

$$\nabla\psi = -\frac{RT}{F}\left\{\frac{D_1 - D_2}{D_1 m_1 + D_2 m_2}\right\}\nabla m_1 \qquad (6.131)$$

which upon resubstitution in equation (6.130) yields

$$J_1 = -\tilde{D}\nabla m_1 \qquad (6.132)$$

with

$$\tilde{D} = \frac{D_1 D_2}{m_1 D_1 + m_2 D_2}(m_1 + m_2) \qquad (6.133)$$

It will be shown in Chapter 8 that a consideration of the kinetics of cation–vacancy exchange leads directly from equation (6.126) to this result without any need to evaluate ψ.

As shown in equation (6.131) there exists a generally non-zero electrostatic field within the solid. Despite the phenomenological correctness of the conception, its physical significance in terms of electrostatics and diffusion models has remained ambiguous. Indeed, investigations along these lines in the chemical literature have often led to paradox and confusion.[19] It is thus essential from the start to examine the fundamentals.[20]

For interdiffusion of initially neutral samples in the infinite couple configuration the field

$$E = -\nabla\psi \qquad (6.134)$$

is entirely self-generated and confined to the diffusion zone. Thus it should always be possible in principle to eliminate E via Poisson's equation which in rationalized SI units is[21]

$$\nabla \cdot E = (\rho + \rho')/\varepsilon_0 \tag{6.135}$$

where ε_0 is the permittivity of free space ($8{\cdot}85 \times 10^{-12} f/m$ in rationalized units), ρ is the real or free change density (cf equation (6.118)) and ρ' is the dipole charge density, and to solve a conventional set of diffusion equations for the concentration profiles C_i in terms of their own boundary and initial conditions.

A particular set of solutions of Poisson's equation is usually assumed to apply to equation (6.130) which is defined by the 'zero net (real) current condition' (6.119). This condition, as applied to binary one-dimensional diffusion of two cations of unequal charge with a common immobile anion, yields Fick-type equations for the countercurrent diffusion of the cations in thermodynamically ideal or Henrian solutions in terms of tracer diffusion coefficients D_i as shown above. In the more general case where $z_1 \neq z_2$, then

$$J_1 = -(D_{11} + D_{22})\ \partial m_1/\partial x$$

and

$$J_2 = -(D_{11} + D_{22})\ \partial m_2/\partial x \tag{6.136}$$

where

$$D_{11} = D_1 D_2 z_2^2 m_2/(D_1 z_1^2 m_1 + D_2 z_2^2 m_2)$$

and

$$D_{22} = D_1 D_2 z_1^2 m_1/(D_1 z_1^2 m_1 + D_2 z_2^2 m_2) \tag{6.137}$$

and the field

$$E_x = -\frac{\partial \psi}{\partial x} = -\frac{RT}{F}\frac{z_1 D_1(\partial m_1/\partial x) + z_2 D_2(\partial m_2/\partial x)}{D_1 z_1^2 m_1 + D_2 z_2^2 m_2} \tag{6.138}$$

which was eliminated from equations (6.130) via equation (6.119). Inferring that for an initially neutral system equation (6.119) implies zero accumulation of real charge ρ, then by equation (6.135) the field must be determined entirely by the dipole charge and its value is

$$E_x = -\frac{RT}{F}\ \frac{z_1(D_1 - D_2)(\partial m_1/\partial x)}{D_1 z_1^2 m_1 + D_2 z_2^2 m_2} \tag{6.139}$$

when the anion concentration as a constant implies

$$z_1\frac{\partial m_1}{\partial x} + z_2\frac{\partial m_2}{\partial x} = 0 \tag{6.140}$$

If we approximate $m_1 + m_2 = m$ as a constant then via equations (6.138) and (6.139) we can integrate across the diffusion zone for countercurrent m_1 and m_2 with $m_1 = m$ at $x = -\infty$ and $m_2 = m$ at $x = -\infty$ and obtain the potential drop

$$\Delta\psi = -\int_{-\infty}^{+\infty} E_x\,\mathrm{d}x = \frac{RT}{F}\frac{z_1(D_1 - D_2)}{(D_1 z_1^2 - D_2 z_2^2)}\ln\frac{D_2 z_2^2}{D_1 z_1^2} \tag{6.141}$$

The diffusion couple is recognized as an electrochemical cell whose potential difference is determined purely by mobility effects. Presumably, this potential difference would be of the order of 1 V. Note that the potential difference cannot be attributed to effective surface charges since the field (equation (6.139)) is not uniform. The excess free energy or negentropy may be associated with unmixing of the velocity distribution (see below). Also a close parallel may be recognized between this structure and that used by Darken in his treatment of the Kirkendall effect. Here, however, the drift velocity is suppressed via charge neutrality to accommodate the static anion lattice and E_x has the analogous role (cf equations (6.113) and (6.139)).

Failure to recognize the pure dipole character of ψ has led a number of investigators to a contradiction between the diffusion formalism and Poisson's equation. The argument starts with the (usually correct) relation

$$\operatorname{div} E = \rho/\varepsilon \tag{6.142}$$

where ρ is the real charge and ε is the dielectric constant. This has the solution in one dimension for $\rho = 0$[22]

$$E_x = \text{constant} \tag{6.143}$$

which for the infinite diffusion couple configuration is clearly in contradiction to equation (6.139).

To resolve the difficulty one must re-derive equation (6.142) from equation (6.135) following Slater and Frank.[21] Note that by definition

$$\rho' = -\operatorname{div} P \tag{6.144}$$

where P is the polarization vector and ρ' is the dipole charge. Thus rearranging equation (6.135) we obtain

$$\operatorname{div}(\varepsilon_0 E + P) = \rho \tag{6.145}$$

Now the polarization is related to E by

$$P = \alpha E \tag{6.146}$$

where α is the polarizability of the medium so equation (6.145) becomes

$$\operatorname{div}(\varepsilon_0 + \alpha) E = \rho \tag{6.147}$$

The dielectric constant ε is defined via the electric displacement

$$D = \varepsilon E \tag{6.148}$$

where

$$\varepsilon = \varepsilon_0 + \alpha \tag{6.149}$$

and provided $\varepsilon \neq 0$, equation (6.147) can be written as (6.142). But herein lies the deadly error, for integrating equation (6.145) for $\rho = 0$ and the boundary condition $E(\pm\infty) = 0$ the unique solution is

$$-\varepsilon_0 E = P \tag{6.150}$$

or $\alpha = -\varepsilon_0$ and

$$\varepsilon = 0 \tag{6.151}$$

which, in equation (6.142), implies division by zero. Clearly, the problem is avoided by focusing attention on the general form of Poisson's equation, equation (6.135). This together with equation (6.144) demonstrates that, subject to equation (6.130) under the boundary condition $E_x(\pm\infty) = 0$, the field is due entirely to diffusion induced dipoles. It also shows that such kinetically induced dipoles with $\alpha = -\varepsilon_0$ bear no relationship to the static dielectric properties of the medium. The latter come into play only when there is a real space charge, when an external field is applied or when there is a permanent or static dipole field (an 'electret'[23]).

Now in consideration of Maxwell's equations, equation (6.151) implies the vanishing of the displacement current, i.e.,

$$J_D = \partial D/\partial t = (\partial/\partial t)(\varepsilon_0 E + P) = 0 \tag{6.152}$$

so the electric dipole current

$$J_d = \partial P/\partial t = -\varepsilon_0(\partial E/\partial t) \tag{6.153}$$

Accordingly, zero net real current implies zero net total current, i.e.

$$I + J_D = 0 \tag{6.154}$$

Figure 6.5 gives a schematic representation of the various functions involved in a diffusion couple ($t > 0$) and offers a summary of the spatial relationships implied by the preceding theoretical equations. Figure 6.5*c* clearly illustrates the pure dipole character of the charge (ρ') distribution (equation (6.144)). Furthermore, we can evaluate the dipole current which gives rise to charge ρ' via equation (6.153) and this is represented schematically in Fig. 6.5*d*. It is matched negatively by the vacuum displacement term which makes the total go to zero. The physics of such a dipole distribution in terms of atom trajectories requires clarification.

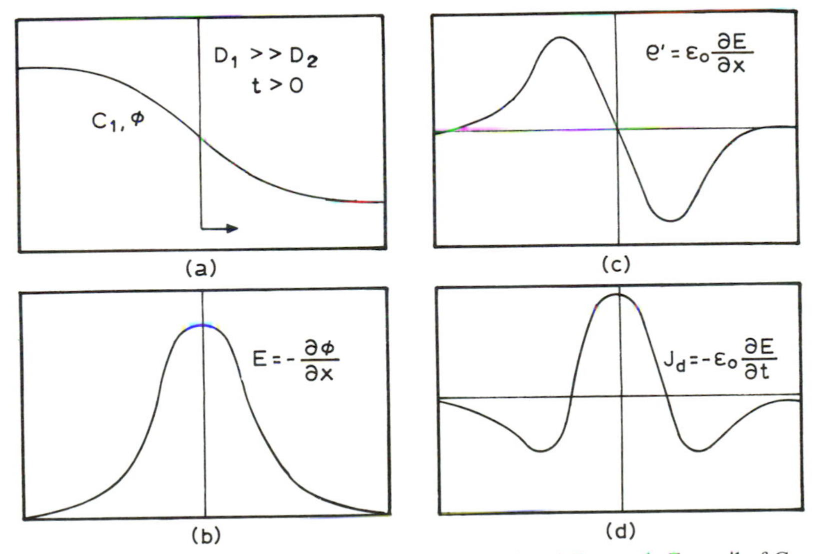

With permission from Canadian Journal of Physics, National Research Council of Canada

Fig. 6.5 Functions derived from the diffusion potential. After Kirkaldy[20]

The 'zero net current' condition is often understood to be a corollary of an assumed 'quasi-steady state' condition. This is because a chemically open and electrically closed system subject to charge conservation necessarily reaches a zero current state at the steady state. Thus an accurate quasi-steady state can be attained in the non-steady state (e.g. an infinite diffusion couple) when the relaxation times for processes moderating the accumulation of charge are very much less than the relaxation time of the driving chemical potential of the neutral species (i.e. the experimental diffusion time). To clarify this general statement it is convenient to again consider a specific model, namely, the countercurrent

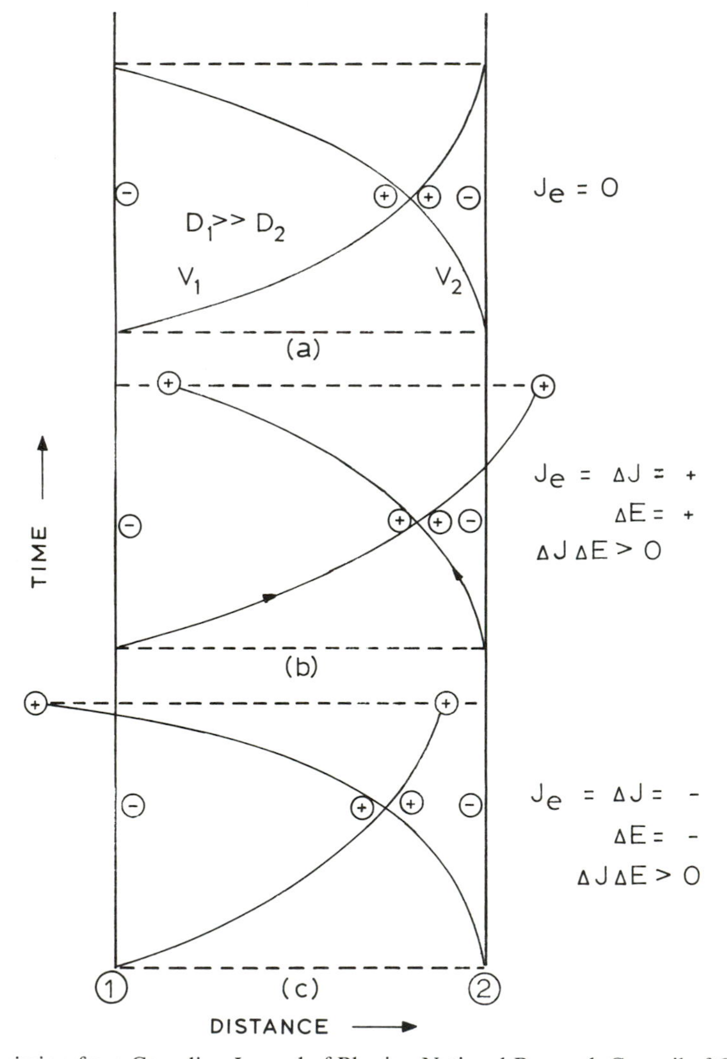

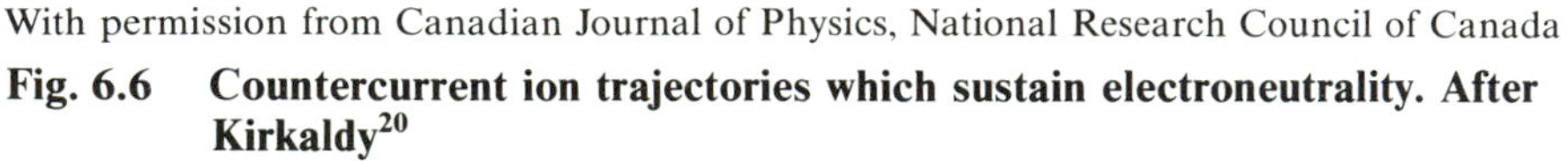

With permission from Canadian Journal of Physics, National Research Council of Canada

Fig. 6.6 Countercurrent ion trajectories which sustain electroneutrality. After Kirkaldy[20]

diffusion of two monovalent cations in a crystal with Schottky defects and immobile anions, and to assume that the conditions leading to equations (6.138) and (6.139) are well-founded.

We note first of all that the unidirectional cation fluxes J_i can be expressed in terms of the mean velocities v_i by

$$J_i = m_i v_i \tag{6.155}$$

Consider next a schematic representation of two average countercurrent trajectories in the zero net-current configuration as in Fig. 6.6*a*. We suppose that all the countercurrent ions 1 and 2 start their trajectories from microscopically separated planes A and B and symmetrically disposed neutral configurations at time $t = 0$. In the instantaneous absence of a field they begin with velocities

$$v_1 = -(D_1/RT)(\partial\mu_1/\partial x) \tag{6.156}$$

and

$$v_2 = -(D_2/RT)(\partial\mu_2/\partial x) \tag{6.157}$$

respectively. Supposing $D_1 \gg D_2$, the mean trajectories will begin as shown in Fig. 6.6*a*, but by the zero net (real) current condition, they must (on the average) reach new neutral positions at exactly the same time. That is to say, through the build-up of mutual electrostatic forces, 1 and 2 must decelerate and accelerate, respectively, as indicated. Their mean configuration during the interchange is clearly that of a non-zero dipole, which when summed and averaged over all particles and configurations must be equivalent to the self-field given by equation (6.139) with the real charge $\rho = 0$. This dipole field is the source of the diffusion potential. Note the necessary null in equations (6.131) and (6.139) when $D_1 = D_2$.

Finally, Figs. 6.4*b* and *c*, representing failures to reach real charge neutrality, indicate how a Le Chatelier principle for neutral stability is to be recognized. Kirkaldy[20] has developed the equivalent quasi-steady principle of minimum entropy production for this state. In the following section we consider a similar problem of ternary multiplicity where cross-interactions appear explicitly within a D-matrix with off-diagonal elements.

6.6 ELECTROSTATIC CROSS-INTERACTIONS IN A TERNARY IONIC SYSTEM

Here a structure is considered which is similar to the foregoing but where two ionic compounds are dissolved in a third forming an anion lattice which can be of non-uniform composition and immobile: to be specific countercurrent diffusion of K^+ and Sr^{++} in a K_2O–SrO–SiO_2 glass.[24, 25] Here the analysis parallels that of the previous section but without the explicit countercurrent constraint $z_1\partial m_1 = -z_2\partial m_2$. In this calculation equation (6.138) is recovered, but the flux equations are

$$J_1 = -D_{11}\frac{\partial m_1}{\partial x} - D_{12}\frac{\partial m_2}{\partial x} \tag{6.158}$$

and

$$J_2 = -D_{21}\frac{\partial m_2}{\partial x} - D_{22}\frac{\partial m_2}{\partial x} \tag{6.159}$$

where D_{11} and D_{22} are given by equations (6.137) and the off-diagonal coefficients take the non-zero values of

$$D_{12} = -(z_2/z_1)\,D_{22} \qquad D_{21} = -(z_1/z_2)\,D_{11} \tag{6.160}$$

It is of particular significance that the determinant of the D-matrix vanishes, i.e.

$$D_{11}D_{22} - D_{12}D_{21} = 0 \tag{6.161}$$

This implies in relation to the diffusion profiles that one of the eigenvalues vanishes and that there is accordingly a permanent discontinuity in m_1 and m_2 at the origin. This arises in turn because of the assumed immobility of the anion sublattice with its static concentration discontinuity in the most general case. This discontinuity in concentration in relation to Fig. 6.7 generates a step function discontinuity in the diffusion potential at the origin and a corresponding Dirac δ-function (cf Appendix 3) superposed upon the E profile which represents a simple dipole at the origin. It is this discrete compensating dipole and its field which, by acceleration or deceleration, forces the cations to produce zero real charge near the origin in the presence of a concentration discontinuity. Note that if the anions possess a non-zero mobility the discontinuity will be moderated accordingly.

Figure 6.7 shows a comparison of the predictions according to the above theory based upon independently measured tracer coefficients. This may be deemed satisfactory in view of the rather poor experimental resolution and precision and the uncertainty in the anion mobility.

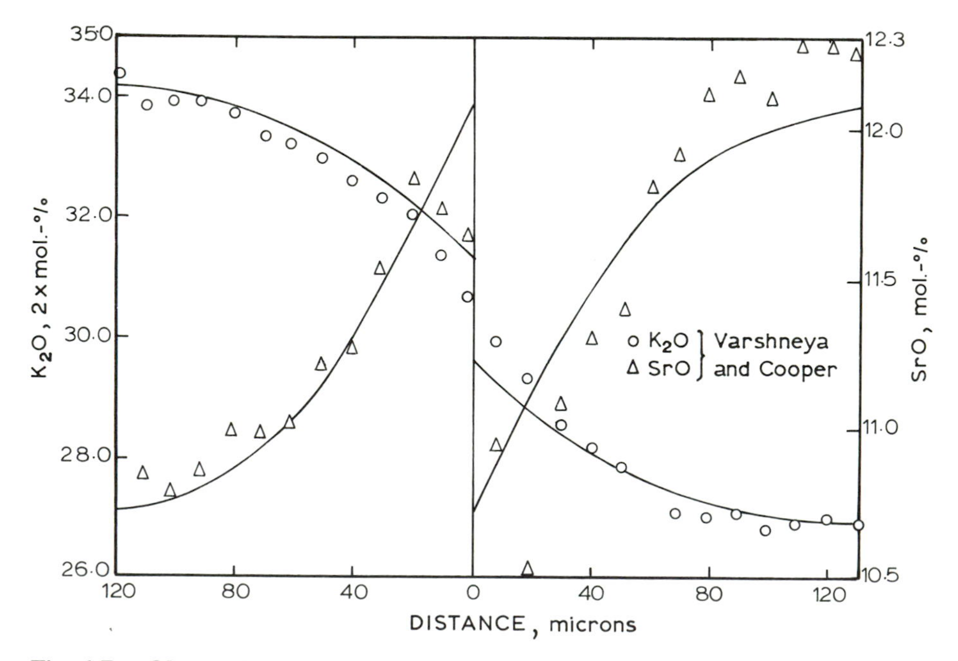

Fig. 6.7 Observed and predicted concentration profiles of K and Sr in a K_2O–SrO–SiO_2 glass after 4·55 h at 798°C. After Okongwu[25]

We conclude by emphasizing that in charged systems a very strong correlation among diffusive flows arises from the electrostatic constraints. These constraints are summarized in the expression of zero net real current

$$\sum_i z_i J_i = 0 \tag{6.162}$$

which is analogous to the expression of matter conservation in an isobaric uncharged system

$$\sum_i \overline{V}_i (J_i)_V = 0 \tag{6.163}$$

Cross-effects therefore arise in essentially the same way. What is different in the charged system, is that the manifestation of a flow compensation mechanism is experimentally inaccessible. Thus while a binary alloy diffusion experiment yields $\tilde{D}$ and v and leads to the estimation of both tracer diffusion coefficients, the analogous experiment on an ionic system without a field determination yields $\tilde{D}$ only. The consequences of this fact are explored in more detail in Chapters 8 and 9.

REFERENCES

1 S. R. DE GROOT AND P. MAZUR: 'Non-equilibrium thermodynamics', 1962, Amsterdam, North-Holland
2 J. S. KIRKALDY: *Can. J. Phys.*, 1958, **36**, 899
3 J. G. KIRKWOOD, R. L. BALDWIN, P. J. DUNLOP, L. J. GOSTING, AND G. KEGELES: *J. Chem. Phys.*, 1960, **33**, 1505
4 J. E. LANE AND J. S. KIRKALDY: *Can. J. Phys.*, 1964, **42**, 1643
5 J. E. LANE AND J. S. KIRKALDY: *ibid.*, 1966, **44**, 2059
6 J. S. KIRKALDY, D. WEICHERT, AND ZIA-UL-HAQ: *ibid.*, 1963, **41**, 2174
7 D. G. MILLER, V. VITAGLIANO, AND R. SARTORIO: *J. Phys. Chem.*, 1986, **90**, 1509
8 J. S. KIRKALDY AND G. R. PURDY: *Can. J. Phys.*, 1969, **47**, 865
9 C. WAGNER: 'Thermodynamics of alloys', 53, 1952, Reading, Mass., Addison-Wesley
10 J. S. KIRKALDY AND G. R. PURDY: *Can. J. Phys.*, 1962, **40**, 202
11 L. C. BROWN AND J. S. KIRKALDY: *Trans. AIME*, 1964, **230**, 223
12 R. E. HOWARD AND A. B. LIDIARD: *Rep. Prog. Phys.*, 1964, **27**, 161
13 J. S. KIRKALDY, D. J. YOUNG, AND J. E. LANE: *Acta Met.*, 1987, **35**, 1273
14 J. R. MANNING: 'Diffusion kinetics for atoms in crystals', 218 *et seq.*, 1968, Princeton NJ, D. Van Nostrand
15 T. HEUMANN: *J. Phys. F (Met. Phys.)*, 1979, **9**, 1997
16 J. S. KIRKALDY: *Scripta Met.*, 1987, **21**, 33
17 M. A. DAYANANDA: *Acta Met.*, 1981, **29**, 1151
18 A. B. LIDIARD: *ibid.*, 1986, **34**, 1487
19 R. R. HAFEMANN: *J. Phys. Chem.*, 1965, **69**, 4226
20 J. S. KIRKALDY: *Can. J. Phys.*, 1979, **57**, 717
21 J. C. SLATER AND N. H. FRANK: 'Electromagnetism', 44 *et seq.*, 1947, New York, McGraw-Hill
22 A. D. MACGILLIVRAY: *J. Chem. Phys.*, 1968, **48**, 2903
23 P. LORRAIN AND D. R. CORSON: 'Electromagnetic fields and waves', 2nd edn, 113, 1970, New York, W. H. Freeman
24 A. K. VARSHNEYA AND A. R. COOPER: *J. Am. Ceram. Soc.*, 1972, **55**, 312
25 D. A. OKONGWU, W-K. LU, A. E. HAMIELEC, AND J. S. KIRKALDY: *J. Chem. Phys.*, 1973, **58**, 777

CHAPTER 7

Ternary metallic solids

7.1 INTRODUCTION

Chapter 2 introduced the Darken equation, which relates the mutual coefficient to the two tracer coefficients and the thermodynamics of a binary substitutional system via transition state theory. The following proceeds in the same fashion by introducing the ternary and higher order transition state theories for exchange and vacancy mechanisms. The ternary equivalents of the Darken equation are then stated. This reveals that the four mutual coefficients are determined to a good approximation by the Onsager reciprocal relations (ORR), the three tracer coefficients, and the thermodynamics. Furthermore, the six intrinsic diffusion coefficients are obtainable from the same data set. In the subsequent section, the structure is specialized to regions of the Gibbs triangle near critical points, where the unique thermodynamic functions are reflected in interesting relations among the mutual coefficients. In the subsequent sections experiments concerned with the *D*-matrix, the ORR and critical points are reviewed.

§§7.2–7.10 deal with the methods available for, and examples of the prediction of ternary coefficients from binary, tracer, and thermodynamic data. Chapters 4, 6, 8, and 9 also deal with this practical problem.

7.2 TRANSITION STATE THEORY FOR THE EXCHANGE MECHANISM

The Onsager coefficients relate the flux to the forces which produce the matter flow, and should therefore be amenable to analysis according to an appropriate kinetic theory. The most general and fundamental theory of rate processes is that of the 'transition state' or 'activated complex' which is described in detail in many texts (for example, Glasstone *et al.*[1]; §2.9). Adopting this approach to mass transfer, diffusion is interpreted as consisting of individual atomic or molecular steps which proceed as activated rate processes. In addition to generating the expected Arrhenius temperature dependence, the methodology has the advantage of being explicitly referred to interparticle interactions. However, it is not normally capable of dealing with certain subtle correlation effects which may be just at the margin of experimental significance (cf §3.11).

It is proposed that the fundamental diffusion step in a condensed phase consists of an interchange between two adjacent constituent 'particles'. If the particles are chemically different and if the net difference between forward and reverse exchange rates is non-zero then matter flow will thereby be effected. The essence of the proposition is therefore that movement of one particle must involve the movement in the reverse direction of another, namely, the mass- or volume-weighted fluxes must satisfy

$$\sum_{i}^{n} \alpha_i J_i = 0 \tag{7.1}$$

This arises, of course, from appreciation of the fact that diffusion of a particular species away from a region does not lead to the creation therein of void space. In this sense it is purely phenomenological and represents no significant addition to the Onsager treatment. Thus it is apparent that the cross-effects will arise in the same way as was described in a macroscopic way in §6.1. However the theoretical structure affords insight into the nature of the diffusional interactions and, when structural details of the material can be specified, it can lead to qualitative predictions as to compositional effects.

Considering diffusion in one dimension, we describe the material as consisting of particles located on planes which are normal to the diffusion direction and separated from each other by a distance λ.[2, 3, 4, 5] This is a number-fixed reference frame. In the transition state theory, the interchanging particles or species i and j form a complex ij, which is considered to be at thermodynamic equilibrium with the surrounding region. It should be noted at this point that one of the participating species in a solid will usually be a vacancy. The equilibrium is described in the usual way by (cf §2.10)

$$a_{ij}^{\dagger}/a_i a_j = K_{ij} = \exp(-\Delta G_{ij}^{\dagger}/RT) \tag{7.2}$$

where a are activities and $\Delta G_{ij}^{\dagger} = \Delta G_{ji}^{\dagger}$ is the standard free energy of formation of the complex from pure i and j.

The rate Γ_{ij} is then given by

$$\Gamma_{ij} = \nu_{ij} X_{ij}^{\dagger} = \nu_{ij} a_{ij}^{\dagger}/\gamma_{ij}^{\dagger} \tag{7.3}$$

where ν_{ij} is the frequency of forward transitions through the activated complex. The forward species flux of species i from a plane (1) to an adjacent plane (2) produced by exchange with species j is therefore given by equations (7.2) and (7.3) as

$$\Gamma_{ij\,(1)\to(2)} = m\lambda \nu_{ij} K_{ij} a_i^{(1)} a_j^{(2)}/\gamma_{ij}^{\dagger} \tag{7.4}$$

Similarly, the rate of the reverse process will be

$$\Gamma_{ij\,(2)\to(1)} = m\lambda \nu_{ji} K_{ji} a_i^{(2)} a_j^{(1)}/\gamma_{ij}^{\dagger} \tag{7.5}$$

The net flux of i due to interchange with j is therefore given by

$$J_i^{i-j} = m\lambda \nu_{ij} K_{ij} \{a_i^{(1)} a_j^{(2)} - a_i^{(2)} a_j^{(1)}\}/\gamma_{ij}^{\dagger} \tag{7.6}$$

where detailed balance requires that $\nu_{ij} K_{ij}/\gamma_{ij}^{\dagger} = \nu_{ji} K_{ji}/\gamma_{ji}^{\dagger}$ (cf §3.8). Taylor series expansion about position (1) and retention of only linear terms leads to

$$J_i^{i-j} = -(m\lambda^2/RT)\, \nu_{ij} K_{ij} a_i a_j \Delta(\mu_i - \mu_j)/\gamma_{ij}^{\dagger} \tag{7.7}$$

In a multicomponent system, the total flux of component i will be the sum of fluxes arising from all such pair-wise exchanges

$$J_i = -\frac{m\lambda^2}{RT}\sum_{j \neq i} \nu_{ij} K_{ij} a_i a_j \nabla(\mu_i - \mu_j)/\gamma_{ij}^{\dagger} \tag{7.8}$$

In the ternary case, the equations are

$$J_1 = -k_{12}a_1a_2\nabla(\mu_1 - \mu_2) - k_{13}a_1a_3\nabla(\mu_1 - \mu_3) \tag{7.9}$$

$$J_2 = -k_{21}a_1a_2\nabla(\mu_2 - \mu_1) - k_{23}a_2a_3\nabla(\mu_2 - \mu_3) \tag{7.10}$$

where

$$k_{ij} = \frac{m\lambda^2}{RT}\nu_{ij}K_{ij}/\gamma_{ij}^{\dagger} \tag{7.11}$$

These equations may be rearranged to yield the standard forms

$$J_1 = -(k_{12}a_1a_2 + k_{13}a_1a_3)\nabla(\mu_1 - \mu_3) + k_{12}a_1a_2\nabla(\mu_2 - \mu_3) \tag{7.12}$$

$$J_2 = k_{12}a_1a_2\nabla(\mu_1 - \mu_3) - (k_{23}a_2a_3 + k_{12}a_1a_2)\nabla(\mu_2 - \mu_3) \tag{7.13}$$

where use has been made of the fact that $k_{12} = k_{21}$. Comparison of equations (7.12) and (7.13) with equations (6.6) and (6.7) yields for the Onsager coefficients

$$\begin{aligned} L_{11} &= k_{12}a_1a_2 + k_{13}a_1a_3 \\ L_{12} &= L_{21} = -k_{12}a_1a_2 \\ L_{22} &= k_{23}a_2a_3 + k_{12}a_1a_2 \end{aligned} \tag{7.14}$$

and it can be seen that this set satisfies the thermodynamic requirements specified by equations (2.86). For a multicomponent system

$$\begin{aligned} L_{ii} &= \sum_{j \neq i}^{n-1} k_{ij}a_ia_j \\ L_{ij} &= -k_{ij}a_ia_j \end{aligned} \tag{7.15}$$

It should be noted that in calculating equations (7.14) and (7.15) the assumption $\overline{V}_i = \overline{V}_m$, which was explicit in equations (6.6) and (6.7), has been used. The results are therefore appropriate to the number-fixed frame which is equivalent to the volume-fixed frame under the above assumption.

It is of interest to consider the ternary results (7.14) for an ideal solution in which it might be supposed that

$$k_{12} = k_{13} = k_{23} \tag{7.16}$$

Substitution of these equalities into equations (7.14) and replacement of activities by mole fractions leads to the ratios

$$\left(\frac{L_{12}}{L_{11}}\right)_V = -\frac{X_2}{1 - X_1} \tag{7.17}$$

$$\left(\frac{L_{21}}{L_{22}}\right)_V = -\frac{X_1}{1 - X_2} \tag{7.18}$$

Converting to the diffusion formalism via equation (4.34) it is readily demonstrated[5] that for an ideal solution

$$(D_{11})_V = (D_{22})_V \tag{7.19}$$

$$(D_{12})_V = (D_{21})_V = 0 \tag{7.20}$$

Here the subscripts V indicate the volume-fixed reference frame. In the more general case, the k_{ij} will be unequal but would be expected to vary only slowly with composition. In dilute solutions the activity coefficients are also expected to be relatively insensitive to concentration changes. To a first approximation then, we expect

$$L_{ii} = K_{ii}X_i \tag{7.21}$$

$$L_{ij} = -K_{ij}X_iX_j \qquad i \neq j \tag{7.22}$$

where the K are slowly varying positive coefficients.[6] Such relations may be expected to be valid and physically significant for alloys in which the Kirkendall drift is small.

Similar expressions have been found by Allnatt[6] using linear response theory. Subsequent extension of that approach by Allnatt[7] has led to a slightly more general description by including processes which involve two particles of the same type. A very general treatment of this approach to the evaluation of [L] for diffusion in liquids has been given by Rice and Gray.[8]

Ternary diffusion coefficients may be calculated from the equations (4.35) using the L_{ij} evaluated above. Writing as a special case via equations (7.12) and (7.13)

$$k'_{ij} = k_{ij}\gamma_i\gamma_j \tag{7.23}$$

and evaluating the γ_i from the Wagner thermodynamic expansion (Appendix 1) making use of the dilute solution statement $X_1, X_2 \ll X_3$ we obtain

$$\begin{aligned} D_{11} &\equiv RTk'_{13}X_3(1 + \varepsilon_{11}X_1) \\ D_{12} &\equiv RT(\varepsilon_{12}k'_{13}X_1X_3 - k'_{12}X_1) \\ D_{21} &\equiv RT(\varepsilon_{12}k'_{23}X_2X_3 - k'_{12}X_2) \\ D_{22} &\equiv RTk'_{23}X_3(1 + \varepsilon_{22}) \end{aligned} \tag{7.24}$$

It is seen that even when thermodynamic cross-effects vanish, diffusional interactions will nonetheless be encountered when the kinetic model is applicable. Diffusional cross-effects vanish only when both ε_{12} and k'_{12} are zero.

7.3 TRANSITION STATE THEORY FOR A VACANCY MECHANISM

Experimental evidence suggests that to a good approximation, atoms in most metals diffuse only by exchange with neighbouring vacancies. The earlier model can be modified simply by identifying the vacancies as the Vth component, and adding the term

$$-\frac{m\lambda^2}{RT}\nu_{iV}\exp\left[-\Delta G^{\ddagger}_{iV}/kT\right]\frac{a_ia_V}{\gamma^{\ddagger}_{iV}}\left(\frac{\partial\mu_i}{\partial x} - \frac{\partial\mu_V}{\partial x}\right) \tag{7.25}$$

to the right-hand side of equation (7.8), the flux J_i now being determined in the lattice-fixed (Kirkendall) frame of reference. A basic postulate of the theory of irreversible processes requires that 'local equilibrium' be maintained throughout the system. This demands that the vacancies be everywhere in equilibrium, so that μ_V is zero, and a_V can be approximated by the equilibrium concentration X_V and incorporated in ν_{iV}. If it is presumed that the frequency of direct atom interchanges is small compared with atom-vacancy interchanges, then, with Bardeen and Herring[9] and LeClaire,[10] the single non-zero flux term is obtained

$$J_i = -\frac{m\lambda^2}{RT}\nu_{iV}\exp\left[-\Delta G^{\dagger}_{iV}/kT\right]\frac{a_i}{\gamma^{\dagger}_{iV}}\frac{\partial\mu_i}{\partial x} \tag{7.26}$$

in the Kirkendall frame of reference. When these fluxes are transformed to the volume- or number-fixed frame the resulting phenomenological coefficients must belong to an L' matrix (cf §6.2). The untransformed matrix is designated by a subscript K to identify it with the Kirkendall frame of reference and defined for a three-component system as

$$(L)_K = \begin{bmatrix} \frac{m\lambda^2}{RT}\nu_{1V}\exp\left[-\frac{\Delta G^{\dagger}_{1V}}{kT}\right]\frac{a_1}{\gamma^{\dagger}_{1V}} & 0 & 0 \\ 0 & \frac{m\lambda^2}{RT}\nu_{2V}\exp\left[-\frac{\Delta G^{\dagger}_{2V}}{kT}\right]\frac{a_2}{\gamma^{\dagger}_{2V}} & 0 \\ 0 & 0 & \frac{m\lambda^2}{RT}\nu_{3V}\exp\left[-\frac{\Delta G^{\dagger}_{3V}}{kT}\right]\frac{a_3}{\gamma^{\dagger}_{3V}} \end{bmatrix} \tag{7.27}$$

Here the diagonal elements are the intrinsic Onsager coefficients. The transformation of this matrix to the volume-fixed or laboratory frame was given in §6.2 so the matrix $[L_V]$ can be evaluated by the substitution of the matrix elements $(L)_K$. Manning, on the other hand, has argued[11, 12, 13] that this matrix should be accorded non-zero off-diagonal elements to account for vacancy wind or correlation effects. Such effects, as in binaries, may be implicit within the diagonal frequencies and, being unlikely to affect the diffusion coefficients by more than 10%, will be ignored.

By transforming $(L)_K$ to the number $N \equiv$ volume-fixed frame of reference V, via the method of Kirkwood *et al.*[14] and Lane and Kirkaldy[5] (cf §6.2), the flux vector for the number-fixed frame of reference is obtained, and by substitution of this into equation (3.45) the rate of entropy production is obtained as

$$T\sigma S = (J_1)_K\frac{\partial\mu_1}{\partial x} + (J_2)_K\frac{\partial\mu_2}{\partial x} + (J_3)_K\frac{\partial\mu_3}{\partial x} \tag{7.28}$$

This assumes that at local equilibrium, $\partial\mu_V/\partial x$, is zero, so that the term $(J_V)_K(\partial\mu_V/\partial x)$ can be added to the right-hand side of the above expression in order to recover the original rate of entropy production, given by substituting the fluxes of equation (7.26) into equation (3.45). Hence, the transformation has fulfilled the requirement that the rate of entropy production be invariant.

From equations (7.7), (7.26) and (7.27) it can be recognized that the terms $(L_{ii})_K$ are functions of the form

$$(L_{ii})_K = \frac{N\lambda^2}{RT}X_iP_i, \tag{7.29}$$

where P_i is the average number of jumps of an i atom per second, and therefore

$$\sum_{i-1}^{3}(L_{ii})_K = \frac{m\lambda^2}{RT}(X_1P_1 + X_2P_2 + X_3P_3) = \frac{m\lambda^2}{RT}P_M \quad (7.30)$$

where P_M is the average number of jumps per atom of the mixture per second. The P_i through the frequencies contain small correlation effects. The off-diagonal coefficients of equation (6.20) may now be written

$$(L_{ij})_N = -X_1X_2(P_1 + P_2 - P_M)\frac{m\lambda^2}{RT} \quad (7.31)$$

Equation (7.31) indicates that, in general, the off-diagonal L coefficients in the number-fixed frame of reference, using a simple vacancy mechanism, are given by

$$(L_{ij})_N = -K_{ij}X_iX_j \quad (7.32)$$

which is qualitatively similar to the result given by the exchange mechanism. The K_{ij} will vary slowly with composition, and will usually be positive. In dilute solutions, they can only take negative values if P_n, where n is the solvent, is greater than the combined sum $\Sigma_{i=1}^{n-1} P_i$; a condition fulfilled if there is a very strong Kirkendall shift towards the solvent-rich end of the couple.

The on-diagonal L coefficients of the matrix (7.27) may be written

$$(L_{ij})_N = \{X_iP_i(1 - 2X_i) + X_i^2P_M\}\frac{m\lambda^2}{RT}, \quad (7.33)$$

which for dilute solutions may be simplified to

$$(L_{ii})_N = K_{ii}X_i \quad (7.34)$$

where K_{ii} is a positive coefficient varying slowly with concentration. Equations (7.31) and (7.33) indicate that the L coefficients for this model are consistent with the thermodynamic requirements (relations (3.10)).

In the special case (comparable with that expressed by the conditions (7.16) for the exchange model), in which

$$P_1 = P_2 = P_3 \quad (7.35)$$

for an ideal ternary solution, we obtain from equations (7.31) and (7.33) the result

$$\frac{(L_{12})_N}{(L_{11})_N} = -\frac{X_2}{1 - X_1}, \quad (7.36)$$

and

$$\frac{(L_{21})_N}{(L_{22})_N} = -\frac{X_1}{1 - X_2} \quad (7.37)$$

The ratios of the L given by equations (7.36) and (7.37) are identical with those given by equations (7.17) and (7.18), obtained for an ideal solution subject to the direct exchange model. The individual Ls are, however, different because the activated rate processes are different. Castleman[15] has explored other features of diffusion in ternary ideal solutions.

For non-ideal solutions, the formal descriptions of the exchange mechanism and the vacancy mechanism (assuming local equilibrium) in the number-fixed frame are identical. The differences in the two models occur entirely in the interpretation of the L coefficients in terms of kinetic parameters. The concentration dependencies for dilute solutions are similar in the two cases. The only qualitative difference is that in the vacancy mechanism the off-diagonal L may conceivably take positive values, whereas in the exchange mechanism they are restricted to negative values. In the following, an approximate method of obtaining the P_i from tracer and binary diffusion data is outlined. By substituting the values of the P_i into equations (7.31) and (7.33), an estimate of the magnitudes of the L coefficients in a multicomponent system may be obtained.

The complete experimental framework for the ternary vacancy mechanism must be more detailed than for the exchange mechanism, as it is necessary to consider an additional flux. These extra data can be obtained by measurements of the Kirkendall shift or by independent tracer experiments.

For dilute multicomponent solutions, the jump probabilities P_i in the vacancy model may be obtained from tracer and binary diffusion data since correlation effects will be negligible. The solvent is designated as component n, and the diffusion of radioactive solvent atoms (or molecules), denoted by a superscript *, is considered in an otherwise pure solvent. As the radioactive species is normally very dilute and the solution very close to ideal, the number-fixed and lattice-fixed reference frames are indistinguishable, so one may write

$$(J_n^*)_\mathrm{N} = -(L_{nn}^*)_\mathrm{K}\frac{\partial \mu_n^*}{\partial x} = -m\,(D_n^*)_\mathrm{N}\frac{\partial X_n^*}{\partial x} \tag{7.38}$$

Also

$$\frac{\partial \mu_n^*}{\partial x} = \frac{RT}{X_n^*}\frac{\partial X_n^*}{\partial x} \tag{7.39}$$

which on substitution into equation (7.38) gives

$$(L_{nn}^*)_\mathrm{K} = D_n^* X_n / RT \tag{7.40}$$

The jump probability P_n is taken to be equal to P_n^*, and is obtained from equations (7.40) and (7.29) as

$$P_n = (D_n^*)_\mathrm{N} / m\lambda^2 \tag{7.41}$$

Note here that the substitution of P_n for P_n^* is only approximate because the frequency of decomposition ν_{nV}^* of the activated complex is dependent to some extent on the mass of the complex, which is different for the radioactive and normal solvent species, and because vacancy associated correlation effects are neglected (refer to § 3.11 and equations (6.117) for the binary case).

Examination of equation (7.26) indicates that the jump probabilities P_i are given by equations of the type

$$P_i = \nu_{iV}^\dagger \exp\left[-\Delta G_{iV}^\dagger / kT\right] \gamma_i / \gamma_{iV}^\dagger \tag{7.42}$$

The term $\exp[-\Delta G^{\dagger}_{iV}/kT]$ is a constant and the ratio $\gamma_i/\gamma^{\dagger}_{iV}$ is not expected to be strongly influenced by the presence of other solutes, but ν_{iV} may show some variation with the composition of a multicomponent system. However, one may proceed with the assumption that the jump probabilities P_i in a dilute multicomponent solution, are approximately independent of composition, and are therefore equivalent to the P_i in the dilute binary solutions. For each binary solution the flux of component i in the number-fixed reference frame is given by

$$(J_i)_N = -(L_{ii})_N \left(\frac{\partial \mu_i}{\partial x} - \frac{\partial \mu_n}{\partial x} \right) = -m(D_i)_N \frac{\partial X_i}{\partial x} \tag{7.43}$$

Removing the term $(\partial \mu_n/\partial x)$ from equation (7.43) by the Gibbs–Duhem relation, and substituting for $(L_{ii})_N$ from equation (7.33) results in

$$\{X_i P_i (1 - X_i)^2 + X_i^2 P_n\} \frac{m\lambda^2}{RT} \frac{1}{X_n} \frac{\partial \mu_i}{\partial x} = (D_i)_N m \frac{\partial X_i}{\partial x} \tag{7.44}$$

Substituting from equation (7.41) for P_n gives

$$P_i = \frac{X_n (D_i)_N \Big/ \left(1 + \dfrac{\partial \ln \gamma_i}{\partial \ln X_i} \right) - X_i (D_n^*)_N}{m\lambda^2 X_n^2} \tag{7.45}$$

In many cases it will be sufficient to use the value of P_i at infinite dilution, in which case equation (7.45) is simplified to

$$\lim_{X_i \to 0} P_i = \lim_{X_i \to 0} (D_i)_N / m\lambda^2 \tag{7.46}$$

By substituting the values of P_n from equation (7.41) and P_i from either equation (7.45) or equation (7.46) into equations (7.31) and (7.33), an estimate of the magnitude of the L coefficients of a multicomponent system, in the number- or volume-fixed frame of reference, may be obtained.

7.4 THE TERNARY EQUIVALENT OF THE DARKEN ANALYSIS AND EQUATION

In this section the L_K-matrix as established in the Kirkendall frame by transition state theory is converted to a mutual D-matrix in the laboratory frame. Consideration is restricted to single-phase substitutional systems, in which possible volume changes of mixing, lateral dimensional changes, and pore formation can be neglected. Indeed, the existence of a rather strict local equilibrium for all chemical species, and for the vacancies as well, will be assumed. Therefore, discussion is restricted to diffusion processes that occur with comparatively small overall composition changes. This should not usually represent a serious restriction, since interest is most often confined to terminal solutions of limited solubility range.

The flux equations are written in terms of the concentration (mol cm^{-3}) gradients or of the chemical potential (J mol^{-1}) gradients in either the number-fixed or the lattice-fixed (Kirkendall) frame of reference. For example, in the number-fixed frame in which component 3 is the solvent,

$$J_1 = -L_{11}\frac{\partial(\mu_1 - \mu_3)}{\partial x} - L_{12}\frac{\partial(\mu_2 - \mu_3)}{\partial x} \tag{7.47}$$

or upon substitution of $\mu_1 = \mu_1(m_1, m_2)$ and $\mu_2 = \mu_2(m_1, m_2)$,

$$J_1 = -D_{11}(\partial m_1/\partial x) - D_{12}(\partial m_2/\partial x) \tag{7.48}$$

The coefficients in these relations are designated as the chemical or mutual Ls and Ds. In the frame of reference moving with the embedded Kirkendall markers (the lattice-fixed frame), the corresponding fluxes are in the most general case

$$J_1' = -L_{11}'\frac{\partial\mu_1}{\partial x} - L_{12}'\frac{\partial\mu_2}{\partial x} - L_{13}'\frac{\partial\mu_3}{\partial x} \tag{7.49}$$

whereby, assuming local equilibrium, the chemical potential of vacancies has been set as $\mu_V = 0$, and

$$J_1' = -D_{11}'\frac{\partial m_1}{\partial x} - D_{12}'\frac{\partial m_2}{\partial x} \tag{7.50}$$

where it is understood that the concentration of component 3 has been chosen as the dependent one. Here the coefficients are designated as intrinsic, understanding that $L' = L_K$ of the previous section. Equations (7.47) and (7.48) are subject to the mass balance

$$J_1 + J_2 + J_3 = 0 \tag{7.51}$$

while equations (7.49) and (7.50) are subject to

$$J_1' + J_2' + J_3' = -J_V' = -(m_1 + m_2 + m_3)\,v = -mv \tag{7.52}$$

where J_V' is the vacancy flux and v is the marker velocity relative to the number-fixed frame. Using the Gibbs–Duhem equation, μ_3 could be eliminated from equations (7.47) and (7.49), but it is preferable to leave these fully expanded for the moment. In this format, the Onsager reciprocal relations are satisfied[5] as

$$L_{12} = L_{21} \tag{7.53}$$

and although the L's are not unique, it is always possible to choose a set such that

$$L_{ik}' = L_{ki}' \tag{7.54}$$

All of these coefficients can be expressed in terms of one another by appropriate linear transformations. The most interesting are those that convert the mutual and intrinsic coefficients. These are obtained[16, 17] by noting that

$$J_i = J_i' + m_i v \tag{7.55}$$

substituting equation (7.50) in (7.52) and solving for v. This yields

$$v = \frac{1}{m_1 + m_2 + m_3}\left\{(D_{11}' + D_{21}' + D_{31}')\frac{\partial m_1}{\partial x} + (D_{12}' + D_{22}' + D_{32}')\frac{\partial m_2}{\partial x}\right\} \tag{7.56}$$

and writing equation (7.55) in terms of Ds and mole fractions X_i gives

$$J_1 = -\{X_2 + X_3)D'_{11} - X_1(D'_{21} + D'_{31})\}\frac{\partial m_1}{\partial x}$$

$$-\{X_2 + X_3)D'_{12} - X_1(D'_{22} + D'_{32})\}\frac{\partial m_2}{\partial x} \quad (7.57)$$

Comparison of equations (7.57) and (7.48) yields for example

$$D_{11} = (X_2 + X_3)D'_{11} - X_1(D'_{21} + D'_{31}) \quad (7.58)$$

and

$$D_{12} = (X_2 + X_3)D'_{12} - X_1(D'_{22} + D'_{32}) \quad (7.59)$$

The determination of the six intrinsic coefficients by chemical diffusion and marker experiments represents a formidable task as described by Kirkaldy and Lane.[18] Fortunately, there is some agreement that the off-diagonal L' coefficients can be approximated by zero in substitutional systems undergoing diffusion by a vacancy mechanism[10,5] (see previous section). Thus, an approximation may be made as in equation (7.27)

$$J'_1 = -L'_{11}\frac{\partial \mu_1}{\partial x} = -D'_{11}\frac{\partial m_1}{\partial x} - D'_{12}\frac{\partial m_2}{\partial x} \quad (7.60)$$

$$J'_2 = -L'_{22}\frac{\partial \mu_2}{\partial x} = -D'_{21}\frac{\partial m_1}{\partial x} - D'_{22}\frac{\partial m_2}{\partial x} \quad (7.61)$$

$$J'_3 = -L'_{33}\frac{\partial \mu_3}{\partial x} = -D'_{31}\frac{\partial m_1}{\partial x} - D'_{32}\frac{\partial m_2}{\partial x} \quad (7.62)$$

Noting that μ_1, μ_2, and μ_3 are all functions of m_1 and m_2

$$D'_{11} = L'_{11}\partial\mu_1/\partial m_1 \quad (7.63)$$

$$D'_{22} = L'_{22}\partial\mu_2/\partial m_2 \quad (7.64)$$

$$D'_{31} = L'_{33}\partial\mu_3/\partial m_1 \quad (7.65)$$

$$D'_{12}/D'_{11} = \frac{\partial \mu_1}{\partial m_2}\Big/\frac{\partial \mu_1}{\partial m_1} \quad (7.66)$$

$$D'_{21}/D'_{22} = \frac{\partial \mu_2}{\partial m_1}\Big/\frac{\partial \mu_2}{\partial m_2} \quad (7.67)$$

$$D'_{32}/D'_{31} = \frac{\partial \mu_3}{\partial m_2}\Big/\frac{\partial \mu_3}{\partial m_1} \quad (7.68)$$

are obtained. Noting with Darken[19] and Le Claire[10] the approximate relationship between the intrinsic coefficient L'_{ii} and the tracer coefficient D^*_{ii} (also deducible from relation (7.27); see §2.14 for the binary analogue)

$$L'_{ii} = m_i D^*_{ii}/RT$$

equations (7.63)–(7.65) can be used to obtain

$$D'_{11} = \frac{m_1 D^*_{11}}{RT} \frac{\partial \mu_1}{\partial m_1} \quad (7.69)$$

$$D'_{22} = \frac{m_2 D^*_{22}}{RT} \frac{\partial \mu_2}{\partial m_2} \quad (7.70)$$

$$D'_{31} = -\frac{D^*_{33}}{RT}\left(m_1 \frac{\partial \mu_1}{\partial m_1} + m_2 \frac{\partial \mu_2}{\partial m_1} \right) \quad (7.71)$$

These latter are accurate only for dilute solutions since vacancy correlation effects are neglected. It can therefore be seen that *a priori* knowledge of the three independent thermodynamic derivatives of the form $\partial \mu_i / \partial m_k$ allows one to calculate the six intrinsic *D*s. In general, however, the appropriate thermodynamic data are not available. Alternatively, a measurement of the three independent mutual *D*s (one is dependent through the Onsager relation (7.53)) would suffice, since the nine independent equations including (7.58), (7.59), and (7.66)–(7.71) can be used to eliminate the three independent thermodynamic derivatives, thus leaving six relations for the determination of the six intrinsic *D*s. Algebraic manipulation leads to

$$D'_{11} = \frac{D^*_{11}}{A} [\{(1 - X_2)D^*_{22} + X_2 D^*_{33}\} D_{11} + \{X_1(D^*_{22} - D^*_{33})\} D_{21}] \quad (7.72)$$

$$D'_{12} = \frac{D^*_{11}}{A} [\{X_1(D^*_{22} - D^*_{33})\} D_{22} + \{(1 - X_2)D^*_{22} + X_2 D^*_{33}\} D_{12}] \quad (7.73)$$

$$D'_{22} = \frac{D^*_{22}}{A} [\{(1 - X_1)D^*_{11} + X_1 D^*_{33}\} D_{22} + \{X_2(D^*_{11} - D^*_{33})\} D_{12}] \quad (7.74)$$

$$D'_{21} = \frac{D^*_{22}}{A} [\{X_2(D^*_{11} - D^*_{33})\} D_{11} + \{(1 - X_1)D^*_{11} + X_1 D^*_{33}\} D_{21}] \quad (7.75)$$

$$D'_{31} = -\frac{D^*_{33}}{A} [\{(1 - X_2)D^*_{22} + N_2 D^*_{11}\} D_{11} + \{(1 - X_1)D^*_{11} + X_1 D^*_{22}\} D_{21}] \quad (7.76)$$

$$D'_{32} = -\frac{D^*_{33}}{A} [\{(1 - X_1)D^*_{11} + X_1 D^*_{22}\} D_{22} + \{(1 - X_2)D^*_{22} + X_2 D^*_{11}\} D_{12}] \quad (7.77)$$

where

$$A = X_1 D^*_{22} D^*_{33} + X_2 D^*_{11} D^*_{33} + X_3 D^*_{11} D^*_{33} \quad (7.78)$$

It can therefore be seen that, as in the binary case, a combination of relatively straightforward tracer and chemical diffusion experiments can yield all of the intrinsic data for the system, and from this both the Kirkendall effect and some thermodynamic data can be calculated. In view of the inherent experimental and analytic difficulties of marker experiments in ternary systems, it is recommended that these be rejected in favor of this tracer-chemical combination.

Alternate analyses of ternary intrinsic behaviour and the Darken structure have been given by Philibert and Guy,[20, 21] Ziebold and Cooper,[22] and Schönert.[23]

7.5 THE KIRKENDALL EFFECT IN RELATIVELY DILUTE TERNARY SOLUTIONS

For many metallic solutions in the dilute range up to about 10 at.-%, the activity coefficients can be adequately represented by Taylor expansions[24]

$$\ln \gamma_1 = \ln \gamma_1^0 + \varepsilon_{11} X_1 + \varepsilon_{12} X_2 \tag{7.79}$$

$$\ln \gamma_2 = \ln \gamma_2^0 + \varepsilon_{21} X_1 + \varepsilon_{22} X_2 \tag{7.80}$$

and

$$\ln \gamma_3 = \varepsilon_{11} X_1^2/2 + \varepsilon_{12} X_1 X_2 + \varepsilon_{22} X_2^2/2 \tag{7.81}$$

subject to the Maxwell symmetry relation

$$\varepsilon_{12} = \varepsilon_{21} \tag{7.82}$$

These interaction parameters are unitless quantities of the magnitude of pair interaction energies divided by kT. Typical values for ε_{12} might be ± 5. If these values are substituted in equations (7.66)–(7.68) and (7.69)–(7.71)

$$D'_{12} = \frac{\varepsilon_{12} X_1}{1 + \varepsilon_{11} X_1} D'_{11} \tag{7.83}$$

$$D'_{21} = \frac{\varepsilon_{12} X_2}{1 + \varepsilon_{22} X_2} D'_{22} \tag{7.84}$$

$$D'_{32} = D'_{31}\{1 + \varepsilon_{12}(X_2 - X_1) + (\varepsilon_{11} X_1 - \varepsilon_{22} X_2)\} \tag{7.85}$$

$$D'_{11} = D^*_{11}(1 + \varepsilon_{11} X_1) \tag{7.86}$$

$$D'_{22} = D^*_{22}(1 + \varepsilon_{22} X_2) \tag{7.87}$$

$$D'_{31} = -D^*_{33}\{1 - (\varepsilon_{11} X_1 + \varepsilon_{12} X_2)\} \tag{7.88}$$

are obtained. Consider a Darken-type diffusion couple[16] for a substitutional solution in which both solutes show an appreciable binary Kirkendall effect. An experimental couple is shown in Fig. 7.1 in which initially uniform tin in copper has been weakly redistributed as a result of a strong superposed zinc gradient.[25] Although the interaction parameter for this system is quite large (~8·5), the redistribution is nonetheless rather weak because of the similarity of values of the mutual diffusion coefficients D_{ii} for the solutes tin and zinc.[26] This is quite common behaviour in substitutional solutions. It means that the Boltzmann-type integral, equation (5.12) expressed in terms of intrinsic coefficients, may be investigated qualitatively for this couple following Kirkaldy and Lane[18] (with Sn = 1, Zn = 2, Cu = 3) by setting $\mathrm{d}m_1/\mathrm{d}\lambda = 0$ and writing (with $m_i(\pm\infty) = m_i^{\pm}$; $m_i(\mathrm{K})$ at marker)

$$\int_{m_2^+}^{m_2(\mathrm{K})} \lambda \, \mathrm{d}m_2 - \lambda_0 m_2 \simeq -2D'_{22} \left.\frac{\mathrm{d}m_2}{\mathrm{d}\lambda}\right|_{\mathrm{K}} \tag{7.89}$$

and

$$\int_{m_3^+}^{m_3(\mathrm{K})} \lambda \, \mathrm{d}m_3 - \lambda_0 m_3 \simeq -2D'_{32} \left.\frac{\mathrm{d}m_2}{\mathrm{d}\lambda}\right|_{\mathrm{K}} \tag{7.90}$$

where $\lambda = x/\sqrt{t}$ and λ_0 is the location of the marker in λ-space. Note that in our approximation $dm_2 \simeq -dm_3$, so that on adding

$$\lambda_0 \simeq \frac{2}{m_2 + m_3}(D'_{22} + D'_{32})\frac{dm_2}{d\lambda}\bigg|_K \tag{7.91}$$

This states what is perhaps obvious, that a ternary diffusion couple with one component that remains substantially uniform has a Kirkendall effect essentially equal to that of the remnant binary combination modified by the concentration dependence of the intrinsic coefficients (cf equations (2.60), (7.85), (7.87), and (7.88)).

If the boundary conditions for components 1 and 2 are interchanged so as to give a second couple with a concentration coordinate common to the first, we have

$$\lambda_0 = \frac{2}{m_1 + m_3}(D'_{11} + D'_{31})\frac{dm_1}{d\lambda}\bigg|_K \tag{7.92}$$

It may often happen that this displacement will have the opposite sign to the other. Consequently, it would be impossible in such a system to find pairs of Darken-type couples (which are the most sensitive to ternary interactions) in terminal solutions to which the Boltzmann analysis for intrinsic coefficients may be applied. This is a further reason for avoiding marker experiments for intrinsic coefficient determinations.

Finally, it is of interest to consider the general ternary relation for the Kirkendall shift in λ-space when both independent gradients are non-zero. This can be evaluated from equation (7.56) and $\lambda = x/\sqrt{t}$ as (cf §2.13)

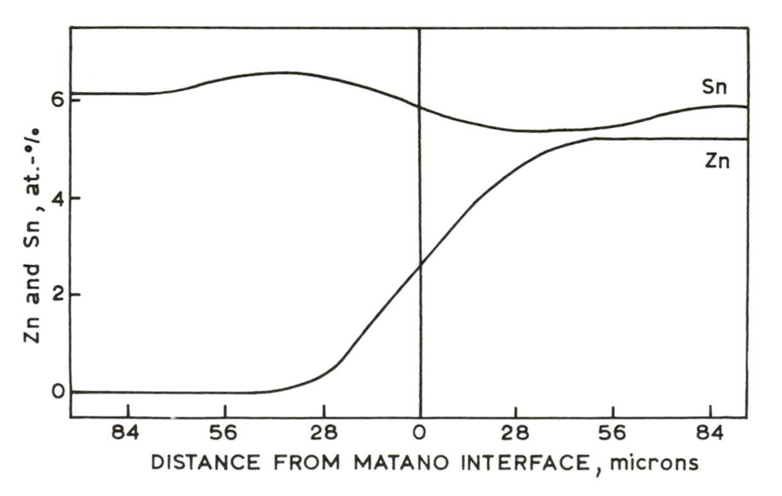

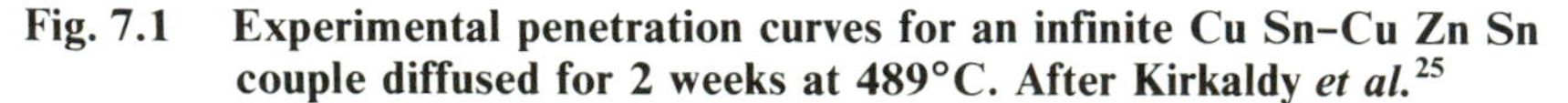

Fig. 7.1 Experimental penetration curves for an infinite Cu Sn–Cu Zn Sn couple diffused for 2 weeks at 489°C. After Kirkaldy *et al.*[25]

$$\lambda_0 = \frac{1}{m_1 + m_2 + m_3}\left[(D'_{11} + D'_{21} + D'_{31})\frac{\mathrm{d}m_1}{\mathrm{d}\lambda}\bigg|_{\mathrm{K}} + (D'_{12} + D'_{22} + D'_{32})\frac{\mathrm{d}m_2}{\mathrm{d}\lambda}\bigg|_{\mathrm{K}}\right] \tag{7.93}$$

Restricting consideration to relatively dilute solutions so that the interaction terms D'_{21} and D'_{12} may be neglected relative to the others, one obtains

$$\lambda_0 \simeq \frac{1}{m_1 + m_2 + m_3}\left\{(D'_{11} + D'_{31})\frac{\mathrm{d}m_1}{\mathrm{d}\lambda}\bigg|_{\mathrm{K}} + (D'_{22} + D'_{32})\frac{\mathrm{d}m_2}{\mathrm{d}\lambda}\bigg|_{\mathrm{K}}\right\} \tag{7.94}$$

We see, therefore, that the observed effect will be simply the sum of the two binary effects weighted according to the relative gradients (and, therefore, in part to the driving composition differences) of the two independent components. Further simplification may be obtained from equation (7.85) which indicates that D'_{31} and D'_{32} are approximately equal in dilute solutions, so that two components with equal diffusivities D'_{11} and D'_{22}, arranged so that their gradients are equal and opposite, would have a zero Kirkendall effect.

The fact that the Kirkendall effect in a ternary solution can be interpreted in terms of binary behaviour suggests that the ternary phenomenon has no unique theoretical interest. Thus, as we have already concluded on grounds of experimental economy, effort should be directed towards the determination of the intrinsic Ds independently of the Kirkendall effect. These do have a fundamental theoretical interest, and their determination as described previously yields as a by-product some thermodynamic data as well as the possibility of calculating the Kirkendall shifts.

A solution of very common metallurgical interest is a ternary solution consisting of one interstitial component (1) and two substitutional components, including the solvent (3).[27,28,29,30] In the treatment of substitutional solutions, the mutual diffusion coefficients and the marker velocity refer to the number-fixed coordinate system. As the partial molar volumes of the substitutional components are very nearly equal, and the solutions dilute, the number-fixed and volume-fixed reference frames are experimentally indistinguishable, so that the previous analysis is equally valid in both coordinate systems. Since a dilute interstitial component contributes negligibly to the volume, the number- and volume-fixed reference frames do not coincide in a system with a non-zero flux of an interstitial component. In the following treatment of this type of system, the volume-fixed rather than the number-fixed frame of reference is used, because the resultant analysis is much simpler. The equations equivalent to (7.52) and (7.53) are

$$J_2 + J_3 = 0 \tag{7.95}$$

and

$$J'_2 + J'_3 = -J'_V = -(m_2 + m_3)v \tag{7.96}$$

Equation (7.47) is altered so that $\partial(\mu_1 - \mu_3)$ becomes $\partial\mu_1$, but the other equations are formally unchanged. The new Darken analysis involving equations (7.95), (7.96), and (7.56) yields modified linear relations between the mutual and intrinsic coefficients, while the Kirkendall shift, given by

$$v = \frac{1}{m_2 + m_3}\{J'_2 + J'_3\} \tag{7.97}$$

is simplified by the absence of explicit terms in the interstitial flux. The effect of the interstitial component appears, therefore, in the concentration dependence of the intrinsic diffusion coefficients and in the cross-terms in the substitutional flux equations. Since interstitial solutions of interest are invariably dilute, these effects will tend to be negligible. Consequently a Kirkendall behaviour that is characteristic of the substitutional binary pair can be expected. Note that an initially uniform distribution of interstitial will act as a distributed marker for the Kirkendall drift.

7.6 DIFFUSION AT AND NEAR TERNARY CRITICAL STATES

The problem of chemical diffusion near critical points has gained relevance through the publication of pertinent data.[31, 32] Since the diffusion matrix is proportional to the product of the positive definite Onsager mobility matrix and the symmetric Hessian of the Gibbs' free energy matrix, which describes the stability character of isothermal, isobaric chemical systems, it follows that the kinetic character of critical states will be determined in part by the equilibrium criteria. It can be demonstrated, in particular, that the determinant of the ternary diffusion matrix is stationary and vanishes in all critical states, and that all elements of the matrix are also stationary and vanish in a particular state designated as a ternary critical point. The consequences in relation to solutions of the diffusion equations in the neighbourhood of critical points are investigated. This problem has been considered by Kirkaldy,[17] Sundelöf,[33] de Fontaine,[34] Kirkaldy and Purdy,[35] and Morral and Cahn.[36]

Treatments of the thermodynamic equilibrium problem have been given many times.[37, 38, 39] A formulation is presented here which is particularly pertinent to the ternary diffusion problem. We assume with the others that the molar Gibbs' function G is differentiable to the third order in isothermal, isobaric critical states. It follows therefore that necessary conditions for critical states are

$$d^2G = 0 \tag{7.98}$$

and

$$d^3G = 0 \tag{7.99}$$

Reference to Appendices 1 and 4 is advised at this juncture.

Consider a system of arbitrary mass whose Gibbs' free energy

$$G' = \mu_1 n_1 + \mu_2 n_2 + \mu_3 n_3 \tag{7.100}$$

where μ are partial molar free energies (chemical potentials) and n are mole numbers. Since G' is a homogeneous function of the mole numbers, the Euler theorem gives

$$n_1 d\mu_1 + n_2 d\mu_2 + n_3 d\mu_3 = 0 \tag{7.101}$$

which is the familiar Gibbs–Duhem equation for isothermal, isobaric systems. Dividing both equations (7.100) and (7.101) by the total moles $n_1 + n_2 + \ldots + n_r$, differentiating equation (7.100), and substituting equation (7.101) yields for the molar free energy in terms of mole fractions, X_i,

$$\begin{aligned} dG &= \mu_1\, dX_1 + \mu_2\, dX_2 + \mu_3\, dX_3 \\ &= (\mu_1 - \mu_3)\, dX_1 + (\mu_2 - \mu_3)\, dX_2 \end{aligned} \tag{7.102}$$

It follows, therefore, that

$$d^2G = d(\mu_1 - \mu_3)\, dX_1 + d(\mu_2 - \mu_3)\, dX_2 + (\mu_1 - \mu_3)\, d^2X_1 + (\mu_2 - \mu_3)\, d^2X_2 \tag{7.103}$$

Without loss in generality, variations along lines can be considered in the composition plane with $d^2X_1 = d^2X_2 = 0$. Hence, d^2G can be written in the quadratic form

$$d^2G = \frac{\partial(\mu_1 - \mu_3)}{\partial X_1} dX_1^2 + \left[\frac{\partial(\mu_1 - \mu_3)}{\partial X_2} + \frac{\partial(\mu_2 - \mu_3)}{\partial X_1}\right] dX_1\, dX_2 + \frac{\partial(\mu_2 - \mu_3)}{\partial X_2} dX_2^2 \tag{7.104}$$

The properties of the solution and its transition points are therefore determined by the properties of the matrix of coefficients. As shown in §6.3

$$\mu_{ik} = \frac{\partial(\mu_i - \mu_3)}{\partial X_k} = \frac{\partial(\mu_k - \mu_3)}{\partial X_i} = \mu_{ki} \tag{7.105}$$

That is to say, the matrix $[\mu_{ik}]$ is also real and symmetric and equation (7.104) has the standard quadratic form. Thus, the solution properties can be expressed in terms of the varying character of the matrix, namely positive definite, positive semidefinite, indefinite, or negative definite (cf Appendix 4).

Let us now investigate the complete set of fluctuations in the composition plane defined by

$$X_2 = \alpha X_1 + b \tag{7.106}$$

or

$$dX_2 = \alpha\, dX_1 \tag{7.107}$$

where α is an arbitrary constant. Let ds be the fluctuation path where

$$ds = [(dX_1)^2 + (dX_2)^2]^{1/2} = dX_1(1 + \alpha^2)^{1/2} \tag{7.108}$$

Equation (7.103) may then be written as

$$(1 + \alpha^2)\frac{d^2G}{ds^2} = \frac{d(\mu_1 - \mu_3)}{dX_1} + \alpha\, \frac{d(\mu_2 - \mu_3)}{dX_1} \tag{7.109}$$

The character of the quadratic form (7.104) is such that it will be positive or zero (positive semidefinite of order 1) when

$$\det\, [\mu_{ik}] = \mu_{11}\mu_{22} - \mu_{21}\mu_{12} = 0 \tag{7.110}$$

If α exists such that equations (7.98) and (7.110) are simultaneously satisfied, then equation (7.110) defines a surface in $X_1 - X_2 - T$ space and equation (7.109) defines a direction α in this surface for which $d^2G = 0$. This is known as the spinodal surface.[38]

Writing equation (7.109) in the form

$$\mu_{11} + \alpha\mu_{12} + \alpha(\mu_{21} + \alpha\mu_{22}) = 0 \tag{7.111}$$

and substituting (7.110), we obtain spinodal relations expressed in terms of that fluctuation direction α for which $d^2G/ds^2 = 0$,

$$\mu_{12} = \mu_{21} = -(1/\alpha)\mu_{11} \tag{7.112}$$

and

$$\mu_{22} = (1/\alpha^2)\mu_{11} \tag{7.113}$$

There might also exist positive semidefinite spinodal states of order 2, for which one (and therefore all) elements of $[\mu_{ik}]$ as well as its determinant are zero. This corresponds to the situation where the fluctuation direction α can take any value (cf equation (7.111)).

Critical states must satisfy equation (7.99) as well as equation (7.98) where from equation (7.109)

$$(1 + \alpha^2)^{3/2}\frac{d^3G}{ds^3} = \frac{d^2(\mu_1 - \mu_3)}{dX_1^2} + \alpha\frac{d^2(\mu_2 - \mu_3)}{dX_1^2} \tag{7.114}$$

Accordingly, there will exist a line of critical states on the spinodal surface defined by the solutions of equations (7.98), (7.99), (7.109), and (7.114), and elimination of α. A positive semidefinite state of order 2 is most likely to exist at the temperature maximum of the line of critical points (the top of the miscibility gap). The normal critical states are designated as 'pseudobinary' since fluctuations are described by a unique α and the latter special state as 'ternary' since critical fluctuations can be in any direction (α is indeterminate). The 'pseudobinary' points are often designated as 'plait points'.[39]

Furthermore, on the basis of analytic continuation it is proposed that, incrementally inside the spinodal surface, the matrix $[\mu_{ik}] = \mu$ is indefinite ($d^2G \lesseqgtr 0$) and that below a ternary critical point it is negative definite ($d^2G < 0$).

The absence of thermodynamic data will usually militate against the direct use of the foregoing relations. However, the values of α which satisfy equations (7.112), (7.113), and (7.114) are often known experimentally since they represent the slope of the tie lines at the limit of the pseudobinary critical points in the set of isotherms of the phase constitution diagram (see Figs. 7.2 and 7.3).

Diffusion in a ternary system in which all components have the same partial molar volume is described in Onsager's formalism by the flux–force equations (4.14) and (4.34). The diffusion coefficient matrix $[D_{ik}]$, in matrix notation, satisfies

$$m[D_{ik}] = [L_{ij}][\mu_{jk}] \tag{7.115}$$

For $n = 3$, the matrix product

$$[L_{ij}][\mu_{jk}] = m[D_{ik}] = \begin{bmatrix} L_{11}\mu_{11} + L_{12}\mu_{21} & L_{11}\mu_{12} + L_{12}\mu_{22} \\ L_{21}\mu_{11} + L_{22}\mu_{21} & L_{21}\mu_{12} + L_{22}\mu_{22} \end{bmatrix} \tag{7.116}$$

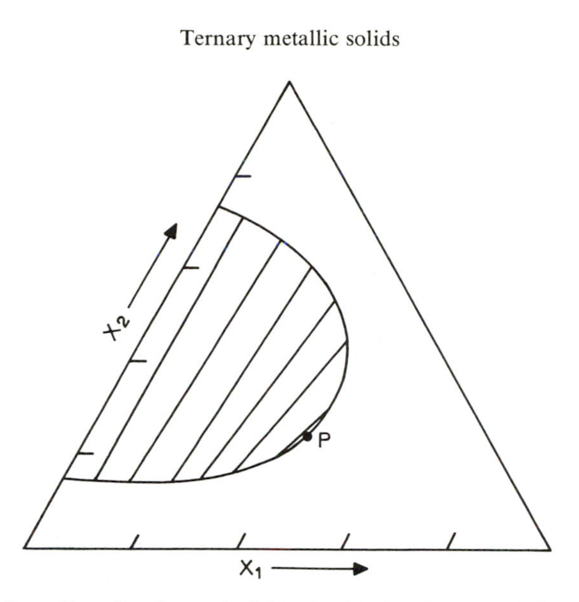

With permission from Canadian Journal of Physics, National Research Council of Canada

Fig. 7.2 Schematic miscibility gap in a ternary isotherm with a pseudobinary critical point *P*. After Kirkaldy and Purdy[35]

Equations (7.112) and (7.113) define a spinodal surface, so upon substitution

$$m[D_{ik}] = \begin{bmatrix} \left(L_{11} - \frac{1}{\alpha}L_{12}\right)\mu_{11} & -\frac{1}{\alpha}\left(L_{11} - \frac{1}{\alpha}L_{12}\right)\mu_{11} \\ \left(L_{21} - \frac{1}{\alpha}L_{22}\right)\mu_{11} & -\frac{1}{\alpha}\left(L_{21} - \frac{1}{\alpha}L_{22}\right)\mu_{11} \end{bmatrix} \tag{7.117}$$

is obtained and we see that the determinant $|D_{ik}|$ vanishes. This result, of course, follows directly from equations (7.110) and (7.115) since the determinant product follows the same multiplication rule as the matrix product and $m|D_{ik}| = |L_{ij}| \cdot |\mu_{jk}|$.[17, 33, 34]

Note also from equation (4.44) that the eigenvalues of the matrix become

$$\lambda_2 = 0 \qquad \lambda_1 = D_{11} + D_{22} > 0 \tag{7.118}$$

For the special case where the limiting tie line parallels the X_2 axis, $\alpha = \infty$ and

$$m[D_{ik}] : \begin{bmatrix} L_{11}\mu_{11}0 \\ L_{21}\mu_{11}0 \end{bmatrix} \tag{7.119}$$

and for a ternary critical point $[\mu_{ik}] = 0$, so

$$m[D_{ik}] = \begin{bmatrix} 0 & 0 \\ 0 & 0 \end{bmatrix} \tag{7.120}$$

That is, all four diffusion coefficients and both eigenvalues of $[D_{ik}]$ vanish.

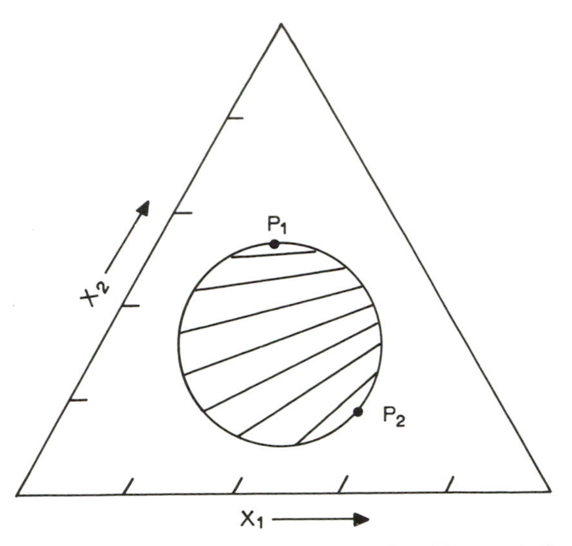

With permission from Canadian Journal of Physics, National Research Council of Canada

Fig. 7.3 Schematic closed miscibility gap in a ternary isotherm with pseudo-binary critical points P_1 and P_2. The coincidence of P_1 and P_2 at higher temperature could conceivably define a ternary critical point. After Kirkaldy and Purdy[35]

Further information can be obtained by investigating the consequences of $d^3G = 0$ (equation (7.99)). Since the eigenvalues of the μ matrix, $\lambda_{i\mu}$, satisfy

$$d^2G = \lambda_{1\mu} dX_1'^2 + \lambda_{2\mu} dX_2'^2 \quad (7.121)$$

where the eigenvectors dX_1' and dX_2' are linear combinations of dX_1 and dX_2, it follows that for variations in the $dX_2' = ds$ direction that

$$\frac{d^2G}{dX_2'^2} = \frac{d^2G}{ds^2} = \lambda_{2\mu} \quad (7.122)$$

It follows, furthermore, from equations (4.44), (7.116), and the binomial expansion that in the vicinity of the critical point

$$\lambda_2 = \frac{|L|(\mu_{11} + \mu_{22})}{m^2(D_{11} + D_{22})}\lambda_{2\mu} \quad (7.123)$$

or

$$\lambda_2 = \frac{|L|(\mu_{11} + \mu_{22})}{m^2(D_{11} + D_{22})}\frac{d^2G}{ds^2} \quad (7.124)$$

This is the ternary analogue of the well-known binary expression (equation (1.183))

$$\widetilde{D} = \frac{L}{m}\frac{d^2G}{dX_2^2} \quad (7.125)$$

To the extent that the elements of the L matrix are slowly varying in the vicinity of the critical point, we may write

$$\frac{d\lambda_2}{ds} = (\text{constant}) \frac{d^3G}{ds^2} \tag{7.126}$$

so at the critical point, in view of $d^2G = d^3G = 0$,

$$\lambda_{2\mu} = \lambda_2 = \frac{d\lambda_{2\mu}}{ds} = \frac{d\lambda_2}{ds} = 0 \tag{7.127}$$

That is to say, one eigenvalue of both $[\mu]$ and $[D]$ is zero and stationary at the critical point. We see also that the determinants of the two matrices are not only zero but stationary at the critical point. Since d^2G is also stationary, it follows that relations (7.112) and (7.113) hold not only at the critical point but also incrementally outside it and therefore matrix (7.117) is valid in the same range. It follows directly that all the elements of μ and D are stationary (minima or maxima) with respect to variations parallel to the limiting tie line.

The foregoing development can yield no information about the temperature variation of the coefficients or of their variation in the composition plane at angles to the limiting tie line since equations (7.98) and (7.99) define the free energy surface in relation to a single direction. One may suppose, however, that plait points on a free energy surface define a positive second derivative ($d^2G > 0$) for directions oblique to the limiting tie line in the composition plane. It may therefore be inferred that the determinant of the diffusion coefficients and all its elements will vary monotonically through the critical point in the composition plane. Furthermore, the necessary shape of the free energy curve at a plait point indicates that d^2G is indefinite in the temperature coordinate below the critical point. Hence $|D_{ik}|$, which vanishes at a plait point, will change from positive to negative on cooling through such a point.

If the relative changes in concentration within a ternary system are small, the diffusion coefficients may be approximated by constants. Solutions may then be sought of the form[40, 41, 27, 17, 42]

$$m_i = a_{i0} + \sum_{k=1}^{2} a_{ik} m^k \tag{7.128}$$

where m^k is a solution of the binary equations (cf §4.6)

$$\frac{\partial m^k}{\partial t} = \lambda_k \frac{\partial^2 m^k}{\partial x^2} \tag{7.129}$$

For the ternary system, the $\lambda_{1,2}$ are given by equation (4.44). To each eigenvalue λ_k corresponds an eigenfunction m^k and a column eigenvector $[a_k]$ which is determined to within a multiplicative constant, the latter being determined by the boundary conditions. The eigenvectors will be mutually orthogonal if, and only if, the D matrix is symmetric.

For a pseudobinary critical point in which the diffusion determinant vanishes, the eigenvalues are given by equations (7.118). Since our solutions (7.128) are only valid for constant D, attention must be restricted to incremental perturbation of the concentration distribution as the initial condition. Such perturbations may be represented by eigenfunctions of the form (cf §4.6)

$$m^k(\lambda_k) = \int_{-\infty}^{\infty} A_k(\beta)\exp(-\beta^2\lambda_k t)\exp(i\beta x)\,d\beta \tag{7.130}$$

where $A_k(\beta)$ is related to the Fourier transform of the initial perturbation. It can therefore be seen that $m^2(\lambda_2)$ is stationary while $m^1(\lambda_1)$ is a function which regresses in time. From the characteristic equation (4.70) and the matrix (7.112), it can be seen that

$$\frac{a_{22}}{a_{12}} = -\frac{D_{11}-\lambda_2}{D_{12}} = -\frac{D_{11}}{D_{12}} = \alpha = \frac{dX_2}{dX_1} \tag{7.131}$$

and

$$\frac{a_{21}}{a_{11}} = -\frac{D_{11}-\lambda_1}{D_{12}} = \frac{D_{22}}{D_{12}} = \frac{L_{21}-\frac{1}{\alpha}L_{22}}{L_{11}-\frac{1}{\alpha}L_{12}} \tag{7.132}$$

That is, in the case of equation (7.131), the eigenvector is parallel to the limiting tie line. In the second case the result is not transparent. However, if the D matrix happened to be symmetric, then

$$\frac{a_{21}}{a_{11}} = -\frac{1}{\alpha} = -\frac{dX_1}{dX_2} \tag{7.133}$$

That is to say the eigenvector (a_{11}, a_{21}) is orthogonal to the limiting tie line. In the usual case, this eigenvector will be oblique to the tie line in the composition plane. This vector corresponds to the direction of fluctuation which will undergo the maximum rate of relaxation of a fluctuation.

For a ternary critical point

$$\lambda_{1,2} = 0 \tag{7.134}$$

so both terms in the solution will be stationary in time. Since the corresponding eigenvectors are indeterminate, perturbations in any direction will remain stationary.

Directly below a pseudobinary critical point in the temperature coordinate, d^2G is indefinite, i.e. the composition point lies on a saddle surface. Equivalently, $|D_{ik}|$ changes from positive to negative, and equation (4.44) yields a positive and a negative eigenvalue. This condition will also hold at subcritical composition points which lie within the spinodal composition–temperature surface swept out by the line $d^2G = 0$.[38, 39] In the absence of gradient or strain energy effects (cf § 12.5; long wavelengths in a liquid) the solutions following a perturbed initial condition are (e.g. λ_2 negative and λ_1 positive)

$$m_1 = a_{10} + \int_{-\infty}^{\infty} a_{11}'A_1(\beta)\exp(-\beta^2\lambda_1 t)\exp(i\beta x)\,d\beta + \int_{-\infty}^{\infty} a_{12}'A_2(\beta)\exp(-\beta^2\lambda_2 t)\exp(i\beta x)\,d\beta \tag{7.135}$$

and

$$m_2 = a_{20} + \int_{-\infty}^{\infty} a_{21}' A_1(\beta) \exp(-\beta^2\lambda_1 t) \exp(i\beta x)\, d\beta$$
$$+ \int_{-\infty}^{\infty} a_{22}' A_2(\beta) \exp(-\beta^2\lambda_2 t) \exp(i\beta x)\, d\beta \quad (7.136)$$

The Fourier coefficients $A_i(\beta)$ and the coefficients $a_{ik} = a_{ik}' A_k(\beta)$ are obtained by the Fourier inversion of

$$m_1(x, 0) - a_{10} = \int_{-\infty}^{\infty} [a_{11}' A_1(\beta) + a_{12}' A_2(\beta)] \exp(i\beta x)\, d\beta \quad (7.137)$$

and

$$m_2(x, 0) - a_{20} = \int_{-\infty}^{\infty} [a_{21}' A_1(\beta) + a_{22}' A_2(\beta)] \exp(i\beta x)\, d\beta \quad (7.138)$$

and application of the characteristic equations and the boundary conditions at infinity.

Equations (7.135) and (7.136) are the solutions for incremental spinodal decomposition of a ternary solution. For each β they contain a decaying composition vector (a_{11}, a_{21}) and a growing vector (a_{12}, a_{22}). It is to be particularly noted from equations (7.137) and (7.138) that, except for fluctuations parallel to (a_{11}, a_{21}), all fluctuations will ultimately grow. The precise direction of the vectors at any temperature below the critical point can only be obtained from a precise knowledge of the free energy surface.

In general, the course of spinodal decomposition must be obtained as a solution of the diffusion equations with variable coefficients, corrected for gradient and strain energy.[43] For uniformly periodic segregation, the mass balance requires that the centres of gravity of symmetrically located positively and negatively segregated zones must lie on a line passing through the composition point in the composition plane. These centers of gravity will in general trace an S-shaped curve terminating at the two ends of the tie line corresponding to the given temperature and composition.

Although it is often assumed in ternary spinodal decomposition that segregation traces a straight line coincident with the equilibrium tie line (that is, the eigenvectors associated with a temperature-dependent λ_2 are constant and parallel to the tie line) there is no theoretical basis for such an assumption. Indeed, all that is required for deviation from a straight line is that $D_{11} \neq D_{22}$ and (or) D_{12} and D_{21} are different from zero.

Directly below a ternary critical point, $d^2G < 0$, and the matrix $[\mu_{ik}]$ becomes negative definite. $[D_{ik}]$ accordingly has negative eigenvalues, i.e.

$$\lambda_1 < 0 \qquad \lambda_2 < 0 \quad (7.139)$$

These are now two mutually oblique growing composition vectors in the solutions. The path of early spinodal decomposition in this regime is indeterminate. However, as one of the eigenvalues necessarily shows a gain in magnitude relative to the other, the segregation path will achieve a bias towards the composition vector corresponding to the greatest eigenvalue. The small eigenvalue will ultimately change sign and a well-defined segregation path will evolve towards the termini of the equilibrium tie line. These conclusions will necessarily undergo modification when strain and gradient energy effects are included (cf § 12.5).

7.7 DIFFUSION IN Cu–Ag–Au ALLOYS

In the following sections the experimental data which relates to the foregoing theoretical sections is reviewed. For certain ternary systems the data is sufficiently accurate and comprehensive that a satisfactory closure with theory can be illustrated. These systems are therefore emphasized. Table 7.1 references all of the papers and their subject matter which have been examined for inclusion in this synthesis.

Ziebold and Ogilvie[31] evaluated the mutal diffusion matrix over the single-phase region of the Gibbs triangle of Cu–Ag–Au at 725°C using infinite diffusion couples and microprobe analysis. Figure 5.1 shows some typical diffusion profiles illustrating the reproducibility of their very careful technique. Note the strong uphill diffusion implied by the silver profile. Figure 7.4 summarizes the

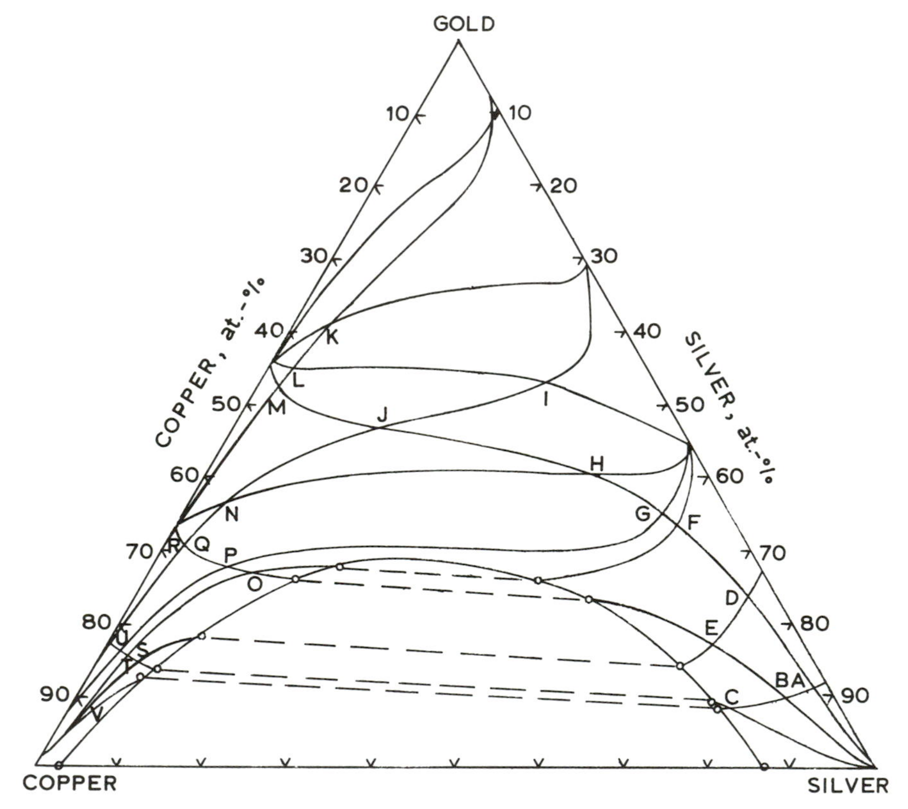

Reprinted with permission from TRANSACTIONS OF THE METALLURGICAL SOCIETY, Vol. 239, p. 9423, (1967), a publication of The Metallurgical Society, Warrendale, Pennsylvania

Fig. 7.4 Composition paths for diffusion at 725°C in the Cu–Ag–Au system; letters indicate intersections used to compute diffusion coefficient. After Ziebold and Ogilvie[31]

full set of diffusion profiles including those passing through the miscibility gap by formation of a phase interface. The latter observations of two-phase tie lines served to provide sufficient thermodynamic data, which combined with binary data allowed the evaluation of the $[\mu_{ij}]$ matrix. The diffusion profiles were analysed using the generalization of the Matano analysis due to Kirkaldy[17] yielding diffusion coefficients at points of crossover of two profiles (cf Fig. 7.4)

Table 7.1 Data on ternary cross-effects in solid metallic systems

System	Composition range	Temperature range, °C	Reference
Fe–C–Si	Fe-rich	1000	Darken,[16] analysed by Kirkaldy[27]
Fe–C–Si	Fe-rich	853–1058	Kirkaldy and Purdy[28]
Fe–C–Ni	Fe-rich	800–1050	Brown and Kirkaldy[30]
Fe–C–Mn	Fe-rich	795–988	Kirkaldy and Purdy[28]
Fe–C–Cr	Fe-rich	1050	Brown and Kirkaldy[30]
Fe–C–Co	Fe-rich	982–1038	Chandhok, Hirth, and Dulis[44]
Fe–Cr–Al	Fe-rich	900	Akuezue[45]
Fe–Cr–Ni	Fe-rich	1000	Pilliar and Kirkaldy[46]
Fe–Ni–Al	β_2 field	1004	Moyer and Dayananda[47]
Fe–Ni–Co	Entire isotherm	1136	Sabatier and Vignes[48]
Fe–Ni–Co	Entire isotherm	1315	Vignes and Sabatier[49]
Fe–Ni–Cr	Entire γ field	1100	Duh and Dayananda[50]
Al–Cu–Mn	Al-rich	556·5	Kirkaldy, Mason, and Slater[51]
Al–Cu–Zn	Al-rich	504	Kirkaldy, Zia-Ul-Haq, and Brown[26]
Ag–Zn–Cd	Ag-rich	600	Carlson, Dayananda, and Grace[52]
Co–Ni–Cr		1300	Guy and Leroy[53]
Cu–Ni–Zn	Entire α field	775	Sisson and Dayananda[54]
Cu–Ni–Zn	Cu-rich	900	Wan and de Hoff[55]
Cu–Ni–Zn–Mn		775	Dayananda[56]
Cu–Sn–Zn	Cu-rich	481–821	Kirkaldy, Brigham, and Weichert[25]
Cu–Sn–Zn	Cu-rich	750	Dayananda, Kirsch, and Grace[57]
Cu–Sn–Mn	Cu-rich	850	Dayananda and Grace[58]
Cu–Zn–Mn	Cu-rich	850	Whittenberger and Dayananda[59]
Cu–Au–Ag	Entire isotherm	725	Ziebold and Ogilvie[31]
V–Zr–Ti		800	Brunsch and Steeb[60]
Ni–Cr–Al	Ni-rich	1100	Nesbitt and Heckel[61]

and optimal points of individual profiles (cf Ag, Fig. 5.1 and §5.3). Table 7.2 shows the record for nine D-matrix points which exhibit uniformly strong cross-effects (D_{12} or D_{21}) relative to the diagonal values.

The Onsager reciprocal relations were tested using the derived formula (cf equation (4.38))

$$\left[g_{22} + \frac{1 - X_2}{X_2} g_{12}\right] D_{11}^3 + \left[\frac{1 - X_1}{X_2} g_{22} + \frac{X_1}{X_2} g_{12}\right] D_{21}^3$$

$$= \left[g_{11} + \frac{1 - X_1}{X_1} g_{21}\right] D_{22}^3 + \left[\frac{1 - X_2}{X_1} g_{11} + \frac{X_2}{X_1} g_{21}\right] D_{12}^3 \qquad (7.140)$$

where 1 = Cu, 2 = Ag and the solvent 3 = Au and

$$g_{ik} = \left(\frac{\partial \ln a_i}{\partial \ln X_k}\right) \qquad (7.141)$$

Table 7.3 shows the comparison of the left and right sides of equation (7.139) for the D-values in Table 7.2. The verification of the ORR can be deemed to be good, considering the precision of the technique and the uncertainty of the thermodynamic data.

As noted in §4.5, Murakami *et al.*[32] have evaluated the D-matrix by a thin-film X-ray method at 242°C for two compositions near the critical point (Ag–50Au–5Cu and Ag–60Au–11Cu; microcouple 2 and 3, respectively in Table 4.2). Table 4.2, microcouple 2 entries are deemed to represent a composition near the critical point where $D_{11}D_{22} - D_{12}D_{21} \rightarrow 0$. Simultaneously $D_{12} \rightarrow D_{11}$ and $D_{21} \rightarrow D_{22}$. Ziebold and Ogilvie found the same trend at 725°C. It is in fact predicted by relation (7.117) due to Kirkaldy and Purdy[35] since for this miscibility gap the limiting or critical tie line direction $\alpha \rightarrow -1$ (Fig. 7.4).

Table 7.2 Measured diffusion coefficients at 725°C: 1 = Cu, 2 = Ag, 3 = Au. After Ziebold and Ogilvie[31]

Couple	Atom composition		cm^2 s^{-1} × 10^{10}			
	$100X_1$	$100X_2$	D_{11}^3	D_{22}^3	D_{12}^3	D_{21}^3
H	14·0	45·7	1·9	2·4	1·1	2·1
I	13·1	34·0	0·99	1·3	0·09	1·7
J	35·1	18·0	2·5	4·0	1·6	2·8
K	35·1	4·2	3·0	2·8	1·7	0·57
L	41·8	3·0	3·1	2·0	0·75	0·29
M	44·6	2·5	3·2	3·3	0·73	0·81
N	58·5	4·8	2·3	3·0	0·53	0·81
O	60·3	12·9	2·3	3·1	1·11	1·8
P	62·9	9·5	2·4	3·6	1·04	1·9

Table 7.3 Test of the Onsager relations: 1 = Cu, 2 = Ag, 3 = Au. After Ziebold and Ogilvie[31]

Couple	Atom composition		Equation (7.140)	
	$100X_1$	$100X_2$	Left side: $aD^3_{11} + bD^3_{21}$	Right side: $cD^3_{22} + dB^3_{12}$
H	14·0	45·7	12·4	9·9
I	13·1	34·0	10·6	5·2
J	35·1	18·0	21·3	21·9
K	35·1	4·2	21·6	21·2
L	41·8	3·0	17·3	12·9
M	44·6	2·5	20·4	18·8
N	58·5	4·8	10·9	13·4
O	60·3	12·9	6·7	12·3
P	62·9	9·5	8·1	14·4

7.8 DIFFUSION IN Fe–Co–Ni ALLOYS

Vignes and Sabatier[48, 49] made a propitious choice of system for detailed study, for the solution Fe–Co–Ni is very nearly ideal in the thermodynamic sense so thermodynamic interactions will be absent. Furthermore, the system shows a strong Kirkendall effect towards the solvent iron ($D^*_{Fe} > D^*_{Co} \sim D^*_{Ni}$) so one may expect strong Onsager cross-effects associated with divergent species mobilities.

Using techniques similar to those of Ziebold and Ogilvie the authors observed the diffusion paths of Fig. 7.5. Note that there is a much better distribution of crossover points as compared to Fig. 7.4, so the Kirkaldy analysis[17] yields a more representative matrix of D over the Gibbs triangle. Other D-values were obtained from extrema in profiles such as in Fig. 5.2.

The complete D-matrix at 1315°C is presented on the Gibbs triangle of Figs. 7.6–7.9. The Onsager L_{ij} coefficients computed assuming an ideal solution are reported in Table 7.4. The confirmation of the symmetry relation is uniformly good over the entire Gibbs triangle. Binary limits are all consistent (§4.5).

The authors also put to the test the theory of Lane and Kirkaldy which relates the L- and D-coefficients to the tracer coefficients. From our equations (7.58) and (7.59) they obtain for the ideal solution case

$$D^k_{ii} = (X_j + X_k)D_i + X_i D_k \tag{7.142}$$

and

$$D^k_{ij} = -X_i(D_j - D_k) \tag{7.143}$$

The tracer diffusion coefficients D_i are known for each component on the three sides of the ternary diagram. On the ij and ik sides, D_i is given either by a radioactive tracer diffusion experiment D_i^* or by a Kirkendall experiment. On the jk side, D_i is equal to $D^0_{i(jk)}$ as determined in the present study (cf §4.5). In the Co–Ni system the Kirkendall shift is so small that it can be inferred that $D_{Ni} \simeq D_{Co}$.

Table 7.4 Phenomenological coefficients L^k_{ij} at 1315°C $L \cdot RT \times 10^{10}$ g cm^{-1} s^{-1}. After Vignes and Sabatier[49]

Composition %$_{Co}$	%$_{Ni}$	L^{Fe}_{CoCo}	L^{Fe}_{NiNi}	L^{Fe}_{CoNi}	L^{Fe}_{NiCo}
9·2	35·6	2·9	13·2	0·56	0·55
2·2	68·3	1·1	31·4	−0·34	−0·13
1·6	72·3	0·95	33·1	−0·70	−0·82
10·3	31·4	2·6	11	0·36	0·49
33·3	30·2	8·54	9	−0·32	−0·54
44·8	27·8	9·7	7·4	−1·55	−1·65
49·7	27·2	9·4	6·6	−2	−1·85
65·7	27·8	8	6·55	−5·3	−5·35
38·5	29	9·3	8·5	−0·84	−0·96
29·2	31·2	8·4	9·4	0·16	0·15
36·5	19·9	8·1	6	0·8	0·55
30·6	40·7	8·3	11·8	−1·36	−1·44
30·3	61·5	8·6	11·6	−6·2	−6·9
68	4·9	8	1·11	0·11	0·13
35·5	35·4	9·2	10·1	−1·6	−1·44
48·3	46·5	9·8	10·4	−8·1	−8·5
54·5	9·8	8·6	2·1	0·62	0·44
67·9	25·7	7·6	6·35	−4·85	−5·15
74	2·9	7·2	0·56	−0·05	0
48	46·5	10	10·3	−8·2	−8·2
25	49	7·65	13·6	−1·75	−1·46
11·5	70·2	5	18·8	−2·9	−3·8
25·5	66·5	7·6	11·7	−6·4	−6·4
48	13·5	8·6	3·7	0·6	0·64
23·7	25·3	6·15	7·6	1·2	0·9
15·2	15·5	3·45	3·2	0·71	0·66

In the Fe–Ni and Fe–Co systems the Kirkendall effects are strong and directed towards the iron-rich part of the couple. Such Kirkendall experiments therefore show that

$$D_{Fe} > D_{Ni} \simeq D_{Co} \tag{7.144}$$

The authors conjecture that such relationships are satisfied throughout the Gibbs triangle and therefore conclude from equations (7.142) and (7.143) that

$$D^{Co}_{FeNi} = -X_{Fe}(D_{Ni} - D_{Co}) \simeq 0 \tag{7.145}$$

and

$$D^{Co}_{NiFe} = -X_{Ni}(D_{Fe} - D_{Co}) < 0 \tag{7.146}$$

over the entire ternary diagram. This is in complete accord with the observations (their Table II). The same theory for an ideal solution leads to the Onsager coefficient

$$\frac{RT}{m} L^{\text{Co}}_{\text{NiFe}} = -X_{\text{Fe}}X_{\text{Ni}}[X_{\text{Fe}}D_{\text{Ni}} + X_{\text{Ni}}D_{\text{Fe}} + X_{\text{Co}}(D_{\text{Fe}} + D_{\text{Ni}} - D_{\text{Co}})] \tag{7.147}$$

Thus according to relations (7.144) this must be negative over the entire diagram, again in accord with the experiments. Finally, they deduce

$$\frac{RT}{m} L^{\text{Fe}}_{\text{NiCo}} = -X_{\text{Ni}}X_{\text{Co}}[X_{\text{Ni}}D_{\text{Co}} + X_{\text{Co}}D_{\text{Ni}} + X_{\text{Fe}}(D_{\text{Ni}} + D_{\text{Co}} - D_{\text{Fe}})] \tag{7.148}$$

Supposing that, consistent with equations (7.144) and a strong Kirkendall effect towards the iron, D_{Fe}-D_{Ni}-D_{Co} is slightly positive, then this coefficient can be expected to change from negative to positive in the approach to the Fe-rich corner of the isotherm (Fig. 7.10). Such an eventuality was predicted by Lane and Kirkaldy.[5] It should be emphasized finally that the diffusion interactions in this system are of entirely kinetic origin (Fig. 5.2), by contrast with interstitial austenites (Figs. 4.2 and 4.4) where the interactions are entirely of thermodynamic origin.

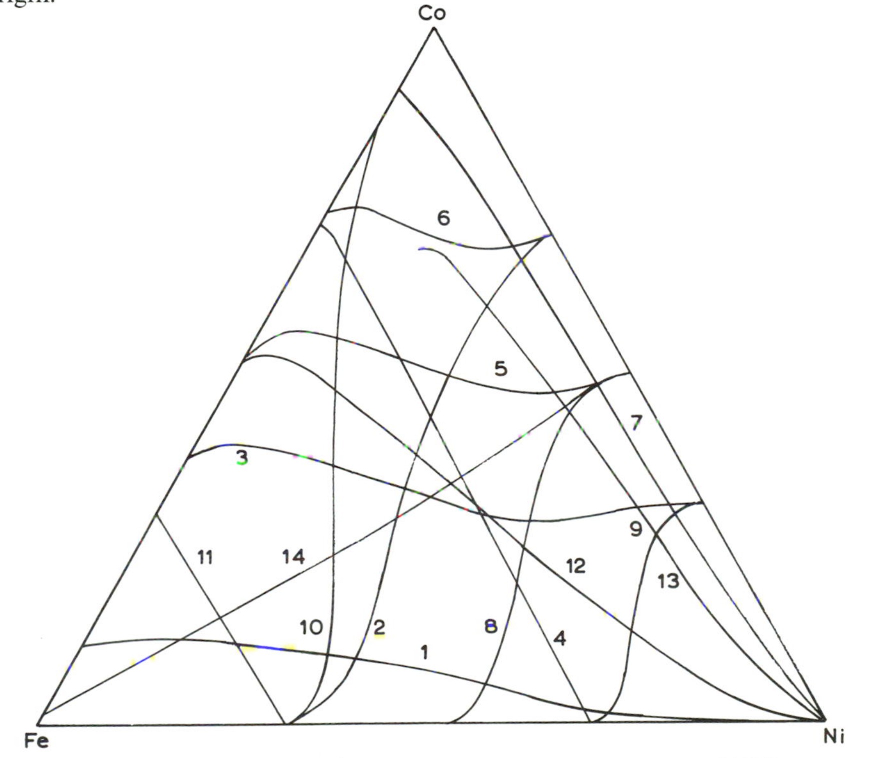

Fig. 7.5 Diffusion paths for diffusion at 1315°C in the Fe–Co–Ni system. Numbers indicate sample number. After Vignes and Sabatier[49]

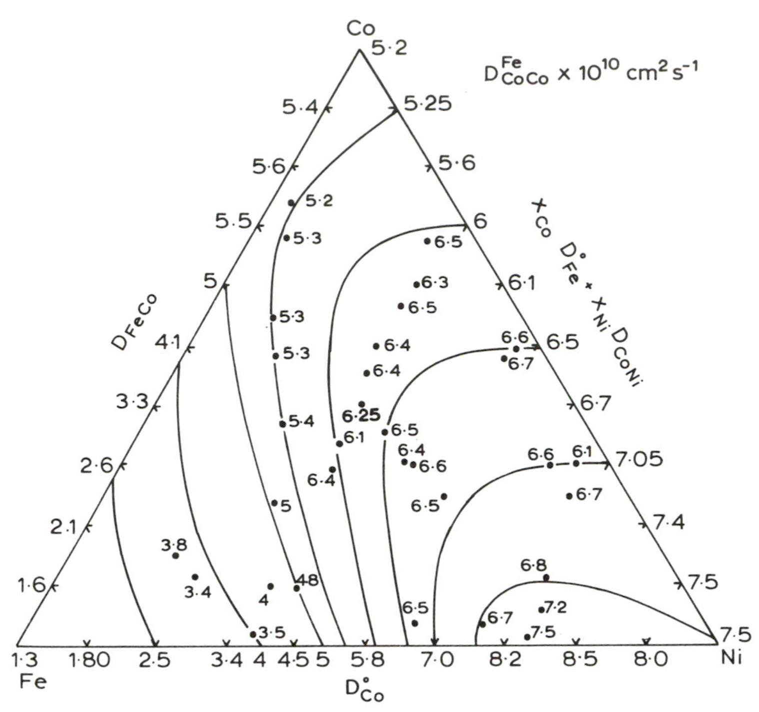

Reprinted with permission from TRANSACTIONS OF THE METALLURGICAL SOCIETY, Vol. 245, p. 1795, (1969), a publication of The Metallurgical Society, Warrendale, Pennsylvania

Fig. 7.6 The direct interdiffusion coefficient $D^{\mathrm{Fe}}_{\mathrm{CoCo}}$ at 1315°C contour map. After Vignes and Sabatier[49]

7.9 DIFFUSION IN DILUTE Fe–C–X ALLOYS

It will be appreciated from the foregoing what a voluminous amount of meticulous experimentation is required to establish the ternary D-matrix for a single isotherm. It is thus essential to deploy every possible theoretical device to reduce the experimentation to manageable proportions. The technologically significant dilute alloys consisting of carbon (or other interstitials) up to the 10 at.-% level and substitutional elements (Si, Mn, Ni, Cr, Mo, and Cu) up to the 4 at.-% level in bcc or fcc iron are a case in point. Since the interstitial and substitutional atoms lie on different sublattices correlations between such species will be weak so the Onsager cross-effects L_{CX}, where subscript X represents any substitutional addition, will tend to be vanishingly small relative to L_{CC} and the thermodynamic interactions will strongly predominate (cf §6.3). Furthermore, since these alloys tend to be dilute, dilute solution thermodynamics is adequate to their analysis. Thus accordingly to §6.3

$$\frac{D_{\mathrm{CX}}}{D_{\mathrm{CC}}} = \frac{\partial \mu_{\mathrm{C}}}{\partial X_{\mathrm{X}}} \bigg/ \frac{\partial \mu_{\mathrm{C}}}{\partial X_{\mathrm{C}}} = \frac{\varepsilon_{\mathrm{CX}} X_{\mathrm{C}}}{1 + \varepsilon_{\mathrm{CC}} X_{\mathrm{C}}} \tag{7.149}$$

and

$$\frac{D_{XC}}{D_{XX}} = \frac{\left(\varepsilon_{CX} + \frac{L_{XC}}{D_{XX}} \frac{1}{X_C}\right) X_X}{1 + \varepsilon_{XX} X_X} \tag{7.150}$$

Although it is not always possible to neglect L_{XC} relative to $L_{XX}X_C$ in equation (7.150) the fact that $D_{CC} \gg D_{XX}$ assures that the two eigenvalues of the matrix become D_{CC} and D_{XX} and the effects of D_{XC} on the concentration profiles becomes negligible (cf equations (4.88) and (4.90)). Knowledge of its precise value thus becomes irrelevant to practical calculations.

Kirkaldy *et al.*[28, 30] have verified equation (7.149) for the systems Fe–C–Si, Fe–C–Ni, Fe–C–Co, Fe–C–Mn and Fe–C–Cr using the quasi-steady diffusion layer couples described in §5.4 and 4.6.6. Figure 7.11 shows the comparison between theory and experiment for fcc iron (austenite) at 1050°C based on the independently determined Wagner interaction parameters of Table 7.5. The small deviations (~10%) could be due to the neglect of correlation effects, but it is more probable that they can be attributed to inaccurate values of the interaction parameters.

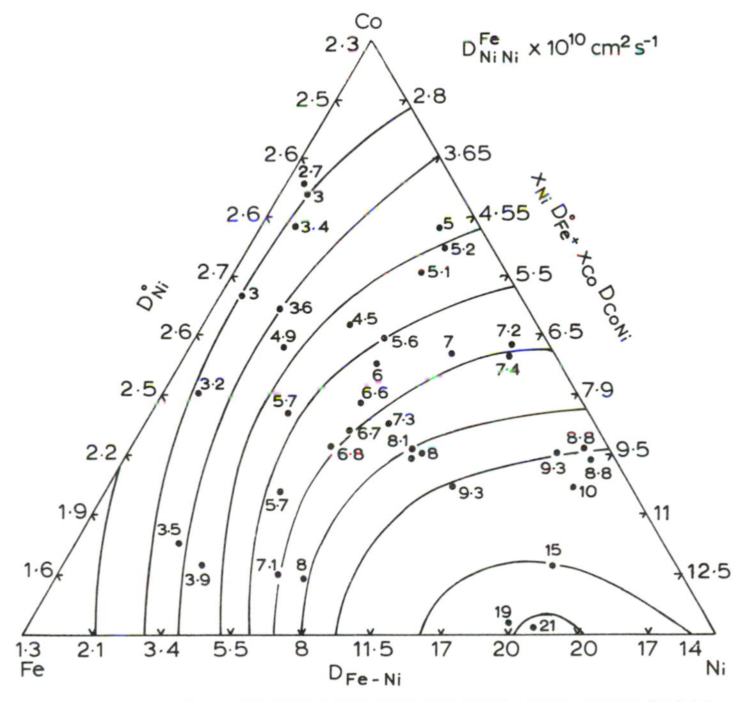

Reprinted with permission from TRANSACTIONS OF THE METALLURGICAL SOCIETY, Vol. 245, p. 1795, (1969), a publication of The Metallurgical Society, Warrendale, Pennsylvania

Fig. 7.7 The direct interdiffusion coefficient D^{Fe}_{NiNi} at 1315°C. After Vignes and Sabatier[49]

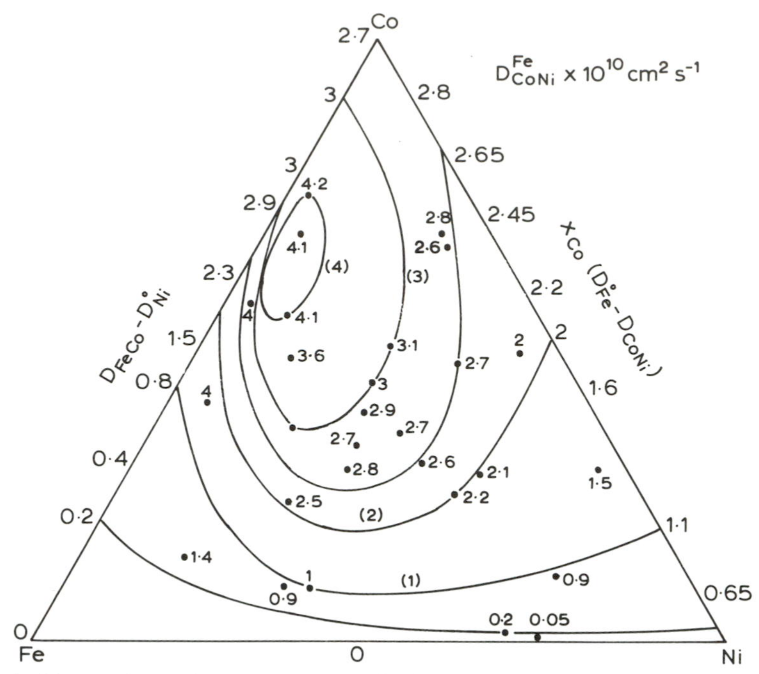

Fig. 7.8 The cross-interdiffusion coefficient D^{Fe}_{CoNi} at 1315°C. After Vignes and Sabatier[49]

It is to be noted that the interaction parameters are generally very slowly varying functions of temperature, so the off-diagonal D_{CX} will have nearly the same temperature dependence as D_{CC}. A regular solution model of such mixtures yields the expression

$$\varepsilon_{CX} = 6\frac{(e_{CX} - e_{FeC})}{kT} \tag{7.151}$$

where the e are pair interaction energies,[28] so the $1/T$ dependence could be used to extend isothermal data over wider temperature ranges. If more than one temperature data point is available we recommend the interpolation formula

$$\varepsilon_{CX} = a/T + b \tag{7.152}$$

7.10 DIFFUSION IN DILUTE Cu–Zn–Sn and Cu–Zn–Mn

Copper-rich alloys with Zn and Sn include many brasses and bronzes. Kirkaldy *et al.*[25] undertook the determination of the D-matrix in the concentration ranges 0–5 at.%Zn, 0–6 at.%Sn by an optimal experimental-theoretical program. A somewhat improved theoretical part is presented here which is appropriate to alloys possessing significant thermodynamic and kinetic cross-effects, and which recognizes from the start that $D_{SnSn} \sim 4D_{ZnZn}$. The summarized theory of Lane and Kirkaldy[5] gives for dilute solutions (equations (7.29) *et seq.*)

$$D_{\mathrm{ZnSn}} = mRT\{(1 + \varepsilon_{\mathrm{ZnSn}})L_{\mathrm{ZnZn}} + L_{\mathrm{ZnSn}}/X_{\mathrm{Sn}}\} \tag{7.153}$$

and

$$D_{\mathrm{SnZn}} = mRT\{(1 + \varepsilon_{\mathrm{ZnSn}})L_{\mathrm{SnSn}} + L_{\mathrm{ZnSn}}/X_{\mathrm{Zn}}\} \tag{7.154}$$

where

$$L_{\mathrm{ZnZn}} \simeq X_{\mathrm{Zn}}D_{\mathrm{ZnZn}}/RT \qquad L_{\mathrm{SnSn}} \simeq X_{\mathrm{Sn}}D_{\mathrm{SnSn}}/RT \tag{7.155}$$

$$L_{\mathrm{ZnSn}} \simeq -X_{\mathrm{Zn}}X_{\mathrm{Sn}}(D^*_{\mathrm{Zn}} + D^*_{\mathrm{Sn}} - D^*_{\mathrm{Cu}})/RT \tag{7.156}$$

and in the limit of infinite dilution

$$D_{\mathrm{ZnZn}} \simeq D^*_{\mathrm{Zn}} \quad \text{and} \quad D_{\mathrm{SnSn}} \simeq D^*_{\mathrm{Sn}} \tag{7.157}$$

From the binary Kirkendall experiment (§2.13)

$$D^*_{\mathrm{Zn}} \simeq 2{\cdot}6D^*_{\mathrm{Cu}}$$

so one can estimate

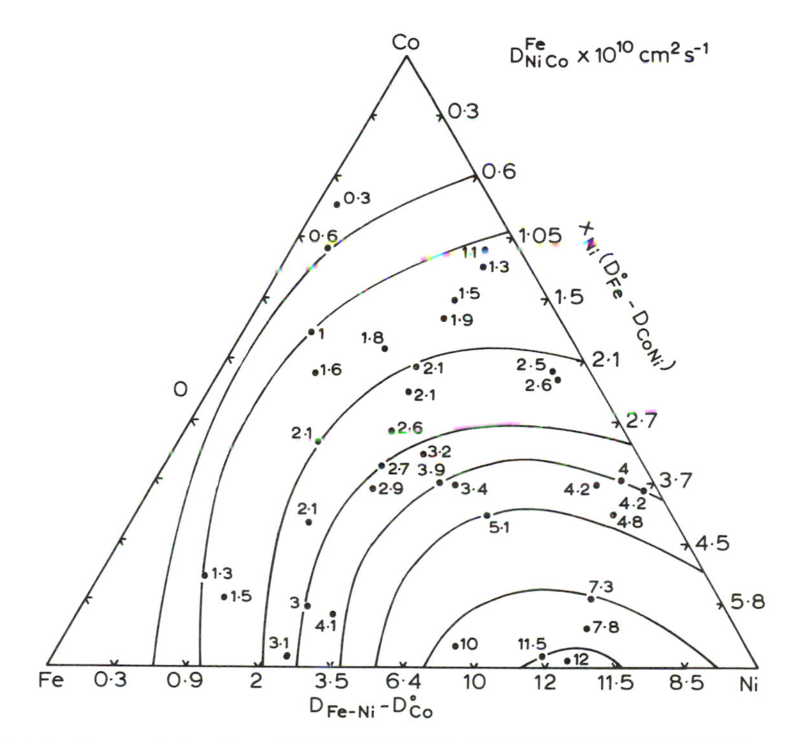

Fig. 7.9 The cross-interdiffusion coefficient $D^{\mathrm{Fe}}_{\mathrm{NiCo}}$ at 1315°C. After Vignes and Sabatier[49]

$$D_{ZnSn}/D_{ZnZn} = (\varepsilon_{ZnSn} - 3{\cdot}75)X_1 \quad (7.158)$$

and

$$D_{SnZn}/D_{SnSn} = (\varepsilon_{ZnSn} - 0{\cdot}2)X_2 \quad (7.159)$$

where ε_{ZnSn} is the thermodynamic cross-interaction parameter. In the experiments using infinite (Fig. 7.1) and finite layered couples (for improved sensitivity, Fig. 5.6) D_{ZnSn}, D_{SnSn} and D_{SnZn} were determined as a function of composition. D_{ZnSn} was inaccessible due to the unfavourable ratio D_{ZnZn}/D_{SnSn}.[25,30] However, the coefficient of X_2 in equation (7.159) was found to be 6·1 at 498°C and 8·8 at 821°C, yielding values of ε_{SnZn} = 6·3 and 9·0, respectively. With this information D_{ZnSn} can be calculated via equation (7.158). Figures 7.12 and 7.13 show the summarizing contours of the *D*-matrix on the Gibbs isotherm at 498 and 821°C. This information was obtained from only six diffusion couples. In principle, it could have been obtained from four, which underlines the importance of experimental design and theoretical exhaustion in an economical approach to data-gathering in ternary diffusion.

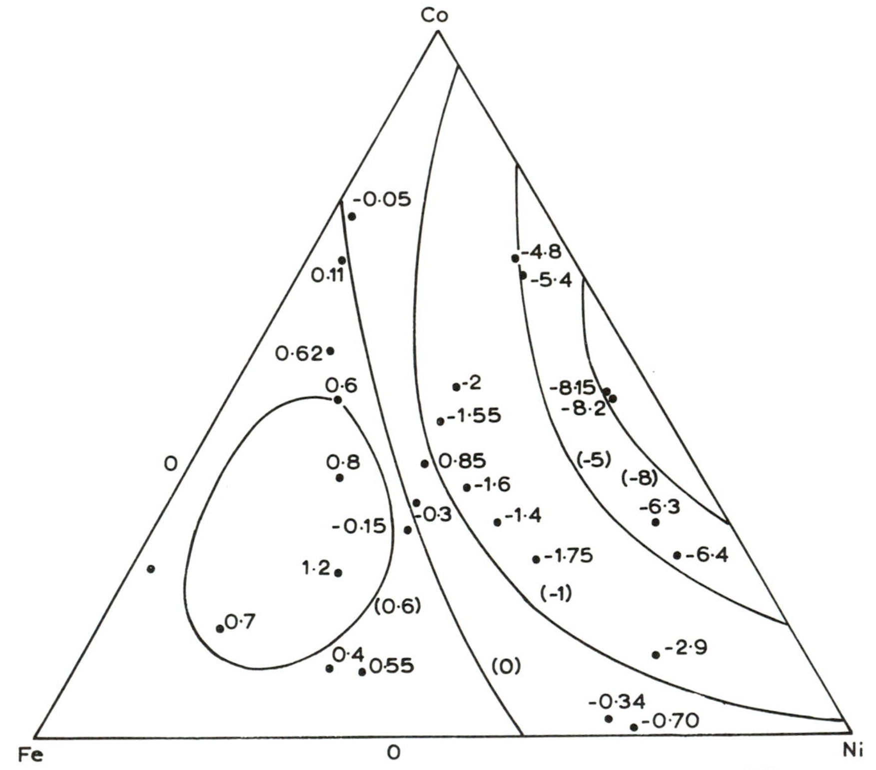

Reprinted with permission from TRANSACTIONS OF THE METALLURGICAL SOCIETY, Vol. 245, p. 1795, (1969), a publication of The Metallurgical Society, Warrendale, Pennsylvania

Fig. 7.10 The cross-phenomenological coefficient L^{Fe}_{CoNi}, L^{Fe}_{NiCo} at 1315°C. ($L \times RT \times 10^{10}$) g cm^{-1} s^{-1}. After Vignes and Sabatier[49]

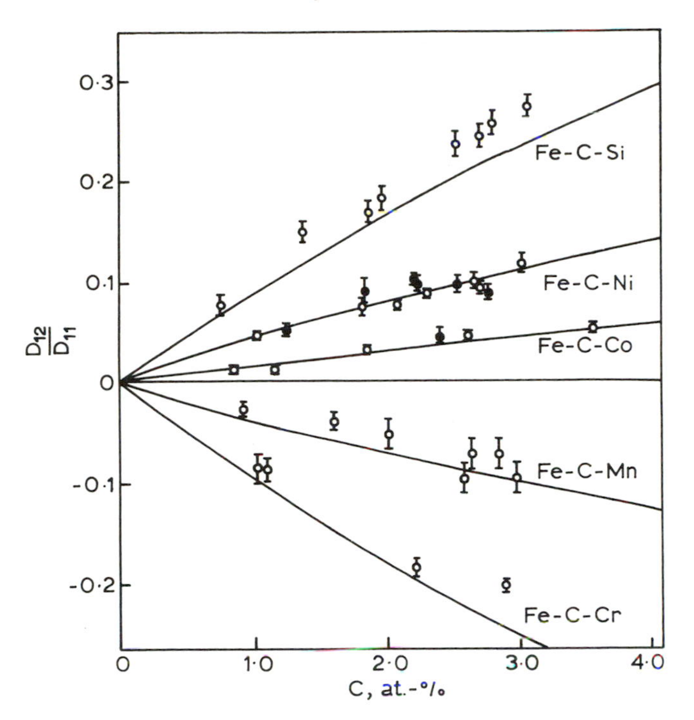

Fig. 7.11 **Variation of D_{12}/D_{11} with carbon concentration (C_1) in the five systems investigated. The solid point on the Fe–Co–C curve was obtained from a couple with a cobalt concentration of 0·040. The solid points on the Fe–Ni–C curve were obtained from couples with nickel concentrations of 0·018 7. The solid lines represent the thermodynamic prediction. After Brown and Kirkaldy[30]**

Table 7.5 Values of interaction parameter in dilute ternary austenites. After Brown and Kirkaldy[30]

Parameter	System	Magnitude
e_{11}	Fe–C	8·90
e_{12}	Fe–Si–C	9·5
	Fe–Ni–C	4·6
	Fe–Co–C	1·7
	Fe–Mn–C	−4·2
	Fe–Cr–C	−10·7

It is worthwhile noting that ε_{ZnSn} as determined here approximately follows a $1/T$ (K) relation as expected on theoretical grounds[28, 30] (cf §7.9).

Dayananda *et al.*[57] have examined Cu–Zn–Sn solid solutions at 750°C in roughly the same composition range using semi-infinite diffusion couples. That

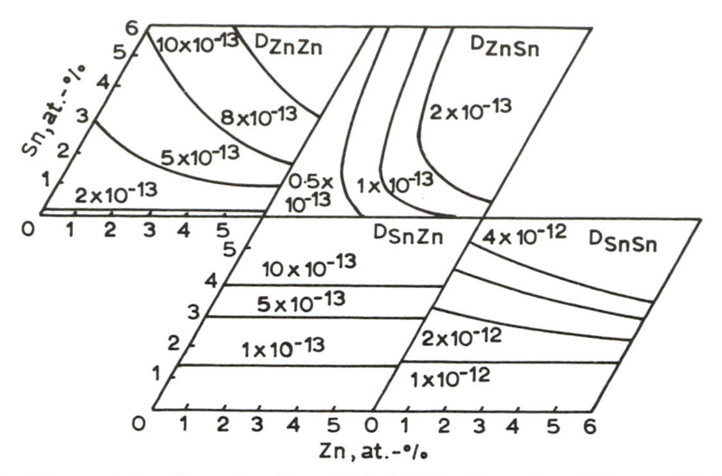

Reprinted with permission from *Acta Met.*, **13**, J. S. Kirkaldy *et al.*, p. 907, copyright 1965, Pergamon Press Ltd

Fig. 7.12 Contour map of the *D* matrix ($cm^2\ s^{-1}$) for diffusion of zinc and tin in copper at 498°C. After Kirkaldy *et al.*[25]

is to say, they inject Zn into a Sn–Cu alloy from a constant pressure vapour phase. The interactions which they record in Fig. 7.14 are completely analogous and quantitatively consistent with those shown in Figs. 7.1 and 5.6. Because matter is being added to the sample this is a moving interface problem as regards the solution of the differential equations. Dayananda and Grace[58] have accordingly generalized a binary solution applicable to Zn–Cu given by Balluffi and Seigle[62] to aid in the analysis.

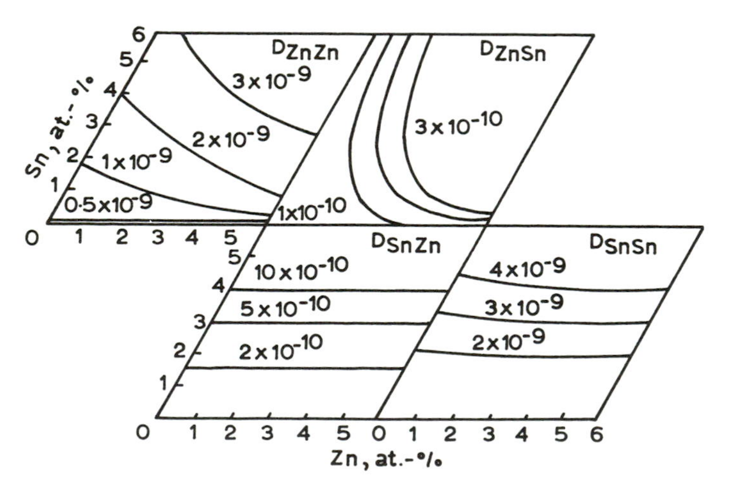

Reprinted with permission from *Acta Met.*, **13**, J. S. Kirkaldy *et al.*, p. 907, copyright 1965, Pergamon Press Ltd

Fig. 7.13 Contour map of the *D* matrix ($cm^2\ s^{-1}$) for diffusion of zinc and tin in copper at 821°C. After Kirkaldy *et al.*[25]

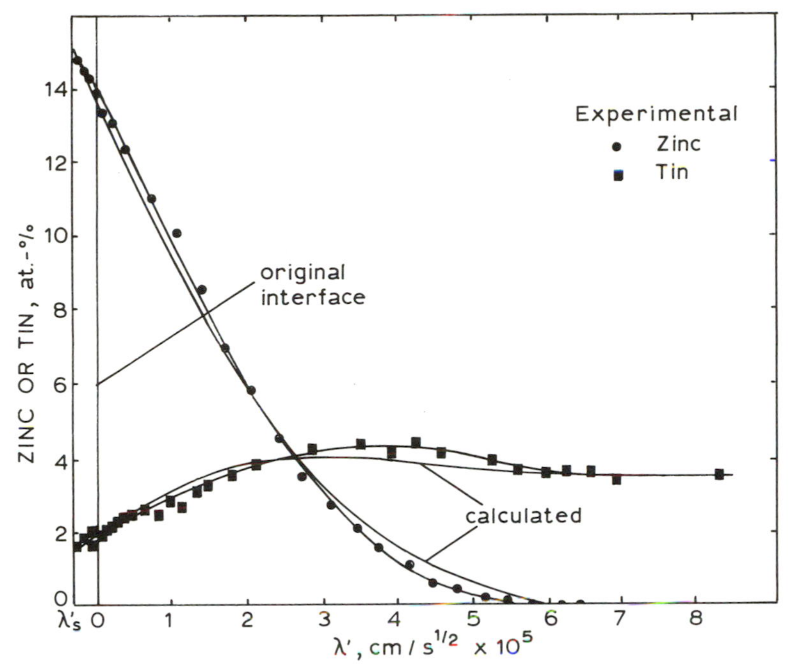

Fig. 7.14 Experimental and calculated concentration profiles for zinc and tin; diffused for 4 days at 750°C. After Dayananda *et al.*[57]

The presence of high vapour pressure constituents makes feasible steady state membrane experiments. Such experiments have the merit of a simpler analysis (cf §§4.6.6 and 5.7). Whittenberger and Dayananda[59] have reported limited results from experiments in Cu–Zn–Mn. Apparently no cross-effects in the form of a non-zero D^{Cu}_{MnZn} or D^{Cu}_{ZnMn} were detected.

7.11 DIFFUSION IN Ag–Zn–Cd

Carlson *et al.*[52] have extended the work of the previous section to a system in which two components (Zn, Cd) can be injected from the vapour phase in a substrate (e.g. Ag). Typical penetration curves for 600°C are shown in Figs. 7.15 and 7.16. On the isotherm, sets of couples between systematically increased vapour pressures and a fixed substrate alloy have paths as shown in Fig. 7.17. Simultaneous observation of outward surface drift and depth of marker burial were made (Fig. 7.18). Analysis of the data according to the method of Philibert and Guy[20,21] has accommodated the evaluation of the six intrinsic diffusion coefficients (cf §7.4). Cooper[63] has examined this data from the point of view of his vector space (matrix) treatment of multicomponent diffusion (see also §§4.6.5 and 7.6). Carlson *et al.*[64] have re-examined this data in relation to the ternary thermodynamics due to Scatchard and Lin.[65] They find some evidence to support the theory of ternary interactions due to Manning.[13]

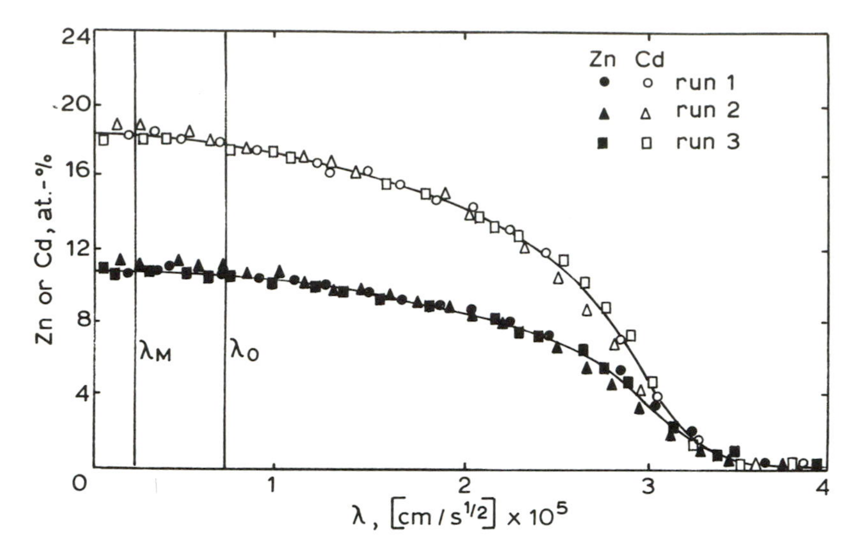

Fig. 7.15 Concentration profiles of zinc and cadmium for the alloy 4/Ag diffusion couple diffused for 3, 5, and 8 days at 600°C. After Carlson *et al.*[52]

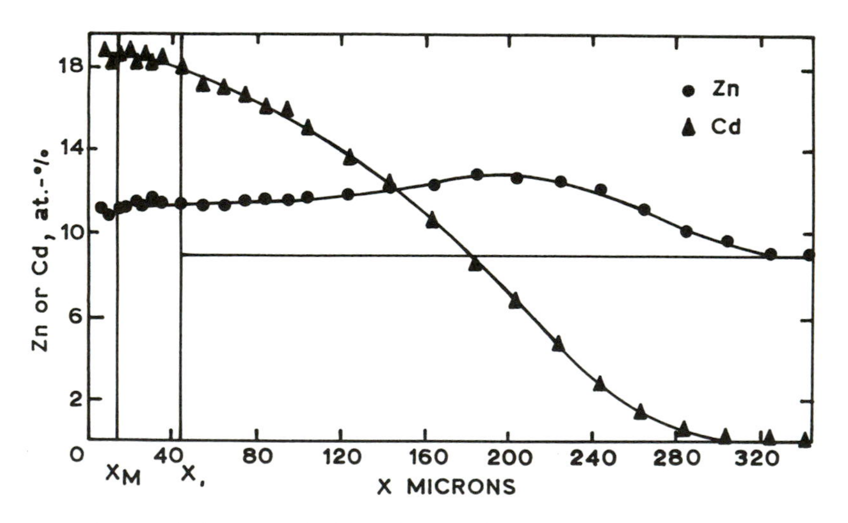

Fig. 7.16 Concentration profiles of zinc and cadmium for the Alloy 4/Alloy B diffusion couple diffused for 5 days at 600°C. After Carlson *et al.*[52]

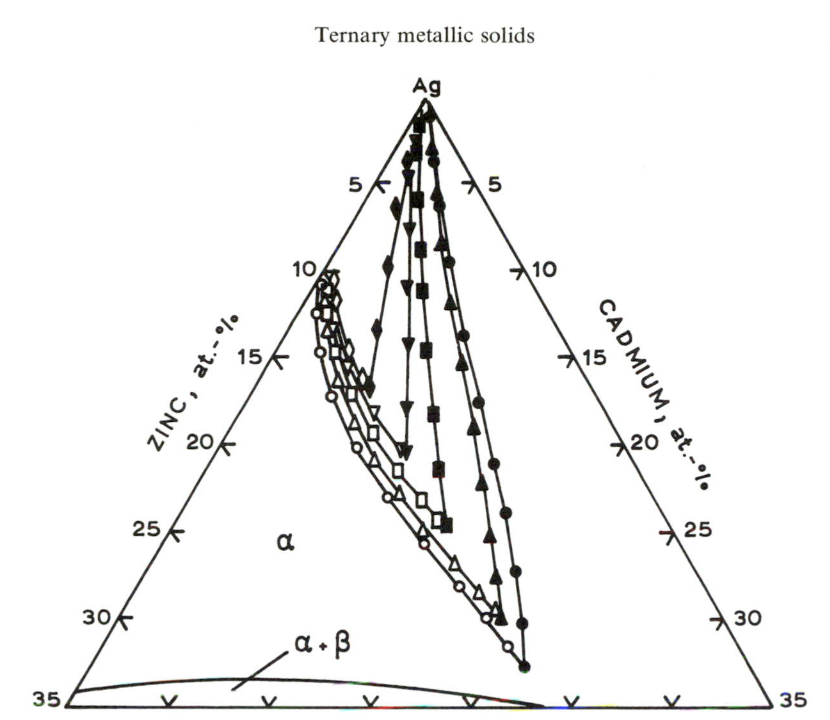

Reprinted with permission from TRANSACTIONS OF THE METALLURGICAL SOCIETY, Vol. 3, p. 189, (1972), a publication of The Metallurgical Society, Warrendale, Pennsylvania

Fig. 7.17 Diffusion paths for runs 4, 5, 21, 22, 23, 24, 1, 6, 25, and 26 at 600°C. After Carlson *et al.*[52]

7.12 DIFFUSION IN Fe-Ni-Cr

Duh and Dayananda[50] have carried out a systematic investigation of ternary diffusion in γ (fcc) alloys of Fe–Ni–Cr at 1100°C. The diffusion paths examined are indicated in Fig. 7.19. Figures 7.20 and 7.21 summarize their deduced chemical coefficients $D^{\mathrm{Fe}}_{\mathrm{NiNi}}$ and $D^{\mathrm{Fe}}_{\mathrm{CrCr}}$. They found in agreement with the earlier observations of Pilliar and Kirkaldy[46] that the off-diagonal coefficients $D^{\mathrm{Fe}}_{\mathrm{NiCr}}$ and $D^{\mathrm{Fe}}_{\mathrm{CrNi}}$ tend to zero. However, their somewhat scattered results suggest a bias towards small negative values. The author's analysis of the thermodynamics (Fig. 7.22) indicates that the system is very nearly ideal so the thermodynamic contribution to the interactions is expected to be small. Furthermore, their inferences from the tracer data of Rothman *et al.*[66] suggest that the tracer coefficients at 1100°C are very nearly equal. In view of the ideal thermodynamics intrinsic coefficients should share the same approximate equality, so the Kirkendall effect, which is the main source of kinetic (Onsager L-) interactions (cf §§6.2 and 7.4), should also be very small for all couples, and the L cross-coefficients should be correspondingly small. This qualitative analysis explains both the vanishing of the off-diagonal D and the fact that the observed paths on the isotherm (Fig. 7.19) are very nearly straight lines. Rather surprisingly, it can be concluded that this very important commercial alloy system is approximately ideal in both the thermodynamic and kinetic senses at points not too close to the phase boundary.

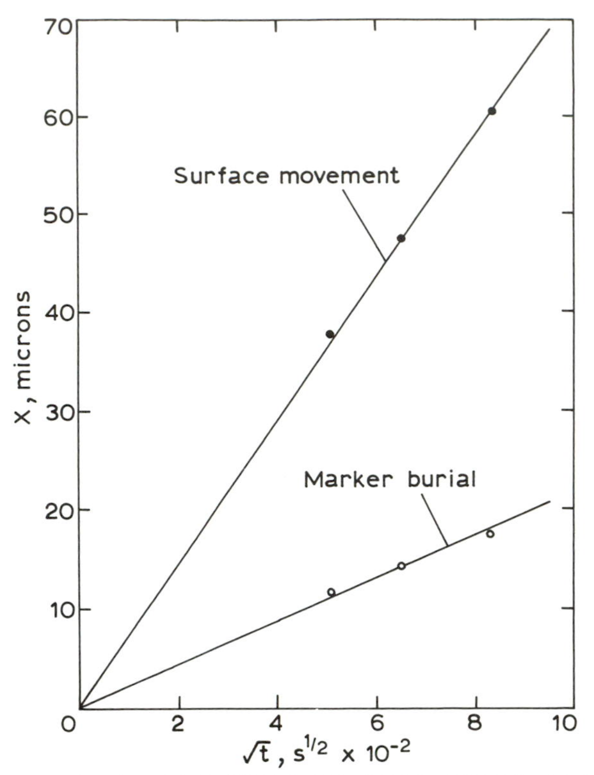

Reprinted with permission from TRANSACTIONS OF THE METALLURGICAL SOCIETY, Vol. 3, p. 189, (1972), a publication of The Metallurgical Society, Warrendale, Pennsylvania

Fig. 7.18 Marker burial and surface movement as functions of square root of diffusion time for the Alloy 4/Ag diffusion couple diffused for 3, 5, and 8 days at 600°C. After Carlson *et al.*[52]

7.13 DIFFUSION IN β_2 Fe-Ni-Al ALLOYS

Moyer and Dayananda[47] have reported on ternary diffusion in the ordered β_2 (CsCl) phase of Fe-Ni-Al alloys at 1004°C. This is located on the phase diagram of Fig. 7.23. The diffusion paths observed are indicated on the isotherm of Fig. 7.24. The interactions are quite dramatic as further indicated by the typical penetration curves of Figs. 7.25 and 7.26. Note in contrast to previous examples (Figs. 7.1 and 5.1) that Al flows up the Ni gradient indicating a thermodynamic attraction between these species. Such a result is not unexpected in an ordering alloy. The analysis of the data finds both on-diagonal chemical coefficients to be positive, taking a minimum value near the centre of the ternary β_2 field, while the off-diagonal coefficients take magnitudes comparable to the on-diagonal ones but change signs in opposite directions as the β_2 field is traversed.

This system and data set deserves further analysis as thermodynamic and tracer data become available.

A summary of data sources accessed for this chapter is presented in Table 7.1.

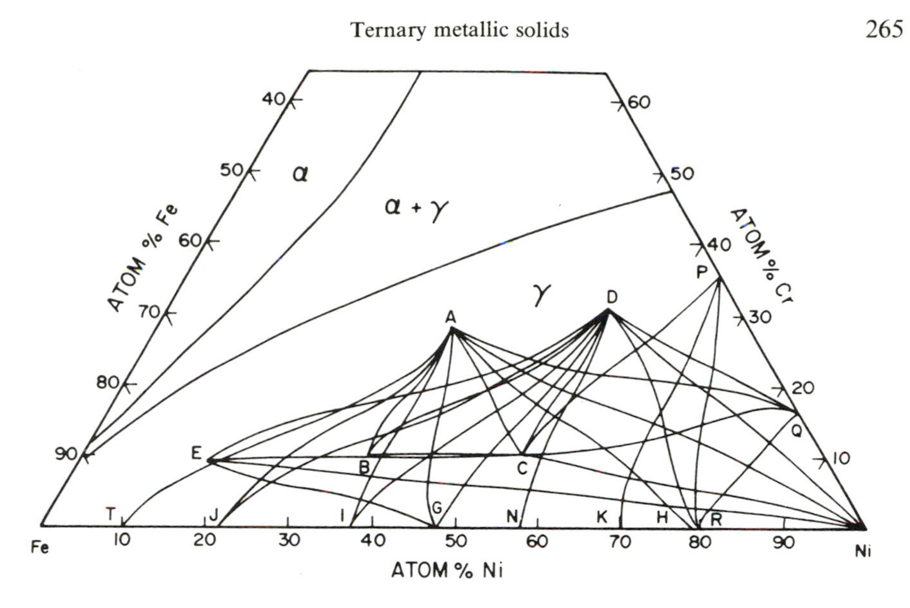

With permission from J. G. Duh and M. A. Dayananda, *Diffusion and Defect Data*, 1985, **39**, 1

Fig. 7.19 Experimental diffusion paths for γ Fe–Ni–Cr diffusion couples annealed at 1100°C for 7 days. After Duh and Dayananda[50]

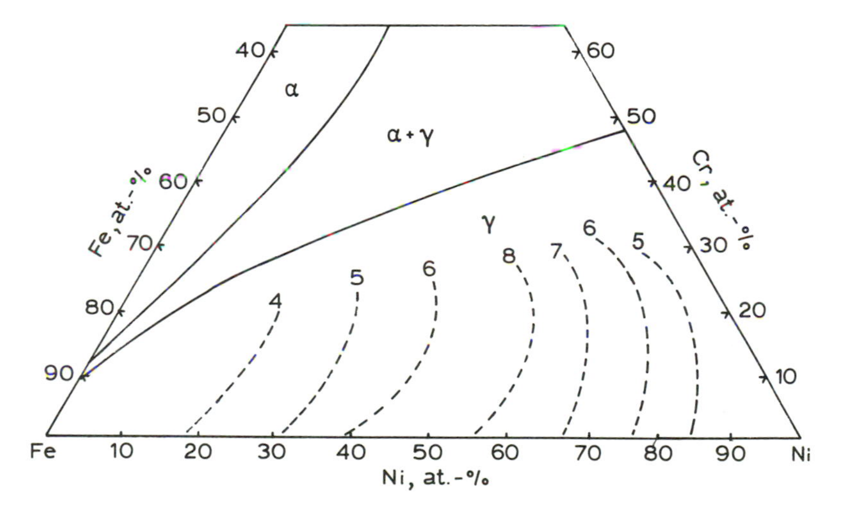

With permission from J. G. Duh and M. A. Dayananda, *Diffusion and Defect Data*, 1985, **39**, 1

Fig. 7.20 Interdiffusion coefficient D^{Fe}_{CrCr} for γ Fe–Ni–Cr alloys at 1100°C (in units of 10^{-11} cm^2 s^{-1}). After Duh and Dayananda[50]

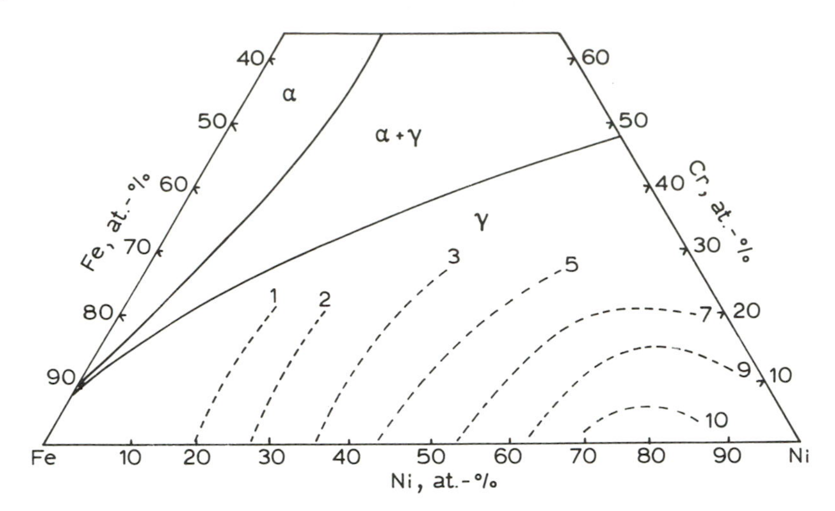

With permission from J. G. Duh and M. A. Dayananda, *Diffusion and Defect Data*, 1985, **39**, 1

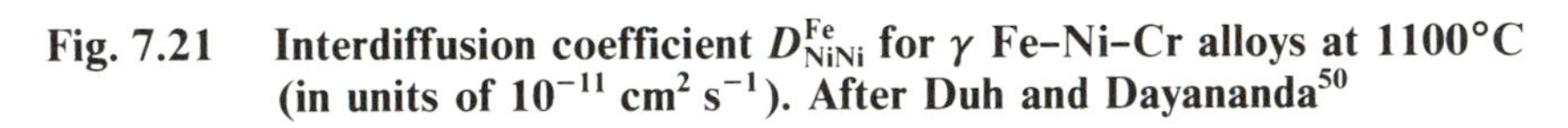

Fig. 7.21 Interdiffusion coefficient D^{Fe}_{NiNi} for γ Fe–Ni–Cr alloys at 1100°C (in units of 10^{-11} cm^2 s^{-1}). After Duh and Dayananda[50]

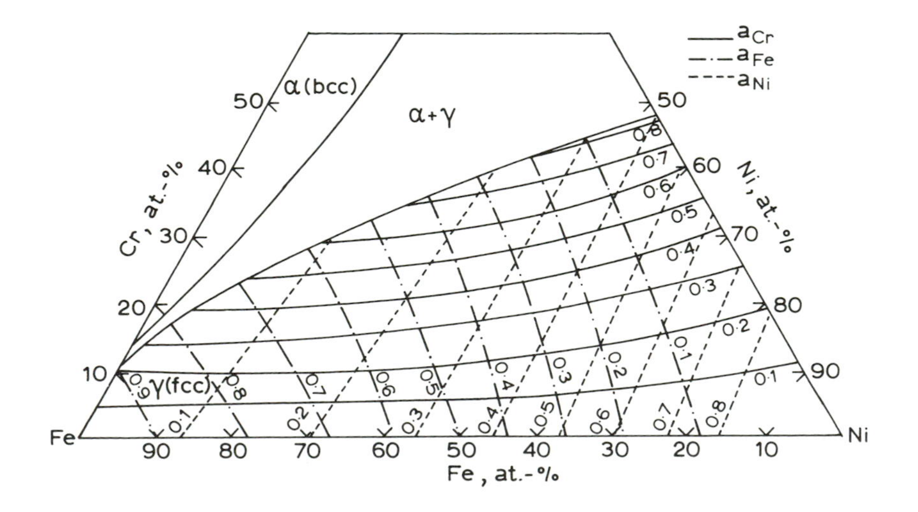

Fig. 7.22 Isoactivity lines for Fe, Ni, and Cr on the Fe–Ni–Cr isotherm at 1200°C based on subregular solution model. After Duh and Dayananda[50]

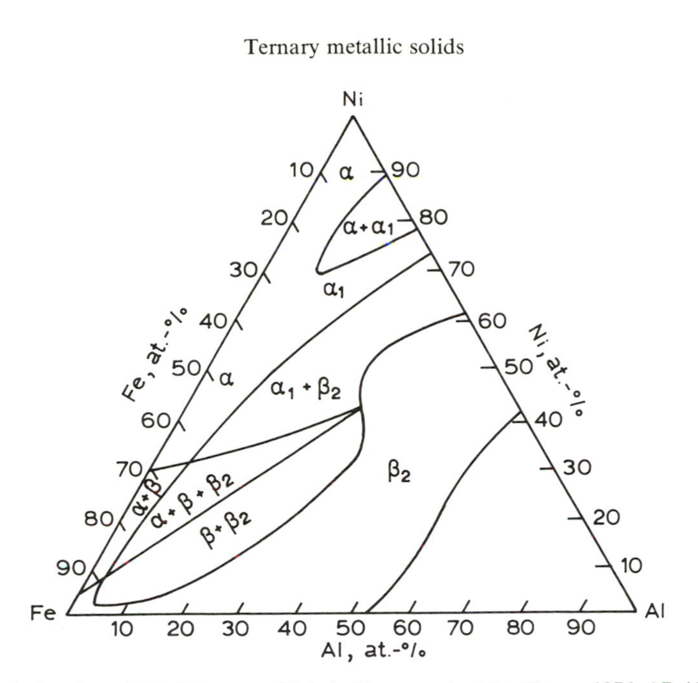

With permission from J. D. Moyer and M. A. Dayananda, *Met. Trans.*, 1976, **A7**, 1035

Fig. 7.23 Fe–Ni–Al constitution diagram. After Moyer and Dayananda[47]

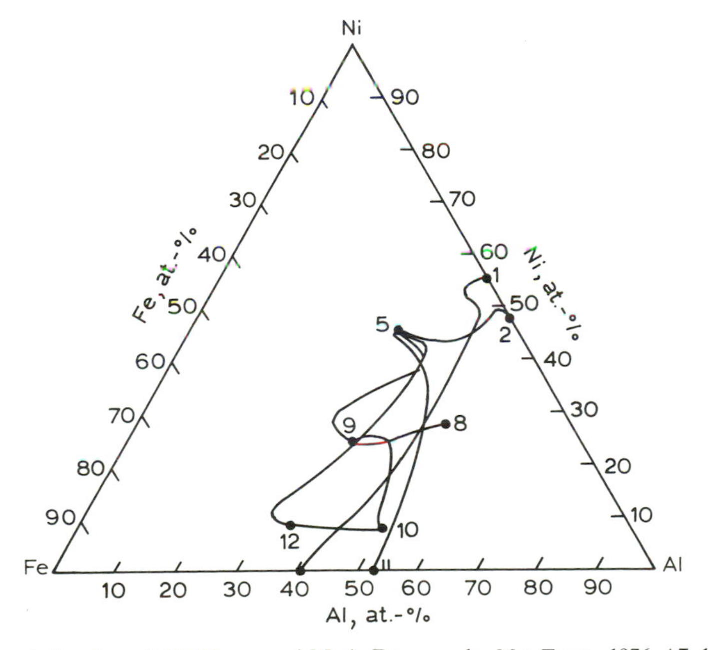

With permission from J. D. Moyer and M. A. Dayananda, *Met. Trans.*, 1976, **A7**, 1035

Fig. 7.24 Diffusion paths for various couples diffused at 1004°C. After Moyer and Dayananda[47]

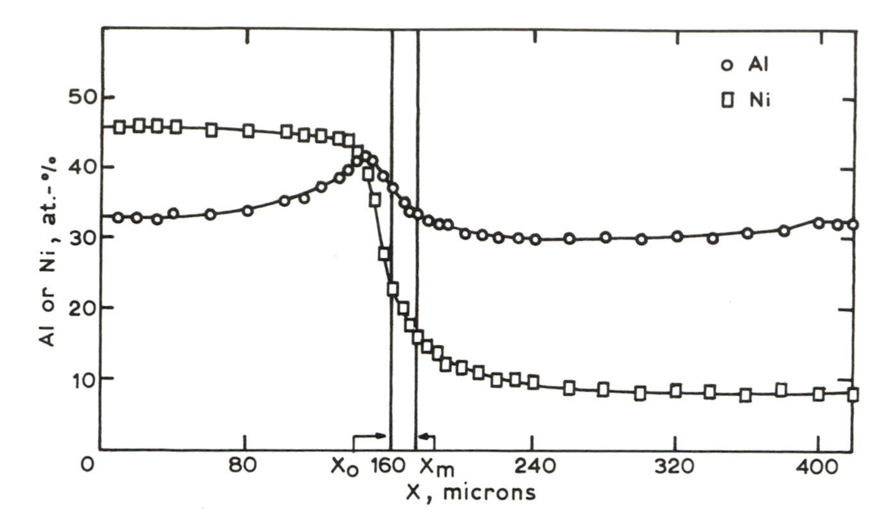

With permission from J. D. Moyer and M. A. Dayananda, *Met. Trans.*, 1976, **A7**, 1035

Fig. 7.25 Concentration profiles of aluminium and nickel for diffusion couple 5/12. After Moyer and Dayananda[47]

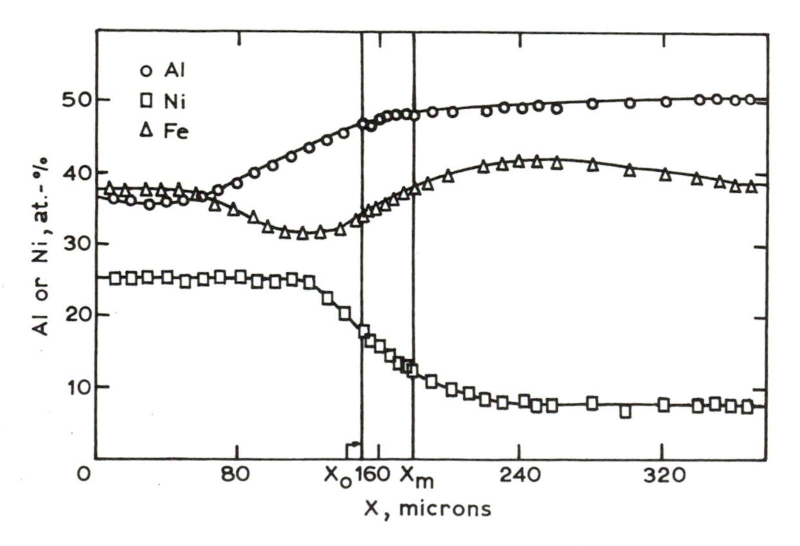

With permission from J. D. Moyer and M. A. Dayananda, *Met. Trans.*, 1976, **A7**, 1035

Fig. 7.26 Concentration profiles of aluminium, iron, and nickel for diffusion couple 9/10. After Moyer and Dayananda[47]

7.14 ZERO FLUX PLANES AND FLUX REVERSALS

Let us refer for the moment to the silver penetration curve in Fig. 5.1 where this component has been pushed up its own gradient due to the mutual repulsion of the copper. Here silver must be flowing to the right from a point near the 200 μm marker and it must be flowing to the left where its own gradient becomes steep. Hence there will be a plane of flux reversal in the couple. This plane has been designated by Dayananda[56] and others as a 'zero flux plane'. Such planes can be formally identified via the Boltzmann integrals (5.12) by recognizing that for a profile at fixed time, t can be factored out of λ and $d\lambda$ to give the flux

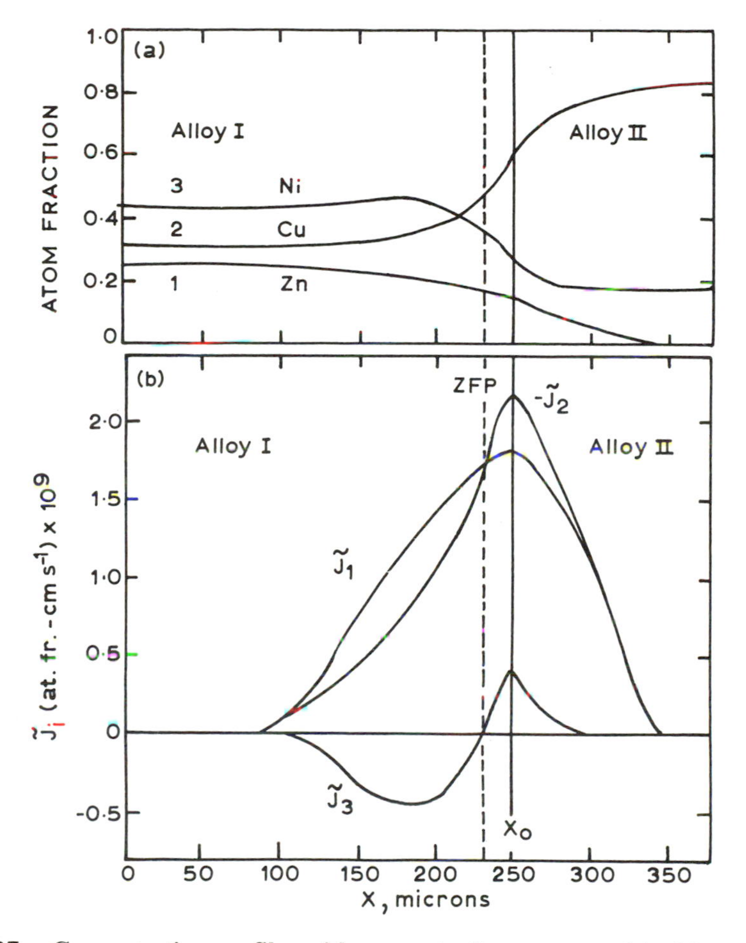

Fig. 7.27 Concentration profiles with concentrations expressed in (*a*) atom fraction for a ternary Cu–Ni–Zn diffusion couple assembled with Alloy I (30·1Cu–44·7Ni–25·2Zn) and Alloy II (80·6Cu–19·4Ni) and diffusion annealed at 775°C for 48 h; (*b*) calculated profiles of interdiffusion fluxes for the components for the couple with the direction from Alloy I to Alloy II taken as positive. After Dayananda[56]

$$J_i = \frac{1}{2t}\int_{m_i^0}^{m_i} x\,\mathrm{d}m_i \qquad (7.160)$$

where the origin of x is established by

$$\int_{m_i^0}^{m_i^1} x\,\mathrm{d}m_i = 0 \qquad (7.161)$$

This implies that the flux at every point in the couple can be evaluated via the appropriate integral over the concentration profiles, and in particular, a zero flux plane will be identified when the integral on the right of equation (7.160) vanishes at a concentration $m_i \neq m_i^1$. Figure 7.27(*b*) shows the flux profiles for a Cu–Ni–Zn diffusion couple obtained from the concentration profiles of Fig. 7.27(*a*) via the integrals (7.160). Herein has been identified a zero flux plane (ZFP) for the Ni component. Zero flux plane identifications have a potential for assisting diffusion diagnostics, at least in special cases. For example, in the system Fe–Co–Ni which was examined in §7.8 and is known to be ideal, the established composition (X_1, X_2) of a zero flux plane for component 1 say, yields the relation via equations (7.12) and (7.14)

$$\frac{X_2}{X_1}\frac{L_{11}^3}{L_{12}^3} = -\frac{\mathrm{d}(X_2/X_3)}{\mathrm{d}(X_1/X_3)} \qquad (7.162)$$

wherein the right hand side can be directly evaluated from the profiles. This also holds for Henrian solutions, that is dilute solutions with weak Wagner interactions. Alternatively, if the tracer coefficients of all species are similar so that Kirkendall drifts are small, implying weak off-diagonal L-coefficients, then for a zero flux plane of component 1

$$\frac{\partial(\mu_1 - \mu_3)}{\partial x} = 0 \qquad (7.163)$$

In general this yields a specific piece of thermodynamic information at the composition of the zero flux plane. If the solution is dilute then this information refers to the Wagner interaction parameters. For an interstitial component, which is necessarily dilute and for which the off-diagonal L always approach zero, equation (7.163) is always valid at its zero flux plane (cf the discussion of quasi-steady experiments in §§6.3 and 7.9).

Dayananda and co-workers have widely explored the nuances and applications of this concept in multicomponent diffusion studies (see Dayananda[56] for a review of the subject).

REFERENCES

1 S. GLASSTONE, K. J. LAIDLER, AND H. EYRING: 'The theory of rate processes', 1941, New York, McGraw-Hill

2 N. F. MOTT AND R. W. GURNEY: 'Electronic processes in ionic crystals', 1940, Oxford University Press

3 F. SEITZ: *Acta Cryst.*, 1950, **3**, 355

4 C. ZENER: in 'Imperfections in nearly perfect crystals', 1952, New York, John Wiley

5 J. E. LANE AND J. S. KIRKALDY: *Can. J. Phys.*, 1964, **42**, 1643

6 A. R. ALNATT: *J. Chem. Phys.*, 1965, **43**, 1855
7 A. R. ALNATT: *J. Phys. C (Solid State Phys.)*, 1982, **15**, 5605
8 S. A. RICE AND P. A. GRAY: 'The statistical mechanics of simple liquids', 535–44, 1965, New York, Interscience
9 J. BARDEEN AND C. HERRING: in 'Atom movements', (ed. J. H. Hollomon), 87, 1951, Cleveland, ASM
10 A. D. LECLAIRE: *Prog. Met. Phys.*, 1953, **4**, 265
11 J. R. MANNING: 'Diffusion kinetics of atoms in crystals', 1968, Princeton NJ, D. Van Nostrand
12 J. R. MANNING: *Can. J. Phys.*, 1968, **46**, 2633
13 J. R. MANNING: *Met. Trans.*, 1970, **1**, 499
14 J. G. KIRKWOOD, R. L. BALDWIN, P. J. DUNLOP, L. J. GOSTING, AND G. KEGELES: *J. Chem. Phys.*, 1960, **33**, 1505
15 L. S. CASTLEMAN: *Met. Trans. A.*, 1983, **14**, 45
16 L. S. DARKEN: *Trans. AIME*, 1949, **180**, 430
17 J. S. KIRKALDY: *Can. J. Phys.*, 1958, **36**, 917
18 J. S. KIRKALDY AND J. E. LANE: *ibid.*, 1966, **44**, 2059
19 L. S. DARKEN: in 'Atom movements', (ed. J. H. Hollomon), 1–25, 1951, Cleveland, ASM
20 J. PHILIBERT AND A. G. GUY: *Compte Rendu*, 1963, **227**, 2281
21 A. G. GUY AND J. PHILIBERT: *Z. Metallk.*, 1965, **56**, 3
22 T. O. ZIEBOLD AND A. R. COOPER: *Acta Met.*, 1965, **13**, 465
23 H. SCHÖNERT: *Z. Phys. Chem.*, 1980, **119**, 53
24 C. WAGNER: 'Thermodynamics of alloys', 53, 1952, Reading, Mass., Addison-Wesley
25 J. S. KIRKALDY, R. J. BRIGHAM, AND D. H. WEICHERT: *Acta Met.*, 1965, **13**, 907
26 J. S. KIRKALDY, ZIA-UL-HAQ, AND L. C. BROWN: *Trans. ASM*, 1963, **56**, 834
27 J. S. KIRKALDY: *Can. J. Phys.*, 1957, **35**, 435
28 J. S. KIRKALDY AND G. R. PURDY: *Can. J. Phys.*, 1962, **40**, 208
29 A. G. GUY AND C. B. SMITH: *Trans. ASM Q.*, 1962, **55**, 1
30 L. C. BROWN AND J. S. KIRKALDY: *Trans. AIME*, 1964, **230**, 223
31 T. O. ZIEBOLD AND R. E. OGILVIE: *ibid.*, 1967, **239**, 942
32 M. MURAKAMI, D. DE FONTAINE, J. M. SANCHEZ, AND J. FODOR: *Thin Solid Films*, 1975, **25**, 465
33 L. O. SUNDELÖF: *Arkiv Kemi*, 1963, **20**, 369
34 D. DE FONTAINE: Ph.D. Thesis, 1966, Northwestern University, Evanston, Ill.
35 J. S. KIRKALDY AND G. R. PURDY: *Can. J. Phys.*, 1969, **47**, 865
36 J. E. MORRAL AND J. W. CAHN: *Acta Met.*, 1971, **19**, 1037
37 J. W. GIBBS: 'Scientific papers, Vol. 1', 1906, New York, Longmans Green
38 I. I. PRIGOGINE AND R. DEFAY: 'Chemical thermodynamics', 1954, New York, Longmans Green
39 J. L. MEIJERING: *Philips Res. Reports*, 1950, **5**, 333
40 L. ONSAGER: *Ann. N.Y. Acad. Sci.*, 1945–46, **46**, 244
41 H. FUJITA AND L. J. GOSTING: *J. Am. Chem. Soc.*, 1956, **78**, 1099
42 J. S. KIRKALDY: *Can. J. Phys.*, 1959, **37**, 30
43 J. W. CAHN: *Acta Met.*, 1961, **9**, 795
44 J. K. CHANDHOK, J. P. HIRTH, AND E. J. DULIS: *Trans. AIME*, 1964, **230**, 223
45 H. C. AKUEZUE: Ph.D. Thesis, 1983, University of California at Berkeley
46 R. M. PILLIAR AND J. S. KIRKALDY: Unpublished research, 1967
47 T. D. MOYER AND M. A. DAYANANDA: *Met. Trans. A*, 1976, **7**, 1035
48 J. P. SABATIER AND A. VIGNES: *Mém. Sci. Rev. Métall.*, 1967, **64**, 225
49 A. VIGNES AND J. P. SABATIER: *Trans. AIME*, 1969, **245**, 1795
50 J. G. DUH AND M. A. DAYANANDA: *Diff. Defect Data*, 1985, **39**, 1
51 J. S. KIRKALDY, G. R. MASON, AND W. J. SLATER: *Trans. Can. IMM*, 1961, **64**, 53
52 P. T. CARLSON, M. A. DAYANANDA, AND R. E. GRACE: *Met. Trans. A*, 1972, **3**, 819

53 A. G. GUY AND V. LEROY: 'Journées internationales des applications du cobalt', June 1964, Brussels

54 R. D. SISSON JR AND M. A. DAYANANDA: *Met. Trans. A*, 1977, **8**, 1949

55 C. C. WAN AND R. J. DEHOFF: *Acta Met.*, 1977, **25**, 287

56 M. A. DAYANANDA: in 'Diffusion in solids: recent developments', (ed. M. A. Dayananda and G. E. Murch), 195, 1985, Warrendale PA, AIME

57 M. A. DAYANANDA, P. F. KIRSCH, AND R. E. GRACE: *Trans. AIME*, 1968, **242**, 885

58 M. A. DAYANANDA AND R. E. GRACE: *ibid.*, 1965, **233**, 1287

59 J. D. WHITTENBERGER AND M. A. DAYANANDA: *Met. Trans.*, 1970, **1**, 3301

60 A. BRUNSCH AND S. STEEB: *Z. Metallk.*, 1974, **65**, 765

61 J. A. NESBITT AND R. W. HECKEL: *Met. Trans. A*, 1987 (in press)

62 R. F. BALLUFFI AND L. L. SEIGLE: *J. Appl. Phys.*, 1954, **25**, 607

63 A. R. COOPER JR: 'Geochemical transport and kinetics', 15, 1974, Carnegie Institute of Washington

64 P. T. CARLSON, M. A. DAYANANDA, AND R. E. GRACE: *Met. Trans. A*, 1975, **6**, 1245

65 G. SCATCHARD AND T-P. LIN: *J. Am. Chem. Soc.*, 1962, **84**, 28

66 S. J. ROTHMAN, L. J. NOWICKI, AND G. E. MURCH: *J. Phys. F*, 1980, **10**, 383

CHAPTER 8

Ternary ionic solids

8.1 INTRODUCTION

This chapter is concerned with diffusion in crystalline solids in which electrostatic effects are important, and the relationship between the electrical conductivity and diffusion coefficient is of interest. Vacancy and interstitial correlation effects are ignored in consideration of the strong electrostatic effects.

In ionic crystals, electrostatic forces are so strong that cations and anions are confined to their own sublattices. Clearly two kinds of vacancies and two kinds of interstitials are possible, at least in principle. Since these point defects carry effective electric charges relative to the perfect lattice, the concentrations of oppositely charged defects are strictly related through the electroneutrality condition. This requirement largely dictates the diffusional behaviour of such substances. The approaches which allow the prediction of ternary coefficients from thermodynamics, tracer and binary coefficients extend to these solid ionic solutions (e.g. §§8.10 and 8.13).

8.2 LATTICE DEFECTS

As discussed in §2.18, an ideal pure stoichiometric compound, MX, will in general contain Schottky and Frenkel defects involving in principle both cation and anion sublattices. In addition, such a compound is susceptible of defect formation by dissolution of impurities and by deviation from stoichiometry. In what follows the defect notation of Kroger and Vink[1] is employed, wherein the lattice species are represented by the symbol $S_M^\times$. Here the subscript represents the normal occupancy in a perfect crystal of the site in question, and the principal symbol represents the species actually occupying the site. The superscript represents the charge of the species relative to normal site occupancy with a prime indicating a negative, a dot positive, and a cross zero charge.

If a solute ion has the same valence as the host lattice species, no additional defect formation accompanies the dissolution process. An example is the formation of magnesiowustite, (Fe, Mg)O by substitutional dissolution

$$Mg + Fe_{Fe}^\times = Mg_{Fe}^\times + Fe$$

However, the introduction of aliovalent (dopant) solute impurities will give rise to charge-compensating defects. Examples are the substitutional dissolution of Cr_2O_3 or Li_2O in NiO

$$2Cr + 3Ni_{Ni}^\times = 2Cr_{Ni}^{\cdot} + V_{Ni}'' + 3Ni \qquad (8.1)$$

$$2\text{Li} + \text{Ni}_{\text{Ni}}^{\times} + \text{V}_{\text{Ni}}'' = 2\text{Li}_{\text{Ni}}' + \text{Ni} \quad (8.2)$$

These reactions will occur simultaneously with the achievement of Schottky equilibrium:

$$\text{zero} = \text{V}_{\text{Ni}}'' + \text{V}_{\text{O}}^{\cdot\cdot} \quad (8.3)$$

and it is seen that both sites and relative charges are conserved. In general, for the compound MX_y

$$\sum_i \text{S}_{\text{M}}^{(i)} = \frac{1}{y}\sum_j \text{S}_{\text{X}}^{(j)} \quad (8.4)$$

$$\sum_i z^{(i)}\text{S}_{\text{M}}^{(i)} = \sum_j z^{(j)}\text{S}_{\text{X}}^{(j)} \quad (8.5)$$

where the $S^{(i)}$ and $S^{(j)}$ are, respectively, the species occupying the cationic and anionic sublattices and $z^{(i)}$ and $z^{(j)}$ are their relative charges.

Real compounds have compositions which may depart from stoichiometry. Consider, for example, a binary oxide in isothermal equilibrium with oxygen. According to the phase rule, the oxide composition varies with oxygen partial pressure. Since the ratio of cation to anion sites is the same whether the compound is stoichiometric or not, any compositional variation implies a change in the concentration of point defects in one or other of the ion sublattices. Thus, for example, an oxide containing a non-stoichiometric excess of metal could accomodate the additional cations as interstitials or by forming compensating anion vacancies

$$\text{MO} = \frac{1}{1+x}(\text{M}_{1+x}\text{O}) + \frac{x}{2}\text{O}_2(\text{g}) \quad (8.6)$$

involving

$$\text{O}_{\text{O}}^{\times} = \tfrac{1}{2}\text{O}_2(\text{g}) + \text{V}_{\text{O}}^{\cdot\cdot} + 2\text{e}' \quad (8.7)$$

or

$$\text{O}_{\text{O}}^{\times} + \text{M}_{\text{M}}^{\times} = \tfrac{1}{2}\text{O}_2(\text{g}) + \text{M}_i^{\cdot\cdot} + 2\text{e}' \quad (8.8)$$

A metal deficit oxide composition can be accommodated by the formation of cation vacancies or anion interstitials

$$\frac{x}{2}\text{O}_2 + \text{MO} = \frac{1}{1-x}(\text{M}_{1-x}\text{O}) \quad (8.9)$$

$$\tfrac{1}{2}\text{O}_2(\text{g}) = \text{O}_{\text{O}}^{\times} + \text{V}_{\text{M}}'' + 2\text{h}^{\cdot} \quad (8.10)$$

$$\tfrac{1}{2}\text{O}_2(\text{g}) = \text{O}_{\text{i}}'' + 2\text{h}^{\cdot} \quad (8.11)$$

Despite deviation from stoichiometry, the conditions of site and charge balance, equations (8.4) and (8.5), are maintained. The various point defects may in principle be ionized to differing degrees, e.g.

$$\text{V}_{\text{M}}^{\times} = \text{V}_{\text{M}}' + \text{h}^{\cdot} \quad (8.12)$$

$$\text{V}_{\text{M}}' = \text{V}_{\text{M}}'' + \text{h}^{\cdot} \quad (8.13)$$

It is seen that semiconducting metal excess oxides contain free electrons, i.e. they are n-type, whereas a metal deficit is associated with the presence of positive holes, i.e. with p-type behaviour.

The thermodynamics of point defects may be handled using the above equations and the methods of statistical mechanics. This methodology has been reviewed by Mott and Gurney,[2] Howard and Lidiard[3] and more recently by Flynn.[4] If, however, there is no requirement to quantitatively evaluate equilibrium constants, the interrelations among the various defects can be described by the application to the various defect equilibria of the law of mass action. This approach is due to Wagner and Schottky[5] and its application has been reviewed many times, notably by Kofstad[6] and Kroger.[7]

As an example the susceptibility of a compound to deviation from stoichiometry and its relationship with the intrinsic disorder of the compound are considered. For a metal oxide subject to Schottky disorder in which neutral vacancies are formed

$$O = V_M^{\times} + V_O^{\times} \tag{8.14}$$

and the equilibrium with the gas phase is represented by

$$\tfrac{1}{2}O_2(g) = O_O^{\times} + V_M^{\times} \tag{8.15}$$

$$O_O^{\times} = \tfrac{1}{2}O_2(g) + V_O^{\times} \tag{8.16}$$

If the composition of the compound is defined as MO_{1+y}, then

$$y = [V_M^{\times}] - [V_O^{\times}] \tag{8.17}$$

$$= K_{15}p_{O_2}^{1/2} - K_{16}p_{O_2}^{-1/2} \tag{8.18}$$

where ideal or Henrian solution behaviour has been assumed.

At the stoichiometric composition, $y = 0$ and the oxygen partial pressure has a particular value $p_{O_2}(0)$. Under these conditions, the defect concentration of the compound is defined by the intrinsic disorder reaction (8.14) which may be compared with equation (8.18) to obtain

$$K_{15}p_{O_2}(0)^{1/2} = K_{16}p_{O_2}(0)^{-1/2} = K_{14}{}^{1/2} \tag{8.19}$$

Resubstitution in equation (8.18) yields

$$y = K_{14}^{1/2}\left\{\left[\frac{p_{O_2}}{p_{O_2}(0)}\right]^{1/2} - \left[\frac{p_{O_2}}{p_{O_2}(0)}\right]^{-1/2}\right\} \tag{8.20}$$

It is apparent how the imposition of a gradient in oxygen potential gives rise to a defect gradient and will lead to diffusion.

Rather than make approximations as to the thermodynamic nature of the solid solution it is clearly preferable to retain an equilibrium description in terms of chemical potentials which are, in any case, the quantities of importance in determining diffusional behaviour. The difficulty lies in incorporating the charge and site balances which are expressed in terms of concentration. This difficulty is avoided by describing the defect structure of the crystal in terms of what are known as building units.

8.3 BUILDING UNITS

The use of equations (8.4) and (8.5) has been explored by Schottky[8] and Kroger *et al.*[9] in arriving at their definition of building units. Building units are groups of lattice species having such a composition that the requirements (8.4) and (8.5) are met when the group is added to the crystal. The nature of these units may be demonstrated by considering the example of a ternary subsitutional oxide (A, B)O containing cation vacancies and positive holes as the only defects. The species constituting such a solid are

$$A_M^\times, B_M^\times, O_O^\times, V_M'', h^\cdot$$

and thus outnumber the thermodynamic compositional variables. Obvious building units made up from this set of species are $\{A_M^\times + O_O^\times\}$ and $\{B_M^\times + O_O^\times\}$. A subset of building units is comprised of relative building units. These consist of the difference between two building units. Thus relative building units represent a change in composition resulting from the replacement of one species with another, e.g. $\{A_M^\times - B_M^\times\}$.

As will shortly be seen such units can be used in combination to form or 'build' a complete structure and therefore to provide a thermodynamic description of the solid. Since relative building units represent compositional change they can be used to describe diffusion. It is clear that a flux of units $\{A_M^\times - B_M^\times\}$ corresponds to interdiffusion of cations A and B via a site-exchange process. Note that since these units can be used to represent both diffusion and equilibrium, they form a harmonious link between the transport properties and local equilibrium state of a solid. Such a description is entirely appropriate to our irreversible thermodynamic treatment.[10]

Since relative building units conserve sites and charge, their movement necessarily satisfies the flux constraints appropriate to diffusion in a field-free system. For one-dimensional diffusion in the oxide (A, B)O under discussion, these constraints are

$$J_A + J_B + J_V = 0 = J_O \tag{8.21}$$

$$2J_V = J_h \tag{8.22}$$

where the fluxes are measured within a solvent-fixed reference frame provided by an immobile anion lattice. It follows that movement of a vacancy must be accompanied by movement of positive holes and is associated with an opposing flux of cations. Relative building units U_i which describe these exchanges are

$$U_1 \equiv \{A_M^\times - V_M'' - 2h^\cdot\} \tag{8.23}$$

$$U_2 \equiv \{B_M^\times - V_M'' - 2h^\cdot\} \tag{8.24}$$

$$U_3 \equiv \{A_M^\times - B_M^\times\} \tag{8.25}$$

A further unit not contributing to diffusion but necessary to complete the structure is

$$U_4 \equiv \{O_O^\times + V_M'' + 2h^\cdot\} \tag{8.26}$$

It is clear that combination in the appropriate proportions of units 1, 2 and 4 yields a solid (A, B)O of any desired degree of substitution and non-stoichiometry.

Thermodynamic meaning is now attached to the relative building units by considering the reactions which lead to the introduction of point defects into the compound.

$$A_{(g)} + V''_M + 2h^{\cdot} = A_M^{\times} \tag{8.27}$$

$$B_{(g)} + V''_M + 2h^{\cdot} = B_M^{\times} \tag{8.28}$$

$$O_O^{\times} + V''_M + 2h^{\cdot} = \tfrac{1}{2}O_2(g) \tag{8.29}$$

These equilibria are described by their corresponding Gibbs equations (3.11) and (3.53) which, under isobaric, isothermal field-free conditions, may be written in terms of molar concentrations m and electrochemical potentials η as

$$\sum_i \eta_i \, dm_i = 0$$

in each case. Since the dm_i are related via the reaction stoichiometry coefficients v_i we may write (cf § 3.4)

$$\sum_i v_i \eta_i = 0 \tag{8.30}$$

whence

$$\begin{aligned} \mu_A &= \eta(A_M) - \eta(V_M) - 2\eta(h) = \mu(U_1) \\ \mu_B &= \eta(B_M) - \eta(V_M) - 2\eta(h) = \mu(U_2) \\ \tfrac{1}{2}\mu_{O_2} &= \eta(O_O) + \eta(V_M) + 2\eta(h) = \mu(U_4) \end{aligned} \tag{8.31}$$

and the potentials of U_1, U_2, and U_4 are seen to be the chemical potentials of the constituent elements μ_i.

The electrochemical potentials of individual lattice species cannot be measured. Moreover, they depend on the local electrostatic potential ψ through the definition

$$\eta(S^z) = \mu(S^z) + zF\psi \tag{8.32}$$

where z is the effective charge of the species and F is the Faraday. The value of ψ is also inaccessible to measurement. It is apparent that appropriate grouping of species leads to the avoidance altogether of the need to consider directly the electrostatic potential or individual species chemical potentials. These quantities are indeterminate within the formalism, just as they are experimentally inaccessible.

Since it is not possible to add, or remove, or diffuse lattice species other than in a way which conserves charge and lattice sites, the use of relative building units is entirely consistent with the fact that the thermodynamics and diffusion kinetics of ionic crystals can always be described in terms of elemental chemical potentials. Relative building units provide a link between the macroscopic thermodynamic/kinetic properties and the point defect structure.

8.4 IRREVERSIBLE THERMODYNAMIC TREATMENT

Under isothermal, isobaric conditions the local time rate of entropy change is given by equation (3.30) which, in the present case, becomes

$$T\dot{s} = -\sum_i \eta_i \dot{m}_i / m \tag{8.33}$$

As shown in Chapter 3, this leads to the evaluation of entropy production per unit volume σ as

$$T\sigma = -\sum_i J_i \nabla \eta_i \tag{8.34}$$

and this entropy source term serves to identify the driving forces which appear in the flux equations

$$J_i = -\sum_j^n L_{ij} \nabla \eta_j \qquad i = 1, 2 \dots n \tag{8.35}$$

Using these results we now proceed to simplify the description for ionic solid defect diffusion.

Taking up once again the example of the p-type oxide (A, B)O, the species fluxes within a solvent-fixed frame specified by $J_O = 0$ may be written

$$J_i = -\sum_j^4 L_{ij} \nabla \eta_j$$

The flux constraints specified by the zero net current and site conservation conditions are

$$\sum_i^4 z_i J_i = 0$$

$$\sum_i^3 J_i = 0$$

which lead to constraints on the coefficients

$$\sum_i^4 z_i L_{ij} = 0$$

$$\sum_i^3 L_{ij} = 0$$

where component 4 has been taken to be the positive hole. Substitution of these relationships, together with the Onsager reciprocal relations, in the flux equations leads to the reduced set

$$J_A = -L_{11}\nabla\{\eta(A_M) - \eta(V_M) - 2\eta(h)\} - L_{12}\nabla\{\eta(B_M) - \eta(V_M) - 2\eta(h)\}$$
$$J_B = -L_{21}\nabla\{\eta(A_M) - \eta(V_M) - 2\eta(h)\} - L_{22}\nabla\{\eta(B_M) - \eta(V_M) - 2\eta(h)\} \tag{8.36}$$

Substitution of relative building unit potentials from equations (8.31) leads to the simple forms ($1 \equiv A, 2 \equiv B$)

$$J_A = -L_{AA}\nabla\mu_A - L_{AB}\nabla\mu_B$$
$$J_B = -L_{BA}\nabla\mu_A - L_{BB}\nabla\mu_B \tag{8.37}$$

Because isothermal diffusion experiments conducted in the absence of any external field disclose nothing but mass transfer rates, the description is complete.

The algebra leading to equations (8.36) is tedious and, in cases involving multiple defect types, can represent an arduous task. Much labour is avoided by using instead the entropy equations (8.33) or (8.34). For the oxide (A, B)O, equation (8.33) may be written explicitly as

$$-Tm\dot{s} = \eta(A_M)\dot{m}(A_M) + \eta(B_M)\dot{m}(B_M) + \eta(V)\dot{m}(V) + \eta(h)\dot{m}(h) + \eta(O_O)\dot{m}(O_O) \quad (8.38)$$

The dm_i may be related to the changes in building unit concentrations via the definitions of equations (8.23), (8.24), and (8.26). Substitution followed by rearrangement leads to

$$\begin{aligned} -Tm\dot{s} &= \{\eta(A_M) - \eta(V) - 2\eta(h)\}\dot{m}_1 \\ &\quad + \{\eta(B_M) - \eta(V) - 2\eta(h)\}\dot{m}_2 \\ &\quad + \{\eta(O_O) + \eta(V) + 2\eta(h)\}\dot{m}_4 \end{aligned} \quad (8.39)$$

which is seen from inspection of equations (8.31) to be equivalent to

$$-Tm\dot{s} = \mu(U_1)\dot{m}_1 + \mu(U_2)\dot{m}_2 + \mu(U_4)\dot{m}_4 \quad (8.40)$$

The flux equations are therefore

$$J_i = -\sum_{j}^{4} L_{ij}\nabla\mu(U_j) \qquad i,j \neq 3 \quad (8.41)$$

Because all three component gradients are used, this set of equations still contains a redundancy, as expressed by the Gibbs–Duhem equation

$$\Sigma m_i \, d\mu_i = 0 \quad (8.42)$$

In the present instance, this is simply removed by adopting the solvent-fixed frame in which $J_O = 0$, and the form (8.37) results.

In proceeding from equation (8.38) it was necessary first to formulate relative building units. This necessity may be avoided by commencing the analysis with the entropy source equation (8.34)

$$-T\sigma = J_{A_M}\nabla\eta(A_M) + J_{B_M}\nabla\eta(B_M) + J_V\nabla\eta(V) + J_h\nabla\eta(h) + J_{O_O}\nabla\eta(O_O) \quad (8.43)$$

Application of the constraints (8.21) and (8.22) to (8.43) leads immediately to

$$-T\sigma = J_{A_M}\nabla\{\eta(A_M) - \eta(V) - 2\eta(h)\} + J_{B_M}\nabla\{\eta(B_M) - \eta(V) - 2\eta(h)\} \quad (8.44)$$

thereby identifying a set of relative building units whose potential gradients define the driving forces in the reduced flux equation set (8.36). It is then a simple matter to devise the equilibrium formulation (8.31) and thus arrive at the flux equation (8.37).

Although the form of equation (8.37) is unique for any compound, the choice of relative building units affords some freedom. Thus for the model compound (A, B)O, the set of units 1, 3 and 4 might instead have been chosen, leading to a superficially different set of reduced flux equations. The representations are, however, equivalent, as unit 3 is equivalent to unit 2 minus unit 1. All three methods of reduction are, of course, equivalent. However, algebraic simplicity is achieved by applying flux constraint equations to the entropy source expression, thereby arriving directly at an appropriate grouping of diffusing lattice species. These groupings reflect the correlations which can exist among the diffusing species. Application of these techniques to a number of different defect solid types is now considered. The early work of Schmalzried and Holt[11, 12] is relevant to these developments.

8.5 COMPOUNDS CONTAINING SCHOTTKY DISORDER

Compounds containing cation and anion vacancies but no electronic defects are considered first. If the anion sublattice is almost immobile and all cations have the same valency, then for zero net charge transfer

$$J_{V_M} = 0$$

Within the oxygen solvent-fixed frame the constraint on the fluxes in a system of n cationic species is

$$\sum_{i}^{n} J_i = 0 \tag{8.45}$$

Application of this constraint to equation (8.34) leads to

$$T\sigma = -\sum_{i}^{n-1} J_i \nabla(\eta_i - \eta_n)$$

and hence

$$J_i = -\sum_{j}^{n-1} L_{ij} \nabla(\eta_i - \eta_n) = -\sum_{j}^{n-1} L_{ij} \nabla(\mu_i - \mu_n) \tag{8.46}$$

The relative building units are seen to be of the type given by equation (8.25) and correspond to diffusion via an exchange mechanism, just as expected for a system in which $J_V = 0$. This does not imply that vacancies are totally uninvolved, for, as noted earlier, building units of this type are equivalent to the difference between units which involve vacancies. In the present instance

$$\{B_M^{\times} - A_M^{\times}\} \equiv \{B_M^{\times} - V_M'' - V_O^{\cdot\cdot}\} - \{A_M^{\times} - V_M'' - V_O^{\cdot\cdot}\}$$

and thus vacancies function as intermediaries. Correlated ionic diffusion arises through the fact that the different cations share the use of the same vacancies. An analogous result was derived by Cooper and Heasley[13] for ternary compounds (A, B)X neglecting the possibility of cross-effects in the L-matrix. This description has been used in the interpretation of interdiffusion in the CaF_2–SrF_2 system by Scheidecker and Berard.[14]

The effects of introducing an aliovalent (dopant) cation to the system are now considered.[15] On the assumption that anion vacancies are practically immobile and very much lower in concentration than dopants, then for a system containing n monovalent cations, and a single divalent dopant D, the entropy source term may be written

$$T\sigma = -\sum_i^n J_i \nabla \eta_i - J_D \nabla \eta(D_M) - J_V \nabla \eta(V_M) \tag{8.47}$$

which is subject to the constraints

$$J_D = J_V \tag{8.48}$$

$$\sum_i^n J_i + J_D + J_V = 0 \tag{8.49}$$

It follows then that

$$J_i = -\sum_j^n L_{ij} \nabla \{\eta_j - \tfrac{1}{2}\eta(D_M) - \tfrac{1}{2}\eta(V_M)\} \tag{8.50}$$

and the relative building units clearly represent the dopant dissolution and the diffusing complex

$$D + 2A_M^\times = D_M^\cdot + V_M' + 2A$$

$$\nabla\{\eta(A_M) - \tfrac{1}{2}\eta(V_M) - \tfrac{1}{2}\eta(D_M)\} = \nabla(\mu_A - \tfrac{1}{2}\mu_D) \tag{8.51}$$

In this case the interchange of two monovalent cations is represented by

$$\{B_M^\times - A_M^\times\} \equiv \{B_M^\times - \tfrac{1}{2}V_M' - \tfrac{1}{2}D_M^\cdot\} - \{A_M^\times - \tfrac{1}{2}V_M' - \tfrac{1}{2}D_M^\cdot\}$$

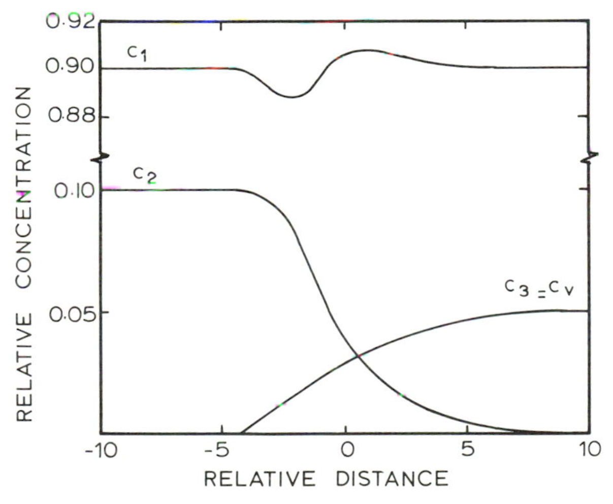

With permission from Canadian Journal of Physics, National Research Council of Canada

Fig. 8.1 Representative countercurrent diffusion profiles calculated for ionic solid containing monovalent solvent cations (1), monovalent solute cations (2), divalent solute cations (3), and cation vacancies (V). After Young *et al.*[15]

and is easily recovered from equation (8.51). Correlated ionic motions in this case arise through the electrostatic constraint expressed in equation (8.48), plus the fact that all cations share the same vacancies. Diffusive interactions calculated in this way are shown in Fig. 8.1. Such a rigorous association between solute and vacancy has been reported by Friaf,[16] Beniere and Chemla[17] and others.

The preceding examples have their solutions predicated largely upon the assumption of an immobile anion lattice. Attention is now directed to the case of a multicomponent substitutional oxide containing as its only defects cation and anion vacancies, both of which are mobile. For a system containing n divalent cations, the entropy source term is given by

$$T\sigma = -\sum_{i}^{n} J_i \nabla \eta_i - J_{V_M} \nabla \eta(V_M) - J_O \nabla \eta(O_O) - J_{V_O} \nabla \eta(V_O) \quad (8.52)$$

which within the lattice-fixed frame is subject to the constraints

$$J_{V_M} = J_{V_O} \quad (8.53)$$

$$\sum_{i}^{n} J_i + J_{V_M} = 0 = J_O + J_{V_O} \quad (8.54)$$

if interchange between the two sublattices does not occur. Application of these constraints to equation (8.52) leads to

$$J_i = -\sum_{j}^{n} L_{ij} \nabla \{\eta_i + \eta(O_O) - \eta(V_M) - \eta(V_O)\} \quad (8.55)$$

and the relative building units are seen to represent correlated cation and anion diffusion via a vacancy mechanism. The corresponding local equilibrium describes the formation of vacancy pairs

$$M_M^\times + O_O^\times = V_M'' + V_O^{\cdot\cdot} + M_{(g)} + \tfrac{1}{2}O_{2(g)}$$

whence

$$\nabla\{\eta(M_M) + \eta(O_O) - \eta(V_M) - \eta(V_O)\} = \nabla\{\mu_M + \tfrac{1}{2}\mu_{O_2}\}$$

If $\nabla \mu_{O_2} = 0$ then, through the application of the Gibbs–Duhem equation (8.42), the flux equations (8.55) reduce to equations (8.46). Thus cocurrent diffusion of all cations is possible only in the presence of an activity gradient in the electronegative species.

8.6 CATION FLUXES IN METAL DEFICIT COMPOUNDS

A multicomponent substitutional oxide in which all cations have the same valency is taken to contain cation vacancies and positive holes as the only defects, both of which are mobile. In the solvent-fixed frame provided by an immobile anion sublattice, the constraints on the fluxes of n divalent cationic species are

$$\sum_{i}^{n} J_i = -J_{V_M} \quad \text{and} \quad 2J_{V_M} = J_h \quad (8.56)$$

where the vacancies are assumed to be doubly ionized. Application of equation (8.56) to equation (8.34) leads to

$$J_i = -\sum_j^n L_{ij}\nabla\{\eta_i - \eta(\mathrm{V_M}) - 2\eta(\mathrm{h})\} \tag{8.57}$$

which is, of course, merely a generalization from equation (8.44). For a binary metal deficit oxide $A_{1-x}O$ the result is simply[18]

$$J_\mathrm{V} = -J_\mathrm{A} = -L_{11}\nabla\{\eta_1 - \eta(\mathrm{V_M}) - 2\eta(\mathrm{h})\}$$

which, with equation (8.31), leads to

$$J_\mathrm{V} = -\frac{L_{11}}{2}RT\nabla\ln(p_{\mathrm{O_2}}/p^{\mathrm{O}}_{\mathrm{O_2}}) \tag{8.58}$$

The application of this result to chemical diffusion in binary metal oxides and sulfides has been examined by a number of workers, e.g. Chen and Peterson[19] and Fryt.[20] In the absence of an oxygen potential gradient there can be no vacancy flux in this system and equation (8.57) reduces to (8.46). Systems of this type have been

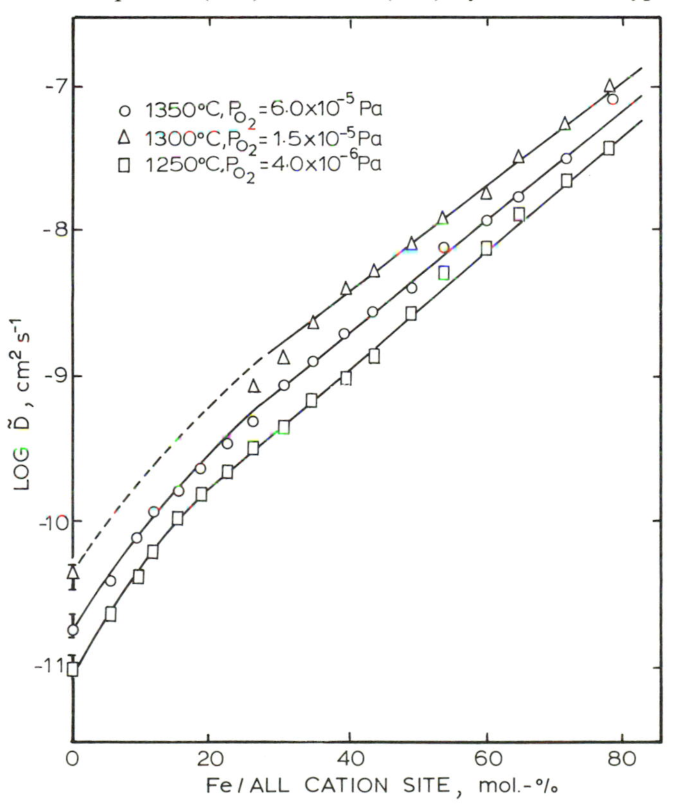

With permission from N. Sato and K. S. Goto, *J. Am. Ceram. Soc.*, 1982, **65**, 158

Fig. 8.2 Interdiffusivity as a function of iron concentration in $Fe_{0.916}O$–MgO diffusion couple. After Sata and Goto[23]

studied by Jones and Cutler[21] and Appel and Pask.[22] However, when the lattice can sustain an oxygen potential gradient then a non-zero vacancy flux is generated. Sata and Goto[23] have demonstrated the effect of p_{O_2} on interdiffusivity in (Mg, Fe)O by carrying out isobaric experiments at a number of different p_{O_2} values. The results of their experiments are shown in Fig. 8.2.

Equations of a form analogous to (8.57), but omitting the cross-terms, have been used by Wagner[24, 25, 26] and by Coates and Dalvi[27] in the description of metal and binary alloy oxidation. In this model it is supposed that a solid scale of (A, B)O grows on the alloy surface at a rate controlled by cation diffusion through the scale. Accordingly the rate of increase in scale thickness x_s is given by

$$\mathrm{d}x_s/\mathrm{d}t = V_{AB}(J_A + J_B)$$

where V_{AB} is the volume of oxide formed per mole of A or B. The use of this description in extracting oxide diffusion data from binary alloy oxidation experiments has been demonstrated for a number of systems and this work has been reviewed by Narita *et al.*[28] An example of the kind of data involved, and the degree of success achieved by the theory, is shown in Fig. 8.3.

If, in addition, anion vacancies exist at a finite concentration and are mobile, then the flux constraints are given by the first of equations (8.56) and

$$J_{O_O} = -J_{V_O} \tag{8.59}$$

$$2J_{V_M} = 2J_{V_O} + J_h \tag{8.60}$$

where the anion vacancies are doubly ionized. Application of equations (8.56), (8.59), and (8.60) to an appropriate form of equation (8.34) (containing $\sum_i^n J_i, J_{V_M}$, J_{O_O}, J_{V_O} and J_h) leads to flux equations relative to the local lattice

$$\begin{aligned} J_i &= -\sum_j^n L_{ij}\nabla\{\eta_j - \eta(V_M) - 2\eta(h)\} - L_{iO}\nabla\{\eta(O_O) - \eta(V_O) + 2\eta(h)\} \\ J_O &= -\sum_j^n L_{ij}\nabla\{\eta_j - \eta(V_M) - 2\eta(h)\} - L_{OO}\nabla\{\eta(O_O) - \eta(V_O) + 2\eta(h)\} \end{aligned} \tag{8.61}$$

The additional relative building unit corresponds to the replacement of an oxygen vacancy by an anion and accompanying positive holes

$$2h^{\cdot} + O_O^{\times} = V_O^{\cdot\cdot} + \tfrac{1}{2}O_2(g)$$

$$\nabla\{\eta(O_O) - \eta(V_O) + 2\eta(h)\} = \tfrac{1}{2}\nabla\mu_{O_2}$$

and the effect of any oxygen gradient on the diffusion of all components is explicit.

8.7 CATION FLUXES IN FRENKEL DISORDER COMPOUNDS

A multicomponent substitutional oxide containing n cations of the same valency is taken to contain as the only defects cation vacancies and interstitial cations, both types of defect being mobile. The flux constraints on such a compound are, for a solvent- (oxygen-) fixed frame

$$\sum_i J_i = -J_{V_M} \tag{8.62}$$

$$\sum_i J'_i = J_{V_M} \tag{8.63}$$

where the prime indicates an interstitial cation and it has been assumed that each species can diffuse both interstitially and via the lattice. Substitution of equations (8.62) and (8.63) into equation (8.34) yields

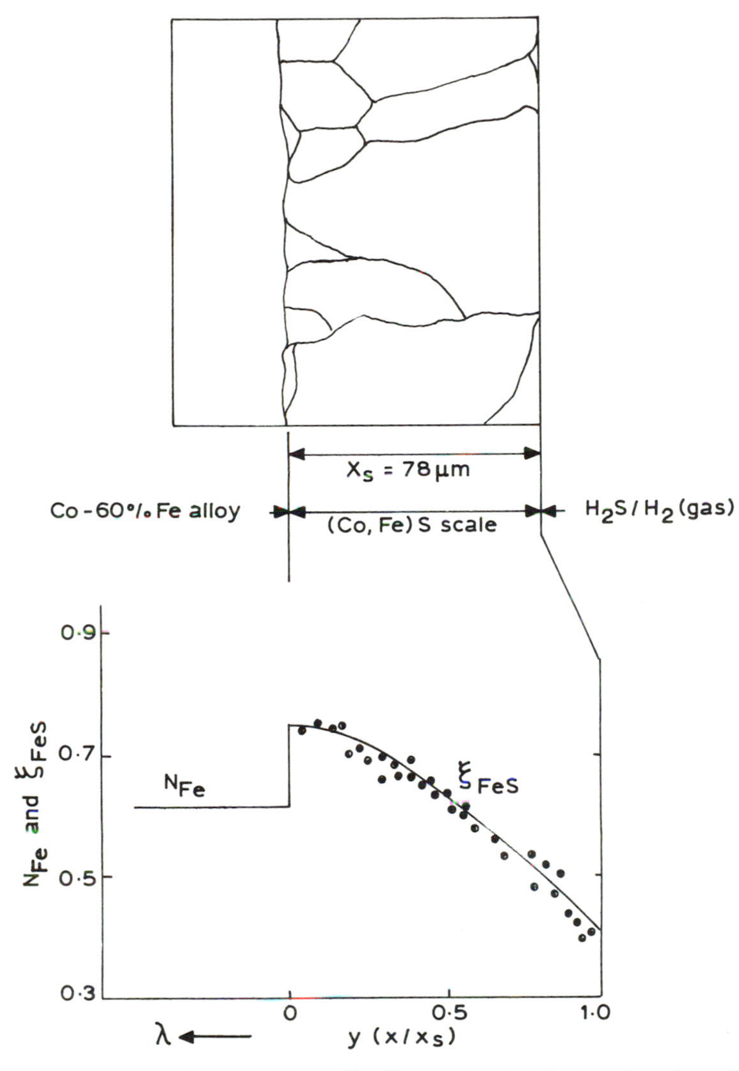

Reprinted by permission of the publisher, The Electrochemical Society, Inc., from D. J. Young, T. Narita, and W. W. Smeltzer, *J. Electrochem. Soc.*, 1980, **127**, 680

Fig. 8.3 (CoFe)S scale formed on Co–60Fe alloy at 600°C and P_{S_2} = 2×10^{-6} atm in 1 h. Compositions of the solid phases were determined from EPMA or Co, Fe, and S concentrations. Micrographic insert is traced. Continuous line represents best fit to diffusion model. After Young *et al.*[29]

$$T\sigma = -\sum_{j}^{n-1} J_j \nabla\{\eta_j(A_M) - \eta_n(A_M)\} - \sum_{j}^{n-1} J'_j \nabla\{\eta_j(A_i) - \eta_n(A_i)\}$$
$$- J_V \nabla\{\eta(V) + \eta_n(A_i) - \eta_n(A_M)\} \tag{8.64}$$

The relative building units involved are seen to correspond to substitutional dissolution on the cation sublattice or the interstitial sublattice, and to the intrinsic disorder reaction

$$A_M^{\times} = A_i^{\cdot\cdot} + V_M''$$

At local equilibrium therefore,

$$\nabla\eta(V_M) + \nabla\eta(A_i) = \nabla\eta(A_M) \tag{8.65}$$

and the third term in equation (8.64) vanishes. Noting that the total flux J^T of a component is given by

$$J_i^T = J_i + J_i'$$

then equation (8.64) is seen to lead to the expected form

$$J_i^T = -\sum_{j}^{n-1} L_{ij} \nabla\{\mu_j - \mu_n\} \tag{8.66}$$

because, for any metals A and B

$$\nabla\{\eta(A_M) - \eta(B_M)\} = \nabla\{\eta(A_i) - \eta(B_i)\} = \nabla\{\mu_A - \mu_B\} \tag{8.67}$$

The form of the correlations among individual species fluxes is readily seen by writing fluxes for a ternary (A, B)O·

$$J_{B_M} = -L_{BB}\nabla\{\eta(B_M) - \eta(A_M)\} - L_{BB_i}\nabla\{\eta(B_i) - \eta(A_i)\}$$
$$J_{B_i} = -L_{B_iB}\nabla\{\eta(B_M) - \eta(A_M)\} - L_{B_iB_i}\nabla\{\eta(B_i) - \eta(A_i)\} \tag{8.68}$$

Substitution into the gradient terms from the equilibrium expression (8.65) makes explicit the preservation of charge neutrality through the correlated motions of interstitials and vacancy–cation pair exchanges. For example

$$J_{B_M} = -L_{BB}\nabla\{\eta(B_M) - \eta(A_i) - \eta(V_M)\} - L_{BB_i}\nabla\{\eta(B_i) - \eta(A_M) + \eta(V_M)\}$$

8.8 CATION FLUXES IN A METAL EXCESS OXIDE

In the presence of free carriers the strict coupling between interstitial and vacancy flows is destroyed, and in addition one must consider reactions of the type

$$M_{(g)} = M_i^{\cdot\cdot} + 2e' \tag{8.69}$$

or

$$O_O^{\times} + M_M^{\times} = \tfrac{1}{2}O_2(g) + M_i^{\cdot\cdot} + 2e' \tag{8.70}$$

both of which lead to non-stoichiometry of the metal excess type. For an oxide containing n divalent cations, all of which are mobile both as interstitial and lattice species, plus doubly ionized cation vacancies and free electrons, the flux constraints within a solvent- (oxide-) fixed reference frame are

$$\sum_i J_i = -J_{V_M} \tag{8.71}$$

$$2\sum_i J_i' = 2J_{V_M} + J_e \tag{8.72}$$

Application of these constraints to equation (8.34) yields

$$T\sigma = \sum_j^{n-1} J_j \nabla\{\eta_j(A_M) - \eta_n(A_M)\} - \sum_j^{n-1} J_j' \nabla\{\eta_j(A_i) - \eta_n(A_i)\} - J_e \nabla\{\eta_e + \tfrac{1}{2}\eta_n(A_i)\} \tag{8.73}$$

where the vacancy flux term has been eliminated through the use of equation (8.65). Component fluxes are then derived with the aid of equation (8.66) as

$$J_i^T = -\sum_j^{n-1} L_{ij} \nabla(\mu_j - \mu_n) - L_{ie} \nabla \mu_n \tag{8.74}$$

where the solid–gas equilibrium (8.69) has been employed to substitute

$$\nabla \mu_M = \nabla \eta(M_i) + 2\Delta\eta(e) \tag{8.75}$$

The three possible correlations are clear from equation (8.73): site conservation imposes a correlation among the fluxes of substitutional lattice species; charge conservation requires that interstitial fluxes be correlated either with each other or with an electron flux. Because of local equilibria among the various defect species, this set of correlations *includes* others such as that between interstitials and vacancies or between electrons and vacancies. These may be recovered explicitly from equation (8.73) by appropriate substitution for the gradient terms.

8.9 EVALUATION OF THE KINETIC COEFFICIENTS

The movement of point defects will be calculated according to the methods of absolute rate theory outlined in §2.10. In applying this theory it is necessary to take into account the effective charge of the moving species and therefore the influence of any local electrostatic field. A method for carrying out such a calculation has been given by Dignam *et al.*[30] for single species and is here adapted to the particular case of interchanges between neighbouring ions and vacancies.

The activity a_{ij} of transition-state complexes (omitting the notation†) is described by their equilibrium with the participating reactants, i and j (cf §2.10)

$$a_{ij}/a_i a_j = K_{ij} = \exp(-\Delta U_{ij}/RT)\exp(\Delta S_{ij}/R) \tag{8.76}$$

where ΔU_{ij} is the internal energy (or enthalpy) of formation of the activated complex and ΔS_{ij} its entropy of formation. It will be assumed that the existence of an electrostatic field leaves ΔS_{ij} unchanged but has an effect on ΔU_{ij}. A profile of the periodic internal energy surface in a direction parallel to that of diffusion is shown in Fig. 8.4. The electrostatic field is supposed to arise solely through the movement of charged species within the solid and is therefore aligned with the diffusion direction. Obviously the height of the energy barrier to diffusion is modified by the presence of a field, the barrier being lower to downfield migration than for upfield migration of an appropriately charged species. If ΔU_{ij} is the height of the barrier in the absence of a field then we may write for the interchange of species i and a vacancy between planes (1) and (2) separated by a distance λ

$$J = m\lambda \nu_{iv} \exp(\Delta S_{iv}/R)\{a_i^{(1)} a_v^{(2)} \exp[-(\Delta U_{iv} - zF[\psi^{(0)} - \psi^{(1)}])/RT] - a_v^{(1)} a_i^{(2)} \exp[-(\Delta U_{iv} - zF[\psi^{(2)} - \psi^{(0)}])/RT]\}/\gamma_{iv} \tag{8.77}$$

where m is the volume concentration of lattice sites, and ν_{iv} is a kinetic frequency term and γ_{iv} is the activity coefficient for the transition state complex. Here z is the effective charge of the vacancy, that of the cation being zero. In the more general case of interchange between any two differently charged species, z is replaced by the difference in effective charges of the participating species. Bracketed superscripts represent the location at which the quantity in question is evaluated.

Equation (8.77) is cleared of common terms and subjected to Taylor series expansion of the terms $a_v^{(2)}$ and $(a_i^{(2)} \exp[-zF\psi^{(2)}/RT])$. Retention of linear terms, in the case where the field is not inordinately high, leads to

$$J = -\frac{m\lambda^2}{RT} \nu_{iv} K_{iv} a_i a_v \{\nabla\mu_i - \nabla\mu_v - zF\nabla\psi\} \tag{8.78}$$

which, upon substituting from equation (8.32), becomes

$$J = -\frac{m\lambda^2}{RT} \nu_{iv} K_{iv} a_i a_v \{\nabla\eta_i - \nabla\eta_v\} \tag{8.79}$$

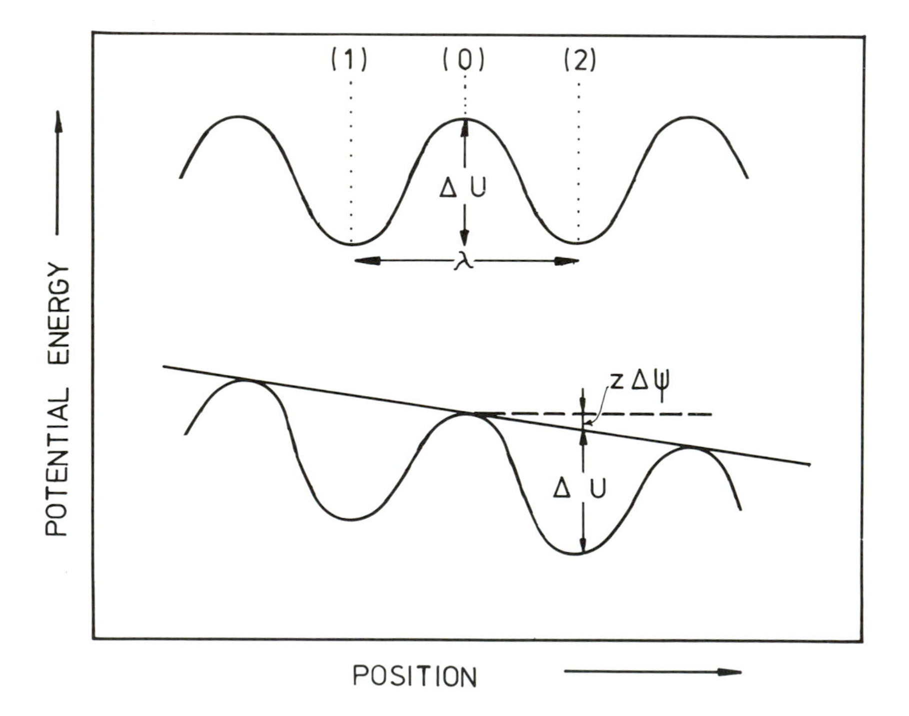

Fig. 8.4 Potential energy profile in diffusion direction: upper curve, no electrostatic field; lower curve showing effect of electrostatic field

In this equation, and in what follows, the constant K_{iv} is redefined to absorb the activity coefficient γ_{iv}. An equation of this form is always found for the interchange of charged species between adjacent sites in an electrostatic field.

8.9.1 Cation diffusion kinetics

The above description is applied first to the ternary solid solution (A, B)X in which cations A and B have the same valence. The anion sublattice will be assumed to be essentially immobile and furthermore it will be supposed that no free carriers exist. Under these conditions no net flux of vacancies can occur.

Since interchanges between nearest neighbours only are considered, it is sufficient to consider two adjacent planes normal to the diffusion direction. The mechanism of site interchange between these planes automatically obeys the site conservation restraint. However each site interchange involves the movement of a vacancy and therefore, according to the zero current constraint equation, countercurrent movement of a vacancy. This balance might be achieved either by the immediate return of the vacancy to its former plane or by the movement of a second vacancy. Since vacancy concentrations are low, the former mechanism is of higher probability. In order for such a mechanism to contribute to diffusion it must involve the two different cation species. The simplest microscopic model which can fulfil these requirements involves three participating sites, one on one plane and two on the other. One of the three sites is, of course, vacant. The successive steps of the only correlated process involving the set $\{1, 2, V\}$ to contribute to diffusion are shown in Fig. 8.5 and will be referred to as mechanism I. Component fluxes due to this mechanism are now evaluated by kinetic methods.

The net species flux from plane (1) to plane (2) in step (a) is found to be

$$J^{(a)} = \frac{m\lambda^2}{RT} \nu_{2v} K_{2v} a_2 a_v \{\nabla\eta_v - \nabla\eta_2\} \tag{8.80}$$

where detailed balance (see Chapter 3) requires that $\nu_{2v} = \nu_{v2}, K_{1v} = K_{v1}$. Similarly it is found that

$$J^{(b)} = \frac{m\lambda^2}{RT} \nu_{1v} K_{1v} a_1 a_v \{\nabla\eta_1 - \nabla\eta_v\} \tag{8.81}$$

In addition, it is possible to formulate an expression for the species flux as a single step correlated transition from the initial to the final state of the

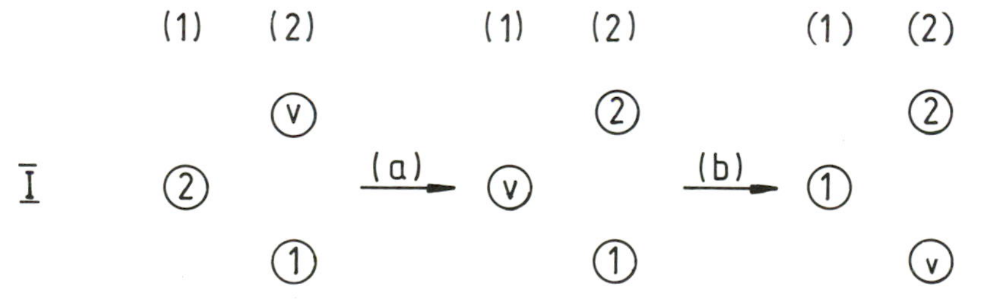

Fig. 8.5 Site exchange model for diffusion in (A, B)O with fixed anion lattice. Species shown are the two different cations and a vacancy

mechanism. In this procedure an activated complex at equilibrium with each of these two states is proposed. Then for species 2,

$$J_2^{\mathrm{I}} = m\lambda \nu_{\mathrm{I}} K_{\mathrm{I}} \{a_2^{(1)} a_{\mathrm{v}}^{(2)} a_1^{(2)} - a_1^{(1)} a_2^{(2)} a_{\mathrm{v}}^{(2)}\} \tag{8.82}$$

which, after Taylor series expansion and retention of linear terms, becomes, on the assumption that $\gamma^{\dagger} \sim 1$,

$$J_2^{\mathrm{I}} = \frac{m\lambda^2}{RT} K_{\mathrm{I}} a_l a_2 a_{\mathrm{v}} \{\nabla \mu_1 - \nabla \mu_2\} \gamma_{\mathrm{I}} \tag{8.83}$$

Because no net charge transfer occurs, the gradients are those of chemical potential.

The quantity ν_{I} is found from the steady state condition

$$J_2^{\mathrm{I}} = J^{(a)} = J^{(b)} \tag{8.84}$$

which, upon substitution from equations (8.80), (8.81), and (8.83) followed by elimination of gradient terms using $\mu_1 = \eta_1$, $\mu_2 = \eta_2$ for this particular model yields

$$\nu_{\mathrm{I}} = \left\{ \frac{K_{\mathrm{I}} a_1}{\nu_{2\mathrm{v}} K_{2\mathrm{v}}} + \frac{K_{\mathrm{I}} a_2}{\nu_{1\mathrm{v}} K_{1\mathrm{v}}} \right\}^{-1} \tag{8.85}$$

Noting that, by definition, ν_{ij} and K_{ij} are independent of concentration, equation (8.85) can be simplified by introducing the new variables m_i', which for Henrian solutions are strictly proportional to the species concentrations, i.e.

$$m_i' = a_i \nu_{i\mathrm{v}} K_{i\mathrm{v}} \tag{8.86}$$

Substitution of equation (8.86) produces the simplified result

$$\nu_{\mathrm{I}} = \frac{\nu_{1\mathrm{v}} \nu_{2\mathrm{v}}}{m_1' + m_2'} \tag{8.87}$$

Combination of equations (8.85)–(8.87) yields

$$J_2^{\mathrm{I}} = \frac{m\lambda^2}{RT} \frac{m_1' m_2'}{m_1' + m_2'} a_{\mathrm{v}} \{\nabla \mu_1 - \nabla \mu_2\} \tag{8.88}$$

Recalling that $J_1 = -J_2$, this kinetic expression is seen to be in accordance with the phenomenological equation (8.46), with

$$L_{11} = \frac{m\lambda^2}{RT} \frac{m_1' m_2'}{m_1' + m_2'} a_{\mathrm{v}} \tag{8.89}$$

There are, of course, no cross-effects in what is in effect a binary system. It is to be noted that even though no vacancy flux exists, the vacancy activity appears explicitly in the phenomenological coefficient because vacancies are required as intermediaries for the process to occur.

The above description may be extended to multicomponent systems using the above model and the methods of §2.10. We consider here the system (A, B, C)X in which cations A, B and C have the same valence. It will be assumed again that the anion lattice is immobile and that no free carriers are present. Under these conditions the system is analogous to one of three electrolytes having a common ion and, as shown by Miller,[31] is, from a diffusion point of view, a ternary system.

As in the previously considered case of (A, B)X, the mechanism of site interchange and the requirement of zero current imply that vacancy movement is correlated to an immediate opposing movement of the vacancy. Again the simplest microscopic model available involves three participating sites, one on one plane and two on the adjacent plane. Two of the sites must be occupied by different cations and the third one vacant. With the three cations designated as species 1, 2 and 3 and the vacancies as V, the three possible correlated processes which can contribute to diffusion are shown in Fig. 8.6. The three mechanisms are designated II, III and IV.

Following the same procedures as for (A, B)X, it is found that

$$J_2^{\mathrm{II}} = \phi_{23}\{\nabla\mu_3 - \nabla\mu_2\} \tag{8.90}$$

$$J_3^{\mathrm{II}} = -J_2^{\mathrm{II}} \qquad J_1^{\mathrm{II}} = 0 \tag{8.91}$$

$$J_1^{\mathrm{III}} = \phi_{13}\{\nabla\mu_3 - \nabla\mu_1\} \tag{8.92}$$

$$J_3^{\mathrm{III}} = -J_1^{\mathrm{III}} \qquad J_2^{\mathrm{III}} = 0 \tag{8.93}$$

$$J_1^{\mathrm{IV}} = \phi_{12}\{\nabla\mu_2 - \nabla\mu_1\} \tag{8.94}$$

$$J_2^{\mathrm{IV}} = -J_1^{\mathrm{IV}} \qquad J_3^{\mathrm{IV}} = 0 \tag{8.95}$$

where

$$\phi_{ij} = \frac{m\lambda^2}{RT}\frac{m_i' m_j'}{m_i' + m_j'} a_{\mathrm{v}} \tag{8.96}$$

Summation of the component fluxes then leads to

$$J_2 = -\phi_{12}\{\nabla\mu_2 - \nabla\mu_1\} + \phi_{23}\{\nabla\mu_3 - \nabla\mu_2\} \tag{8.97}$$

$$J_3 = -\phi_{23}\{\nabla\mu_3 - \nabla\mu_2\} - \phi_{13}\{\nabla\mu_3 - \nabla\mu_1\} \tag{8.98}$$

which upon rearrangement yield

$$\begin{aligned} J_2 &= -(\phi_{12} + \phi_{23})\{\nabla\mu_2 - \nabla\mu_1\} + \phi_{23}\{\nabla\mu_3 - \nabla\mu_1\} \\ J_3 &= \phi_{23}\{\nabla\mu_2 - \nabla\mu_1\} - (\phi_{13} + \phi_{23})\{\nabla\mu_3 - \nabla\mu_1\} \end{aligned} \tag{8.99}$$

and thus

$$\begin{aligned} L_{22} &= \phi_{12} + \phi_{23} \\ L_{23} &= L_{32} = -\phi_{23} \\ L_{33} &= \phi_{13} + \phi_{23} \end{aligned} \tag{8.100}$$

The Onsager reciprocal relations are explicit and it is noted that, as invariably found in practice, the relative importance of the off-diagonal terms depends on the relative concentrations of the three cations and on the choice of which two fluxes appear in the reduced representation (8.99).

The coefficients derived above are now used to calculate diffusion coefficients. If the solution is ideal or Henrian, then equation (8.46) applied to the system (A, B, C)O yields

$$J_2 = -RT\left\{\frac{L_{22}}{m_2} + \frac{L_{22}}{m_1} + \frac{L_{23}}{m_1}\right\}\nabla m_2 - RT\left\{\frac{L_{23}}{m_3} + \frac{L_{23}}{m_1} + \frac{L_{22}}{m_1}\right\}\nabla m_3$$

$$J_3 = -RT\left\{\frac{L_{32}}{m_2} + \frac{L_{32}}{m_1} + \frac{L_{33}}{m_1}\right\}\nabla m_2 - RT\left\{\frac{L_{33}}{m_3} + \frac{L_{33}}{m_1} + \frac{L_{32}}{m_1}\right\}\nabla m_3 \qquad (8.101)$$

Assuming further that the solution is dilute

$$m_2/m_1 \ll 1 \gg m_3/m_1$$

and substituting for L_{ij} from equation (8.100) one obtains

$$\frac{J_2}{m\lambda^2} = -\frac{a_{\rm v}}{m_2}\left\{\frac{m_1'm_2'}{m_1' + m_2'} + \frac{m_2'm_3'}{m_2' + m_3'}\right\}\nabla m_2 + \frac{a_{\rm v}}{m_3}\left\{\frac{m_2'm_3'}{m_1' + m_3'}\right\}\nabla m_3$$

$$\frac{J_3}{m\lambda^2} = \frac{a_{\rm v}}{m_2}\frac{m_2'm_3'}{m_2' + m_3'}\nabla m_2 - \frac{a_{\rm v}}{m_3}\left\{\frac{m_1'm_3'}{m_1' + m_3'} + \frac{m_2'm_3'}{m_2' + m_3'}\right\}\nabla m_3 \qquad (8.102)$$

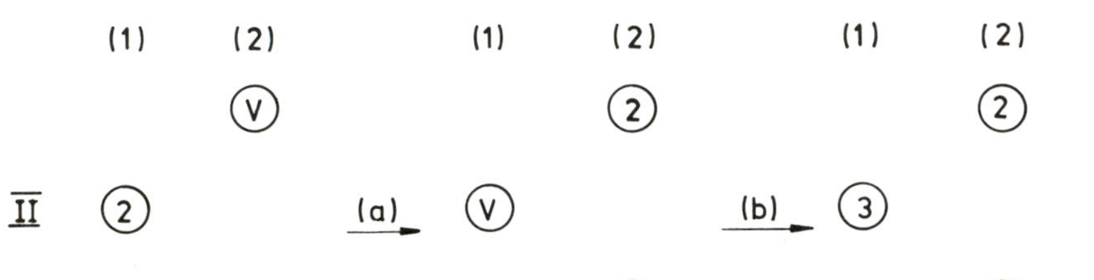

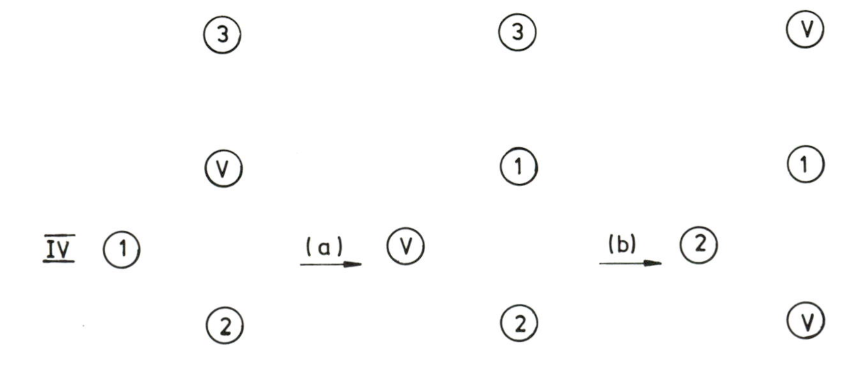

Fig. 8.6 Three mechanisms for diffusion in (A, B,C)O with fixed anion lattice

It is evident that the sufficient stability conditions (from equations (4.46)–(4.51))

$$D_{22},\ D_{33} > 0 \qquad D_{23}D_{32} > 0 \qquad D_{22}D_{33} - D_{23}D_{32} > 0$$

are all obeyed.

8.9.2 Cation fluxes with aliovalent solute

A more interesting situation arises when one of the solute cations has a different valence. This situation was examined in §8.5 and is exemplified by a system containing univalent solvent cation (1), univalent solute cation (2), divalent solute cation (3), cation vacancies (V) analysed on the assumptions of an immobile anion sublattice and absence of free carriers. Following Young *et al.*[15] we consider ion–vacancy interchanges between two adjacent planes which are normal to the diffusion direction. The site conservation constraint is thereby obeyed. However, each site interchange involves movement of a vacancy and therefore, according to the zero current equation (8.48), the cocurrent movement of species 3. Further, in order to satisfy the local charge neutrality condition, the microscopic model must contain in each of the adjacent planes being considered at least two participating sites. The simplest process capable of simultaneously satisfying the above constraints involves four species (one vacancy, one species 3 and two other species) distributed two to a plane. Higher order processes will be ignored. Recalling that species 1 is the solvent, then, in decreasing order of probability, the possible sets of configurations are {1, 1, 3, V}, {1, 2, 3, V}, {3, V, 3, V}, and {2, 2, 3, V}. The latter two will be ignored because of their low probability of occurrence.

The successive steps of the only correlated process involving the set {1, 1, 3, V} to contribute to diffusion are shown in Fig. 8.7. Following the same procedures as before, the component fluxes due to this mechanism, denoted by a superscript V, are found to be

$$J_1^{\mathrm{V}} = \phi\{\nabla\mu_3 + \nabla\mu_{\mathrm{V}} - 2\nabla\mu_1\} \tag{8.103}$$

$$J_3^{\mathrm{V}} = J_{\mathrm{V}}^{\mathrm{V}} = -\tfrac{1}{2}J_1^{\mathrm{V}} \qquad J_2^{\mathrm{V}} = 0 \tag{8.104}$$

where

$$\phi = \frac{m\lambda^2}{RT}\frac{m_1' m_3'}{m_1' + 2m_3'}a_{\mathrm{V}} \tag{8.105}$$

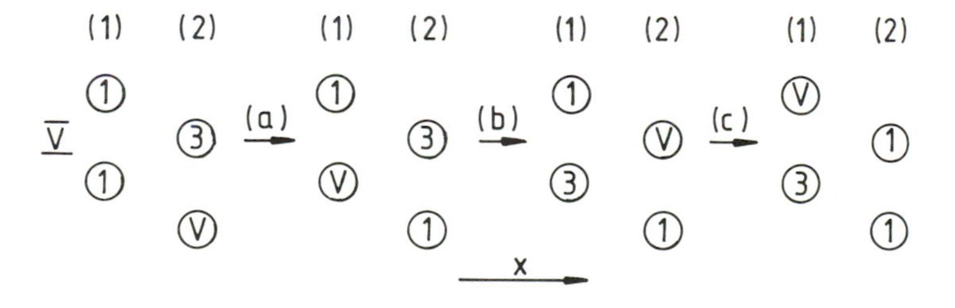

Fig. 8.7 Site exchange model for cation sublattice containing solvent monovalent species 1, divalent species 3 and vacancies

and the m_i' are defined by equation (8.86). Use has been made of the equalities

$$\nabla\mu_1 = \nabla\eta_1 \qquad \nabla\mu_2 = \nabla\eta_2$$

$$\nabla\mu_3 + \nabla\mu_V = \nabla\eta_3 + \nabla\eta_V$$

Once again, the electrostatic potential vanishes from the microscopic description as it did from the phenomenological equations. This is an immediate consequence of the fact that both levels of description have been formulated in such a way as to explicitly forbid any net charge transfer.

The species set {1, 2, 3, V} can be distributed in pairs among two adjacent planes in six distinguishable ways. Each of these states can be obtained from one of the others simply by interchanging the two planes. It is therefore appropriate to designate three of the stages as initial and three as final, as is shown in Fig. 8.8. The resulting three mechanisms are identified as VI, VII and VIII. Following the same procedures as before

$$J_1^{VI} = \phi'\{\nabla\mu_3 + \nabla\mu_V - \nabla\mu_1 - \nabla\mu_2\} \tag{8.106}$$

$$J_1^{VI} = J_2^{VI} = -J_3^{VI} = -J_V^{VI} \tag{8.107}$$

$$J_1^{VII} = -\phi'\{\nabla\mu_1 + \nabla\mu_V - \nabla\mu_2 - \nabla\mu_3\} \tag{8.108}$$

$$J_1^{VII} = J_V^{VII} = -J_2^{VII} = -J_3^{VII} \tag{8.109}$$

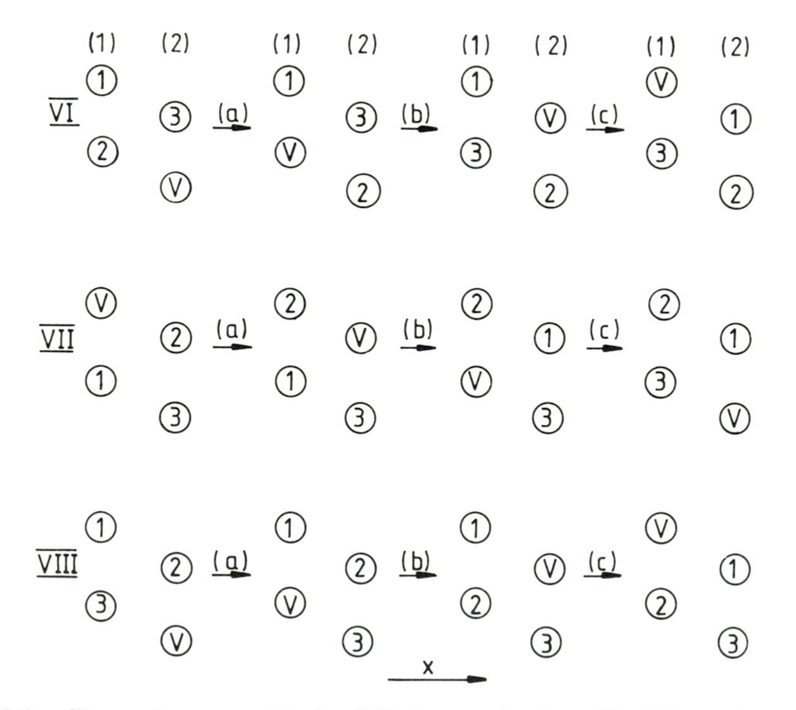

Fig. 8.8 Site exchange models for diffusion mechanisms VI, VII, and VIII

$$J_1^{VIII} = -\phi'\{\nabla\mu_1 + \nabla\mu_3 - \nabla\mu_2 - \nabla\mu_V\} \quad (8.110)$$

$$J_1^{VIII} = J_3^{VIII} = -J_2^{VIII} = -J_V^{VIII} \quad (8.111)$$

where

$$\phi' = \frac{m\lambda^2}{RT}\frac{m_1'm_2'm_3'}{m_1'm_2' + m_2'm_3' + m_1'm_3'}a_V \quad (8.112)$$

Additional contributions to the fluxes of species 1 and 2 can arise from the contributions of mechanism IV, i.e. by an exchange process involving no net vacancy flux. Summation of the component fluxes from mechanism IV to VIII, followed by rearrangement using equation (8.51) yields

$$J_1 = -(\phi_{12} + 2\phi + 3\phi')\nabla\{\mu_1 - \tfrac{1}{2}\mu_V - \tfrac{1}{2}\mu_3\} + (\phi_{12} + \phi')\Delta\{\mu_2 - \tfrac{1}{2}\mu_V - \tfrac{1}{2}\mu_3\} \quad (8.113)$$

$$J_2 = (\phi_{12} + \phi')\nabla\{\mu_1 - \tfrac{1}{2}\mu_V - \tfrac{1}{2}\mu_3\} - (\phi_{12} + 3\phi')\nabla\{\mu_2 - \tfrac{1}{2}\mu_V - \tfrac{1}{2}\mu_3\}$$

These expressions have the same form as the phenomenological equations (8.50), comparision with which yields the coefficients

$$L_{11} = \phi_{12} + 2\phi + 3\phi'$$

$$L_{22} = \phi_{12} + 3\phi'$$

$$L_{12} = L_{21} = -(\phi_{12} + \phi')$$

These expressions can now be used to evaluate the multicomponent diffusion coefficients[15] in the same way as was done earlier for the compound (A, B, C)X. Figure 8.1 shows the result of a representative numerical integration of the non-linear diffusion equations to illustrate an uphill diffusion effect.

8.9.3 Cation fluxes with free carriers

Consider first a metal deficit oxide $M_{1-x}O$ containing divalent cation vacancies and positive holes as the only defects. Since the anion sublattice is immobile, we need consider only the cation sublattice in the microscopic kinetic model, which is shown in Fig. 8.9. The planes which constitute the sublattice are taken formally to contain both cation vacancies and holes. Mechanism IX is seen to obey both the zero net current constraint and the local charge neutrality condition through the correlated motion of matching numbers of vacancies and positive holes.

From our standard procedures, the component fluxes due to this mechanism are found to be

$$J_V^{IX} = \omega_1\{\nabla\mu_1 - 2\nabla\mu_h - \nabla\mu_V\} \quad (8.114)$$

$$J_V^{IX} = \tfrac{1}{2}J_h^{IX} = -J_1^{IX} \quad (8.115)$$

where

$$\omega_1 = \frac{m\lambda^2}{RT}\frac{m_1'm_h'}{m_h' + 2m_1'a_V} \quad (8.116)$$

and

$$m_h' = a_h\nu_h K_h \quad (8.117)$$

and the relations

$$\mu_1 = \eta_1 \qquad \mu_V + 2\mu_h = \eta_V + 2\eta_h$$

have been employed. The above kinetic result is seen to be of the form of the phenomenological equation (8.57), whence it is found that

$$L_{VV} = \omega_1 \tag{8.118}$$

The above description may be extended to a ternary metal deficit oxide $(A, B)_{1-x}O$ in which cation vacancies and positive holes are the only defects. A microscopic kinetic model for the motion of the second cation is shown as mechanism X in Fig. 8.9. By analogy with the case described by equation (8.114), we may write

$$J_V^X = \omega_2\{\nabla\mu_2 - 2\nabla\mu_h - \nabla\mu_V\} \tag{8.119}$$

$$J_V^X = \tfrac{1}{2}J_h^X = J_2^X \qquad J_1^X = 0 \tag{8.120}$$

It is obvious that mechanisms IX and X give rise to no cross-effect between the cations. Such an effect arises, nonetheless, through the operation of mechanism I. Summation of the component fluxes due to mechanisms I, IX and X leads to

$$J_1 = -\omega_1\{\nabla\mu_1 - \nabla\mu_V - 2\nabla\mu_h\} - \phi_{12}\{\nabla\mu_1 - \nabla\mu_2\} \tag{8.121}$$

$$J_2 = -\omega_2\{\nabla\mu_2 - \nabla\mu_V - 2\nabla\mu_h\} - \phi_{12}\{\nabla\mu_2 - \nabla\mu_1\} \tag{8.122}$$

which may be rearranged to yield

$$J_1 = -(\omega_1 + \phi_{12})\{\nabla\mu_1 - \nabla\mu_V - 2\nabla\mu_h\} + \phi_{12}\{\nabla\mu_2 - \nabla\mu_V - 2\nabla\mu_h\} \tag{8.123}$$

$$J_2 = \phi_{12}\{\nabla\mu_1 - \nabla\mu_V - 2\nabla\mu_h\} - (\omega_2 + \phi_{12})\{\nabla\mu_2 - \nabla\mu_V - 2\nabla\mu_h\} \tag{8.124}$$

These are of the form of the phenomenological equation (8.57) whence

$$L_{11} = \omega_1 + \phi_{12} \qquad L_{22} = \omega_2 + \phi_{12}$$

$$L_{12} = L_{21} = -\phi_{12}$$

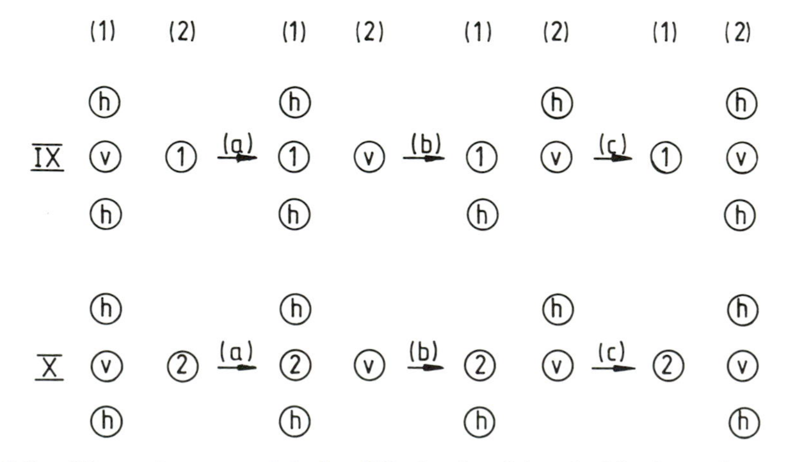

Fig. 8.9 Site exchange models for diffusion involving doubly charged vacancies associated with positive holes

The correlation between the cation fluxes is seen to arise through site conservation whereas the correlation between total (net) cation flux and positive hole flux is due to electrostatic coupling.

8.9.4 Kinetics of interstitial diffusion

In a solid containing an immobile anion sublattice and no free carriers, charge neutrality is preserved by the correlated motions of interstitials and vacancy-cation pair exchanges. Kinetic mechanisms of this type which contribute to mass transport in a ternary compound are shown in Fig. 8.10. In these models, the planes which constitute the crystal are supposed formally to contain both lattice and interstitial sites. Applying the usual procedures to these models and adding the contribution from mechanism I (A–B interchange without net vacancy flux) leads to the result

$$J_A = \kappa_{AB_i}\{\nabla\eta(V) + \nabla\eta(B_i) - \nabla\mu(A)\} + \kappa_{A_iB}\{\nabla\mu(B) - \nabla\eta(A_i) - \nabla\eta(V)\} + \phi_{12}\{\nabla\mu_B - \nabla\mu_A)\} \tag{8.125}$$

where

$$\kappa_{AB_i} = \frac{m\lambda^2}{RT}\frac{m'_A m'_{B_i}}{m'_{B_i} + m'_A a_V} a_V$$
$$\kappa_{A_iB} = \frac{m\lambda^2}{RT}\frac{m'_{A_i} m'_B}{m'_{A_i} + m'_B a_V} a_V \tag{8.126}$$

If the Frenkel defect equilibrium equation (8.65) is employed, then it is easily shown that

$$J_A = (\kappa_{AB_i} + \kappa_{A_iB} + \phi_{AB})\nabla\{\mu_B - \mu_A\} \tag{8.127}$$

which is of the expected form for a pseudobinary system. This treatment is easily extended to multicomponent systems. As an example we consider the quaternary compound (A, B, C)X. Kinetic mechanisms such as those shown in Fig. 8.10, but involving component C, lead to additional contributions to the flux of component A

$$J_A = (\kappa_{AB_i} + \kappa_{A_iB} + \phi_{AB})\nabla\{\mu_B - \mu_A\} + (\kappa_{AC_i} + \kappa_{C_iA} + \phi_{AC})\nabla\{\mu_C - \mu_A\}$$

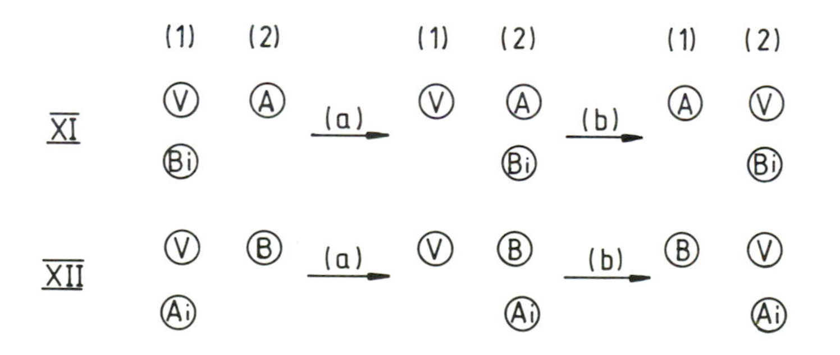

Fig. 8.10 Site exchange models for countercurrent cation-vacancy flow accompanied by interstitial cation movement

which may be regrouped to read

$$J_A = -(\alpha + \beta)\nabla\{\mu_A - \mu_C\} + \beta\nabla\{\mu_B - \mu_C\} \tag{8.128}$$

where

$$\beta = \kappa_{AB_i} + \kappa_{A_iB} + \phi_{AB} \qquad \alpha = \kappa_{AC_i} + \kappa_{A_iC} + \phi_{AC}$$

Similarly, it is found that

$$J_B = \beta\nabla\{\mu_A - \mu_C\} - (\gamma + \beta)\nabla\{\mu_B - \mu_C\} \tag{8.129}$$

where

$$\gamma = \kappa_{BC_i} + \kappa_{B_iC} + \phi_{BC}$$

Once again, the Onsager reciprocal relations are explicit.

The correlations revealed thus far are a consequence of the strict electrostatic coupling of interstitial and vacancy flows. In the presence of free carriers this coupling is destroyed, and interstitial species are free to move uncoupled from the lattice species. As interstitial mobilities can be much higher than those of lattice species it will frequently be sufficient to consider only the interstitial cations and accompanying electrons. At low concentrations, the only coupling is electrostatic: countercurrent interstitial flows or cocurrent interstitial plus electron flows, corresponding to the second and third terms of equation (8.73). It is easily demonstrated that the appropriate kinetic equations for divalent interstitial cations are

$$J_{A_i} = -\varepsilon_i\nabla\{\eta(A_i) + 2\eta(e)\} - \sum_j \phi'_{ij}\nabla\{\mu_i - \mu_j\} \tag{8.130}$$

where

$$\varepsilon_i = \frac{m\lambda^2}{RT}\frac{m'_i m'_e}{2m'_i + m'_e}$$

$$\phi'_{ij} = \frac{m\lambda^2}{RT}\frac{m'_i m'_j}{m'_i + m'_j}$$

The kinetic mechanisms proposed above are not capable of absolute numerical predictions. However, they indicate the expected concentration dependencies of the phenomenological coefficients and the direction in which correlations effect individual ionic fluxes. They are therefore of use in designing multicomponent diffusion experiments.

8.10 DEFECT PAIRING

Point defects may be able to interact strongly with one another,[32, 33] forming complexes which are relatively immobile within the host lattice. An example of this additional kind of local equilibrium is pair formation between two oppositely charged species, e.g. donors and acceptors in a semiconductor (cf §8.12) represented by the chemical equation

$$D^{\cdot} + A' = DA$$

for which we may write, assuming ideal or Henrian solution behaviour

$$X_{DA}/X_{D^{\cdot}}X_{A'} = K \tag{8.131}$$

In this case the concentrations of species actually free to move, X_D and $X_{A'}$, differ from the analytically measured concentrations of these components, X_1 and X_2, i.e.

$$X_{D^{\cdot}} = X_1 - X_{DA} \qquad X_{A'} = X_2 - X_{DA} \tag{8.132}$$

Since diffusion experiments on solids invariably rely on analytical measurements of total component concentration, the diffusion equations must be couched in these terms. It is necessary, therefore, to describe the thermodynamics of the solution in terms of total component concentrations.

Whilst the free species $D^{\cdot}$ and A', if dilute, will be Henrian, it is most improbable that the solute components 1 and 2 can be so regarded. This is an immediate consequence of the fact that X_1, X_2 are not direct measures of $X_{D^{\cdot}}$ and $X_{A'}$ because of the formation of pairs. Consequently activities referred to the pure solutes are expected to deviate markedly from their values in the absence of pair formation. One particular solution model is now proposed and its consequences for the diffusion equations explored.

If the two solutes have negligible solubility for the solvent and each other, then the phase boundaries of the solution may be defined, at constant temperature, as constant activity lines (Fig. 8.11)

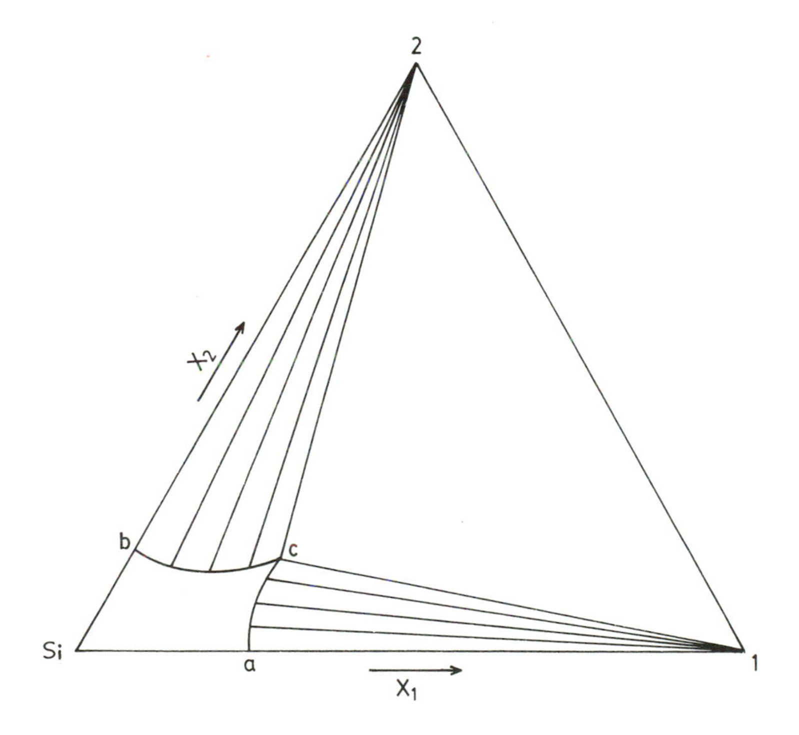

Fig. 8.11 Isothermal phase diagram for system Si–1–2 showing the Si-based solution in equilibrium with almost pure component 1 (along line a–c) and with almost pure component 2 (along line b–c)

$$\gamma_1 X_1 = 1 \quad \text{and} \quad \gamma_2 X_2 = 1 \tag{8.133}$$

for, respectively, the solution–component 1 and solution–component 2 boundaries. If the further condition is applied that unassociated solutes display Henrian solution behaviour, then, for Henrian activity coefficients $h_{A'}, h_{D^{\cdot}}$,

$$h_{D^{\cdot}} X_{D^{\cdot}} = 1 \quad \text{and} \quad h_{A'} X_{A'} = 1 \tag{8.134}$$

from which it follows that the phase boundaries are defined by constant concentrations of these species

$$X_{D^{\cdot}} = X^0_{D^{\cdot}} \quad \text{and} \quad X_{A'} = X^0_{A'} \tag{8.135}$$

respectively. The component activity coefficients γ_1, γ_2 can now be calculated.

Substitution from equations (8.132) and (8.135) into equation (8.131) followed by rearrangement yields for the pair concentration

$$X_{DA} = \frac{KX^0_D}{1 + KX^0_D} X_2 \tag{8.136}$$

which applies along the solution–component 1 boundary. The concentration of X_1 along this boundary is then found from equations (8.132) and (8.136) as

$$X_1 = X^0_D + \frac{KX^0_D}{1 + KX^0_D} \cdot X_2$$

whence, through equation (8.133)

$$\ln \gamma_1 \simeq \ln \left(\frac{1 + KX^0_D}{X^0_D} \right) - KX^0_D - KX_2 \tag{8.137}$$

Similarly, it is easily shown that along the solution–component 2 boundary

$$\ln \gamma_2 \simeq \ln \left(\frac{1 + KX^0_A}{X^0_A} \right) - KX^0_A - KX_1 \tag{8.138}$$

These results are seen to continue analytically to Wagner's expressions (see Appendix 1) for activity coefficients in dilute ternary solutions, namely

$$\begin{aligned} \ln \gamma_1 &= K_1 + \beta_1 X_1 + \beta X_2 \\ \ln \gamma_2 &= K_2 + \beta X_1 + \beta_2 X_2 \end{aligned} \tag{8.139}$$

with $\beta_1 = \beta_2 = \beta = -K$ (cf Dorward[33]).

Flux equations in the solvent-fixed frame for the system under discussion can be written as

$$J_1 = -L_{11} \nabla \mu_1 - L_{12} \nabla \mu_2$$

$$J_2 = -L_{21} \nabla \mu_1 - L_{22} \nabla \mu_2$$

Calculation of the chemical potential gradients using (8.137) and (8.138) then leads to the following evaluation of the D_{ij}

$$\begin{aligned} D_{11} &= RT(L_{11}/X_1 - KL_{12}) \simeq RTL_{11}/X_1 \\ D_{12} &= RT(-KL_{11} + L_{12}/X_2) \simeq -RTL_{11}K \\ D_{21} &= RT(L_{21}/X_1 - KL_{22}) \simeq -RTL_{22}K \\ D_{22} &= RT(-KL_{21} + L_{22}/X_2) \simeq RTL_{22}/X_2 \end{aligned} \quad (8.140)$$

the approximations being validated for dilute X_1 and X_2 ($L_{12} = L_{21} \simeq 0$). Since for strong complexing K can approach $1/X_1$ or $1/X_2$ (cf equation (8.131)) the off-diagonal Ds can approach the magnitude of the on-diagonal Ds despite the diluteness of the solution. Note that the negative off-diagonal Ds imply that diffusion will be biased in a countercurrent experiment, reflecting the strong attraction between the two species. This is equivalent to a negative interaction parameter ε_{12} in a system of uncharged species. In a Darken-type experiment we would anticipate behaviour similar to Fig. 5.2 (Co → Si, Ni → donor, and Fe → acceptor). A number of examples of the phenomenon have been reported by Reiss *et al.*[34] An evaluation of other aspects of this complex problem is presented in §8.12.

8.11 RELATIONSHIP BETWEEN DIFFUSIVITY AND CONDUCTIVITY

In the situation where only one charged species is mobile, the relationship between its diffusion coefficient and electrical mobility is described by the Einstein equation in molar units (cf §2.18)

$$u_i/D_i = z_iF/RT \quad (8.141)$$

where z_i represents the *real* charge of the moving species. The electrical mobility u_i is related to the general mobility of a species, B_i, defined as velocity per unit driving force, following

$$J_i = m_iv_i = L_{ii}X_i \quad (8.142)$$

as

$$B_i = L_{ii}/m_i \quad (8.143)$$

Substitution for this quantity in equation (8.142) and expansion of X_i then yields for an ideal or Henrian solution

$$J_i = -B_iRT\nabla m_i - m_iB_iz_iF\nabla\psi \quad (8.144)$$

which is recognized to have the form of the Nernst–Planck equation

$$J_i = -D_i\nabla m_i - m_iu_i\nabla\psi \quad (8.145)$$

Here the definition of electrical mobility by the equation

$$J_i = m_iu_iE \quad \text{for} \quad \nabla\mu_i = 0$$

with E being the electrostatic field strength, has been employed. Comparison of equations (8.144) and (8.145) shows that (cf equation (2.116))

$$u_i = B_iz_iF = D_iz_iF/RT$$

more generally

$$D_i = \frac{L_{ii}RT}{m_i}\frac{\partial \ln a_i}{\partial \ln m_i}$$

from which the Einstein relation follows immediately if the species behaves as an ideal or Henrian solute. Unfortunately, this simple relation has no general applicability in the situation where more than one species is mobile.

We consider first the simple situation in which two monovalent cations move on the same sublattice via neutral vacancies in the absence of any other mobile defect. Since sites are conserved, the general diffusion equation is

$$J_1 = -L_{11}\nabla\{\eta_1 - \eta(\mathrm{V})\} - L_{12}\nabla\{\eta_2 - \eta(\mathrm{V})\}$$
$$J_1 = -J_2 \tag{8.146}$$

and the diffusion equations reduce to

$$J_1 = -L_{11}\nabla\{\eta_1 - \eta_2\} = -L_{11}\nabla\{\mu_1 - \mu_2\} \tag{8.147}$$

Since $\mathrm{d}m_1 = -\mathrm{d}m_2$, this result may be expressed for an ideal or Henrian solution as

$$J_1 = -L_{11}RT\left\{\frac{1}{m_1} + \frac{1}{m_2}\right\}\nabla m_1 \tag{8.148}$$

and thus

$$D_1 = L_{11}RT\frac{m_1 + m_2}{m_1 m_2} \tag{8.149}$$

In the conductivity experiment the constraint (8.146) does not apply (if appropriate electrodes are provided) but we have

$$\nabla\mu_1 = 0 = \nabla\mu_2$$

In this case

$$J_1 = -(z_1 L_{11} + z_2 L_{12})F\nabla\psi \tag{8.150}$$

and, therefore

$$u_1 = (z_1 L_{11} + z_2 L_{12})F/m_1 \tag{8.151}$$

Comparison of equations (8.149) and (8.151) yields in place of the usual Einstein relationship

$$\frac{u_1}{D_1} = \frac{z_1 L_{11} + z_2 L_{12}}{RTL_{11}} \cdot \frac{m_2}{m_1 + m_2} \cdot F \tag{8.152}$$

It is apparent that only if component 1 is dilute and the cross-effect small, does the simpler relationship (8.141) apply.

In a ternary oxide (A, B)O containing divalent vacancies, positive holes, and an immobile anion sublattice, the general diffusion equation (8.57) is

$$J_1 = -L_{11}\nabla\{\eta_1 - \eta(\mathrm{V}) - 2\eta(\mathrm{h})\} - L_{12}\nabla\{\eta_2 - \eta(\mathrm{V}) - 2\eta(\mathrm{h})\} \tag{8.153}$$

If the vacancy concentration is very small, then to a first approximation this expression reduces to equation (8.147) which, for an ideal or Henrian solution leads to the estimate of D_1 provided by equation (8.149). The form of equation (8.153) results from the existence of the constraints appropriate to a diffusion experiment

$$J_1 + J_2 + J_V = 0$$

$$J_1 + J_2 + 2J_h = 0$$

In a conductivity experiment, the latter constraint no longer applies. Application of the remaining constraint to the entropy source expression yields as the transport equation

$$J_1 = -L_{11}\nabla\{\eta_1 - \eta(\mathrm{V})\} - L_{12}\nabla\{\eta_2 - \eta(\mathrm{V})\} - L_{13}\nabla\eta(\mathrm{h}) \tag{8.154}$$

Since, in the conductivity experiment,

$$\nabla\mu_1 = \nabla\mu_2 = \nabla\mu_h = \nabla\eta_V = 0$$

we have

$$J_1 = -(z_1L_{11} + z_2L_{12} + z_hL_{13})F\nabla\psi \tag{8.155}$$

and therefore,

$$u_1 = (z_1L_{11} + z_2L_{12} + z_hL_{13})F/m_1 \tag{8.156}$$

Comparison of (8.149) and (8.156) yields

$$\frac{u_1}{D_1} = \frac{z_1L_{11} + z_2L_{12} + z_hL_{13}}{RTL_{11}} \cdot \frac{m_2}{m_1 + m_2} \cdot F \tag{8.157}$$

It is evident that a different relationship between mobility and diffusion coefficient will be found for each different set of mobile species. These relationships are easily found by applying the appropriate constraints to the entropy source expression and comparing the resulting diffusion and transport equations.

8.12 DIFFUSION IN A DOUBLY DOPED SEMICONDUCTOR

In silicon containing two fully ionized dopants, one a donor and one an acceptor, the individual species and imperfections are

$$\mathrm{Si}_{\mathrm{Si}}^{\times},\ \mathrm{V}_{\mathrm{Si}}^{\times},\ \mathrm{V}_{\mathrm{Si}}',\ \mathrm{V}_{\mathrm{Si}}^{\cdot},\ \mathrm{D}_{\mathrm{Si}}^{\cdot},\ \mathrm{A}_{\mathrm{Si}}',\ \mathrm{e}',\ \mathrm{h}^{\cdot}$$

where multiple vacancy charge states have been allowed for.[35] This very practical problem is explored to indicate the theoretical limitations of current kinetic models. Additional complications will arise through defect pair formation (see §8.10).[33,34] The flux constraints in such a system are

$$J_{\mathrm{Si}^\times} + J_{\mathrm{V}^\times} + J_{\mathrm{V}^\cdot} + J_{\mathrm{V}'} + J_{\mathrm{D}^\cdot} + J_{\mathrm{A}'} = 0 \tag{8.158}$$

$$J_{\mathrm{V}'} + J_{\mathrm{A}'} + J_{\mathrm{e}'} = J_{\mathrm{V}^\cdot} + J_{\mathrm{D}^\cdot} + J_{\mathrm{h}^\cdot} \tag{8.159}$$

There are 24 possible pairs of species fluxes which can be eliminated from the entropy source term. Choosing the pair $J_{\mathrm{Si}^\times}$ and $J_{\mathrm{e}'}$, application of equations (8.158) and (8.159) to (8.34) leads to

$$\begin{aligned} -T\sigma = {} & J_{\mathrm{V}^\times}\nabla\{\eta(\mathrm{V}_{\mathrm{Si}}^\times) - \eta(\mathrm{Si}_{\mathrm{Si}}^\times)\} + J_{\mathrm{V}'}\nabla\{\eta(\mathrm{V}') - \eta(\mathrm{Si}_{\mathrm{Si}}^\times) - \eta(\mathrm{e}')\} \\ & + J_{\mathrm{V}^\cdot}\nabla\{\eta(\mathrm{V}_{\mathrm{Si}}^\cdot) - \eta(\mathrm{Si}_{\mathrm{Si}}^\times) + \eta(\mathrm{e}')\} + J_{\mathrm{D}^\cdot}\nabla\{\eta(\mathrm{D}^\cdot) - \eta(\mathrm{Si}_{\mathrm{Si}}^\times) + \eta(\mathrm{e}')\} \\ & + J_{\mathrm{A}}\nabla\{\eta(\mathrm{A}') - \eta(\mathrm{Si}_{\mathrm{Si}}^\times) - \eta(\mathrm{e}')\} + J_{\mathrm{h}^\cdot}\nabla\{\eta(\mathrm{h}^\cdot) + \eta(\mathrm{e}')\} \end{aligned} \tag{8.160}$$

The first three relative building units correspond, in the equilibrium sense, to intrinsic vacancy formation

$$Si_{Si}^{\times} = V_{Si}^{\times} + Si_{(g)}$$

$$e' + Si_{Si}^{\times} = V_{Si}' + Si_{(g)}$$

$$Si_{Si}^{\times} = V_{Si}^{\cdot} + e' + Si_{(g)}$$

whence

$$-\nabla\mu_{Si} = \nabla\{\eta(V_{Si}^{\times}) - \eta(Si_{Si}^{\times})\} \tag{8.161}$$

$$= \nabla\{\eta(V_{Si}') - \eta(Si_{Si}^{\times}) - \eta(e')\} \tag{8.162}$$

$$= \nabla\{\eta(V^{\cdot}) - \eta(Si_{Si}^{\times}) + \eta(e')\} \tag{8.163}$$

The appearance of their potential gradients in the entropy source expression, and hence in the flux equations, corresponds to diffusion of Si via a vacancy mechanism. It shows explicitly how correlation between the movements of free carriers and lattice defects arise through the electrostatic constraint expressed in equation (8.159).

The remaining building units in equation (8.160) correspond to extrinsic defect formation

$$D_{(g)} + Si_{Si}^{\times} = D_{Si}^{\cdot} + e' + Si_{(g)}$$

$$A_{(g)} + Si_{Si}^{\times} + e' = A_{Si}' + Si_{(g)}$$

and intrinsic electron–hole equilibrium

$$e' + h^{\cdot} = 0$$

Then from the Gibbs equations for these equilibria

$$\nabla(\mu_D - \mu_{Si}) = \nabla\{\eta(D_{Si}^{\cdot}) - \eta(Si_{Si}^{\times}) + \eta(e')\} \tag{8.164}$$

$$\nabla(\mu_A - \mu_{Si}) = \nabla\{\eta(A_{Si}') - \eta(Si_{Si}^{\times}) - \eta(e')\} \tag{8.165}$$

and, if the Fermi level is constant,

$$\nabla\{\eta(e') + \eta(h^{\cdot})\} = 0 \tag{8.166}$$

The gradient terms in equations (8.164) and (8.165) correspond to diffusion via an exchange mechanism correlated with an appropriate free carrier movement. This does not imply that vacancies are not involved, since building units of this type are equivalent to the difference between units which involve vacancies

$$\{D_{Si}^{\cdot} - Si_{Si}^{\times} + e'\} = \{D_{Si}^{\cdot} - V_{Si}^{\times} + e'\} - \{Si_{Si}^{\times} - V_{Si}^{\times}\} \tag{8.167}$$

It is apparent that vacancies function as intermediaries and correlated diffusion arises through the fact that the different species share the use of the same vacancies.

The entropy source expression may now be expressed in terms of constitutional variables. Noting that the total vacancy flux, J_V^T, is given by

$$J_V^T = J_{V^{\times}} + J_{V'} + J_{V^{\cdot}}$$

and that

$$J_V^T + J_{Si} + J_D + J_A = 0$$

then from equations (8.160)–(8.165) it is found that

$$-T\sigma = J_{Si}\nabla\mu_{Si} + J_D\nabla\mu_D + J_A\nabla\mu_A \qquad (8.168)$$

In the (silicon) solvent-fixed frame

$$\begin{aligned} -J_D &= L_{11}\nabla\mu_D + L_{12}\nabla\mu_A \\ -J_A &= L_{21}\nabla\mu_D + L_{22}\nabla\mu_A \end{aligned} \qquad (8.169)$$

and the usual ternary diffusion equations result.

In an isothermal, field-free diffusion experiment, equations (8.169) provide a unique and complete description. However, the set of driving forces identified by equation (8.160) resulted from a particular choice of which species fluxes were to be eliminated from the entropy source expression. It is obviously not unique but serves nonetheless to identify some of the possible correlations among the diffusing species and imperfections. The complete range of possible correlations for any particular species can be determined by identifying all relative building units containing that species. As an example, the correlations available to the solute species $D^{\cdot}$ are now enumerated.

Relative building units containing $D^{\cdot}$ are most easily found by examining the application of equations (8.158) and (8.159) to the entropy source expression (8.34). As before, we proceed by eliminating two species fluxes by expressing them in terms of, *inter alia*, J_D. Substitution of these expressions into equation (8.34) leads to identification of the required relative building units in the form of equation (8.160). Inspection of equations (8.158) and (8.159) reveals that there are 18 distinguishable pairs of species not involving $D^{\cdot}$ which may be simultaneously eliminated in this way. The relative building units thereby revealed are as follows

$$\{D^{\cdot}_{Si} - V^{\cdot}_{Si}\} \quad \{D^{\cdot}_{Si} - V^{\times}_{Si} - h^{\cdot}\} \quad \{D^{\cdot}_{Si} - V^{\times}_{Si} + e'\} \quad \{D^{\cdot}_{Si} - 2V^{\times}_{Si} + V'_{Si}\}$$

$$\{D^{\cdot}_{Si} - V'_{Si} + 2e'\} \qquad \{D^{\cdot}_{Si} - V'_{Si} + 2h^{\cdot}\}$$

$$\{D^{\cdot}_{Si} - Si^{\times}_{Si} + e'\} \quad \{D^{\cdot}_{Si} - Si^{\times}_{Si} - h^{\cdot}\} \quad \{D^{\cdot}_{Si} - 2Si^{\times}_{Si} + V'\}$$

$$\{D^{\cdot}_{Si} - A'_{Si} + 2e'\} \qquad \{D^{\cdot}_{Si} - A'_{Si} - 2h^{\cdot}\}$$

$$\{D^{\cdot}_{Si} + A'_{Si} - 2V^{\times}_{Si}\} \qquad \{D^{\cdot}_{Si} + A'_{Si} - 2Si^{\times}_{Si}\}$$

Only 13 distinguishable units result because all six units involving $V^{\cdot}_{Si}$ reduce to $\{D^{\cdot}_{Si} - V^{\cdot}\}$. It should be noted that these units reflect the simplest forms of correlation, as any linear combination of the units would obviously satisfy all thermodynamic and conservative requirements.

The relative building units involving $D^{\cdot}_{Si}$ necessarily reflect the flux constraints of equations (8.158) and (8.159). The insertion of these constraints into the entropy source expression imposes the requirements that movement of $D^{\cdot}_{Si}$ be balanced by an opposite movement of another lattice species, and that any net charge flow thereby arising be neutralised. Thus four principal correlations are seen to exist: countercurrent donor–vacancy movement, countercurrent donor–solvent movement, countercurrent donor–acceptor movement and cocurrent donor–acceptor movement. The first group of units corresponds, in the equilibrium sense, to donor dissolution

$$D_{(g)} + V^{\cdot}_{Si} = D^{\cdot}_{Si}$$

$$D_{(g)} + V^{\times}_{Si} + h^{\cdot} = D^{\cdot}_{Si}$$

$$D_{(g)} + V^{\times}_{Si} = D^{\cdot}_{Si} + e' \qquad \text{etc.}$$

It is readily demonstrated that the electrochemical potential gradient of each of these units is equal to $\nabla\mu_D$. Units in the second group reflect substitutional dissolution of the donor

$$D_{(g)} + Si^{\times}_{Si} = D^{\cdot}_{Si} + e' + Si_{(g)} = D^{\cdot}_{Si} - h^{\cdot} + Si_{(g)}$$

$$D_{(g)} + 2Si^{\times}_{Si} = D^{\cdot}_{Si} + V_{Si} + 2Si_{(g)}$$

and their electrochemical potential gradients are seen to be equal to either $\nabla(\mu_D - \mu_{Si})$ or $\nabla(\mu_D - 2\mu_{Si})$. The two units in the third group correspond to substitution of donor for acceptor and their electrochemical potential gradients are equal to $\nabla(\mu_D - \mu_A)$. Units in the last group reflect simultaneous dissolution of both dopants with and without substitution. Their gradients are accordingly given by either $\nabla(\mu_D + \mu_A - 2\mu_{Si})$ or $\nabla(\mu_D + \mu_A)$.

The relative importance of these various possible correlations will be determined by the probabilities of finding the required species in adjacent positions within the solid (i.e. of locating the required relative building unit). Thus the most likely of the various possible donor–vacancy exchange mechanisms will be determined by the vacancy ionization equilibria. All correlations involving the simultaneous movement of two constituents require the presence of a vacancy. If the vacancy is explicit in the corresponding relative building unit, then net vacancy transport also occurs. If the unit contains no vacancy, then the vacancies function only as intermediaries as shown in equation (8.167). It follows then that donor–solvent countercurrent diffusion can occur with and without net vacancy movement. The favoured mechanism will depend on the relative abundance of V'_{Si} and $V^{\times}_{Si}$. Cocurrent donor–acceptor diffusion can also occur both with and without a vacancy flux. Given that the probability of finding a $Si^{\times}_{Si}$ species approaches unity, the unit $\{D^{\cdot}_{Si} + A'_{Si} - 2Si^{\times}_{Si}\}$ adjacent to an additional $V^{\times}_{Si}$ is more probable than $\{D^{\cdot}_{Si} + A'_{Si} - 2V^{\times}_{Si}\}$ alone. The zero vacancy flux situation therefore seems more probable. Countercurrent donor–acceptor diffusion cannot involve a net vacancy flux but must involve an electronic current. The relative importance of cocurrent and countercurrent donor–vacancy correlations will depend on kinetic factors and on the relative availability of neutral vacancies and free carriers. These considerations together with defect pairing have been further explored by Young and Kirkaldy.[36]

8.13 CHEMICAL AND TRACER DIFFUSION IN IONIC SOLIDS

The kinetic description proposed in this chapter will now be used to arrive at relationships between diffusion in a multicomponent solid, and tracer diffusion in pure compounds. As noted in §6.6, this question can be approached by calculating the 'diffusion potential' and thereby estimating $\tilde{D}$. However, if cross-effects are not to be ignored, such a calculation assumes a complexity of considerable proportions for a multicomponent solid. First the situation dealt with in §6.6, counterdiffusing cations associated with an immobile anion sublattice, for which equation (8.46) applies, will be considered. Geometrical correlation effects (see §2.7) will be neglected.

A tracer diffusion experiment involves interchange of labelled and unlabelled cations in the absence of any potential gradient in the electronegative crystal constituent. Consequently no vacancy flux is generated, and equation (8.46) applies to this situation also, i.e.

$$J_{1*} = -L\nabla(\mu_{1*} - \mu_1) \tag{8.170}$$

The coefficient is evaluated through equation (8.89) to yield

$$J_{1*} = -\frac{m\lambda^2}{RT}\frac{m'_1 m'_{1*}}{m'_1 + m'_{1*}} a_V \nabla(\mu_{1*} - \mu_1) \tag{8.171}$$

For the ideal solution behaviour expected of a tracer, this becomes

$$J_{1*} = -m\lambda^2 \frac{m'_1 m'_{1*}}{m'_1 + m'_{1*}} a_V \frac{m_1 + m_{1*}}{m_1 m_{1*}} \nabla m_{1*} \tag{8.172}$$

Recalling the definition of m' in equation (8.86) and noting that in the tracer case, when correlations associated with the vacancies are omitted, then

$$\nu_{1V} K_{1V} = \nu_{1*V} K_{1*V}$$

D_{1*} follows from equation (8.172) as

$$D_{1*} = \lambda^2 \nu_{1V} K_{1V} a_{1V} \tag{8.173}$$

and a corresponding expression is found for D_{2*}. Thus the expected relationship between D^* and D_V (equation (6.94)) is found.

Turning now to interdiffusion of two different cations, denoted by 1 and 2, equation (8.89) leads to an evaluation of the kinetic coefficient to yield

$$J_1 = -\frac{m\lambda^2}{RT}\frac{m'_1 m'_2}{m'_1 + m'_2} a_V \nabla(\mu_1 - \mu_2) \tag{8.174}$$

Since $z_1 = z_2$ it follows that $\nabla m_1 = -\nabla m_2$ and, if the solution is ideal or Henrian, equation (8.174) becomes

$$J_1 = -m\lambda^2 \frac{m'_1 m'_2}{m'_1 + m'_2} a_V \frac{m_1 + m_2}{m_1 m_2} \nabla m_1 \tag{8.175}$$

Substitution into equation (8.175) for m'_i from equations (8.86) and (8.173) for ideal solutions yields

$$J_1 = -\frac{D_{1*} D_{2*}}{m_1 D_{1*} + m_2 D_{2*}}(m_1 + m_2)\nabla m_1 \tag{8.176}$$

whence

$$\tilde{D} = \frac{D_{1*} D_{2*}}{m_1 D_{1*} + m_2 D_{2*}}(m_1 + m_2) \tag{8.177}$$

This is precisely the result (equation (6.133)) found earlier by calculating the diffusion potential. The need to evaluate this potential is avoided in the present calculation by virtue of the fact that the relative building unit formulation ensures that the various species diffuse in charge-neutral groups.

It should be noted that the assumption of Henrian solution behaviour employed in arriving at equation (8.177) may be in serious error. The thermodynamic correction term $(1 + \partial \ln \gamma / \partial \ln m)$ may be the most important element in the relationship between $\tilde{D}$ and D^*, as has been reported by Osenbach *et al.*[37, 38] for Cr-doped MgO crystals.

In the ternary case, the kinetic terms are evaluated in equations (8.97)–(8.100). Proceeding from equation (8.98) on the basis of an assumed ideal or Henrian solution, but without other approximations, one finds

$$J_2 = -\{(m_1 + m_2)\tilde{D}_{12} + m_3\tilde{D}_{23}\}\nabla m_2 - \{m_2\tilde{D}_{12} - m_2\tilde{D}_{23}\}\nabla m_3$$

and

$$J_3 = -\{(m_3\tilde{D}_{13} - m_3\tilde{D}_{23}\}\nabla m_2 - \{(m_1 + m_3)\tilde{D}_{13} + m_2\tilde{D}_{23}\}\nabla m_3 \tag{8.178}$$

where

$$\tilde{D}_{ij} = \frac{D_i^* D_j^*}{m_i D_i^* + m_j D_j^*} \tag{8.179}$$

It is seen that cross-terms vanish only if $\tilde{D}_{12} = \tilde{D}_{23}$ and $\tilde{D}_{13} = \tilde{D}_{23}$. As expected, the relative importance of the cross-terms depends on the concentrations of solute species.

REFERENCES

1 F. A. KROGER AND H. J. VINK: *Solid State Physics*, 1956, **3**, 307
2 N. F. MOTT AND R. W. GURNEY: 'Electronic processes in ionic crystals', 1940, Oxford, Clarendon Press
3 R. E. HOWARD AND A. B. LIDIARD: *Rep. Prog. Phys.*, 1964, **27**, 161
4 C. P. FLYNN: 'Point defects and diffusion', 1972, Oxford, Clarendon Press
5 C. WAGNER AND W. SCHOTTKY: *Z. Phys. Chem. B*, 1931, **11**, 163
6 P. KOFSTAD: 'Nonstoichiometry, diffusion and electrical conductivity in binary metal oxides', 1972, New York, Wiley–Interscience
7 F. A. KROGER: 'The chemistry of imperfect crystals', 2nd edn, 1974, Amsterdam–New York, North-Holland–Elsevier
8 W. SCHOTTKY: in 'Halbleiterprobleme, Vol. 4', (ed. W. Schottky), 1958, Braunschweig, Fr. Viewig
9 F. A. KROGER, F. H. STIELTJES, AND H. J. VINK: *Philips Res. Reports*, 1959, **14**, 557
10 D. J. YOUNG AND J. S. KIRKALDY: *J. Phys. Chem. Solids*, 1984, **45**, 781
11 H. SCHMALZREID: *Prog. Solid State Chem.*, 1965, **2**, 265
12 H. SCHMALZREID AND J. B. HOLT: *Z. Phys. Chem. Neue Folge*, 1968, **60**, 220
13 A. R. COOPER AND J. H. HEASLEY: *J. Am. Ceram. Soc.*, 1966, **42**, 280
14 R. SCHEIDECKER AND M. F. BERARD: *ibid.*, 1973, **56**, 204
15 D. J. YOUNG, E. DELAMOTTE, AND J. S. KIRKALDY: *Can. J. Phys.*, 1979, **57**, 722
16 R. J. FRIAF: *J. Phys. Chem. Solids*, 1969, **30**, 429
17 F. M. BENIERE AND M. CHEMLA: *J. Chem. Phys.*, 1972, **56**, 549
18 D. J. YOUNG: *Scripta Met.*, 1975, **9**, 159
19 W. K. CHEN AND N. L. PETERSON: *J. Phys. Chem. Solids*, 1975, **36**, 1097
20 E. FRYT: *Oxid. Met.*, 1982, **18**, 83
21 J. T. JONES AND I. B. CUTLER: *J. Am. Ceram. Soc.*, 1971, **54**, 335
22 M. APPEL AND J. A. PASK: *ibid.*, 152

23 N. SATA AND K. S. GOTO: *ibid.*, 1982, **65**, 158
24 C. WAGNER: *Z. Phys. Chem. B*, 1932, **21**, 25
25 C. WAGNER: *ibid.*, 1936, **32**, 447
26 C. WAGNER: *Corros. Sci.*, 1969, **9**, 91
27 D. E. COATES AND A. D. DALVI: *Oxid. Met.*, 1970, **2**, 331
28 T. NARITA, K. NISHIDA, AND W. W. SMELTZER: *J. Electrochem. Soc.*, 1982, **129**, 209
29 D. J. YOUNG, T. NARITA, AND W. W. SMELTZER: *ibid.*, 1980, **127**, 679
30 M. J. DIGNAM, D. J. YOUNG, AND D. W. G. GOAD: *J. Phys. Chem. Solids*, 1973, **34**, 1227
31 D. G. MILLER: *J. Phys. Chem.*, 1959, **63**, 570
32 I. PRIGOGINE AND R. DEFAY: 'Chemical thermodynamics', 409 *et seq.*, 1954, New York, Longmans Green
33 R. C. DORWARD: Ph.D. Thesis, 1967, McMaster University
34 H. REISS, C. S. FULLER, AND F. J. MORIN: *Bell Tech. J.*, 1956, **35**, 535
35 D. SHAW: 'Atomic diffusion in semiconductors', 1973, London, Plenum
36 D. J. YOUNG AND J. S. KIRKALDY: Unpublished research, 1987
37 J. W. OSENBACH, W. R. BITTER, AND V. S. STUBICAN: *J. Phys. Chem. Solids*, 1981, **42**, 509
38 J. W. OSENBACH, W. R. BITTER, AND V. S. STUBICAN: *ibid.*, 1982, **43**, 413

CHAPTER 9

Multicomponent diffusion in liquids

9.1 INTRODUCTION

An enormous literature on multicomponent diffusion in liquids has accrued since the first accurate work of Dunlop and Gosting.[1] This body of work[2, 3, 4, 5, 6, 7] serves to demonstrate the validity of the Onsager reciprocal relations for both non-electrolyte and electrolyte solutions. It will be recalled that the experimentally determined D-matrix reflects both the kinetic and thermodynamic terms

$$[D] = [L][\mu] \tag{9.1}$$

In liquid solutions it may be that the thermodynamic effects are overwhelming, notably in electrolyte/non-electrolyte solutions where salting in or salting out effects can be dramatic.[8, 9, 10] several examples of this situation are identified by Cussler.[5] Such effects have been discussed in a general way in § 6.3 and will not be pursued further in this chapter. The application of binary data and thermodynamics to the prediction of ternary coefficients is dealt with in § § 9.4–9.6.

Our discussions of diffusion in metals and in ionic solids have derived much of their insight from a preliminary consideration of the structure of the diffusion medium. As indicated in § 2.16, such an examination is exceptionally complex for the liquid state, and accordingly no entirely satisfactory theory exists. In the absence of a comprehensive theory, the understanding of diffusion is only partial and most 'predictive' calculations in fact rest on correlations. Of these, the most time-honoured is the Stokes–Einstein equation[11, 12, 13, 14, 15] which has been applied with varying degrees of success to diffusion in non-electrolyte solutions.

We here reiterate the statement made in § 1.29 concerning the tendency for practical processes to involve diffusion and convection in complex juxtaposition. The general subject of mass transfer deals quantitatively and semi-empirically with such composite processes.[15] This chapter is primarily concerned with theories of pure multicomponent diffusion and their tests against accurate experimentation. However, diffusion in molten slags, glasses and salts and in capillaries and fissures, as in membranes, physiological and geological structures is often free of convection so diffusion data is directly relevant to process rates.

9.2 DILUTE NON-ELECTROLYTE SOLUTIONS

For a binary system observed in the volume-fixed frame, diffusion is phenomenologically described by

$$J_1 = -L_{11}\nabla\mu_1 = -D_{12}\nabla m_1 \tag{9.1}$$

or

$$J_1 = m_1 v_1 \tag{9.2}$$

where D_{12} ($\equiv \widetilde{D}$ within the solid state literature) is the chemical or mutual diffusion coefficient for the flow of component 1 relative to the mixture. The velocity of the solute may also be expressed[12, 13, 14] as

$$v_1 = \frac{1}{N_A \xi} \nabla \mu_1 = \frac{kT}{m_1 \xi} \left(\frac{\partial \ln a_1}{\partial \ln m_1} \right) \nabla m_1 \tag{9.3}$$

where k is the Boltzmann constant and ξ is the viscous resistance per molecule. Activity is here defined in units of m. From equations (9.1)–(9.3) it is apparent that

$$D_{12} = \frac{kT}{\xi} \frac{\partial \ln a_1}{\partial \ln m_1} \tag{9.4}$$

To this point the argument is phenomenological and as such can produce no useful conclusion. To proceed, it is necessary to evaluate ξ in terms of system parameters. Such an evaluation was attempted by Stokes (see Bird *et al.*[15]) for the case of a spherical particle of radius r_1 moving in a continuous medium of viscosity Γ_2. The diffusion of such a particle is then found by a modification of Stoke's law[11] to be described by

$$D_{12} = \frac{kT}{\alpha \pi \Gamma_2 r_1} \frac{\partial \ln a_1}{\partial \ln m_1} \tag{9.5}$$

with

$$\alpha = \frac{6(1 + 2\Gamma_2/\beta r_1)}{(1 + 3\Gamma_2/\beta r_1)} \tag{9.6}$$

where β is a slip coefficient. The slip coefficient can vary between 0 and ∞ corresponding to variations in α between 4 and 6. For a large spherical molecule diffusing in a solvent of low molecular weight the Stokes particle diffusion model could conceivably apply and equation (9.5) becomes

$$D_{12} = \frac{kT}{6\pi \Gamma_2 r_1} \frac{\partial \ln a_1}{\partial \ln m_1} \tag{9.7}$$

which is the well-known Stokes–Einstein equation. As discussed in §2.17, this relationship works well in the case of large polymeric molecules diffusing in low molecular weight solvents.

There are several limitations to the above description: solute molecules are not invariably hard spheres and they frequently tend to solvate, strict limitation to dilute solutions is required and in many cases of interest, solute molecules are of the same order in size as the solvent, or even smaller. The failure of the Stokes–Einstein relationship under various of these circumstances has been amply documented by Evans *et al.*[16] Particularly dramatic instances of the failure have been reported. Hiss and Cussler[17] showed that for a small solute diffusing in high viscosity, high molecular weight solvents

$$D_{12}\Gamma^{2/3} = \text{constant} \tag{9.8}$$

Davies *et al.*[18] reported for the diffusion of CO_2 in various solvents that

$$D_{12}\Gamma^{0.45} = \text{constant} \tag{9.9}$$

Thus the predicted exponent of unity is not observed in these cases and accordingly the Stokes–Einstein equation does not have broad applicability outside the rather narrowly defined regime for which it was derived. In particular, the use of the equation to describe tracer self-diffusion involving identical sized molecules is quite misplaced.

Because viscosity is easier to measure than a diffusion coefficient, and because the physical processes which determine Γ are presumed also to play a part in determining D, correlations between the two quantities are still sought. Many such correlations have been summarized and assessed by Reid *et al.*[19] Various approaches to the description of tracer self-diffusion have been reviewed by Ertl *et al.*[20] and Ertl and Dullien.[21]

Absolute rate theory of diffusion by vacant sites has been used to arrive at an alternative derivation of the Stokes–Einstein equation.[22] It may be shown that

$$\Gamma = \frac{h}{\lambda^3} \exp(\Delta G_\Gamma^\dagger/RT) \tag{9.10}$$

and

$$D_{12} = \frac{kT}{h}\lambda^2 \exp(-\Delta G_D^\dagger/RT) \tag{9.11}$$

where h is Planck's constant and λ is the average 'jump distance' of a moving species, and has here been set equal to the average intermolecular distance. Combination of these two equations leads to a result of the form of equation (9.7) for the ideal case ($\partial \ln a/\partial \ln m = 1$) on the assumption $\Delta G_\Gamma^\dagger = \Delta G_D^\dagger$. Alternately, it has been suggested by Hiss and Cussler[17] that for a small solute molecule in solvents of large molecular weight, the quantity λ is more properly associated with the solvent molecule. In that case, for a series of different solvents they obtain

$$D_{12}\Gamma^{2/3} = kT\,h^{1/3} \exp(-\Delta G_D^\dagger/3RT) \tag{9.12}$$

It should be noted, however, that the applicability of absolute rate theory to diffusion in non-electrolytes is by no means universally accepted.

Hildebrand[23] has trenchantly criticized the use of absolute rate theory in describing diffusion (or viscous flow) in non-electrolytes. The argument is based largely on the experimental findings that both fluidity (≡ reciprocal viscosity) and diffusion coefficients are *linearly* dependent on temperature. The linear coefficients were shown to be quantitatively consistent with changes in relative free volume of solvent, i.e.

$$\frac{1}{\Gamma} = B(\bar{V} - \bar{V}_0)/\bar{V}_0 \tag{9.13}$$

where B is a constant and $\bar{V}_0$ is the intrinsic volume, that is, the molar volume at which fluidity and diffusion rates are zero. Thus temperature effects arise simply through the thermal expansion of a liquid, and Hildebrand's view of molecular motion in a liquid is that of a van der Waals fluid. This view receives support from a number of computer simulations of the molecular dynamics of hard sphere models (cf §2.16 and the following).

The description of diffusion in a fluid composed of hard spheres starts with Enskog's kinetic theory[24] based on energy and momentum transfer via successive uncorrelated binary collisions. Computer simulation[25, 26] allowed the calculation of the position and velocity of each of a large number of hard spheres at the time of each successive collision and hence the determination of free path lengths. The results showed that the assumption that only uncorrelated collisions contribute to transport was wrong. Appropriate corrections to the Enskog description have been calculated from the molecular dynamics approach by Alder *et al.*[27] Additional corrections must be made for the fact that collisions may cause the transfer of translational to rotational momentum.[28] Calculations of this sort show that mean free paths of diffusing molecules are substantially less than molecular diameters. This is consistent with the view adopted by Hildebrand and quite inconsistent with diffusion by vacant sites within absolute rate theory. The calculations have now been tested against experimental data for tracer diffusion in a wide range of solvents[16] and for chemical diffusion in binary liquid mixtures.[29, 30] These tests have generally shown the calculations to be accurate, but some systems were not satisfactorily described. Interestingly, Ruby *et al.*[31] in an X-ray and Mossbauer study of diffusion of Fe^{2+} in $H_3PO_4 + 3{\cdot}27H_2O$, while verifying that jumps the size of lattice dimensions do not occur, find nonetheless that Arrhenius behaviour obtains over four decades with an activation energy of 0·55 eV. It may be therefore that absolute rate theory retains validity in some average sense.

It seems then that the molecular dynamics approach has become well established in the description of non-electrolyte diffusion. Unfortunately the technique has apparently not completely revealed the nature of the correlations in microscopic molecular motions. The calculations have not yet been extended to ternary systems and no insight into the nature of kinetic cross-effects is thereby available.

9.3 CROSS-EFFECTS IN NON-ELECTROLYTE SOLUTIONS

A number of studies have shown that off-diagonal L-coefficients are in general non-zero (see, for example, Dunlop[32] and Ellerton and Dunlop[8, 9]). A knowledge of how to predict these coefficients is, however, lacking. This theoretical gap can be illustrated by considering the simple case of binary solutions, for which the Darken relationship

$$D_{12} = \{X_1 D_{2*} + X_2 D_{1*}\} \frac{\partial \ln a}{\partial \ln m} \tag{9.14}$$

subject to weak correlation effects in liquids might be expected to apply (cf §3.11).

Vignes[33] has used experimental data for a number of binary systems to demonstrate that equation (9.14) does not apply to non-electrolyte mixtures. Table 2.2, due to Tyrrell and Harris,[7] illustrates the same point. In obtaining equation (9.14) a relationship between D_1 and D_{1*} is necessary and the form of equation (2.61) has been used. Since this involves a tracer coefficient which is, by definition, inconsistent with the Stokes–Einstein equation, it is perhaps not surprising that this incorrectly derived version of the Darken equation should fail. A more self-consistent procedure was followed by Hartley and Crank[34] and also by Carman and Stein[35] using equation (9.5) to evaluate the D. The resulting expression is

$$\Gamma D_{12} = \{X_1 \Gamma_1 D_{21}^{(0)} + X_2 \Gamma_2 D_{12}^{(0)}\} \frac{\partial \ln a}{\partial \ln m} \tag{9.15}$$

where the $D^{(0)}$ are diffusion coefficients at infinite dilution:

$$D_{12}^{(0)} = \lim_{m_1 \to 0} D_{12}$$

This equation has also been tested by Vignes[33] and shown to fail. Recalling that the Stokes–Einstein equation is thought to be applicable to a large solute molecule in a 'continuum' of small solvent molecules, it is clear that if the model is valid for species 2 as solute in solvent 1 then it cannot be valid for species 1 as solute in solvent 2. Thus the failure of equation (9.15) is only to be expected.

Vignes[33] has proposed an empirical correlation of the form

$$D_{12} = (D_{12}^{(0)})^{X_2} (D_{21}^{(0)})^{X_1} \tag{9.16}$$

and demonstrated that it applies to a considerable number of systems in which molecular association (dimerization, etc.) does not occur. The correlation has been further tested by Dullien[36] and found to be reasonably successful. Imaginative approaches to the application of absolute rate theory to liquid diffusion[37,38] have led to proposed justifications of equation (9.16) but, as discussed above, this tactic is unlikely to be universally successful.

Given that we are unable to describe binary interdiffusion, it is hardly surprising that attempts to calculate ternary cross-effects have not been highly successful. Cussler and Lightfoot[39] have proposed that hydrodynamic correlations arise through perturbations (by solutes) of the solvent liquid flow. Their derivation leads, however, to an L-matrix which is not symmetric. Lane and Kirkaldy[40,41,42] have proposed that liquids be modelled as quasi-crystalline materials in which diffusion occurs via the activated jumps of particles, either by direct exchange of moving species or their exchange with vacant sites. The theory predicts cross-effects which are symmetric and obey the requirements for stability of the D-matrix. Moreover, for the exchange model it succeeds in predicting off-diagonal terms which are negative, in agreement with experimental observations. Order of magnitude agreement with the experimental diagonal and cross-interactions for the non-electrolyte system dilute sucrose-mannitol in water was obtained.[8] Ellerton and Dunlop's choice of system was a salutary one in which the intrinsic diffusion rate for the solvent is about five times that for the solutes. This is precisely the extreme condition under which the predicted off-diagonal coefficients change from negative to positive within the vacant site model[42] (§7.3). Since the observed coefficients are strongly negative, this model is unequivocally ruled out (cf Table 9.1), in agreement with the above conclusions pertaining to binary systems. Note from this table that the experiments were not sufficiently accurate to unequivocally confirm the Onsager reciprocal relations.

It is concluded that our present knowledge of multicomponent diffusion in non-electrolyte solutions has not yet reached the desired state of predictive ability, even for dilute solutions, and concentrated solutions present even more formidable problems.[43]

Table 9.1 Experimental and calculated phenomenological coefficients $(L_{ij})_V$ for sucrose, mannitol, water, and for the volume frame of reference

m_1 (mol l^{-1})	0·25	0·25	0·50	0·50
m_2 (mol l^{-1})	0·25	0·50	0·25	0·50
$[(L_{11})_V \times 10^{20}]$exptl	3·47	3·04	5·07	4·31
$[(L_{11})_V \times 10^{20}]$vac	3·87	3·40	6·13	5·29
$[(L_{11})_V \times 10^{20}]$exch	4·86	4·73	9·16	8·90
$[(L_{12})_V \times 10^{20}]$exptl	−0·172	−0·334	−0·269	−0·460
$[(L_{12})_V \times 10^{20}]$vac	0·038	0·068	0·062	0·110
$[(L_{12})_V \times 10^{20}]$exch	−0·027	−0·054	−0·054	−0·108
$[(L_{21})_V \times 10^{20}]$exptl	−0·108	−0·314	−0·148	−0·276
$[(L_{21})_V \times 10^{20}]$vac	0·038	0·068	0·062	0·110
$[(L_{21})_V \times 10^{20}]$exch	−0·027	−0·054	−0·054	−0·108
$[(L_{22})_V \times 10^{20}]$exptl	4·67	8·14	3·71	6·32
$[(L_{22})_V \times 10^{20}]$vac	4·91	8·67	3·84	6·64
$[(L_{22})_V \times 10^{20}]$exch	6·19	11·98	5·86	11·31

The phenomenological coefficients $(L_{ij})_V$ are in mol^2 J^{-1} cm^{-1} s^{-1} and the diffusion coefficients D in cm^2 s^{-1}. 1 ≡ sucrose, 2 ≡ mannitol. The following limiting binary diffusion coefficients were used in computing the $(L_{ij})_V$: $D^0_{\text{sucrose}} = 0{\cdot}522\,6 \times 10^{-5}$ and $D^0_{\text{mannitol}} = 0{\cdot}664 \times 10^{-5}$. The self-diffusion coefficient of water was taken to be $2{\cdot}27 \times 10^{-5}$. Experimental entries are after Ellerton and Dunlop.[8] Calculated entries 'vac' and 'exch' are based on the predictions of Lane and Kirkaldy (1966).[8, 42]

9.4 DESCRIPTION OF DIFFUSION IN BINARY ELECTROLYTES

In dilute aqueous electrolyte solutions the ions are sufficiently far apart that long range coulomb forces are predominant, with the consequence that one may assume that the flux of an ion is proportional to the gradient of its own thermodynamic potential only. In this approximation one can derive the Nernst–Hartley limiting expression for the diffusion coefficient which is the analogue of the solid state relation (6.133) (see also below). In strong solutions, specific ion effects are increasingly important, activity coefficients become concentration dependent above 0·1 M and both the additivity of conductances and the Nernst–Hartley equation fail. Ion fluxes must then be influenced by other ions and their gradients.

A binary electrolyte, consisting of a dissociated salt in a non-dissociated solvent (e.g. H_2O), is one of the simplest systems for which Onsager cross-effects are non-trivial. This is because one can recognize a pair of independent forces and fluxes in the presence of an external electric field. If with Miller[44] we choose the solvent-fixed (SF) reference frame ($J_0 = 0$ in Miller's notation) the transport equations for the cation (1) and anion (2) can be written as

$$J_1 = -l_{11}\nabla\eta_1 - l_{12}\nabla\eta_2$$

and

$$J_2 = -l_{21}\nabla\eta_1 - l_{22}\nabla\eta_2 \qquad (9.17)$$

where the electrochemical potentials $\eta_i = \mu_i + z_iF\psi$ and $l_{12} = l_{21}$. Since we are here assuming an external electric field there will be a non-zero electric current and J_1 and J_2 may therefore be regarded as independent. The generalization to ternary and higher order solutions is obvious. It is to be noted that the gradients of the chemical potentials of individual ionic species are inaccessible to experiment. Moreover, the electrostatic potential gradient is expressible in terms of these gradients in an isothermal diffusion experiment. An early step in a discussion of electrolyte diffusion is therefore to dispose of these operationally undefined gradients together with the diffusion potential gradient. As in the case of ionic solids (Chapter 8) the procedure used is to group individual ionic species according to the chemical equilibria in effect.

An electrolyte solution consists of a solvent, which we will consider to be non-ionized, and a number of salts, each of which dissolves according to a dissociative equilibrium

$$C_{\nu_{1c}}A_{\nu_{1a}} = \nu_{1c}C^{z_1} + \nu_{1a}A^{z_2} \qquad (9.18)$$

Here z_1 and z_2 are the valences of the cation and anion respectively. The electroneutrality condition for the dissolution of one such salt is therefore

$$z_1\nu_{1c} + z_2\nu_{1a} = 0 \qquad (9.19)$$

and the condition for a solution containing a number of electrolytes is simply the concentration weighted sum of the several corresponding expressions.

The dissociative equilibrium for salt dissolution is specified by the Gibbs equation

$$\mu_s\,dm_s = \eta_{1c}\,dm_{1c} + \eta_{1a}\,dm_{1a} \qquad (9.20)$$

where μ_s is the chemical potential of the neutral salt. Since $dm_{1c} = \nu_{1c}\,dm_s$ and $dm_{1a} = \nu_{1a}\,dm_s$, this statement may be rewritten as

$$\mu_s = \nu_{1c}\eta_{1c} + \nu_{1a}\eta_{1a} \qquad (9.21)$$

which, upon expanding the electrochemical potentials, becomes

$$\mu_s = \nu_{1c}\mu_{1c} + \nu_{1c}z_1F\psi + \nu_{1a}\mu_{1a} + \nu_{1a}z_2F\psi \qquad (9.22)$$

Application of the charge neutrality condition (9.19) then leads immediately to

$$\mu_s = \nu_{1c}\mu_{1c} + \nu_{1a}\mu_{1a} \qquad (9.23)$$

Since it is not possible to add or remove ions other than in the form of electroneutral salts, it is always possible to describe the thermodynamics of an electrolyte solution in terms of neutral salt chemical potentials. This construction is closely analogous to the 'building unit' construction developed in Chapter 8.

Since the charge neutrality constraint which leads to equation (9.23) applies on a local basis as well as on the macroscopic scale, this result may be used in evaluating the local rate of entropy change (cf equation (3.30))

$$T\frac{\mathrm{d}s}{\mathrm{d}t} = -\sum_{l}^{2n-1} \eta_l \frac{\mathrm{d}X_l}{\mathrm{d}t} \tag{9.24}$$

where the species l are $2n - 2$ ions or solvent. Substitution from equation (9.20) yields

$$T\sigma = -\sum_{i} \mu_i \frac{\mathrm{d}X_i}{\mathrm{d}t} \tag{9.25}$$

where the species i are now electroneutral salts or solvent. This leads in the usual way to the flux equation (cf §3.6)

$$J_i = -\sum_{j}^{n} L_{ij}\nabla\mu_j \tag{9.26}$$

whereby the flows of $(n - 1)$ electroneutral salts and their solvent are described. Since in a diffusion experiment no other flow is possible, the description is complete. Note that the L_{ij} are necessarily functions of the l_{ij} (see below).

The set $\nabla\mu_j$ are related via the Gibbs–Duhem equation, and the J_i are also related in a way determined by the choice of reference frame. These dependencies are eliminated using the methods of §4.2. For electrolyte diffusion experiments we again choose with Miller[44] to use the solvent-fixed reference frame specified by $J_0 = 0$. Proceeding in this way we obtain for the $(n - 1)$ solutes

$$J_i = -\sum_{j}^{n-1} L_{ij}\nabla\mu_j \tag{9.27}$$

Because the equation (9.27) has been properly formulated, the Onsager reciprocal relations apply to the L_{ij}. The corresponding diffusion matrix is given by equation (4.35). For the case of a single salt s dissolved in water, diffusion is characterized by a single coefficient,

$$J_s = -L_{ss}\nabla\mu_s = -L_{ss}\frac{\partial\mu_s}{\partial m_s}\nabla m_s = -D_s\nabla m_s \tag{9.28}$$

even though the salt is dissociated into two oppositely charged ions, each of which diffuses. This simplification is an immediate consequence of the fact that charge separation cannot occur to any significant extent in the absence of an external field. Thus the diffusion of cation and anion are correlated in a simple, but rigorous way.

Of course in this formulation, the electrically conductive properties of a solution are algebraically inaccessible. The matter is of importance if one wishes to examine the mechanism whereby the movement of a cation influences that of an anion and to apply electrical measurements to the evaluation of the salt diffusion coefficient D_s. One accordingly seeks the relationship between L_{ss} and the coefficients l_{ij} in equations (9.17). This is achieved by eliminating the diffusion potential and in so doing transforming the force combination in equations (9.17) to that required by equation (9.25) for insertion in equation (9.27). The two formalisms can be rigorously matched at the zero electrical current condition, i.e.

$$z_1 J_1 + z_2 J_2 = 0 \tag{9.29}$$

Upon substitution from equations (9.17) this yields

$$-F\frac{\partial\psi}{\partial x} = \left[\frac{(z_1 l_{11} + z_2 l_{21})\dfrac{\partial\mu_1}{\partial x} + (z_1 l_{12} + z_2 l_{22})\dfrac{\partial\mu_2}{\partial x}}{z_1^2 l_{11} + z_1 z_2(l_{12} + l_{21}) + z_2^2 l_{22}}\right] \tag{9.30}$$

and

$$\nabla\eta_1 = -\left(\frac{z_1 l_{12} + z_2 l_{22}}{z_1 l_{11} + z_2 l_{21}}\right)\nabla\eta_2 \tag{9.31}$$

Now differentiation of equation (9.21) yields

$$\nabla\mu_s = \nu_1\nabla\eta_1 + \nu_2\nabla\eta_2 \tag{9.32}$$

which can be used with equation (9.31) together with (9.19) to obtain expressions for the electrochemical potential gradients in terms of $\nabla\mu_s$. The results are

$$\nabla\eta_1 = \frac{z_2}{\nu_1}\left\{\frac{z_1 l_{12} + z_2 l_{22}}{z_1^2 l_{11} + z_2 z_1(l_{12} + l_{21}) + (z_2)^2 l_{22}}\right\}\nabla\mu_s$$

$$\nabla\eta_2 = \frac{z_1}{\nu_2}\left\{\frac{z_1 l_{11} + z_2 l_{21}}{z_1^2 l_{11} + z_2 z_1(l_{12} + l_{21}) + (z_2)^2 l_{22}}\right\}\nabla\mu_s \tag{9.33}$$

Substitution for $\nabla\eta_1$, $\Delta\eta_2$ in equation (9.17) leads to

$$J_1 = -\frac{z_2 z_1}{\nu_2}\left\{\frac{l_{21}l_{12} - l_{11}l_{22}}{z_1^2 l_{11} + z_2 z_1(l_{12} + l_{21}) + z_2^2 l_{22}}\right\}\nabla\mu_s \tag{9.34}$$

The electrolyte flux is given by

$$J_s = \frac{J_1}{\nu_1} = \frac{J_2}{\nu_2} \tag{9.35}$$

From (9.28), (9.34), and (9.35) it is seen in agreement with Miller[44] that

$$L_{ss} = \frac{z_2 z_1}{\nu_2\nu_1}\left\{\frac{l_{21}l_{12} - l_{11}l_{22}}{z_1^2 l_{11} + z_2 z_1(l_{12} + l_{21}) + z_2^2 l_{22}}\right\} \tag{9.36}$$

As made clear in the above the l_{ij} are uniquely determined only when electrical as well as diffusion measurements are made. In a conductance experiment carried out at uniform chemical potentials the current

$$I = (z_1 J_1 + z_2 J_2)F = -\lambda(\partial\psi/\partial x) \tag{9.37}$$

where λ is the specific conductance, so

$$\lambda = F^2[z_1^2 l_{11} + z_1 z_2(l_{12} + l_{21}) + z_2^2 l_{22}] \tag{9.38}$$

The Hittorf transference number t_i^h, or the fraction of the current carried by the ith ion relative to the solvent in a solution of uniform composition is defined by

$$t_i^h = z_i J_i F/I \tag{9.39}$$

whence

$$t_1^{\mathrm{h}} = (z_1^2 l_{11} + z_1 z_2 l_{12})F^2/\lambda \tag{9.40}$$

and

$$t_2^{\mathrm{h}} = (z_1 z_2 l_{21} + z_2^2 l_{22})F^2/\lambda = 1 - t_1^{\mathrm{h}} \tag{9.41}$$

The cell or emf transference numbers t_i^{c} are defined by

$$-F\frac{\partial \psi}{\partial x} = \frac{t_1^{\mathrm{c}}}{z_1}\frac{\partial \mu_1}{\partial x} + \frac{t_2^{\mathrm{c}}}{z_2}\frac{\partial \mu_2}{\partial x} \tag{9.42}$$

and are thus extractable from equation (9.30). The Hittorf and cell transference numbers are equal on account of the Onsager reciprocal relations. Thus the three independent l_{ij} are uniquely available from λ, one of the transference numbers, and the salt diffusion coefficient D_s. The reverse transformation can be simply expressed[44] as

Table 9.2 Ionic transport coefficients l_{ij} for binary electrolyte solutions at 25°C. After Miller[44]

System	m	$10^{12}\, l_{11}/m^a$	$10^{12}\, l_{12}/m$	$10^{12}\, l_{22}/m$
H_2O–HCl	0·001	37·34	0·109	8·148
	0·5	33·40	0·940	7·235
	1·0	31·10	1·096	6·781
	2·0	28·85	1·355	6·108
	3·0	23·09	1·564	5·573
H_2O–LiCl	0·001	4·103	0·073	8·112
	0·5	3·316	0·700	6·827
	1·0	2·942	0·699	6·286
	2·0	2·406	0·622	5·475
	3·0	1·954	0·500	4·718
H_2O–NaCl	0·001	5·325	0·081	8·123
	0·5	4·613	0·840	7·121
	1·0	4·311	0·911	6·772
	2·0	3·812	0·911	6·035
	3·0	3·366	0·858	5·393
H_2O–KCl	0·001	7·826	0·086	8·129
	0·5	7·143	1·038	7·478
	1·0	7·077	1·214	7·364
	2·0	6·866	1·362	7·160
	3·0	6·734	1·440	6·929
H_2O–KCl	0·001	7·826	0·086	8·129
	0·5	7·143	1·038	7·478
	1·0	7·077	1·214	7·364

[a]Units of l_{ij} are mol^2 J^{-1} cm^{-1} s^{-1}

$$l_{ij} = \frac{t_i^{\rm h} t_j^{\rm c} \lambda}{z_1 z_2 F^2} + \nu_i \nu_j L_{\rm s} \tag{9.43}$$

Miller and co-workers[44, 45] have analysed the available data for many binary salt solutions to yield the l_{ij} in tabular and graphical form. Such data are very useful for estimating diffusion coefficients in ternary and higher order systems (cf Table 9.2).

9.5 DESCRIPTION OF DIFFUSION IN TERNARY ELECTROLYTES

Miller[3] has generalized the binary treatment of the previous section to ternary aqueous solutions with two cations and a common anion, specified as $C_{\nu_{1c}}A_{\nu_{1a}}$ and $D_{\nu_{2c}}A_{\nu_{2a}}$. There are nine coefficients in the analogues of equations (9.17) which by application of the Onsager reciprocal relations are reduced to six independent ones. These are demonstrated to be uniquely determined by the three observable independent salt mobility coefficients (including the cross term), the conductance, and the two independent transference numbers. Miller gives explicit relations analogous to equation (9.43) together with the diffusion mobilities in terms of the ion mobilities analogous to equation (9.36)

$$L_{ij} = \frac{\sum_k^3 \sum_l^3 z_k z_l (l_{ij} l_{kl} - l_{il} l_{kj})}{\nu_{ic} \nu_{jc} \sum_k^3 \sum_l^3 z_k l_{kl} z_l} \qquad i,j = 1, 2 \tag{9.44}$$

where species 3 is the anion. It is easily seen as a consequence of $l_{ij} = l_{ji}$ that $L_{12} = L_{21}$. The expressions for higher order systems are correspondingly more complex. The common anion case has been analysed by Miller.[4] The application of this approach to a ternary isotopic system has been presented by Anderson and Paterson.[46]

A means of predicting ternary coefficients from measured binary data is highly desirable. Wendt[47] proposed that because in the dilute limit ionic interactions would be negligible, they might be treated as such even in non-dilute solution. Furthermore, it was suggested that the mobilities of the ions are the same at reasonable concentrations as at infinite dilution and can therefore be related to the infinitely dilute ionic conductances. This model was later applied by Woolf[48] to an electrolyte in an ionizing solvent. These assumptions are not borne out by experimental tests.[49] In fact the l_{ij} cross-coefficients are 5–30% of the on-diagonal coefficients for 0·5–3 M alkali halide solutions. Examples of experimentally evaluated l_{ij} are shown in Table 9.2.

A more successful scheme for predictive calculation of the six ternary l_{ij} from the three binary ones has been proposed by Miller[3] and further tested by Kim *et al.*[50]

For two electrolytes, 1 and 2, having a common anion we define equivalent concentrations for cations, N_1 and N_2, and the anion, N_3, as follows

$$N_1 = \nu_{1c} z_1 m_1$$

$$N_2 = \nu_{2c} z_2 m_2$$

$$N_3 = -z_3 (\nu_{1a} m_1 + \nu_{2a} m_2)$$

and the total concentration of equivalents N is equal to N_3. The ternary on-diagonal cationic coefficients are given by

$$l_{ii}/N = X_i(l_{ii}^{(b)}/N)_{N_i} \qquad i = 1, 2 \tag{9.45}$$

where $X_i = N_i/(N_1 + N_2)$ and $l_{ii}^{(b)}$ is the cationic coefficient for the binary electrolyte i evaluated at the same concentration as that of *component i* in the ternary mixture, as indicated by the subscript N_i. The anion coefficient is evaluated from

$$l_{33}/N = X_1(l_{aa}^{(b)}/N)_{1,N} + X_2(l_{aa}^{(b)}/N)_{2,N} \tag{9.46}$$

where $l_{aa}^{(b)}$ is the anionic coefficient in a binary solution (1 or 2) evaluated at the same concentration as that of *total* electrolyte, as indicated by the subscript N. Cross-terms are evaluated from

$$l_{i3}/N = X_i(l_{i3}^{(b)}/N)_N \qquad i = 1, 2 \tag{9.47}$$

$$l_{12}/N = X_1X_2(l_{13}l_{23}/N^2)^{1/2} \tag{9.48}$$

where $l_{i3}^{(b)}$ is the cation i–anion 3 cross-term in the $i3$ binary, evaluated at the same total equivalent concentration N as in the ternary.

These calculation methods can be extended to quaternary and higher orders systems.[4] It is to be emphasized that these equations are semi-empirical, but quite successful.

The binary l_{ij} have been critically tabulated for about ten systems based on diffusion and electrical measurements[44,45] (Table 9.2) so many more ternary systems are subject to accurate prediction of their diffusion parameters. Table 9.3 shows a comparison between theory and experiment for NaCl and KCl in H_2O according to the Wendt dilute solution approximation and Miller's calculation scheme. The dilute limit for the on-diagonal L corresponds to the Nernst–Hartley binary formula and is generally well-verified. The Miller scheme clearly performs best for all mobilities at the higher concentrations.

Rather than measure the various binary quantities at each concentration, it would be preferable to be able to calculate the effect of varying concentration. Onsager and Fuoss[51] and Onsager and Kim[52] sought to calculate the l_{ij} in terms of the limiting (infinite dilution) parameters and solution ionic strength. This theory is very successful at relatively low concentrations. Pikal[53] has extended the theory of the l_{ij} for dilute solutions based on the early Debye–Hückel thermodynamic theory[54] and Onsager–Fuoss theory[51] taking account of short and long range electrostatic interactions. Figure 9.1 (after Pikal[53]) shows a comparison of theory with experiment for dilute HCl-H_2O at 25°C. While the magnitude of the off-diagonal l_{12} is relatively small in this dilute range and vanishes with C as expected, it reaches the magnitude of the on-diagonal coefficients for rich solutions. This theory appears to have increased substantially the concentration range accessible to calculation. Further progress in this direction will probably come from the use of cluster expansion methods and the reader is referred to the monograph by Friedman.[55] Meanwhile the equation due to Pikal appears to offer the best means of calculation from a 'limiting law' approach.[56]

Diffusion in an electrolyte solution, as in any other medium, can also in principle be described using the methods of non-equilibrium statistical mechanics. This 'linear response' approach seeks to evaluate velocity correlation

Table 9.3 Ternary phenomenological coefficients (L_{ij}) for H_2O–NaCl–KCl at 25°C*

	Experimental	Wendt[a]	Miller[b]	Lane and Kirkaldy[c]	Lane and Kirkaldy[d]
0·5 M NaCl, 0·25 M KCl					
$10^8 RT(L_{11})_V$	0·472	0·495	0·465	0·492	0·472
$(L_{12})_V$	−0·110[e]	−0·128	−0·108	−0·120	−0·115
$(L_{22})_V$	0·378	0·395	0·380	0·393	0·377
0·25 M NaCl, 0·5 M KCl					
$10^8 RT(L_{11})_V$	0·280	0·293	0·273	0·289	0·283
$(L_{12})_V$	−0·105[e]	−0·120	−0·102	−0·113	−0·111
$(L_{22})_V$	0·620	0·624	0·621	0·629	0·617
0·5 M NaCl, 0·5 M KCl					
$10^8 RT(L_{11})_V$	0·509	0·541	0·497	0·533	0·516
$(L_{12})_V$	−0·162[e]	−0·189	−0·160	−0·173	−0·168
$(L_{22})_V$	0·681	0·700	0·683	0·701	0·679
1·5 M NaCl, 1·5 M KCl					
$10^8 RT(L_{11})_V$	1·289	1·587	1·293	1·51	1·31
$(L_{12})_V$	−0·494[e]	−0·642	−0·521	−0·49	−0·43
$(L_{22})_V$	1·743	1·985	1·865	1·99	1·72

*Data from Miller,[3] units of LRT are mol cm^{-1} s^{-1}; (*a*) predicted by Miller using Wendt's approximation; (*b*) predicted by Miller's empirical method, equations (9.44)–(9.48); (*c*) predicted by Lane and Kirkaldy's ion-solvent interchange model, equations (9.69)–(9.70); (*d*) predicted by Lane and Kirkaldy's interchange model with relative viscosity correction; (*e*) average of experimental L_{12}, L_{21}.

coefficients (time integrals of particle velocities averaged over the entire ensemble of particles, cf §2.9). These coefficients are related to the phenomenological coefficients.[57, 58] Because momentum transfer between the constituent particles is involved, a mass-fixed frame is employed where

$$\sum_i J_i N_i = 0 \qquad J_i = -\sum_j^n L_{ij} \nabla \eta_j \tag{9.49}$$

$$L_{ij} = \frac{1}{3kTV} \int_0^\infty \langle J_i(0)\, J_j(t) \rangle \, \mathrm{d}t \tag{9.50}$$

and

$$J_i = \frac{1}{N_A} \sum_{l=\alpha}^{N_i} N_i v_{il} \tag{9.51}$$

Here V is the volume, N_i is the number of particles of component i, neglecting isotopic differences, and v_{il} is the velocity of particle l of that component measured with respect to the local centre of mass. The angle brackets indicate an average over the equilibrium ensemble. Defining the velocity correlation coefficient by

$$f_{ij} = \int_0^\infty \langle v_{i\alpha}(0)\, v_{j\alpha}(t) \rangle \, dt \qquad i \neq j \tag{9.52}$$

then from the above equations it follows that

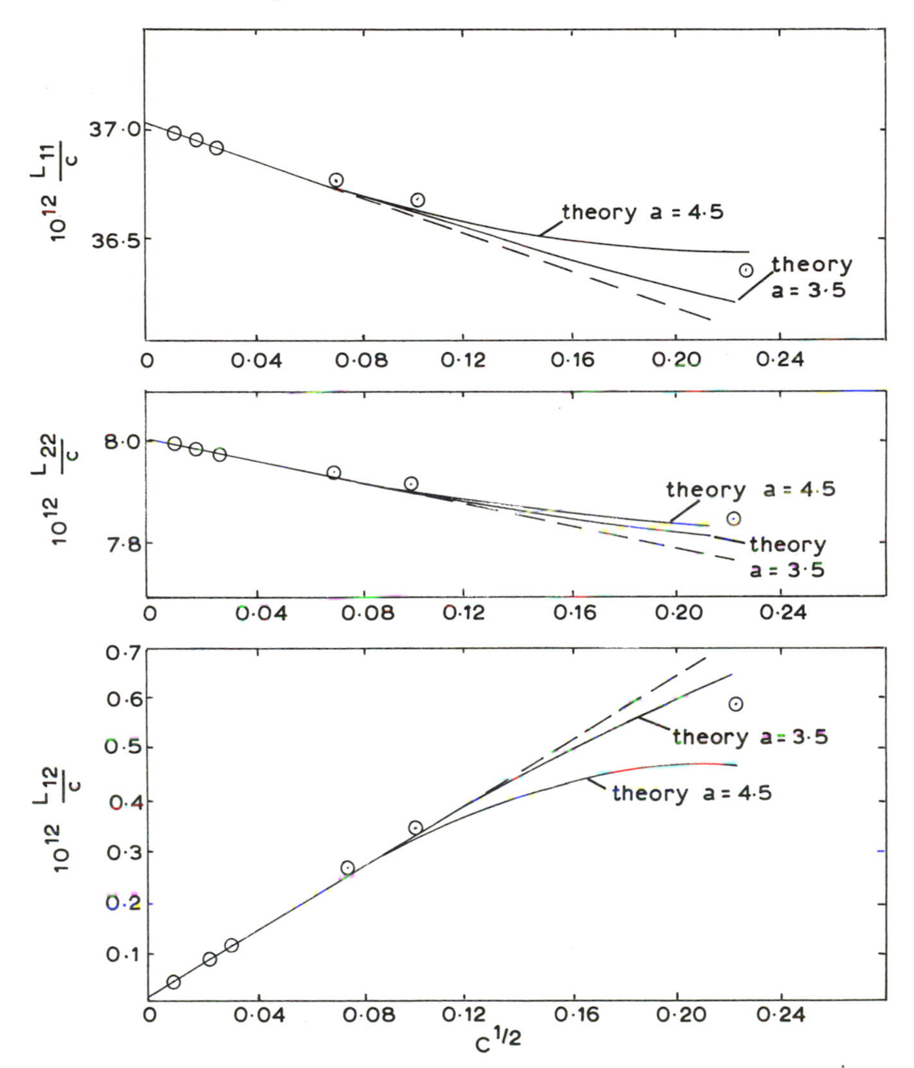

Reprinted with permission from M. J. Pikal, *J. Phys. Chem.*, 1971, **75**, 3124. Copyright 1971 American Chemical Society

Fig. 9.1 l_{ij}/c for HCl–H_2O at 25°C. Comparison of theory with experiment: O experimental. After Pikal[53]

$$L_{ij} = \frac{Vm_i m_j}{3kT} f_{ij} \tag{9.53}$$

$$L_{ii} = \frac{Vm_i^2}{3kT}\left\{\frac{f_{i\alpha\alpha}}{N_i} + f_{i\alpha\beta}\right\} \tag{9.54}$$

The term $f_{i\alpha\alpha}$ is the autocorrelation function and the term $f_{i\alpha\beta}$ is a measure of the correlation between different particles of the same species.

The velocity correlation coefficients have been related to the various ionic transport coefficients (l_{ij}) by Hertz[59] and Woolf and Harris.[60] Calculations of the different velocity correlation coefficients for a number of systems have been presented by Woolf[61] and their limiting behaviour at infinite dilution has been investigated by Miller.[56] These calculations derive from experimentally measured ionic transport parameters and contribute no new knowledge. *A priori* calculation of the f_{ij} via the statistical approach is the only way in which our understanding can be advanced. Unlike the non-electrolyte case, however, interparticle interactions in electrolyte solutions have not, as yet, been successfully described. In the absence of suitable potential functions for ion–ion and ion–solvent interactions there can be no progress. Indeed, Miller[56] suggests that, because of their behaviour at infinite dilution, velocity correlation functions may be inherently unsuitable for computer simulation. An investigation of the correlation functions of some other dynamical variable might be profitable.

9.6 TRANSITION STATE THEORY APPLIED TO ELECTROLYTE DIFFUSION

An elementary predictive methodology has been developed by Lane and Kirkaldy[41, 42] (cf § 7.2) on the basis of a quasi-crystalline model for the solution.[62] These authors begin with the assumption that each solute ion has as its nearest ionic neighbours species of opposite charge and that all species — anions, cations, salts and solvent molecules — may be described as residing on an integral number of equivalent lattice points. Furthermore, the model makes explicit use of the assumption that diffusion is effected through particle–particle interchanges which may be conceived of as equivalent to 'jumps' over a fixed distance.

It is further argued that in the lattice-fixed frame only ion–solvent interchanges need be considered. This is reasonable for any dilute solution, but particularly so for an electrolyte. Thus the existence of adjacent ions of like sign is improbable because of the high repulsive energy involved. Adjacent ions of opposite charge are much more likely to exist, but their exchange would lead to the location of each ion at the centre of an ionic atmosphere, most of which has a similar charge to its own. It seems most likely that such a high energy configuration would decay by a reversal of the original interchange. Such a process is argued to contribute nothing to diffusion and may be ignored.

Proceeding therefore on the basis that ion–solvent interchanges are the only processes which need be considered, we use the methods of § 7.2 to write for the ionic species i in the lattice-fixed frame[41, 42, 43]

$$J_i = -k_{i0} m_i m_0 \{\nabla \eta_i - \nabla \mu_0\} \tag{9.55}$$

where the subscript 0 denotes solvent and for the present illustration $\overline{V}_i = \overline{V}_0$ is assumed. The model is first applied to the case of a binary electrolyte (equation (9.18)) to illustrate the methodology and its limitations. The rate coefficients k_{i0} are estimated in terms of the partial ionic conductances at infinite dilution as

$$k_{i0} = k_{i0}^0 = \lambda_i^0/10^3 m_o^0 |z_i| F^2 \tag{9.56}$$

where superscript 0 indicates pure solvent. Individual cation and anion fluxes J_c and J_a are given by

$$J_c = -k_{c0} m_c m_0 \{\nabla \eta_c - \nabla \mu_0\}$$
$$J_a = -k_{a0} m_a m_0 \{\nabla \eta_a - \nabla \mu_0\} \tag{9.57}$$

where μ_0 is the solvent chemical potential. The salt flux is given in the volume-fixed frame (which is here identical with the lattice frame) by

$$J_s = -L_{ss}\{\nabla \mu_s - (\nu_c + \nu_a)\nabla \mu_0\} = -L_{ss}\left\{\nabla \mu_s - \frac{\overline{V}_s}{\overline{V}_0}\nabla \mu_0\right\} \tag{9.58}$$

since $\nabla \mu_s = \nu_c \nabla \eta_c + \nu_a \nabla \eta_a = \nu_c \nabla \mu_c + \nu_a \nabla \mu_a$. Here use has been made of the assumption $\overline{V}_i = \overline{V}_0 = 1$ to arrive at the ratio

$$\overline{V}_s/\overline{V}_0 = \nu_c + \nu_a \tag{9.59}$$

Graf *et al.*[63] point out that the zero current condition is strictly applicable only in the solvent-fixed frame. To apply it here we restrict consideration to infinite dilution where all frames are equivalent. Thus we write

$$J_s = \frac{J_c}{\nu_c} = \frac{J_a}{\nu_a} \tag{9.60}$$

to yield, after elimination of gradient terms

$$L_{ss} = \frac{k_{c0} k_{a0} m_c m_a m_0}{\nu_c^2 k_{a0} m_a + \nu_a^2 k_{c0} m_c} \tag{9.61}$$

Recalling that $\nu_a z_a = -\nu_c z_c$, it is readily demonstrated that this result is equivalent to equation (9.36) for

$$l_{11} = k_{c0} m_c m_0 \qquad l_{22} = k_{a0} m_a m_0 \tag{9.62}$$

and at infinite dilution

$$l_{12} = l_{21} = 0 \tag{9.63}$$

It follows from equation (9.56) and the foregoing that equation (9.61) is equivalent to the Nernst–Hartley equation

$$D_s = \frac{(\nu_a + \nu_c)\lambda_c^0 \lambda_a^0}{F^2 \nu_c |z_c| (\lambda_c^0 + \lambda_a^0)} \frac{\partial \mu_s}{\partial m_s} \tag{9.64}$$

where the λ_i^0 are the limiting values of the partial ionic conductances at infinite dilution.

For non-dilute solutions this model implies non-zero values of the solvent-fixed l_{ij} via a simple transformation of the lattice-fixed diagonal matrix defined by equations (9.57) to the solvent-fixed frame. Graf *et al.*[63] have carried out the transformation and offer the formula (see equation (A4.9) for δ_{ij})

$$l_{ij} = k_{i0}m_1m_0\left(\delta_{ij} + \frac{m_s}{m_0}\nu_j\right) + \frac{m_i}{m_0}\sum_{k=1}^{2} k_{k0}m_km_0\left(\delta_{kj} + \frac{m_s}{m_0}\nu_k\right) \tag{9.65}$$

which is accurately approximated by

$$l_{11} = k_{10}m_1m_0 \qquad l_{22} = k_{20}m_2m_0 \tag{9.66}$$

and

$$l_{12} = l_{21} = (k_{10} + k_{20})m_1m_2 \tag{9.67}$$

The latter, like the diagonal elements, are always positive. Figure 9.2 shows a comparison of predicted on-diagonal coefficients with Miller's compilation (Table 9.2). Equation (9.65) is seen to be successful at low concentrations. For simple salts like NaCl we can write

$$\frac{l_{12}}{l_{11} + l_{22}} = \frac{m_2}{m_0} \tag{9.68}$$

which allows a rapid check on the relative off-diagonal magnitudes. Table 9.4 compares this predicted ratio with that tabulated for rich solutions of HCl, LiCl,

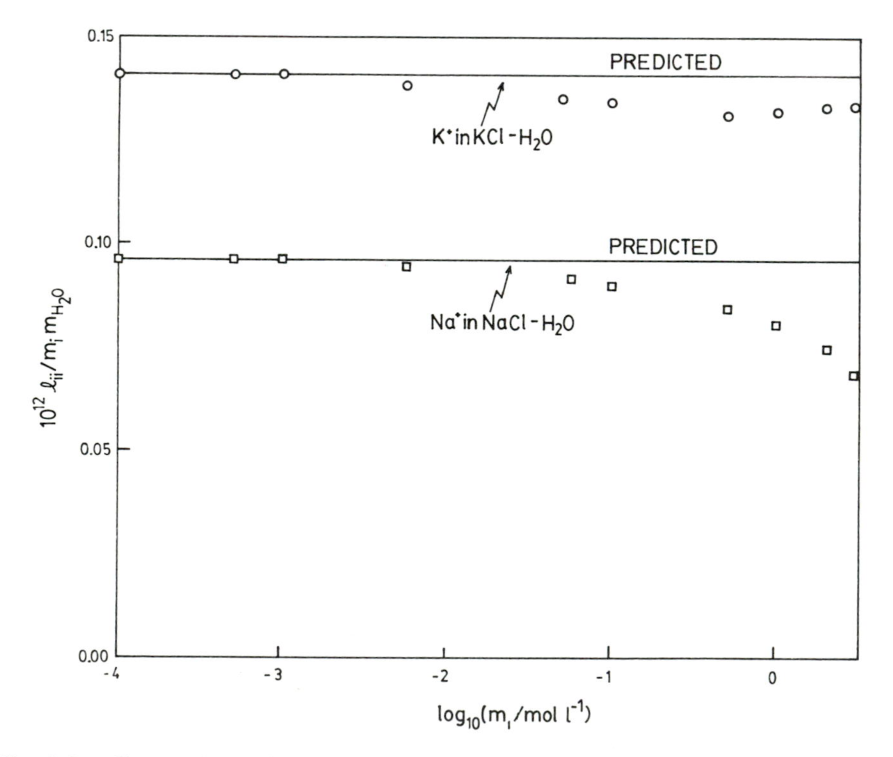

Fig. 9.2 **Comparison of observed and predicted concentration dependence of the on-diagonal solvent-fixed Onsager coefficients with the quasi-crystalline theory of Lane and Kirkaldy**

Table 9.4 Predicted and observed mobility ratios*

Molarity	HCl		LiCl		NaCl		KCl	
	$\frac{l_{12}}{(l_{11}+l_{22})}$	$\frac{m_2}{m_0}$	$\frac{l_{12}}{(l_{11}+l_{22})}$	$\frac{m_2}{m_0}$	$\frac{l_{12}}{(l_{11}+l_{22})}$	$\frac{m_2}{m_0}$	$\frac{l_{12}}{(l_{11}+l_{22})}$	$\frac{m_2}{m_0}$
0·5	0·023	0·009	0·069	0·009	0·072	0·009	0·071	0·009
1	0·029	0·018	0·076	0·018	0·082	0·018	0·084	0·018
2	0·041	0·036	0·079	0·036	0·092	0·036	0·097	0·036
3	0·055	0·054	0·075	0·054	0·098	0·054	0·106	0·054

*Values of l_{ij} from Miller.[44]

NaCl, and KCl in H_2O. The predictions are all of the right sign and magnitude, becoming relatively poorer with decreasing concentration. This latter defect, however, has a minimal effect on the prediction of diffusion coefficients. Graf *et al.*[63] have studied these deviations systematically through D_s^0, seeking to better understand the model and to amend it. In this way they are developing very useful systematics for diffusion in ground water, a matter of great importance.

The Lane–Kirkaldy model[41] generalizes trivially to multicomponent systems via equations like (9.57). For mixtures of strong electrolytes like NaCl and KCl the mobilities are given by

$$L_{ii} = k_{10}m_1m_0(z_3)^{-2}\left\{1 - (z_i)^2k_{i0}m_i/\sum_{j=1}^{3}(z_j)^2k_{j0}m_j\right\} \tag{9.69}$$

and

$$L_{12} = L_{21} = -k_{10}k_{20}m_1m_2m_0z_1z_2/(z_3)^2\sum_{j=1}^{3}(z_j)^2k_{j0}m_j \tag{9.70}$$

The predictions of these equations for NaCl–KCl have been entered in the second to last column of Table 9.3 for comparison with the observations and Miller's calculations. In making the latter comparison it is to be emphasized that appreciably more empirical information goes into Miller's formula. It is for this reason that Graf and co-workers, who deal with complex natural aqueous solutions, seek to perfect a model which requires a minimal input of empirical data.

Lane and Kirkaldy[42] offered a second model based on ion exchange with vacant sites and this was also quite successful in predicting the L-matrices, particularly when a relative viscosity correction was included. In view of the failure of this class of models in the case of ternary non-electrolytes (cf §9.3) we recommend that the vacancy model be dropped from further consideration. However, there may be some merit in amending the exchange model via the relative viscosity correction (see also Graf *et al.*[63]). The final column of Table 9.3 includes this amendment. While such a correction is somewhat *ad hoc* it is of the form expected from a consideration of non-electrolyte mass transfer processes (cf Lane and Kirkaldy[42]).

The kinetic models of Lane and Kirkaldy are idealized and, on a molecular level, somewhat unrealistic. Their success in predicting both the sign and magnitude of the L coefficients derives from two considerations. Firstly, the models amount to simple place exchange processes which automatically conserve volume and the flows are strongly coupled so as to preserve charge neutrality. These two physical effects are, of course, the origins of correlated ionic motions. Secondly, individual ionic 'kinetic parameters' are in fact calculated from macroscopic properties (the conductances) which represent average ionic motions in the same way as do the phenomenological L coefficients.

9.7 IONIC MELTS

Ionic melts may be regarded as highly concentrated electrolytic fluids. Not surprisingly, they are quite highly ordered and a quasi-crystalline model is often

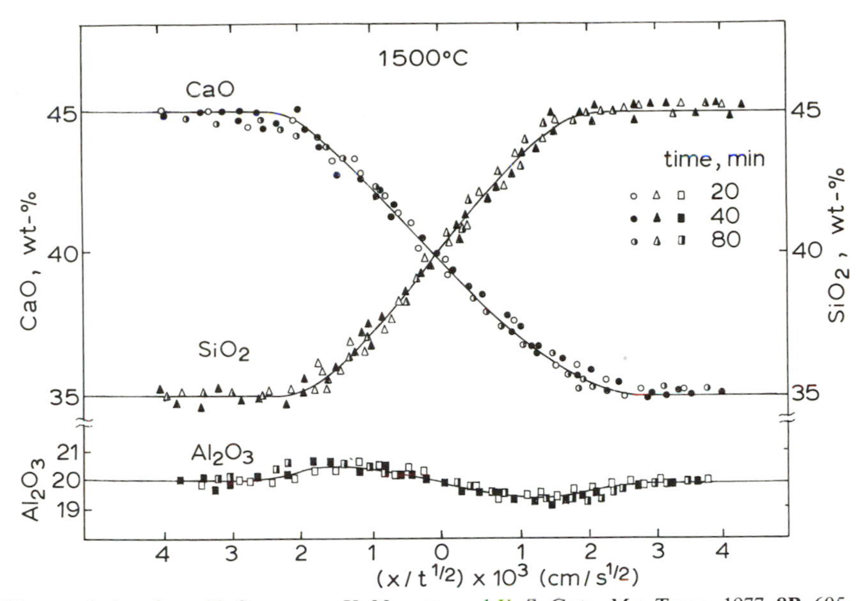

With permission from H. Suguwara, K. Nagata, and K. S. Goto, *Met. Trans.*, 1977, **8B**, 605

Fig. 9.3 Relation between the concentrations of lime, alumina, and silica and distance divided by square root of time for three diffusion times of 20, 40, and 80 min at 1773 K. After Sugawara *et al.*[69]

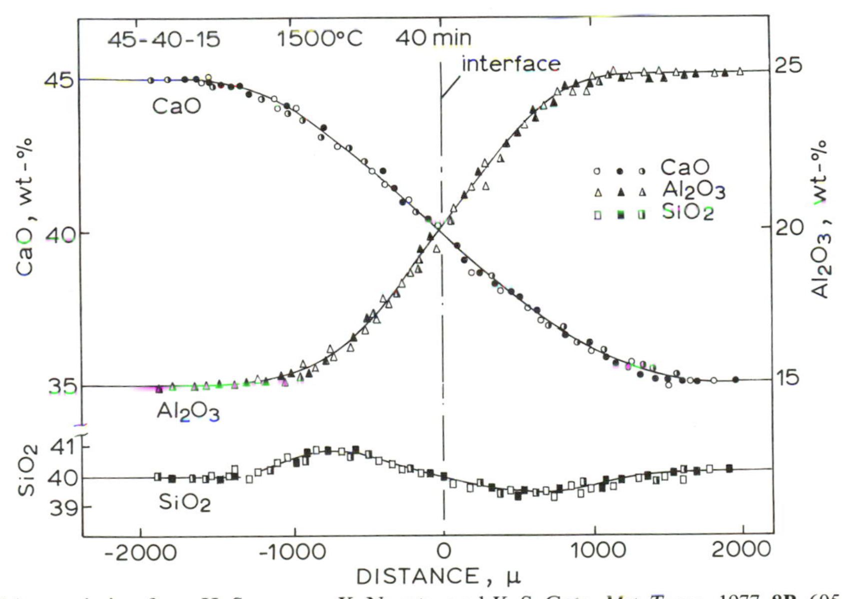

With permission from H. Suguwara, K. Nagata, and K. S. Goto, *Met. Trans.*, 1977, **8B**, 605

Fig. 9.4 Concentration penetration curves from three independent diffusion runs for 40 min at 1773 K with constant silica content. After Sugawara *et al.*[69]

employed as an approximate model description. The use of computer simulations in the study of ionic liquid structures has been recently reviewed by Adams and Hill.[64] These studies show that the transport properties of melts may be discussed within the framework of a highly disordered crystal model.[65]

The phenomenology of transport in ionic melts has been reviewed by Laity[66] and Nagata and Goto.[67] The procedure is precisely the same as was developed in Chapter 8 for ionic solids or §9.4 for electrolyte solutions (in the absence of a solvent). In describing isothermal diffusion, the electrostatic potential is eliminated from the formalism by grouping ionic species in charge-neutral sets which correspond to real components, possessed of measurable chemical potentials. Thus ionic transport equations of the form $J_i = -\Sigma l_{ij} \nabla \eta_j$ are reduced to salt (or component) transport equations of the form (9.28). The relationships between the ionic phenomenological coefficients l_{ij} and the electrical conductivity, transport numbers, and diffusion coefficients are the same as for an electrolyte solution §§9.4 and 9.5. Using these relationships, Nagata and Goto calculated the ternary diffusion coefficients for a CaO–SiO_2–Al_2O_3 slag, obtaining fairly good agreement with penetration curves measured for this system by Oishi.[68] A later direct determination of the ternary $\tilde{D}$ values from the penetration curves of Figs. 9.3 to 9.5 also showed good agreement.[69] It appears to be generally true from these figures and their analysis that for these ionic oxide melts the phenomenological cross-terms, D_{ij} and L_{ij}, are significant.[70]

Diffusion in oxide melts is expected to be similar to diffusion in solid ionic oxides in that the oxygen partial pressure should, in part, determine the rates of

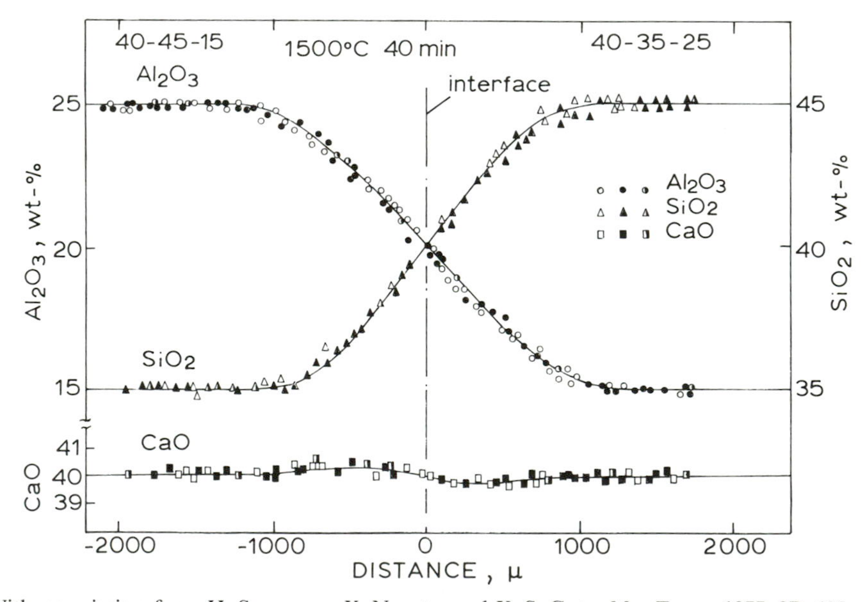

With permission from H. Suguwara, K. Nagata, and K. S. Goto, *Met. Trans.*, 1977, **8B**, 605

Fig. 9.5 Concentration penetration curves from three independent diffusion runs for 40 min at 1773 K with constant lime content. After Sugawara *et al.*[69]

cation diffusion (cf § 8.6). Such an effect has been demonstrated for $CaO-SiO_2-Fe_xO$ melts by Goto *et al.*[71]

9.8 OXIDE GLASSES

In the sense that glasses are highly viscous fluids, it is appropriate that their diffusion behaviour be discussed in this chapter. Attention will be confined to the technologically important class of oxide glasses. Much of the literature has been reviewed by Doremus[72] who describes diffusion in terms of a simple structural model.

Oxide glasses are composed of network-forming oxides, such as SiO_2, Al_2O_3 or B_2O_3, and, additionally, strongly ionic oxides such as those of alkali and alkali earth metals which partially disrupt the covalent oxide network. These latter oxides are known as network modifiers. Their presence is manifested by the existence of metal cations associated with oxygen anions which are covalently bonded to the oxide network. Not surprisingly, the mobilities of the 'free' cations are much greater that those of the network forming elements.

Because the oxygen anions are essentially immobile at relatively low temperatures, the only diffusing species are cations, and electrostatic cross-effects arise through the condition

$$\sum_i z_i J_i = 0$$

assuming no electron transport. The situation is thus analogous to a multicomponent electrolyte solution possessing a common anion at a uniform concentration. The phenomenological analysis is identical and will not be repeated here. An explicit treatment has been given by Cooper[73] for the case in which no cross-effects, other than electrostatic, exist in the intrinsic reference frame. The treatment has been extended by Okongwu *et al.*[74] (cf § 6.6) to the situation in which the separate halves of a diffusion couple have different oxide ion concentrations. The consequence of this asymmetry in the couple is the appearance of discontinuities in the cation concentration profiles at the interface (cf Fig. 6.5).

No kinetic model is available for diffusion in glasses. It seems reasonable to suppose, however, that cation diffusion occurs via an exchange mechanism. In this way the requirement of local charge neutrality is met, the cations having their equilibrium positions associated with the immobile oxygen anions. It is also probable that vacancies exist in the cation array as a consequence of mixing metal oxides of different valences. Cation exchange would therefore be expected to occur via their agency. Thus cation diffusion in glasses can be visualized in precisely the same way as in the case of ionic crystals containing immobile anions. The finding that cation mobility in multicomponent glasses increases with decreasing cation charge is thought to support this view.[75]

Because of its technological importance in surface toughening of glasses, cation diffusion between ionic melts and oxide glasses has attracted some attention.[76] The process is usually accelerated by the application of an electrostatic field. Transport behaviour under these conditions has been analysed for binary ion exchange.[77, 78]

REFERENCES

1 P. J. DUNLOP AND L. G. GOSTING: *J. Am. Chem. Soc.*, 1955, **77**, 5238
2 D. G. MILLER: *Chem. Rev.*, 1960, **60**, 15
3 D. G. MILLER: *J. Phys. Chem.*, 1967, **71**, 616
4 D. G. MILLER: *ibid.*, 3588
5 E. L. CUSSLER: 'Multicomponent diffusion', 1976, New York, Elsevier
6 E. L. CUSSLER: 'Diffusion: mass transfer in fluid systems', 1984, Cambridge University Press
7 H. J. V. TYRRELL AND K. R. HARRIS: 'Diffusion in liquids', 1984, London, Butterworths
8 H. D. ELLERTON AND P. J. DUNLOP: *J. Phys. Chem.*, 1967, **71**, 1291
9 H. D. ELLERTON AND P. J. DUNLOP: *ibid.*, 1538
10 V. VITAGLIANO AND K. SARTORIO: *ibid.*, 1970, **74**, 2949
11 G. G. STOKES: 'Mathematical and physical papers, Vol. 3', (1850), 1903, Cambridge University Press
12 A. EINSTEIN: *Ann. Phys. (Leipzig)*, 1905, **17**, 549
13 A. EINSTEIN: *ibid.*, 1906, **19**, 371
14 W. SUTHERLAND: *Phil. Mag.*, 1905, **9**, 781
15 R. B. BIRD, W. E. STEWART, AND E. N. LIGHTFOOT: 'Transport phenomena', 571, 1960, New York, John Wiley
16 D. F. EVANS, T. TOMINAGA, AND H. J. DAVIS: *J. Chem. Phys.*, 1981, **74**, 1298
17 T. G. HISS AND E. L. CUSSLER: *J.A.I.Ch.E.*, 1973, **19**, 693
18 G. A. DAVIES, A. B. PONTER, AND K. CRAINE: *Can. J. Chem. Eng.*, 1967, **45**, 372
19 R. C. REID, J. M. PRAUZNITZ, AND T. K. SHERWOOD: 'The properties of liquids and gases', 3rd edn, 1977, New York, McGraw-Hill
20 H. ERTL, R. K. GHAI, AND F. A. L. DULLEIN: *J.A.I.Ch.E.*, 1974, **20**, 1
21 H. ERTL AND F. A. L. DULLEIN: *ibid.*, 1973, **19**, 1215
22 S. K. GLASSTONE, K. J. LAIDLER, AND H. EYRING: 'Theory of rate processes', 1941, New York, McGraw-Hill
23 J. L. HILDEBRAND: 'Viscosity and diffusion', 1977, New York, John Wiley
24 S. CHAPMAN AND T. G. COWLING: 'The mathematical theory of non-uniform gases', 1939, New York, Cambridge University Press
25 B. J. ALDER AND T. E. WAINWRIGHT: *Phys. Rev. A*, 1970, **1**, 18
26 B. J. ALDER, D. M. GLASS, AND T. E. WAINWRIGHT: *J. Chem. Phys.*, 1970, **53**, 3813
27 B. J. ALDER, W. E. ALLEY, AND J. H. DYMOND: *ibid.*, 1974, **61**, 1415
28 D. CHANDLER: *Acc. Chem. Res.*, 1974, **7**, 246
29 S. J. BERTUCCI AND W. F. FLYGARE: *J. Chem. Phys.*, 1975, **63**, 1
30 J. H. DYMOND AND L. A. WOOLF: *J. Chem. Soc. Faraday I*, 1982, **78**, 991
31 S. L. RUBY, J. C. LOVE, P. A. FLINN, AND B. J. ZABRANSKY: *Appl. Phys. Lett.*, 1975, **27**, 320
32 P. J. DUNLOP: *J. Phys. Chem.*, 1957, **61**, 1619
33 A. VIGNES: *Ind. Eng. Chem. Fundamentals*, 1966, **5**, 189
34 G. S. HARTLEY AND J. CRANK: *Trans. Faraday Soc.*, 1949, **45**, 801
35 P. C. CARMAN AND L. H. STEIN: *Trans. Faraday Soc.*, 1956, **52**, 619
36 F. A. L. DULLEIN: *Ind. Eng. Chem. Fundamentals*, 1971, **10**, 41
37 H. T. CULLINAN AND M. R. CUSICK: *ibid.*, 1967, **6**, 72
38 J. LEFFLER AND H. T. CULLINAN: *ibid.*, 1970, **9**, 84
39 E. L. CUSSLER AND E. N. LIGHTFOOT: *J. Phys. Chem.*, 1965, **69**, 2875
40 J. E. LANE AND J. S. KIRKALDY: *Can. J. Phys.*, 1964, **42**, 1643
41 J. E. LANE AND J. S. KIRKALDY: *Can J. Chem.*, 1965, **43**, 1812
42 J. E. LANE AND J. S. KIRKALDY: *ibid.*, 1966, **44**, 477
43 E. L. CUSSLER: *J.A.I.Ch.E.*, 1980, **26**, 43
44 D. G. MILLER: *J. Phys. Chem.*, 1966, **70**, 2639
45 D. G. MILLER, J. A. RARD, L. B. EPPSTEIN, AND R. A. ROBINSON: *J. Sol. Chem.*, 1980, **9**, 467
46 J. ANDERSON AND R. PATERSON: *J. Chem. Soc. Faraday I*, 1975, **71**, 1335

47 R. P. WENDT: *J. Phys. Chem.*, 1965, **69**, 1227
48 L. A. WOOLF: *ibid.*, 1972, **76**, 1166
49 R. P. WENDT AND M. SHAHIM: *ibid.*, 1970, **74**, 2770
50 H. KIM, G. REINFELDS, AND L. J. GOSTING: *ibid.*, 1973, **77**, 934
51 L. ONSAGER AND R. M. FUOSS: *ibid.*, 1932, **36**, 2689
52 L. ONSAGER AND S. K. KIM: *ibid.*, 1957, **61**, 215
53 M. J. PIKAL: *ibid.*, 1971, **75**, 3124
54 P. DEBYE AND E. HÜCKEL: *Z. Phys.*, 1923, **24**, 185, 305
55 H. L. FRIEDMAN: 'Ionic solution theory, based on cluster expansion methods', 1962, New York, Interscience
56 D. G. MILLER: *J. Phys. Chem.*, 1981, **85**, 1137
57 D. W. MCCALL AND D. C. DOUGLASS: *ibid.*, 1967, **71**, 987
58 D. C. DOUGLASS AND H. L. FRISCH: *ibid.*, 1969, **73**, 3039
59 H. G. HERTZ: *Ber. Bunsenges*, 1977, **81**, 656
60 L. A. WOOLF AND K. R. HARRIS: *J. Chem. Soc. Faraday I*, 1978, **74**, 933
61 L. A. WOOLF: *J. Phys. Chem.*, 1978, **82**, 959
62 E. A. GUGGENHEIM: 'Thermodynamics: an advanced treatment for chemists and physicists', 5th edn, 1967, Amsterdam, North-Holland
63 D. L. GRAF, D. E. ANDERSON, AND J. B. WOODHOUSE: *Acta Geochim. Cosmochim.*, 1983, **47**, 1985
64 D. ADAMS AND G. HILLS: in 'Ionic liquids', (ed. D. Inman and D. G. Lovering), 1981, New York, Plenum
65 J. RICHTER: *ibid.*
66 R. W. LAITY: 'The structure and properties of ionic melts', 1962, Aberdeen University Press
67 K. NAGATA AND K. S. GOTO: *J. Electrochem. Soc.*, 1976, **123**, 1814
68 Y. OISHI: Japan–USA joint symp. on 'Ceramics', 1972
69 H. SUGAWARA, K. NAGATA, AND K. S. GOTO: *Met. Trans. B*, 1977, **8**, 605
70 K. S. GOTO, M. SASABE, AND M. KAWAKAMI: *Trans. Iron Steel Inst. Jpn*, 1977, **17**, 212
71 K. S. GOTO, T. KURAHASHI, AND M. SASABE: *Met. Trans. B*, 1977, **8**, 523
72 R. H. DOREMUS: 'Glass science', 1973, New York, Wiley–Interscience
73 A. R. COOPER: *Phys. Chem. Glasses*, 1965, **6**, 55
74 D. A. OKONGWU, W-K. LU, A. E. HAMIELEC, AND J. S. KIRKALDY: *J. Chem. Phys.*, 1973, **58**, 777
75 A. K. VARSHNEYA AND A. R. COOPER: *J. Am. Ceram. Soc.*, 1972, **55**, 220
76 A. JAMBON AND J. P. CARRON: *Acta Geochim. Cosmochim.*, 1976, **40**, 897
77 M. ABOU-EL-LEIL AND A. R. COOPER: *J. Am. Ceram. Soc.*, 1979, **62**, 390
78 E. E. SHAISHA AND A. R. COOPER: *ibid.*, 1981, **64**, 278

CHAPTER 10

Multicomponent transport in polymers, molecular sieves and membranes

10.1 INTRODUCTION

In §2.17 the subject of transport in gels, or systems which are structurally disperse, was introduced, and binary interactions in polymers, zeolites and membranes were considered. It was reported that transport in polymers and zeolites could be at least qualitatively understood in terms of Fickian diffusion kinetics and according to mechanisms analogous to those which apply in simple liquids and solids. Such should also be the case for systems of three or more components. The following two sections explore this conjecture and illustrate its limitations in terms of a specific example for a ternary polymeric system and a ternary zeolite system.

By contrast, transport in membranes, although closely related to that in polymers, is greatly complicated by strong association or complexing which occurs within the walls. This is particularly the case for biological membranes and the commercially important ion exchange membranes. Such specificity can be understood in a broad way through an irreversible thermodynamic treatment of coupled diffusion-reaction processes. It thus emerges that solutes can diffuse against their own gradients through association with mobile species (carrier transport) or by the conventional diffusion interactions, or both. §10.4 presents a number of models and results pertaining to this generally complex process.

10.2 TERNARY KINETICS IN THE DYEING OF POLYMERS

The kinetics of mixing of Benzopurpurine 4B and Sky Blue 6B into swollen cellophane presents a classic problem in multicomponent diffusion. Sekido and Morita[1] prepared cellophane sheet cut into pieces 0·06 m wide and 0·55 m long, washed with distilled water and wound tightly round a glass tube. After dyeing in an extensive medium the film roll was rinsed, then the dyed cellophane was unrolled and dried. The sample was then cut, layer by layer, for analysis. The mean thickness of the swollen sheet was $4{\cdot}66 \times 10^{-5}$ m.

The experiment may be analysed as a semi-infinite couple, the surface dye concentration being held constant throughout the diffusion. The ternary diffusion analysis was carried out using the solutions of Fujita and Gosting[2] (cf §4.6). Typical penetration curves for ternary uptake are given in Figs. 10.1 and

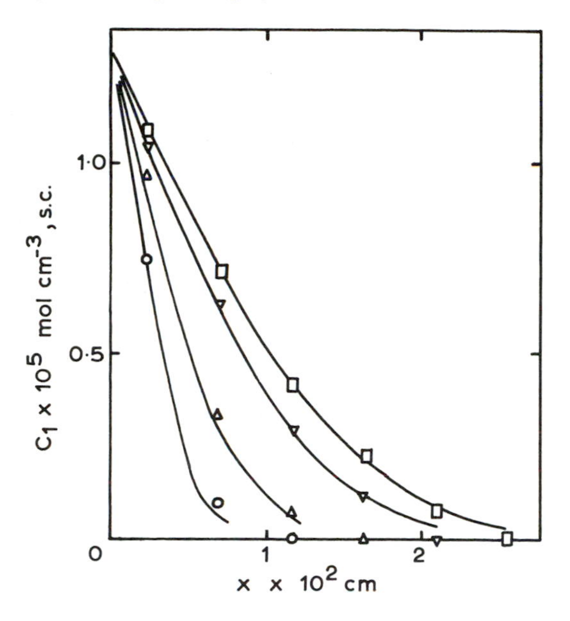

Fig. 10.1 The distributions of Benzopurpurine 4B at various times of mixture dyeing: ○ 187; △ 540; ▽ 1448; □ 2130 min
— theoretical curve
each dye concentration = $5{\cdot}0 \times 10^{-5}$ mol l^{-1}; 90°C; 0·03 mol NaCl l^{-1}. After Sekido and Morita[1]

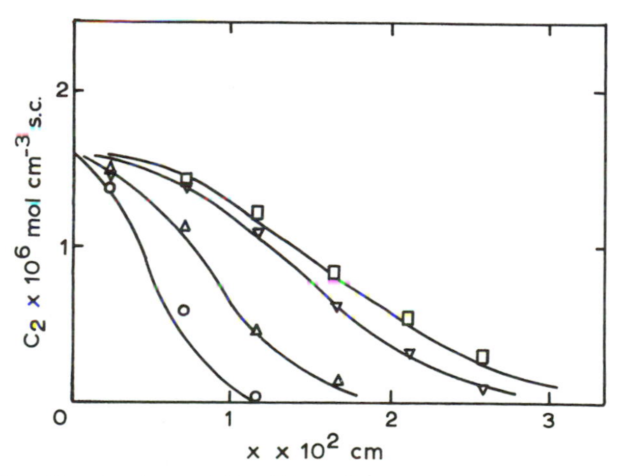

Fig. 10.2 The distributions of Sky Blue 6B at various times of mixture dyeing: ○ 187; △ 540; ▽ 1448; □ 2130 min
— theoretical curve
each dye concentration = $5{\cdot}0 \times 10^{-5}$ mol l^{-1}; 90°C; 0·03 mol NaCl l^{-1}. After Sekido and Morita[1]

Table 10.1 Four diffusion coefficients and surface concentrations in mixture dyeing (at 90°C). After Sekida and Morita[1]

D_{11}	D_{22}	D_{12}	D_{21}	$C_1^0 \times 10^5$	$C_2^0 \times 10^6$
	($\times 10^8$ cm^2/min.)			(mol./cm^3)	
2·9	2·0	−0·01	0·10	0·49	0·38
3·1	2·3	−0·01	0·40	0·97	0·96
3·3	3·0	−0·01	0·80	1·3	1·6
3·3	4·0	−0·09	1·1	1·5	2·2

10.2. The analysed diffusion matrix is given in Table 10.1. One notes, in accordance with dilute solution theory of multicomponent diffusion (cf § §6.3 and 7.9), that D_{11} and D_{22} are approximately constant and that the average D_{21} is substantially smaller than the diagonal Ds and increasing linearly with concentration C_2. The fact, however, that $D_{12} \sim 0$ is not in accord with what we would expect in an interstitial, ternary metallic solution. The thermodynamics of the gel are evidently more complex.

10.3 ADSORPTION KINETICS IN UPTAKE OF A BINARY GAS BY A ZEOLITE

Kärger and Bülow[3] have developed a model for binary adsorption gas kinetics which incorporates the D-matrix of the generalized Fick equation (4.67). They assume a highly different adsorption behaviour of the two components, negligible Onsager cross-coefficients and applicability of the Langmuir model for the equilibria. They thus obtain an analytic expression for the sorption kinetics which they claim accords with observations in the n–neptane–benzene/NaX–zeolite system.

The thermokinetic assumptions are very similar to those explored in § §6.3 and 7.9, and particularly to those of interstitial solutions in solids, whereby the Onsager coefficients

$$L_{12} = L_{21} = 0 \tag{10.1}$$

the tracer coefficients

$$D_i = RTL_{ii}/X_i \tag{10.2}$$

and the chemical coefficients

$$D_{ij} = D_i \frac{X_i \, \partial \ln a_i}{X_j \, \partial \ln X_i} \qquad i, j = 1, 2 \tag{10.3}$$

Although the thermodynamics of the solution was expressed by Kärger and Bülow in terms of a Langmuir model, discussion of the thermodynamics will be deferred until the experimental record has been examined. However, as with those authors, the special case $D_1 \gg D_2$ will be examined, which greatly simplifies the differential equations (cf §4.6.1) and accords with their choice of experimental system. The boundary value problem is developed for zeolite spheres of radius R

assuming a surface concentration at equilibrium which is to be established by the ultimate saturation conditions observed experimentally. The appropriate Fickian differential equations are (cf equation (4.67))

$$\frac{\partial X_1}{\partial t} = \frac{1}{r^2}\frac{\partial}{\partial r}\left[r^2\left(D_{11}\frac{\partial X_1}{\partial r} + D_{12}\frac{\partial X_2}{\partial r}\right)\right] \tag{10.4}$$

and

$$\frac{\partial X_2}{\partial t} = \frac{1}{r^2}\frac{\partial}{\partial r}\left[r^2\left(D_{21}\frac{\partial X_1}{\partial r} + D_{22}\frac{\partial X_2}{\partial r}\right)\right] \tag{10.5}$$

Now since $D_1 \gg D_2$, within spherical symmetry the quasi-steady approximation for X_1 can be assumed, so that

$$\frac{\partial X_1}{\partial r} \simeq -\frac{D_{12}}{D_{11}}\frac{\partial X_2}{\partial r} \tag{10.6}$$

for the conditions following an initial transient, whence

$$\frac{\partial X_2}{\partial t} \simeq \frac{1}{r^2}\frac{\partial}{\partial r}\left[r^2\frac{D}{D_{11}}\frac{\partial X_2}{\partial r}\right] \tag{10.7}$$

where

$$D = D_{11}D_{22} - D_{12}D_{21} \tag{10.8}$$

Next the results of the authors' macroscopic uptake experiments[4] with sorption from an azeotropic mixture of n–heptane (n–C_7H_{16}) and benzene (C_6H_6) into zeolite Na-X at 85°C (Fig. 10.3) are considered. Clearly the faster diffusing n–heptane reaches approximately seven times its average saturation concentration near the end of the initial transient (approximately t_0 = 20 min). There is accordingly a very strong pumping of the n–heptane down the initially steep benzene gradient indicating that D_{12}/D_{11} reaches very large positive values.

We can estimate this ratio by reference to Fig. 10.4 and the results of Fig. 10.3. We can then write the mass balance for the effective diffusion coefficient in equation (10.7) using the mass ratio ~10/100, realized at the time t_0 when the n–heptane maximum is reached

$$\frac{\frac{1}{2}X_2 \cdot 2(Dt_0/D_{11})^{1/2}\, 4\pi R^2}{X_1 \cdot \frac{4}{3}\pi R^3} \simeq \frac{10}{100} \tag{10.9}$$

Referring to equation (10.6) and Fig. 10.4 and combining with equation (10.9) we estimate

$$\frac{X_2}{X_1} \simeq -\frac{\Delta X_2}{\Delta X_1} \simeq \frac{1}{3}\cdot\frac{10}{100}\cdot\frac{R}{(Dt_0/D_{11})^{1/2}} = \frac{D_{11}}{D_{12}} \tag{10.10}$$

Now if t_s is the time to saturation we can estimate

$$(Dt_s/D_{11})^{1/2} \sim 3R \tag{10.11}$$

whence

$$D_{12}/D_{11} \sim 100(t_0/t_s)^{1/2} \simeq 20 \tag{10.12}$$

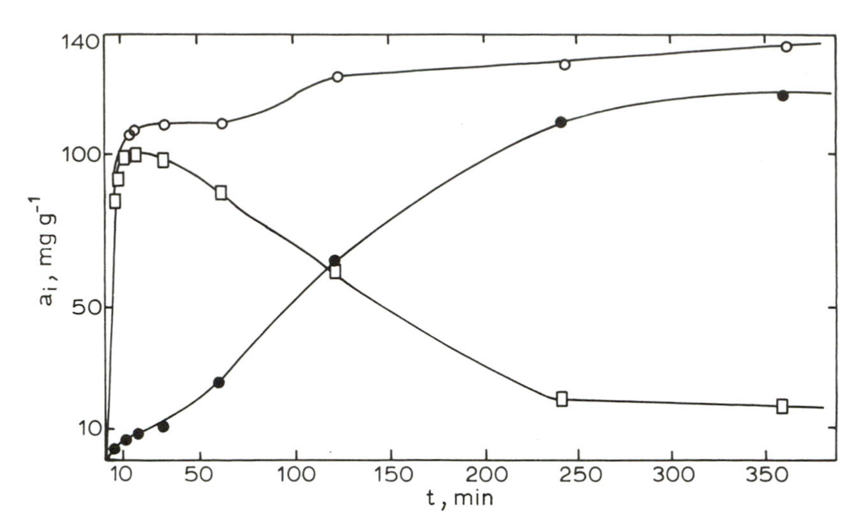

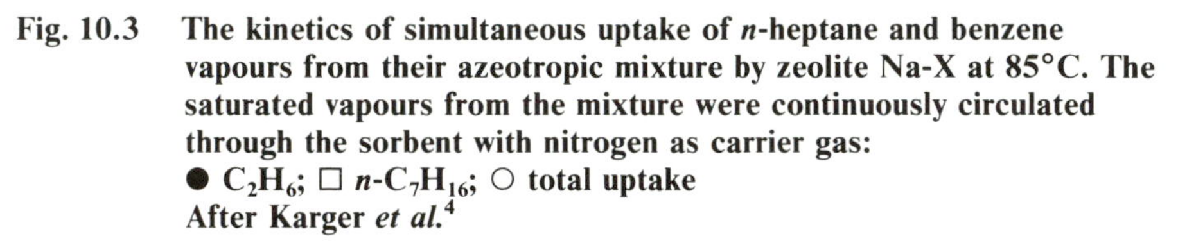

Fig. 10.3 The kinetics of simultaneous uptake of *n*-heptane and benzene vapours from their azeotropic mixture by zeolite Na-X at 85°C. The saturated vapours from the mixture were continuously circulated through the sorbent with nitrogen as carrier gas:
● C_2H_6; □ *n*-C_7H_{16}; ○ total uptake
After Karger *et al.*[4]

which is very large indeed. Clearly, this could not be accommodated by Wagner-type dilute solution thermodynamics (§ §6.3 and 7.9) where

$$\frac{D_{12}}{D_{11}} = \varepsilon_{12}X_1 \simeq \frac{X_1}{X_2}\frac{D_{21}}{D_{22}} \tag{10.13}$$

Furthermore, from equation (10.13) it can be seen that with such *D*-ratios the radical in the above relations becomes imaginary for feasible concentrations. Indeed, it would appear that, at the high chosen gas concentrations which lead to apparent interim supersaturations for n-heptane of 700%, an n-heptane-rich phase may be forming a chemical association or precipitating in analogy to internal oxidation or Liesegang precipitation, then redissolving as the initial gradients moderate. Therefore it may be supposed that the experiments should be repeated at gas concentrations of less than 10% of those used previously, if a simple, one-phase diffusion analysis is to be mounted and verified.

10.4 MEMBRANES

A membrane may be regarded as a thin layer of solution in contact at each side with bulk solutions or reservoirs (refer to §2.17.3). A diagrammatic representation of a planar membrane is shown in Fig. 10.5. In a diffusion experiment the compositions of the two solutions, which the membrane separates, differ. Furthermore the composition within the membrane will vary across its thickness whereas the compositions of the reservoirs may usually be regarded as homogeneous. The transport of material from one reservoir to the other involves dissolution into the membrane, diffusion across its width, and exsolution at the far side. Since the experimentalist has access to the reservoirs but not to the interior of the membrane, it is necessary first to relate membrane fluxes to the states of the reservoirs. This is done using a steady state description.

10.4.1 Flux equations for unreactive membranes

In the steady state

$$\operatorname{div} \boldsymbol{J}_i = 0$$

and, therefore, for the one-dimensional case under consideration

$$J_i = \text{constant} \tag{10.14}$$

Thus a component flux is the same across each membrane surface and has the same value at all points within the interior. Steady state diffusion is achieved by maintaining the reservoir compositions constant.

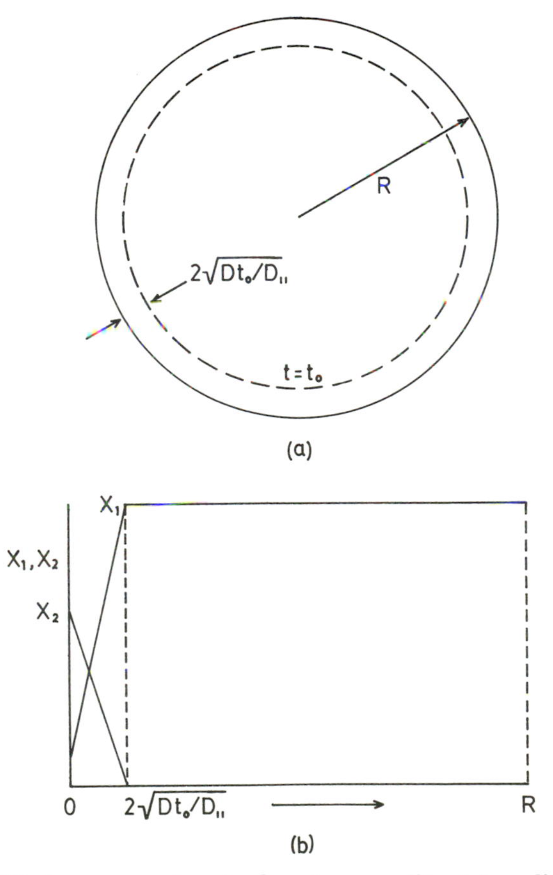

Fig. 10.4 Diffusion parameters in charging a zeolite: (*a*) zeolite sphere dimensions; (*b*) concentration relations at $t \simeq t_0$, the time to the *n*-heptane maximum in Fig. 10.3

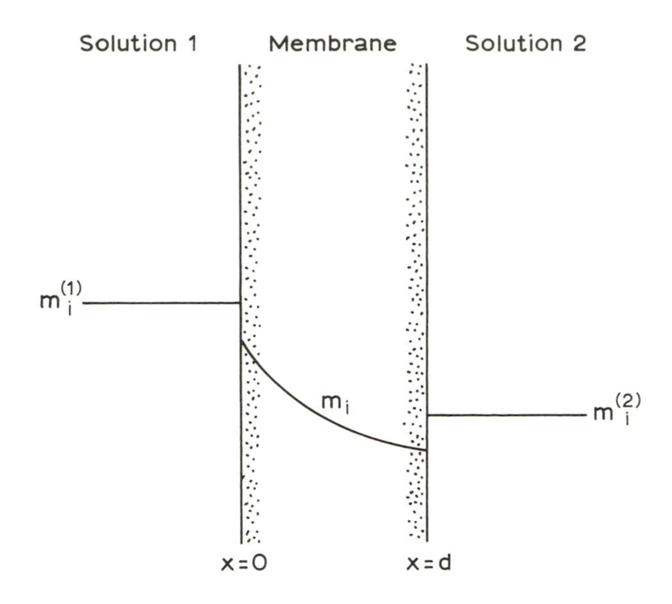

Fig. 10.5 Cross-sectional view of a membrane showing equilibration of solutes between reservoirs and membrane

For isothermal diffusion, the dissipation rate within a volume element of the membrane is given by

$$T\sigma = -\sum_{i}^{n} J_i \nabla\mu_i \tag{10.15}$$

Since the fluxes are constant, this expression may be integrated across the membrane thickness to yield

$$T\int_0^d \sigma \, \mathrm{d}x = \Phi = \sum_{i}^{n} J_i(\mu_i^0 - \mu_i^d)$$

$$= \sum_{i}^{n} J_i \Delta\mu_i \tag{10.16}$$

where Φ is the dissipation function per unit area of membrane. Katchalsky and Curran[5] have applied this result to membrane diffusion and showed how it can be transformed to an equivalent expression in terms of hydrostatic and osmotic pressure changes. For a binary system

$$\Phi = J_v \Delta p + J_D \Delta\pi \tag{10.17}$$

where J_v is the volume flow, J_D is the flow of solute relative to the solvent, Δp is the hydrostatic pressure, and π is the osmotic pressure. The validity of the Onsager reciprocal relations in this description has been questioned[6, 7] and certainly their range of applicability is limited to the linear regime of thermodynamics. Within this regime however, the formulation appears to be quite accurately in accord with experiment.[8]

It is frequently possible to ignore hydrostatic pressure effects and consequently phrase the flux equations in the usual diffusion formalism. From equation (10.16) we write

$$J_i = -\sum_j L_{ij}\Delta\mu_j/d \tag{10.18}$$

where we have chosen to make explicit the effect of membrane thickness. For a membrane containing a single uncharged solute which is ideal or Henrian, equation (10.18) may be rewritten, by expanding the chemical potential, as

$$J_1 = -D_1\Delta m_1/d \tag{10.19}$$

where

$$D_1 = L_{11}RT/m_1 \tag{10.20}$$

The concentration difference appearing in equation (10.19) refers to the membrane interior and does not have the same value as the concentration difference between the reservoirs, $m_1^{(2)} - m_1^{(1)}$.

Solute is exchanged between a reservoir and the membrane across the interphase boundary. If the rates of forward and backward solute movement are large compared to the difference between them, then the process may be described approximately as being at equilibrium. The partition, or distribution, of solute between reservoir and membrane is therefore given by

$$\beta_i^{(1)} = \frac{m_i(0)}{m_i^{(1)}} \bigg/ \beta_i^{(2)} = \frac{m_i(d)}{m_i^{(2)}} \tag{10.21}$$

In general β_i is expected to be a function of concentration, but for small concentration differences we adopt the approximation

$$\beta_i^{(1)} \approx \beta_i^{(2)}$$

whereupon equation (10.19) becomes

$$J_1 = -P_1(m_1^{(2)} - m_1^{(1)}) \tag{10.22}$$

where the permeability P_1 is given by

$$P_1 = D_1\beta_1/d \tag{10.23}$$

Thus membrane permeability depends on the diffusion coefficient, the distribution coefficient and membrane thickness.

Electrolytes diffusing in membranes may be treated in the same way as non-electrolytes by expressing fluxes in terms of neutral salts. Thus a solution of two salts in water is described for the common anion case by

$$C_{\nu_{1c}}A_{\nu_{1a}} = \nu_{1c}C^{z_1} + \nu_{1a}A^{z_3} \tag{10.24}$$

$$D_{\nu_{2c}}A_{\nu_{2a}} = \nu_{2c}D^{z_2} + \nu_{2a}A^{z_3} \tag{10.25}$$

When such a solution diffuses through an unreactive membrane, the entropy source term is given by

$$T\sigma = -\sum_i^4 J_i \cdot \nabla\eta_i \tag{10.26}$$

the four diffusing species being cations C^{z_1} and D^{z_2}, the anion A^{z_3} and water. This description is simplified by adopting the solvent-fixed reference frame

$$J_4 = 0 \tag{10.27}$$

and by applying the zero net current constraint

$$\sum_i^3 z_i J_i = 0 \tag{10.28}$$

Combination of equations (10.24), (10.25), and (10.26), together with the identities

$$z_1 \nu_{1c} + z_3 \nu_{1a} = 0 = z_2 \nu_{2c} + z_3 \nu_{2a} \tag{10.29}$$

yields

$$T\sigma = \frac{J_1}{\nu_{1c}}(\nu_{1c}\nabla\eta_1 + \nu_{1a}\nabla.\eta_3) + \frac{J_2}{\nu_{2c}}(\nu_{2c}\nabla.\eta_2 + \nu_{2a}\nabla\eta_3) \tag{10.30}$$

By writing the Gibbs equations for the isothermal equilibria (10.24) and (10.25) it is seen that

$$\nabla\mu_{1s} = \nu_{1c}\nabla\eta_1 + \nu_{1a}\nabla\eta_3 \tag{10.31}$$

$$\nabla\mu_{2s} = \nu_{2c}\nabla\eta_2 + \nu_{2a}\nabla\eta_3 \tag{10.32}$$

where the subscripts 1s and 2s denote each of the two salts. Noting that through stoichiometry

$$J_{1s} = \frac{J_1}{\nu_{1c}} \qquad J_{2s} = \frac{J_2}{\nu_{2c}} \tag{10.33}$$

it follows from equations (10.30), (10.31), and (10.32) that

$$T\sigma = J_{1s}\nabla\mu_{1s} + J_{2s}\nabla\mu_{2s} \tag{10.34}$$

Application of the steady state approximation allows integration of equation (10.34) across the membrane thickness yielding

$$\Phi = J_{1s}\Delta\mu_{1s} + J_{2s}\Delta\mu_{2s} \tag{10.35}$$

whence

$$J_{1s} = -L_{11}\Delta\mu_{1s} - L_{12}\Delta\mu_{2s} \tag{10.36}$$

$$J_{2s} = -L_{21}\Delta\mu_{1s} - L_{22}\Delta\mu_{2s} \tag{10.37}$$

The phenomenological coefficients L_{ij} may be calculated in terms of individual ionic l_{ij} according to the methods of §§9.4 and 9.5. A recent example of the application of Miller's method (equations (9.45)–(9.48)) for calculating these coefficients has been described by Konturri *et al.*[9] for electrolyte diffusion through an inert membrane. Diffusion coefficients in membranes are lower than in free solution because of the tortuous diffusion paths. Thus, for example, the self-diffusion coefficient of Cl^- in a cation-exchange resin membrane was measured by Mackay and Meares[10] as 7×10^{-6} cm^2 s^{-1}, compared to the value of $1{\cdot}5 \times 10^{-5}$ cm^2 s^{-1} found in aqueous solution at room temperature.

Transformation of the phenomenological equations (10.36) and (10.37) into generalized permeability equations yields

$$J_i = - \Sigma P_{ij}\{m_j^{(2)} - m_j^{(1)}\} \qquad (10.38)$$

where

$$P_{ij} = D_{ij}\beta_j/d \qquad (10.39)$$

Obviously cross-effects can arise through selective dissolution within the membrane. A particularly transparent case is provided by the selective dissolution of a common ion. Consider the permeation of the salt CA both alone and in the presence of a second salt DA, through a membrane in which the solubilities of cation C and anion A are high but the solubility of cation D is low. The effect of adding DA to the reservoir which acts as source to the membrane, is

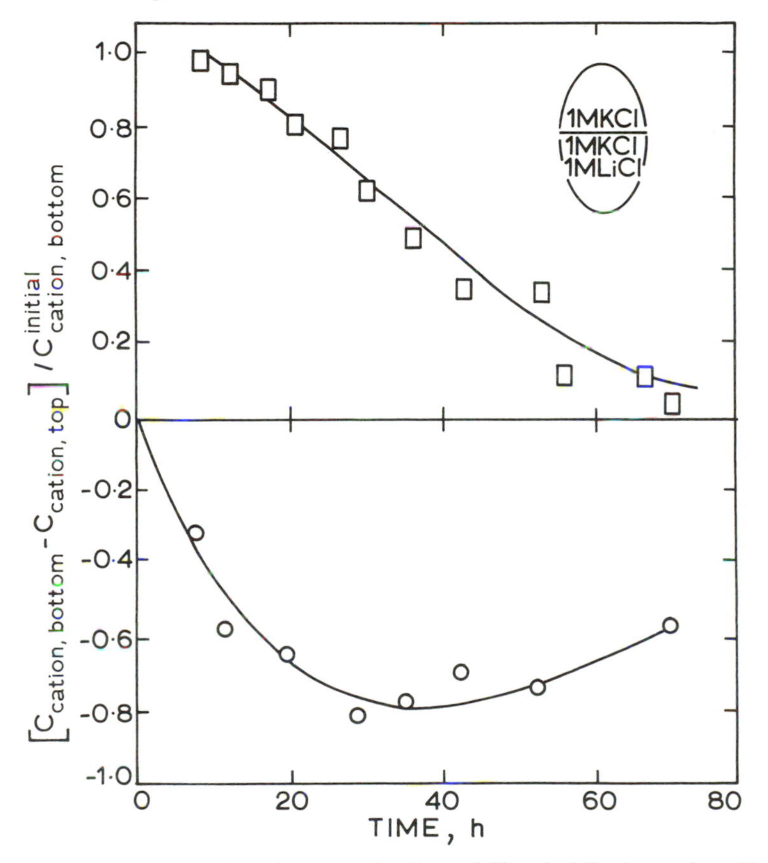

Reproduced by permission of the American Institute of Chemical Engineers from F. Caracciola, E. Cussler, and D. F. Evans, *J.A.I.Ch.E.*, 1975, **21**, 163

Fig. 10.6 Uphill diffusion of potassium cations resulting from common ion effect. Solubility of K^+ at the bottom side of the membrane is enhanced as a result of the higher Cl^- concentration. After Carriocola *et al.*[11]

to increase the membrane anion content. This in turn enhances the solubility of cation C and hence the permeability of the membrane to the salt CA. The effect is not a diffusive cross-effect; it is due simply to an enhancement of the concentration of diffusing species within the membrane. An experimental observation of this effect due to Caracciolo *et al.*[11] is shown in Fig. 10.6.

10.4.2 Electrostatic effects in unreactive membranes

In many situations it is desirable to express membrane permeation rates in terms of the electrostatic potential difference across the membrane (see Appendix 1, Donnan equilibrium). This requires integration of the flux equation, which contains the potential gradient. Since this was first done by Planck[12, 13] several alternate approaches embodying different simplifying approximations have followed. Many have been reviewed by Sten-Knudsen[14] and Schultz.[15] All involve the often unjustified approximation that the transport of each ion is governed only by its own electrochemical potential gradient, i.e. cross-coefficients in the L-matrix are zero. A similar problem arises with thin solid films and its treatment has been reviewed by Fromhold.[16] In this case the flux equation for ion i is given by the Nernst–Planck equation rewritten here as

$$J_i = -D_i\left\{\frac{\mathrm{d}m_i}{\mathrm{d}x} + \frac{m_i z_i}{RT}F\frac{\mathrm{d}\psi}{\mathrm{d}x}\right\} \tag{10.40}$$

Multiplying this equation by $\exp(z_i F\psi/RT)$ leads, after rearrangement, to

$$J_i \exp\left(\frac{z_i F\psi}{RT}\right) = -D_i\left\{\frac{\mathrm{d}}{\mathrm{d}x}\left[m_i \exp\left(\frac{z_i F\psi}{RT}\right)\right]\right\} \tag{10.41}$$

If D_i is independent of position, the above equation may be integrated across the membrane thickness under steady state conditions to yield

$$J_i \int_0^d \exp\left(\frac{z_i F\psi}{RT}\right)\mathrm{d}x = -D_i\left[m_i^{(\mathrm{d})}\exp\left(\frac{z_i F\Delta\psi}{RT}\right) - m_i^{(0)}\right] \tag{10.42}$$

The necessary integration requires a knowledge of the functional dependence

$$\psi = \psi(x)$$

The most common assumption is that because the concentration of locally uncompensated charge is low, then through Poisson's equation the electric field $(-\mathrm{d}\psi/\mathrm{d}x)$ is constant:

$$-\frac{\mathrm{d}^2\psi}{\mathrm{d}x^2} = \frac{\mathrm{d}E}{\mathrm{d}x} = \frac{F}{\varepsilon}\sum_i z_i m_i \approx 0 \tag{10.43}$$

where ε is the electrostatic permittivity. In this case we have

$$\psi = \frac{x}{d}\Delta\psi \tag{10.44}$$

if the zero of potential is defined to be at one side of the membrane. This approximation was first applied to membrane diffusion by Goldman[17] and had been used earlier in treating transport through thin solid films. Substitution from equation (10.44) into (10.42) allows integration which yields

$$J_i = -\frac{D_i z_i F\Delta\psi}{RTd}\left[\frac{m_i^{(d)}\exp(z_iF\Delta\psi/RT) - m_i^{(0)}}{\exp(z_iF\Delta\psi/RT) - 1}\right] \tag{10.45}$$

The concentrations appearing in the above expression are boundary values of membrane concentrations and the potential difference $\Delta\psi$ is the difference appearing across the membrane. These quantities are not measurable. Hodgkin and Katz[18] suggested that the distribution coefficients defined by equation (10.21) be assumed independent of concentration, in which case the liquid-junction potential differences developed across the two membrane–reservoir interfaces are equal and opposite. In this case $\Delta\psi$ is equal to the emf developed between the two reservoirs, $\Delta\psi'$. Equation (10.45) may be written for this case as

$$J_i = -\frac{P_i z_i F\Delta\psi'}{RT}\left[\frac{m_i^{(1)} - m_i^{(2)}\exp(z_iF\Delta\psi'/RT)}{1 - \exp(z_iF\Delta\psi'/RT)}\right] \tag{10.46}$$

where P_i is defined by equation (10.23). This equation is commonly referred to in the membrane literature as the Goldman–Hodgkin–Katz equation. These authors extended the description to multicomponent electrolyte solutions and, on the assumption that cross-effects are zero, showed that equation (10.46) applies to each participating ion when the electrostatic field is constant (equation (10.43)). Clearly the two assumptions are self-consistent as they both require dilute solutions.

In an isothermal diffusion experiment the potential difference $\Delta\psi'$ appearing in equation (10.46) is established by the separation of the diffusing ions themselves. It is therefore calculable by the methods of §§6.5 and 9.4.

One of the centrally important assumptions used in arriving at the Goldman–Hodgkin–Katz equation and other integrations of the Nernst–Planck equation was that the membrane plays no role other than to define geometrically a region through which diffusion occurs. Since in many cases of interest this is not so, it is desirable to develop an experimental method of determining whether membrane permeation is due to simple diffusion. Such a test is provided by the ratio between the tracer fluxes of a species across a membrane in each of the two directions, $J^{1\rightarrow2}$ and $J^{2\rightarrow1}$. This ratio was first calculated for the simple diffusion case by Behn[19] and applied by Ussing.[20]

Tracer j of solute i has its reservoir concentrations held at $m_j^{(1)} \neq 0$ and $m_j^{(2)} = 0$. Then its rate of diffusion is given by equation (10.42) as

$$J_j^{1\rightarrow2}\int_0^d \exp(z_iF\psi/RT)\,\mathrm{d}x = D_i m_j^{(0)} \tag{10.47}$$

on the assumption that $D_j = D_i$. If a second tracer k of solute i has the reservoir concentrations held at $m_k^{(1)} = 0$ and $m_k^{(2)} \neq 0$, its rate of diffusion from (2) to (1) is given by

$$J_k^{2\rightarrow1}\int_0^d \exp(z_iF\psi/RT)\,\mathrm{d}x = D_i m_k^{(d)}\exp(z_iF\Delta\psi/RT) \tag{10.48}$$

assuming $D_k = D_i$. Elimination of the integral between the two equations leads to the ratio

$$\frac{J_j^{1\to2}}{J_k^{2\to1}} = \frac{m_j^{(0)}}{m_k^{(d)}} \exp(z_i F \Delta\psi/RT) \tag{10.49}$$

Incorporation of the liquid junction potentials

$$m_j^{(0)} = m_j^{(1)} \exp[-z_i F(\psi^{(1)} - \psi^{(0)})/RT]$$

$$m_k^{(d)} = m_k^{(2)} \exp[-z_i F(\psi^{(2)} - \psi^{(d)})/RT] \tag{10.50}$$

leads to

$$\frac{J_j^{1\to2}}{J_k^{2\to1}} = \frac{m_j^{(1)}}{m_j^{(2)}} \exp(z_i F \Delta\psi'/RT) \tag{10.51}$$

For ideal tracers

$$\frac{J_j^{1\to2}}{m_j^{(1)}} = \frac{J_i^{1\to2}}{m_i^{(1)}} \qquad \frac{J_k^{2\to1}}{m_k^{(2)}} = \frac{J_i^{2\to1}}{m_i^{(2)}} \tag{10.52}$$

and therefore,

$$\frac{J_i^{1\to2}}{J_i^{2\to1}} = \frac{m_i^{(1)}}{m_i^{(2)}} \exp[z_i F \Delta\psi'/RT) \tag{10.53}$$

Whilst this treatment avoids the necessity for specifying the potential profile within the membrane it is still dependent on the assumption that cross-effects are negligible and therefore restricted in application to relatively dilute solutions.

If the fluxes of an ion can be described by equation (10.53) over a range of concentrations and electric potential differences, then it may be concluded that transport is due entirely to electrochemical potential differences. The converse is, however, not necessarily true.

Several commonly observed features of membrane diffusion cannot be described with the transport equations derived from integration of the Nernst–Planck equation. These features are

(*a*) fluxes vary linearly with change in electrochemical potential difference but reach a maximum as $\Delta\eta$ increases. This is the 'saturation' effect

(*b*) even though $\Delta\eta$ for a given solute is maintained unchanged, the addition of a second solute can suppress the diffusion of the first. This cross-effect is known as 'competitive inhibition'

(*c*) fluxes can be much larger and much more strongly coupled than expected from liquid diffusion data.

It is clear from reference to equation (10.46) that saturation and competitive inhibition can be described only if P_i, the permeability coefficient, can be regarded as a function of concentration, both of the diffusing species and of the other species present. For this to be so, we must abandon our assumption that membranes serve simply as solvents of defined physical extent, and admit the possibility of specific interaction between solute and membrane species. We consider first interactions at fixed sites within the membrane.

10.4.3 Membranes with fixed reactive sites

Because the reactive sites within the membrane are spatially located, it is possible to define their electrostatic potentials with respect to the membrane surfaces. By making generalized statements as to the location of these sites it becomes possible to apply rate theory to the processes of solute transfer to and from these sites, just as was done for ionic solids (§ 8.9). The application of this analysis to membrane diffusion was first suggested by Eyring *et al.*[21] and developments since then have been reviewed by Hille and Schwarz.[22, 23] Analogous treatments have been applied to thin solid films (for example, Young and Dignam[24]).

Consider a membrane containing a low density of sites capable of binding a diffusing ion. Figure 10.7 shows a profile of the internal energy surface along a diffusion path which traverses the membrane and contains one such site S located a distance λ from one side of the membrane. An ion bound to this site is seen to be metastable with respect to the reservoirs and the membrane is therefore permeable to the ion. The energy barriers shown in Fig. 10.7 are modified by the electrostatic potential difference $\Delta\psi$ which must exist across the membrane when diffusion occurs. Fluxes between adjacent positions in the absence of an electric field may be calculated from elementary rate theory. Thus for example, the forward flux of species i from the left side of the membrane to binding sites is proportional to the species concentration and to the probability of finding an empty site

$$J_i^{0\to s} = k_{0s}\lambda m_i^{(0)}[S]/[S_T] \tag{10.54}$$

where $[S]$ is the concentration per unit membrane area of empty active sites and $[S_T]$ is the total concentration of such sites. The rate constant k_{0s} reflects the magnitude of the activation barrier and a frequency term. Incorporating the electrostatic potential effects,

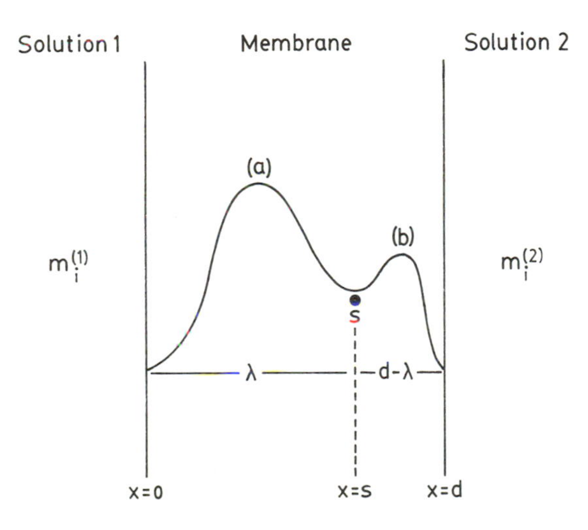

Fig. 10.7 Internal energy profile for membrane diffusion path containing one binding site, *S*. The profile shape is appropriate to zero electrostatic potential difference across the membrane

$$J_i^{0\to s} = \lambda m_i^{(0)} \frac{[S]}{[S_T]} k_{0s} \exp[z_i F(\psi^{(a)} - \psi^{(0)})/RT] = m_i^{(0)} \frac{[S]}{[S_T]} k'_{0s}$$

$$J_i^{s\to 0} = [i|S] k_{s0} \exp[-z_i F(\psi^{(s)} - \psi^{(a)})/RT] = [i|S] k'_{s0} \tag{10.55}$$

$$J_i^{s\to d} = [i|S] k_{sd} \exp[z_i F(\psi^{(b)} - \psi^{(s)})/RT] = [i|S] k'_{sd}$$

$$J_i^{d\to s} = (d - \lambda) m_i^{(d)} \frac{[S]}{[S_T]} k_{ds} \exp[-z_i F(\psi^{(d)} - \psi^{(b)})/RT] = m_i^{(d)} \frac{[S]}{[S_T]} k'_{ds} \tag{10.56}$$

is obtained. Here the concentration of sites with bound ions is denoted by $[i|S]$. Since sites are conserved

$$[S_T] = [S] + [i|S] = \text{constant} \tag{10.57}$$

If the constant field approximation is adopted, the potential at each of the indicated points within the membrane can be evaluated in terms of $\Delta\psi$ by means of equation (10.44).

Under steady state conditions, the net flux of i across the membrane is given by

$$J_i = k'_{0s} m_i^{(0)} \frac{[S]}{[S_T]} - k'_{s0}[i|S] \tag{10.58}$$

$$J_i = k'_{sd}[i|S] - k'_{ds} m_i^{(d)} \frac{[S]}{[S_T]} \tag{10.59}$$

The solution of equations (10.55)–(10.59) for J_i is

$$J_i = [S_T] \left\{ \frac{m_i^{(0)} k'_{0s} k'_{sd} - m_i^{(d)} k'_{ds} k'_{s0}}{[S_T] k'_{sd} + [S_T] k'_{s0} + m_i^{(0)} k'_{0s} + m_i^{(d)} k'_{ds}} \right\} \tag{10.60}$$

When one reservoir contains none of the solute, e.g. $m_i^{(d)} = 0$, then

$$J_i = [S_T] \frac{m_i^{(0)} k'_{0s} k'_{sd}}{[S_T] k'_{sd} + [S_T] k'_{s0} + m_i^{(0)} k'_{0s}} \tag{10.61}$$

Obviously J_i saturates with increasing $m_i^{(0)}$ if $\Delta\psi$ is constant. The saturation in the flux is due to the fact that $[i|S]$ saturates at a finite level. It is also clear that the presence of another species with an affinity for S will lead to competitive interactions which will alter the observed value of J_i. It is evident that such an effect can formally be described as a cross-effect.

Notwithstanding the fact that this system deviates strongly from the unreactive membrane case, it can be demonstrated by substituting equations (10.55), (10.56), and (10.61) into (10.47) and (10.48) that the flux-ratio equation (10.53) is nonetheless obeyed.

10.4.4 Membranes with mobile reactive centres

In a number of cases, diffusion across membranes is found to proceed very much faster than would be predicted from simple diffusion theory. In addition, strong cross-effects are found to be capable of inducing 'uphill' diffusion, i.e. transport of a species in a direction contrary to that indicated by its own potential gradient. In order to explain these effects, the existence of mobile carriers within the membrane has been postulated. These carriers are reactive species which form a bound complex with the diffusing species at one membrane surface, diffuse across the membrane 'carrying' with them the solute, and then release the solute to the adjacent reservoir at the far surface. The process is described as 'facilitated' or 'carrier-mediated' transport. Although no carrier species appears to have been unequivocally identified, let alone isolated for study, in biological membranes a number of chemically well-defined carrier-containing membranes have now been produced.[7] Many of the suggested mechanisms involving mobile carriers have been reviewed by LeFevre[25] who summarizes a great deal of the experimental evidence.

One particular model of carrier–solute interaction known as 'active transport' is due to Kedem and Caplan[26, 27] and is based on the postulate that coupling exists between diffusion fluxes and chemical reaction flows. Such a postulate is inconsistent with irreversible thermodynamics[28] because, according to the Curie theorem, coupling is possible only between fluxes having tensorial characters which are identical or differ by a multiple of two. Coupling between a scalar reaction flow and vector diffusive flow obviously falls outside this definition and will not be considered further here. Such phenomena are to be understood in terms of diffusion with complexing (§ 8.10 and equation (8.140)).

We turn now to a consideration of the kinetics of carrier–solute complex formation and transport. A description of binary (single solute) carrier-mediated transport is first developed and then extended to two ternary cases of particular interest, countertransport and cotransport.

10.4.5 Binary carrier-mediated transport

A general kinetic scheme is shown diagrammatically in Fig. 10.8 where it is seen that in addition to dissolution and exsolution of the diffusing species, reaction between that species and the mobile carriers must be considered. A general solution is available under the steady state description of chemical kinetics (see Schultz *et al.*[29]) but for present purposes it is sufficient to adopt the equilibrium approximation for the reaction

$$\text{solute}(i) + \text{carriers}(\mathrm{S}) \rightleftarrows \text{complex}(i\,|\,\mathrm{S})$$

expressed in volume concentrations as

$$m_{i|\mathrm{S}} = K m_i m_\mathrm{S} \tag{10.62}$$

The concentrations $m_i^{(0)}$ and $m_i^{(d)}$ are related to reservoir concentrations by equation (10.21). Thus it is being assumed that diffusion in the membrane is very much slower than the rates of the complex formation and dissociation reactions. Diffusion of an uncharged solute is now considered.

Under steady state condition the net flux of complex is given by

$$J_{i|\mathrm{S}} = D_\mathrm{c}(m_{i|\mathrm{S}}^{(0)} - m_{i|\mathrm{S}}^{(d)})/d = D_\mathrm{S}(m_\mathrm{S}^{(d)} - m_\mathrm{S}^{(0)})/d \tag{10.63}$$

where D_c and D_S are, respectively, the diffusion coefficients of the bound complex

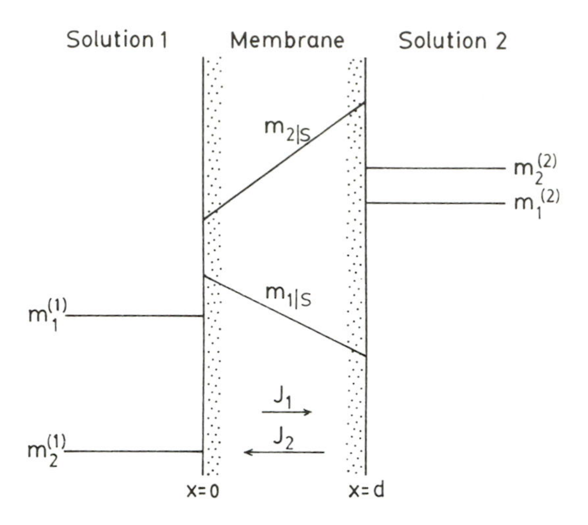

Fig. 10.8 Idealized concentration profiles for carrier-solute complexes, 1 | S and 2 | S, during steady state countertransport under diffusion control

and of the uncomplexed carrier. Carriers are presumed to exist only within the membrane and their total concentration is therefore constant

$$[S_T] = \int_0^d (m_{i|S} + m_S)\,\mathrm{d}x = \text{constant} \qquad (10.64)$$

In the steady state, the gradients of $m_{i|S}$ and m_S are constant and equation (10.64) integrates to yield

$$[S_T] = \mathrm{d}\{m_{i|S}^{(0)} + m_{i|S}^{(d)} + m_S^{(0)} + m_S^{(d)}\}/2 \qquad (10.65)$$

Equations (10.63) and (10.65), together with equation (10.62) evaluated at each of the interfaces, may then be solved to yield

$$J_{i|S} = \frac{2[S_T]}{d^2}\frac{D_S D_c K(m_i^{(0)} - m_i^{(d)})}{(1 + Km_i^{(0)})(D_S + D_c K m_i^{(d)}) + (1 + Km_i^{(d)})(D_S + D_c K m_i^{(0)})} \qquad (10.66)$$

The functional form of this relationship is investigated by setting the solute concentration in one reservoir at zero, say $m_i^{(2)} = 0$, then the unidirectional flux becomes

$$J_{i|S}^{1\to 2} = \frac{2[S_T]}{d^2}\frac{D_S D_c K m_i^{(0)}}{D_S(1 + Km_i^{(0)}) + (D_S + D_c K m_i^{(0)})} \qquad (10.67)$$

It is obvious that the flux saturates as $m_i^{(0)}$ is increased, because the membrane becomes saturated with bound complexes. The total flux of solute i is given by the sum of equations (10.22) and (10.67). A particularly well-known example of this behaviour is the facilitation of oxygen diffusion by haemoglobin, data for which is shown in Fig. 10.9. Diffusive fluxes of oxygen are compared for membranes containing haemoglobin, which complexes oxygen, and methemoglobin which does not. The difference between the two fluxes, representing the carrier-mediated flux, is seen to increase rapidly and then saturate, as described in equation (10.67). The presence of a back pressure of oxygen at the outlet side of the membrane exercises a strong effect on the carrier-mediated portion of the flux as shown in Fig. 10.9(*b*). Presumably the carrier becomes saturated throughout the membrane and can no longer support a significant gradient in complex concentration.

It is clear from equation (10.67) that carrier-mediated transport is strongly dependent on the value of the equilibrium constant describing the complex formation process. Thus highly selective membrane permeabilities are to be expected and separation processes are based on this feature. An example of the selective recovery of Cu^{2+} from a Cu^{2+}/Ni^{2+} solution is shown in Fig. 10.10 (Lee *et al.*[32]). The steady state fluxes of the cations through a supported liquid membrane of oximes are seen to strongly favour removal of Cu^{2+} from the parent solution.

The above analysis describes the limiting case in which diffusion across the membrane is the rate-limiting process, the interfaces of the membrane being essentially at equilibrium with the adjoining reservoirs. Another special case which can be simply described[34,35] is the fast diffusion case in which the membrane interfacial mass transfer processes are rate-controlling. The overall process is, in this case, not controlled by diffusion and will not be discussed here.

10.4.6 Countertransport

Countertransport is so called because one solute is transported against its concentration difference by a flux of a second solute moving in the opposite direction. As will be seen, the phenomenon is simply a consequence of the two solutes differing in their reactivity towards complex formation. The process is shown diagrammatically in Fig. 10.8 where it has been assumed that diffusion is very much slower than complex formation and dissociation which are assumed to be at equilibrium. It is seen from the figure that no diffusional interaction need be postulated to explain countertransport, which is simply an artifact produced by the competitive equilibria

$$\begin{aligned} \text{solute 1 (1)} + \text{carrier (S)} &= \text{complex } (1\,|\,\text{S}) \\ \text{solute 2 (2)} + \text{carrier (S)} &= \text{complex } (2\,|\,\text{S}) \end{aligned}$$

Thus the large concentration difference in component 2 between the reservoirs ensures a large concentration gradient in the complex (2 | S) across the membrane. This in turn leads to an inverse gradient in complex (1 | S). The size of the effect is determined by the relative magnitude of the equilibrium constants

$$K_1 = \frac{m_{1|S}^{(0)}}{m_1^{(0)} m_S^{(0)}} = \frac{m_{1|S}^{(d)}}{m_1^{(d)} m_S^{(d)}} \tag{10.68}$$

$$K_2 = \frac{m_{2|S}^{(0)}}{m_2^{(0)} m_S^{(0)}} = \frac{m_{2|S}^{(d)}}{m_2^{(d)} m_S^{(d)}} \tag{10.69}$$

where ideality has been assumed. Obviously the flux of component 1 opposes that of component 2.

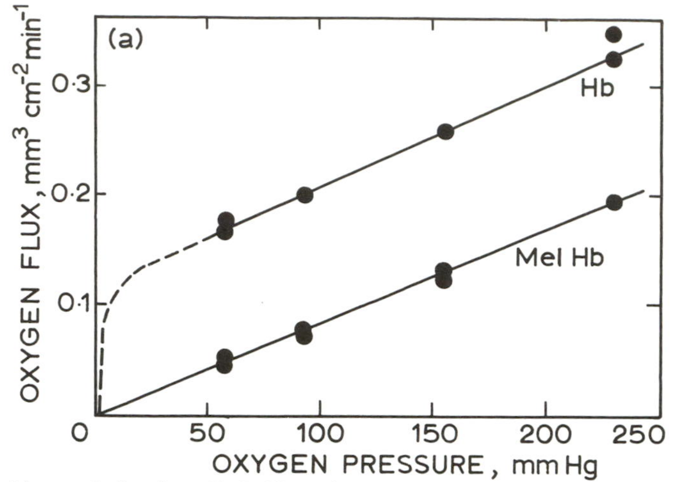

Reprinted with permission from E. A. Hemmingsen, *Science*, 1960, **132**, 1379. Copyright 1960 by the AAAS

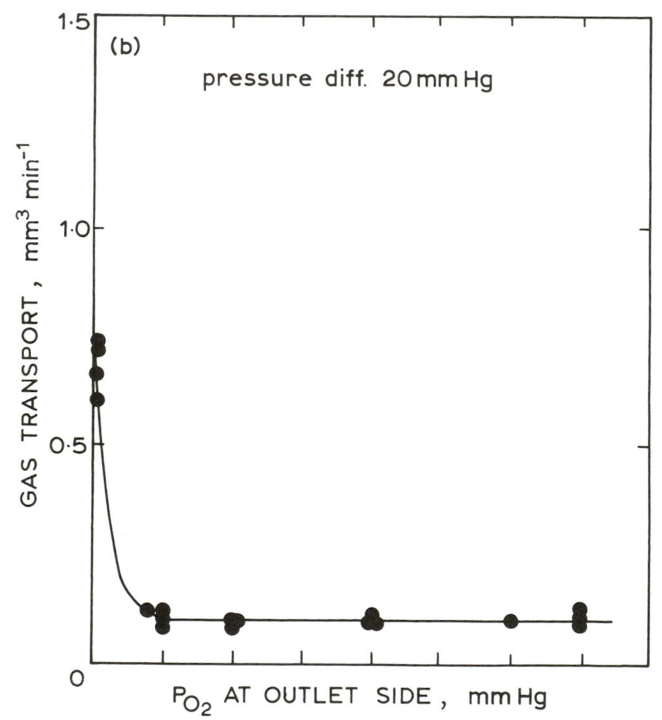

Reprinted with permission from E. A. Hemmingsen and P. F. Scholander, *Science*, 1962, **135**, 733. Copyright 1962 by the AAAS

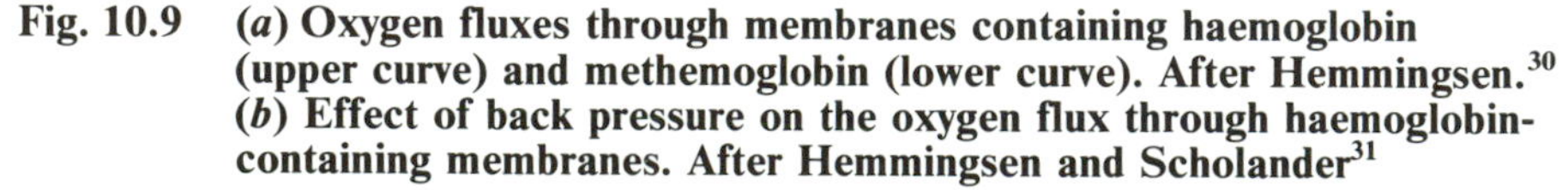

Fig. 10.9 **(*a*) Oxygen fluxes through membranes containing haemoglobin (upper curve) and methemoglobin (lower curve). After Hemmingsen.[30] (*b*) Effect of back pressure on the oxygen flux through haemoglobin-containing membranes. After Hemmingsen and Scholander[31]**

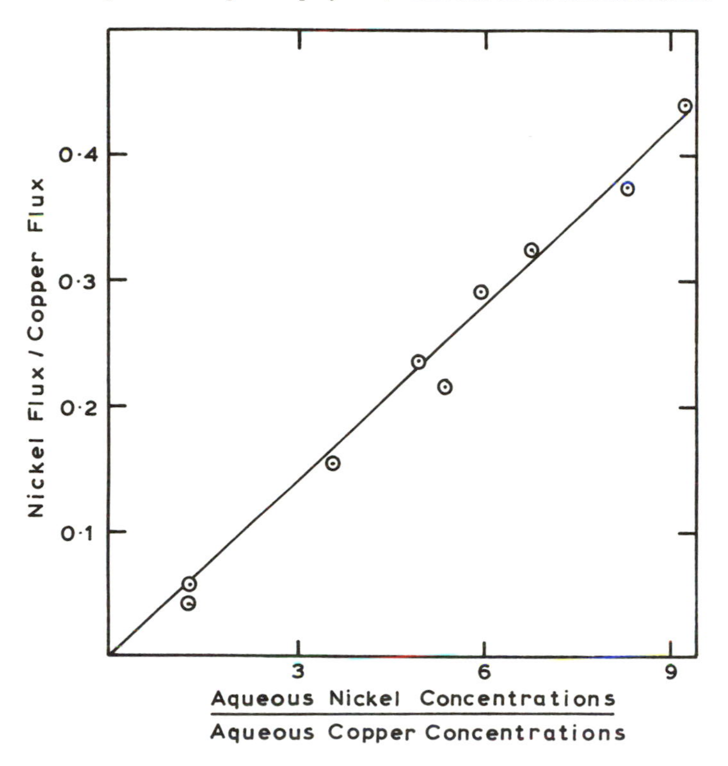

Fig. 10.10 Selectivity of supported liquid membranes for copper over nickel

The analysis proceeds in the same way as for binary carrier-mediated transport, that is to say diffusional cross-effects are ignored and the steady state description is employed. Since carriers are presumed to exist only within the membrane, in analogy with equation (10.65)

$$[S_T] = d\{m^{(0)}_{1|S} + m^{(d)}_{1|S} + m^{(0)}_{2|S} + m^{(d)}_{2|S} + m^{(0)}_S + m^{(d)}_S\}/2 \quad (10.70)$$

The flux equations are, for ideal or Henrian behaviour

$$\begin{aligned} J_{1|S} &= D_{1|S}(m^{(0)}_{1|S} - m^{(d)}_{1|S})/d \\ J_{2|S} &= D_{2|S}(m^{(0)}_{2|S} - m^{(d)}_{2|S})/d \\ J_S &= D_S(m^{(0)}_S - m^{(d)}_S)/d \\ \Sigma J &= 0 \end{aligned} \quad (10.71)$$

and diffusion of unbound solutes 1 and 2 is ignored. Solution of the equations (10.68)–(10.71), for the case $D_{1|S} = D_{2|S} = D_S = D$ leads to the simple result

$$J_1 = \frac{DK_1[S_T]}{d^2}\left\{\frac{m_1^{(0)}}{1 + K_1m_1^{(0)} + K_2m_2^{(0)}} - \frac{m_1^{(d)}}{1 + K_1m_1^{(d)} + K_2m_2^{(d)}}\right\} \quad (10.72)$$

where the boundary values $m_i^{(0)}$, $m_i^{(d)}$ are given by equation (10.21).

In countertransport,

$$m_1^{(d)} > m_1^{(0)} \tag{10.73}$$

because

$$m_1^{(2)} > m_1^{(1)}$$

According to equation (10.72), a positive flux of component 1 will nonetheless be generated provided that

$$\frac{m_1^{(0)}}{m_1^{(d)}} > \frac{1 + K_2 m_2^{(0)}}{1 + K_2 m_2^{(d)}} \tag{10.74}$$

This requirement is consistent with the countertransport definition (10.73) if

$$m_2^{(d)} > m_2^{(0)}$$

It is by now clear that although the chemical potential difference in component 1 between the two reservoirs is opposed to transport, a favourable gradient of the complex 1 | S develops within the membrane as a result of competitive interaction between the two equilibrium processes. The carrier transports one solute in one direction across the membrane to the interface where it is freed. It is then preferentially bound to the outer solute which it then transports back across the membrane in the opposite direction.

Equation (10.72) may be reformulated with the aid of equation (10.21) as

$$J_1 = \frac{-DK_1[S_T]}{d^2}\left\{\frac{\beta_1(1 + K_2\beta_2\bar{m}_2)\Delta m_1 - K_2\beta_1\beta_2\bar{m}_1\Delta m_2}{(1 + K_1\beta_1 m_1^{(1)} + K_2\beta_2 m_2^{(2)})(1 + K_1\beta_1 m_1^{(2)} + K_2\beta_2 m_2^{(2)})}\right\} \tag{10.75}$$

where $\bar{m}_1$, $\bar{m}_2$ are the average reservoir solute concentrations

$$\bar{m}_1 = (m_1^{(1)} + m_1^{(2)})/2$$

$$\bar{m}_2 = (m_2^{(1)} + m_2^{(2)})/2$$

As it is possible to vary the reservoir concentration differences Δm_i independently of the averages, this procedure is legitimate. This result is seen to possess the form of a ternary membrane diffusion equation

$$J_1 = -P_{11}\Delta m_1 - P_{12}\Delta m_2 \tag{10.76}$$

where

$$P_{11} = \frac{P_1[S_T]}{d}\,\frac{(1 + K_2\beta_2\bar{m}_2)}{(1 + K_1\beta_1 m_1^{(1)} + K_2\beta_2 m_2^{(1)})(1 + K_1\beta_1 m_1^{(2)} + K_2\beta_2 m_2^{(2)})}$$

$$P_{12} = \frac{-P_1[S_T]}{d}\,\frac{K_2\beta_2\bar{m}_1}{(1 + K_1\beta_1 m_1^{(1)} + K_2\beta_2 m_2^{(1)})(1 + K_1\beta_1 m_1^{(2)} + K_2\beta_2 m_2^{(2)})}$$

and similarly for component 2. The apparent uphill diffusion represented by the negative coefficient P_{12} is seen to result from the competitive equilibria represented by equations (10.68) and (10.69). Diffusion within the membrane is in fact described by simple binary expressions (10.71) containing no cross-terms and is, of course, a downhill process.

The countertransport phenomenon is well illustrated by its use in concentrating metals in solution. This application has been reviewed in some detail by Danesi.[33] Countercurrent transport of metal and hydrogen ions through a supported liquid membrane as illustrated in Fig. 10.11.[36] The metals diffuse from the high pH to the low pH side of the membrane, against their own concentration gradients. Because one metal is much more strongly complexed than the other, it is preferentially extracted from the feed solution, as shown in Fig. 10.12.

10.4.7 Cotransport

In cotransport, two solutes are transported together in the same direction. Because transport involves a ternary complex

$$\text{solute 1 (1)} + \text{solute 2 (2)} + \text{carrier (S)} = \text{complex } (1|2|S) \tag{10.77}$$

one of the components can be transported against its own chemical potential difference by virtue of a large driving force provided for the other. The process is shown diagrammatically in Fig. 10.13. A fairly general treatment has been given by Schultz and Curran[37] but we consider here only the simple case where the ternary complex and the uncomplexed carrier are the only species mobile within the membrane.

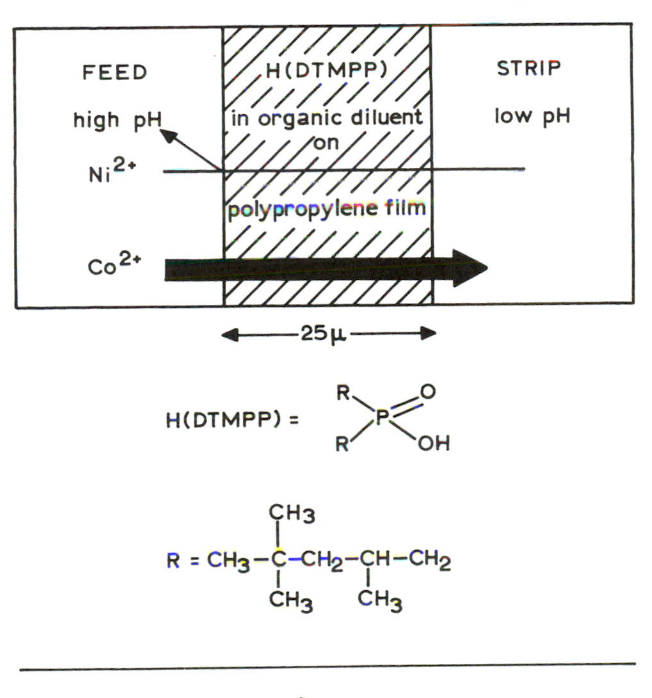

Fig. 10.11 **Schematic of supported liquid membranes used to separate Co^{2+} from Ni^{2+}. Organic diluent is 67% decalin-33% diisopropylbenzene. After Danesi[33]**

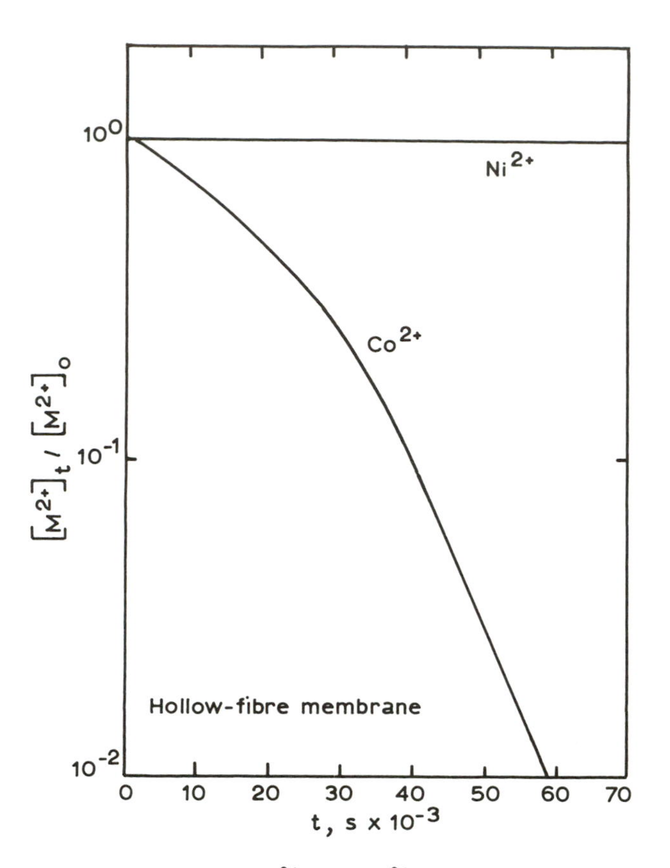

Fig. 10.12 Time dependence of Ni^{2+} and Co^{2+} concentrations in solution treated with membrane shown in Fig. 10.11. Feed solution has initial concentrations $[Co^{2+}]_0 = [Ni^{2+}]_0 = 0.1M$. After Danesi[33]

Using the same methods as for countertransport we write the complex formation equilibrium

$$K = \frac{m_{12S}^{(0)}}{m_1^{(0)} m_2^{(0)} m_S^{(0)}} = \frac{m_{12S}^{(d)}}{m_1^{(d)} m_2^{(d)} m_S^{(d)}} \tag{10.78}$$

Carriers are assumed to exist only within the membrane

$$[S_T] = d\{m_{12S}^{(0)} + m_{12S}^{(d)} + m_S^{(0)} + m_S^{(d)}\}/2 \tag{10.79}$$

and the steady state flux equations are

$$\begin{aligned} J_1 &= J_2 = J_{12S} \\ &= D_{12S}\{m_{12S}^{(0)} - m_{12S}^{(d)}\}/d \end{aligned} \tag{10.80}$$

$$J_S = D_S\{m_S^{(0)} - m_S^{(d)}\}/d \tag{10.81}$$

$$J_S = -J_{12S} \tag{10.82}$$

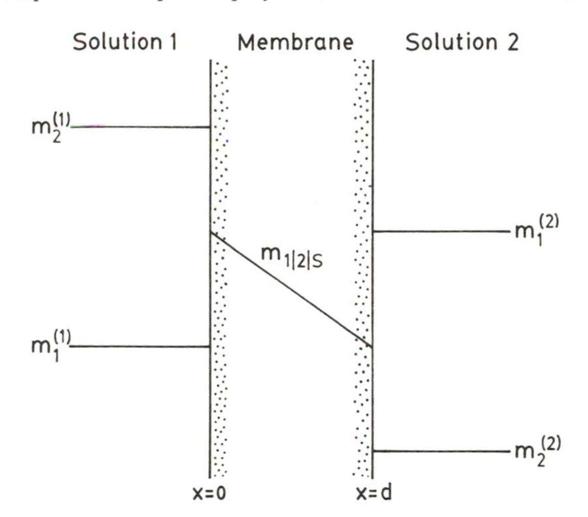

Fig. 10.13 Idealized concentration profiles for carrier-solute complex during steady state co-transport under diffusion control

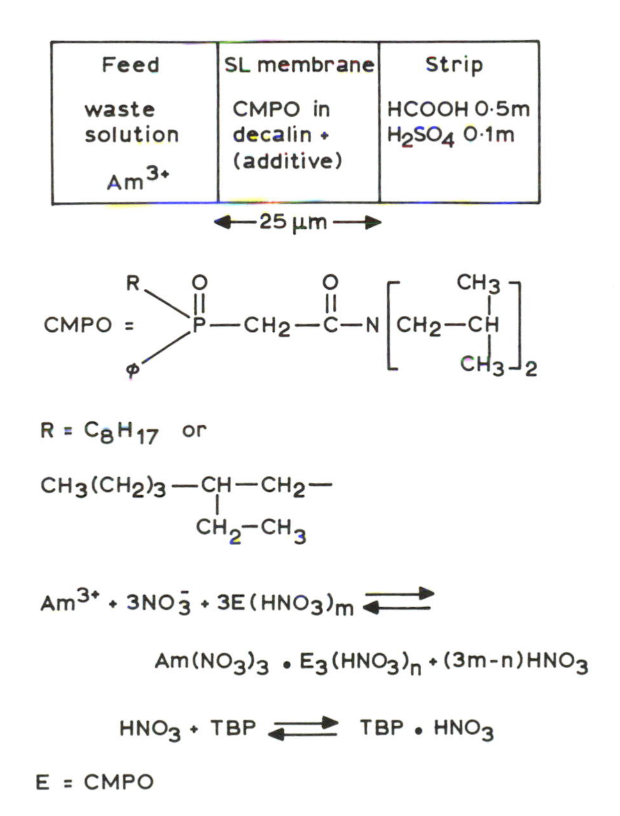

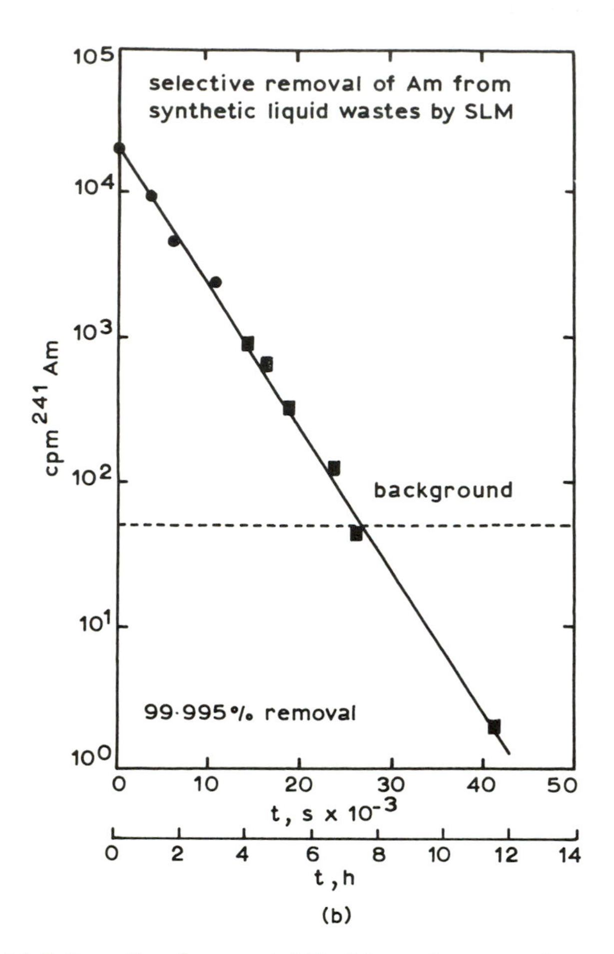

Fig. 10.14 **(*a*) Schematic of supported liquid membrane used to selectively remove Am^{3+} and other actinides from synthetic liquid waste, showing the chemical reactions responsible for co-transport of Am^{3+} and HNO_3. (*b*) Decay in Am^{3+} concentration as it is selectively removed through the membrane. After Horwitz *et al.*[39]**

under the assumptions that diffusional cross-effects are zero. Solution of equations (10.78)–(10.82) for the case $D_S = D_{12S} = D$ yields

$$J_1 = \frac{DK[S_T]}{d^2}\left\{\frac{m_1^{(0)}m_2^{(0)}}{1 + Km_1^{(0)}m_2^{(0)}} - \frac{m_1^{(d)}m_2^{(d)}}{1 + Km_1^{(d)}m_2^{(d)}}\right\} \tag{10.83}$$

which may be rewritten as

$$J_1 = -P_{11}\Delta m_1 - P_{12}\Delta m_2 \tag{10.84}$$

where

$$P_{11} = P_1 \frac{[S_T]}{d} \bar{m}_2/(1 + Km_1^{(0)}m_2^{(0)})(1 + Km_1^{(d)}m_2^{(d)})$$

$$P_{12} = P_{11}\frac{\bar{m}_1}{\bar{m}_2} \qquad (10.85)$$

and $\bar{m}_1$, $\bar{m}_2$ are the averages of the boundary value concentrations. Equation (10.84) is seen to have the form of a ternary membrane diffusion equation but once again this is due entirely to the complex formation reaction. Diffusion is in fact described by simple binary expressions (10.80) and is of course a downhill process.

A number of cotransport membrane systems have been described by Schultz and Curran,[37] Schiffer *et al.*,[38] and Danesi.[33] An example involving the co-transport of Am^{3+} and HNO_3 is shown in Fig. 10.14.[39]

The observation of rapid 'uphill' transport is generally taken to imply that membrane diffusion is carrier mediated. The reasons for this are, firstly, that while negative correlations may be predicted in phases containing no carriers, they are generally weak and would not be expected to produce the strong effects noted in membranes. Secondly, although membranes containing fixed reactive sites exhibit competitive inhibition effects, these sites are simultaneously accessible to solutes from both reservoirs and uphill transport is impossible within the framework of this description. The observation of uphill transport can readily be accounted for with the aid of the mobile carrier model and is therefore taken as evidence for its reality.

REFERENCES

1 M. SEKIDO AND A. MORITA: *Bull. Chem. Soc. Jpn*, 1962, **35**, 1375
2 H. FUJITA AND L. J. GOSTING: *J. Am. Chem. Soc.*, 1956, **78**, 1099
3 J. KÄRGER AND M. BÜLOW: *Chem. Eng. Sci.*, 1975, **30**, 893
4 J. KÄRGER, M. BÜLOW, AND W. SCHIRMER: *Z. Phys. Chem. (Leipzig)*, 1975, **256**, 144
5 A. KATCHALSKY AND P. F. CURRAN: 'Nonequilibrium thermodynamics in biophysics', 1965, Cambridge, Mass., Harvard University Press
6 E. H. BRESLER AND R. P. WENDT: *J. Phys. Chem.*, 1969, **73**, 264
7 E. L. CUSSLER: 'Multicomponent diffusion', 1976, Amsterdam, Elsevier
8 T. FOLEY AND P. MEARES: *J. Chem. Soc. Faraday Trans. I*, 1976, **72**, 1105
9 K. KONTURRI, P. FORSSELL, AND A. SIPILA: *ibid.*, 1982, **78**, 3613
10 D. MACKAY AND P. MEARES: *Trans. Faraday Soc.*, 1959, **55**, 122
11 F. CARACCIOLO, E. L. CUSSLER, AND D. F. EVANS: *J.A.I.Ch.E.*, 1975, **21**, 160
12 M. PLANCK: *Ann. Phys. Chem.*, 1890, **39**, 161
13 M. PLANCK: *ibid.*, **40**, 56
14 O. STEN-KNUDSEN: in 'Membrane transport in biology, Vol. 1', (ed. G. Giebisch, D. C. Tosteson, and H. H. Ussing), 1978, Berlin, Springer-Verlag
15 S. G. SCHULZ: 'Basic principles in membrane transport', 1980, Cambridge University Press
16 A. T. FROMHOLD: 'Theory of metal oxidation, Vol. 1', 1976, Amsterdam, North-Holland
17 D. E. GOLDMAN: *J. Gen. Physiol.*, 1943, **27**, 37
18 A. L. HODGKIN AND B. KATZ: *J. Physiol. (London)*, 1949, **108**, 37
19 U. A. R. BEHN: *Ann. Phys. (Leipzig)*, 1897, **62**, 54
20 H. H. USSING: *Acta Physiol. Scand.*, 1949, **19**, 43
21 H. EYRING, R. LUMRY, AND J. W. WOODBURY: *Rec. Chem. Prog.*, 1949, **10**, 100

22 B. HILLE: in 'Membranes — a series of advances' (ed. G. Eisenman), 1975, New York, Marcel Dekker
23 B. HILLE AND W. SCHWARZ: *J. Gen. Physiol.*, 1978, **72**, 409
24 D. J. YOUNG AND M. J. DIGNAM: *J. Phys. Chem. Solids*, 1972, **34**, 1235
25 P. LEFEVRE: in 'Current topics in membranes and transport', (ed. F. Bronner and A. Kleingeller), 1975, New York, Academic Press
26 O. KEDEM: in 'Membrane transport', (ed. A. Klingeller and A. Kotyk), 1961, Prague, Czech. Acad. Sciences
27 O. KEDEM AND S. R. CAPLAN: *Trans. Faraday Soc.*, 1965, **61**, 1897
28 S. R. DE GROOT AND P. MAZUR: 'Non-equilibrium thermodynamics', 1962, Amsterdam, North-Holland
29 J. S. SCHULZ, J. D. GODDARD, AND S. R. SUCHDEO: *J.A.I.Ch.E.*, 1974, **20**, 417, 625
30 E. A. HEMMINGSEN: *Science*, 1962, **135**, 733
31 E. A. HEMMINGSEN AND P. F. SCHOLANDER: *ibid.*, 1960, **132**, 1379
32 K-H. LEE, D. F. EVANS, AND E. L. CUSSLER: *J.A.I.Ch.E.*, 1978, **24**, 860
33 P. R. DANESI: *Sep. Sci. Technol.*, 1985, **19**, 857
34 W. J. WARD: *J.A.I.Ch.E.*, 1970, **16**, 405
35 J. D. GODDARD, J. S. SCHULTZ, AND S. R. SUCHDEO: *ibid.*, 1974, **20**, 625, 831
36 P. R. DANESI, L. REICHLER-YINGER, C. CIANETTI, AND P.-G. RICHERT: *Solv. Extr. Ion Exch.*, 1984, **2**, 781
37 J. S. SCHULTZ AND P. F. CURRAN: *Physiol. Rev.*, 1970, **50**, 637
38 D. K. SCHIFFER, A. M. HOCHHAUSER, D. F. EVANS, AND E. L. CUSSLER: *Nature*, 1974, **250**, 484
39 E. P. HORWITZ, H. DIAMOND, AND D. KALINA: in 'Plutonium chemistry', (ed. W. T. Carnella and G. Choppin), ACS Symp. Series No. 216, p. 433, 1983

CHAPTER 11

Stable and unstable interfaces in ternary diffusion

11.1 INTRODUCTION

The prediction of the diffusion-controlled motions of phase interfaces in ternary systems is a straightforward generalization of the binary case provided thermodynamic conditions obtain for a stable, planar interface. It is exceedingly important to the understanding of commercial processes such as superalloy coating and galvanizing. But here, as with the binary solidification process (§1.19), the interaction of the two independent transport fields may be such that regions of supersaturation obtain along the diffusion path near the phase interface, implying that such an interface is thermodynamically unstable and therefore prone to develop protuberances or isolated precipitates. Such planar unstable paths, if predictable, can be designated as *virtual paths*.[1] Grain boundaries which represent nucleation sites, will naturally influence the morphologies which subsequently develop. If the diffusion paths are contrived to cross complicated Gibbs isotherms containing two and three phase regions, as often is the case with alloy oxidation and sulphidation, the morphologies in the diffusion zone can become very complex indeed. Here it can be assumed with some validity that a one-dimensional path, however complex, will be iso-activity on the lateral cross-section. The path, which due to infinite boundary conditions will tend to be parabolic, can therefore be represented diagnostically on the isotherm as a unique iso-activity path. Except in the simplest cases, however, such paths cannot be predicted. A large part of this chapter will therefore be concerned with diagnostics rather than predictions.

The cases which can be dealt with analytically are associated with weak perturbations of a planar interface or with fine distributions of isolated precipitates (e.g. internal oxidation). The latter may form stable or periodic distributions, the periodic case being known as the Liesegang phenomenon.[2] In all cases of morphological change, surface tension plays a role in the moderation of structure. This important effect will not be introduced explicitly until Chapter 12.

11.2 THE MOTION OF PLANAR PHASE INTERFACES

The general system of one-dimensional diffusion equations for multicomponent diffusion in a substitutional solid solution alloy is given in §4.4. These equations have the parametric solutions (see also Seith[3])

$$C_i = C_i(\lambda) \qquad \lambda = x/\sqrt{t}, \tag{11.1}$$

provided the initial conditions can be expressed as

$$C_i(+\infty) = C_{i1}$$

and

$$C_i(-\infty) = C_{iN}, \tag{11.2}$$

i.e., the conditions for the infinite diffusion couple.

As the following analysis will demonstrate, these same parametric solutions exist in one-dimensional multiphase systems provided very close to constant concentrations obtain at all interfaces, irrespective of the number of phases which appear in the diffusion zone of the couple. That these constant concentrations must be equilibrium values follows from the fact that as $t \rightarrow \infty$ and the diffusion gradients approach zero, each increment in the zone, including the interface increment, must approach local equilibrium. Accordingly, any diffusion measurement that confirms relation (11.1) demonstrates uniquely that very close to interfacial equilibrium is maintained throughout the diffusion period. The only preconditions to be imposed on the terminal alloys in this description are that they be homogeneous solid solutions, that diffusion be structure independent, and that the phase constitution allows diffusion to proceed from them without any compositions appearing in the zone which require more than two phases in equilibrium. For example, the two alloys A and B marked on the 775°C section of the copper–nickel–zinc phase diagram of Fig. 11.1 would probably meet the conditions. In this case, a third phase β would appear in the diffusion zone between the terminal α and γ phases.

Figure 11.2 is a diagrammatic penetration curve for an n-component alloy involving a total of N phases. Each continuous region of the curve is described by the solution of the ordinary equations (4.77)[4, 5] which fits the appropriate phase boundary compositions. In general the diffusion coefficients are functions of all the concentrations, but in the present calculation they are assumed to be an

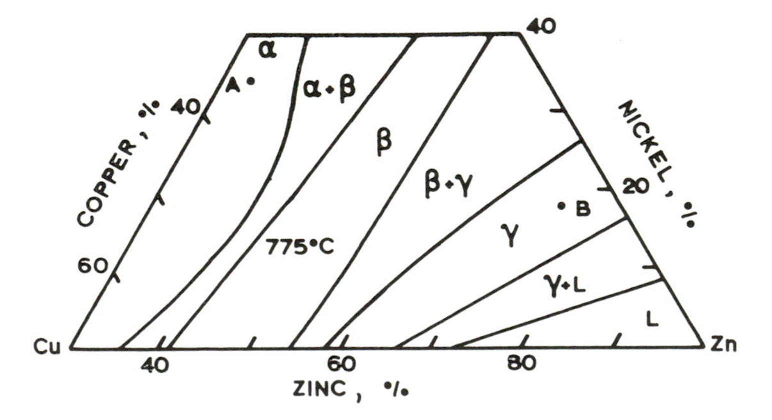

With permission from Canadian Journal of Physics, National Research Council of Canada

Fig. 11.1 775°C section of the copper–nickel–zinc constitution diagram. After Kirkaldy[4]

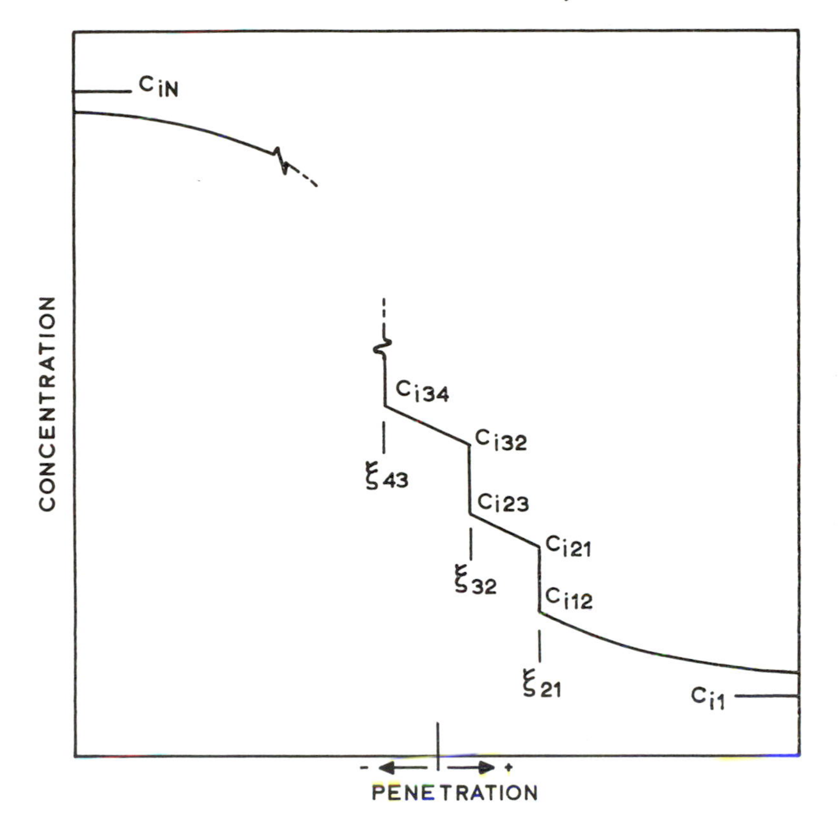

With permission from Canadian Journal of Physics, National Research Council of Canada

Fig. 11.2 Schematic penetration curve for the *i*th component of a multi-component diffusion couple with intermediate phases. After Kirkaldy[4]

appropriate average value for each phase so that an analytic solution can be obtained.

The coordinates of the interfaces ξ are determined simultaneously with the interface concentrations by application of interface continuity relations of the form

$$-(C_{i21} - C_{i12})\xi_{21}/2\sqrt{t} = J_{i12} - J_{i21} \tag{11.3}$$

and the equilibrium conditions. In this evaluation we have $2(N-1)n$ interface concentrations plus $N-1$ phase boundary coordinates as unknowns, and these are uniquely determined by the $(n-1)(N-1)$ independent relations of type (11.3), the $n(N-1)$ equilibrium chemical potential relations of the form (Appendix 1)

$$\mu_{i1}(C_{112}, C_{212}, \ldots, C_{n-1,12}) = \mu_{i2}(C_{121}, C_{221}, \ldots, C_{n-1,21}), \tag{11.4}$$

and the $2(N-1)$ concentration balances of the form

$$\sum_{i=1}^{n} C_{i12} \simeq \text{constant} \tag{11.5}$$

The isothermal constitution diagram is equivalent to the latter two sets of equations.

The required solutions take for constant coefficients the same form as those presented in §4.6.2, namely

$$C_i = a_{i0} + \sum_{k=1}^{n-1} a_{ik} \int_{\lambda}^{\xi_{ij}} \exp\left[-\lambda^2/4u_k\right] \mathrm{d}\lambda \tag{11.6}$$

subject to the characteristic equations

$$u_k = \sum_{j=1}^{n-1} \frac{a_{jk}}{a_{ik}} D_{ij} \tag{11.7}$$

and the appropriate phase boundary values. In the evaluation of the coefficients it is convenient to normalize the integrals by letting

$$a_{ik} = d_{ik} \Big/ \int_{\xi_{i+1,j+1}}^{\xi_{ij}} \exp\left[-\lambda^2/4u_k\right] \mathrm{d}\lambda \tag{11.8}$$

where the d_{ik} are functions of the boundary Cs and Ds as determined by equation (11.7). The complete numerical calculation involves as well the solution of the simultaneous non-linear set of equation (11.3) with the phase diagram equivalent of sets (11.4) and (11.5). These solutions are analogous to the binary ones given in §1.20.

The foregoing suggests some results of a general nature which are worth summarizing.

1. For an infinite diffusion couple consisting of an arbitrary number of components and an arbitrary number of parallel intermediate phases, the diffusion penetration depth will tend to follow a parabolic ($x = a\sqrt{t}$) law. This statement holds for both constant and variable diffusion coefficients and is a result of the fact that approximate interface local equilibrium is attained in a very short time. Exceptions, if such appear, could be due to an unusually high activation energy for atom transfer across one or more of the phase interfaces and would be exemplified by deviations from the parabolic law near the time origin.
2. Intermediate phases with a 'zero' solubility range will have 'zero' width in the diffusion zone.
3. Intermediate phases with a small solubility range and/or relatively small diffusion coefficients will have correspondingly small widths in the diffusion zone.
4. Intermediate phases with a large solubility range and/or relatively large diffusion coefficients will have a correspondingly large width in the diffusion zone.
5. The accuracy of the phenomenological description will be prejudiced to a greater or lesser extent by interface inhibitions (films or high transfer

activation energies), grain boundary diffusion, high defect densities, anisotropic diffusion, and porosity and lateral dimensional changes due to the Kirkendall effect.

The theoretical and experimental description of multicomponent diffusion with the appearance of intermediate phases has applications in the understanding of metal to metal bonding and in the determination of alloy phase diagrams.[4,6]

Obviously, the problem becomes much more complicated if resistance associated with surface reactions is involved, for then the boundary concentrations are a function of time[7] (cf § 1.25). Bückle[8] has studied this problem for the binary case. An even greater complication is identified when one or more of the intermediate phases exhibits delayed nucleation.[9]

11.3 AN APPROXIMATE SOLUTION FOR A PLANAR PRECIPITATE

In the following we develop in detail a special case of the general solution in § 11.2 which concerns a planar grain boundary precipitate. Here one of the independent elements diffuses much faster than the other. In particular, we analyse the precipitation of β (bcc) in a γ (fcc) iron alloy containing carbon and a slow-diffusing substitutional element (Mn). For constant coefficients, subject to the boundary conditions at infinity and at the interface in λ space ($\lambda = \alpha$) we state the formal conditions[10]

$$C_i(\lambda = \infty) = C_{i0} \qquad C_i(\lambda = \alpha+) = C_{i1} \qquad C_i(\lambda = \alpha-) = C'_{i1} \qquad (i = 1, 2) \tag{11.9}$$

Quantities denoted by a single prime refer to the daughter phase (ferrite). Unprimed quantities refer to the parent phase (austenite). The solutions are

$$C_i = a_{i0} + \sum_{k=1}^{n-1} a_{ik} \operatorname{erfc}(\lambda/2\sqrt{u_k}) \qquad (i = 1, 2) \tag{11.10}$$

where the coefficients a_{ik} and u_k remain to be determined in terms of the diffusion coefficients and boundary conditions. The boundary conditions at ∞^+ lead to

$$\Delta_i = \sum_{k=1}^{n-1} d_{ik} \tag{11.11}$$

where

$$\Delta_i = C_{i1} - C_{i0} \quad \text{and} \quad d_{ik} = a_{ik} \operatorname{erfc}(\alpha/2\sqrt{u_k})$$

The d_{ik} are given by

$$d_{11} = \left\{D_{12}\Delta_2 + [(D_{11} - D_{22}) + D]\frac{\Delta_1}{2}\right\}/D$$

$$d_{21} = \left\{D_{21}\Delta_1 - [(D_{11} - D_{22}) - D]\frac{\Delta_2}{2}\right\}/D$$

$$d_{12} = \Delta_1 - d_{11}$$

$$d_{22} = \Delta_2 - d_{21}$$

where as usual (cf § 4.4)

$$D = [(D_{11} - D_{22})^2 + 4D_{12}D_{21}]^{1/2}$$

$$u_1 = (D_{11} + D_{22} + D)/2$$

and

$$u_2 = (D_{11} + D_{22} - D)/2$$

The interfacial mass balances are

$$(C'_{i1} - C_{i1})v = -J_i(\alpha) \tag{11.12}$$

where the J_i are given by equation (4.34) and the interface velocity is

$$v = \alpha/2\sqrt{t} \tag{11.13}$$

The parameter α can therefore be used as an index of reaction rate.

In general, the growth-rate coefficient α must be found by simultaneous solution of the two flux balances. If, as in the binary case, the interfacial concentrations were fixed, no solution would exist, since the two flux balances would yield two different interface velocities. There is, however, a degree of freedom afforded by the unknown tie line (in the two-phase region) specifying the equilibrium interfacial concentrations. The two mass balances can, therefore, be simultaneously satisfied by the selection of an appropriate tie line.

In the special case of ternary austenites (fcc), the diffusion coefficient of carbon (D_{11}) is usually much greater than that of a substitutional solute (component 2). If, in addition, the constitution diagram is similar to that of the Fe–C–Mn system, the mass balances demand that for high supersaturations the change in concentration of the substitutional element between the bulk austenite and ferrite (bcc) must be negligible, i.e. there must be negligible manganese partition.

This may be demonstrated by referring to the schematic Fe–Mn–C constitution diagram and the penetration curves of Fig. 11.3*a*. The fact that $D_{11} \gg D_{22}$ is indicated by a relatively short manganese penetration in the austenite. Let the composition of the bulk austenite be represented by a point A on the ternary isotherm, which represents rather high supersaturation. The 'zero manganese partition' tie line is then given by the full line (B–B′). This tie line represents the extreme of the possible tie line solutions, and all other possible tie line solutions must lie below this, as represented by the dashed line (G–G′). Now all the corresponding values of the carbon build-up (Δ_1) will be in excess of the value corresponding to the zero-partition tie line, which is relatively large to begin with. This implies that all possible solutions must show a large carbon build-up in the austenite and therefore that α must also be large (i.e. the growth rate must be relatively large). If the manganese mass balance is to be simultaneously satisfied, then the material (i.e. the area) in the very thin manganese spike must be balanced by a correspondingly small manganese depletion, $(C_{20} - C'_{21})\alpha$, of the ferrite. Since this small depletion is spread over a relatively large distance α, it must correspond to a negligible concentration difference $(C_{20} - C'_{21})$. That is, without reference to the diffusion solutions, the mass balance requires that there be negligible manganese partition for compositions such as A, and, therefore, that the limiting tie line B–B′ must be very close to the correct one. It is evident that the same argument holds true for all compositions A lying along the line from B′

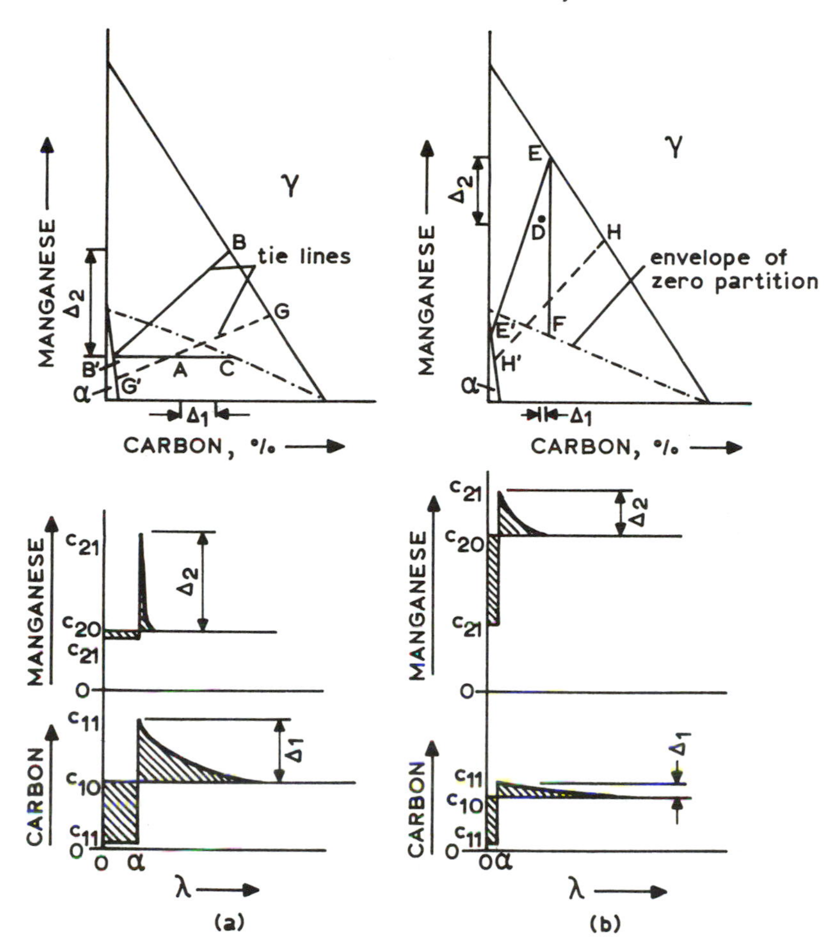

Reprinted with permission from TRANSACTIONS OF THE METALLURGICAL SOCIETY, Vol. 230, p. 1025, (1964), a publication of The Metallurgical Society, Warrendale, Pennsylvania

Fig. 11.3 (*a*) and (*b*) Schematic penetration curves in λ space for ferrite growth in Fe–C–Mn austenites. The mass balances are represented by equal shaded areas on either side of the interface ($\lambda = \alpha$). The system is transforming in each case with interfacial concentrations given by the tie lines designated *B–B′* and *E–E′*. After Purdy *et al.*[10]

almost up to C. The line B′–C therefore represents a spectrum of bulk austenite compositions which all transform with interfacial concentrations given approximately by the tie line B–B′. Needless to say, the exact solutions of the diffusion equations and the mass balances confirm this qualitative inference.

If we restrict attention to these regions of high supersaturation and note that manganese on account of its low mobility cannot partition, then equation (11.10) in combination with the mass balances (11.12) may be simplified to yield an approximate expression for the growth parameter α or the related parameter $\beta_1 = \alpha/2\sqrt{u_1} \simeq \alpha/2\sqrt{D_{11}}$, ie.

$$\beta_1 = \frac{\Delta_1}{\sqrt{\pi}C_{11}}\left(1 + \frac{D_{12}}{D_{11}}\frac{\Delta_2}{\Delta_1}\right)\frac{\exp(-\beta_1^2)}{\operatorname{erfc}\beta_1} \tag{11.13}$$

It is significant to note that only carbon diffusion coefficients appear. This simply says that all we need to know about the manganese coefficients is that they are relatively very small.

Let us now consider an alloy D (Fig. 11.3*b*) which lies in a region of relatively low supersaturation. The limiting tie line is given by the solid line (E–E′), and all other possible tie lines must again lie below this as indicated by the dashed line (H–H′) (otherwise the austenite would show a deficit in carbon rather than a build-up). Thus, in this case, the manganese must partition. The only way that both mass balances can be simultaneously satisfied under this condition is if Δ_1 is small, or, equivalently, if the limiting tie line E–E′ is very close to the correct one. This would imply that α (and therefore the growth rate) is very small and that therefore $(C_{20} - C'_{21})$ in the manganese depletion term $\alpha(C_{20} - C'_{21})$ must be significant. That is, alloys of composition *D* must show an appreciable partition of manganese and the growth rates are circumscribed to very small values corresponding to the low manganese diffusion rates. The same arguments hold for all alloys D lying between the points E and F. The approximate solution for $\beta_2(= \alpha/2\sqrt{u_2} \simeq \alpha/2\sqrt{D_{22}})$ analogous to equation (11.13) and corresponding to these conditions is given by

$$\beta_2 = \frac{1}{\sqrt{\pi}}\frac{\Delta_2}{(C_{21} - C'_{21})}\frac{\exp(-\beta_2^2)}{\operatorname{erfc}\beta_2} \tag{11.14}$$

These considerations suggest a simple construction for approximately determining the tie line which gives the interfacial concentrations. The composition of the bulk austenite, A or D, must lie on either of two mutually perpendicular sides of a right triangle, the hypotenuse of which is a tie lie and the remaining sides of which are parallel to the carbon- and manganese-concentration axes. If the austenite composition lies on the side parallel to the manganese-concentration axis, the transformation will be essentially controlled by the diffusion of manganese; if it lies on the side parallel to the carbon axis, the transformation will proceed with negligible partition of manganese. This approximation will be more accurate the greater the difference between D_{11} and D_{22}. A number of these triangles were used to construct the 'envelopes of zero partition' shown dot-dashed in Figs. 11.3*a* and *b*. These envelopes therefore give a precise definition of our distinction between high and low supersaturation.

Qualitatively, we see that two classes of transformation exist: one in which the major effect of manganese additions is to alter the boundary conditions for carbon diffusion (a lesser effect is due to carbon diffusion on the manganese gradient in advance of the interface) and one in which the gross redistribution of manganese is required — where manganese-diffusion control is predominant. In this latter case, a drastic reduction in the transformation rate is to be expected.

Applying these considerations to the case of the planar growth of incoherent

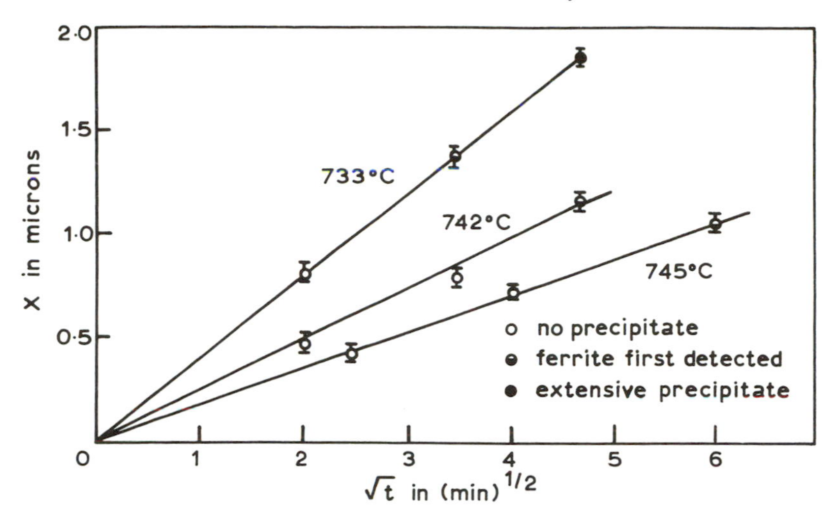

Reprinted with permission from TRANSACTIONS OF THE METALLURGICAL SOCIETY, Vol. 230, p. 1025, (1964), a publication of The Metallurgical Society, Warrendale, Pennsylvania

Fig. 11.4 Ferrite growth data for the 0·285%C, 3·16%Mn diffusion couples. After Purdy *et al.*[10]

grain boundary ferrite in Fe–C–Mn austenites, we may infer that, for alloy compositions which allow a transition from low to high supersaturation with decreasing temperature, the growth rate of ferrite must show a sharp inflection at the transition. At temperatures above this point, growth rates will be relatively small. At lower temperatures the rate will increase rapidly with decreasing temperatures.

Purdy *et al.*[10] carried out experiments in Fe–C–Mn with boundary conditions similar to relations (11.9) and compared the results with predictions based on the Fe–C–Mn phase diagram and ternary diffusion data given in §7.9. The raw parabolic growth data is represented by Fig. 11.4 while the summary in comparison with theoretical predictions is given in Fig. 11.5. Figure 11.6 demonstrates the drastic effect of slow-diffusing manganese on the transformation rate of Fe–C (plain carbon steel) alloys. Such effects are very important in heat treatment of alloy steels. Hillert[11] and Popov and Mikhalev[12] have arrived at similar conclusions using pseudobinary arguments. Gilmour *et al.*[13] and Sharma and Kirkaldy[14] have extended this analysis to more realistic boundary conditions.

11.4 NON-PLANAR INTERFACES IN TWO-PHASE TERNARY DIFFUSION COUPLES

To introduce a very complex subject we begin with some qualitative arguments and observations pertaining to the system Cu–Zn–Sn.[15] The extra degree of freedom introduced by a second independent concentration in a ternary system gives rise to the possiblity of unstable planar interfaces in infinite diffusion layer couples. Diffusion paths which pass continuously in and out of two-phase regions of the phase diagram (Fig. 11.7) but which are not coincident with tie lines

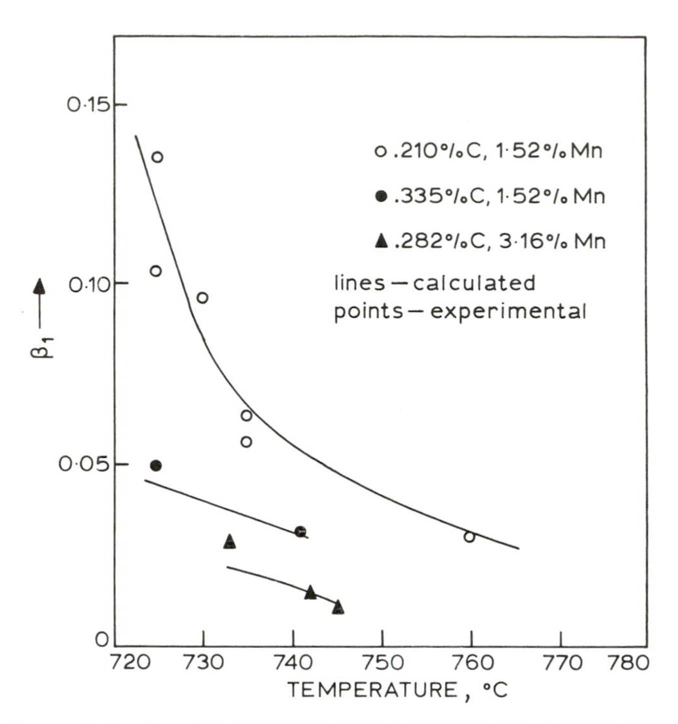

Reprinted with permission from TRANSACTIONS OF THE METALLURGICAL SOCIETY, Vol. 230, p. 1025, (1964), a publication of The Metallurgical Society, Warrendale, Pennsylvania

Fig. 11.5 Summary of ferrite growth data for all ternary compositions studied. The points are experimental. After Purdy *et al.*[10]

give rise to supersaturated zones which may be either adjacent to or isolated from planar phase interfaces, with corresponding non-isolated or isolated precipitates. Non-isolated precipitates are equivalent to morphological development of the interface. Selected couples in the ternary system Cu–Zn–Sn were examined metallographically and both isolated and non-isolated precipitates were observed.[15] The diffusion path in one of the couples showing interface morphological development was examined using electron probe microanalysis.

Ignoring the spurious effects of preferential grain boundary diffusion and anisotropic crystal growth, it can be unequivocally stated that it is impossible for a binary isothermal diffusion couple constructed of two unsaturated phases to sustain a non-planar phase interface. This is because surface tension will always change the activity and therefore the diffusion conditions at the tip of protuberances in such a way as to moderate the perturbations which led to the protuberances. In ternary systems the same effect is operative, but forces associated with supersaturation can be strong enough to overcome the surface energy constraint (cf § 12.3).

Consider a ternary system having the constitution of Fig. 11.7 and suppose that we have calculated diffusion path X–Y by solution of the appropriate diffusion

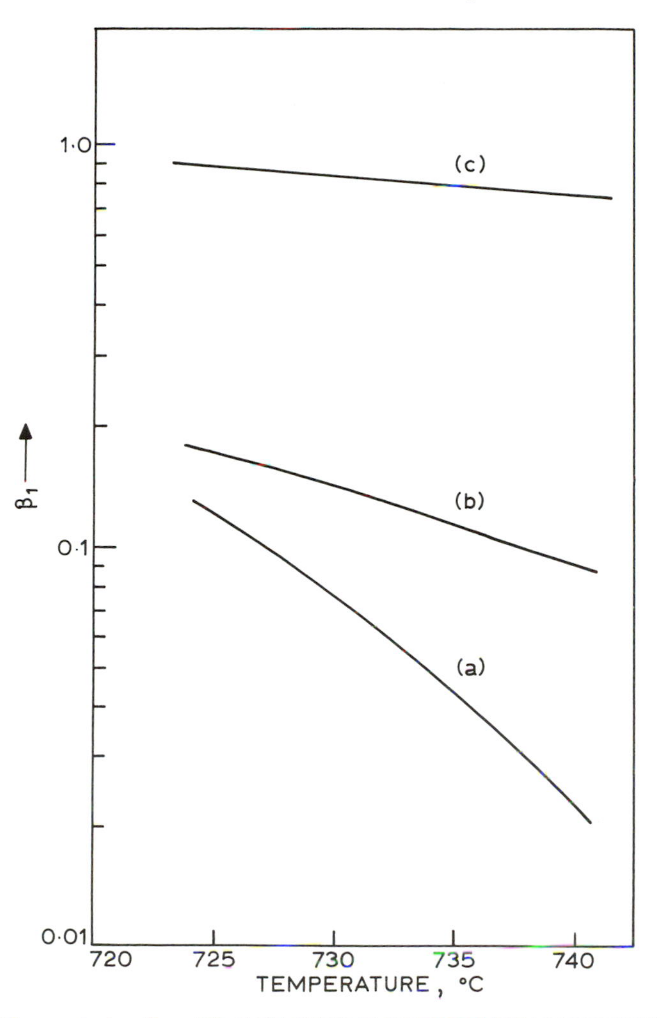

Fig. 11.6 Calculated variation of rate parameter β_1 with temperature: (*a*) complete calculation for the grain boundary reaction; (*b*) assuming a manganese constitutional effect, but neglecting ternary-diffusion effects ($D_{12} = 0$); (*c*) assuming that the alloy contains no manganese. After Purdy *et al.*[10]

equations and mass balances by assuming initially that the interface is flat. The only general restriction on the form of such a theoretical path is that it should cross the line joining its termini at least once. If it does not do so, then the mass balance is necessarily violated. This will be called a virtual path, since the distinct possibility appears that the calculated path dips into the two-phase region of the

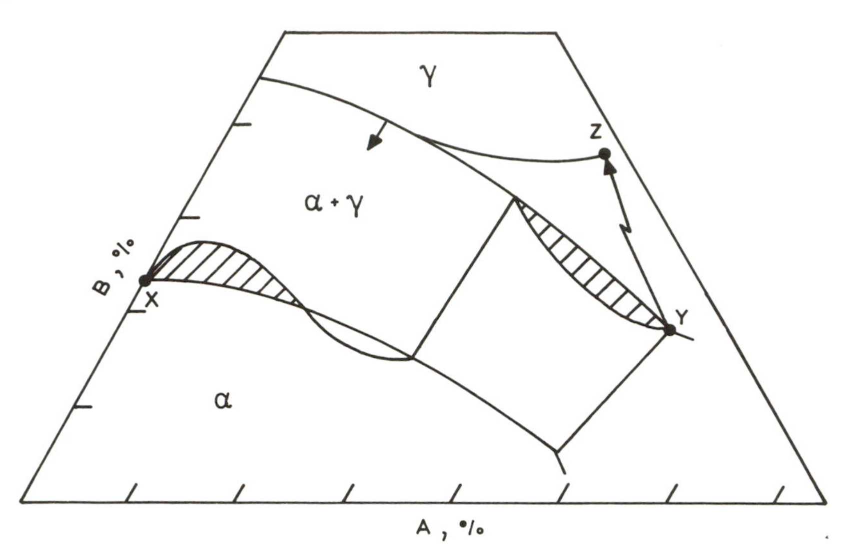

Reprinted with permission from TRANSACTIONS OF THE METALLURGICAL SOCIETY, Vol. 224, p. 490, (1962), a publication of The Metallurgical Society, Warrendale, Pennsylvania

Fig. 11.7 Schematic isothermal section of a ternary constitution diagram showing a representative diffusion path X–Y. Hatch marks indicate regions of supersaturation. After Kirkaldy and Fedak[15]

diagram producing a condition of supersaturation within the couple which promotes nonplanar morphologies. Two such cases are indicated schematically in Fig. 11.7. At the Y end of the couple the region of supersaturation appears in contact with the phase boundary (i.e. with the tie line) while at the X end it appears isolated from the interface. In the latter case one would anticipate the possibility of an isolated precipitate of γ appearing within the α end of the couple while in the first case one might anticipate both isolated and non-isolated precipitates of α appearing in the γ-phase. The non-isolated precipitates are of course equivalent to a non-planar morphology.

One can clearly see the existence of a critical condition for the onset of a non-planar morphology as a function of the terminal compositions. For example, as the terminal Y changes to Z (or in the extreme, to some binary γ alloy on the B-axis) the region of non-isolated supersaturation would be expected to vanish. Thus, the critical point can be designated by the coincidence of the diffusion path and the constitution diagram phase boundary (see, however, § 11.5).

A close analogy exists between the description of this phenomena and the steady state growth of alloy crystals from the melt. In that configuration a region of supersaturation appears in the liquid ahead of a virtual flat interface, provided certain critical conditions are satisfied or exceeded (cf Fig. 1.23). These conditions are defined by the coincidence of the actual liquid temperature distribution and the liquidus trace within the melt. This supersaturation makes possible the development of non-planar morphologies such as cells or dendrites. In both

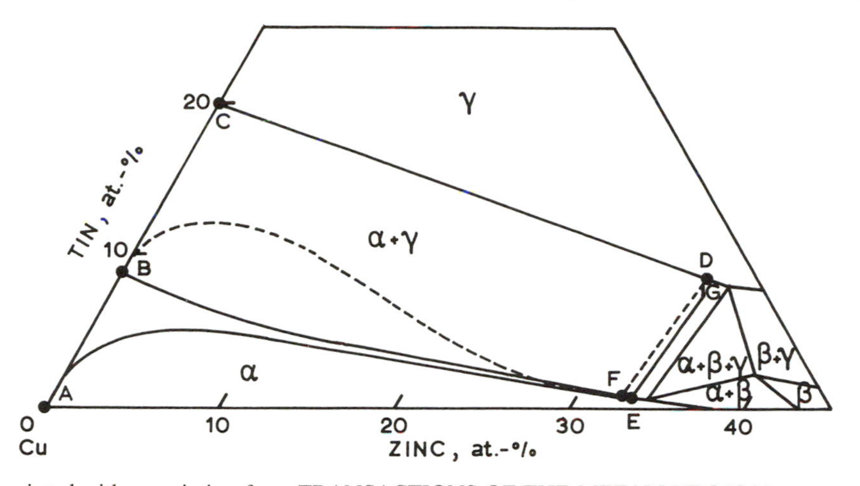

Reprinted with permission from TRANSACTIONS OF THE METALLURGICAL SOCIETY, Vol. 224, p. 490, (1962), a publication of The Metallurgical Society, Warrendale, Pennsylvania

Fig. 11.8 500°C isotherm of the Cu–Zn–Sn constitution diagram showing experimental diffusion paths AD and BD. In preparing this section the $\alpha + \gamma$ field has been altered from that tentatively given in the ASM Handbook (1948) to agree with our own metallographic observations. After Kirkaldy and Fedak[15]

phenomena the critical conditions as stated do not take into account the fact that the surface tension provides a barrier to the formation of morphological units. Hence we anticipate that appreciable supersaturation will be required before non-planar morphologies can actually appear.

It must be emphasized that the conditions for the existence of non-planar interfaces noted above are necessary but not in general sufficient, since the appearance of a flat interface under these same conditions need not violate the kinetic equations and the internal constraints. Should a morphological unit be formed, a new kinetic or order parameter enters into the description (e.g. the length of the unit) and with it an internal degree of freedom. In the absence of detailed knowledge of the cooperative phenomena which control a morphological transition this degree of freedom can only be removed by invoking thermodynamic optimization methods (cf Chapter 12).

In the research of Kirkaldy and Fedak[15] on Cu–Zn–Sn a series of couples was examined at 500°C. The pertinent constitution diagram is shown in Fig. 11.8. Alloys B, C, D, E were prepared from the melt and extensively homogenized. Various couples (BC, CE, BD, and AD) were prepared by machining and polishing alloy samples and lightly clamping pairs.

As expected, the binary couple BC produced a flat interface. Surprisingly, it was found that couple CE also sustained a flat interface. Evidently such a couple does not introduce a supersaturated region. On the other hand, couple BD must have produced a region of isolated supersaturation since a zone of isolated γ precipitate appears in the α phase, as shown in Fig. 11.9. The dashed line BFD on

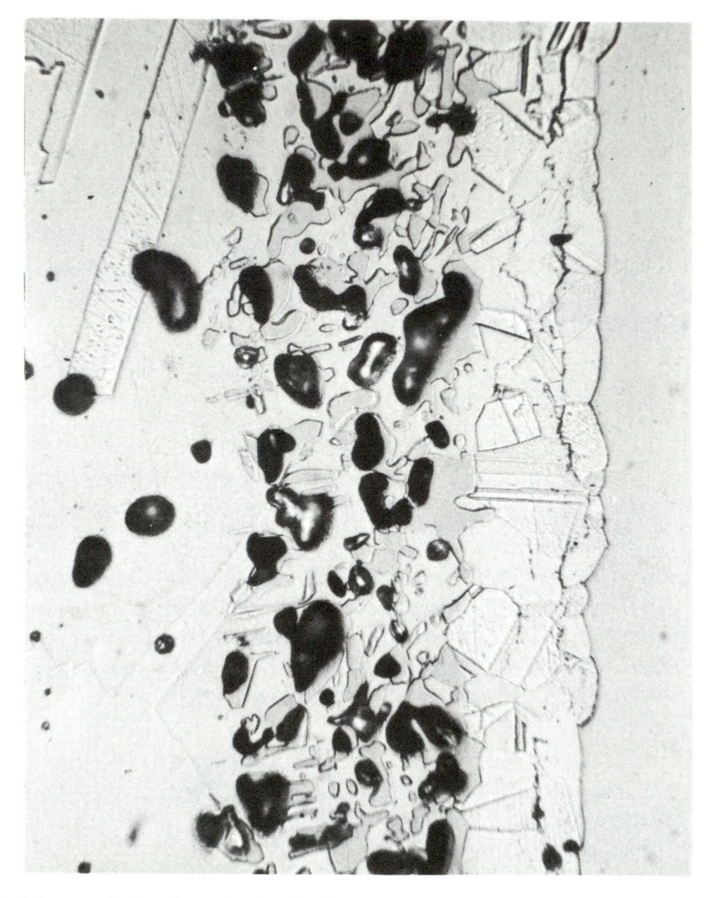

Fig. 11.9 Micrograph of couple BD, (see Fig. 11.8) annealed for 120 h at 500°C and cooled at 50°C per min. Note the isolated γ- precipitate within the α-phase. The apparent extensive porosity is due to an unfortunate etching effect (×530). Reduced approximately 10% for reproduction. After Kirkaldy and Fedak[15]

Fig. 11.8 gives a schematic diffusion path which satisfactorily represents the metallographic observations. In drawing this path the authors were influenced somewhat by microprobe measurements on an adjacent path as described below. This type of precipitation appears to be analogous to the internal oxidation which occurs in binary alloys (cf § 11.7).

Couple AD gave rise to a non-isolated region of supersaturation on the γ side of the phase boundary which led to both isolated grain boundary and non-isolated precipitates. The projecting Widmanstätten-type figures are needles with a roughly rectangular cross-section, as indicated by Fig. 11.10. Combination of the electron probe scan, Fig. 11.11, with rather uncertain constitution data allows us to construct a semiquantitative diffusion path for this couple. This path is shown as a line joining compositions A E G D in Fig. 11.8. The part of the path near D with the small jog records the quantitative metallographic aspects of Fig. 11.10. It also recognizes that some tin must be transported out of the γ-phase by providing a small (experimentally undetected) tin gradient. One may think of this jog as the remnants of a region of non-isolated supersaturation in the corresponding virtual diffusion path.

An interesting feature of the recorded diffusion path of Fig. 11.11 is that it lies very close to the zero tin axis for a considerable distance within the α-phase. This means that the accumulation of tin that appears well out in the α-phase has been delivered by the thermodynamic force provided by the zinc gradient. It is therefore inferred that zinc appreciably increases the activity of tin in copper in accord with §7.10. Another feature of this path is that the weld markers have shifted from the Matano interface towards the zinc-rich side of the couple (cf §2.13). The amount of motion is somewhat less than observed by Smigelskas and Kirkendall[16] in homogeneous brasses, but this is not surprising considering the multiphase, multicomponent nature of our couple.

The appearance of both isolated and non-isolated precipitates within the diffusion zone of Cu–Zn–Sn diffusion couples confirms the postulated existence

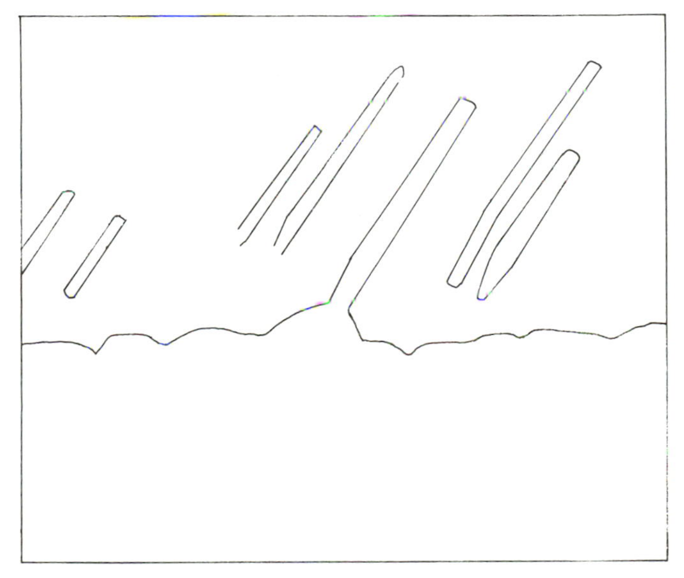

Fig. 11.10 Tracing of high power micrograph of α-Widmanstätten figures obtained in couple AD demonstrating their acicular nature (×860). Reduced approximately 25% for reproduction. After Kirkaldy and Fedak[15]

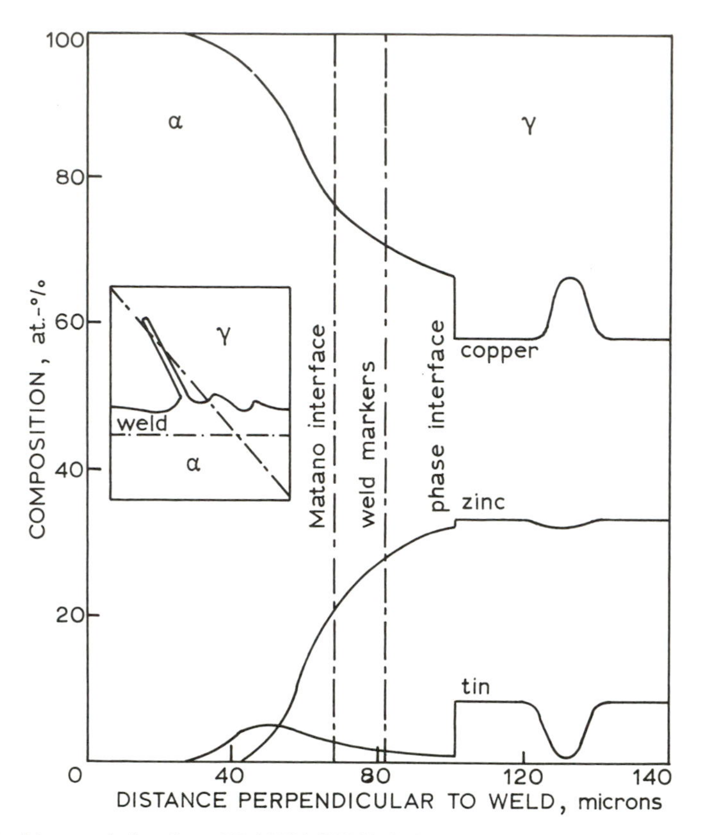

Reprinted with permission from TRANSACTIONS OF THE METALLURGICAL SOCIETY, Vol. 224, p. 490, (1962), a publication of The Metallurgical Society, Warrendale, Pennsylvania

Fig. 11.11 Electron probe scan of diffusion zone of couple AD. Insert shows scanning path on couple. Note that the spike has a central composition which is very little different from than at its α-phase root. Diffusion time 120 h. After Kirkaldy and Fedak[15]

of diffusion paths which intercept two phase regions of the phase diagram in such a way as to produce regions of supersaturation. The description of this behaviour in terms of the steady state virtual paths on the ternary diagram is precisely analogous to be conventional description of steady state alloy crystal growth. This analogy provides an obvious key to a synthetic description of morphological development as explored in the following section.

Practical investigations pertaining to galvanizing of steel have been reported by Urednicek and Kirkaldy.[17, 18]

11.5 PERTURBATION ANALYSIS OF INTERFACE INSTABILITY IN TERNARY SYSTEMS

Perturbation analysis in relation to phase interfaces was first introduced by Wagner[19, 20] and perfected by Mullins and Sekerka[21, 22] (see also § 12.3). Here we adopt the linear perturbation theory to analyse the morphological stability of moving planar phase interfaces in isothermal ternary systems.[23] In order to attain generality with respect to constitution, capillarity has been excluded from the boundary conditions. This implies that the derived stability criterion applies only to long wavelength perturbations. This criterion has been tested through comparison with the results of earlier treatments of related problems and is utilized to establish criteria for (i) interfaces in diffusion bonded materials, (ii) interfaces associated with isothermal systems in which solid alloys are in contact with liquid metals, (iii) oxide–metal interfaces during oxidation of alloys, and (iv) precipitate–matrix interfaces in ternary systems.

In considering the unidirectional steady state solidification of dilute binary alloys, Rutter and Chalmers[24] proposed that the transition from a planar to cellular growth morphology is associated with the onset of constitutional supercooling (CSC) in the liquid layer adjacent to the moving phase interface. This concept was made quantitative by Tiller *et al.*[25] who developed the well-known CSC stability criterion (cf § § 1.19 and 12.9). Mullins and Sekerka[22] have re-examined the problem of solid–liquid interface stability in terms of linear perturbation theory with the inclusion of capillarity and have derived a more complete stability criterion.

Coates and Kirkaldy[26] employed an analogue with the concept of CSC to obtain a quantitative stability criterion for this phenomenon. We shall henceforth call this the elementary criterion. Consider the isothermal system A–B–C shown schematically in Fig. 11.12. A and B have been selected as the independent components and the original ternary phase diagram has been replotted onto an orthogonal coordinate system in which the ordinate and abscissa are the concentrations of A and B, respectively. Depending on the values of the diffusion coefficients, this virtual diffusion path might enter the two-phase field in either or both phases, as is indicated by the dotted lines PR and SQ. If this is the case, it indicates that a region of supersaturation would be established adjacent to a planar interface. Thus the latter shape is unstable and could break down to a non-planar morphology in an attempt to eliminate the supersaturation. By considering the tangency of the diffusion path and phase boundary lines as the marginal state for instability one concludes as the analogue of the CSC criterion that for any isothermal ternary system the stability criterion is

$$m^{\alpha} G^{\alpha}_{\mathrm{BF}} - G^{\alpha}_{\mathrm{AF}} > 0 \quad \text{and/or} \quad m^{\gamma} G^{\gamma}_{\mathrm{BF}} - G^{\gamma}_{\mathrm{AF}} > 0 \quad \text{(unstable)} \tag{11.15}$$

where $G^{k}_{i\mathrm{F}}$ (i = A, B; $k = \alpha, \gamma$) is the interfacial gradient of component i in the k-phase and m^{α} and m^{γ} are the slopes (including sign) of the phase boundary lines ($\alpha/\alpha + \gamma$ and $\alpha + \gamma/\gamma$ respectively) at the points corresponding to planar interface compositions (i.e. points R and S in Fig. 11.12). The above criterion is subject to the following conditions: (*a*) component A must be the ordinate and B the abscissa, (*b*) the naming of components and phases for a given configuration must conform to a certain sign convention given by Coates and Kirkaldy[26] and (*c*) the positive direction is from α to γ. With Coates and Kirkaldy[23] we now examine this interfacial stability problem using the linear perturbation methods developed by Wagner[19, 20] and Mullins and Sekerka[21, 22] and compare the results with the elementary approach just described. Throughout the treatment it is assumed that local equilibrium (as defined by the ternary phase diagram) is

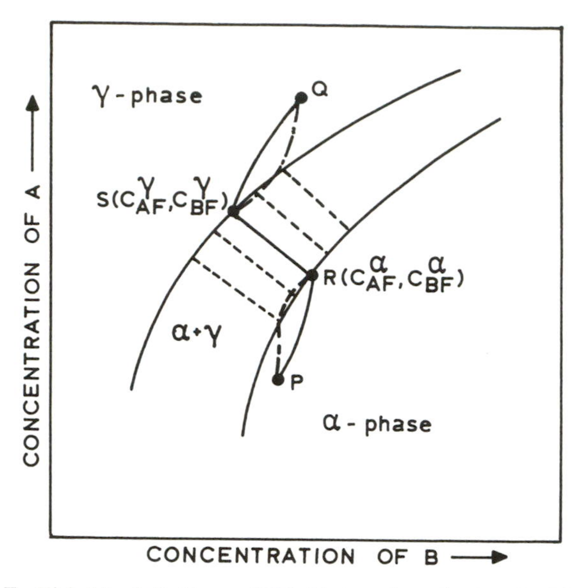

From Coates, D. E., Kirkaldy, J. S., *Trans. ASM*, **62**, American Society for Metals, Metals Park, OH 44073, USA, p. 426, 1969

Fig. 11.12 Schematic isothermal ternary phase diagram. Points *P* and *Q* define the terminal compositions of a diffusion couple and points *R* and *S* define the interface concentrations, (C_{AF}, C_{BF}) and (C_{AF}, C_{BF}) respectively, corresponding to solution of the diffusion equations for a planar interface. For local equilibrium at the interface, *R* and *S* must define a tie line. The solid and dotted lines *PRSQ* represent four possibilities for the calculated diffusion path, i.e. $\overline{PRSQ}$, $\overline{P}\,\overline{RSQ}$, $\overline{PRS}\,\overline{Q}$, and $\overline{P}\,\overline{RS}\,\overline{Q}$. After Coates and Kirkaldy[23]

maintained at phase interfaces, that effects of grain boundary and surface diffusion are negligible, that the strength of off-diagonal diffusional interactions is negligible and the on-diagonal (ordinary) diffusion coefficients are concentration independent. Furthermore, we assume that capillarity effects are negligible. This means that our stability criteria will be precisely applicable only to later time parabolic growth or to long wavelength perturbations. No assumptions are made about the constitution of the systems being considered.

A two-phase (α–γ) infinite ternary diffusion couple, in which the interface is planar and moving at the instantaneous velocity v, is considered at time t. At this instant the interface is coincident with the x-axis of a stationary coordinate system having its z-axis directed towards the γ-phase. Accordingly, the velocity is positive if the interface moves towards the γ-phase. A sinusoidal perturbation of infinitesimal amplitude δ is now imposed on the interface such that the equation of the latter becomes

$$z = \delta(t) \sin \omega x = \phi(x, t) \tag{11.16}$$

where $\omega = 2\pi/\lambda$ is the wavenumber of the perturbation and λ is the wavelength. A calculation of the sign of the time derivative of the amplitude, $\dot{\delta} = \mathrm{d}\delta/\mathrm{d}t$, establishes whether the perturbation grows or decays and hence whether or not the originally planar interface is stable. Because the field equations are linear and the boundary conditions can be linearized for perturbations of small amplitude, the behaviour of an infinitesimal perturbation of arbitrary shape can be conveniently treated by expanding it into sinusoidal Fourier components and applying the superposition principle. Thus no loss of generality is suffered by considering the stability of the interface with respect to a sinusoidal perturbation of arbitrary wavelength. The concentration distributions of the two independent components A and B in each phase must first be obtained in the vicinity of the perturbed interface. These can conveniently be written in the form

$$\begin{aligned} C_i^k(x,z,t) &= F_i^k(zt^{-1/2}) + C_{i\mathrm{p}}^k(x,z,t) \\ &= C_{i\mathrm{F}}^k + G_{i\mathrm{F}}^k z + C_{i\mathrm{p}}^k \qquad i = \mathrm{A, B} \quad k = \alpha, \gamma \end{aligned} \tag{11.17}$$

where the F_i^k are the known solutions for a flat interface (cf §4.6.2) and the $C_{i\mathrm{p}}^k$ are the differences in the solutions for a perturbed interface and the solutions for a flat interface. Because the C_i^k are required only in the vicinity of the interface, the F_i^k can be expanded to the first order in z. In order that mathematically tractable solutions may be obtained, it is assumed the average interface advances at a slow enough velocity v and the perturbation grows or decays slowly enough that the $C_{i\mathrm{p}}^k$ can be approximated by solutions of the time-independent Laplace equations

$$\left(\frac{\partial^2}{\partial x^2} + \frac{\partial^2}{\partial z^2}\right) C_{ip}^k = 0 \qquad i = \mathrm{A, B} \qquad k = \alpha, \gamma \tag{11.18}$$

A necessary condition for use of this approximation is that the following inequalities hold

$$\frac{v}{2D_i^k} \ll \omega \qquad i = \mathrm{A, B} \quad k = \alpha, \gamma \tag{11.19}$$

where the D_i^k are diffusion coefficients. In treating the stability of precipitate-matrix interfaces in binary systems, Sekerka[27] has demonstrated that Laplacian and complete time-dependent solutions lead to the same stability criterion for diffusion controlled growth. This of course is not proof that in the present more complicated situation the same will hold true. However, it does suggest that use of Laplacian solutions will lead to a good approximation.

The following boundary conditions apply:

1. The $C_{i\mathrm{p}}^k$ vanish with distance from the perturbed interface
2. The C_i^k are subject to local equilibrium at every position on the perturbed interface. That is, for each phase the independent concentrations at a given position along the interface must define a point on the corresponding phase boundary of the ternary phase diagram and the two points thus defined must be joined by a tie line. Because the amplitude of the perturbation is infinitesimal, these points are infinitesimally removed from the points corresponding to a flat interface (i.e. R and S in Fig. 11.12). It is valid therefore to approximate the phase boundaries in this vicinity of the phase diagram by their tangents at the points defined by the concentrations at a flat interface. Accordingly, one can write

$$C^{\alpha}_{A\phi} = m^{\alpha} C^{\alpha}_{B\phi} + b^{\alpha}$$
$$C^{\gamma}_{A\phi} = m^{\gamma} C^{\gamma}_{B\phi} + b^{\gamma} \qquad (11.20)$$

where $C^{\alpha}_{A\phi}$, $C^{\alpha}_{B\phi}$, $C^{\gamma}_{A\phi}$, and $C^{\gamma}_{B\phi}$ are the equilibrium concentrations at the perturbed interface, $z = \phi(x, t)$, m^{α} and m^{γ} are, as before, the slopes of the tangents to the phase boundaries, and b^{α} and b^{γ} are the corresponding intercepts. Equations (11.20) will ensure that the interface concentrations in each phase define points on the corresponding phase boundaries. However, a further condition is necessary to ensure that these points are joined by a tie line of the phase diagram. Accordingly, the following condition is imposed

$$C^{\alpha}_{B\phi} = m^{\alpha\gamma} C^{\gamma}_{B\phi} + b^{\alpha\gamma} \qquad (11.21)$$

where $m^{\alpha\gamma}$ is defined as the slope of the tangent, at the point defined by the planar interface compositions, to a plot of the equilibrium interface concentration of B in the α-phase versus B in the γ-phase and $b^{\alpha\gamma}$ is the corresponding intercept. It can be proven that $m^{\alpha\gamma}$ is related to the slope m of the given tie line by

$$m^{\alpha\gamma} = \frac{m^{\gamma} - m}{m^{\alpha} - m} \qquad (11.22)$$

All three of equations (11.20) and (11.21), taken together, define the local equilibrium boundary condition. The parameters involved in these relationships are obtained directly from the ternary phase diagram in question, after a calculation of the unique diffusion path for a planar interface. The formulation of the local equilibrium boundary condition in this empirical manner allows consideration of all conceivable ternary systems and releases one from any assumptions relating to dilute solution limits and ideal or regular solution behaviour. Furthermore, this formulation yields results which can be related to the elementary approach previously described.

3. The following mass balance boundary conditions apply to every position along the perturbed interface

$$v(x) = \frac{1}{(C^{\alpha}_{A\phi} - C^{\gamma}_{A\phi})}\left[D^{\gamma}_{A}\left(\frac{\partial C^{\gamma}_{A}}{\partial z}\right)_{\phi} - D^{\alpha}_{A}\left(\frac{\partial C^{\alpha}_{A}}{\partial z}\right)_{\phi}\right]$$
$$= \frac{1}{(C^{\alpha}_{B\phi} - C^{\gamma}_{B\phi})}\left[D^{\gamma}_{B}\left(\frac{\partial C^{\gamma}_{B}}{\partial z}\right)_{\phi} - D^{\alpha}_{B}\left(\frac{\partial C^{\alpha}_{B}}{\partial z}\right)_{\phi}\right] \qquad (11.23)$$

where $v(x)$ is the velocity of the perturbed interface in the z direction at a given position defined by the x coordinate. It should be noted that for a planar interface $v(x) = V$ and equations (11.23) reduce to

$$V(C^{\alpha}_{AF} - C^{\gamma}_{AF}) = D^{\gamma}_{A} G^{\gamma}_{AF} - D^{\alpha}_{A} G^{\gamma}_{AF}$$
$$V(C^{\alpha}_{BF} - C^{\gamma}_{BF}) = D^{\gamma}_{B} G^{\gamma}_{BF} - D^{\alpha}_{B} G^{\alpha}_{BF} \qquad (11.24)$$

Following Mullins and Sekerka,[22] the interface concentrations are first written in the form

$$C^{k}_{i\phi} = C^{k}_{iF} + a^{k}_{i}\delta \sin \omega w \qquad i = A, B \quad k = \alpha, \gamma \qquad (11.25)$$

where the four constants a^{k}_{i} are to be determined from the boundary conditions.

Solutions of equations (11.18), which satisfy the boundary condition that they vanish at a distance from the perturbed interface, are introduced into equation (11.17) to give complete solutions which, to the first order in δ, reduce to equations (11.25) at the interface, $z = \delta \sin \omega x$. The complete solutions are

$$\left.\begin{aligned} C_i^\alpha &= C_{iF}^\alpha + G_{iF}^\alpha z + \delta(a_i^\alpha - G_{iF}^\alpha)\sin(\omega x)\exp(wz) \\ C_i^\gamma &= C_{iF}^\gamma + G_{iF}^\gamma z + \delta(a_i^\gamma - G_{iF}^\gamma)\sin(\omega x)\exp(-wz) \end{aligned}\right\} \quad i = \mathrm{A, B} \qquad (11.26)$$

Substituting equations (11.25) into equations (11.20) and (11.21) and equating coefficients of $\sin \omega x$ in each of the latter equations leads to

$$\begin{aligned} a_\mathrm{A}^\alpha &= m^\alpha m^{\alpha\gamma} a_\mathrm{B}^\gamma \\ a_\mathrm{B}^\alpha &= m^{\alpha\gamma} a_\mathrm{B}^\gamma \\ a_\mathrm{A}^\gamma &= m^\gamma a_\mathrm{B}^\gamma \end{aligned} \qquad (11.27)$$

which leaves a_B^γ as the remaining unknown constant. On substituting $z = \delta \sin \omega x$ into the z derivative of equations (11.25) and expanding to the first order in δ, expressions for the concentration gradients at the perturbed interface are obtained. These, along with equations (11.25) and (11.27), are introduced into equations (11.23). Coefficients of $\sin \omega x$ in the latter equations are equated to give an expression for the remaining constant a_B^γ. Rearrangement of this expression with equations (11.24) and application of conditions (11.19) leads to

$$a_\mathrm{B}^\gamma = \frac{(C_\mathrm{AF}^\alpha - C_\mathrm{AF}^\gamma)(D_\mathrm{B}^\gamma G_\mathrm{BF}^\gamma + D_\mathrm{B}^\alpha G_\mathrm{BF}^\alpha) - (C_\mathrm{BF}^\alpha - C_\mathrm{BF}^\gamma)(D_\mathrm{A}^\gamma G_\mathrm{AF}^\gamma + D_\mathrm{A}^\alpha G_\mathrm{AF}^\alpha)}{(C_\mathrm{AF}^\alpha - C_\mathrm{AF}^\gamma)(m^{\alpha\gamma} D_\mathrm{B}^\alpha + D_\mathrm{B}^\gamma) - C_\mathrm{BF}^\alpha - C_\mathrm{BF}^\gamma)(m^\alpha m^{\alpha\gamma} D_\mathrm{A}^\alpha + m^\gamma D_\mathrm{A}^\gamma)} \qquad (11.28)$$

The derivative of the amplitude of the perturbation is given by

$$v(x) = V + \dot{\delta} \sin \omega x \qquad (11.29)$$

Equation (11.28) can be equated to either one of equations (11.23); the last is arbitrarily chosen. Accordingly, interfacial gradients from equations (11.26) and components of equations (11.25) and (11.27), together with equation (11.28), are introduced into equation (11.23). Then the coefficient of $\sin \omega x$ is equated to $\dot{\delta}$ in equation (11.29) to give the following result after application of conditions (11.19) and rearrangement with equations (11.24)

$$\dot{\delta}/\delta = v\omega\eta/\varepsilon$$

where

$$\eta = (D_\mathrm{A}^\gamma G_\mathrm{AF}^\gamma + D_\mathrm{A}^\alpha G_\mathrm{AF}^\alpha)(m^{\alpha\gamma} D_\mathrm{B}^\alpha + D_\mathrm{B}^\gamma) - (D_\mathrm{B}^\gamma G_\mathrm{BF}^\gamma + D_\mathrm{B}^\alpha G_\mathrm{BF}^\alpha)(m^\alpha m^{\alpha\gamma} D_\mathrm{A}^\alpha + m^\gamma D_\mathrm{A}^\gamma)$$

and

$$\varepsilon = (D_\mathrm{A}^\gamma G_\mathrm{AF}^\gamma - D_\mathrm{A}^\alpha G_\mathrm{AF}^\alpha)(m^{\alpha\gamma} D_\mathrm{B}^\alpha + D_\mathrm{B}^\gamma) - (D_\mathrm{B}^\gamma G_\mathrm{BF}^\gamma - D_\mathrm{B}^\alpha G_\mathrm{BF}^\alpha)(m^\alpha m^{\alpha\gamma} D_\mathrm{A}^\alpha + m^\gamma D_\mathrm{A}^\gamma) \qquad (11.30)$$

The problem is to ascertain under what conditions the above expression becomes positive (i.e. the perturbation grows, indicating instability). To this end, mean diffusivities for each component are defined in the following manner

$$\overline{D}_A = \tfrac{1}{2}(m^\alpha m^{\alpha\gamma} D_A^\alpha + m^\gamma D_A^\gamma)$$
$$\overline{D}_B = \tfrac{1}{2}(m^{\alpha\gamma} D_B^\alpha + D_B^\gamma) \qquad (11.31)$$

and weighted interfacial concentration gradients can be defined such that

$$G_{iF}^k = \frac{D_i^k G_{iF}^k}{\overline{D}_i} \qquad i = A, B \quad k = \alpha, \gamma \qquad (11.32)$$

Substitution of equations (11.31) into equation (11.30), division of numerator and denominator by $2\overline{D}_A\overline{D}_B$, introduction of equations (11.32), and then slight rearrangement leads to

$$\frac{\dot{\delta}}{\delta} = v\omega \left\{ \frac{(G_{AF}^\gamma - G_{BF}^\gamma) + (G_{AF}^\alpha - G_{BF}^\alpha)}{(G_{AF}^\gamma - G_{BF}^\gamma) - (G_{AF}^\alpha - G_{BF}^\alpha)} \right\} \qquad (11.33)$$

Noting the similarity between terms in the numerator and denominator, the latter are multiplied by the denominator to obtain

$$\frac{\dot{\delta}}{\delta} = v\omega \left\{ \frac{(G_{AF}^\gamma - G_{BF}^\gamma)^2 - (G_{AF}^\alpha - G_{BF}^\alpha)^2}{[(G_{AF}^\gamma - G_{BF}^\gamma) - (G_{AF}^\alpha - G_{BF}^\alpha)]^2} \right\} \qquad (11.34)$$

Thus only the numerator need be considered to determine the sign of the above expression. Since ω is necessarily positive, the appropriate stability criterion is

$$v\{|G_{AF}^\gamma - G_{BF}^\gamma| - |G_{AF}^\alpha - G_{BF}^\alpha|\} > 0 \qquad (11.35)$$

Coates and Kirkaldy[23] show that a number of commonly recognized cases submit to this analysis (see also Whittle *et al.*[28]). Furthermore, they demonstrate that it is only in very special cases that the elementary criterion (relation (11.15)) agrees with relation (11.35) and that in general the former is neither a necessary nor a sufficient condition for instability to set in. Because of the great complexity of relation (11.35) workers have not been greatly tempted to seek a closure with experiment (cf Mullins and Sekerka[22]). Thus the qualitative diagnostic methods introduced in §11.4 persist as the main pathway to understanding and to technological applications.

11.6 DIAGNOSTICS FOR COMPLEX DIFFUSION PATHS ON THE ISOTHERM

This section deals with a series of statements or theorems concerning the microstructural paths of ternary multiphase diffusion couples, expanding the earlier work of Rhines[29] and Clark and Rhines.[30] Here the review article of Kirkaldy and Brown[1] is followed.

First a number of theorems of both positive and negative nature are stated which pertain to the representation of the diffusion paths for single-phase infinite diffusion couples upon an isothermal section of the ternary constitution diagram. While such representations are prevalently used and are convenient from some points of view, they incompletely describe the kinetics of the situation. A complete representation must include the penetration curves for the two independent concentrations.

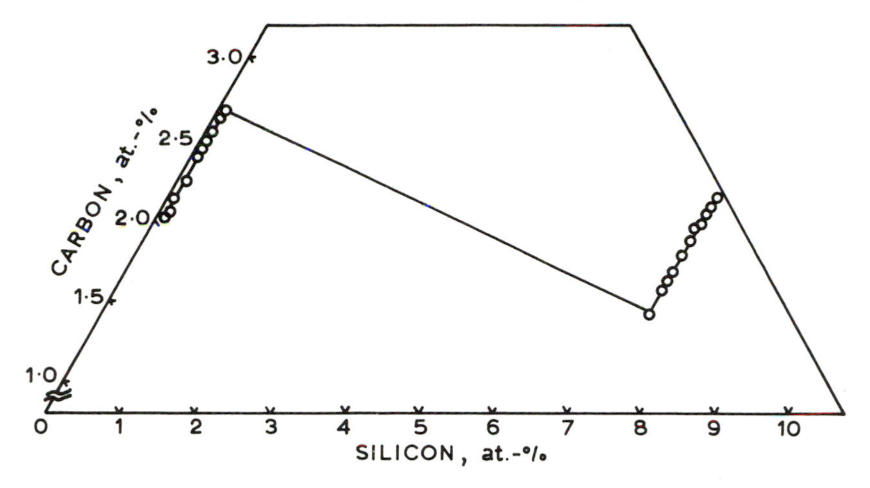

Reprinted with permission from *Can. Met. Q.*, **2**, J. S. Kirkaldy and L. C. Brown, p. 89, copyright 1963, Pergamon Press Ltd

Fig. 11.13 Map of the penetration curve of Fig. 4.2 on the ternary isotherm. After Kirkaldy and Brown[1]

1. *Diffusion penetration curves can be mapped onto the ternary isotherm for all times as stationary lines*

Figure 11.13 shows the result of mapping the penetration curves of Fig. 4.2 on the Fe–C–Si 1050°C isotherm. This time-invariant mapping is a direct consequence of the fact that both solute concentrations can be expressed as unique functions of the same parameter λ. However, in discarding λ, and with it the metric of the penetration curves, we discard most of the kinetic information contained therein.

2. *Calculated paths on ternary isotherms remain invariant as the four diffusion constants are varied in direct proportion*

This is simply because the quantities involving D, like $\sqrt{t}$, act as scale factors for x, and so the solutions of the diffusion equations remain in proportion spatially under these proportionate variations.

3. *Diffusion paths cannot be mapped back into the C_1–C_2–λ space without the reintroduction of diffusion data*

This statement reaffirms the closing statement of Theorem 1. Even if one combines the stationary path with the mass balance there still remain an infinite number of curves in C_1–C_2–λ space which correspond to the stationary path.

4. *A diffusion path for an infinite couple on the ternary isotherm must cross the straight line joining the terminal compositions at least once*

This statement, due to Meijering,[31] is a simple consequence of the mass balance. Considering an infinite couple as a whole, it is clear that the mean composition must lie along the line joining the termini. Now if the diffusion path were to lie all on one side of this line (such as ACB in Fig. 11.14), then the mean composition of

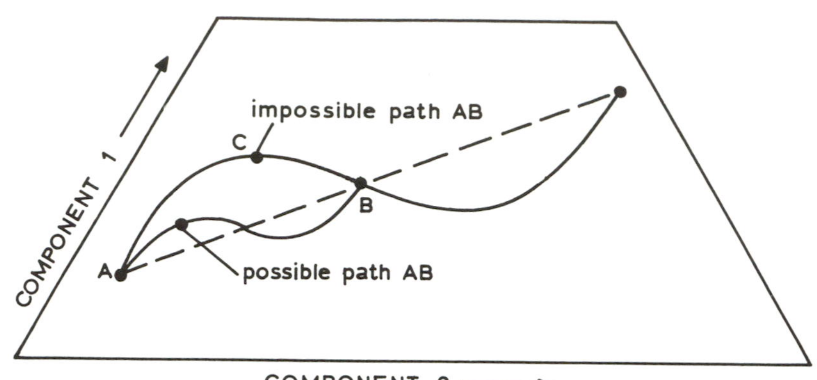

Reprinted with permission from *Can. Met. Q.*, **2**, J. S. Kirkaldy and L. C. Brown, p. 89, copyright 1963, Pergamon Press Ltd

Fig. 11.14 Schematic representation of possible and impossible paths on the ternary isotherm. After Kirkaldy and Brown[1]

the couple would have to be off the line, in contradiction to the original mass conservation requirement.

5. *The diffusion path for a system with constant diffusion coefficients crosses the straight line joining the terminal compositions once and only once at the mid-point of this line*

The proof of this is given by Kirkaldy and Brown[1] as is that for the following statement.

6. *The diffusion path for a system with constant coefficients will be a straight line when*

$$D_{21}(C_{10} - C_{11})^2 - D_{12}(C_{20} - C_{21})^2 = (D_{11} - D_{22})(C_{10} - C_{11})(C_{20} - C_{21}) \quad (11.36)$$

7. *A path in a single-phase region in which the off-diagonal coefficients are negligible will always be S-shaped*

The degree of eccentricity will increase with increasing difference between the two diffusion on-diagonal coefficients. In the case where the coefficients are equal, the curve will degenerate into a straight line (see Theorem 6). If the two terminal compositions lie on adjacent sides of the ternary phase diagram, then the diffusion path will bend away from the most rapidly diffusing component, as noted by Clark and Rhines.[30] However, with other boundary conditions and non-zero off-diagonal coefficients, such a rule fails (see, for example, Fig. 11.13).

8. *A diffusion path on the ternary isotherm is defined uniquely only by its terminal compositions*

While this appears obvious from the form of the diffusion equations, workers have made the contrary statement that *any* two points on a path uniquely define it. This maintains, in effect, that the path of a couple with termini that lie upon the path of another couple will have a coincident path. This is clearly incorrect, for path ACB in Fig. 11.14 between compositions A and B is an impossible one by Theorem 4.

9. *There is no theoretical restriction that prevents paths radiating from one terminal composition from crossing*

If the on-diagonal diffusion coefficients are constant, paths radiating from a single composition to the isotherm boundaries will not cross. However, when the off-diagonal coefficients are non-zero and all coefficients are concentration-dependent, or when the locus of the moving terminus does not coincide with the isotherm boundary, cross-overs are bound to occur (Fig. 11.15).

10. *Paths radiating from a single point to points along any line in the isotherm will reach all points lying between the extreme paths at least once*

This follows from the continuity of solutions of the Fick equations with changes in one of the boundary concentrations and from Theorem 9.

Turning next to diffusion paths in multiphase regions we refer to Figs. 11.9, 11.10, 11.16, and 11.17 which present metallographically some examples of multiphase paths in the system Cu–Sn–Zn. Because multiphase paths through a ternary diffusion couple contain a number of lengths of single-phase path, many of the theorems or rules stated above will apply to these as well.

11. *Theorems 1, 2, 3, 4, 8, and 9 for single-phase paths apply to multiphase paths containing all flat interfaces, while Theorems 4, 8, and 9 apply to paths containing two-phase regions (non-planar interfaces) as well*

It has been noted that an exact parametric solution (λ-dependent) does not exist for systems with non-planar interfaces. Accordingly it is not possible to represent

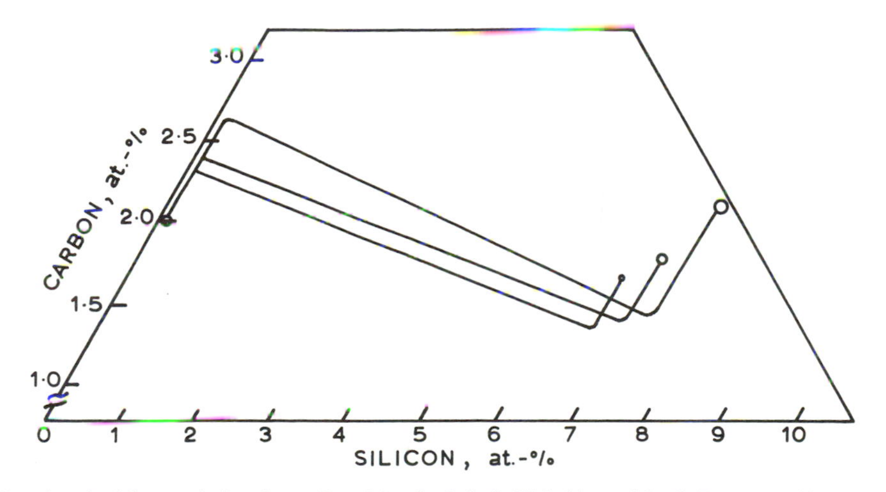

Reprinted with permission from *Can. Met. Q.*, **2**, J. S. Kirkaldy and L. C. Brown, p. 89, copyright 1963, Pergamon Press Ltd

Fig. 11.15 Calculated diffusion paths in the 1 000°C isotherm of the Fe–C–Si diagram, which radiate from a point but nonetheless cross over. $D_{11} = 4{\cdot}8 \times 10^{-7}$ cm^{2}s^{-1}, $D_{22} = 4{\cdot}6 \times 10^{-11}$ cm^{2}s^{-1}, $D_{12}/D_{11} = 12{\cdot}4\, X_c$, $D_{21} = 0$. After Kirkaldy and Brown[1]

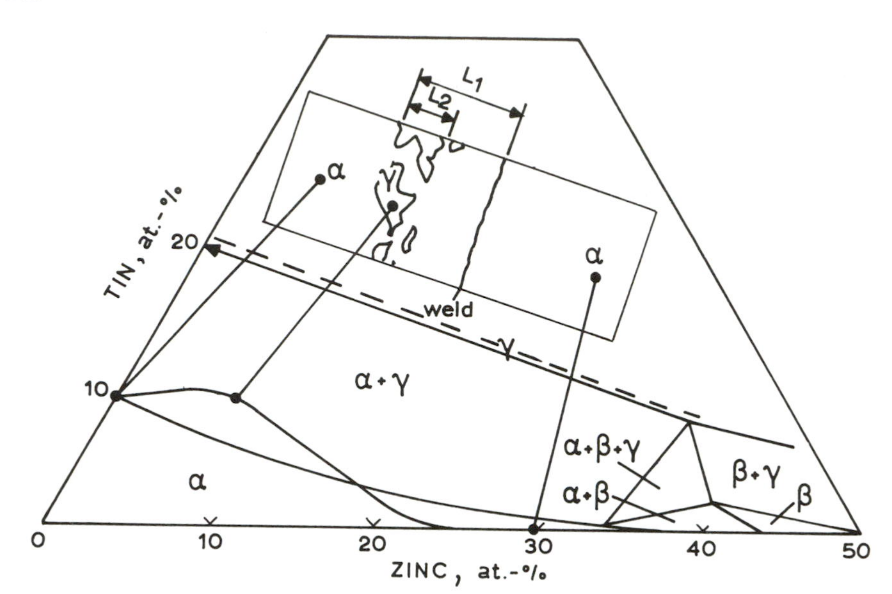

Reprinted with permission from *Can. Met. Q.*, **2**, J. S. Kirkaldy and L. C. Brown, p. 89, copyright 1963, Pergamon Press Ltd

Fig. 11.16 Two-phase diffusion path in the 500°C isotherm of the Cu–Zn–Sn diagram. Micrographic inset is traced. Virtual diffusion path plotted for $D_{11} = 9{\cdot}2 \times 10^{-12}$ cm²s⁻¹, $D_{22} = 4{\cdot}0 \times 10^{-12}$ cm²s⁻¹, $D_{12} = +9{\cdot}2 \times 10^{-12}$ cm²s⁻¹, $D_{21} = +2{\cdot}3 \times 10^{-12}$ cm²s⁻¹. Magnification of inset micrograph: about 520×. After Kirkaldy and Brown[1]

such paths by a single stationary line on the ternary isotherm. It is thus necessary to qualify Theorem 1 by Theorem 12.

12. *To the extent that lateral diffusion and non-uniformity of layer interfaces can be ignored or averaged out, the diffusion paths involving two-phase regions may be approximated by a stationary path connecting a continuous series of local equilibria*

A roughly equivalent statement is that the thermodynamic activity is approximately constant on every cross-section and the activity profile regresses in a parabolic fashion. The results of Rhines[29] and Clark and Rhines[30] experimentally bear this out and record a parabolic growth law within the experimental accuracy. Clark and Rhines originally suggested that each parallel layer of the couple is in equilibrium across the whole couple, with no lateral diffusion. However, some lateral diffusion must occur, since the diffusion equations cannot otherwise accommodate two-phase growth. The very ragged nature of the observed columnar layers and their necessary change in dispersion with time bear out the existence of lateral diffusion.

Assuming then, that a path through a two-phase region can be represented to a fair approximation by a stationary line on the ternary isotherm, the possible configurations for such lines and their interpretation must be considered.

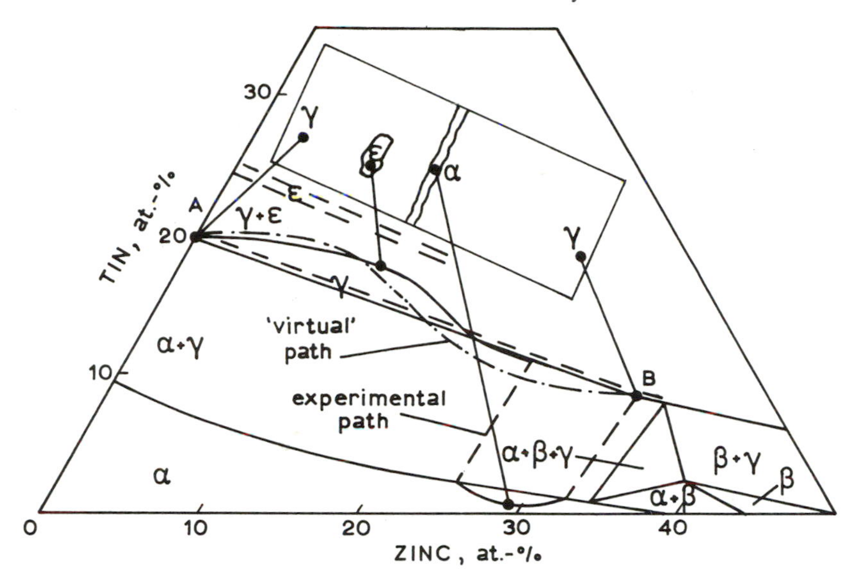

Fig. 11.17 Three-phase diffusion path in the 500°C isotherm of the Cu–Zn–Sn diagram. Micrographic inset is traced. Virtual path calculated for $D_{ZnZn} = 2{\cdot}3 D_{SnSn}$ and neglecting off-diagonal coefficients. Magnification of inset micrograph: about 46×. After Kirkaldy and Brown[1]

13. *A diffusion path that passes through a two-phase region coincident with a tie line contains a planar interface whose local equilibrium specification is given by that tie line*

As a convention, the section of path coincident with a tie line should be dashed to indicate that it corresponds to zero spatial extent in the couple.

14. *A diffusion path that passes into a two-phase region from a single-phase one at an angle to the tie lines and returns immediately to that same single phase describes a region of isolated precipitation*

Figures 11.9 and 11.16 show examples of this behaviour along with corresponding calculated penetration curves. Bear in mind that the path lying within the two-phase region designates the lever rule quantities of the two phases at each point. The actual diffusion paths are not designated, but must conform at phase interfaces to the local equilibrium concentrations given by the tie line projected out from the drawn path. In support of Rule 12, it is evident from the complex random nature of these diffusion zones that expansion of the precipitate cannot occur without simultaneous dissolution and three-dimensional diffusion. Figure 11.18 shows the growth plot corresponding to the couple of Fig. 11.16. In agreement with the observations of Clark and Rhines, this is approximately parabolic except for a possible slight delay in the leading edge of the precipitate near the time origin.

15. *A diffusion path that passes into a two-phase region from a single-phase one at an angle to the tie lines but exits into another phase represents a columnar or a columnar-plus-isolated-precipitate two-phase zone*

Figure 11.10 shows a microstructural example of this behaviour. Here the path on the phase diagram enters the two-phase region at an angle to the tie lines but exits coincident with one of them. The cuspoidal point on the path represents a discontinuity at which the precipitating phase jumps from a fraction to 100% of the cross-section. Note that the mixed zone of this couple contained both isolated (not shown) and non-isolated precipitates, the latter columnar precipitates being rooted in the parent phase.

16. *A diffusion path in a two-phase region may not reverse its order of crossing of the tie lines*

The activities of both independent components of the phases at the opposite extremes of a two-phase region must be monotonic functions of position along the phase boundary. This is because an optimum in the activity along this direction would imply entry into another two-phase region. Thus, a path that reverses its order within a two-phase region implies an optimum in activity of both components in both phases at some point along the diffusion path (i.e. when the path parallels the tie line). Such optima are thermodynamically unstable, and if they were to occur in a diffusion couple they would immediately moderate into a section of tie line plus a section of skew path, as in Fig. 11.10.

On occasion, skew paths in two-phase regions will encounter the boundaries of a three-phase triangle and emerge either into a single-phase region at a vertex or into a two-phase region along a second skew path. To the extent that such skew two-phase paths can be approximated by a single line their intersection with the triangle boundary will be in local equilibrium with the intersection point of any

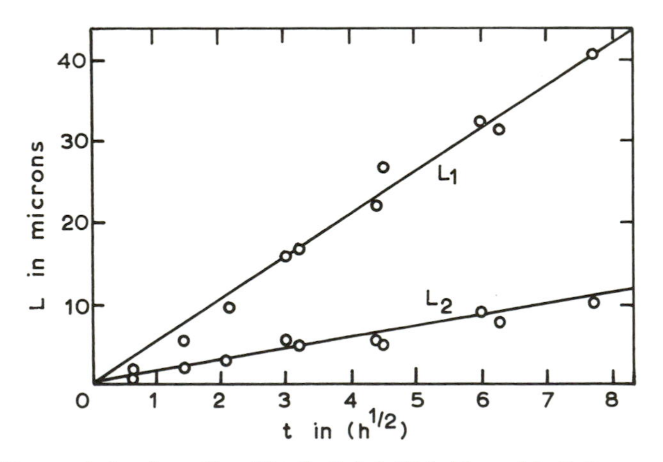

Reprinted with permission from *Can. Met. Q.*, **2**, J. S. Kirkaldy and L. C. Brown, p. 89, copyright 1963, Pergamon Press Ltd

Fig. 11.18 Parabolic growth plots for the leading and trailing precipitate edges corresponding to the couple of Fig. 11.16. After Kirkaldy and Brown[1]

other path. Such path pairs must accordingly meet at a planar interface (a consequence of the Gibbs phase rule). The convention should be adopted that such path pairs be joined by a dashed straight line to indicate that this section of the path corresponds to zero spatial extent in the couple (cf Rule 13). Clark and Rhines observed planar interfaces of this type in the two-phase *vs* one-phase and the two-phase *vs* two-phase combinations.

17. *Paths that pass through three-phase triangles must do so along a straight line representing at its extremes the local equilibrium existing at a planar multiphase interface*

A path may not enter and exit on the same face of the triangle because of Theorem 16.

The problem of calculation of paths that contain non-planar interfaces or precipitate layers appears to be impossible of exact solution. However, the concept of the virtual path in some cases allows one to predict whether or not a path will have planar or non-planar interfaces or precipitate layers. A virtual path is calculated by guessing an order of appearance of the phases between the termini which satisfies Theorem 4 and attempting to calculate the diffusion path assuming that all interfaces are flat. For such a calculation, diffusion data will be extrapolated out of single-phase regions as required. If a solution is found which duplicates the assumed order and satisfies local equilibrium at all interfaces (i.e. the path crosses all two-phase regions coincident with tie lines) then this may be considered as a possible solution. If this calculated path does not at any point enter a two-phase field other than at a point of crossing, then this is the observable path. If, on the other hand, the calculated path dips into and out of two-phase regions in one or more places without crossing them, then the calculated path is unstable and indeed 'virtual'.

Figure 11.7 shows a schematic path having this possible character. At the X end of the couple the path dips in and out of the $\alpha + \gamma$ field from the α side at points removed from the interface, producing a region of supersaturation. One therefore expects isolated precipitation to appear here. Figures 11.9 and 11.16 show experimental examples of such behaviour. The diffusion paths in these figures were calculated as virtual paths using available diffusion constants. The calculations have in both cases successfully predicted the appearance of isolated precipitates. The parabolic growth plot in Fig. 11.18 for the diffusion couple of Fig. 11.16 gives a value for the rate constant of the leading edge of the precipitate $8{\cdot}7 \times 10^{-4}$ mm^2 s^{-1}. The rate constant for the diffusion front calculated using the α-phase diffusion constant is 11×10^{-4} mm^2 s^{-1}. This agreement indicates that virtual diffusion paths calculated by extrapolating diffusion coefficients for single-phase regions into adjoining two-phase areas are useful in making estimates of actual path behaviour.

At the Y end of the diffusion path in Fig. 11.7 the region of supersaturation is in contact with the interface (the tie line) so that one might expect both isolated and non-isolated precipitates to occur. Naturally, when and if precipitation occurs, the virtual path will vanish and a complex real one takes its place.

Figure 11.17 represents the result of a particularly significant experiment wherein the possible path is non-unique or degenerate (cf Chapter 12). This is a diffusion couple constructed of terminal samples of the same intermetallic phase but of widely different composition. Since the phase extent on the isotherm is very narrow, it is to be expected that a virtual path will necessarily swing into two-phase regions. The dashed path is a virtual one calculated on the basis that $D_{ZnZn} = 2{\cdot}3D_{SnSn}$ (as in the α-phase). This path suggests four distinct physical possibilities: (i) the path will remain metastable, (ii) ε-phase will precipitate on the A end but the path will remain metastable on the B end, (iii) α-phase will precipitate on the B end but the path will remain metastable on the A end, or (iv) precipitation will occur on both ends removing the supersaturation. The high density of lattice defects as nucleation sites rule out cases (i)–(iii), but as far as the diffusion equation and boundary conditions are concerned all cases are acceptable solutions. It was indeed a surprise to find experimentally that the actual solution is none of these. As the micrograph and solid path A–B show, the path dips in and out of the α–phase along tie lines to produce a uniform band of α rather than a scattered precipitate layer. It is experimentally found that this band thickens parabolically.

There is a hint in this result as to the basis of nature's choice of path from among the mathematically possible ones. Brown and Kirkaldy had inferred from the relative widths of diffusion zones that diffusion of both Zn and Sn is much faster in the γ than in the α-phase. Hence, nature's interjection of the high resistance α-layer into the diffusion zone of Fig. 11.17 effectively slows down the rate of dissipation of free energy within that couple. Furthermore, in the couples of Figs. 11.9 and 11.16 the overall series resistance is maintained high by formation of discontinuous precipitates of the low-resistance γ-phase in the high-resistance α-phase. It is very likely that the ε-phase has an even lower resistance than the γ-phase since it lies much closer to the liquidus. Therefore, in the couple of Fig. 11.17 the overall series resistance of the couple is again maintained high by formation of discontinuous ε-precipitates within the γ matrix.

Although these couples are all non-constrained, unsteady systems, the results appear to manifest in some qualitative way the principle of minimum entropy production. Noting the confirmatory statement of Clark and Rhines that 'the interface type of two-phase structure (corresponding to parallelism of the composition path and a two-phase tie line) occurs with unexpected frequency', we are inclined to think that high-resistance configurations are preferred as a manifestation of a thermodynamic imperative. This thermodynamic principle is to be ultimately derived from irreversible thermodynamics. Such a principle must remove not only the topographical degree of freedom but also that associated with the choice of degree of dispersion. The problems of degeneracy and pattern formation are discussed more fully in Chapter 12. Further applications of the foregoing virtual path concepts are to be found in articles by Smeltzer and co-workers,[32, 33, 34] Dalvi and Coates,[35] Young *et al.*,[36] Hofmann and Politis,[37] and Olander.[38]

11.7 INTERNAL PRECIPITATION AND SPONTANEOUS EMULSIFICATION

An example of internal precipitation brought on by interdiffusion was offered in Fig. 11.9. A more common situation occurs during the oxidation or sulphidation of a binary alloy from the gas phase wherein a zone of fine oxide or sulphide particles penetrates from the surface in a more or less parabolic fashion.[29, 39]

Fig. 11.19 Spontaneous emulsification of oil containing 2% oleic acid (×14). After Mansfield[41]

Analogously, when a liquid consisting of 10% methyl alcohol and 90% toluene is placed quietly upon water in a test tube, the water remains clear but water droplets appear in the upper liquid forming a turbid emulsion.[40] Both phenomena involve ternary diffusion, which through an appropriate constitution creates a growing zone of supersaturation and parallel precipitation, as in Fig. 11.9, or growing protuberances ('streamers' in the liquid case) as in Fig. 11.10. The emulsification problem is usually complicated by buoyancy effects and violent turbulence caused by changes in surface tension with concentration (the Marangoni effect; cf McBain and Wu,[40] Mansfield,[41] Sternling and Scriven[42]) so there has not been a great deal of progress in experimental quantification of the phenomena. In solids, however, there are fewer artifacts, so considerable progress has been made on a closure between theory and experiment.

As in the previous section a parabolic isothermal ternary diffusion interaction, as commonly found in alloy oxidation, can be usefully described as a path on the isotherm. See, for example, Clark and Rhines,[30] Kirkaldy and Brown,[1] and Coates and Dalvi.[35] Some possible diffusion paths for oxidation are shown on the schematic isotherm of Fig. 11.20. This isotherm is for a standard-state pressure of component 1 (sulphur or oxygen) for which no liquid phases appear. Since the condensed-state phase boundaries are virtually unaffected by changes in the gas-phase pressure, the same diagram can be used in discussions about interactions at all pressures less than that for the standard state. For given reacting alloys f, b, d, h, and j and $p_1 < p_1^0$, there will be unique outer compositions of the oxides (MO or NO) at points like e, a, c, k, g, and i, respectively. Although the specified pressure uniquely defines only a line of constant activity in the diagram, there is always a further condition which restricts the outer composition to a point on that line. For

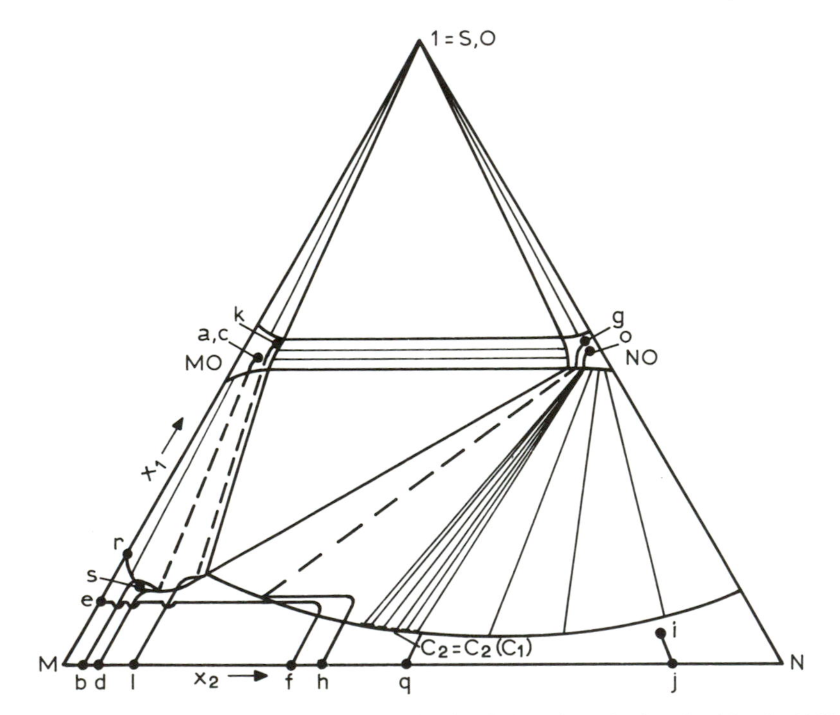

From J. S. Kirkaldy, 'Oxidation of Metals and Alloys', American Society for Metals, 1971. With permission

Fig. 11.20 Schematic diffusion paths on the isotherm corresponding to alloy oxidation. After Kirkaldy[43]

example, path ij which lies entirely in the metal phase (no oxide formation) must coincide with a component ray from point 1 since no significant relative motion of M and N is implied or possible. Thus point i is determined as the intersection of the component ray through j with a constant activity line defined by p_1.

Path ef describes the case where component N is evaporating from the couple in countercurrent motion to component 1 diffusing inward. C_{20} is therefore zero so p_1 at the binary limit uniquely defines point e.

Paths ab, cd, kl and gh imply the formation of an outer oxide layer. If the cations are diffusing outward, the layer front advancing at a velocity v from the original surface defines two relations of the form

$$vC_{20} = J_{20}(C_{20}, C_{20}) \tag{11.37}$$

and

$$vC_{30} = J_{30}(C_{20}, C_{30}) \tag{11.38}$$

where C are the outer concentrations and J are predictable values of the fluxes of metal feeding the reaction front. By eliminating v, a relation of the form

$$F(C_{20}, C_{30}) = 0 \tag{11.39}$$

is obtained which, combined with the iso-activity line defined by p_1, uniquely defines the outer boundary conditions. On the other hand, if the main process is inward diffusion of anions, then the oxide diffusion path must be on a component ray which, with the inner interface mass balance, is sufficient constraint to uniquely define the path.

It will be noted in Fig. 11.20 that paths ab, kl, ef and gh, in contra-distinction to stable paths ij and cd, all cross tie lines in two-phase regions and thus imply the existence of supersaturation in the metallic diffusion zone.

It must be finally remarked that only one of the diffusion paths presented in Fig. 11.20 explicitly contains ternary diffusion cross-effects as indexed by a non-zero value of D_{12}. The reversal of concentration C_1 along path ab could only arise through a positive value of D_{12}. The pronounced knees in some of the other curves in the metal phase are simply a consequence of the implicit assumption that $D_1 > D_2$. The latter condition is by far the most common situation leading to diffusion instability.

An actual path in the metal, such as oq in Fig. 11.20, will now be quantitatively described. If the precipitation is occurring from an approximately uniform solution of composition (C_1, C_2) to a precipitate of composition (C_{10}, C_{20}) then the precipitate volume fraction p evolves according to Kirkaldy[43]

$$(C_1 - C_{10})\frac{\partial p}{\partial t} = (1 - p)\frac{\partial C_1}{\partial t} \tag{11.40}$$

and

$$(C_2 - C_{20})\frac{\partial p}{\partial t} = (1 - p)\frac{\partial C_2}{\partial t} \tag{11.41}$$

In a non-uniform solution, on the other hand, there is a net flow of solutes 1 and 2 in or out of each volume element. Equations (11.40) and (11.41) can be corrected simply by adding to the right-hand side the divergence terms $\partial J_{1x}/\partial x$ and $\partial J_{2x}/\partial x$, respectively, where J_{1x} and J_{2x} are the total fluxes in the metal crossing unit area. Neglecting ternary interactions, these will be

$$J_{1x} = -f(p)D_1\frac{\partial C_1}{\partial x} \tag{11.42}$$

and

$$J_{2x} = -f(p)D_2\frac{\partial C_2}{\partial x} \tag{11.43}$$

where $f(p)$ is the average fraction of the cross-sectional area not blocked by the precipitates. This fraction is taken to be

$$f(p) \simeq 1 - p \tag{11.44}$$

which is a poor approximation. However, since we will seek a solution for $p \ll 1$ the error is inconsequential. Hence equations (11.40) and (11.41) become

$$(C_1 - C_{10})\frac{\partial p}{\partial t} = (1-p)\frac{\partial C_1}{\partial t} - (1-p)D_1\frac{\partial^2 C_1}{\partial x^2} + D_1\frac{\partial p}{\partial x}\frac{\partial C_1}{\partial x} \tag{11.45}$$

and

$$(C_2 - C_{20})\frac{\partial p}{\partial t} = (1-p)\frac{\partial C_2}{\partial t} - (1-p)D_2\frac{\partial^2 C_2}{\partial x^2} + D_2\frac{\partial p}{\partial x}\frac{\partial C_2}{\partial x} \tag{11.46}$$

Provided p is not too large (say $<0{\cdot}1$), the non-linear right-hand term may be discarded and the equations written as

$$\frac{1}{1-p}\frac{\partial p}{\partial t} = \frac{1}{(C_1 - C_{10})}\frac{\partial C_1}{\partial t} - \frac{D_1}{(C_1 - C_{10})}\frac{\partial^2 C_1}{\partial x^2} \tag{11.47}$$

and

$$\frac{1}{1-p}\frac{\partial p}{\partial t} = \frac{1}{(C_2 - C_{20})}\frac{\partial C_2}{\partial t} - \frac{D_2}{(C_2 - C_{20})}\frac{\partial^2 C_2}{\partial x^2} \tag{11.48}$$

Now if the equation of the solubility curve is

$$C_2 = C_2(C_1) \tag{11.49}$$

equations (11.47) and (11.48) are combined and the left-hand side eliminated to give finally

$$\frac{\partial C_1}{\partial t} = D_1\frac{\left(1 - \dfrac{D_2}{D_1}\dfrac{m}{n}\right)}{\left(1 - \dfrac{m}{n}\right)}\frac{\partial^2 C_1}{\partial x^2} - \frac{D_2}{n-m}\frac{\mathrm{d}m}{\mathrm{d}C_1}\left(\frac{\partial C_1}{\partial x}\right)^2 \tag{11.50}$$

where

$$m = \frac{\mathrm{d}C_2}{\mathrm{d}C_1} < 0 \quad \text{and} \quad n = \frac{C_2 - C_{20}}{C_1 - C_{10}} \tag{11.51}$$

Again discarding the non-linear right-hand term as higher order ($n \neq m$)

$$\frac{\partial C_1}{\partial t} = D_1 \frac{\left(1 - \frac{D_2}{D_1}\frac{m}{n}\right)}{\left(1 - \frac{m}{n}\right)} \frac{\partial^2 C_1}{\partial x^2} \tag{11.52}$$

is obtained. One can also write down an analogous expression for C_2. Note that, although the non-linear terms have been discarded in the derivatives, the equation has not necessarily been linearized because m and n can still be strong functions of concentration.

In principle, it remains as a boundary value problem to integrate for C_1 and C_2, finally solving for the precipitate fraction p by integrating equations (11.47) or (11.48). This procedure has been carried out by Kirkaldy[43] in the analysis of the sulphidation of Fe–Mn alloys. A rather simpler, more transparent and very common case where $D_1 \gg D_2$ has been based on equation (11.52)[43,44] by Laflamme and Morral[45] and Ohriner and Morral.[46] Their theoretical construction is paraphrased in the following. They identify two limiting cases of subscale formation. One is a familiar case associated with small solubility products where the amount of precipitate is constant across the subscale zone and precipitation occurs at the subscale matrix interface.[39,47] In the other case precipitation occurs continuously through the subscale as the subscale forms. This case is treated in detail for situations when atoms from the gas phase have a higher mobility than solute atoms in the base metal, and it is found that the amount of precipitate follows an error function across the subscale zone.

Laflamme and Morral discuss the two limiting cases of subscale formation in terms of the ratio $K/C_{1S}C_{20}$, in which K is the solubility product for the formation of a compound, BC; C_{1S} is the initial concentration of the gaseous element at the surface of the material. They recognize that the subscales can form when the above ratio falls in the range $0 < K/C_{1S}C_{20} < 1$ and, accordingly, that the two limiting cases fall at the extremes of this range. If the ratio is greater than one the system is not saturated after reaction with the gas phase and precipitation will not result. Figures (11.21) and (11.22) illustrate how the two limiting cases can be distinguished on phase diagrams from the resulting microstructure. In the first case, for the limit as $K/C_{1S}C_{20} \rightarrow 0$ (Fig. 11.21) species C = 1 diffuses to the subscale interface at X where it reacts with solute B = 2 to form the compound BC. In Fig. 11.21*b* one can see that as the concentration of C varies from C_{1S} to near zero there is a negligible change in the lever arm length which represents the amount of BC present. Therefore, the amount of BC remains constant across the subscale zone.

In the other limit as $K/C_{1S}C_{20} \rightarrow 1$ (Fig. 11.22) species C reacts with B to form BC all across the subscale zone until at the subscale interface X its concentration drops below the saturation value. One can see in Fig. 11.22*b* that the lever arm length changes gradually from a maximum value to zero as the concentration varies between C_{1S} at the surface and C_{1i} at the subscale interface. Therefore, the amount of BC will change continuously across the subscale zone, from a maximum value at the surface to zero at the subscale interface. This variation of precipitate amount with distance is a characteristic and required feature of the second limiting case.

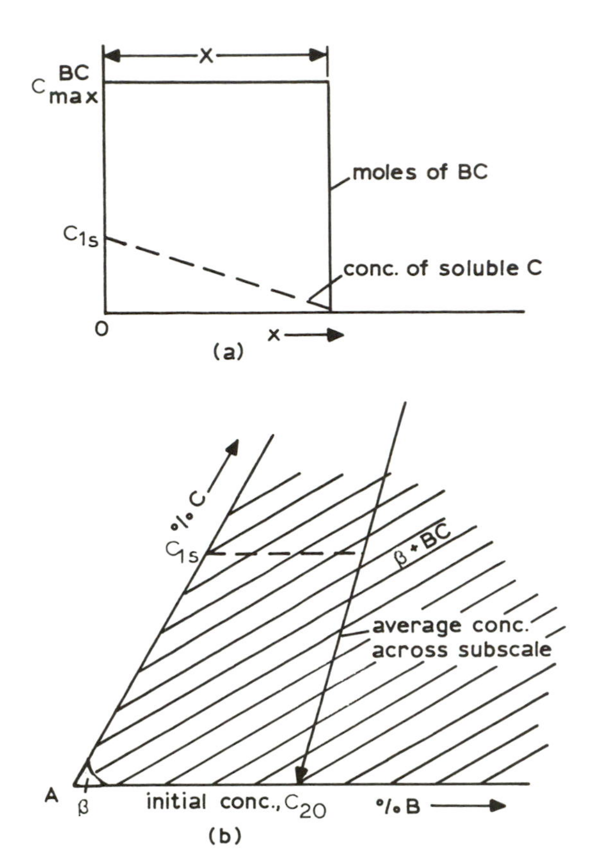

Reprinted with permission from *Acta Met.*, **26**, G. R. Laflamme and J. E. Morral, p. 1791, copyright 1980, Pergamon Press Ltd

Fig. 11.21 First limiting case. In (*a*) the moles of BC, C^{BC} is constant as a function of distance, *x*, across the subscale zone to *X*, the subscale-matrix interface. In (*b*), an isothermal section of the ternary phase diagram, the lever arm length is seen to b· constant until the concentration of C (*c*) approaches vanishingly small values at the interface. After Laflamme and Morral[45]

The first, and very common experimental limiting case has been treated theoretically in detail by Wagner, and reviews of its history and current status have been given by Rapp[47] and Swisher.[48] The proof that it is a special case solution of Kirkaldy's equation (11.52) is due to Morral and co-workers.[45,46] Proceeding from equation (11.52) the limiting cases can be demonstrated by setting $D_2 \ll D_1$ where

$$\frac{\partial C_1}{\partial t} = D_1 \frac{\partial^2 C_1}{\partial x^2} \bigg/ \left(1 - \frac{m}{n}\right) \tag{11.53}$$

where $C = C_1$ is the gas concentration, and $D = D_1$ is the diffusion coefficient of

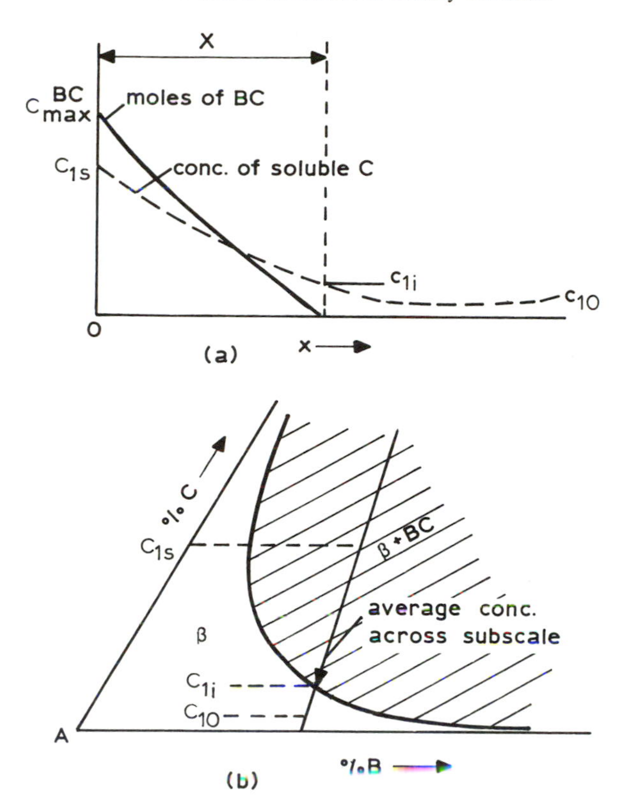

Fig. 11.22 Second limiting case. The same as Fig. 11.21 except the moles of BC varies with concentration of C, starting at a maximum value at the surface and dropping to zero at C_{1i}, the saturation value of the concentration. In (*b*), it can be seen that the lever arm varies continuously with C and goes to zero at C_{1i}. After Laflamme and Morral[45]

the gas species. A standard solubility product within the precipitate zone can now be assumed

$$C_1 C_2 = K \tag{11.54}$$

and the diffusion equation manipulated to the form

$$\frac{\partial C_1}{\partial t} = \frac{D_1}{K} n C_1^2 \frac{\partial^2 C_1}{\partial x^2} \tag{11.55}$$

Now since there is no long-range diffusion of B the mole fraction of precipitate can be written via the lever rule

$$f(x, t) = \frac{C_{20} - C_2(x, t)}{C_{2p}} \tag{11.56}$$

where C_{2p} is the concentration of B in the precipitate ($C_2 \ll C_{2p}$). Now if f_s is the mole fraction of precipitate at the sample surface, i.e.

$$f_s = (1 - K/C_{20}C_{1S})C_{20}/C_{2p} = \alpha C_{20}/C_{2p} \tag{11.57}$$

and the fraction of precipitate relative to the surface value is defined as

$$r = \frac{f}{f_s} \tag{11.58}$$

C_1 can be evaluated in terms of r as

$$C_1 = (K/C_{20})/(1 - \alpha r) \tag{11.59}$$

and thus equation (11.55) be converted to

$$\frac{\partial r}{\partial t} = \frac{\partial}{\partial x}\left[\frac{D_1(1-\alpha)}{q(1-\alpha r)^2}\right]\frac{\partial r}{\partial x} \tag{11.60}$$

where

$$q = C_{20}/nC_{1S} \tag{11.61}$$

A parabolic solution is sought by letting $r = r(\eta)$ where

$$\eta = \frac{x}{2(Dt(1-\alpha)/q)^{1/2}} \tag{11.62}$$

which reduces equation (11.60) to the ordinary equation

$$\frac{\mathrm{d}}{\mathrm{d}\eta}\left[\frac{(1-\alpha)}{(1-\alpha r)^2}\frac{\mathrm{d}r}{\mathrm{d}\eta}\right] + 2\eta\frac{\mathrm{d}r}{\mathrm{d}\eta} = 0 \tag{11.63}$$

Since $r = 0$ at $x = \eta = \infty$ and unity at the surface the initial and boundary conditions can be written uniquely in terms of η, i.e.

$$r(\eta = \infty) = 0 \qquad \text{at} \quad t = 0,\ x > 0$$

and

$$r(\eta = 0) = 1 \qquad \text{at} \quad t > 0,\ x = 0 \tag{11.64}$$

and the precipitate distribution is then uniquely described by the solution

$$r = r(\alpha, \eta) \tag{11.65}$$

Figure 11.23 shows the solution for r as a function of α and the normalized depth parameter

$$y = \frac{x}{2(D_1 t/q)^{1/2}} \tag{11.66}$$

When $\alpha \rightarrow 1$, the original case treated by Wagner,[39] the precipitate fraction is constant up to the subscale–matrix interface, which occurs at $y = 1/\sqrt{2}$. The second limiting case of $\alpha \rightarrow 1$ treated by Laflamme and Morral[45] and Ohriner and Morral,[46] which is simplest for an initially saturated alloy, yields the solution

$$r = \operatorname{erfc} y \tag{11.67}$$

Kirkaldy treated a more general case in which diffusion of M was not set equal to

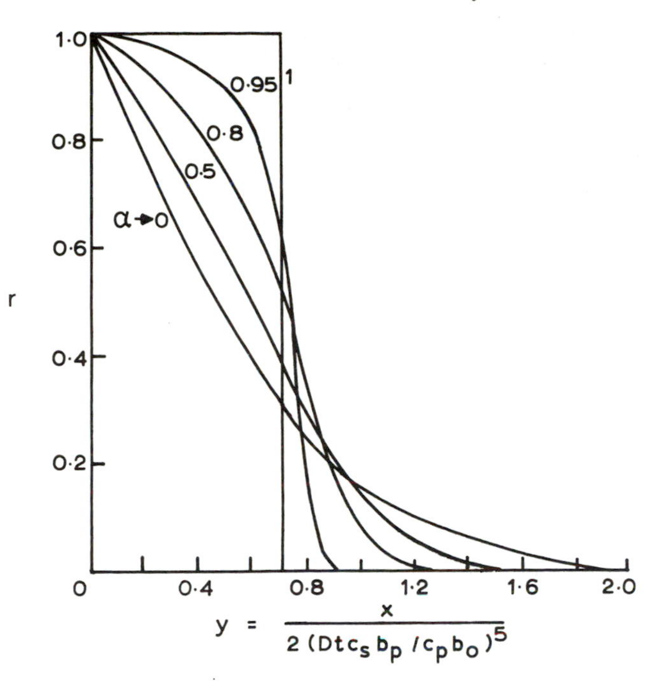

Reprinted with permission from *Scripta Met.*, **13**, E. K. Ohriner and J. E. Morral, p. 7, copyright 1979, Pergamon Press Ltd

Fig. 11.23 The dependence of the relative precipitate fraction, r, on y is shown for several values of the solubility parameter α. After Ohriner and Morral[46]

zero.[43] This also led to an erf-shaped solution and to a good prediction of the precipitate distribution and penetration of MnS precipitates in Fe.[49] Indeed, the mean frontal penetration of the sulphide was found experimentally to follow

$$\xi \simeq 6{\cdot}9 \times 10^{-7}\, t^{1/2}\,\mathrm{m} \qquad (11.68)$$

and predicted to follow

$$\xi \simeq 2(D_s t)^{1/2} \simeq 6{\cdot}6 \times 10^{-7}\, t^{1/2}\,\mathrm{m} \qquad (11.69)$$

REFERENCES

1 J. S. KIRKALDY AND L. C. BROWN: *Can. Met. Q.*, 1963, **2**, 89
2 R. E. LIESEGANG: *Naturwiss. Wochenschr.*, 1886, **11**, 353
3 W. SEITH: 'Diffusion in metallen', 157 *et seq.*, 1955, Berlin, Springer-Verlag
4 J. S. KIRKALDY: *Can. J. Phys.*, 1958, **36**, 917
5 G. V. KIDSON: *J. Nucl. Mater.*, 1960, **3**, 21
6 A. D. ROMIG JR AND J. I. GOLDSTEIN: *Met. Trans. A*, 1983, **14**, 1224
7 L. S. CASTLEMAN AND L. L. SEIGLE: *Trans. AIME*, 1957, **209**, 1173
8 H. BÜCKLE: *Metallforschung*, 1946, **1**, 175
9 R. C. DORWARD AND J. S. KIRKALDY: *Trans. AIME*, 1968, **242**, 2055
10 G. R. PURDY, D. H. WEICHERT, AND J. S. KIRKALDY: *ibid.*, 1964, **230**, 1025

11 M. HILLERT: *Trans. Iron Steel Inst. Jpn*, 1971, **11**, 1153
12 A. A. POPOV AND M. S. MIKHALEV: *Phys. Met. Metallogr. (U.S.S.R.)*, 1959, **7**, 36
13 J. B. GILMOUR, G. R. PURDY, AND J. S. KIRKALDY: *Met. Trans.*, 1972, **3**, 1455
14 R. C. SHARMA AND J. S. KIRKALDY: *Can. Met. Q.*, 1973, **12**, 391
15 J. S. KIRKALDY AND D. G. FEDAK: *Trans. AIME*, 1962, **224**, 490
16 A. D. SMIGELSKAS AND E. O. KIRKENDALL: *Trans. AIME*, 1947, **171**, 130
17 M. URIDNECEK AND J. S. KIRKALDY: *Z. Metallk.*, 1973, **64**, 419
18 M. URIDNECEK AND J. S. KIRKALDY: *ibid.*, 899
19 C. WAGNER: *J. Electrochem. Soc.*, 1954, **101**, 225
20 C. WAGNER: *ibid.*, 1956, **103**, 571
21 W. W. MULLINS AND R. F. SEKERKA: *J. Appl. Phys.*, 1963, **34**, 323
22 W. W. MULLINS AND R. F. SEKERKA: *ibid.*, 1964, **35**, 444
23 D. E. COATES AND J. S. KIRKALDY: *Trans. ASM*, 1969, **62**, 426
24 J. W. RUTTER AND B. CHALMERS: *Can. J. Phys.*, 1953, **31**, 15
25 W. A. TILLER, J. W. RUTTER, K. A. JACKSON, AND B. CHALMERS: *Acta Met.*, 1953, **1**, 428
26 D. E. COATES AND J. S. KIRKALDY: *J. Cryst. Growth*, 1968, **3**, **4**, 549
27 R. F. SEKERKA: *J. Phys. Chem. Solids*, 1967, **28**, 983
28 D. P. WHITTLE, D. J. YOUNG, AND W. W. SMELTZER: *J. Electrochem. Soc.*, 1976, **123**, 1073
29 F. N. RHINES: *Metals Technology*, Feb. 1940, p. 1
30 J. B. CLARK AND F. N. RHINES: *Trans. ASM*, 1959, **51**, 199
31 J. L. MEIJERING: discussion to Clark and Rhines, 1959
32 W. W. SMELTZER AND A. D. DALVI: *J. Electrochem. Soc.*, 1974, **121**, 386
33 W. W. SMELTZER AND D. P. WHITTLE: *ibid.*, 1978, **125**, 1116
34 W. W. SMELTZER AND P. C. PATNAIK: *ibid.*, 1985, **132**, 1233
35 A. D. DALVI AND D. E. COATES: *Oxid. Met.*, 1972, **5**, 113
36 D. J. YOUNG, T. NARITA, AND W. W. SMELTZER: *J. Electrochem. Soc.*, 1980, **127**, 679
37 P. HOFMANN AND C. POLITIS: *J. Nucl. Mater.*, 1979, **87**, 375
38 D. R. OLANDER: *ibid.*, 1983, **115**, 271
39 C. WAGNER: *Z. Electrochem.*, 1959, **63**, 772
40 J. W. MCBAIN AND T-M. WOO: *Proc. R. Soc.*, 1937, **A163**, 182
41 W. W. MANSFIELD: *Aust. J. Res.*, 1952, **5A**, 331
42 C. V. STERNLING AND L. E. SCRIVEN: *J.A.I.Ch.E.*, 1959, **5**, 514
43 J. S. KIRKALDY: *Can. Met. Q.*, 1969, **8**, 35
44 J. S. KIRKALDY: in 'Oxidation of metals and alloys', 101, 1971, Ohio, ASM
45 G. R. LAFLAMME AND J. E. MORRAL: *Acta Met.*, 1978, **26**, 1791
46 E. K. OHRINER AND J. E. MORRAL: *Scripta Met.*, 1979, **13**, 7
47 R. A. RAPP: *Corrosion*, 1965, **21**, 382
48 J. H. SWISHER: in 'Oxidation of metals and alloys', 210, 1971, Ohio, ASM
49 H. NAKAO: M.Eng. Thesis, 1967, McMaster University

CHAPTER 12

Self-organizing structures with diffusion-reaction control

12.1 INTRODUCTION

In previous chapters the existence and evolution of ordered or patterned structures within a purely diffusive or dissipative (entropy-producing) milieu was recognized. While some have detected in this a contradiction to the second law of thermodynamics, such does not in fact exist. Firstly, a rich macroscopic pattern need not possess high order in the (neg) entropic sense. Secondly, a misunderstanding arises from a failure to recognize that the universal entropy increase applies only to isolated systems, and even within the latter, a local entropy decrease is not denied within average entropy increase. This was emphasized by Lotka in the early years of this century[1,2] and reiterated by von Bertalanffy[3] and Prigogine[4] in later years. Prigogine's group has adopted the phrase 'dissipative structures' to describe certain evolutionary processes of order from chaos, including animate and inanimate diffusion-controlled, pattern-forming processes. The descriptive phrase 'self-organising structures' is adopted here, since by Prigogine's definition the term 'dissipative structures' is overly restrictive (cf § 12.7). Irrespective of these fundamental problems, this chapter represents a survey of some of the more complex initial and boundary value problems in combined diffusion and chemical reaction. At this time the optimizations offered may seem speculative to some readers.

12.2 THE PHYSICAL NATURE OF SPONTANEOUS PATTERN FORMATION

In the most elementary pattern-forming process such as exemplified by Fig. 5.2 the dissipative mixing of cobalt and iron has led to the unmixing of nickel in the ternary solution and to the corresponding stationary, skew-symmetric concentration pattern in $\lambda(=x/\sqrt{t})$ space. The self-organization is attributable to 'feedback' through the Onsager cross-interaction quantified by off-diagonal terms in the D-matrix. The term 'feedback' has been chosen here, for it is used in the same cooperative or collective sense as in Wiener's control theory[5] and in the Turing class of autocatalytic models of diffusion-reaction morphogenesis.[6,7]

It has often been emphasized[7] that dissipative structures occur in non-linear systems which are far from equilibrium. Actually, many interesting self-organizing systems which are amenable to theoretical analysis and experimental observation satisfy the local equilibrium criterion (cf Fig. 1.33) and evolve subject to weak driving forces. What seems to characterize many systems which are rich in pattern are singular yet loose internal constraints (analogous to Onsager reciprocity in the continuum) which allow nucleate 'bifurcations' of initial value structures and thus lead to a degeneracy or multiplicity of equivalent solutions in the steady or oscillatory state limit. The analogous internal constraints in the continuum are often represented by autocatalytic terms in Lotka–Volterra type growth–death coupled differential equations[8] (cf Chapter 12.11). These are in contrast to Gibbs-type constitutional constraints at internal surfaces, including capillarity effects, which identify free boundaries and their corresponding kinetic indeterminacy (cf Figs. 1.24 and 1.33). Since with free internal boundaries there are generally a finite or infinite number of stationary states defined or attainable, the problem of 'pattern selection' becomes paramount. In some cases different initial states lead to different final states within the same boundary conditions, a characteristic designated as 'hysteresis'. Often only chaotic or turbulent states are attained while in other cases a unique average stationary state is attained via a fluctuation search over the feasible states. Such fluctuations generally occur at imperfect internal and external surfaces, and if sufficiently frequent, but nonetheless benign so as to avoid chaos, allow the system to locate a unique average patterned state which is invariant in the presence of such fluctuations. It is this restricted class of Gibbs-constrained systems which particularly concerns us here (see, however, § 12.12.1).

While it is sometimes feasible in principle to predict the stable patterns for such systems via an appropriate perturbation analysis (an example is given in § 12.8) or via the computational modelling of the fluctuation process invoking renormalization group theory,[9] a much simpler approach to pattern selection offers itself through the hypothesis that the set of feasible stationary states coupled by fluctuations, as in the foregoing, represents a conventional statistical phase space which evolves towards a state of maximum probability at the stable stationary state. As we shall demonstrate in § 12.7, this corollary to the second law of thermodynamics implies that a stable but weakly fluctuating, steady-patterned state is characterized by a maximum or minimum (or both) in the entropy production rate. This represents at least a partial theoretical answer to the question raised by Nicolis and Prigogine[7] as to how one can 'find a set of thermodynamic properties characterizing the [dissipative] structure as uniquely as possible' and provides an operational criterion which has proven effective in practice.

12.3 INSTABILITY OF A SPHERICAL PRECIPITATE IN A SUPERSATURATED SOLUTION

The perturbation approach to the problem of interface stability was mooted by Wagner,[10,11] extended by Mullins and Sekerka,[12] and applied by many others (cf § 11.5). It involves the investigation of the growth and decay of perturbations applied to an interface and thus presents a pradigm for dendritic processes. The method is analogous to a technique widely used for testing hydrodynamic stability.[13] Since general perturbation methods are mathematically difficult as applied to the problems of interest here, the method will be introduced simply by considering the elementary problem of diffusion controlled precipitation of a sphere in a dilute supersaturated solution subject to small amplitude perturbations

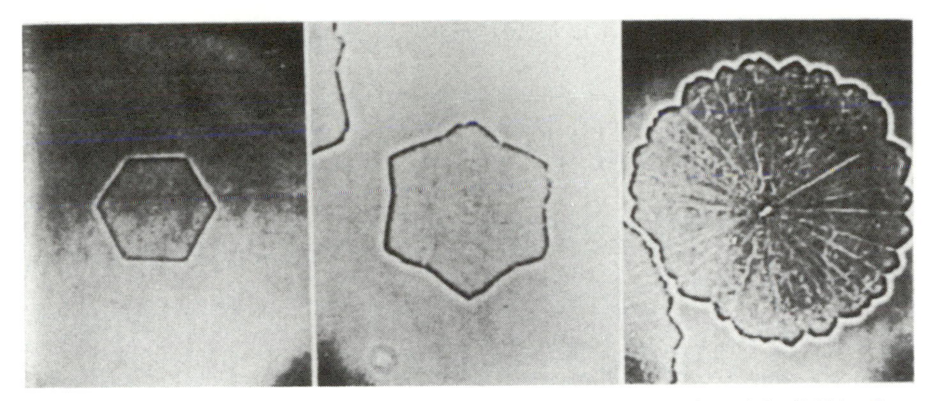

Fig. 12.1 Perturbation development on a hexagonal crystal of isotactic polystyrene. After Keith and Padden[14]

(cf § 1.23 for the stable solutions). Such a calculation is aimed at understanding morphological pattern development as in Fig. 12.1 (see also § 11.5).

In the method of Mullins and Sekerka[12] the assumption is made that the precipitating particle is much richer in solute than the solution from which it derives. In such a case the average interface moves a much shorter distance than the surrounding diffusion field and so to a good approximation the latter may be regarded as stationary. Furthermore the spherical symmetry in association with a quasi-stationary interface assures that the diffusion field in the neighbourhood of the interface may be accurately obtained as a solution of Laplace's equation (see Appendix 2). This approximation has been widely used in consideration of the precipitation of water and snow droplets from supersaturated air (see, for example Kirkaldy[15]).

A nearly spherical particle of radius R growing in a supersaturated solution of concentration C_∞ under the conditions noted above is considered and the behaviour of a small distortion of shape is investigated. The equation of the distorted sphere may be taken as

$$r = \rho(\theta, \psi) = R + \delta Y_{lm}(\theta, \psi) \tag{12.1}$$

where δ is a small amplitude of displacement and Y_{lm} is a single spherical harmonic (see Appendix 3). For a dilute solution, the equilibrium concentration $C_s(\theta, \psi)$ is given by the approximate Gibbs–Thomson equation (A1.48) as

$$C_s = C_0 + C_0 \Gamma K \tag{12.2}$$

where C_0 is the equilibrium concentration for a flat interface, $\Gamma = \sigma V/RT$, σ is the surface tension, and K is the mean curvature (equal to $2/R$ for a sphere).

The curvature of a slightly distorted sphere is given as

$$K(\theta, \psi) = \frac{2}{R}\left[1 - \frac{\delta Y_{lm}}{R}\right] - \frac{\delta \Lambda Y_{lm}}{R^2} \tag{12.3}$$

where Λ is the angular part of the Laplacian operator ∇^2, i.e.

$$\Lambda = \frac{1}{\sin\theta}\frac{\partial}{\partial\theta}\left(\sin\frac{\partial}{\partial\theta}\right) + \frac{1}{\sin^2\theta}\frac{\partial^2}{\partial\psi^2} \tag{12.4}$$

From the well-known fact that

$$\Lambda Y_{lm} = -l(l+1)Y_{lm} \tag{12.5}$$

and combining equations (12.3) and (12.2) we obtain for the equilibrium concentration along the interface

$$C_s(\theta,\psi) = C_0\left[1 + \frac{2\Gamma}{R}\right] + (l+2)(l+1)\frac{\Gamma\delta Y_{lm}}{R^2} \tag{12.6}$$

The first term contains the usual Gibbs–Thomson capillarity correction for a sphere, whereas the second term represents a small correction for deviations from sphericity. A solution of Laplace's equation is therefore sought which reduces to equation (12.6) at the interface and satisfies the boundary condition at infinity, C_∞. A sufficiently general solution is

$$C(r,\theta,\psi) = \frac{A}{r} + \frac{B\delta Y_{lm}}{r^{l+1}} + C_\infty \qquad l \geqslant 1 \tag{12.7}$$

where the first term corresponds to a spherical harmonic of order zero ($l = 0$). The second term represents the perturbation of the diffusion field due to the surface fluctuation. To the first order in δ from equations (12.1) and (12.7) give the approximation

$$C(r,\theta,\psi) = \frac{A}{R}\left[1 - \frac{\delta Y_{lm}}{R}\right] + \frac{B\delta Y_{lm}}{R^{l+1}} + C_\infty \tag{12.8}$$

It therefore remains to determine the coefficients A and B by evaluating equation (12.8) at the interface and equating coefficients with those in equation (12.6). This results in the solution

$$C(r,\theta,\psi) = \frac{(C_0 - C_\infty)R + 2C_0\Gamma}{r} + \frac{[(C_0 - C_\infty)R^l + C_0\Gamma R^{l-1}l(l+1)]\,\delta Y_{lm}}{r^{l+1}} + C_\infty \tag{12.9}$$

The radial velocity is given by

$$v = \frac{\mathrm{d}r}{\mathrm{d}t} = \frac{\mathrm{d}R}{\mathrm{d}t} + Y_{lm}\frac{\mathrm{d}\delta}{\mathrm{d}t} \tag{12.10}$$

Furthermore, the interface mass balance yields

$$v(C_p - C_s) = D\left.\frac{\partial C}{\partial r}\right|_{r=\rho} \tag{12.11}$$

Here C_p is the concentration of the precipitate and D is the diffusion coefficient in the solution. Thus,

$$v = \frac{D}{C_p - C_s}\left\{\frac{C_\infty - C_R}{R} + \left[(l-1)\frac{C_\infty - C_0}{R^2} - \frac{C_0}{R^3}[l(l+1)^2 - 4]\right]\delta Y_{lm}\right\} \tag{12.12}$$

where $C_R = C_0[1 + 2\Gamma/R]$ is the concentration on an undistorted sphere. Equating coefficients of Y_{lm} in equations (12.10) and (12.12) and noting that $C_p - C_R \gg (C_\infty - C_R)$

$$\dot{\delta}_l = \frac{d\delta}{dt} = \frac{D(l+1)}{(C_p - C_R)R}\left[\frac{C_\infty - C_R}{R} - \frac{C_0\Gamma}{R^2}(l+1)(l+2)\right]\delta_l \tag{12.13}$$

is obtained for the lth harmonic. It can be seen that $\dot{\delta}_l$ is composed of two terms — the first a gradient term which favours the growth of the harmonic, and the second a capillarity term which inhibits it. The question of stability therefore reduces to a determination of which term dominates.

The fundamental harmonic for $l = 1$ is

$$Y_{10} = (3/4\pi)^{1/2}\cos\theta \tag{12.14}$$

and this corresponds to a simple translation of the centre of the sphere a distance δ and so does not contribute to the growth. It is for this reason that $\dot{\delta}_l$ vanishes in equation (12.13). The fundamental harmonic for $l = 2$ is

$$Y_{20} = (5/4\pi)(\tfrac{3}{2}\cos^2\theta - \tfrac{1}{2}) \tag{12.15}$$

which corresponds to an ellipsoid of eccentricity $2\delta/R$ to the first order in δ. According to equation (12.13) this will have a positive growth rate provided

$$(l+1)(l+2) + 2 < R(C_\infty - C_0)/\Gamma C_0 \tag{12.16}$$

and the corresponding critical radius for growth of the harmonic $l = 2$ will be

$$R_c(l) = 7R^* \tag{12.17}$$

where

$$R^* = 2\Gamma C_0/(C_\infty - C_0) \tag{12.18}$$

is the critical nucleation radius corresponding to $C_s = C_R = C_\infty$, above which the sphere itself begins to grow (cf equation (12.12)). That is to say, the sphere becomes unstable at a radius seven times that at which it may grow. In a typical case of 10% supersaturation, $R_c \sim 0{\cdot}1\ \mu$m.

It is interesting to examine the shape change of a particle by examining the shape parameter

$$\frac{\dot{\delta}_l/\delta_l}{\dot{R}/R} = (l-1)\left[1 - \frac{R_c(l)}{R}\right]\left[1 - \frac{R^*}{R}\right]^{-1} \tag{12.19}$$

It can be seen that for $R \gg R_c(l)$ and $l = 2$ the shape parameter does not change with time, so that the shape of a sphere perturbed by this harmonic will be preserved.

Turning to the harmonic $l = 3$, an instability is found which leads to a change of shape. In this case the fundamental harmonic is

$$Y_{30} = (7/4\pi)^{1/2}(\tfrac{5}{2}\cos^3\theta - \tfrac{3}{2}\cos\theta) \tag{12.20}$$

The condition is determined whereby the shape parameter

$$-\frac{\dot{\delta}_3/\delta_3}{\dot{R}/R} > 1 \tag{12.21}$$

Further examination of equation (12.13) shows that there will be a harmonic of mean wavelength λ_M which grows faster than the others. This may be estimated by optimization for large l as

$$\lambda_M \approxeq \pi(6RR^*)^{1/2} \tag{12.22}$$

It might therefore be conjectured with Mullins and Sekerka that an arbitrary perturbation containing such a wavelength component will ultimately grow at the expense of the others and produce a dimpled surface on the scale of λ_M. Figure 12.1 shows some observations by Keith and Padden[14] which appear to correspond to this situation (see also Townsend and Kirkaldy[16]). It must, however, be emphasized that this optimization argument to establish the pattern dimensions is based within the linear, low amplitude regime and so can yield at best an estimate. A theory which deals with finite amplitude, non-linear morphological development as in Fig. 12.1 must draw upon more powerful optimization procedures to remove the evident free boundary degeneracy (cf § 12.7).

12.4 OSTWALD RIPENING: THE GROWTH OF DISPERSED PRECIPITATES

In § § 1.24 and 1.25 the subject was introduced pertaining to the precipitation of a collection of spherical particles via diffusion and/or reaction control without reference to the effects of surface tension. It was implicitly recognized that a free boundary problem is involved for it was necessary to introduce an experimental final mean radius ρ_0 (cf equation (1.155)) for a complete closure between theory and experiment. In the following treatment of the problem due to Wagner[17] simultaneous diffusion and reaction control is dealt with together with a non-uniform distribution of particle radii and the explicit effect through surface tension of such radii on the diffusive boundary conditions. In particular, high curvature raises the surface chemical potential or concentration so solute tends to flow from small to large particles and the latter to grow at the expense of the former. This process is known as Ostwald ripening.[18]

Following Wagner the precipitation is considered of a pure spherical crystal in a solution of vanishing solubility C_0, so according to equation (A1.45) the surface concentration in the matrix is (see Appendix 1)

$$C_r = C_0 \exp(2\sigma v^\beta/rRT) \tag{12.23}$$

which can be approximated for not too small particles as

$$C_r = C_0[1 + (2\sigma v^\beta/rRT] \tag{12.24}$$

To describe a multidispersed system of particles a distribution function is introduced and defined as

$$f(r,t) = \lim_{\Delta r \to 0} \frac{N(r, r + \Delta r, t)}{\Delta r} \tag{12.25}$$

where N is the number of particles per unit volume with radius lying between r and $r + \Delta r$ at time t. The number of particles which reach or surpass a certain radius per unit time is equal to

$$f\frac{\mathrm{d}r}{\mathrm{d}t} = f\dot{r} \tag{12.26}$$

so the time rate of change of the distribution function can be defined as the negative divergence of $f\dot{r}$, i.e.

$$\frac{\partial f}{\partial t} = -\frac{\partial(f\dot{r})}{\partial r} \tag{12.27}$$

The total number of particles is obtained via

$$z = \int_0^\infty f(r, t)\,\mathrm{d}r \tag{12.28}$$

The loss of z per unit time is given by the number of particles per unit time which converge to $r = 0$ per unit time, i.e.

$$-\frac{\mathrm{d}z}{\mathrm{d}t} = -\frac{\mathrm{d}}{\mathrm{d}t}\int_0^\infty f(r, t)\,\mathrm{d}r = -\lim_{r\to 0} f\dot{r} \tag{12.29}$$

The volume of dispersed particles is given by

$$V = \int_0^\infty \frac{4\pi r^3}{3} f(r, t)\,\mathrm{d}r \tag{12.30}$$

and at least for late times this is stationary in time. Following the arguments and approximations of § § 1.24 and 1.25 one can evaluate $\dot{n}$, the mean rate of change of solute concentration in the matrix, as

$$\dot{n} = -\frac{4\pi r^2 GD}{Gr + D}(C_r - C) \tag{12.31}$$

where D is the diffusion coefficient, G is the reaction rate constant, and C is the mean concentration of the matrix far from the particles. Since

$$\frac{\mathrm{d}}{\mathrm{d}t}\left(\frac{4\pi r^3}{3}\right) = 4\pi r^2 \dot{r} = \dot{n}\,v^\beta \tag{12.32}$$

it follows that (cf equation (1.164))

$$\dot{r} = -\frac{GDv^\beta}{Gr + D}(C_r - C) \tag{12.33}$$

Upon integration this yields equation (1.165).

To solve for the concentration C it is recognized that

$$\frac{\partial C}{\partial t} = -\int_0^\infty \dot{n}(r)\, f(r, t)\,\mathrm{d}r \tag{12.34}$$

Substituting equation (12.24) into (12.31) and the latter into (12.34) yields

$$\frac{C}{C_0} = 1 + \frac{2\sigma v^\beta}{RT}\,\frac{\displaystyle\int_0^\infty f(r, t)\frac{r\,\mathrm{d}r}{Gr + D}}{\displaystyle\int_0^\infty f(r, t)\frac{r^2\,\mathrm{d}r}{Gr + D}} \tag{12.35}$$

Manipulating equations (12.24), (12.35) and (12.33) then yields

$$\dot{r} = \frac{2\sigma C_0(v^\beta)^2}{rRT}\frac{GD}{Gr+D}\left[1 - \frac{r\int_0^\infty f(r,t)\frac{r\,\mathrm{d}r}{GR+D}}{\int_0^\infty f(r,t)\frac{r^2\,\mathrm{d}r}{GR+D}}\right] \tag{12.36}$$

That particle size r^* which is stationary with respect to the environment ($\dot{r} = 0$) can be evaluated as

$$r^* = \frac{\int_0^\infty f(r,t)\frac{r^2\,\mathrm{d}r}{Gr+D}}{\int_0^\infty f(r,t)\frac{r\,\mathrm{d}r}{Gr+D}} \tag{12.37}$$

Thus equation (12.36) can be written as

$$\dot{r} = \frac{2\sigma C_0 v^\beta}{RT}\cdot\frac{GD}{Gr+D}\left(\frac{1}{r^*} - \frac{1}{r}\right) \tag{12.38}$$

Finally the distribution function defined by equation (12.27) can be evaluated as a solution of the differential equation

$$\frac{\partial f(r,t)}{\partial t} = -\frac{2\sigma C_0(v^\beta)^2}{r^*RT}\frac{\partial}{\partial r}\left[\frac{GD}{Gr+D}(r-r^*)\frac{f(r,t)}{r}\right] \tag{12.39}$$

For the case of pure diffusion control ($Gr \gg D$) equation (12.38) becomes

$$\dot{r} = \frac{2\pi C_0 v^\beta}{RTr^*}\cdot\frac{D}{r}\left(1 - \frac{r^*}{r}\right) \tag{12.40}$$

a form first obtained by Greenwood[19] by less rigorous methods. In his derivation r^* is the mean particle radius.

One immediately recognizes the emergence of a free boundary problem, for an initial condition for $f(r,t)$ is required if a unique solution is to be found. But physically this is not defined within the model for it is intimately tied in with the fluctuation spectrum of particle sizes above and below the critical nucleation radius, the latter particles having been implicitly excluded from the analysis. (Note that r^* or $\bar{r}$ turn out to be not much in excess of the critical nucleation radius.) Wagner was therefore forced to proceed on the basis of an *ad hoc* initial distribution function which he chose to be Gaussian. Compare this constraint with the imposition of an empirical final mean radius ρ_0 in the elementary process discussed in § § 1.24 and 1.25 and the preamble to this section. Wagner remarks that this work has great relevance to the development of creep resistant alloys, but will only attain full utility when re-expressed in a multi-component formalism. To our knowledge this has not as yet been carried out.

Greenwood encountered the same fundamental problem of indeterminacy in relation to equation (12.40). He assumed that the distribution rapidly attains a quasi-steady state in which the distribution of radii is independent of initial conditions and conjectured by analogy with Zener,[20] Hillert[21] and others that the steady state optimum lies at a maximum in $\dot{r}$ with respect to r, which yields $r = 2r^*$

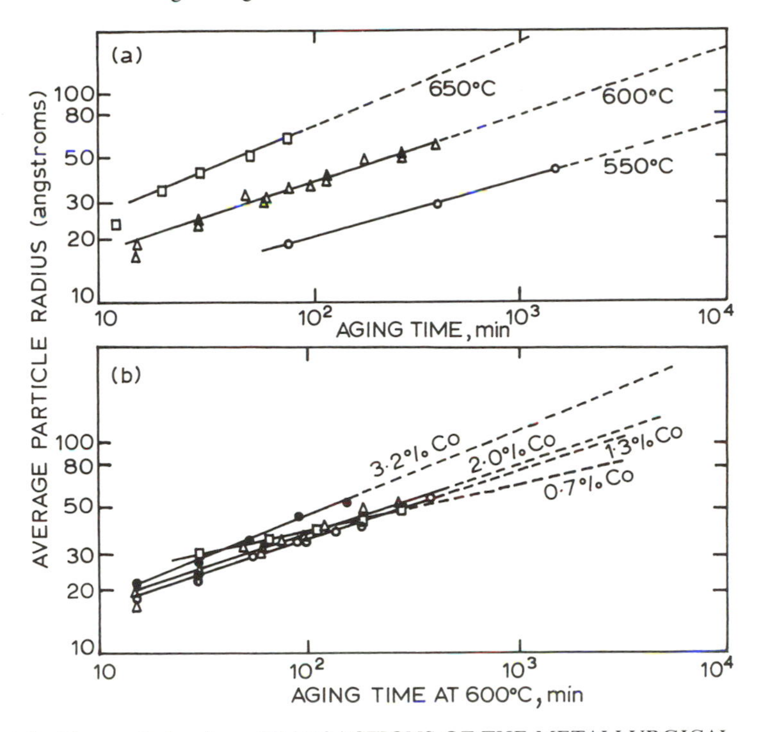

Reprinted with permission from TRANSACTIONS OF THE METALLURGICAL SOCIETY, Vol. 215, p. 556, (1959), a publication of The Metallurgical Society, Warrendale, Pennsylvania

Fig. 12.2 **(*a*) Change in average particle radius (determined magnetically) with aging time at various temperatures for an alloy containing 2 wt-% Co. (*b*) Change in average particle radius with aging time at 600°C for alloys containing varying percentages of cobalt. The average slopes are in reasonable agreement with $r^{*3} - r_0^3 = (\text{const})t$. After Livingston[23]**

(cf § 1.26 and equation (1.181)) and $r^{*3} - r_0^3 = (\text{const})t$, where r_0 represents the first observation. That is to say, particles with radius greater than r^* are growing with increasing rates, and at any instant particles of radius twice the arithmetic mean are growing at the fastest rate. This, of course, intimates that some sort of instantaneous dissipation principle is operative (cf § 12.7).

Both the Greenwood and Wagner theories have been expanded upon (see, for example, Ardell[22]) and tested against experiment with reasonable success[23] (Fig. 12.2). The problem continues to attract the attention of researchers.

Böhm and Kahlweit[24] have applied the method to the process of internal oxidation (cf § 11.7). It is to be appreciated that this latter application pertains to a ternary or higher order system where the phase constitution is more complex and ternary cross-effects may sometimes be important. In the following a ternary precipitation phenomenon is explored which submits to quite rigorous mathematical methods.

12.5 SPINODAL DECOMPOSITION IN TERNARY SYSTEMS INCORPORATING STRAIN AND GRADIENT ENERGY TERMS

In § § 1.27 and 7.6 some elementary aspects of the thermodynamics and kinetics of spinodal decomposition for binary alloys were explored. Notable contributions to the theory of ternary spinodal decomposition have been made by de Fontaine,[25] Kirkaldy and Purdy,[26] and Morral and Cahn,[27] and these are synthesized in the following. There remain important gaps in the knowledge of this important diffusion-controlled phase transformation.

The general thermodynamic and kinetic properties of the diffusion matrix which pertains to a purely chemical multicomponent spinodal decomposition were first explored by Kirkaldy *et al.*[15, 28, 29] and Sundelöf.[30] As emphasized earlier, the D-matrix can be expressed as the product of a symmetric, positive definite mobility L-matrix and a symmetric thermodynamic μ-matrix

$$\rho D = L\mu \tag{12.41}$$

where ρ is the density. For ternary substitutional solid solutions the thermodynamic matrix can be adequately represented by

$$\mu_{ik} = \frac{\partial(\mu_i - \mu_3)}{\partial X_k} = \frac{\partial(\mu_k - \mu_3)}{\partial X_i} = \mu_{ki} \tag{12.42}$$

where 3 designates the solvent, μ_i are chemical potentials and Xs are mole fractions. Since determinants follow the same rule of multiplication as matrices

$$D = D_{11}D_{22} - D_{12}D_{21} = (L_{11}L_{22} - L_{12}^2)(\mu_{11}\mu_{22} - \mu_{12}^2)/\rho^2 \tag{12.43}$$

The determinant enters explicitly into the eigenvalues of the D-matrix

$$u_1 = \tfrac{1}{2}[D_{11} + D_{22} + ((D_{11} + D_{22})^2 - D)^{1/2}] \tag{12.44}$$

and

$$u_2 = \tfrac{1}{2}[D_{11} + D_{22} - ((D_{11} + D_{22})^2 - D)^{1/2}] \tag{12.45}$$

and it is the signs of these quantities which determine whether the diffusion solutions contain growing or decaying eigenfunctions (cf § § 4.6 and 4.6.5). Clearly $u_2 = 0$ when $D = 0$ and goes negative as D goes negative, indicating that the corresponding eigenfunction will have a growing amplitude. The chemical spinodal surface is thus defined by the vanishing of the diffusion determinant. This is attributable in turn to the vanishing of the determinant μ, since the determinant of the Onsager L-matrix is generally positive.

In defining the initial value problem for ternary systems which involve strain and gradient energy Morral and Cahn[27] write the diffusion equations generalized from equation (1.188) as

$$\frac{\partial C_1}{\partial t} = D_{11}^{(1)}\nabla^2 C_1 + D_{12}^{(1)}\nabla^2 C_2 - D_{11}^{(2)}\nabla^4 C_1 - D_{12}^{(2)}\nabla^4 C_2 \tag{12.46}$$

and

$$\frac{\partial C_2}{\partial t} = D_{21}^{(2)}\nabla^2 C_1 + D_{22}^{(1)}\nabla^2 C_2 - D_{21}^{(2)}\nabla^4 C_1 - D_{22}^{(2)}\nabla^4 C_2 \tag{12.47}$$

where $D^{(1)}$ is a generalization of the D-matrix containing strain energy as well as thermodynamic terms and $D^{(2)}$ pertains to gradient energy effects. Explicit expressions for the two D-matrices were first given by de Fontaine.[25] If we assume that the strains due to the two independent solute accumulations are additive and that the gradient energies are represented by a standard, symmetric quadratic form in the gradients we can follow Cahn's procedure for binaries[31] which led to equation (1.187) to evaluate the matrix elements for isotropic materials as

$$\rho D_{11}^{(1)} = L_{11}(\mu_{11} + \phi\eta_1^2) + L_{12}(\mu_{22} + \phi\eta_1\eta_2)$$
$$\rho D_{12}^{(1)} = L_{11}(\mu_{12} + \phi\eta_1\eta_2) + L_{12}(\mu_{22} + \phi\eta_2^2)$$
$$\rho D_{22}^{(1)} = L_{21}(\mu_{11} + \phi\eta_1^2) + L_{22}(\mu_{21} + \phi\eta_1\eta_2)$$
$$\rho D_{22}^{(1)} = L_{21}(\mu_{12} + \phi\eta_1\eta_2) + L_{22}(\eta_{22} + \phi\eta_2^2) \quad (12.48)$$
$$\rho D_{11}^{(2)} = 2(L_{11}\chi_{11} + L_{12}\chi_{21}) \qquad \rho D_{21}^{(2)} = 2(L_{21}\chi_{11} + L_{22}\chi_{21})$$
$$\rho D_{12}^{(2)} = 2(L_{11}\chi_{12} + L_{12}\chi_{22}) \qquad \rho D_{22}^{(2)} = 2(L_{21}\chi_{12} + L_{22}\chi_{22}) \quad (12.49)$$

where $\phi = 2Y(1 - \pi)$, Y is Young's Modulus, π is the Poisson ratio, η_1 and η_2 are the strain energy parameters[31] and the mathematically symmetric matrix elements χ_{ik} are the gradient energy expansion coefficients. The strain energy parameters are directly related to the Vegard's law coefficients for dilute solutions,[32] i.e., the linear lattice expansion which usually occurs with alloy additions of greater or lesser atomic radii are reflected in proportionate local lattice strains when composition modulations occur. There is little or no quantitative information on the gradient energy matrix. Hillert[33, 34] and Cahn and Hilliard[35] have provided solution models for χ in the binary case only. Evidently, the ternary elements must possess at least one interdependency other than symmetry which may be associated with the net balance of forces in the solution. Since these parameters depend on pair interactions, χ is obviously related to μ. The matrix $D^{(2)}$ can be written as the product

$$\rho D^{(2)} = L\chi \quad (12.50)$$

Since L is positive definite and χ is symmetric, and probably positive definite as well near a pseudo-binary critical temperature (solvent X_3 dilute), $D^{(2)}$ will have a positive determinant and eigenvalues so its effect in the solutions should be towards a decay in modulation amplitudes. The matrix $D^{(1)}$ remains as the product of the positive definite L matrix and a symmetric matrix of elements, $\mu'_{ik} = \mu_{ik} + \phi\eta_i\eta_k$, so can be written

$$\rho D^{(1)} = L\mu' \quad (12.51)$$

The determinant of μ' is

$$\mathrm{Det}\,\mu' = \mu_{11}\mu_{22} - \mu_{12}^2 + \phi(\eta_1^2\mu_{22} - 2\mu_{12}\eta_1\eta_2 + \eta_2^2\mu_{11}) \quad (12.52)$$

When $\mathrm{Det}\,\mu'$ vanishes, the right-hand term is recognized as a perfect square, $\phi(\eta_1\sqrt{\mu_{22}} - \eta_2\sqrt{\mu_{11}})^2$, which indicates that excess chemical supersaturation must be provided for onset of an instability when strain energy effects are non-zero. Recognizing that the strain energy coefficients have been assumed to be relatively independent of concentration but not of temperature, the equation

$$\mathrm{Det}\,\mu' = 0 \quad (12.53)$$

according to equation (12.52) defines the long wavelength spinodal surface over concentration and temperature (see also §4.6.5).

Turning now to the early stages of decomposition and focusing on the binary solutions of the equation (cf equation (1.187))

$$\frac{\partial C}{\partial t} = D^{(1)}\nabla^2 C - D^{(2)}\nabla^4 C \tag{12.54}$$

the general solution pertaining to arbitrary initial perturbations of an infinite uniform solution in one dimension is written[31]

$$C = a_0 + a\int_{-\infty}^{\infty} A(\beta) \exp -(u\beta^2 + v\beta^4) \exp(i\beta x)\, \mathrm{d}\beta \tag{12.55}$$

where β is the magnitude of the wavenumber and x is the space dimension. Substituting equation (12.55) into (12.54) it is found that

$$u = D^{(1)} \quad \text{and} \quad v = D^{(2)} \tag{12.56}$$

The coefficient $aA(\beta)$ is obtained via

$$C(\mathbf{r}, 0) - a_0 = \int_{-\infty}^{\infty} aA(\beta) \exp(i\beta x)\, \mathrm{d}\beta \tag{12.57}$$

i.e., as the Fourier transform of the initial condition (Appendix 2).

This structure generalizes to ternary systems rather simply for the case where the gradient energy terms are ignored, as shown by Kirkaldy and Purdy[26] (cf §4.6.5). One can identify their treatment as the long wavelength approximation in a liquid. However, for the case where $D^{(2)}$ is non-zero the obvious generalization fails due to the fact that two independent matrices within the same linear equation set cannot be simultaneously brought to the diagonal.[36, 37] Indeed, consider the generalizations

$$C_1 - a_{10} = a_{11}\int_{-\infty}^{\infty} A_1(\beta) \exp[-(u_1\beta^2 + v_1\beta^4)t] \exp(i\beta x)$$
$$+ a_{12}\int_{-\infty}^{\infty} A_2(\beta) \exp[-(u_2\beta^2 + v_2\beta^4)t] \exp(i\beta x)\, \mathrm{d}\beta \tag{12.58}$$

and

$$C_2 - a_{20} = a_{21}\int_{-\infty}^{\infty} A_1(\beta) \exp[-(u_1\beta^2 + v_1\beta^4)t] \exp(i\beta x)$$
$$+ a_{22}\int_{-\infty}^{\infty} A_2(\beta) [\exp -(u_2\beta^2 + v_2\beta^4)t] \exp(i\beta x)\, \mathrm{d}\beta \tag{12.59}$$

where u_1, u_2 and v_1, v_2 are the eigenvalues of the $D^{(1)}$ and $D^{(2)}$ matrices, respectively. By substitution in equations (12.46) and (12.47) we obtain the characteristic equations

$$(u_2 - D_{11}^{(1)})/D_{12}^{(1)} = D_{21}^{(1)}/(u_2 - D_{22}^{(1)}) = \frac{a_{22}}{a_{12}} = (v_2 - D_{11}^{(2)})/D_{12}^{(2)} = D_{21}^{(2)}/(v_2 - D_{22}^{(2)}) \quad (12.60)$$

and

$$(u_1 - D_{22}^{(1)})/D_{21}^{(1)} = D_{12}^{(1)}/(u_1 - D_{11}^{(1)}) = \frac{a_{11}}{a_{21}} = (v_1 - D_{22}^{(2)})/D_{21}^{(2)} = D_{12}^{(2)}/(v_1 - D_{11}^{(2)}) \quad (12.61)$$

Here the difficulty is recognized, for if u and v are the eigenvalues as defined, the right-hand equalities involve only $D^{(2)}$ matrix elements while the left-hand side equalities involve only $D^{(1)}$ matrix elements. This means that in general solutions cannot be found in terms of elementary functions and Fourier decompositions. Solutions have been given for analogous but lower order diffusion equations with certain restrictions on the coefficient matrices.[38, 39] Aifantis, for example, considers one diagonal matrix and the other with a vanishing determinant.

A conditional solution, adequate for illustrative purposes can be developed according to forms (12.58) and (12.59). One notes that the overdetermination of the coefficients is removed when the right-hand ratios of equations (12.60) and (12.61) become indeterminate, i.e. when

$$v_1 - D_{11}^{(2)} = v_2 - D_{11}^{(2)} = D_{12}^{(2)} = D_{21}^{(2)} = v_1 - D_{22}^{(2)} = v_2 - D_{22}^{(2)} = 0 \quad (12.62)$$

or when

$$v_1 = v_2 = D_{11}^{(2)} = D_{22}^{(2)} \quad \text{and} \quad D_{12}^{(2)} = D_{21}^{(2)} = 0 \quad (12.63)$$

which is to say, a diagonal $D^{(2)}$ matrix with equal diagonal elements $D_{ii}^{(2)}$, and it is this case that will be dealt with in one dimension. As indicated above this should be a realistic representation near a pseudo-binary critical temperature (X_3 dilute).

To apply a Fourier integral analysis of initial conditions for equations (12.58) and (12.59), single integrals on the right must be related to initial value functions of x on the left. Thus, multiplying equation (12.58) by a_{21} and equation (12.59) by a_{11}, subtracting and inserting ratios from equations (12.60) and (12.61) on the left, we obtain for even initial conditions (cf §1.27)

$$\Delta C_1^\circ = C_1^\circ - a_{10} \quad (12.64)$$

and

$$\Delta C_2^\circ = C_2^\circ - a_{20} \quad (12.65)$$

with

$$\frac{(u_1 - D_{22}^{(1)})}{u_1 - u_2}\left(\Delta C_2^\circ - \frac{D_{21}^{(1)}}{u_1 - D_{22}}\Delta C_1^\circ\right) = a_{22}\int_{-\infty}^{\infty} A_2(\beta)\cos\beta x \, d\beta \quad (12.66)$$

and symmetrically

$$\frac{(u_2 - D_{11}^{(1)})}{u_2 - u_1}\left(\Delta C_1^\circ - \frac{D_{12}^{(1)}}{u_2 - D_{11}}\Delta C_2^\circ\right) = a_{11}\int_{-\infty}^{\infty} A_1(\beta)\cos\beta x \, d\beta \quad (12.67)$$

Further, from equations (12.60) and (12.61)

$$\frac{(u_1 - D_{22}^{(1)})}{(u_2 - D_{11}^{(1)})}\frac{D_{12}^{(1)}}{(u_1 - u_2)}\left(\Delta C_2^\circ - \frac{D_{21}^{(1)}}{u_1 - D_{22}}\Delta C_1^\circ\right) = a_{12}\int_{-\infty}^{\infty} A_2(\beta)\cos\beta x \, \mathrm{d}\beta \quad (12.68)$$

and

$$\frac{(u_2 - D_{11}^{(1)})}{(u_1 - D_{22}^{(1)})}\frac{D_{21}^{(1)}}{(u_2 - u_1)}\left(\Delta C_1^\circ - \frac{D_{12}^{(1)}}{u_1 - D_{22}^{(1)}}\nabla C_2^\circ\right) = a_{21}\int_{-\infty}^{\infty} A_1(\beta)\cos\beta x \, \mathrm{d}\beta \quad (12.69)$$

To complete the solution, a specification of the eigenvalues is required. Equation (12.63) is a reminder of the assumption

$$v_1 = v_2 = D_{11}^{(2)} = D_{22}^{(2)} = D^{(2)} \quad (12.70)$$

As an example, a shallow spinodal decomposition is explored where (cf Appendix 4)

$$|4 \det D^{(1)}| \ll (\mathrm{tr}\, D^{(1)})^2$$

Hence

$$u_1 = \mathrm{tr}\, D^{(1)} \quad \text{and} \quad u_2 = \det D^{(1)}/\mathrm{tr}\, D^{(1)} < 0 \quad (12.71)$$

A kinetically unsymmetric case is further specified with $D_{11} \simeq 10D_{22} > 0$; $D_{21}/D_{11} \simeq 0{\cdot}1$ and $D_{12}/D_{11} \simeq 1$. The positive signs of D_{21} and D_{12} express the net repulsive forces between components 1 and 2.

It is convenient to choose for the one-dimensional initial conditions

$$\frac{\sin\beta_0 x}{\beta_0 x} = \Delta C_1^\circ = \mp\Delta C_2^\circ \quad (12.72)$$

The function on the left is the Fourier transform of a double step function which is $1/2\beta_0$ for $-\beta_0 < \beta < \beta_0$ and zero elsewhere. Although this does not satisfy the mass balance it can be supposed that the initial perturbation arises in part from a minute local inhomogeneity. The set of solutions is accordingly

$$\Delta C_2 = -\frac{1}{2\beta_0}\int_{-\beta_0}^{\beta_0} \exp\left[-\beta^2(\det D^{(1)} + \beta^2 D^{(2)})t\right]\cos\beta x \, \mathrm{d}\beta \quad (12.73)$$

with $\Delta C_2 = -\Delta C_1$, and

$$\Delta C_2 = \frac{1}{2{\cdot}2\beta_0}\left[\int_{-\beta_0}^{\beta_0} 0{\cdot}2 \exp\left[-\beta^2(\mathrm{tr}\, D + \beta^2 D^{(2)})t\right]\right.$$

$$\left. + \, 0{\cdot}9 \exp\left[-\beta^2(\det D^{(1)} + \beta^2 D^{(2)})\right]\cos\beta x \, \mathrm{d}\beta \right] \quad (12.74)$$

with $\Delta C_1 = \Delta C_2$, where it is understood that $\det D^{(1)} < 0$. Thus the eigenvectors on the right-hand side will grow within a certain range of β and may reach a maximum rate at the value of β for which the exponent is a maximum, e.g.

$$\beta_{\mathrm{m}} = (\det D^{(1)}/2D^{(2)})^{1/2} \quad (12.75)$$

This wavenumber disposition is shown in Fig. 1.38. In the first case (equation

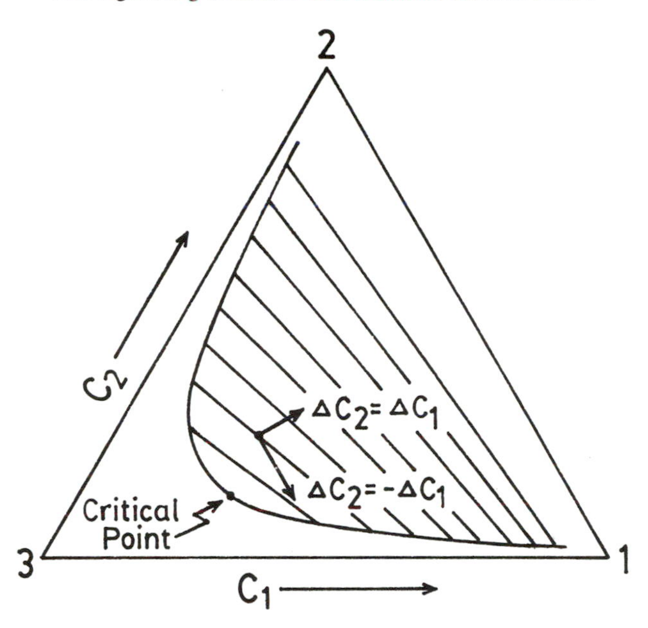

Fig. 12.3 Schematic decomposition paths during ternary spinodal decomposition

(12.73)), the initial perturbation from the uniform solution at each location in the sample can begin and continue growth along the same vector defined by the initial condition (Fig. 12.3). In the second case (equation (12.74)) the first eigenvector ($\sim a_{11}, a_{21}$) may dominate at first so for each location there will be a decay in amplitude before the second growing eigenvector with the larger coefficient becomes predominant (Fig. 12.3). This will occur in a time

$$\tau \sim 3[\beta^2(\operatorname{tr} D + \beta^2 D^{(2)})]^{-1} \tag{12.76}$$

In any case, it is apparent that in the low amplitude range of constant diffusion coefficients an initial perturbation in which $\Delta C_2 \propto \Delta C_1$ will at every location retain its initial vector on the Gibbs triangle. In due course, the phenomenological coefficients start to vary with time as the amplitude of decomposition becomes significant and the average vector direction is forced to rotate in the Gibbs triangle toward parallelism with the tie line. Indeed, there is already a hint of such an effect, predicated upon our thermodynamically consistent set of parameters, whereby the first sample fluctuation ($\Delta C_2 = -\Delta C_1$) considered above, which is tending towards parallelism with the expected tie line direction, is free of the damping term and its dominance is thereby favoured (Fig. 12.3).

The foregoing eventualities are predicated upon the requirement that the wavenumber β_0 in the initial fluctuation not be appreciably smaller than β_m, for then the strong selective amplification of wavenumber β_m cannot occur. To put this another way, growth of fluctuations with $\beta_0 \geqslant \beta_m$ will be favoured in this process.

In applying this theory quantitatively, one will find it necessary to specify a realistic value of β_0 for the initial value problem. This problem has been discussed in some detail for the binary system by Huston *et al.*[40] If we hypothesize that the appropriate fluctuation spectrum is that associated with a temperature above the critical value where preprecipitation or clustering appears (cf Aalders *et al.*[41,42]) and from which the sample was quenched with $\beta_0 \ll \beta_m$, and since the latter value is not expected to be contained within the spectrum, it accordingly cannot be reached through exponential amplification (cf integral limits). This is evidently not a viable concept. The fluctuation spectrum which rapidly emerges at the reaction temperature must accordingly be focused upon. This would evidently be dominated by pulses for which $\beta_0 \simeq \beta_m$ and the amplification will have already been completed before an appropriate initial value problem can be mounted. Should this be correct reasoning then it remains as an important free boundary value problem to specify the precise constant of proportionality between β_0 and β_m which is thermokinetically optimal.

12.6 PERIODIC PRECIPITATION: THE LIESEGANG PHENOMENON

The Liesegang phenomenon, which pertains to the evolution of concentric rings or ordered layers of precipitates in ternary or higher order systems, has been the subject of a century of study.[43,44,45] The most comprehensive and widely quoted theory is due to Wagner,[46] as amended by Kahlweit[47,48] and this has been adapted for application to a number of recent experimental studies.[49,50,51] Figure 12.4 shows the Liesegang bands observed in 9Cd1Mg Ag after oxidation in air (see also Osinski *et al.*,[52] and Patnaik[53]). Generally, the spacing Δx_n between bands at a distance x_n from the surface can be represented by the geometric series

$$\frac{x_n + \Delta x_n}{x_n} = \frac{x_{n+1}}{x_n} = k$$

or in terms of a constant spacing coefficient[51]

$$S = k - 1 = \Delta x_n / x_n \tag{12.77}$$

Reprinted with permission from *Acta Met.*, **23**, V. A. Van Rooijen *et al.*, p. 987, copyright 1975, Pergamon Press Ltd

Fig. 12.4 Liesegang bands in AgCdMg (9 at-% Cd, 1 at-% Mg) oxidized in air for 18 h at 800°C. After van Rooijen *et al.*[51]

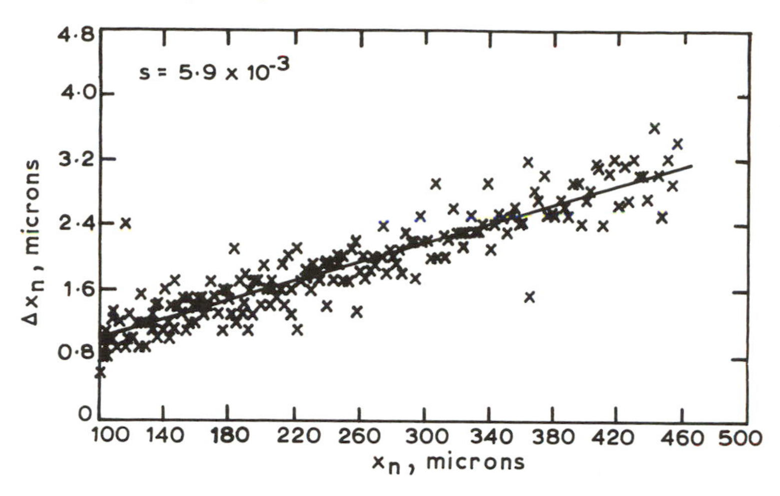

Reprinted with permission from *Acta Met.*, **23**, V. A. Van Rooijen *et al.*, p. 987, copyright 1975, Pergamon Press Ltd

Fig. 12.5 Band spacing *vs* penetration depth. AgCdAl (9 at-% Cd, 0·4 at-% Al) oxidized for 18 h in air at 800°C. After van Rooijen *et al.*[51]

This is known as the Jablczynski relation.[54] Hedges[44] remarks that almost any theory seems to be consistent with this gradual increase in spacing. Figure 12.5 shows the experimental plot for AgCdAl whereby a value of $S = 5{\cdot}9 \times 10^{-3}$ was determined.

In spite of its wide acceptance the Wagner theory is in some ways unsatisfactory for it contains a number of *ad hoc* steps required by the discontinuous nucleation and growth process conjectured. Furthermore, its assumptions are in conflict with certain observations on Liesegang-banded crystallization in gels where the structure proves to be a single dendritic crystal of continuously variable volume fraction. It seems worthwhile therefore to seek a continuum theory of the process analogous to spinodal decomposition.

A possible framework for such a theory has already been presented and remarked upon in §11.7. It was noted in our discussion of internal oxidation that the parameters within the effective diffusion coefficients were such that the latter might under certain circumstances take negative values. Accordingly, the concentrations and precipitate fractions could undergo spatial instabilities which are mathematically analogous to those in spinodal decomposition.

Consider then the diffusion problem defined by the dissolution of compound AB into an alloy of concentration X_B in solvent C as defined by the virtual path ab (cf §11.6) in Fig. 12.6. Here it is assumed that ternary interactions are negligible and $D_A \gg D_B$. The Fick equations for the concentrations of A and B between the precipitates are

$$\frac{\partial C_A}{\partial t} = D_{A\mathrm{eff}} \frac{\partial^2 C_A}{\partial t^2} \tag{12.78}$$

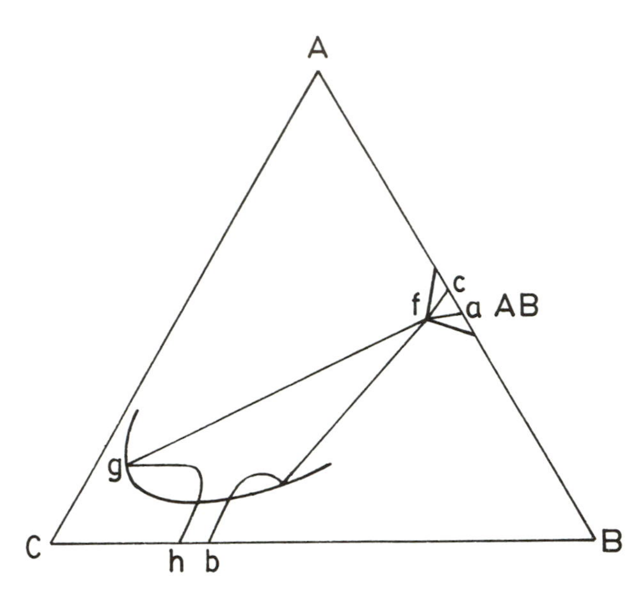

Fig. 12.6 Virtual paths and Gibbs triangle

and

$$\frac{\partial C_B}{\partial t} = D_{\text{Beff}} \frac{\partial^2 C_B}{\partial t^2} \tag{12.79}$$

where

$$D_{\text{Aeff}} = D_A \left(1 - \frac{D_B}{D_A} \cdot \frac{dX_B}{dX_A}\right) \Big/ \left(1 - \frac{dX_B}{dX_A}\right) \tag{12.80}$$

and

$$D_{\text{Beff}} = D_B \left(1 - \frac{D_A}{D_B} \cdot \frac{dX_A}{dX_B}\right) \Big/ \left(1 - \frac{dX_A}{dX_B}\right) \tag{12.81}$$

Here the lever rule constraint used in § 11.7 has been dropped, so incomplete precipitation is general within the diffusion zone. Now in accordance with Fig. 12.6 and the previous assumption, the case will be considered that

$$\frac{D_A}{D_B} \gg \frac{dX_B}{dX_A} \gg 1 \tag{12.82}$$

so that

$$D_{\text{Beff}} = D_{\text{Aeff}} \simeq -D_A \cdot dX_A/dX_B < 0 \tag{12.83}$$

Hence an analogue of spinodal type decompositions is possible within this model (path ab in Fig. 12.6). However, the required phase constitution and other conditions seem to be too specialized to be generally pertinent to the Liesegang phenomenon. Indeed, virtual paths such as cfgh in Fig. 12.6 are much more appropriate, and since $dX_B/dX_A < 0$ equations (12.81) and (12.82) deny the possibility of a spinodal type instability in this mode.

We have found that the key to generalization lies in the inclusion of capillarity within the precipitation model as in § 12.4 (see also Ortolova[55, 56]). Here we consider a Henrian solution X_A, X_B in contact with stoichiometric AB precipitates subject to the capillarity-dependent solubility product (cf Appendix 1).

$$X_A X_B = K \exp\left(\frac{2\sigma v}{RT \cdot r}\right) \tag{12.84}$$

It is here assumed that the molar volume V is the same as that of the matrix. To relate the precipitate radius r to the precipitate mole fraction nucleation site saturation on n available nucleation sites per unit volume is assumed and

$$p = \frac{4}{3}\pi r^2 n = \frac{4}{3}\pi n \left(\frac{2\sigma v}{RT}\right)^3 \bigg/ \left(\ln \frac{X_A X_B}{K}\right)^3 \tag{12.85}$$

obtained. For the kinetic relations to follow

$$\frac{1}{2}\dot{p} = -P_A \frac{\partial X_A}{\partial t} - P_B \frac{\partial X_B}{\partial t} \tag{12.86}$$

will be required, where

$$P_A = \frac{1}{2}\frac{\partial p}{\partial X_A} = \frac{2\pi n}{X_A}\left(\frac{2\sigma v}{RT}\right)^3 \bigg/ \left(\ln \frac{X_A X_B}{K}\right)^4 \tag{12.87}$$

and

$$P_B = \frac{1}{2}\frac{\partial p}{\partial X_B} = \frac{2\pi n}{X_B}\left(\frac{2\sigma v}{RT}\right)^3 \bigg/ \left(\ln \frac{X_A X_B}{K}\right)^4 \tag{12.88}$$

Assuming an infinitesimal volume fraction of precipitates, equal molar volumes of precipitate and matrix and negligible ternary interactions, we obtain from equations (11.47) and (11.48) the Fick-type equations

$$-D_A \frac{\partial^2 X_A}{\partial x^2} + \frac{\partial X_A}{\partial t} = -\frac{\dot{p}}{2} \tag{12.89}$$

and

$$-D_B \frac{\partial^2 X_B}{\partial x^2} + \frac{\partial X_B}{\partial t} = -\frac{\dot{p}}{2} \tag{12.90}$$

Combining these with equation (12.86) yields

$$\frac{\partial X_A}{\partial t} = \frac{1 - P_B}{1 - P_A - P_B} D_A \frac{\partial^2 X_A}{\partial x^2} + \frac{P_B}{1 - P_A - P_B} D_B \frac{\partial^2 X_B}{\partial x^2} \tag{12.91}$$

and

$$\frac{\partial X_B}{\partial t} = \frac{P_A}{1 - P_A - P_B} D_A \frac{\partial^2 X_A}{\partial x^2} + \frac{(1 - P_A)}{(1 - P_A - P_B)} D_B \frac{\partial^2 X_B}{\partial x^2} \tag{12.92}$$

These have the standard form of the ternary diffusion equations (cf Chapter 4) but through equations (12.87) and (12.88) are strongly non-linear. Note in particular that the determinant of coefficients is

$$|D| = \frac{D_A D_B}{1 - P_A - P_B} \tag{12.93}$$

which in view of the positive signs of P_A and P_B and their composition dependencies can become negative for an appropriate high density n indicating the onset of spinodal-like instabilities. Suppose as a special case that $D_A \gg D_B$ and in the reaction zone $X_A \gg X_B$ such that via equations (12.87) and (12.88) $P_B \gg 1 \gg P_A$, whence $|D| < 0$ and the diffusion equations approximate to

$$\frac{\partial X_A}{\partial t} \simeq D_A \frac{\partial^2 X_A}{\partial t^2} \tag{12.94}$$

and

$$\frac{\partial X_B}{\partial t} \simeq -\frac{D_B}{\bar{P}_B} \frac{\partial^2 X_B}{\partial t} \tag{12.95}$$

Thus in boundary conditions similar to virtual path efgh in Fig. 12.5 the fast diffuser X_A follows a normal error function profile while the slow diffuser undergoes a profile oscillation of the approximate form (cf §1.27)

$$\Delta X_B = \Delta X_B^\circ \exp\left(\frac{D_B}{\bar{P}_B} \cdot \frac{4\pi^2}{\lambda^2} \cdot t\right) \cos \frac{2\pi}{\lambda} x \tag{12.96}$$

where the bar represents an average of P_B over one wavelength λ, which in turn varies slowly with distance along the X_A profile. Now it is noted that a point of half amplitude of X_A penetrates roughly as the distance

$$x_A \simeq 2(D_A t)^{1/2} \tag{12.97}$$

whereas a point of half amplitude of ΔX_B is reached in a time when $t\partial\Delta X_B/\partial t \simeq \Delta X_B/2$ or from equation (12.96)

$$\lambda = 2\sqrt{2}\pi\sqrt{(D_B/\bar{P}_B)t} \tag{12.98}$$

Since for a stable solution these amplitudes must synchronize on a continuous space and time basis at the precipitation front it follows that

$$\frac{\Delta x_i}{x_i} \simeq \frac{\lambda}{x_A} \simeq \sqrt{2}\pi\sqrt{(D_B/\bar{P}_B D_A)} \simeq \text{constant} \ll 1 \tag{12.99}$$

which is the Jablczynski relation. This is the scaling law which removes the degeneracy in this quasi-stationary free boundary problem. With other boundary conditions numerical methods must necessarily be used. In undertaking such a numerical exercise, however, it will immediately become apparent that there is a fundamental difficulty in assigning the initial condition. This is a manifestation of the free boundary nature of the problem as discussed in previous sections of this Chapter. The evident singularities in $|D|$ suggest that there will also be computational difficulties.

It is worth noting that Liesegang structures appear spontaneously in nature. In particular they can be recognized in geological formations[44, 56] and in the

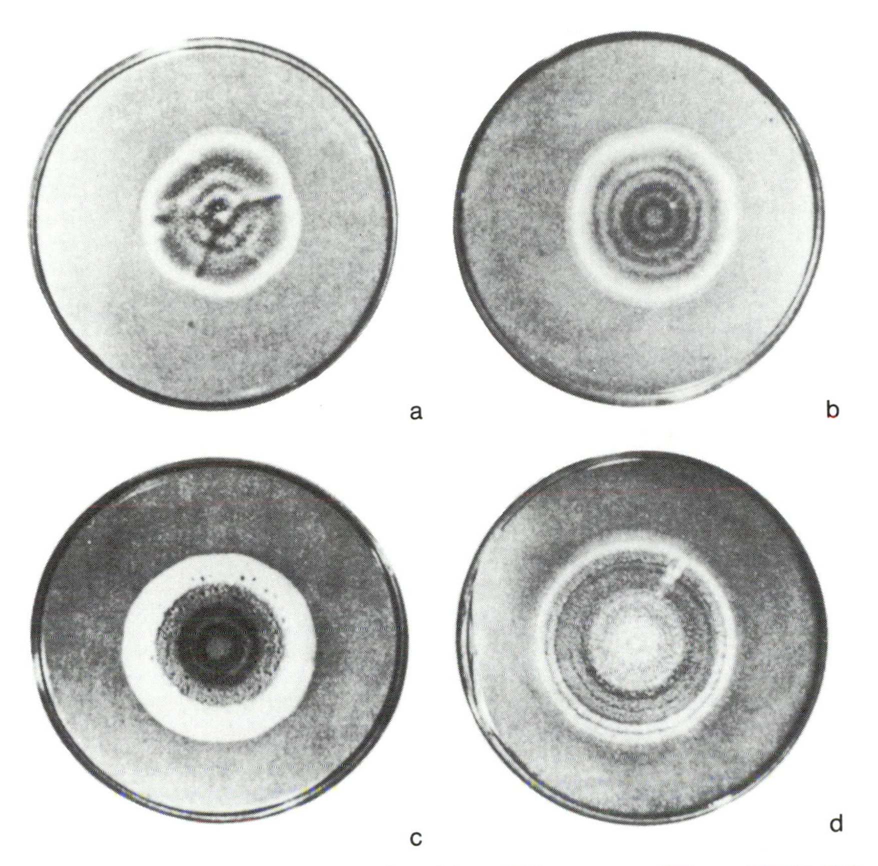

Fig. 12.7 ***Penicillium commune*** **series: (*a*) and (*b*)** ***commune*** **Thom, NRRL 890; (*c*) and (*d*)** ***lanoso-griseum*** **Thom, NRRL 894. After Raper and Thom[57]**

zonation of fungi such as *Penicillium urticae* grown in a nutrient gel in a petri dish (Fig. 12.7[57]). The scale of the rings in Fig. 12.7 is very close to that obtained with crystals grown in supersaturated gels, as might be expected on the basis of the controlling liquid diffusion coefficients (cf Ullschek[58] and Hedges[44]). A further remarkable analogy is to be recognized in the fact that both the crystalline and fungal structures can by accident grow as spirals rather than rings, as indicated in Figs. 12.8[45] and 12.9.[59] This configuration represents a soluble free-boundary problem in the sense of § 12.7, for analysis shows that the solutions of the controlling differential equations are degenerate, yielding an infinity of related values of the spacing and rotation frequency.[60] We will return to this problem following the establishment of an optimizing principle (§ 12.11).

12.7 MAXIMUM PATH PROBABILITY AND THE MINIMAX PRINCIPLE

The science of morphogenesis originated prior to the twentieth century. Lotka recognized in 1922 that the internal degree of freedom or degeneracy associated

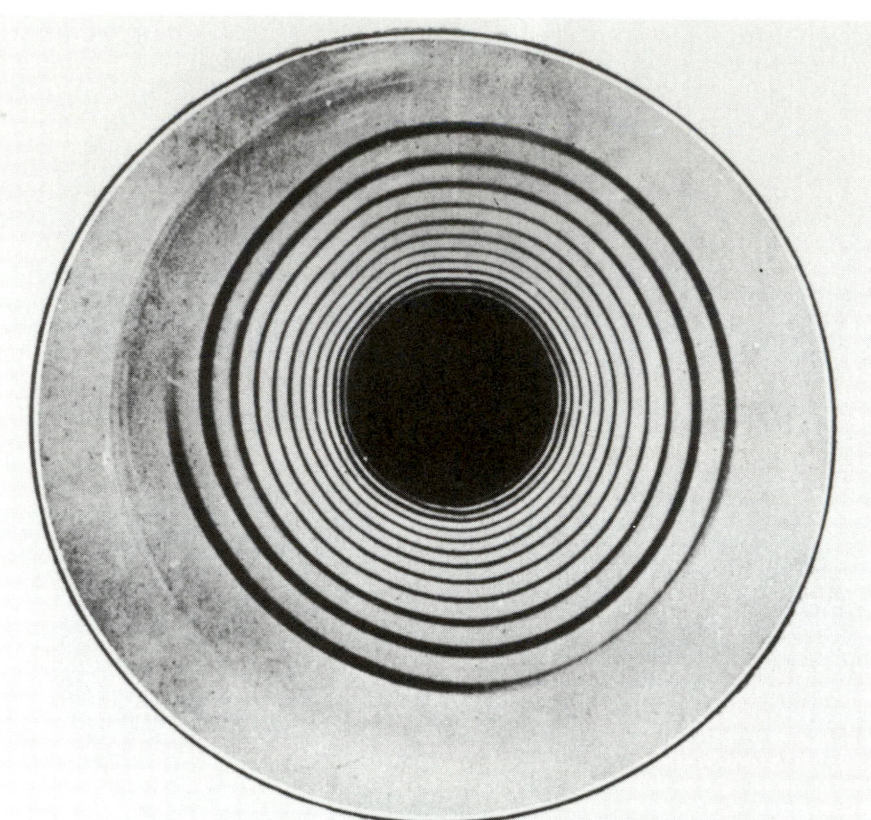

With permission from S. Veil, 'Les phénomènes périodiques de la chimie', 1934, Hermann, Paris

Fig. 12.8 Precipitation of silver chromate by reaction of $AgNO_3$ and $K_2Cr_2O_7$ as a simple spiral. After Veil[45]

with differentiation in both biological individuals and populations could be removed by invoking a thermodynamic principle which went beyond the current understanding of the second law of thermodynamics, and proposed a maximum power principle. This is now recognized for isothermal systems to be equivalent to a maximum in the dissipation. In 1945 he summarized an evolving dichotomy of viewpoints by noting that the understanding of development in the biological world required the recognition of opposing maximal and minimal dissipation principles, a formulation which can now be expressed as a minimax in the entropy production rate. This principle is very closely related to the minimax principle of Game Theory which pertains to the dynamics of economic competition.[61, 62, 63]

Onsager[64] anticipated this induction as an aside to his Minimum Dissipation principle (the Prigogine and de Groot version of this elementary theorem is invoked in §12.10). The latter was inclusive of the minimum dissipation principles perceived by Lord Kelvin[65] and Lord Rayleigh[66] and popularized by de Groot[67] and Prigogine[4] but it also noted that spontaneously generated internal

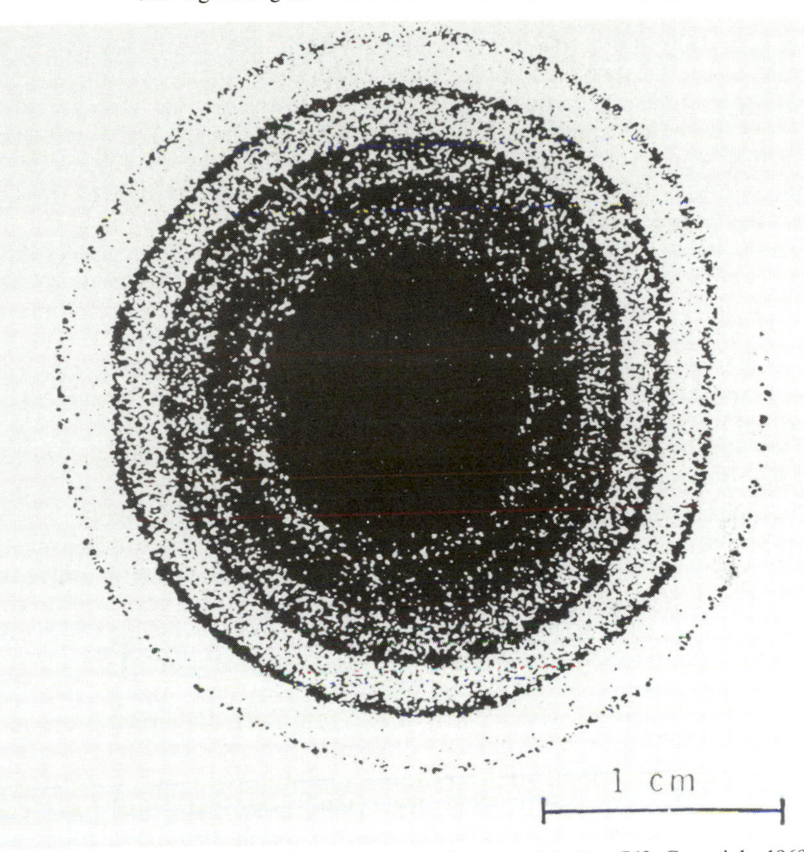

Fig. 12.9 Archimedean spiral produced by culture zonation in *Penicillium diversum*. After Bourret *et al.*[59]

flux constraints would necessarily reduce the dissipation. It therefore followed that the relaxation of such spontaneous constraints (for example, Onsager cited a stress-induced crack in a crystal conducting heat) would ultimately lead back to a state of maximum dissipation. Because Onsager and his successors have based their arguments upon such processes within the rather restrictive linear formalism of irreversible thermodynamics, applications of the optimal principles to morphogenesis have been misinterpreted and often criticized. Other authors have failed to distinguish between nondegenerate dissipative systems for which variational principles are redundant, and therefore trivial, and degenerate dissipative systems for which optimal principles or scaling laws are essential.

While Prigogine in his early work surmised that the principle of minimum entropy production was a useful determinant in biological development,[4] during the intervening years he and his associates have emphasized the exploration and mapping of bifurcated solutions of model differential equations, thus generating a complex of deep mathematical problems in non-linear pattern formation. He has not, however, forsaken the idea that thermodynamic principle may be essential to the unique solution of certain non-linear problems. To quote from the recent monograph by Nicolis and Prigogine[1] 'Assuming that a dissipative structure has been formed (order through fluctuations) we would like to find a set of thermodynamic properties characterizing the structure as uniquely as possible ... A general solution of this important problem is not yet available'.

While the latter statement seems to suggest the existence of a thermodynamic potential which selects for stable patterns, Prigogine and co-workers are very emphatic that such potentials do not in general exist outside the thermodynamic or near-equilibrium 'branch' where the principle of minimum entropy production obtains.[7] However, they do admit that if we eschew interest in sustained thermokinetic oscillations Thom's theory of catastrophes,[68] with its explicit inclusion of a potential and parametric description of morphologies, has great utility. We believe the approach to be described in the following falls within the same rubric because the non-oscillatory systems of main interest here lie in or near the thermodynamic branch as defined by Nicolis and Prigogine.

There remains considerable ambiguity as to what situations are to be assigned the terms order, pattern or dissipative structure. In § 12.2 the term order was assigned to the unmixing of nickel in the curves of Fig. 5.2 and the term pattern to the antisymmetric shape of the Ni curve in Fig. 5.2, whereas Prigogine and Nicolis have offered a proof of the 'impossibility of ordered behavior in the linear range of irreversible processes.[7] Furthermore, we regard the highly patterned morphology which appears in Fig. 1.33 as a dissipative structure, yet this would be ruled out by the definition of Nicolis and Prigogine, since the low supersaturation structure lies within the thermodynamic (local equilibrium) branch. Furthermore, the patterns of Figs. 1.33 and 12.9 can be developed spontaneously within an isolated system, seemingly contradicting the conventional wisdom that dissipative structure is a property of open systems. Indeed this possibility is essential to the minimax theorem which follows.

The development of the minimax theorem requires a number of preliminaries, including the irreversible thermodynamic framework. The derivation focuses on a compartmentalized discontinuous system as in Fig. 12.10*a*, the uniform upper reservoir with instantaneous entropy $S_1(t)$, the lower reservoir with entropy $S_2(t)$, the intermediary small subsystem 3 with entropy $S_2(t)$, and the total entropy being the sum

$$S(t) = S_1(t) + S_2(t) + S_3(t) \tag{12.100}$$

Since the system as a whole is isolated, the internal entropy production rate according to the second law of thermodynamics

$$\dot{S}_i = \dot{S} = \frac{dS}{dt} > 0 \tag{12.101}$$

Since, furthermore, the reservoirs are uniform, all of the entropy production is to be attributed to irreversible processes within the small subsystem. The entropy changes of the reservoirs are to be associated with heat $\dot{q}$ transferred per unit time isothermally across their boundaries (of magnitude $\dot{q}/T$) to or from the small subsystem. We propose now to investigate fluctuations in the system entropy

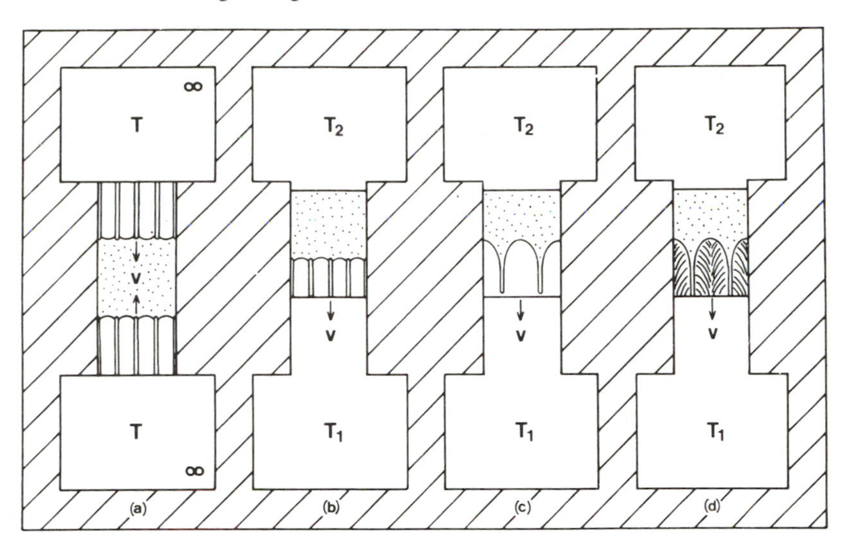

Fig. 12.10 Schematic representation of dissipative systems (corresponding to Figs. 1.33, 12.14, 1.24, and 12.32) arranged to achieve idealized quasi-steady states within isolated systems consisting of very large, perfectly mixed heat baths in contact with relatively small reacting subsystems. The drift velocities *v* in (*b*), (*c*), and (*d*) (by gravity, say) contribute negligibly to the dissipation. These idealizations are analogous to the reversibility ideal in equilibrium thermodynamics. After Kirkaldy[69]

about a mean path $S(t)$ associated with macroscopic pattern fluctuations which are initiated by microscopic fluctuations in the subsystem entropy S_3.

The ideas of a kinetic phase space, a related statistical ensemble, and a transient state of maximum path probability were introduced by Onsager.[64] While he focused on linear processes in isolation, there is nothing in the theoretical structure which excludes non-linear systems of the kind discussed here. Kikuchi,[67] for example, deals with non-linear diffusive and ordering systems within this paradigm. The Boltzmann relation (which Onsager noted is 'exceedingly general') can be used to define a thermodynamic probability W for a pair of path states separated by a time increment τ and expressed as a path entropy

$$\mathcal{S} = k \ln W = \tfrac{1}{2}[S(t) + S(t + \tau)] - g \qquad (12.102)$$

Here g is a positive semidefinite function of the fluctuation coordinates and expresses the effects of correlated motions within the distribution of kinetic states. In the linear theory, g is a quadratic expression as a function of the fluctuation coordinates and possesses a positive definite matrix of coefficients (Onsager reciprocity). In our more general case we need only recognize that when time-dependent or non-integrable relations obtain between coordinates (correlated motions), the number of statistical degrees of freedom is reduced in accord with the number of such relations and W as obtained by combinatorics is reduced accordingly. For maximum path probability the mean quantity $\overline{\mathscr{S}}$ will necessarily tend toward a maximum. Onsager demonstrated that this optimization of W generates the 'principle of minimum dissipation' in the linear case.

In this presentation we follow Onsager by defining the statistical ensemble over which an average is ultimately to be taken as a large number of related patterned and irreversible states which can originate from the same initial thermodynamic state. This ensemble is to be constructed out of an array of theoretically perfect states, defined by one or more order parameters p, together with a set of imperfect patterned states covering the same range of average p values.

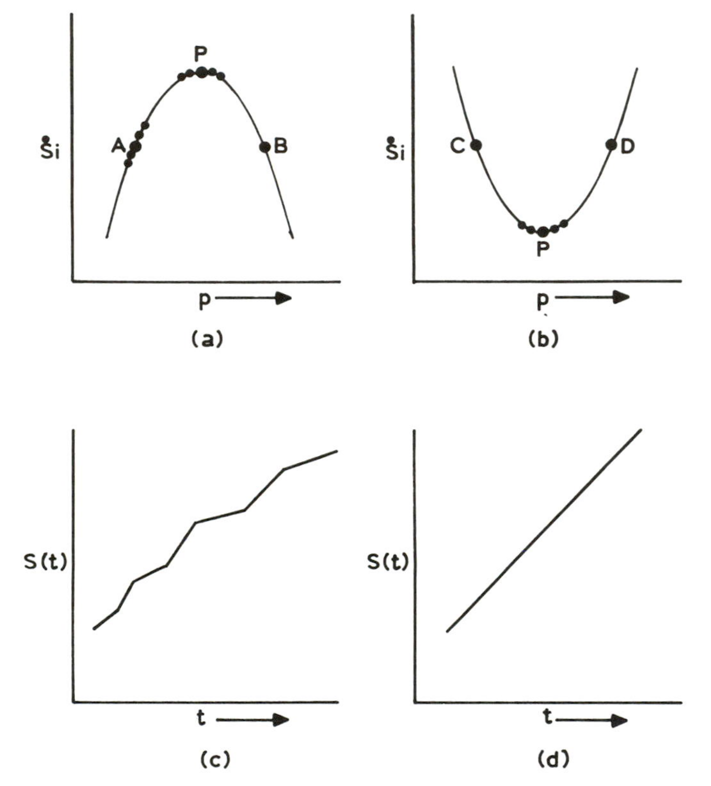

Fig. 12.11 Schematic representation of the dissipation rate as a function of order parameter p; (a) a maximal path and (b) a minimal path. Also the effect of macroscopic subsystem pattern fluctuation on the enclosure entropy: (c) off the optimum in $\dot{S}_i$, e.g. point A; (d) at an optimum in $\dot{S}_i$, e.g. points P. After Kirkaldy[69]

One now undertakes the evaluation of the ensemble averages for the right-hand side of equation (12.102) with fixed average p and then to seek the maximum with respect to average p. The argument proceeds in reference to Fig. 12.11. The key step is to characterize the correlation function g. The correlations of interest here obtain between the macroscopic pattern changes instantaneously generated by microscopic fluctuations in the small subsystem and the subsequent system entropy changes associated with ordering in the dissipative subsystem and arising from the changes in heat transfer rate to the reservoirs. In this macroscopic theory these are conventionally indexed by the time correlation moment[71]

$$C = \Delta\dot{S}_i(t)\,\Delta S(t+\tau) \tag{12.103}$$

$\Delta\dot{S}_i$ measures the instantaneous entropy production increment following a pattern change and $\Delta S = \Delta(S_1 + S_2 + S_3)$ designates the subsequent change in the system entropy over and above some base $S(t)$ which is self-consistently defined. This is clearly monotonic with the strength of the correlations and vanishes as the correlations vanish. On account of the assumed small scale of the subsystem and the required macroscopic lifetimes of interesting patterns, the total system response can be taken to be prompt so that during the lifetime τ of the new pattern we can assume that $\dot{S}_i$ is constant and from equations (12.100) and (12.101) evaluate the steady state integral for the subsequent system entropy change as

$$\Delta S = \int_t \Delta\dot{S}_i\,\mathrm{d}t = \tau\Delta\dot{S}_i \tag{12.104}$$

This implies a perfect correlation within the idealized phase space of quasi-steady states. The initiating fluctuation δS_3 is assumed to be microscopic so need not be entered into the global entropy accounts.

Proceeding, we deduce from equations (12.103) and (12.104) that the correlation product is positive semidefinite, i.e.

$$C = \tau(\Delta\dot{S}_i)^2 \tag{12.105}$$

Since g is also positive semidefinite it follows that $g = g(C)$ and is positive monotone in C, with $g(0) = 0$. The appropriate average over the deviations from the path entropy is obtained via equation (12.102). Through Taylor expansion and averaging via equations (12.104) and (12.105) the average path entropy for a given trial state is

$$\begin{aligned}\bar{\mathscr{S}} &= \frac{1}{2}\overline{[S(t) + S(t+\tau)]} - g(\bar{C}) \\ &= S(t) + \overline{\frac{\tau}{2}\Delta\dot{S}_i} - \overline{g(\tau[\Delta\dot{S}_i]^2)}\end{aligned} \tag{12.106}$$

Because of the inferred relation between g and C it is immaterial whether we use $\overline{g(C)}$ or $g(\bar{C})$ in the optimization.

Next, in reference to Fig. 12.11a, we explore a trial stable state on the degenerate entropy production curve (point A) recognizing the empirical requirement that analysable patterns must not fluctuate too widely over the order parameters in the stable state (i.e. in accord with the tight distribution of dots about the hypothetical steady state A). Finally it is assumed that the fluctuating τ is uncorrelated with $\Delta\dot{S}_i(p)$ and that in the quasi-steady mode with a vanishing regression rate $\Delta\dot{S}_i = \Delta S/\tau$ takes either sign with equal probability about the mean (Fig. 12.11*a*). Thus we can set $\overline{\tau\Delta\dot{S}_i/2} = 0$ for each trial mean p and identify the state of maximum path probability by

$$g = \overline{\tau[\Delta\dot{S}_i]^2} = 0 \tag{12.107}$$

Thus we conclude that

$$\Delta\dot{S}_i = 0 \tag{12.108}$$

for all fluctuations over p at the stable state, which is to say that the entropy production rate $\dot{S}_i$ is a maximum or a minimum (or both) over one or more order parameters p (points P rather than A in Fig. 12.11*a* and *b*).

In view of equations (12.108) and (12.104) it is seen that all macroscopic fluctuations in the entropy vanish ($\Delta S = 0$; Figs. 12.11*c* and *d*). This implies that the second derivative of the entropy with respect to the time is macroscopically defined and zero in value, as required for a rigorous quasi-steady state in isolation. Equilibrium stability theory based on the second law takes it for granted that d^2S is macroscopically defined for non-equilibrium states, so we could have reversed the argument using equation (12.104) and $d^2S = \Delta S = \tau\Delta\dot{S}_i = 0$ to prove the minimax principle. Evidently, we can thereby claim that the minimax principle for degenerate systems is a corollary of the second law.

The present result is not inconsistent with Onsager's linear theory with its principle of minimum dissipation, but it is clearly a weaker statement since the character of the optimum is not specified. The two principles merge only when and if the degeneracy vanishes. This is a salutary and important result for it explains why maximal and minimal states seem to appear in nature willy nilly. The principle is not inclusive of non-linear, non-degenerate steady state systems which have no internal order parameters and thus a restricted kinetic phase space. However, if such parameters are artificially created an analogous minimax emerges.[62] A number of applications follow. For a more fundamental exposition of this problem refer to Kirkaldy.[63, 69] Minimal and maximal aspects of a generalized dissipation principle have been explored by Tykodi,[72, 73] Paltridge,[74] and Sawada[75] (see also § 12.11).

12.8 THE GROWTH OF LAMELLAR PEARLITES: THE SPACING PROBLEM

There is a large group of isothermal interface reactions of the form

$$\text{Phase I} \rightarrow \text{Phase II} + \text{Phase III} \tag{12.109}$$

which spontaneously form lamellar or hexagonally spaced rod arrays of the product phases with a unique average spacing at each level of undercooling ΔT or supersaturation. These are variously designated as eutectoid (solid $\rightarrow$ solid), eutectic (liquid to solid) and for the lamellar eutectoid(ic) case are adequately described by Figs. 1.32–1.34. The same morphologies can result when an externally supplied reactant diffuses into and reacts with an alloy to form two

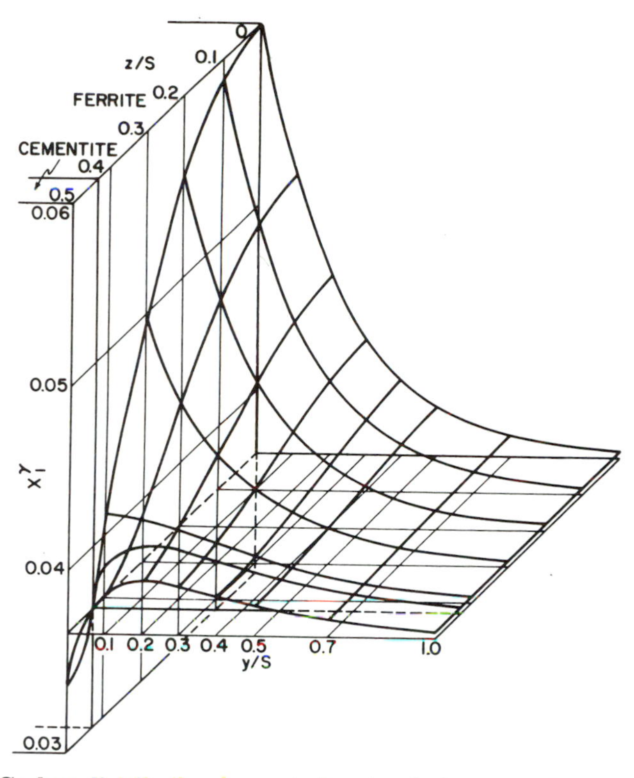

Fig. 12.12 Carbon distribution in austenite ahead of the pearlite boundary at 660°C. After Hashiguchi and Kirkaldy[82]

product phases[76, 77, 78, 79] or in the form where Phase II is a metastable derivative of Phase I (discontinuous precipitation[80, 81]). Figure 12.12, due to Hashiguchi and Kirkaldy[82] shows the calculated carbon diffusion profile ahead of the advancing structure interface corresponding to these Figs. 1.32–1.34. The preliminary discussion of Chapter 1 was concerned primarily with the diffusive boundary value problem defined by the phase constitution of Fig. 1.32 and the lamellar morphologies of Figs. 1.33 and 1.34. We turn now to the removal of the solution degeneracy recognized in § 1.26, a problem which has intrigued materials scientists for half a century. This kinetic indeterminacy appears within the derived velocity-spacing relation (cf equation (1.181)) of the form

$$v = v(\Delta T, s) \tag{12.110}$$

which implies an infinite number of steady state patterns over s for each undercooling. To establish a unique solution as observed, the theorist is pressed to find a perturbation argument which specifies the stability of a unique state from among the spectrum allowed by equation (12.110) or a potential function which optimizes at the same average pattern (cf § 12.7). In the following both procedures are followed and the necessary equivalence is demonstrated.

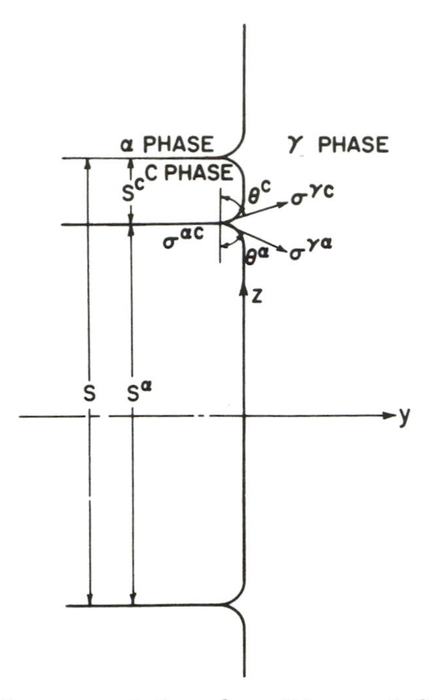

Fig. 12.13 Schematic representation of pearlite growth front. After Hashiguchi and Kirkaldy[82]

An experimental test is found in the model of Hashiguchi and Kirkaldy[82] for the Fe–C eutectoid (pearlite) which is developed according to the parameters in Fig. 12.13. The model of § 1.26 is generalized by the inclusion of parallel volume and phase boundary diffusion of C in accordance with current intelligence.[83, 84, 85]

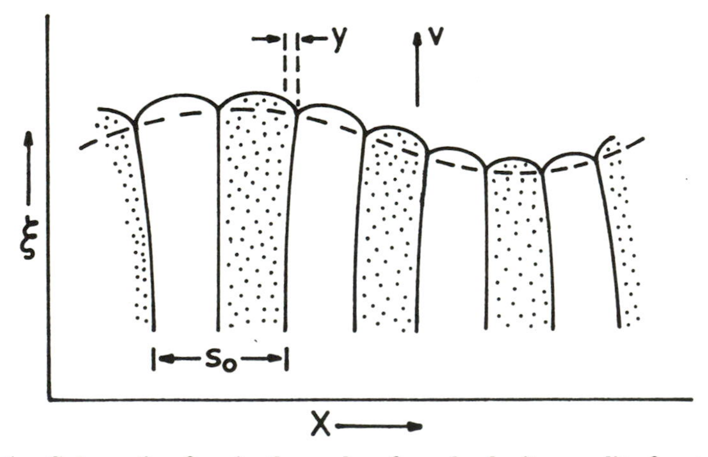

Fig. 12.14 Schematic of an isothermal or forced velocity pearlite front which is evolving following a spatial and/or velocity perturbation. After Langer[87]

The resulting $v(s)$ formula is in precise accordance with the limiting volume diffusion (§ 1.26) and phase boundary diffusion formulas of Hillert.[21,86]

The perturbation methodology is due to Langer[87] as adapted by Kirkaldy.[88,89] In relation to Fig. 12.14 and long wavelength perturbations one connects the displacement coordinates $y(x, t)$ and $\xi(x, t)$ by the first order continuity relation in terms of a base mean velocity v_0, i.e.

$$\frac{\partial v}{\partial t} = -v_0 \frac{\partial \xi}{\partial x} \tag{12.111}$$

This is a strongly intuitive scaling law over the shape coordinates. Langer has also indicated how the discrete local lamellar spacing should be expanded analytically about some mean value s_0, i.e.

$$s(x, t) \simeq s_0(1 + \mathrm{d}y/\mathrm{d}x) \tag{12.112}$$

Both expressions are valid only in the limit of long wavelength, low amplitude perturbations in ξ and s. Note the discontinuous sequence for Fig. 12.14

$$\begin{aligned} s_0 &= s_0 \\ s_1 &= s_0 + y_1 = s_0[1 + y_1/s_0] \\ s_n &= s_0 + (y_n - y_{n-1}) = s_0[1 + (y_n - y_{n-1})/s_0] \end{aligned} \tag{12.113}$$

which confirms the continuous relation (12.112). Implicit within this analytic description is the assumption that there exists a prompt and not too disruptive mechanism for isothermal spacing change (e.g. lamellar faults).

The perturbation argument proceeds by developing a non-linear differential equation which relates the temporal evolution of the spacing of a perturbed system to the curvature of the front. Firstly, differentiation of equation (12.112) yields

$$\frac{\partial (s/s_0)}{\partial x} = \frac{\partial^2 y}{\partial x^2} \tag{12.114}$$

and

$$\frac{\partial (s/s_0)}{\partial t} = \frac{\partial^2 y}{\partial t\, \partial x} = \frac{\partial^2 y}{\partial x\, \partial t} \tag{12.115}$$

where $\partial y/\partial t$ is the lateral velocity at the front. Differentiating equation (12.111) yields through equation (12.115)

$$-v_0 \frac{\partial^2 \xi}{\partial x^2} = \frac{\partial (s/s_0)}{\partial t} \tag{12.116}$$

This consequence of equations (12.111) and (12.112) implies that any spacing distribution existing at a frontal boundary of zero curvature is stationary to the first order in the time, or alternatively that there is a spontaneous trend for the frontal surface energy to be minimized.

Differentiating once more we obtain

$$-v_0 \frac{\partial^2}{\partial x^2}\left(\frac{\partial \xi}{\partial t}\right) = \frac{\partial^2 (s/s_0)}{\partial t^2} \tag{12.117}$$

Since the local forward velocity v can be expressed as

$$v = v_0 + \frac{\partial \xi}{\partial t} \tag{12.118}$$

Equation (12.118) can be written as

$$-v_0 \frac{\partial^2 v}{\partial x^2} = \frac{\partial^2 (s/s_0)}{\partial t^2} \tag{12.119}$$

In this treatment of the isothermal case we assume that the instantaneous forward velocity can also be expressed by (cf § 1.26)

$$v = v[s(x,t)] \simeq \frac{\Delta T}{\alpha s^n}\left(1 - \frac{s_c}{s}\right) \tag{12.120}$$

where $n = 2$ for volume diffusion and $n = 3$ for phase boundary diffusion. That is to say in the limit of perturbations marginal to each member of the spectrum of accessible steady states the local v is accurately approximated in terms of local s

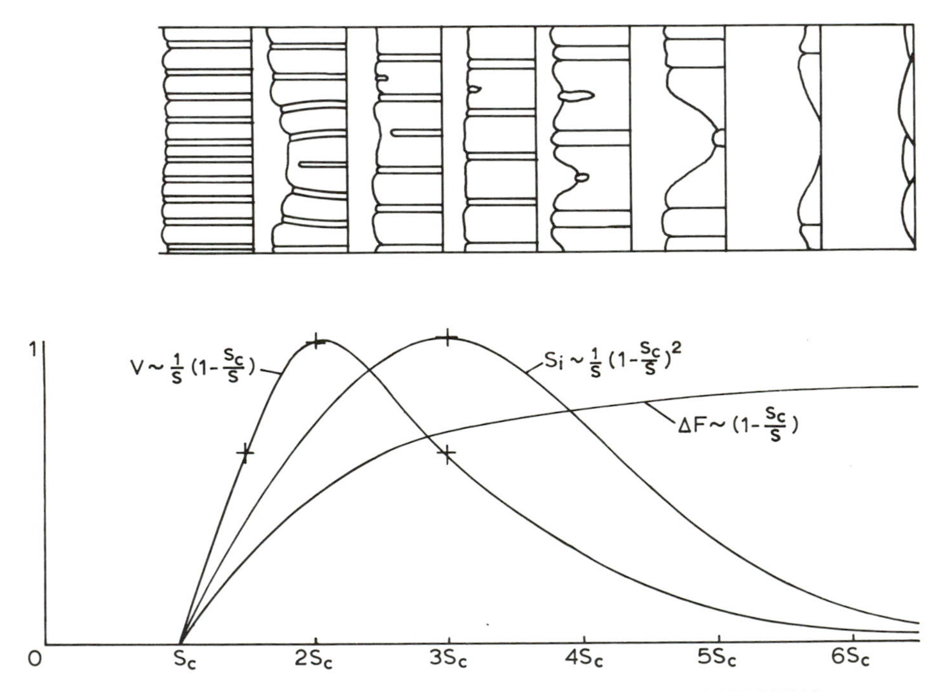

Reprinted with permission from TRANSACTIONS OF THE METALLURGICAL SOCIETY, Vol. 16A, p. 1781, (1985), a publication of The Metallurgical Society, Warrendale, Pennsylvania

Fig. 12.15 Schematic microstructures and matched controlling functions for isothermal lamellar eutectoids (pearlites) corresponding to a volume diffusion mechanism. The local interface shapes shown are close to those given by theory. After Kirkaldy[69]

by its steady state Zener–Hillert expression. Since we will be primarily concerned in the following with spacing perturbations at a flat interface there can be no doubt that equation (12.120) is an excellent approximation. $v_0 = v_0(s_0)$, of course, represents the special case of a uniform unperturbed state.

Substituting equation (12.120) into equation (12.119), keeping in mind that v is not explicit in t, gives the second order, non-linear regression equation

$$\frac{\partial^2 s}{\partial t^2} = -s_0 v_0 \left[\frac{\mathrm{d}v}{\mathrm{d}s}\right] \frac{\partial^2 s}{\partial x^2} - s_0 v_0 \left[\frac{\mathrm{d}^2 v}{\mathrm{d}s^2}\right] \left(\frac{\partial s}{\partial x}\right)^2 \tag{12.121}$$

where the square-bracketed derivatives remain as functions of $s = s(x, t)$. Consider then a base state at the velocity optimum (Fig. 12.15) and a low amplitude linear perturbation of the spacing ($\partial^2 s/\partial x^2 \sim 0$) at zero curvature which, via equation (12.116), implies first order stationarity at the initial state, that is

$$-v_0 \frac{\partial^2 \xi}{\partial x^2} = \frac{\partial (s/s_0)}{\partial t} = 0 \tag{12.122}$$

According to Fig. 12.15 and equation (12.120) ($n = 1$)

$$\frac{\mathrm{d}v}{\mathrm{d}s} \simeq 0 \qquad \frac{\mathrm{d}^2 v}{\mathrm{d}s^2} < 0 \tag{12.123}$$

at the maximum of v, so from equation (12.121), $\partial^2 s/\partial t^2 > 0$ at regions of both augmented and deficit spacing ($s >$ or $< s_0$). Since $\partial s/\partial t = 0$ this implies that spacings to the right and left as well as at the optimum are unstable and all tend spontaneously to a spacing that is greater than that at this coordinate. This result was first argued qualitatively by Frank and Puttick.[90]

Next consider the same planar state ($\partial s/\partial t = 0$) with an initial weak linear spacing distribution near the inflection point of the velocity curve (Fig. 12.15; $\mathrm{d}v/\mathrm{d}s < 0$; $\mathrm{d}^2v/\mathrm{d}s^2 \approxeq 0$) and identify the inflection point by the coordinate s_{opt}. It is immediately seen that for all $s_0 < s_{\mathrm{opt}}$, $\mathrm{d}^2v/\mathrm{d}s^2 < 0$ and $\mathrm{d}^2s/\mathrm{d}t^2 > 0$, so the state is unstable, spontaneously regressing toward s_{opt}. Similarly, for all $s_0 > s_{\mathrm{opt}}$, $\mathrm{d}^2v/\mathrm{d}s^2 > 0$ and $\mathrm{d}^2s/\mathrm{d}t^2 < 0$ and regression is again toward s_{opt}. The latter accordingly defines the stability point uniquely to the first and second order in the time, and is expressed by

$$\frac{\mathrm{d}^2 v}{\mathrm{d}s^2} = 0 \tag{12.124}$$

Finally, consider a sinusoidal perturbation in the wavelength with s_0 at the inflection point of the velocity curve and refer to equation (12.121) and Fig. 12.15. We see that perturbed regions where $\partial^2 s/\partial x^2 < 0$ corresponding to $s > s_0$ are subject to $\partial^2 s/\partial x^2 < 0$, so they spontaneously regress toward s_0. Similarly, regions where $\partial^2 s/\partial t^2 > 0$ corresponding to $s < s_0$ are subject to $\partial^2 s/\partial t^2 > 0$ so also regress toward s_0. To put this in broader terms, equation (12.121) is a non-linear wave equation which describes a perfectly damped wave at the inflection point only (note the signs of the coefficients). Equation (12.121) defines the same state of stability as the entropy production maximum (cf Fig. 12.15 and below). For volume diffusion this identifies the point $s = 3s_c$, and for boundary diffusion it is $s = 2s_c$.

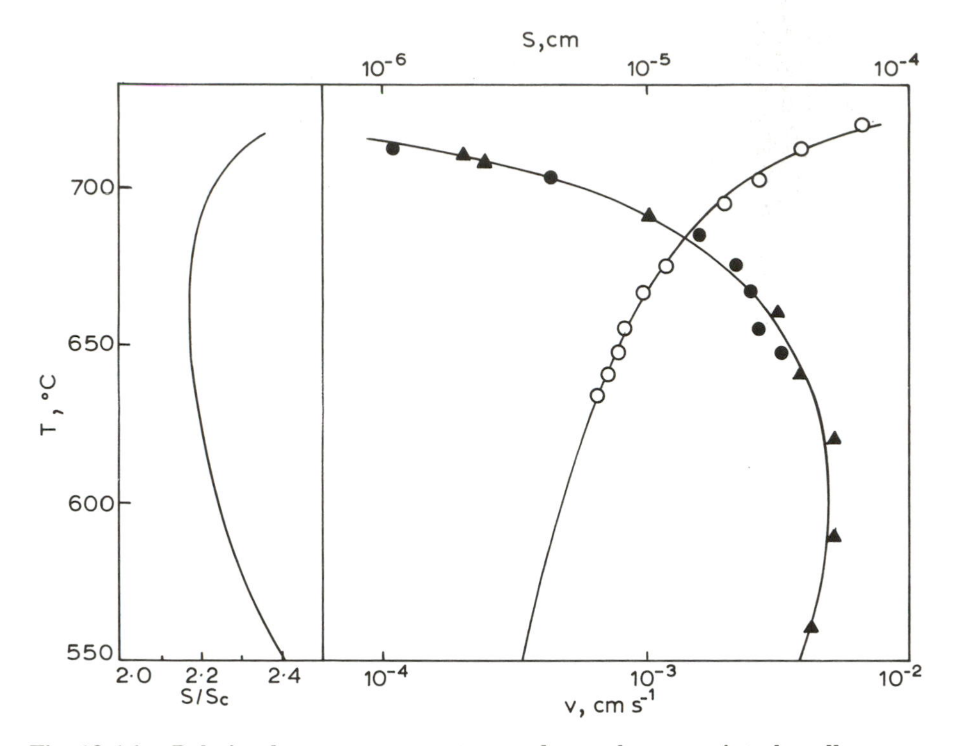

Fig. 12.16 Relation between temperature and growth rate, *v*, interlamellar spacing, *s*, and ratio of *s* to critical spacing, s_0. Points are experimental and lines are calculated. After Hashiguchi and Kirkaldy[82]

As Kirkaldy and Sharma[91] have demonstrated, this stability point also defines a state of minimum global frontal curvature, which explains why point-nucleated pearlite cells tend to grow as spheres.

Hashiguchi and Kirkaldy[82] have analysed the Fe–C eutectoid data due to Brown and Ridley.[92] Their optimization of the theoretical fit to the known material parameters and the independently evaluated velocities and spacings (Fig. 12.16) according to the maximum entropy production principle identifies a dominant phase boundary diffusion process, a boundary diffusion coefficient which is plausible and a stable ratio $s/s_c \rightarrow 2$ in accord with a dominant phase boundary diffusion process (cf the inset of Fig. 12.16).

The forced velocity eutectic growth problem identifies the undercooling ΔT as the free variable. The stabilization, however, remains at a maximum in the dissipation and the system invariants $s\Delta T$ and vs^2 are unchanged.[89]

Bolze *et al.*[93] have modelled the ternary eutectoid reaction for a volume diffusion process. An approximate ternary theory due to Sharma *et al.*[94] has been tested with fair results for the system Fe–Cr–C. The maximum dissipation principle has been applied with reasonable success in a number of other investigations.[74, 95, 96, 97] It is reasonable to conjecture that it is determinant in the morphogenesis of biological membranes as in Figs. 2.24–2.26.

12.9 CELLULAR SOLIDIFICATION OF A BINARY ALLOY

In § 1.19 the boundary conditions for the present problem of interface instability were defined, so reference is made to Figs. 1.23 and 1.24. The latter shows a micrograph of a two-dimensional cellular structure which evolves after breakdown of an initial planar migrating single crystal solid interface. The trailing edges of the liquid protrusions on the right terminate in a cylindrical shape with a curvature which approaches the capillarity nucleation limit ($\sim 0{\cdot}1\ \mu m$ in this case; cf Fig. 12.17). The stable or stationary cellular structure appears only in a narrow range of the effective supersaturation. Above this range the cells evolve side branches becoming dendrites, and below they enter a wavy, non-stationary milieu.[98] Because of the capillarity near-singularity at the trailing edge (the root) the cell walls must be out of local equilibrium if the cell tips and root are in local equilibrium. These cell walls tend to seek low surface energy configurations so at most crystal orientations the cell walls are not parallel to the growth direction even though the line of cell tips remains perpendicular to it. In experiments which relate the more or less unique cell lengths and spacings to the growth parameters it is essential to avoid the skew orientations.

The present model based on Fig. 12.17 makes the following assumptions with respect to the boundary conditions[99]

(1) local equilibrium exists at the cell tips
(2) local equilibrium modified by capillarity exists at the cell roots. Having chosen the first two constraints on the grounds that these interfacial locations are known to be irrational it follows within our model that it is impossible for precise local equilibrium to exist elsewhere
(3) diffusion penetration in the solid is negligible except in the vicinity of the

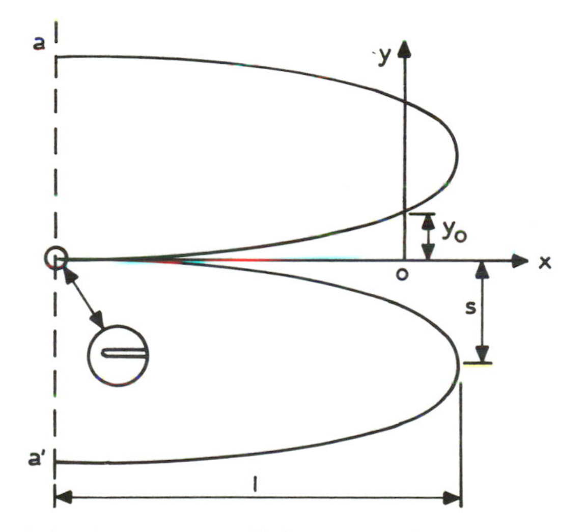

Reprinted with permission from *Acta Met.*, **32**, D. Venugopalan and J. S. Kirkaldy, p. 893, copyright 1984, Pergamon Press Ltd

Fig. 12.17 Schematic interface shape used in our calculations. After Venugopalan and Kirkaldy[99]

root, where it can be represented by a diffusion field analogous to that used in the theory of Widmanstätten plates.[20, 21]

(4) the non-equilibrium cell walls are stabilized by kinetic and crystallographic anisotropy effects over a limited range of growth parameters. Extremes of the parameters lead to dendritic breakdown

(5) the temperature gradients in the solid and the liquid are the same. This condition decouples the heat transfer problem from the solute diffusion problem and corresponds to the Jackson and Hunt[100] thin film experiment with which we are concerned here

(6) the shape of the cells can be represented by a parametric elliptic conic attached to a cylindrical root (Fig. 12.17). This experimentally justified condition substitutes for our present lack of knowledge of non-equilibrium interface constraints. This constraint is not as strong as might appear at first sight for the anisotropy requirement for cell formation specifies *a priori* that a substantial length of the walls must be planar and nearly parallel

Assumptions 1, 2, and 6 were invoked in part to reduce the degeneracy of the free boundary problem and the related optimization procedure to manageable proportions. Even so, there remain two internal degrees of freedom.

The steady state two-dimensional diffusion equation for the liquid in a moving frame of reference (stationary with respect to the interface) is the two-dimensional form of equation (1.110), i.e.

$$\frac{\partial^2 X}{\partial x^2} + \frac{\partial^2 X}{\partial y^2} + \frac{v}{D}\frac{\partial X}{\partial x} = 0 \tag{12.125}$$

A general solution to this equation, subject to the conditions $\partial X/\partial y = 0$ at $y = 0$ and s, is obtained via separation of variables as

$$X = X_0' + \sum_{n=0}^{\infty} B_n' \exp(-\eta_n x) \cos \lambda_n y + \sum_{n=0}^{\infty} A_n' \exp(\mu_n x) \cos \lambda_n y \tag{12.126}$$

where

$$\lambda_n = \frac{n\pi}{s} \qquad \mu_n = \frac{1}{2}\left\{\left[\left(\frac{v}{D}\right)^2 + 4\lambda_n^2\right]^{1/2} - \frac{v}{D}\right\}$$

$$\eta_n = \frac{1}{2}\left\{\left[\left(\frac{v}{D}\right)^2 + 4\lambda_n^2\right]^{1/2} + \frac{v}{D}\right\} \qquad n = 0, 1, 2, \ldots \tag{12.127}$$

and $2s$ is the cell spacing.

Because of the discontinuity assumed in the interface mass balance to accommodate the discretely localized solid diffusion field (condition 3 above) the steady state boundary conditions cannot be satisfied with the single set of coefficients designated in relation (12.126). Consider that near the root, where the groove is of negligible width with nearly parallel walls (i.e. limit of the elliptical conic), there can be no lateral gradient, so the diffusion solution must be accurately of the form

$$X = a + b \exp(-vx/D) \qquad (12.128)$$

provided the origin for x is chosen to be at the root. The value of the leading coefficient a is predicated upon the approximate mass balance at the root ($x = 0$) which was expressed by Kirkaldy[101] as

$$v\Delta X_L = v(1 - k_0)X \Big|_{\text{root}} = a_0 D_s \frac{\Delta X_S}{r} - D\frac{\partial X}{\partial x}\Big|_{\text{root}} \qquad (12.129)$$

where $a_0 \sim 1$, r is the radius of the root, ΔX_L is the step in liquid and ΔX_S the step in solid concentration. Substitution of equation (12.128) into (12.129) and rearrangement yields

$$a_0 D_s \frac{\Delta X_S}{r} = v(a - k_0 X_r) \qquad (12.130)$$

In the cellular configuration near the root, solute segregation is always such that $k_0 X_r > X_0$ with $k_0 < 1$. Furthermore the root capillarity is such that the diffusion in the solid must be in a direction away from the interface ($\Delta X_s > 0$). Thus in equation (12.131)

$$a > k_0 X_r > X_0 \qquad (12.131)$$

which is to say that the $x = \infty$ asymptote for the one-dimensional solute profile deep within the groove is greater than the original alloy content, X_0. In contradistinction, the $x = \infty$ asymptote for the solute profile ahead of the tips is X_0, as indicated in equation (12.126). It follows therefore that this latter relation and choice of origin cannot represent the solution over the entire liquid volume, and in particular at the root. The approximate procedure, following an analogous problem in heat transfer,[102] is to construct two solutions so as to independently satisfy the boundary condition limits and to match up the results via continuity of the concentration and the flux (the gradient) at an origin ($x = 0$) located an unspecified distance x_0 behind the tips. This quantity is to be regarded henceforth as a free internal order parameter. This manipulation can be achieved because solution (12.126) defines two distinct classes of solution with $X'_0 = X'_{01}$ and $A'_n = 0$ or $X'_0 = X'_{02}$ and $B'_n = 0$, μ_n and η_n being different as expressed by equation (12.127).

To proceed, the x origin is set such that x_0 is the adjustable distance to the cell tips and the diffusion solutions are written behind and in front of the origin, respectively, as

$$X = X_0 + A_0 + b \exp(-vx/D) + \sum_{n=1}^{\infty} A_n \exp(\mu_n x) \cos \lambda_n y \qquad x < 0 \quad (12.132)$$

and

$$X = X_0 + B_0 \exp(-vx/D) + \sum_{n=1}^{\infty} B_n \exp(-\eta_n x) \cos \lambda_n y \qquad x > 0 \quad (12.133)$$

In the choice of terms the necessary asymptotic limits and the requirement for matching at $x = 0$ have been recognized. Indeed, the unique matching of these two solutions in value and gradient at $x = 0$ demands the interrelations for the coefficients[98]

$$B_0 = A_0 + b \tag{12.134}$$

and

$$A_n = B_n = -2\frac{v}{D}\frac{A_0 \sin \lambda_n y_0}{(\mu_n + \eta_n)\lambda_n(s - y_0)} \tag{12.135}$$

where y_0 is the half-width of the groove at $x = 0$. Note that the combined equations represent an exact partial solution to the boundary value problem for the liquid. The appearance of the order parameter x_0 is one indicator of the transformation to a free boundary problem. Of course, A_0, b, and x_0 remain to be determined by other conditions on the problem.

The remaining boundary conditions are as follows:

(*a*) At the cell tips local equilibrium yields the tip temperature in terms of the melting point of the solvent, the liquidus slope and the tip (t) composition

$$T_t = T_0 + mX_r \tag{12.136}$$

The root temperature can similarly be written as

$$T_r = (T_0 + mX_r)\left(1 + \frac{\sigma}{Hr}\right) \tag{12.137}$$

for a cylindrical geometry where σ is the surface tension and H is the latent heat of fusion. In these, T_t and T_r are related by

$$T_t = T_r + Gl \tag{12.138}$$

Thus equations (12.136), (12.137), and (12.138) yield the condition

$$mX_t = mX_r + (T_0 + mX_r)\frac{\sigma}{Hr} + Gl \tag{12.139}$$

(*b*) Local equilibrium and the mass balance at the tip yields

$$v(1 - k_0)X_t = -D\left.\frac{\partial X}{\partial x}\right|_t \tag{12.140}$$

(*c*) Mass conservation at the root is given by equation (12.129), which leads to the approximation

$$\Delta X_S = k_0 \Delta X_L = k_0 \frac{\sigma V}{RT_r r(1 - k_0)} \tag{12.141}$$

and $a_0 \simeq \frac{1}{2}$ for cylindrical geometry.[21]

(*d*) The steady state overall solute balance requires that the average composition of the solid behind the interface be equal to X_0. This condition is expressed mathematically as

$$\int_0^s X_S \, dy = sX_0 \tag{12.142}$$

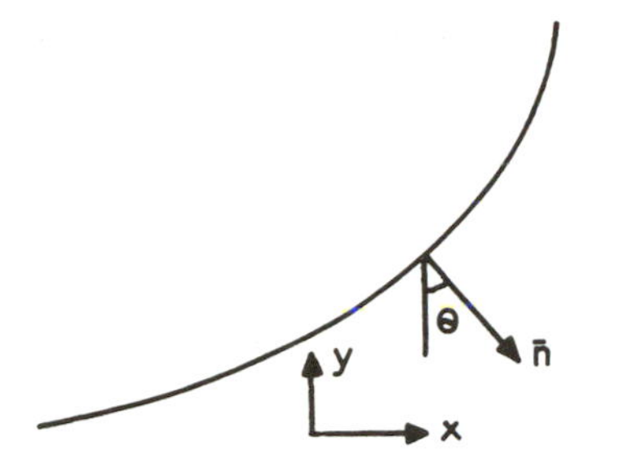

Reprinted with permission from *Acta Met.*, **32**, D. Venugopalan and J. S. Kirkaldy, p. 893, copyright 1984, Pergamon Press Ltd

Fig. 12.18 Relationship between the normal vector and the coordinate axes. After Venugopalan and Kirkaldy[99]

If due to low gradients and the small diffusion coefficients we neglect diffusion in the solid everywhere except adjacent to the root, then the solid composition $X_S(y)$ at any point along a–a′ in Fig. 12.17 is equal to the interface solid composition for the same value of y. Evaluation proceeds via mass conservation at the interface (neglecting diffusion in the solid) yielding (Fig. 12.18)

$$v_n(X_L - X_S) = -D\frac{\partial X}{\partial n} \tag{12.143}$$

where $\partial X/\partial n$ is the gradient normal to the interface. Since, according to our model, all but two points of the cellular interface are out of equilibrium, phenomenological relations between X_L and X_S cannot be easily expressed so equation (12.144) is used for evaluation of X_S. This procedure may be mounted only if the interface is known *a priori*. Thus to make the analysis tractable, and following the practise of earlier workers, we construct a parametric interface shape in reasonable accordance with experiment. From an examination of experimental cellular interfaces we choose to operate upon a parametric elliptic conic with minor axis $2s$ and major axis $2l$ which terminates in a semicircle of radius r (Fig. 12.17). This is expressed as

$$[x + (l - x_0) - r]^2 + y^2 = r^2 \qquad 0 < y < r \tag{12.144}$$

and

$$\left[\frac{x + l - x_0 - r}{l - r}\right]^p + \left[\frac{s - y}{s - r}\right]^p = 1 \qquad r < y < s \tag{12.145}$$

where the index p by observation lies between 2 and 4. Further, from Fig. 12.18

$$v_n = v\sin\theta \qquad \frac{\partial X}{\partial n} = \frac{\partial X}{\partial x}\sin\theta + \frac{\partial X}{\partial y}\cos\theta \qquad \tan\theta = \frac{\mathrm{d}y}{\mathrm{d}x} \tag{12.146}$$

Equations (12.143)–(12.146) are used to calculate X_S everywhere except behind the root, where it is given simply as k_0X_r, thus enabling us to apply the integral (12.142) as a global constraint.

To summarize, the four global boundary conditions, equations (12.129),

(12.139), (12.140), and (12.142) have been articulated. For a complete solution of the problem we are required to evaluate the parameters A_0, b, x_0, l, s, and r. Hence the free boundary problem has two degrees of freedom. These may be identified with x_0 and r at the fundamental level or with the cell length and spacing (l and s) at the observational level.

The two degrees of freedom or degeneracies in the free boundary problem described above must be eliminated to obtain a unique solution. Resolution of this problem demands the use of a thermokinetic principle or a scaling law. Following an early suggestion of Kirkaldy,[103] now quantified by Venugopalan and Kirkaldy[99] (cf § 12.7) we focus on the dissipation rate.

For the cell problem this quantity at the quasi-steady state can be evaluated as either an integral over squared gradients (cf § 12.10) or via the net entropy fluxes at the boundaries. It is convenient to focus on the latter which yields for unit cross-sectional area of the system[103, 104]

$$\Delta \dot{S}_i = v \, \Delta s_V \tag{12.147}$$

where s_V is the average entropy per unit volume of the solid. It is assumed that the fixed thermal environment, as obtained in the thin film experiment, allows us to attribute the fluctuating entropy production rate entirely to diffusion and chemical reaction effects. The entropy s_V can then be related to the chemical segregation distribution in the solid. If $X_S(y)$ is the solid concentration profile behind the interface, then recalling that solid diffusion is significant only at the root, the average entropy per unit volume along a plane passing through the root is approximated as (cf Appendix 1)

$$s_V = \frac{-R \int_0^{2s} [X_s \ln X_s + (1 - X_s) \ln (1 - X_s)] \, dy}{2s} \tag{12.148}$$

and the optimal point of this integral is the same as for equation (12.147). This optimum is then sought with respect to the identified order parameters — cell spacing and length. A unique bivariate minimum is in fact obtained for growth conditions very close to those established in the experiments (see below).

In the computation procedure, for a given alloy content, temperature gradient and velocity, an arbitrary set for spacing, length and $y_0(x_0)$ (position of origin) is assumed. The remaining three variables (A_0, b, and r) are evaluated using the three boundary conditions expressed by equations (12.139), (12.140), and (12.141). Equation (12.142) is then used to check if the choice of y_0 satisfies the overall mass balance in the solid. The entire calculation is repeated until a stable y_0 is found which satisfies all the boundary conditions for the chosen spacing and length. Next, keeping the spacing fixed, the length is varied and the entire procedure is repeated until an optimum is found with respect to the length (at fixed spacing) in the optimizing function (entropy flux). The same procedure is repeated at different spacing values until an optimum with respect to spacing is also found.

The model and optimization applied to a dilute solution of salol in the transparent organic succinonitrile predicts that the spacing and length decrease as the velocity increases at fixed alloy content and temperature gradient. Increasing the alloy content decreases the spacing and increases the length.

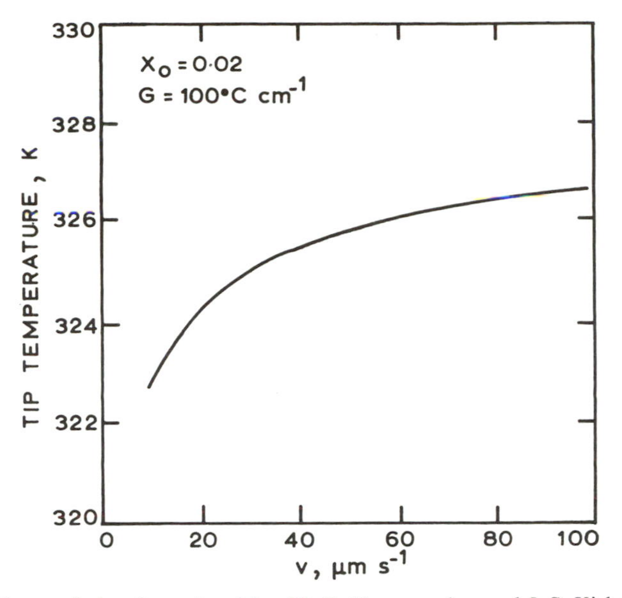

Fig. 12.19 Typical variation of tip temperature with velocity for fixed X_0 and G. After Venugopalan and Kirkaldy[99]

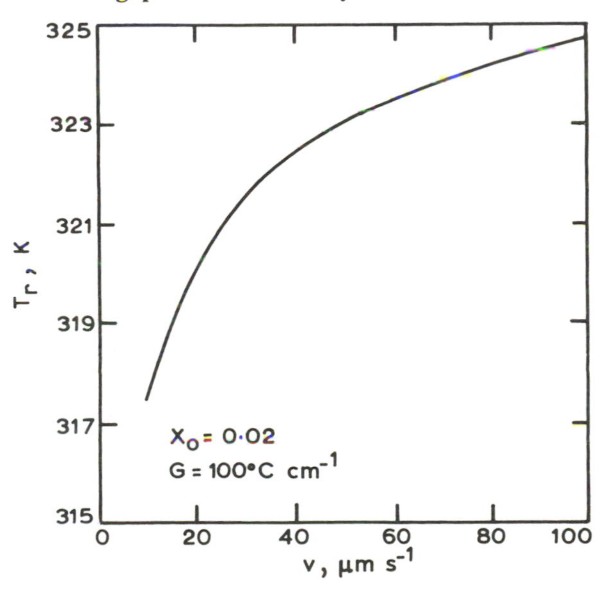

Fig. 12.20 Typical variation of root temperature with velocity for fixed X_0 and G. After Venugopalan and Kirkaldy[99]

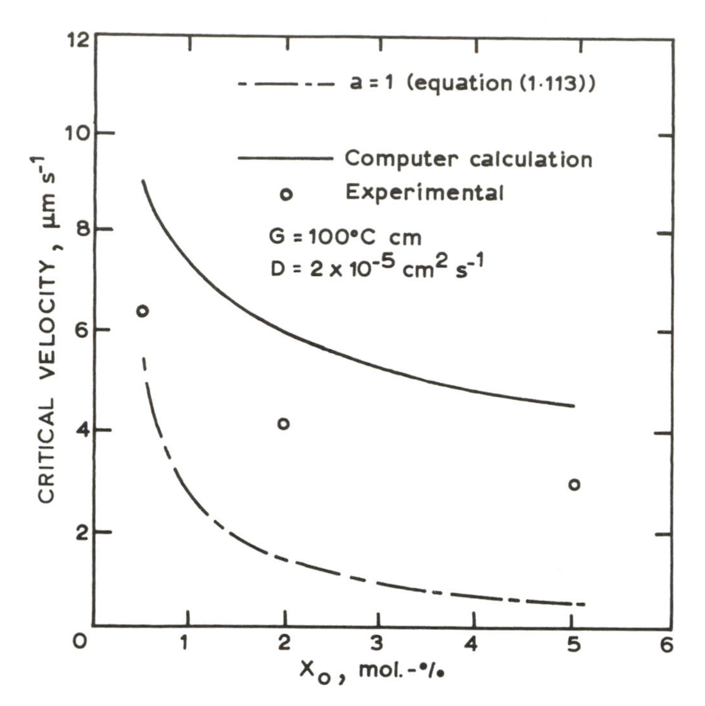

Reprinted with permission from *Acta Met.*, **32**, D. Venugopalan and J. S. Kirkaldy, p. 893, copyright 1984, Pergamon Press Ltd

Fig. 12.21 Critical velocity for the onset of stable cells. The CS criterion (α = 1) is given for comparison. After Venugopalan and Kirkaldy[99]

Increasing the temperature gradient decreases both the spacing and length. Some of the results of the calculations are shown in Figs. 12.19–12.22. The tip temperature increases with growth rate at fixed X_0 and G as does the root temperature (Figs. 12.19 and 12.20). Both curves indicate asymptotic limits in the high velocity regime.

The root radius is a function of the growth condition and lies between 2 and 6 times r^*, where r^* is the critical radius for a liquid nucleus of the root composition at root temperature. It decreases with increasing velocity, X_0 and decreasing G.

A salient feature of the theoretical result is the absence of steady-state solutions at very low velocities. Below a critical velocity the solution algorithm does not in all cases converge and it is not possible to satisfy all the boundary conditions. This is an important theoretical result as it predicts an unstable regime lying between the breakdown of the planar interface and the onset of stationary cells. The critical velocity is a function of the alloy content as shown in Fig. 12.21. The theory thus predicts the experimentally observed gap in supersaturation between the critical velocity and equation (1.113) in which stable cells are not obtained. The theory also predicts that stable cells are considerably longer than the

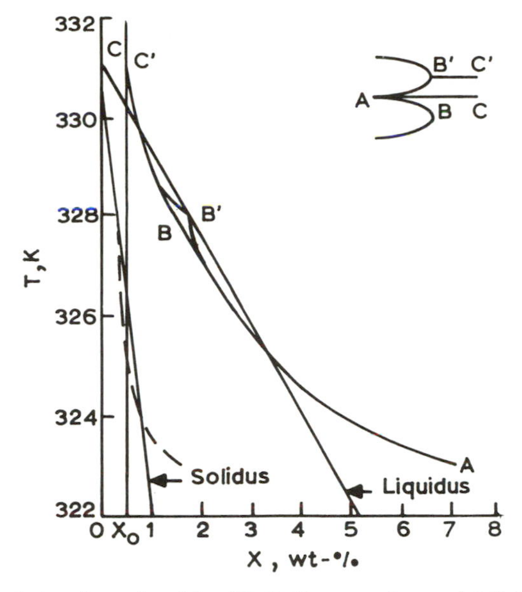

Fig. 12.22 Liquid concentration profile plotted on the phase diagram (X_0 = 0·005, G = 150°C cm^{-1}, v = 20 μm s^{-1}). Notice the approximately even division of supersaturation between the solid (superheat) and the liquid (supercooling). After Venugopalan and Kirkaldy[99]

diffusion length (D/v). In fact for lengths less than about 1·5 D/v, steady state solutions could not be obtained. This behaviour is also in agreement with Kirkaldy's analytic model[101] and with experiments, as very short stationary cells are not obtained. The short length limit increases as the temperature gradient decreases.

In Fig. 12.22 a typical predicted composition profile is superposed upon the phase diagram. For nearly half the cell length the cell walls are superheated and for the other half supercooled. This partitioning of the supersaturation is an implicit manifestation of the principle of minimum entropy production rate, which requires that the integral over the squares of gradients and chemical potential differences be minimized. As the velocity is increased, the cells become shorter and the extent of supersaturation increases along the cell walls. Thus at very high velocities dendritic breakdown of cell walls may be anticipated, as in such cases the chemical driving force is able to overcome the kinetic and crystallographic stabilizing effects.[105]

As a conclusive test of the theoretical structure we refer to the comparisons of observations and predictions based on independent data for the material properties (Fig. 12.23 and 12.24). Other workers[106, 107, 108] have also tested the minimum dissipation principle with variable results. In current research Kirkaldy and Venugopalan[109] have made some progress in developing the scaling law for this problem.

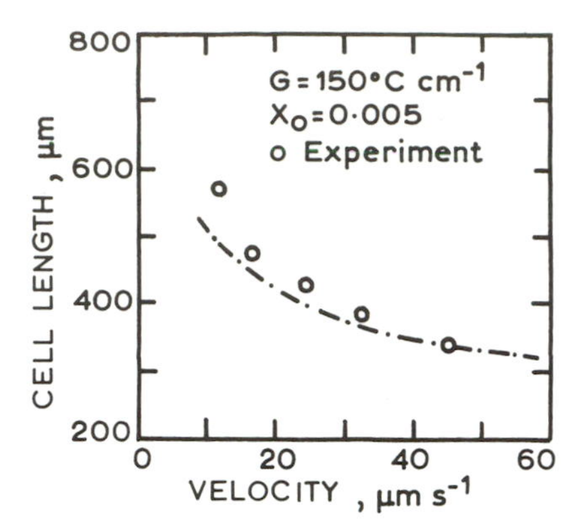

Reprinted with permission from *Acta Met.*, **32**, D. Venugopalan and J. S. Kirkaldy, p. 893, copyright 1984, Pergamon Press Ltd

Fig. 12.23 Call length *vs* velocity at fixed composition and gradient in a salol-succinonitrile solution. $D = 10^{-5}$ cm^2 s^{-1}. After Venugopalan and Kirkaldy[99]

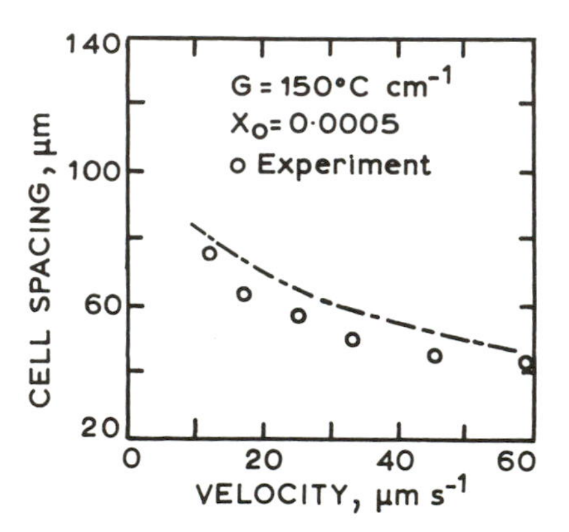

Reprinted with permission from *Acta Met.*, **32**, D. Venugopalan and J. S. Kirkaldy, p. 893, copyright 1984, Pergamon Press Ltd

Fig. 12.24 Cell spacing *vs* velocity at fixed composition and temperature gradient. $D = 10^{-5}$ cm^2 s^{-1}. After Venugopalan and Kirkaldy[99]

12.10 COUPLED CHEMICAL REACTION AND DIFFUSION AT A SOLIDIFICATION INTERFACE

In the previous section, it was recognized that stable patterns occur only with the cell walls out of local equilibrium and where the differential chemical potentials along the wall interfaces arrange themselves in the direction of minimum entropy production (cf Fig. 12.22 and associated text). This draws our attention to a related but elementary 'free boundary' problem concerning the explicit establishment of the stable disposition of chemical potential differences across a planar interface which drives the steady state interfacial chemical reaction. In principle, we are concerned here with a synthesis of the reaction and diffusion processes discussed in §§3.4 and 3.5. This problem has been discussed among others by Kirkaldy,[103, 104] Baker and Cahn,[110] and Langer and Sekerka.[11] The latter authors deal with a non-steady state two-phase solid diffusion couple spanning a miscibility gap (cf Figs. 1.35 and 1.36) and attempt to define the thermokinetic properties of the interface zone uniquely by analytic continuation from the early stages of a spinodal reaction which occurs in the same thermodynamic regime. In this way, they establish a unique disposition of the non-local equilibrium surface concentrations in terms of the gradient energy coefficient (cf §§1.27 and 12.5) and model parameters. They make very clear, however, that this procedure is not applicable to 'singular interfaces, where the crystal symmetry changes abruptly across a single lattice spacing'. Here a singular, steady state solid–liquid interface with non-zero velocity is considered in particular. This has two degrees of degeneracy and is thus properly classified as a free boundary problem. Naturally, the problem carries practical significance only if the interface resistance and/or the velocity are high enough that significant deviations from the local equilibrium interface concentrations occur. Thus the problem bears particular significance in relation to faceted interfaces[100] and it carries general significance in relation to rapid solidification processes and the formation of metallic glasses. The problem is also theoretically important because its conditions lie at the margin between the thermodynamic and dissipative structure branches as defined by Prigogine (cf §12.7) and therefore the process provides a bridge between the principle of minimum entropy production (PMEP) in linear irreversible thermodynamics, Prigogine's general evolution criterion[7] or the minimax theorem in non-linear irreversible thermodynamics.

We start with an expression for the interface steady state isothermal entropy production rate in planar, (A–B) solidification processes due to Kirkaldy[103] (cf §3.4)

$$T\sigma = -mv[(1 - X_S^B)\Delta\mu^A + X_S^B\Delta\mu^B] \tag{12.149}$$

where v is the reaction velocity, X_S are solid mole fractions, μ are chemical potentials, and the fluxes and forces are

$$J_A = mv(1 - X_S) \qquad J_B = mvX_S \tag{12.150}$$

where m is the average molar density, and

$$X_A = -\nabla\mu_A \qquad X_B = -\nabla\mu_B \tag{12.151}$$

These identifications are in accordance with the mass balances and the entropy balance derived from the Gibbs energy equation, and so are a suitable basis for a rigorous treatment of the problem.[4, 67]

It is to be noted that this problem is structurally rather different from that associated with the two force, discontinuous systems which are cited in discussions of Onsager reciprocity and the principle of minimum entropy production (PMEP).[4, 67] Here the direct constraint is upon the thermal boundary values producing a somewhat constrained thermal dissipation, and this is reflected through the thermodynamics in an implicit constraint upon variations of the chemical fluxes and forces which relax to define the steady state configuration. The present situation is therefore more complex than for systems in which the constraint is applied to one of the forces, as for example with a materially closed thermal diffusion cell. For this latter situation we have

$$T\sigma = J_1X_1 + J_2X_2 \qquad (12.152)$$

where

$$J_1 = L_{11}X_1 + L_{12}X_2 \qquad (12.153)$$

and

$$J_2 = L_{21}X_1 + L_{22}X_2 \qquad (12.154)$$

and where X_1 and X_2 are the appropriate thermal and chemical potential gradients defined via the entropy balance. In Prigogine's theorem of 1955[4] X_1 is supposed to be fixed and J_1, J_2 and X_2 are allowed to relax until the steady state defined by $J_2 = 0$ and $L_{12} = L_{21}$ is reached. Within this global, near-equilibrium approximation the steady state is also defined by the PMEP ($2J_2 = \mathrm{d}\sigma/\mathrm{d}X_2 = 0$) or by an equivalent Le Chatelier principle of the form

$$\delta J_2 \delta X_2 > 0 \qquad (12.155)$$

Kirkaldy,[104] in amplifying an example given by Prigogine,[4] has demonstrated that the PMEP applies in the linear regime even though the constraints are implicit. This argument is based upon the equivalence of the Le Chatelier principle and the PMEP as outlined by de Groot[67] for systems analogous to the thermal diffusion example reviewed above. Forced velocity, non-local equilibrium solidification represents a non-trivial application of these stability principles. A detailed validation of the PMEP based on the Le Chatelier principle for the configuration at hand will be discussed later.

We write the most general form of the reaction equations as

$$J_A = L_{AA}X_A + L_{AB}X_B \qquad (12.156)$$

and

$$J_B = L_{BA}X_A + L_{BB}X_B \qquad (12.157)$$

It is understood that J and X are implicitly connected through their common definitional roots in the mass conservation relations and the entropy balance, and are therefore expressible in terms of Onsager α-variables (cf §3.8). Provided these pairs of J and X are independent, then the L-matrix is uniquely symmetric, due to Onsager reciprocity. Should the pairs J and X prove to be dependent, symmetry can be arbitrarily assigned without loss in generality.[112]

Besides equations (12.156) and (12.157) a term for the coupled diffusion field must be included which defines the chemical potentials adjacent to the interface in the liquid. We will develop this term in the form consisting of functionals

$$J_D = L_D X_D \qquad (12.158)$$

There are no cross-terms acting between equations (12.156), (12.157), and (12.158) since the two processes share only a boundary value, taking place within different volumes. There is also conceivably a boundary value coupled term associated with heat conduction. This, however, makes a zero contribution to the isothermal variation in the entropy production rate which, in the first instance, we emphasize here[103] (see below).

We have found it convenient to write $X_S^B = X_0$,

$$X_A = -(\mu_S^A - \mu_L^A) = -\{(\mu_S^A - \mu_{eq}^A) - (\mu_L^A - \mu_{eq}^A)\} = -(u - w) \tag{12.159}$$

which defines u and w, and

$$X_B = -(\mu_S^B - \mu_L^B) = -\{(\mu_S^B - \mu_{eq}^B) - (\mu_L^B - \mu_{eq}^B)\} = -\left\{\frac{k - X_0}{X_0} w - \frac{1 - X_0}{X_0} u\right\} \tag{12.160}$$

where eq designates the equilibrium values, where the local equilibrium chemical potentials are defined for $X_S^B = X_0$ and $X_L^B = X_0/k$, where k is the equilibrium partition coefficient and X_0 is the alloy composition (cf §1.19). In formulating equation (12.160) we have applied the Gibbs–Duhem equation to the small differences $\mu_S^B - \mu_{eq}^B$ and $\mu_L^B - \mu_{eq}^B$ and introduced the non-equilibrium variational approximation $X_S^B \simeq kX_L^B = X_0$.

The entropy production term for unit area of the diffusion field has been given by Kirkaldy[103] for a dilute solution as

$$(T\sigma)_D = DmRT \int_0^\infty \frac{(\nabla X_L^B)^2}{X_L^B} \, dx \tag{12.161}$$

where D is the liquid diffusion coefficient and the concentration distribution is

$$X_L^B(x) = X_0 + [X_L^B(0) - X_0] \exp(-vx/D) \tag{12.162}$$

Approximating the integral subject to $X_0 < k \ll 1$ (i.e. $X_0 \ll X_L^B - X_0$),

$$(T\sigma)_D \simeq RTmv(X_L^B - X_0) = RTmv(X_{Leq}^B - X_0)\left(1 + \frac{\Delta X_L^B}{X_{Leq}^B - X_0}\right)$$

$$\simeq RTmvX_0 \frac{(1-k)}{k}\left(1 + \frac{1}{2}\frac{k}{1-k}\frac{\Delta X_L^B}{X_0}\right)^2 \tag{12.163}$$

where ΔX_L^B is the deviation of $X_L^B(0)$ from its local equilibrium value. From this can be selected the appropriate functionals for inclusion in the linear relation (12.158), i.e.

$$X_D = -RT\left(1 + \frac{1}{2}\frac{k}{1-k}\frac{\Delta X_L^B}{X_0}\right) \quad \text{and}$$

$$J_D = -vX_0 \frac{(1-k)}{k}\left(1 + \frac{1}{2}\frac{k}{1-k}\frac{\Delta X_L^B}{X_0}\right) = L_D X_D \tag{12.164}$$

These will be required in application of the Le Chatelier principle.

For the dilute Henrian solutions under consideration

$$w = \mu_L^A - \mu_{eq}^A = RT_0 \ln \frac{X_L^A}{X_{Leq}^A} = RT\frac{\Delta X_L^A}{X_{Leq}^A} \simeq -RT_0 \Delta X_L^B \tag{12.165}$$

so $(T\sigma)_D$ is linear in w. Hence for the total entropy production rate write

$$(T\sigma)_t = L_{AA}(u - w)^2 - 2L_{AB}(u - w)[(1 - X_0)u/X_0 - (k - X_0)w/X_0]$$
$$+ L_{BB}[(1 - X_0)u/X_0 - (k - X_0)w/X_0] - mvw + \text{constant} \tag{12.166}$$

Now assuming in the first instance that u and w are independent the isothermal steady state optimum is evaluated via

$$\frac{\partial(T\sigma)_t}{\partial u} = 2L_{AA}(u-w) - 2L_{AB}[(1-X_0)u/X_0 - (k-X_0)w/X_0 + (u-w)(1-X_0)/X_0] + 2L_{BB}[(1-X_0)u/X_0 - (k-X_0)w/X_0](1-X_0)/X_0 = 0 \quad (12.167)$$

and

$$\frac{\partial(T\sigma)_t}{\partial v} = -2L_{AA}(u-w) + 2L_{AB}[(1-X_0)u/X_0 - (k-X_0)w/X_0 + (u-w)(k-X_0)/X_0] - 2L_{BB}[(1-X_0)u/X_0 - (k-X_0)w/X_0](k-X_0)/X_0 - mv = 0 \quad (12.168)$$

Next, from the steady state flux equations,

$$(L_{AA}L_{BB} - L_{AB}^2)(u-w) = [-L_{BB}(1-X_0) + L_{AB}X_0]mv \quad (12.169)$$

and

$$(L_{AA}L_{BB} - L_{AB}^2)[(1-X_0)u/X_0 - (k-X_0)w/X_0] = [L_{AA}X_0 - L_{AB}(1-X_0)]mv \quad (12.170)$$

Combining these with equation (12.167) the unenlightening result that $0 = 0$ is obtained. However, combining with equation (12.168) and cross-multiplying the determinant the significant and apparently contradictory result is that

$$(L_{AA}L_{BB} - L_{AB}^2)/(L_{AA}L_{BB} - L_{AB}^2) = \tfrac{1}{2}(1-k) \quad (12.171)$$

Thus for acceptable solutions of the problem the determinant

$$L_{AA}L_{BB} - L_{AB}^2 = 0 \quad (12.172)$$

which shows in turn what should have been realized from the start; the average fluxes are rigidly cocurrent and thus perfectly correlated at the steady state (equations (12.151)).

Combining equation (12.172) with (12.169) and (12.170) allows expression of all the coefficients in terms of one, which is chosen to be L_{BB}, i.e.

$$L_{AA} = \left(\frac{1-X_0}{X_0}\right)^2 L_{BB} \qquad L_{AB} = L_{BA} = \frac{(1-X_0)}{X_0} L_{BB} \quad (12.173)$$

Then taking second differentials according to equations (12.167) and (12.168) and substituting equation (12.173) it is shown that the u-variation constrained by the steady state solution lies on a line ($\partial^2(T\sigma)/\partial u^2 = 0$ for all u) while the w-variation defines a minimum in the entropy production. Finally, having properly removed the 0/0 from equation (12.168) by adjusting the right-hand side of (12.171) to unity, equation (12.168) can be evaluated according to equation (12.173) to give

$$w = \frac{mv}{L_{BB}} \cdot \frac{X_0^2}{(1-k)} = \frac{J_B X_0}{L_{BB}(1-k)} > 0 \quad (12.174)$$

and the stable entropy production rate associated with the surface reaction

$$(T\sigma)_R = \frac{L_{BB}w^2(1-k)^2}{X_0^2} = \frac{(mvX_0)^2}{L_{BB}} = \frac{J_B^2}{L_{BB}} > 0 \tag{12.175}$$

Substituting from equations (12.157) and (12.173) for J_B verifies that we have correctly dealt with the indeterminacy (0/0) in equation (12.168). Necessarily, both quantities w and $(T\sigma)_R$ vanish as the interface mobility $L_{BB} \rightarrow \infty$. Note also via equation (12.165) and $L_{BB} > 0$ that

$$\Delta X_L^B < 0 \tag{12.176}$$

This implies solute trapping, a characteristic which Baker and Cahn[110] regard as essential to a viable theory.

Interestingly, the quantity $u = \mu_B^A - \mu_{eq}^A$ is indeterminate within the formalism, its particular value having no effect on the stable entropy production rate. This arises mathematically because of the absence of an optimum along this variational direction and physically because it was not anticipated that the interface environs could undertake temperature as well as composition excursions in establishing the steady state stability point. If the interface were to remain at its local equilibrium temperature then $u = 0$ since the solid chemical potential would be constrained by the fixed steady state composition X_0. There is, however, no such constraint on the temperature so μ_S^A can migrate in accordance with the relaxation of the thermal field and an appropriate subsidiary optimal principle. It will be conjectured here and argued below that for overall stability the solute B (rather than the solvent A) must relax within the kinetic degree of freedom to a constrained local equilibrium at the steady state, i.e.

$$X_B = -\Delta\mu_B = 0 \tag{12.177}$$

The stable temperature is accordingly evaluated via

$$\mu_S^A(T, X_0) - \mu_S^A[T, X_{Seq}^A(T)] = -RT(X_0 - X_{Seq}^B) = u = \frac{k - X_0}{1 - X_0}v \tag{12.178}$$

For $k > X_0$ this implies $X_{Seq}^B > X_0$ so with a typical terminal phase diagram of constant negative liquidus slope $-\alpha$

$$\Delta T = T - T_0 = \frac{\alpha}{k}(X_0 - X_{Seq}^B) \simeq -\frac{\alpha m v}{2L_{BB}RT}\,\frac{k - X_0}{k}\,\frac{X_0^2}{(1-k)} \simeq \alpha\Delta X_L^B \tag{12.179}$$

Proceeding with the alternate $X_A = 0$, we obtain $\Delta T \simeq -\alpha X_L^B/k$, which represents a much greater undercooling. An intuitive stability principle in problems of this kind is 'minimum undercooling' which favours equation (12.178) as conjectured. It can be shown that this corresponds approximately to a maximum in the entropy production (cf 12.8) so the stability point is a minimax.

The correctness of the foregoing analysis depends on a perturbation validation of the PMEP in this particular application. Following de Groot[67] and Kirkaldy[104] there is need only to demonstrate that the PMEP generates the Le Chatelier principle of moderation involving mass-coupled fluxes and forces of the form

$$\delta J \delta X > 0 \tag{12.180}$$

In view of constraints (12.177) or (12.178), which fix the interface temperature, isothermal variations in w can henceforth be attributed, in accordance with equation (12.165) with u fixed. Since together with equation (12.177), J_A and J_B are completely constrained by X_A, the latter as a function of w can be taken as the independent force.

To phrase the Le Chatelier principle, the variational equivalent of equation (12.168) is written in accordance with the Kirkaldy procedure[104] as

$$J_A \delta X_A + J_B \delta X_B + J_D \delta X_D = 0 \tag{12.181}$$

and from equation (12.156)

$$\delta J_A = L_{AA} \delta X_A + L_{AB} \delta X_B \tag{12.182}$$

Combining equations (12.156) and (12.157) with these and noting the vanishing of the determinant (equation (12.172)) for the structure based on equations (12.168) or (12.181),

$$\delta J_A \delta X_A = L_{AB} (\delta X_A)^2 \cdot \left(- \frac{J_D}{J_B} \right) \cdot \left(\frac{\delta X_D}{\delta X_A} \right) > 0 \tag{12.183}$$

The sign can be checked by considering a perturbation of the negatively relaxed ΔX_L^B in relation to equations (12.159), (12.164), (12.165), and (12.150) and deducing that each of the product quantities on the right-hand side of equation (12.183) is positive. Similary $\delta J_B \delta X_B > 0$. Thus the Le Chatelier principle for a conserved quantity such as mass is validated. Expressions such as equation (12.183) state simply that a mass accumulation due to an increased flux is moderated by an increased force which tends to counter the initiating flux.

12.11 SPIRAL ZONATION OF *Penicillium diversum*

Fungi, and in particular colonies of *Penicillium*, growing at or near the surface of a sugar-enriched rigid agar–agar or gelatinous medium exhibit many of the features of growth of crystals from supersaturated solution, namely, dendritic growth (of the root system or mycelium), diffusion depletion of nutrients (in regions of colony competition), circular morphology of the dendritic front within circular boundary conditions and zonation (Figs. 12.7 and 12.9). These similarities suggest that many of the morphological features of these organisms may be explained qualitatively and/or quantitatively on the basis of a simple diffusional growth model and without reference to detailed biochemical models and the genetic complement (see the discussion of the Liesegang phenomenon, § 12.6).

In the following attention is restricted to boundary conditions for growth of a single, central, submerged root system of a fungal colony growing on a thin layer of rigid agar–agar based nutrient medium in a standard 100 mm diameter petri dish (cf Figs. 12.7, 12.9 and 12.25). It should be emphasized at the outset that dilute agar media are in fact water-based solutions, their rigidity being provided by gel membranes or networks which occupy an insignificant fraction of the total volume (cf Chapter 10). The diffusion properties are therefore nearly identical to those of water solutions at the temperature of interest.

As a preliminary we discuss the overall growth of a disc-shaped colony limited by diffusion of sugar from the direction of the dish periphery at a concentration

C^∞ to a ring of average composition C^s at radius r_{av}. Thus as boundary conditions for the diffusion problem we take

$$C = C^\infty \quad \text{at} \quad r = 50 \text{ mm} \tag{12.184}$$

and

$$C = C^s \quad at \quad r = r_{av} \tag{12.185}$$

and apply the quasi-steady state approximation to Fick's equation (1.47) with constant D so that in cylindrical coordinates we must solve

$$\frac{d^2C}{dr^2} + \frac{1}{r}\frac{dC}{dr} = 0 \tag{12.186}$$

This has the solution

$$C = k_1 \ln r + k_2 \tag{12.187}$$

where

$$k_1 = \frac{C^\infty - C^s}{\ln(50/r_{av})} \quad \text{and} \quad k_2 = \left[C^\infty - \frac{(C^\infty - C^s)}{\ln(50/r_{av})}\ln 50\right] \tag{12.188}$$

Noting that the front velocity v is dr_{av}/dt the approximate mass balance in differential form can be written

$$(C^a - C^s)\frac{dr_{av}}{dt} = D\frac{(C^\infty - C^s)}{r_{av}\ln(50/r_{av})} \tag{12.189}$$

Estimating $C^s \simeq 0$ and $C^a \sim C^\infty$ (cf Fig. 12.15, $C \rightarrow C_2$) we obtain the differential equation

$$v = \frac{dr_{av}}{dt} = \frac{D}{r_{av}\ln(50/r_{av})} \tag{12.190}$$

Now the interesting thing about this result for a petri dish is that for most of the growth regime (say r_{av} = 10–30 mm) the quantity $r_{av}\ln(50/r_{av}) \simeq 16$ mm, which is a constant. Hence we predict that the velocity at intermediate times will be constant. This non-trivial result is in good agreement with experiment.[57] Indeed for this radial range equation (12.190) reduces to

$$v(\text{mm s}^{-1}) = \frac{D(\text{mm}^2\text{ s}^{-1})}{16(\text{mm})} \tag{12.191}$$

Since typical nutrient sugars (sucrose, dextrose) have diffusion coefficients in H_2O in the range 5–7 × 10^{-4} mm^2 s^{-1} at room temperature, this relation predicts colony growth velocities of 27–38 mm in 10 days. A brief perusal of the monograph by Raper and Thom[57] will verify that the majority of *Penicillium* colonies do in fact grow at rates lying in and bracketing this range.

The theoretical treatment to be given refers to zonation of a submerged fungal root or mycelial network. We follow the hypothesis of Ullscheck[58] which requires that the organism reject a certain by-product of the degradation of sugar to the environment thus creating a countercurrent diffusive flow of sugar and the by-product. This combination with the solvent H_2O, in the presence of the expanding

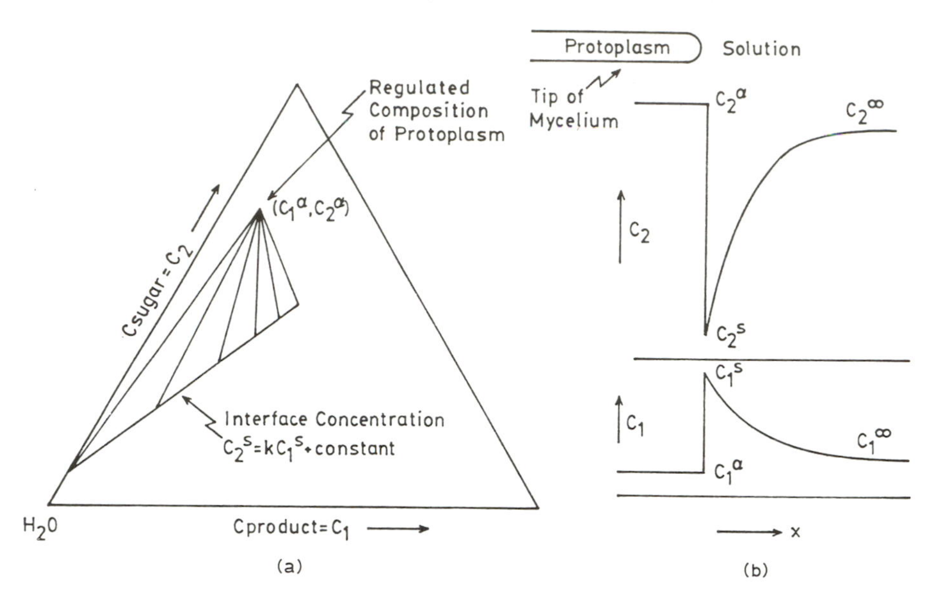

Fig. 12.25 Boundary conditions for spiral zonation of *Penicillium diversum*: (*a*) schematic 'phase diagram'; (*b*) average concentration distribution in advance of the dendritic mycelial front. A single mycelial membrane (hyphal) tip is indicated in upper frame right.

mycelium represents a ternary membrane equilibrium which has inherent instabilities. Let Fig. 12.25 describe the constitutional and kinetic relationships of the reactants and products in relation to the mycelial protoplasm concentration. This model assumes that the sugar and by-product interface concentrations C_2^s and C_1^s are forced by osmotic law to adjust according to the straight line relation

$$C_2^s = kC_1^s + K \qquad k > 0 \tag{12.192}$$

to maintain the protoplasm concentration constant at (C_1^a, C_2^a). It implies that if the surface concentration of nutrient C_2^s is raised, this tends to raise the protoplasm concentration C_2^a. To regulate this, more by-product will have to be produced, and to regulate C_1^a more C_1 will have to appear in the environment and C_1^s must rise, all of which is chemically consistent if not quantitative. It must be assumed that the growth of the mycelium occurs not only at the interface but also as some yet to be determined fine-grained distribution behind the frontal interface. The concentrations of C_1 and C_2 will represent the average distributions lying between the growing root (hyphal) tips as well as in the untouched environment. If the volume fraction of the mycelium is small (cf § 12.6) then the differential equations for diffusion between the mycelial branches (the hyphae) can be written as

$$-D_1\nabla^2 C_1 + \frac{\partial C_1}{\partial t} = -\dot{p} \tag{12.193}$$

and

$$-D_2\nabla^2 C_2 + \frac{\partial C_2}{\partial t} = -n\dot{p} \tag{12.194}$$

where $\dot{p}$ is the rate of accumulation of component 1 per unit volume within the hyphae, $n = C_2^\alpha/C_1^\alpha$ and $n\dot{p}$ is the rate of accumulation of component 2 within the hyphae (cf § 12.6).

By combining equations (12.192)–(12.194) and eliminating $\dot{p}$ the standard Fick equation which is applicable within the mycelial network is obtained

$$\frac{\partial C_1}{\partial t} = \left[\frac{nkD_1 - D_2}{nk - 1}\right]\nabla^2 C_1 \tag{12.195}$$

with an effective diffusion coefficient

$$D_{\text{eff}} = \left[\frac{nkD_1 - D_2}{nk - 1}\right] \tag{12.196}$$

When D_{eff} is positive the solutions of equation (12.195) are stable. If, on the other hand, D_{eff} is negative, then the solutions of equation (12.195) are unstable. For example, if $n = 10$, $k = 1$, $D_1 = 10\,D_2$, $D_{\text{eff}} \simeq D_1$, or if $n = 10$, $k = 1$, $D_1 = 0{\cdot}01\,D_2$, $D_{\text{eff}} \simeq -10\,D_1$. It is this latter case where 1 is the byproduct and 2 is the sugar which will be explored in the following calculations. In accordance with earlier considerations the unstable solutions in a ring mode are periodic in space with wavelengths related to the period by (cf § 12.6) $\tau\,\partial\Delta C_1/\partial t \sim \Delta C_1$ or

$$\lambda \sim 2\pi(|D_{\text{eff}}|)\,\tau)^{1/2} \tag{12.197}$$

The concentration of hyphal material $p = \int \dot{p}\,\mathrm{d}t$ will also be periodic with wavelength λ. The spacial oscillations should have only a slight effect on the average growth rate of a colony since the main resistance to growth is provided by the long diffusion path of nutrient up to the average front. For example, the zonated colony in *Penicillium diversum* of Fig. 12.9 has a growth rate of 25 mm in 10 days (cf equation (12.191)), a wavelength of 2·5 mm and a period of a day.[59]

If it is assumed that there is no fundamental or quantitative distinction between ring and spiral zonation (and the experiments bear this out[59]) then this data for *Penicillium diversum* gives an estimate of $|D_{\text{eff}}| \sim 2 \times 10^{-6}$ mm^2 s^{-1}, which in view of the previous estimate for an unstable case that $D_{\text{eff}} \sim -10D_1$, suggests that the by-product molecules are appreciably larger than the sugar molecules ($D_1 \ll D_2$).

The foregoing general description of zonation is applicable to spiral as well as ring zonation. This is made evident through the observation that equation (12.195) has a set of approximate solutions which generate an Archimedean spiral in the concentration or in the mycelium root density.

Writing equation (12.195) in cylindrical coordinates for $i = 1, 2$

$$D_{\text{eff}}\left[\left(\frac{\partial^2 C_i}{\partial \rho^2}\right)_{\phi,t} + \frac{1}{\rho}\left(\frac{\partial C_i}{\partial \rho}\right)_{\phi,t} + \frac{1}{\rho^2}\left(\frac{\partial^2 C_i}{\partial \phi^2}\right)_{\rho,t}\right] = \left(\frac{\partial C_i}{\partial t}\right)_{\phi,r} \tag{12.198}$$

Linear transformation to a frame of reference in which $C_i = C_i(r, \phi')$ can be achieved for a location near the growth front via the transformations

$$r = \rho - \frac{D_2}{16}t \tag{12.199}$$

and

$$\phi' = \phi - \omega t \tag{12.200}$$

where $D_2/16$ is the average growth rate of the colony (equation (12.191)) and ω is an angular frequency yet to be defined. Now equations (12.199) and (12.200) result in

$$\left[\frac{\partial C_i(r,\phi')}{\partial \rho}\right]_{\phi,t} = \left(\frac{\partial C_i}{\partial r}\right)_{\phi'} \quad \left[\frac{\partial C_i(r,\phi')}{\partial \phi}\right]_{\rho,t} = \left(\frac{\partial C_i}{\partial \phi'}\right)_r$$

$$\left[\frac{\partial C_i(r,\phi')}{\partial t}\right]_{\rho,\phi} = -\frac{D_2}{16}\left(\frac{\partial C_i}{\partial r}\right)_{\phi'} - \omega\left(\frac{\partial C_i}{\partial \phi'}\right)_r \tag{12.201}$$

$$\left[\frac{\partial^2 C_i(r,\phi')}{\partial \rho^2}\right]_{\phi,t} = \left(\frac{\partial^2 C_i}{\partial r^2}\right)_{\phi'} \quad \left(\frac{\partial^2 C_i}{\partial \phi^2}\right)_{\phi,t} = \left(\frac{\partial^2 C_i}{\partial \phi'^2}\right)_r \tag{12.202}$$

Thus in the moving coordinate system, equation (12.198) becomes

$$D_{\text{eff}}\left[\frac{\partial^2 C_i}{\partial r^2} + \frac{1}{\rho}\frac{\partial C_i}{\partial r} + \frac{1}{\rho^2}\frac{\partial^2 C_i}{\partial \phi'^2}\right] = -\frac{D_2}{16}\frac{\partial C_i}{\partial r} - \omega\frac{\partial C_i}{\partial \phi'} \tag{12.203}$$

Since $\rho = \rho(t)$ has not been eliminated it is clear that exact time-independent solutions of equation (12.198) do not exist. However, if attention is focused on a region near the growth front where ρ is a slowly varying function and $\rho \simeq \rho_{\text{av}} \simeq$ constant, the approximate time-independent equation can be obtained.

$$D_{\text{eff}}\left[\frac{\partial^2 C_i}{\partial r^2} + \left\{\frac{1}{\rho_{\text{av}}} + \frac{D_2}{16\,D_{\text{eff}}}\right\}\frac{\partial C_i}{\partial r}\right] = -\omega\frac{\partial C_i}{\partial \phi'} - \frac{D_{\text{eff}}}{\rho_{\text{av}}^2}\cdot\frac{\partial^2 C_i}{\partial \phi'^2} \tag{12.204}$$

Now variables can be separated by the substitution

$$C_{1,2} = R(r)\,\Phi(\phi') \tag{12.205}$$

to obtain

$$\frac{D_{\text{eff}}}{\rho_{\text{av}}^2\omega R}\left[\frac{\partial^2 R}{\partial\left(\frac{r}{\rho_{\text{av}}}\right)^2} + \left\{\frac{1}{\rho_{\text{av}}} + \frac{D_2}{16\,D_{\text{eff}}}\right\}\rho_{\text{av}}\frac{\partial R}{\partial\left(\frac{r}{\rho_{\text{av}}}\right)}\right] \tag{12.206}$$

$$= \frac{1}{\Phi}\left[-\frac{\partial \Phi}{\partial \phi'} - \frac{D_{\text{eff}}}{\omega\rho_{\text{av}}^2}\frac{\partial^2\Phi}{\partial \phi'^2}\right] = m^2$$

where m^2 is a non-negative constant. This gives two ordinary locally applicable differential equations, one in r and one in ϕ'. For component 1 ($C_1(r = \infty) = 0$) the latter has the azimuthal solution

$$\Phi = \Phi_0 \exp\left[-\tfrac{1}{2}\{(\alpha^2 - 4m^2\alpha)^{1/2} + \alpha\}\phi'\right]$$

where

$$\alpha = \frac{\omega \rho_{av}^2}{D_{eff}} \qquad \Phi_0 = \text{constant} \tag{12.207}$$

and with α negative this properly decays in the ϕ' direction. The first equation has a particular local solution for $1/\rho_{av} \ll D_2/16|D_{eff}|$ $(1/30 \ll 3)$

$$R = R_0 \exp\left[-\frac{\beta}{2}\frac{r}{\rho_{av}}\right] \cos\theta\left[\frac{r}{\rho_{av}}\right] \tag{12.208}$$

(cf equation (12.96)) where

$$\beta = \frac{D_2 \rho_{av}}{16\, D_{eff}} \quad \text{and} \quad \theta = \tfrac{1}{2}(-4\alpha m^2 - \beta^2)^{1/2} \tag{12.209}$$

where R_0 is a constant. Note that since the wave disturbance must start near the origin and be amplified by the positive exponent, which in turn causes the disturbance to spread, the average amplitude must decay with distance. However, there is nothing within the truncated local solution (12.208) to indicate this. Fortunately, there is no need for the extended solution.

Clearly, when the product $R\,.\,\Phi$ is transformed back to the laboratory frame via equations (12.199) and (12.200), the concentration distribution C_1 (and therefore the mycelial distribution function p) has the form of an Archimedian spiral.

Now the parameter θ is related to the wavelength of the radial function by

$$\theta\left(\frac{r}{r_{av}}\right) = \frac{2\pi r}{\lambda} \tag{12.210}$$

and since

$$\lambda = v\tau = \frac{D_2}{16}\tau \tag{12.211}$$

where v is the front velocity and τ is the period of rotation of the spiral, and further since

$$\tau = 2\pi/\omega \tag{12.212}$$

where ω is the angular frequency, then

$$\theta = 16\rho_{av}\omega/D_2 \tag{12.213}$$

Thus equations (12.209) and (12.213) give

$$m^2 = -\frac{D_2^2}{4(16)^2 \omega D_{eff}} - \frac{16^2 D_{eff}\omega}{D_2^2} > 0 \tag{12.214}$$

Now in developing a steady state solution we were forced to detach the concentration distribution from the initial conditions. As a consequence the parameter m^2 has become kinetically indeterminate, in complete analogy with the steady state free boundary problems discussed in §12.8 and 12.9. The optimum for this problem can be deduced directly from the principle of maximum path probability, for the positive parameter m^2 is a monotone measure of the degree of correlation between radial and azimuthal motions of the molecules (cf equation

(12.206) and § 12.7). The optimum is accordingly defined by a minimum in m^2 with respect to ω. This yields the parameter values

$$\omega = \frac{D_2^2}{2(16)^2|D_{\text{eff}}|} \tag{12.215}$$

$$\tau = \frac{2\pi}{\omega} = \frac{4\pi(16)^2|D_{\text{eff}}|}{D_2^2} \tag{12.216}$$

and

$$\lambda = v\tau = 4\pi(16)\frac{|D_{\text{eff}}|}{D_2} = 2\pi^{1/2}(|D_{\text{eff}}|\tau)^{1/2} \tag{12.217}$$

together with $m^2 = 1$. For the *Penicillium diversum* spirally zonated colonies previously discussed and using $D_2 = 0{\cdot}5 \times 10^{-3}\,\text{mm}^2\,\text{s}^{-1}$ and the observed $\lambda = 2{\cdot}5$ mm, equation (12.217) gives

$$|D_{\text{eff}}| \approxeq 6 \times 10^{-6}\,\text{mm}^2\,\text{s}^{-1} \tag{12.218}$$

This is of the same magnitude as that obtained from an analysis of ring zonations. This equivalence of the two mechanisms is in fact a prediction of the theory, for if equation (12.216) is substituted into estimate (12.197)

$$\lambda = 2\pi(16)(4\pi)^{1/2}\frac{|D_{\text{eff}}|}{D_2} \tag{12.219}$$

(in millimetres) is obtained, which is only a factor of $\sqrt{\pi}$ different from estimate (12.217). This can be taken as the error in the intuitive estimate (12.197).

A more rigorous thermodynamic model could be constructed via § 12.6 by drawing upon the close relationship between osmotic and capillarity effects (cf Appendix 1). However, this would not significantly affect the foregoing specific conclusions and the generalities which accrue.

The global equivalence of ring and spiral zonation, which is established experimentally, draws attention to the connection between the intuitive scaling law which was invoked for the ring case in § 12.6 and a maximum in the path probability or minimum in molecular path correlations which was invoked for the spiral case. This intelligence concerning the delicate laws of collective molecular action will be used in the following section, which seeks to express the greatest generality in dealing with self-organizing systems.

12.12 PATTERN FORMATION IN INITIALLY CONTINUOUS SYSTEMS

In the kinetic description of most of the self-organizing systems of this and the preceding chapter the initial condition was assumed to be heterogenous, morphogenesis pertaining to the change of shape of already existing interfaces. The significant exception is spinodal decomposition, to which we will return in due course. This restricted set of initial conditions evidently depopulates the phase space which must be searched for stable solutions and therefore greatly simplifies the mathematical techniques for establishing the persistence or stability of periodic, steady or quasi-steady endpoints, if they in fact exist.

The search for generality in the investigation of the stability of initially continuous systems has spawned a large, and powerful mathematical literature, but has not been notably successful in producing quantitative (i.e. numerical) predictions of patterns (based on independently accessible rate parameters) which can be compared with experiments. This is in contrast to the methods of § § 12.1–12.11 where, at worst, order of magnitude predictions are validated by experiment and, at best, a very good closure with experiment is obtained, The exception to these statements may be the Belousov–Zhabotinskii and related reactions[113, 114, 115] (see Figs. 12.28 and 12.29 below) which have attracted increasing experimental attention during the past two decades.

The initiation and evolution of this problem may be traced to two chains of thought and their ultimate synthesis. The first is attributable to Turing[6] of computer science's 'Turing machine' fame). He was the first to realize the importance of non-linearity due to autocatalysis and to adopt for a marginal perturbation analysis the linear sets of diffusion-reaction equations for two component concentrations X and Y in the dimensionless form

$$\frac{\partial X}{\partial \tau} = a(X-h) + b(Y-k) + \frac{\partial^2 X}{\partial \theta^2} \tag{12.220}$$

$$\frac{\partial Y}{\partial \tau} = c(X-h) + d(Y-k) + \frac{\partial^2 Y}{\partial \theta^2} \tag{12.221}$$

where θ is a dimensionless space coordinate and τ is dimensionless time, and to demonstrate that the marginal solutions imply the evolution of symmetry-breaking stationary spatial waves as well as oscillatory solutions (cf § § 12.6, 12.11, and 12.12.2). Turing's most general conjectural model extends to concentration-dependent coefficients a, b, c, and d and to multicomponent concentrations X. It can be noted here that these can be converted to the equations of spinodal decomposition by substituting fourth derivatives with respect to θ for the terms linear ln X and Y. Since the spinodal terms apparently always dampen the amplitudes in time there should be no temporal oscillations (cf § 12.5).

The second important element is due to Prigogine and co-workers[7] who undertook the extension of the principle of minimum entropy production to far-from-equilibrium continuous systems. They found a generally non-integrable 'evolution criterion' which was inclusive of non-dissipative, 'velocity-dependent' processes. Furthermore they were able to demonstrate on this basis that a 'local potential' analogous to the entropy production could be defined for a steady state, given the kinetic coordinates of the steady state. This potential was demonstrated to have application to the variational determination of non-linear steady states in certain heat transfer and fluid dynamics problems.[116]

The structure of the local potential and its Taylor expansion coefficients is such as to give a strong intimation of symmetry-breaking instabilities. Prigogine and co-workers have demonstrated further that a fluctuation theory consistent with the local potential implies large fluctuations at a marginal point of instability analogous to a second order phase transition and these fluctuations must moderate the symmetry-breaking transition from the 'thermodynamic branch' to the 'bifurcated' branch of 'dissipative structures' along a path defined by an appropriate state parameter(s).

Prigogine and co-workers[7] have also reconstructed the Turing theory within

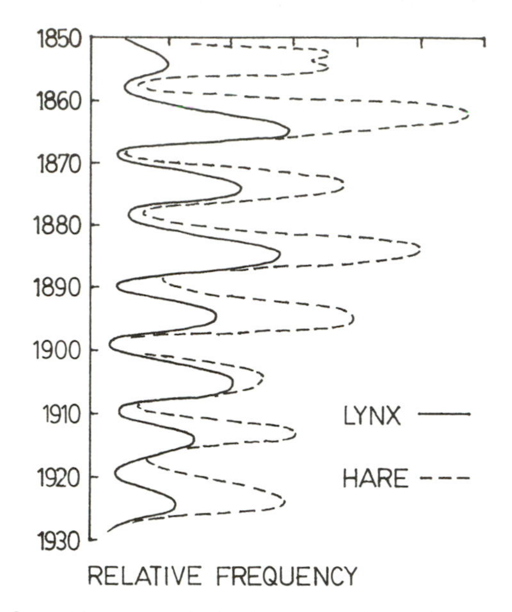

Fig. 12.26 **The 9 or 10 year cycle in the population of hares is closely followed by the lynx. When the hares have reached a peak, the lynxes, with abundant prey, are still building up. Thus when the hares drop off because of overcrowding or disease, many lynxes starve. With fewer predators and enough food the hares again multiply. Graphs based on pelts handled by the Hudson's Bay Company show cycles repeated over eight decades**

the framework of the mathematical theory of non-linear diffusion-reaction differential equations[117, 68] and have investigated various conjectured autocatalytic, homogeneous reaction models with a view to studying the branching of solutions under more or less tractable initial and boundary conditions. They note, in particular, that there often exists more than one patterned solution, which again raises the central problem of nature's mode of pattern selection. Furthermore, they remark that 'the multiplicity of solutions corresponds to a gradual acquisition of autonomy from the environment', the ultimate independence being expressed by the interpretation given here of the minimax principle (§ 12.7).

In the following, a pedagogical approach to the theory of the Belousov–Zhabotinskii reactions is presented which trades upon historically-established elementary (non-diffusive) oscillatory models in anticipation of the rather difficult diffusive, autocatalytic models reviewed by Nicolis and Prigogine.[7]

12.12.1 The Lotka–Volterra models of competitive reactions

Lotka was the first to attempt the quantification of population dynamics by analogy with chemical reaction rate equations.[1, 8] Indeed, he considered equations equivalent to Turing's equations (12.220) and (12.221) with variable coefficients but with suppression of the diffusion term on the right. The analysis is accordingly irrelevant to the examination of spatial patterns.

Temporal oscillations in populations are common in harsh environments.[118, 119] The periodic variation of the population of snowshoe hares and their predators, the Canada lynx, is representative of such ecological instabilities. Figure 12.26 shows data based on Hudson's Bay Company records. The cycle is seen to be 9 or 10 years and the lynx cycle lags the hare cycle slightly.

A simple ecology such as that involving sub-artic vegetation, hares and lynxes can be modelled to a first approximation as a system of coupled, competing first order chemical reactions.[120]

Let X_1, X_2, X_3 be the instantaneous volume concentrations of vegetation, herbivores and predators, respectively, in the environment. We can then write

$$\frac{\mathrm{d}X_1}{\mathrm{d}t} = k_1 X_1 - k_{12} X_2 \tag{12.222}$$

and

$$\frac{\mathrm{d}X_2}{\mathrm{d}t} = k_{21} X_1 - k_3 X_3 \tag{12.223}$$

where k are positive 'reaction rate' constraints related to the behaviour of the participating species. Equation (12.222) states that the rate of growth of vegetation is proportional to the amount of vegetation at any instant minus the rate of absorption of the vegetation by the herbivores. The first term will be strictly valid only in the early growth stages of an ecology whose initial conditions are such that $X_1, X_2, X_3 = 0+$. More generally, non-linearity must be introduced to deal with saturation of the space and nutrients for vegetation growth (see below). Equation (12.223), on the other hand, states that the rate of growth of herbivores is proportional to the concentration of vegetation, decreased by a factor proportional to the concentration of predators. A linear term proportional to the mating rate X_2 is ignored in equation (12.223).

In order to decouple these equations we need a relation between X_2 and X_3. Let us, therefore, assume that the concentrations of predators and herbivores are proportional, i.e.

$$X_3 = K_4 X_2 \tag{12.224}$$

where $K_4 = X_3/X_2$ is the steady state food chain ratio.

By minor manipulation and differentiation it can be shown that X_1 and X_2 both satisfy the differential equation

$$\frac{\mathrm{d}^2 X_{1,2}}{\mathrm{d}t^2} - (k_1 - k_3 K_4)\frac{\mathrm{d}X_{1,2}}{\mathrm{d}t} + (k_{12}k_{21} - k_1 k_3 K_4)X_{1,2} = 0 \tag{12.225}$$

These have the solutions

$$X_{1,2} = K_{1,2}\,[\exp(r_1 t) + \exp(r_2 t)] \tag{12.226}$$

where the eigenvalues

$$r_{1,2} = \tfrac{1}{2}[(k_1 - k_3 k_4) \pm \{(k_1 - k_3 K_4)^2 - 4(k_{12}k_{21} - k_1 k_3 K_4)\}^{1/2}] \tag{12.227}$$

and where henceforth we represent the discriminant by $\{\surd\}$. Now if the initial conditions are X_1, X_2, X_3 equal to some small positive value, then equation (12.226) becomes

$$X_{1,2} = K_{1,2} \exp [\tfrac{1}{2}(k_1 - k_3K_4)t]\{\exp [+\tfrac{1}{2}\{\surd\}t] - \exp [-\tfrac{1}{2}\{\surd\}t]\} \qquad (12.228)$$

and the solution will describe a stable or growing ecology provided that

$$k_1 - k_3K_4 \geqslant 0 \qquad (12.229)$$

$$k_1k_3K_4 > k_{12}k_{21} \qquad (12.230)$$

Now a persistent oscillatory instability will be defined by an imaginary discriminant, provided at the same time equation (12.229) is satisfied. Hence

$$4(k_{12}k_{21} - k_1k_3K_4) > (k_1 - k_3K_4)^2 \qquad (12.231)$$

It will first be noted that the presence of a carnivore population is not essential to the instability for the structure of the solutions is not changed essentially by setting $k_3 = 0$. What is highly important is that k_{12} be large (heavy depredation of the vegetation by the herbivores) and or k_{21} be large (high consumption and fecundity of the herbivores) and k_1 be small (low recovery rate of the vegetation). All of these would appear to be satisfied by these populations in the sub-artic. With $k_3 = 0$ and taking the oscillation as marginal then

$$k_1^2 - 4k_{12}k_{21} \sim -k_1^2 \qquad (12.232)$$

and the period is

$$\tau = 4\pi/k_1 \qquad (12.233)$$

If the half-period for recovery of the vegetation after depredation is $1/k_1 \simeq 1$ year then the oscillation period is about the right magnitude of 10 years. Note, however, that this oscillation is superposed upon an initially growing ecology (equation (12.228)). Were we to consider an oscillation about a stable state of zero overall growth defined via predation by the carnivore ($k_1 = k_3K_4$) then the period is

$$\tau = 2\pi | k_{12}k_{21}k_1k_3 - K_4 |^{-1/2} \qquad (12.234)$$

which could be quite different from the foregoing estimate. The latter estimate, however is likely to be the most defective since this state comes about by neglecting the effect of the strong non-linear seli-damping terms on the vegetation growth rate. Evidently a thorough approach to the problem must be through non-linear rate equations.

Lotka[1] offered the non-linear set

$$\frac{\mathrm{d}X}{\mathrm{d}t} = k_1AX - k_2XY \qquad (12.235)$$

$$\frac{\mathrm{d}Y}{\mathrm{d}t} = k_2XY - K_3BY \qquad (12.236)$$

and these were later studied in detail by Volterra[121] and Kerner.[122] Nicolis and Prigogine[7] point out that these equations correspond to the irreversible, homogeneous chemical reaction set

$$\begin{aligned} A + X &\xrightarrow{k_1} 2X \\ X + Y &\xrightarrow{k_2} 2Y \\ Y + B &\xrightarrow{k_3} E + B \end{aligned} \qquad (12.237)$$

where the first two are autocatalytic and the concentrations A and B are constants, implying that this is an open system. On an ecological interpretation the decrement in the vegetation growth rate is proportional to its own concentration (self-competition) and the herbivore concentration, and the increment in the herbivore growth rate is proportional to its own concentration (mating pairs) and the vegetation concentration (food availability). Note the antisymmetric relation assumed for the cross terms ($k_2 = -k_{12} = k_{21}$). The non-trivial steady state solution (X_0, Y_0) is given by

$$X_0 = \frac{k_0 B}{k_2} \qquad Y_0 = \frac{k_1 A}{k_2} \qquad (12.238)$$

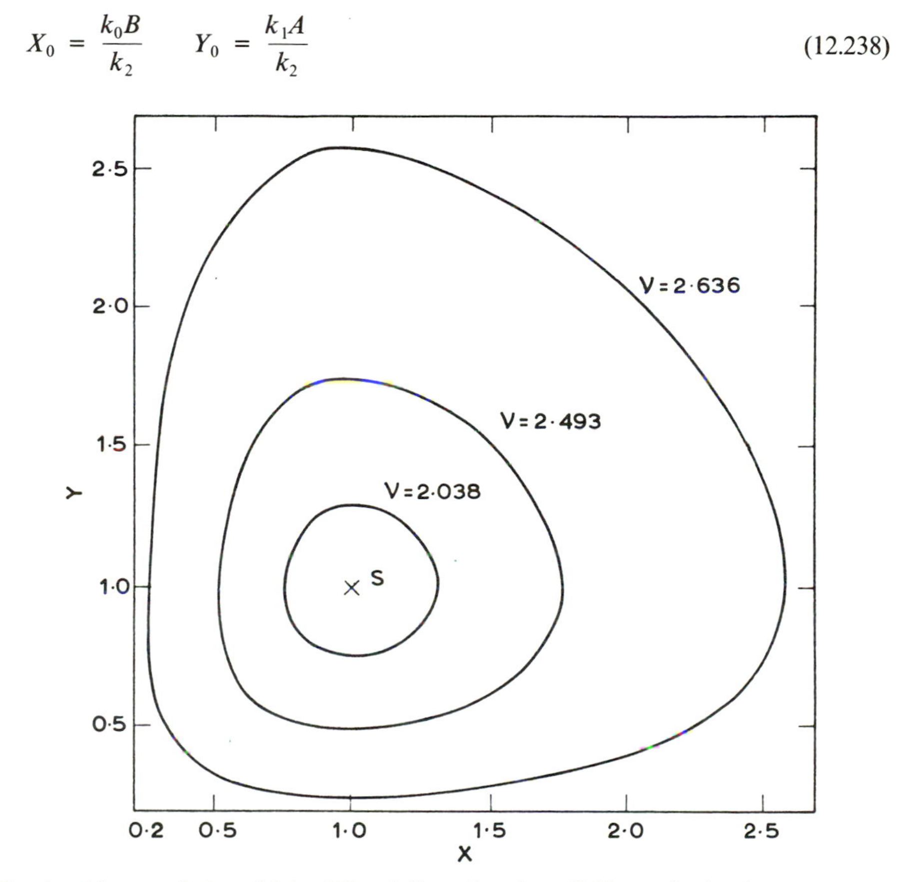

Reprinted by permission of John Wiley & Sons, Inc., from 'Self-organization in nonequilibrium systems', G. Nicolis and I. Prigogine, copyright © 1977, John Wiley & Sons

Fig. 12.27 Lotka–Volterra cycles obtained for successively higher values of constant of motion V and for $k_3B/k_2 = k_1A/k_2 = 1$. S denotes steady state. After Nicolis and Prigogine[7]

A linear (low amplitude) perturbation analysis at this point reveals that small perturbations are periodic with a universal period[7]

$$\tau = 2\pi(k_1 k_2 AB)^{-1/2} \tag{12.239}$$

Now the frequency depends explicitly on the vegetation nutrient availability (A) and the carnivore biomass load (B) as might be expected. However, further analysis of large amplitude motions about X_0, Y_0 reveals that there are actually an infinite number of oscillations of different amplitudes and correspondingly different frequencies (Fig. 12.27) which depend ultimately on different initial conditions. This degeneracy is somewhat analogous to that deduced for the spiral growth problem of the previous section. Nicolis and Prigogine take this continuous spectrum of possible trajectories as implying that the system must switch orbits as a result of small perturbations so that there will be no average preferred orbit. They give this as an argument against the use of such mathematical structures for modelling the oscillations observed in nature. In the contrary view given here the acceptable models for nature are provided by observations in nature, and the fact that deterministic constructions in the form of differential equations possess solutions with certain unsavoury properties cannot invalidate the model or the differential equations. Rather we must learn how to interpret the solutions as appropriate to the circumstances. In the present case Nicolis and Prigogine have defined for us a macroscopic phase space of short-lived orbits which can mutually transform via internal fluctuations or external perturbations (for example fluctuating coefficients A and B in equations (12.235) and (12.236). Because a mean stable orbit actually exists, and the theory is plausible, the associated phase space of perfect and imperfect orbits must surely possess a most probable mean, i.e. a state of maximum path probability. While we have not as yet explored this question of oscillatory stability in any detail we conjecture that the stable frequency will be defined by the minimum in a correlation moment like (cf § § 12.1 and 12.11)

$$C = \frac{\overline{XY}}{\overline{X} \cdot \overline{Y}} \tag{12.240}$$

and will be not too far removed from the marginal value of equation (12.239).

The temporal models which are favoured by the Prigogine school are those which possess multiple but separated time-periodic solutions with limit cycles. Generally speaking, the concentration dependencies of the rates must be at least to the third order in equations like (12.220) and (12.221). The limit cycles define end points of unique period and amplitude. These authors apparently suppose that such deterministic solutions with endpoints independent of initial conditions are the only ones which nature can use in its self-regulatory processes. This is a powerful idea which has spawned many applications of the theory in the biological and medical sciences.[7, 123] Such solutions only exist for systems sufficiently far from equilibrium (as measured in terms of state parameters λ) that some critical value λ_c has been reached or surpassed and where the solutions on a continuum thermodynamic or near equilibrium branch become singly or multiply bifurcated. It is apparently only those solutions which lie on the non-thermodynamic branch which are to be associated with self-organization and the designation 'dissipative structures'. It must now be concluded that *all* of the pattern-forming systems previously discussed are to be excluded from the Prigogine definition of 'self-organizing' systems. Most of these are excluded because a continuous thermodynamic branch in terms of continuous state variables λ cannot be defined, and in the two cases where that is possible (spinodal decomposition and the Lotka–Volterra model) no bifurcation point is identifiable. Evidently, unless we can find a further synthesis, we are faced with the adoption of two meanings for the term self-organization. Before proceeding to this general question we offer a synopsis of the theory of Noyes and co-workers[124, 125] describing the dissipative structures which arise from the Belousov–Zhabotinskii reaction–diffusion reactions (Figs. 12.28 and 12.29).

12.12.2 The Oregonator and the Belousov–Zhabotinskii reactions

As Figs. 12.28 and 12.29[126, 127] and our own observations[128] have demonstrated, uniform dilute solutions of the key substances $HBrO_2 = X$, $Br^- = Y$ and

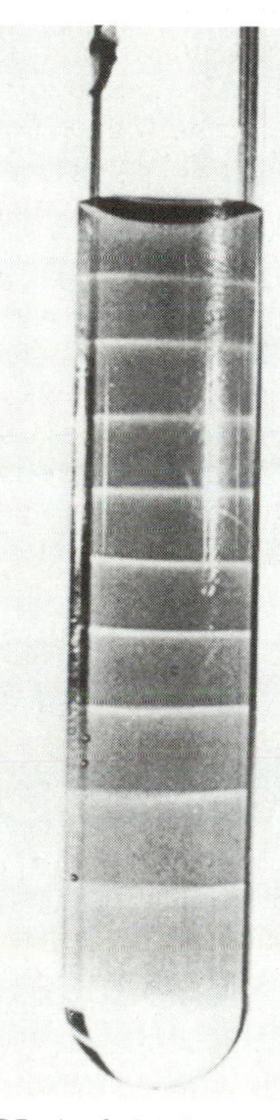

From M. Herschkowitz-Kaufman, *C.R. Acad. Aci. Sec. C*, 1970, **270**, 1049, by permission of Académie des Sciences de Paris

Fig. 12.28 Horizontal bands in the Belousov–Zhabotinskii reaction. Light strips correspond to regions where oxidation steps are predominant. After Herschkowitz-Kaufman[126]

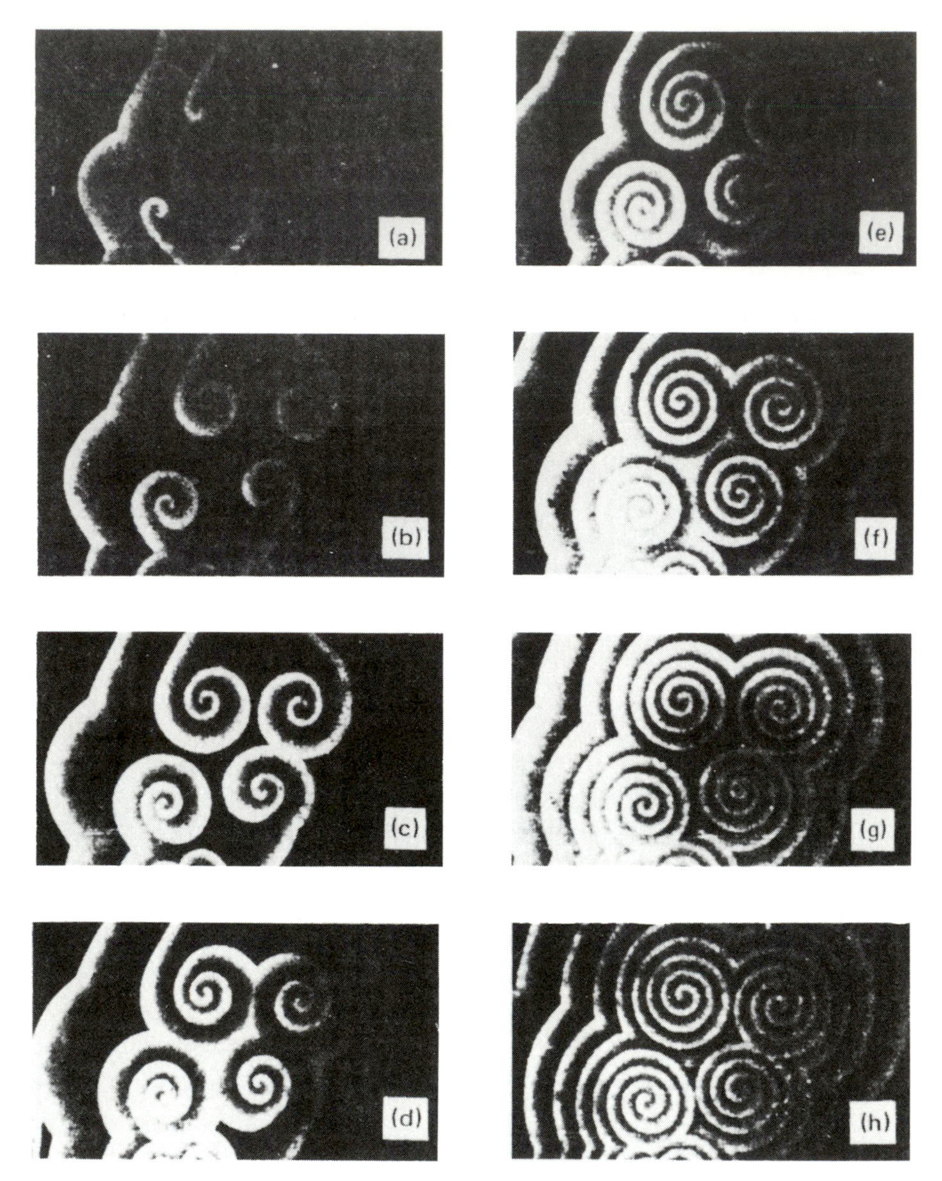

Fig. 12.29 Spiral wave activity in Belousov–Zhabotinskii reaction carried out in petri dish. After Zhabotinskii[127]

$2Ce^{4+} = Z$ together with constant concentrations of $BrO_3^- = A = B$ and waste product concentrations P, Q show instabilities which lead to both spatial and temporal oscillations in the apparent absence of phase change (cf Figs. 12.28 and 12.29). The rotating spiral forms of the reaction in a shallow layer of a petri dish are particularly dramatic[127, 128] for they combine the temporal and spatial periodicities. The important reaction steps have been identified as[124]

$$
\begin{aligned}
&A + Y \xrightarrow{k_1} X \\
&X + Y \xrightarrow{k_2} P \\
&B + X \xrightarrow{k_{34}} 2X + Z \\
&2X \xrightarrow{k_5} Q \\
&Z \xrightarrow{k_6} fY
\end{aligned}
\tag{12.241}
$$

where f is a stoichiometric coefficient. The third and fourth reactions represent the autocatalytic steps. There are other non-controlling steps which assure the charge and mass balances. When diffusion (spatial periodicity) is suppressed the differential equations expressed in terms of unitless quantities become

$$\frac{\mathrm{d}x}{\mathrm{d}\tau} = s(y - xy + x - qx^2) \tag{12.242}$$

$$\frac{\mathrm{d}y}{\mathrm{d}\tau} = \frac{1}{s}(-y - xy + fz) \tag{12.243}$$

$$\frac{\mathrm{d}z}{\mathrm{d}\tau} = w(x - z) \tag{12.244}$$

where $q \simeq 8{\cdot}4 \times 10^{-6}$, $s = 77{\cdot}3$ and $w = 0{\cdot}16\,k_6$ as established by independent experiments. It has been proven mathematically that these generate at least one finite-amplitude, periodic trajectory.[129] A limit cycle has been established by computer simulations and the predicted frequency is in agreement with observations for reasonable values of f and k_6.

The inclusion of the second order diffusion term greatly complicates the analysis. Nonetheless, mathematical analysis has indicated that spatial periodicities are represented among the solutions of such equations provided there are more than two variables.[129,7] Such would also be the case for spinodal processes if we could identify imaginary eigenvalues for the $D^{(2)}$ matrix (cf § 12.5).

DeSimone *et al.*[130] and Tyson[131] have examined the solutions of a special case of the linear Turing equations (12.220) and (12.221) in cylindrical coordinates for a pair of coupled chemical reactions in concentrations C_1 and C_2, i.e.

$$\frac{\partial C_1}{\partial t} = aC_1 + bC_2 \tag{12.245}$$

and

$$\frac{\partial C_2}{\partial t} = D\left(\frac{\partial C_2}{\partial r^2} + \frac{1}{r}\frac{\partial C_2}{\partial r} + \frac{1}{r^2}\frac{\partial^2 C_2}{\partial \theta^2}\right) + cC_1 + dC_2 \tag{12.246}$$

Their solution proceeds analogously to that presented in § 12.11 for the fungal Liesegang phenomenon. For arbitrary wave number $k = 2\pi/\lambda$ they find an asymptotic ($r \to \infty$) stationary solution in the laboratory frame of reference

$$C_2 \simeq \frac{2}{\sqrt{\pi k r}}\cos\left[kr - \frac{(2m+1)\pi}{4} \pm (\omega t \pm m\theta)\right]; \qquad m = 0, 1, 2, \ldots \tag{12.247}$$

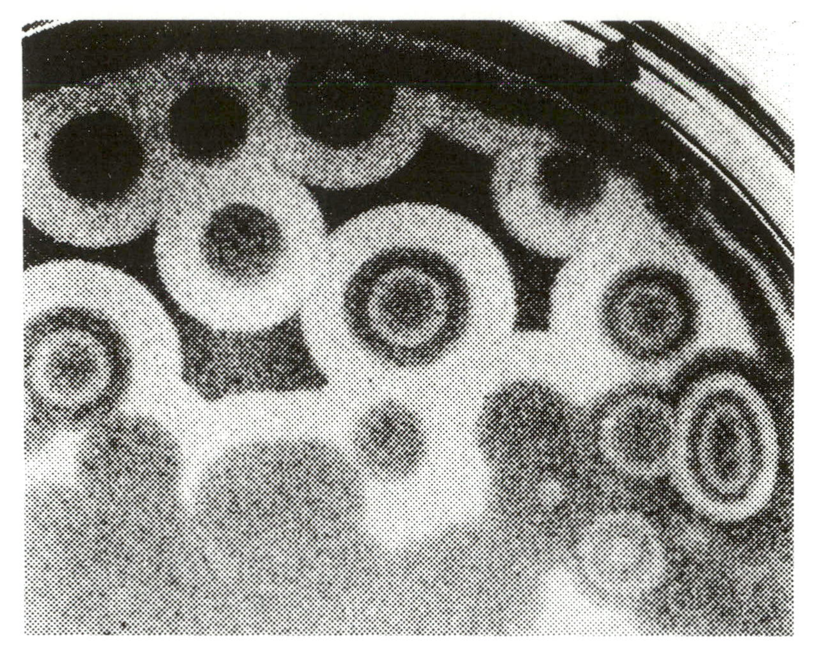

Fig. 12.30 Spatial wave pattern observed at 5°C in a thin (2 mm) layer of reactive solution with initial composition $[CH_2(COOH)_2]$ = 0·0033 M, [NaI] = 0·09 M. $[NaClO_2]$ = 0·1 M, $[H_2SO_4]$ = 0·0056 M, and starch as indicator

where

$$\omega^2 = -a^2 - bc > 0 \qquad (12.248)$$

and the autocatalytic condition

$$k^2D - (a + d) > 0 \qquad (12.249)$$

hold for stationarity. This properly decays with radius r and for $m = 0$ describes concentric rings (the target pattern) which expand with time. For $m = 1$ a point of constant C_2 generates an Archimedes spiral of pitch $1/k$, and for $m = 2$ a double spiral, and so on. These figures are not unlike the Zhabotinskii rings and spirals of Figs. 12.29 and 12.30. Note that k is indeterminate within the formalism on account of the non-analyticity of the solution at the origin[131] so analogously to the Liesegang structure a subsidiary principle is required to establish the unique configuration as usually observed. We must make clear, however, that the nature of the instability and the figures are different in the two cases (D here is positive, the instability being effected by the add-on chemical terms). Also here the entire figure changes continuously with time whereas Liesegang figures change only at the periphery (a spatial instability only).

During the past decade many new chemical oscillators have been discovered,[123] including a chlorite oscillator which represents an analogue of Liesegang rings (Fig. 12.30). This important area of research in natural pattern formation is sufficiently rich as to be approaching a discipline in magnitude of effort. The group activity, however, lacks a synthetic framework, and this we will attempt to provide in our closing section.

12.12.3 A synthetic view of self-organization in chemical reaction–diffusion systems
A survey of the pattern forming, irreversible systems examined in Chapters 11 and 12 suggests that a synthesis can be created by adopting a greatly extended meaning for the equilibrium concept of a phase transition.[132, 133] The concept must indeed be much broader than that based on symmetry-breaking through analogy with second order transitions (such as order–disorder reactions) and seemingly favoured by Turing and by Prigogine and co-workers, for in a general change of pattern one cannot in general attribute discrete symmetry elements quantified by mathematical groups or groupoids to initial and final states. In fact, all that one can attribute to macroscopic patterns in general is that they consist of distinct sets of objects or shapes within time and space, the spectral set being recognized in the previous section as of prime interest. A 'phase transition' in the broadest sense is accordingly a transition of one recognizable set (including subsets) into another. This is inclusive of highly ordered structures transforming into chaotic ones (turbulence) and vice versa. In logic a linguistic interpretation of a set is a predicate statement, and a transition between predicate statements is an inference. Self-organization as exemplified by pattern formation is accordingly equivalent to an inferential statement. Thus in comprehending pattern formation in the sense of self-organization we are really talking about elementary logic.[63, 134, 135] Huberman and Hogg[136] have in fact modelled adaptive computing structures which are 'analogous to the behavior of a dissipative dynamical system with many degrees of freedom'. We thus adopt the general definition of a phase transition as the transformation of one distinct set of macroscopic thermokinetic objects or forms into another distinct set of objects (cf Langer[137]).

Clearly, our definition is inclusive of both first and second order equilibrium transitions for the mathematical groups which are used to characterize crystal structure or symmetry changes are also sets. Furthermore, all the previously described reactions which convert homogeneous or heterogeneous supersaturated structures in open or closed systems into recognizable patterns fall within our elementary rubric. We can recognize analogues of first and second order phase transitions in kinetic (constrained non-equilibrium) phase space. For example, the onset of the Rayleigh–Benard instability in thermal-hydraulics or the Oregonator transition can be seen as simple symmetry-breaking transitions on the macroscopic scale. On the other hand, the transition of a laser-pumped, supersaturated exciton gas in germanium to a steady state patterned cloud of electron–hole drops has all the features of a spinodal, near-equilibrium first order phase transition.[97, 139]

These analogies are interesting, but in fact misleading in respect to the great variety of transitions of both discrete and gradual character which do not fall within conventional classifications. For example, the sequence of binary solidification patterns of Fig. 12.31, which can be reached via a smooth variation of the experimental control parameters, pass through planar, weakly turbulent; cellular, cellular-dendritic (oscillatory) and multidendritic modes.[99] Such structures appear under conditions only slightly removed from local equilibrium.

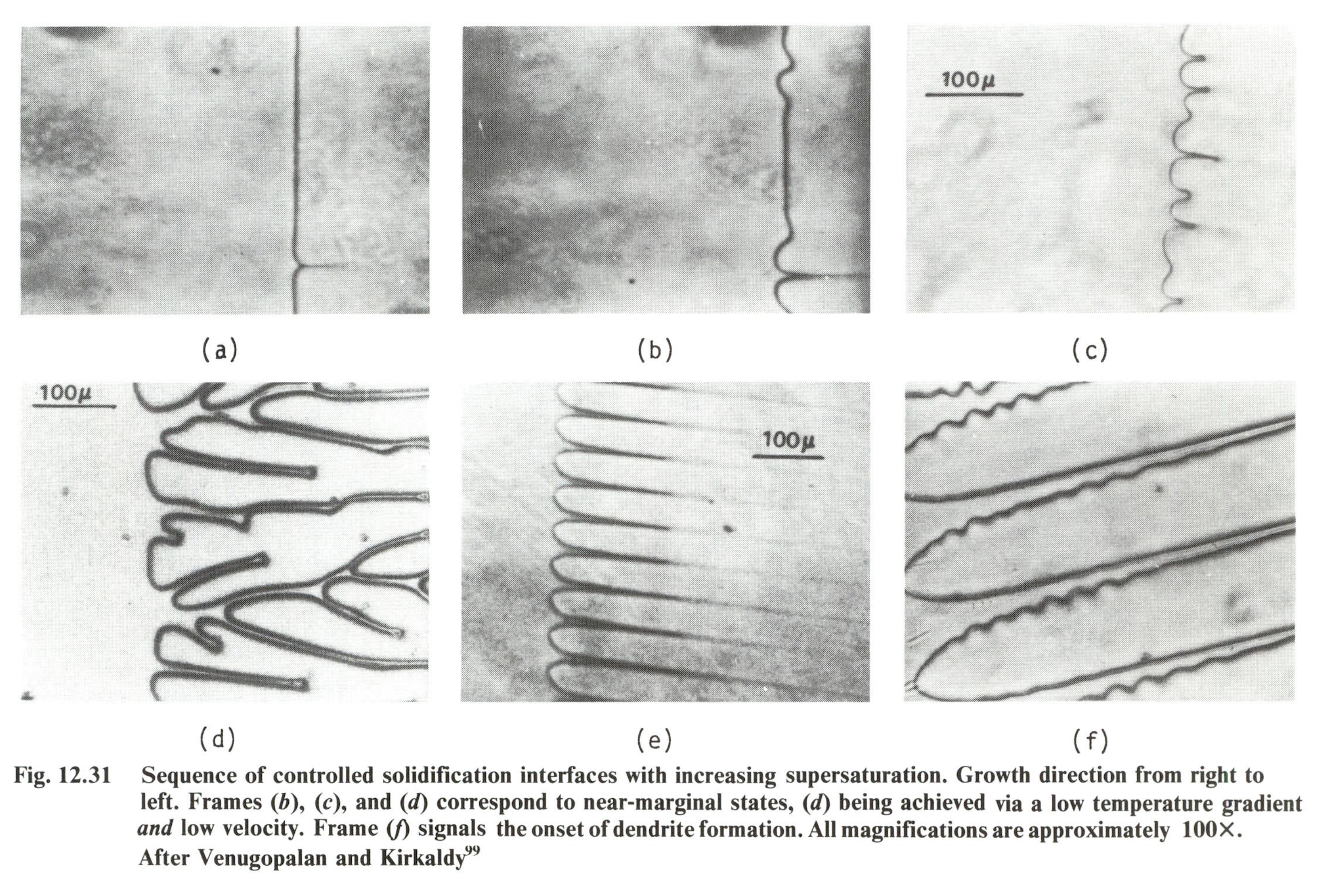

Fig. 12.31 Sequence of controlled solidification interfaces with increasing supersaturation. Growth direction from right to left. Frames (*b*), (*c*), and (*d*) correspond to near-marginal states, (*d*) being achieved via a low temperature gradient *and* low velocity. Frame (*f*) signals the onset of dendrite formation. All magnifications are approximately 100×. After Venugopalan and Kirkaldy[99]

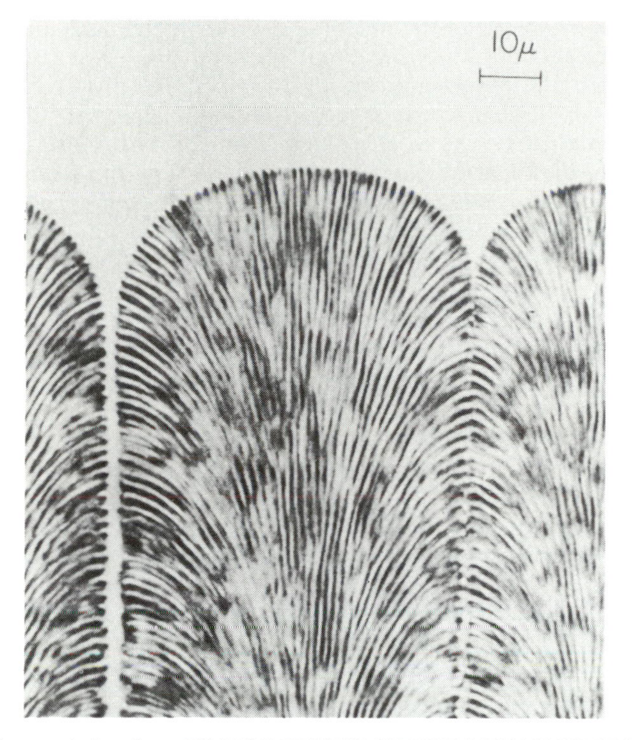

Fig. 12.32 Superposed cellular and eutectic growth in solidification of a ternary solution. This demonstrates very well the thermokinetic uniqueness of the eutectic lamellar spacing, for new lamellae must be regularly created at the cell front to sustain the optimum spacing and dissipation. After Jackson and Hunt[100]

In passing, it is remarked that while the 'weakly turbulent' pattern of Fig. 12.31*d* has negligible translational or rotational symmetry and apparent high entropy it possesses many equivalence sets consisting of randomly ordered segments of fixed curvatures or inflection points and partial elliptic surfaces. In this sense it is quite highly 'ordered' and therefore pattern-rich. Indeed, the chaotic patterns known as 'fractals'[140] possess considerable order. This conception extends to the dynamics of growth as observed under the microscope, where the patterns of 'pulsation' replace those of periodicity in the case of Figs. 12.31*f* and 12.32.

Within this general conception wherein phase transitions are defined as pattern changes in the most general sense the dissipative structures of Prigogine and co-workers pertain to a very special case, namely, the temporal and spatial decomposition of a dilute, highly supersaturated, autocatalytic solution of simultaneously reacting and diffusing chemical constituents. If there are two controlling rate equations then the concentration terms must be to the third order, or if the concentration terms are second order there must be at least three controlling rate equations if bifurcations leading to unique or discretely multiple limit cycles or spatial periodicities are to be attained. The proven existence of such universally asymptotic states is a very strong and important statement about macroscopic nature. It says in effect that a generalized phase space possesses kinetic singularities. On the other hand, classical first and second order phase transitions also specify singularities in an equilibrium phase space. The current distinction lies only in the fact that we know and accept a potential determining the latter whereas Prigogine and co-workers express the doubt that such a potential exists for dissipative structures.

At the same time they clearly regard such a state as perturbation stable. If this is the case the local fluctuation phase space certainly possesses a conditional state of maximum path probability, which we have seen is a directly applicable stabilization principle in certain cases (spiral zonation). In many other cases this principle is equivalent to a minimax in the dissipation which in turn is equivalent to a scaling law. In the applications of their theory to primordial chemical evolution[132] they make clear that initial evolution is towards higher dissipation with a final (interim) minimal regression towards a stable state. Are they not admitting that the determinate trajectories defined by their autocatalytic models lie on a saddle surface, the saddle point being inaccessible only because of an unrealistic determinacy of the trajectories (cf Kirkaldy,[134, 141, 142] Chambers[62])? Can it be that continuum homogeneous dilute solution models fail (do not analytically continue) when forced within the range of transition to dissipative structure except as they pertain to hysteritic states? Should that be the case then all unique, observable patterned structures are of a kind, representing a state of maximum path probability. Maximum path probability is manifested via a variety of operations which depend on whether the terminal state is steady, quasi-steady, oscillatory or chaotic. Operational equivalences which can be identified are perturbation regression, dissipation optimization, intuitive scaling, convergent trajectories (limit cycles), minimization of correlated motions and renormalization group methodology.[143, 144] This latter difficult statistical mechanical methodology, which deals accurately with fluctuating systems with 'many scales of length', is the fundamental route available for the demonstration of such equivalences.

REFERENCES

1 A. J. LOTKA: *Proc. Nat. Acad. Sci.*, 1920, **6**, 420

2 A. J. LOTKA: *ibid.*, 1922, **8**, 147

3 L. VON BERTALANFFY: *Science*, 1950, **111**, 23

4 I. PRIGOGINE: 'Introduction to thermodynamics of irreversible processes', 1955, Springfield, Ill., Charles C. Thomas

5 N. WIENER: 'Cybernetics', 1948, New York, John Wiley

6 A. M. TURING: *Phil. Trans.*, 1952, **B237**, 37

7 G. NICOLIS AND I. PRIGOGINE: 'Self-organization in nonequilibrium systems', 1977, New York, John Wiley

8 A. J. LOTKA: 'Mathematical biology', 1956, New York, Dover Publications

9 L. KADANOFF: private communication, 1986

10 C. WAGNER: *J. Electrochem. Soc.*, 1954, **101**, 225

11 C. WAGNER: *ibid.*, 1956, **103**, 571

12 W. W. MULLINS AND R. F. SEKERKA: *J. Appl. Phys.*, 1963, **34**, 323
13 S. CHANDRASEKHAR: 'Hydrodynamic and hydromagnetic stability', 1961, Oxford, Clarendon Press
14 H. D. KEITH AND F. J. PADDEN: quoted by W. A. Tiller in *Science*, 1964, **146**, 871
15 J. S. KIRKALDY: *Can. J. Phys.*, 1958, **36**, 446
16 R. D. TOWNSEND AND J. S. KIRKALDY: *Trans. ASM*, 1968, **61**, 605
17 C. WAGNER: *Z. Electrochem.*, 1961, **65**, 581
18 W. OSTWALD: *Z. Phys. Chem.*, 1900, **34**, 495
19 G. W. GREENWOOD: *Acta Met.*, 1956, **4**, 243
20 C. ZENER: *Trans. AIME*, 1946, **167**, 550
21 M. HILLERT: *Jernkontorets Ann.*, 1957, **141**, 757
22 A. J. ARDELL: *Acta Met.*, 1972, **20**, 61
23 J. D. LIVINGSTON: *Trans. AIME*, 1959, **215**, 566
24 G. BOHM AND M. KAHLWEIT: *Acta Met.*, 1964, **12**, 641
25 D. DE FONTAINE: Ph.D. Thesis, 1967, Northwestern University, Evanston, Ill.
26 J. S. KIRKALDY AND G. R. PURDY: *Can. J. Phys.*, 1969, **47**, 865
27 J. E. MORRAL AND J. W. CAHN: *Acta Met.*, 1971, **19**, 1037
28 J. S. KIRKALDY: *Can. J. Phys.*, 1958, **36**, 899
29 J. S. KIRKALDY, D. WEICHERT, AND ZIA-UL-HAQ: *ibid.*, 1963, **41**, 2166
30 L. O. SUNDELOF: *Arkiv Kemi*, 1963, **20**, 369
31 J. W. CAHN: *Acta Met.*, 1961, **9**, 795
32 W. B. PEARSON: 'A handbook of lattice spacings and structure of metals and alloys', 1967, New York, Wiley-Interscience
33 M. HILLERT: D.Sc. Thesis, 1956, MIT, Cambridge, Mass.
34 M. HILLERT: *Acta Met.*, 1961, **9**, 525
35 J. W. CAHN AND J. HILLIARD: *J. Chem. Phys.*, 1959, **31**, 688
36 S. MACHLUP AND L. ONSAGER: *Phys. Rev.*, 1953, **91**, 1512
37 H. TOOR: *Chem. Eng. Sci.*, 1965, **20**, 941
38 E. AIFANTIS: *Acta Met.*, 1979, **27**, 683
39 A. I. LEE AND J. M. HILL: *J. Math. Anal. Applications*, 1982, **89**, 530
40 E. L. HUSTON, J. W. CAHN, AND J. E. HILLIARD: *Acta Met.*, 1966, **14**, 1053
41 J. AALDERS, C. VAN DIJK, AND S. RADELAAR: *J. Phys F (Met. Phys.)*, 1984, **14**, 2801
42 J. AALDERS, C. VAN DIJK, AND S. RADELAAR: *Phys. Rev. B*, 1984, **30**, 1646
43 R. E. LIESEGANG: *Naturwiss. Wochenschr.*, 1896, **11**, 353
44 E. S. HEDGES: 'Liesegang rings and other periodic structures', 1932, London, Chapman and Hall
45 S. VEIL: 'Les phénomenes périodiques de la chimie', 1934, Paris, Herman et Cie
46 C. WAGNER: *J. Colloid. Sci.*, 1950, **5**, 85
47 M. KAHLWEIT: *Z. Phys. Chem.*, 1962, **32**, 1
48 M. KAHLWEIT: *Prog. Chem. Solids*, 1965, **2**, 134
49 R. L. KLUEH AND W. W. MULLINS: *Acta Met.*, 1969, **17**, 59, 69
50 Y. S. SHEN, E. J. ZDANUK, AND R. H. KROCK: *Met. Trans.*, 1971, **2**, 2839
51 V. A. VAN ROOIJEN, E. W. VAN ROYEN, J. VRIGEN, AND S. RADELAAR: *Acta Met.*, 1975, **23**, 987
52 K. OSINSKI, A VRIEND, G. F. BASTIN, AND F. J. J. VAN LOO: *Z. Metallk.*, 1982, **73**, 258
53 P. C. PATNAIK: Ph.D. Thesis, 1984, McMaster University
54 K. JABLCZYNSKI AND S. KOBRYNER: *Bull. Soc. Chim.*, 1926, **39**, 383
55 P. ORTOLOVA: in 'Theoretical chemistry, Vol. IV', (ed. H. Eyring), 1978, New York, Academic Press
56 P. ORTOLOVA: in 'Chemical instabilities', (ed. G. Nicolis and F. Baras), NATO ASI series, 1984, Dordrecht, Holland, D. Reidel
57 K. B. RAPER AND C. THOM: 'A manual of penicillia', 1968, New York, Hafner
58 F. ULLSCHEK: *Bot. Arch.*, 1928, **23**, 289
59 J. A. BOURRET, R. G. LINCOLN, AND B. H. CARPENTER: *Science*, 1969, **166**, 763

60 J. S. KIRKALDY: unpublished research, 1985
61 J. VON NEUMANN AND O. MORGENSTERN: 'Theory of games and economic behaviour', 1944, Princeton University Press
62 D. B. CHAMBERS: Ph.D. Thesis, 1973, McMaster University
63 J. S. KIRKALDY: *Phys. Rev. A*, 1985, **31**, 3376
64 L. ONSAGER: *Phys. Rev.*, 1931, **37**, 405; **38**, 2265
65 LORD KELVIN: *Math. Phys. Paper*, 1882, **1**, 232
66 LORD RAYLEIGH: *Proc. Math. Soc.*, 1873, **4**, 357
67 S. R. DE GROOT: 'Thermodynamics of irreversible processes', 1952, Amsterdam, North-Holland
68 R. THOM: 'Stabilité structurelle et morphogénèse', 1972, New York, Benjamin
69 J. S. KIRKALDY: *Met. Trans. A*, 1985, **16**, 1781
70 R. KIKUCHI: *Phys. Rev.*, 1961, **124**, 1682, 1691
71 A. MUNSTER: 'Statistical thermodynamics, Vol. 1', 59–62, 396, 1969, Berlin, Springer-Verlag
72 R. J. TYKODI: 'Thermodynamics of steady states', 1967, New York, Macmillan
73 R. J. TYKODI: *J. Chem. Phys.*, 1984, **80**, 1652
74 G. W. PALTRIDGE: *Q.J.R. Met. Soc.*, 1981, **107**, 531
75 Y. SAWADA: *Prog. Theor. Phys.*, 1981, **66**, 68
76 D. J. YOUNG, W. W. SMELTZER, AND J. S. KIRKALDY: *J. Electrochem. Soc.*, 1976, **123**, 1758
77 J. FRIDBERG AND M. HILLERT: *Acta Met.*, 1977, **25**, 19
78 A. D. TOMSETT, D. J. YOUNG, AND M. S. WAINWRIGHT: *J. Electrochem. Soc.*, 1984, **131**, 2476
79 V. S. BHIDE AND W. W. SMELTZER: *ibid.*, 1983, **130**, 2483
80 K. TASHIRO: Ph.D. Thesis, 1986, McMaster University
81 K. TASHIRO AND G. R. PURDY: submitted *Acta Met.*, 1986
82 K. HASHIGUCHI AND J. S. KIRKALDY: *Scand. J. Metall.*, 1984, **13**, 240
83 J. W. CAHN AND W. C. HAGEL: in 'Decomposition of austenite by diffusional processes', (ed. V. F. Zackay and H. I. Aaronson), 131, 1962, New York, Interscience
84 B. E. SUNDQUIST: *Acta Met.*, 1968, **16**, 1413
85 B. E. SUNDQUIST: *ibid.*, 1969, **17**, 967
86 M. HILLERT: *ibid.*, 1971, **19**, 769
87 J. S. LANGER: *Phys. Rev. Lett.*, 1980, **44**, 1023
88 J. S. KIRKALDY: *Scripta Met.*, 1981, **15**, 1255
89 J. S. KIRKALDY: *Phys. Rev. B*, 1984, **30**, 30
90 F. C. FRANK AND K. E. PUTTICK: *Acta Met.*, 1956, **4**, 206
91 J. S. KIRKALDY AND R. C. SHARMA: *ibid.*, 1980, **28**, 1009
92 D. BROWN AND N. RIDLEY: *J.I.S.I.*, 1969, **207**, 1223
93 G. BOLZE, M. P. PULS, AND J. S. KIRKALDY: *Acta Met.*, 1972, **20**, 73
94 R. C. SHARMA, G. R. PURDY, AND J. S. KIRKALDY: *Met. Trans. A*, 1979, **10**, 1129
95 W. V. R. MALKUS AND G. VERONIS: *J. Fluid Mech.*, 1958, **4**, 225
96 T. D. HAMILL AND K. J. BAUMEISTER: 'Proc. 3rd international heat transfer congress', 59, IV, 1966, A.I.Ch.E.
97 J. S. KIRKALDY AND L. R. B. PATTERSON: *Phys. Rev. A*, 1983, **28**, 1612
98 D. VENUGOPALAN: Ph.D. Thesis, 1982, McMaster University
99 D. VENUGOPALAN AND J. S. KIRKALDY: *Acta Met.*, 1984, **32**, 893
100 K. A. JACKSON AND J. D. HUNT: *Trans. AIME*, 1966, **236**, 1129
101 J. S. KIRKALDY: *Scripta Met.*, 1980, **14**, 739
102 V. S. ARPACI: 'Conduction heat transfer', 211, 1966, Reading, Mass., Addison-Wesley
103 J. S. KIRKALDY: *Can. J. Phys.*, 1959, **37**, 739
104 J. S. KIRKALDY: *ibid.*, 1960, **38**, 1343
105 P. G. SHEWMON: *Trans. AIME*, 1965, **233**, 736
106 I. JIN AND G. R. PURDY: *J. Cryst. Growth*, 1974, **23**, 29, 37
107 B. BILLIA, A. L. COULET, AND L. CAPELLA: *ibid.*, 1976, **35**, 201
108 B. BILLIA AND L. CAPELLA: *ibid.*, 1978, **44**, 235

109 J. S. KIRKALDY AND D. VENUGOPALAN: unpublished research, 1986
110 J. C. BAKER AND J. W. CAHN: *Acta Met.*, 1969, **17**, 575
111 J. S. LANGER AND R. F. SEKERKA: *ibid.*, 1975, **23**, 1225
112 S. R. DE GROOT AND P. MAZUR: 'Non-equilibrium thermodynamics', 1962, Amsterdam, North-Holland
113 B. B. BELOUSOV: *Sb. Ref. Radiat. Med.*, 1958, Moscow
114 A. M. ZHABOTINSKII: *Biofizika*, 1964, **9**, 306
115 A. M. ZHABOTINSKII: *Nature*, 1970, **225**, 535
116 P. GLANSDORFF AND I. PRIGOGINE: 'Thermodynamics of structure, stability and fluctuations', 1971, New York, Wiley–Interscience
117 A. M. LIAPOUNOV: Problème général de la stabilité du mouvement', 1949, Princeton University Press
118 C. S. ELTON: *J. Exp. Biol.*, 1924, **2**, 119
119 F. A. PITELKA: in 'Artic biology', 153, 1957, Oregon State University Press
120 J. S. KIRKALDY AND P. N. SMITH: unpublished research, 1968
121 V. VOLTERRA: 'Leçons sur la theorie mathematique de la lutte sur la vie', 1936, Paris, Gauthier-Villars
122 E. KERNER: *Bull. Math. Biophys.*, 1957, **19**, 121
123 'Chemical instabilities', (ed. G. Nicolis and F. Baras), 1984, Dordrecht, D. Reidel
124 R. M. NOYES AND R. J. FIELD: *Am. Rev. Phys. Chem.*, 1974, **25**, 95
125 R. J. FIELD AND R. M. NOYES: *J. Chem. Phys.*, 1974, **60**, 1877
126 M. HERSCHKOWITZ-KAUFMAN: *C.R. Acad. Sci. Sec. C*, 1970, **270**, 1049
127 A. M. ZHABOTINSKII: 'Concentrated auto-oscillations', 1974, Moscow, Nauka
128 A. T. WINFREE: *Sci. Am.*, 1974, **230**, 82
129 J. STANSHINE: Ph.D. Thesis, 1975, MIT
130 J. A. DESIMONE, D. L. BEIL, AND L. E. SCRIVEN: *Science*, 1973, **180**, 946
131 J. J. TYSON: 'The Belousov–Zhabotinskii reaction', 102, 1976, Berlin, Springer-Verlag
132 I. PRIGOGINE, G. NICOLIS, AND A. BABLOYANTZ: *Physics Today*, 1972, **25**, (11), 23; (12), 38
133 J. S. KIRKALDY: in 'Decomposition of austenite by diffusional processes', (ed. V. Zackay and H. Aaronson), 78, 1962, New York, Interscience
134 J. S. KIRKALDY: 'Life, logic and bootstrap physics', 1978, Ancaster, Canada, JASAK
135 D. D'HUMIÈRES AND B. A. HUBERMAN: *J. Stat. Phys.*, 1984, **34**, 361
136 B. A. HUBERMAN AND T. HOGG: *Phys. Rev. Lett.*, 1984, **52**, 1048
137 S. K. LANGER: 'Symbolic logic', 3rd edn, 1967, New York, Dover Publications
138 E. L. KOSCHMIEDER: *Adv. Chem. Phys.*, 1975, **32**, 109
139 J. S. KIRKALDY: *Scripta Met.*, 1983, **17**, 115
140 B. B. MANDELBROT: 'The fractal geometry of nature', 1977, New York, W. H. Freeman
141 J. S. KIRKALDY: *Can. J. Phys.*, 1964, **42**, 1447
142 J. S. KIRKALDY: *J. Biophys.*, 1965, **5**, 965, 981
143 K. G. WILSON: *Rev. Mod. Phys.*, 1975, **47**, 773
144 K. G. WILSON: *Sci. Am.*, 1979, **241**, 158

APPENDIX 1

Solution thermodynamics including the effects of capillarity and osmosis

A1.1 THE GIBBS-DUHEM EQUATION

In the following, upper case extensive variables V, G, S, etc., refer to total quantities, while lower case v, g, s, etc., refer to molar quantities. The total Gibbs free energy of an arbitrary system is defined as

$$G = E + pV - TS = H - TS \tag{A1.1}$$

where V is the volume, E is the internal energy, S is the entropy and H is the enthalpy. The reversible Gibbs energy equation from the second law is

$$\mathrm{d}E = T\,\mathrm{d}S - p\,\mathrm{d}V + \sum_{i=1}^{r} \mu_i\,\mathrm{d}n_i \tag{A1.2}$$

where n_i is the number of moles of component i and

$$\mu_i = \left(\frac{\partial E}{\partial n_i}\right)_{S,V,n_{j\neq i}} \tag{A1.3}$$

is called the chemical potential. Its evaluation will be discussed later. Combining the differential of equation (A1.1) with equation (A1.2) yields the differential form of the Gibbs free energy

$$\mathrm{d}G = V\,\mathrm{d}p - S\,\mathrm{d}T + \sum_{i=1}^{r} \mu_i\,\mathrm{d}n_i \tag{A1.4}$$

By the optimal part of the second law of thermodynamics it can be proved that $\mathrm{d}G = 0$ (minimum) for processes at constant T and p. It follows from equation (A1.4) that the partial molar free energy

$$\overline{G}_i = \left(\frac{\partial G}{\partial n_i}\right)_{p,T,n_{j\neq i}} = \mu_i \tag{A1.5}$$

the chemical potential. Also, integrating equation (A1.4) at constant p and T a second expression for G is obtained, namely

$$G = \sum_{i=1}^{r} \mu_i n_i \tag{A1.6}$$

Further, integrating equation (A1.2) at constant T, p, and μ_i yields

$$E = TS - pV + \sum_{i=1}^{r} \mu_i n_i \tag{A1.7}$$

Differentiating once more and subtracting equation (A1.2) yields

$$\sum_{i=1}^{r} n_i \, \mathrm{d}\mu_i = V \, \mathrm{d}p - S \, \mathrm{d}T \tag{A1.8}$$

This can be divided by the total number of moles

$$n = \sum_{i=1}^{r} n_i \tag{A1.9}$$

to yield the molar form, using X_i to denote mole fraction

$$\sum_{i=1}^{r} X_i \, \mathrm{d}\mu_i = v \, \mathrm{d}p - s \, \mathrm{d}T \tag{A1.10}$$

which is known as the Gibbs–Duhem equation. The existence of such a relation is a special case of the Euler theorem for homogeneous equations. Note that in contrast to the energy equation in which the independent variables are all extensive, the independent variables are all intensive. The integral of this expresses the chemical potentials as functions of pressure and temperature.

A1.2 PHASE EQUILIBRIA

Consider a constant p and T system containing phases I, II, etc. at equilibrium and virtual transfers of mass of the form $\mathrm{d}n_i^{\mathrm{I}} = -\mathrm{d}n_i^{\mathrm{II}} = -\mathrm{d}n_i^{\mathrm{III}}$, etc. It follows from equation (A1.4) with $\mathrm{d}G = 0$ that

$$\mu_i^{\mathrm{I}} = \mu_i^{\mathrm{II}} = \mu_i^{\mathrm{III}} = \ldots \tag{A1.11}$$

Thus the chemical potential has the same role for chemical equilibrium as p and T have for mechanical and thermal equilibrium. It is these relations which allow us to evaluate the chemical potential of arbitrary phases by comparison with those for ideal gases.

A1.3 STANDARD STATES, THE ACTIVITY AND THE ACTIVITY COEFFICIENT

Consider a pure ideal gas for which

$$pv = RT \tag{A1.12}$$

and choose a standard state pressure p^0 (say 1 atm). Now integrating equation (A1.10) from p^0 to p ($X_i = 1, r = 1$) yields

$$\mu = \mu^0 + RT \ln \frac{p}{p^0} \tag{A1.13}$$

where μ^0, the integration constant, is called the standard chemical potential (which for a pure material is equal to the standard Gibbs free energy per mole, g^0;

cf equation (A1.6)). For a mixture of ideal gases obeying Dalton's law of partial pressures

$$p = \sum_{i=1}^{r} p_i \quad \text{or} \quad \mathrm{d}p = \sum_{i=1}^{r} \mathrm{d}p_i \tag{A1.14}$$

and

$$p_i v = X_i RT \tag{A1.15}$$

Equations (A1.10) and (A1.15) yield for constant T

$$\sum_{i=1}^{r} X_i \,\mathrm{d}\mu_i = \sum_{i=1}^{r} X_i RT \frac{\mathrm{d}p_i}{p_i} \tag{A1.16}$$

so a chemical potential can be consistently defined for each component via

$$\mathrm{d}\mu_i = RT \frac{\mathrm{d}p_i}{p_i} \tag{A1.17}$$

or

$$\mu_i = \mu_i^0 + RT \ln \frac{p_i}{p_i^0} \tag{A1.18}$$

The ratio p_i/p_i^0 is designated as the activity

$$a_i = \frac{p_i}{p_i^0} \tag{A1.19}$$

of the ith component of the solution and is always unity in the standard state. If perchance p_i^0 were chosen as a fixed value p for each i then by equations (A1.14) and (A1.15)

$$a_i = X_i \tag{A1.20}$$

Alternatively, if ideal vapours in equilibrium with a liquid or solid solution of the same components in an isothermal enclosure are considered, and the standard state for each component is defined as the vapour pressure over the corresponding pure condensed phase, then μ_i^0 is the Gibbs free energy of that pure phase and the chemical potential of the ith component in that phase is, by equations (A1.11)

$$\mu_i = \mu_i^0 + RT \ln \frac{p_i}{p_i^0} \tag{A1.21}$$

These observed ratios p_i/p_i^0 for the equilibrated gases, which are a function of solution composition, can be designated as the activities for the condensed phase solution, and the chemical potentials written in general as

$$\mu_i = \mu_i^0 + RT \ln a_i \tag{A1.22}$$

The vapour pressure method is widely used for establishing chemical potentials numerically. If perchance the experiments yield

$$a_i = X_i \qquad \text{(A1.23)}$$

then the condensed phase solution is designated as ideal. For dilute solutions of i it is experimentally found that

$$a_i = \gamma_i^0 X_i \qquad \text{(A1.24)}$$

where γ_i^0 is a constant. This is known as Henry's law and is derivable within statistical mechanics. From this it can be proven via the Gibbs–Duhem equation that, for almost pure component i, equation (A1.23) is satisfied and thus expresses ideality. This is known as Raoult's law, first discovered experimentally via the partial pressures of rich solutions. The standard states can be chosen such that $\gamma_i^0 = 1$, in which case the activities are said to be Henrian. Generally

$$a_i = \gamma_i(X_k) X_i \qquad \text{(A1.25)}$$

where the activity coefficient γ_i designates deviations from ideality.

The quantity $\ln \gamma_i$ can be expanded as a Taylor expansion about a state of infinite dilution and to the first order has the form

$$\ln \gamma_i = \ln \gamma_i^0 + \sum_{k=1}^{r} \varepsilon_{ik} X_k + \ldots \qquad \text{(A1.26)}$$

where the γ_i^0 are the Henry's law coefficients. The Maxwell relation for exact differentials yields the symmetry relations

$$\frac{\partial \ln \gamma_i}{\partial X_k} = \varepsilon_{ik} = \varepsilon_{ki} = \frac{\partial \ln \gamma_k}{\partial X_i} \qquad \text{(A1.27)}$$

which greatly reduces the experimentation involved in evaluating the ε matrix. This has been evaluated for many important dilute solutions and plays an important role in ternary diffusion theory.

A1.4 REGULAR AND BRAGG-WILLIAMS SOLUTIONS

A binary solution which possesses an ideal entropy of mixing

$$s_{\mathrm{m}}^{\mathrm{id}} = -R(X_1 \ln X_1 + X_2 \ln X_2) \qquad \text{(A1.28)}$$

is called a regular solution. For such a condensed solution the enthalpy of mixing is given by

$$\Delta h_{\mathrm{m}} = RT(X_1 \ln \gamma_1 + X_2 \ln \gamma_2) \qquad \text{(A1.29)}$$

A special symmetric case of this structure based on statistical mechanics and pair interaction energies is much favoured by physicists (the Bragg–Williams or quasi-chemical approximation), and expresses

$$\ln \gamma_i = \frac{zw}{kT}(1 - X_i)^2 \qquad i = 1, 2 \qquad \text{(A1.30)}$$

where z is the coordination number and

$$w = \left(e_{12} - \frac{e_{11} + e_{22}}{2}\right) \qquad \text{(A1.31)}$$

is the excess pair interaction energy. If w is positive or $\gamma_i > 1$ then the unlike pair energy is higher than the mean of the like pair energies implying a net repulsion and a tendency towards phase separation. The free energy $g(X_2)$ and the phase diagram of this Bragg–Williams or quasi-chemical solution have been discussed in relation to Figs. 1.35 and 1.36. If w is positive there is a miscibility gap and a negative sign of d^2g/dX^2 between the spinodes s implying uphill diffusion and spontaneous decomposition (equation (1.36)). If w is negative, implying attraction between unlike atoms, then the evolution will be towards ordering or compound formation. It is to be noted in relation to Fig. 1.36 that the common tangent construction (line AB) which expresses the equilibrium minimum of g also expresses the equality of chemical potentials. Furthermore, it is recognized in general from the graph that the tangent at any point of $g(X_2)$ is

$$\frac{dg}{dX_2} = \mu_2 - \mu_1 \tag{A1.32}$$

This general result is relevant to the reduction of the mobility (L) matrix to the diffusion (D) matrix in §1.6.

The foregoing transformation is the paradigm for 'first order' transformations where there is a discontinuity in extensive variables. Transformations which occur continuously in extensive variables but show a discontinuity in the first derivative of extensive variables are designated as 'second order'. Many order–disorder transformations are of this type.

A1.5 THE EQUILIBRIUM CONSTANT

For a general chemical reaction, which includes phase transitions, we can write the equilibrium of reactants and products as

$$aA + bB + \ldots \rightleftarrows cC + dD + \ldots \tag{A1.33}$$

where upper case represents species and lower case represents stoichiometric coefficients. The free energy change of the reaction is

$$\Delta G = cg^0_C + dg^0_D + \ldots - ag^0_A - bg^0_B \ldots + cRT \ln a_C + dRT \ln a_D + \ldots - aRT \ln a_A - bRT \ln a_B \tag{A1.34}$$

At equilibrium $\Delta G = 0$ and this can be written as

$$\Delta G^0(T) = -RT \ln \frac{a^c_C a^d_D \ldots}{a^a_A a^b_B \ldots} = -RT \ln K(T) \tag{A1.35}$$

$K(T)$ is called the equilibrium constant and its existence implies a unique concentration relation through a at every temperature, and thus a phase diagram constraint. It is, in fact, equivalent to the Gibbs–Duhem equation together with equal chemical potentials between phases. Its use is sometimes more convenient, particularly in relation to defect and association equilibria in solutions (cf §1.17).

More comprehensive treatments are to be found in Darken and Gurry[1] and Prigogine and Defay.[2]

A1.6 THE EFFECT OF SURFACE CURVATURE ON THE CHEMICAL POTENTIAL

Many problems of mass transfer or phase transformation involve finely divided phases, bubbles or emulsions so the surface tension and its action through the curvature upon the chemical potential at the surfaces of such phases cannot be ignored. This effect is most easily comprehended as a pressure or stress

$$p = \sigma\left(\frac{1}{r_1} + \frac{1}{r_2}\right) = \sigma(\Gamma_1 + \Gamma_2) \tag{A1.36}$$

where Γ are the principal curvatures and p acts towards the centre of curvature, where σ is the surface free energy (or tension) and r_1, r_2 are the principle radii of curvature. For a sphere of radius r this yields

$$p = 2\sigma/r \tag{A1.37}$$

and for a cylinder

$$p = \sigma/r \tag{A1.38}$$

If the curvature is variable then for a surface $y = y(x)$

$$\Gamma = \frac{d^2y}{dx^2}\bigg/\left[1 + \left(\frac{dy}{dx}\right)^2\right]^{3/2}$$

The connection with the surface chemical potential or alloy concentration is established via an integral of the Gibbs–Duhem equation.

The problems of sintering, or thermal densification of porous material, and grain boundary grooving involve solid–gas interfaces of pure or alloy materials. For pure crystals the Gibbs–Duhem equation can be integrated

$$d\mu = v\,dp \tag{A1.39}$$

from zero curvature at constant molar volume v to the value Γ of interest, i.e.

$$\Delta\mu = v(p - 0) = v\sigma\Gamma > 0 \tag{A1.40}$$

It will here be appreciated that the integral will only be valid for condensed phases where v will be accurately constant with variable p.

There are two integrals for binary solutions and one for a ternary solution which are of special interest. First, referring to the phase diagram of Fig. A1.1, the precipitation of pure β spheres in an α solution of concentration X_B is considered. The Gibbs–Duhem equations are written for the two phases at equal chemical potentials

$$X_A^\alpha\,d\mu_A + X_B^\alpha\,d\mu_B = v^\alpha\,dp^\alpha \tag{A1.41}$$

and

$$X_A^\beta\,d\mu_A + X_B^\beta\,d\mu_B = v^\beta\,dp^\beta \tag{A1.42}$$

then integrated from zero curvature of the sphere assuming there is no pressure change in the parent α phase. Hence subtraction gives

$$(X_B^\beta - X_B^\alpha)(\Delta\mu_B - \Delta\mu_A) = v^\beta\Gamma\sigma \tag{A1.43}$$

Now the A solution as $X_B \to 0$ is ideal while the B solution is Henrian so $\Delta\mu_A$ vanishes relative to $\Delta\mu_B$ and

$$\Delta\mu_B = v^\beta\Gamma\sigma > 0 \tag{A1.44}$$

or

$$X_B^\alpha(r) = X_B^\alpha(r = \infty)\exp\left(v^\beta\Gamma\sigma/RT\right) \tag{A1.45}$$

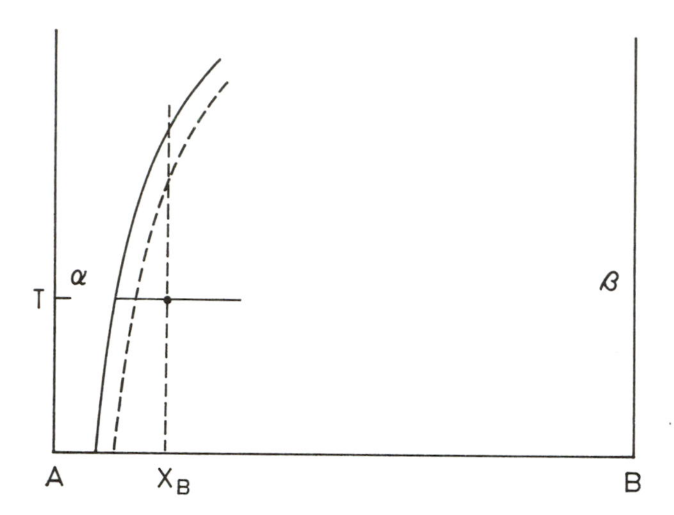

Fig. A1.1 Schematic phase diagram involving precipitation of pure crystals β in dilute α-phase. Effect of curvature on β-precipitate is to shift solubility limit to right

This is known variously as the Gibbs–Thomson or Ostwald–Freundlich equation. It implies that the solubility limit is increased by the curvature (—— → ---- in Fig. A1.1). Indeed, if the curvature is great enough a spherical particle must dissolve rather than grow (cf § 1.23).

Next the phase diagram of Fig. A1.2 is considered, where both solutions α and β are relatively dilute and therefore reasonably approximated by Henrian

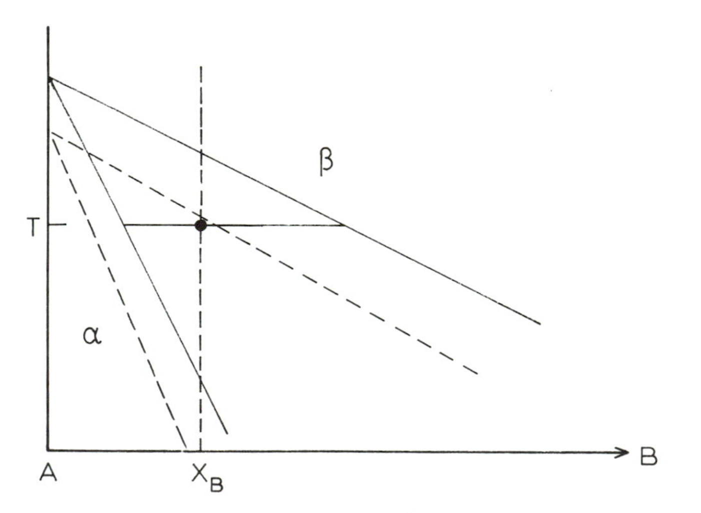

Fig. A1.2 Schematic phase diagram involving precipitation of dilute α-phase in dilute β-phase. Effect of curvature on α-precipitate is to shift both solubility lines to left

behaviour. Whilst equations (A1.41) and (A1.42) remain valid, the α-phase is here identified as the spherical precipitate and it is recognized that (cf § 1.18)

$$X_B^\alpha \simeq k_0 X_B^\beta \tag{A1.46}$$

Equation (A1.43) becomes

$$(X_B^\beta - X_B^\alpha)\,\Delta\mu_B = -v^\alpha \Gamma\sigma \tag{A1.47}$$

or

$$X_B^\beta(r) = X_B^\beta(r=\infty)\exp(-v^\alpha\Gamma\sigma/[RT(1-k_0)]) \tag{A1.48}$$

with

$$X_B^\beta(r) \simeq k_0 X_B^\beta(r) \tag{A1.49}$$

Note that the negative exponent in equation (A1.48) implies that both phase boundaries in Fig. 3.2 shift towards the $X_B = 0$ axis. This implies as before that if the curvature is sufficiently large the α particles will dissolve.

Finally, a ternary analogue of equation (A1.45) is considered where a precipitate compound *AB* has a small solubility product in a dilute solution X_A, X_B. This gives approximately

$$X_A X_B = K\exp[2v^{AB}\sigma/RT] \tag{A1.50}$$

where *K* is the solubility product for a planar phase AB.

For such corrections to the condensed state phase diagrams to become significant, radii of curvatures of $<10\,\mu m$ must usually be attained. At radii of 0·1 μm, which is commonly the case with solid state precipitates, sols and non-settling emulsions, pressures of $10^9\,N\,m^{-2}$ (10 000 atm) are reached. At such pressures even solids suffer significant volume changes so the foregoing relations exhibit corresponding errors.

A1.7 OSMOSIS AND DONNAN EQUILIBRIUM

There are two generalizations of the foregoing discussions of phase equilibria which stem from the existence of semipermeable membranes (cf § 2.17.3 and Chapter 10). The tendency for mobile species, either ionic or non-ionic, to equilibrate across such membranes invites an extension of the foregoing arguments applied to the effects of surface tension. For example, if in Fig. A1.3 one considers a large spherical membrane M immersed in pure water (α) which can sustain pressure over atmospheric, and containing a dilute solution of non-ionizable sugar (β) say, which cannot penetrate the membrane (i.e. a non-dializable species) then one can apply the Gibbs–Duhem equations (A1.41) and (A1.42) subject to the conditions for essentially flat membranes

$$dp^\alpha = X_S^\alpha = d\mu_A = 0 \tag{A1.51}$$

via equation (A1.42) to obtain the differential equation for solution β

$$X_S^\beta\,d\mu_S = v^\beta\,dp^\beta \tag{A1.52}$$

Integrating this by adding sugar S to the chamber within Henrian dilution limits with

$$d\mu_S = RT\,d(\ln kX_S) \tag{A1.53}$$

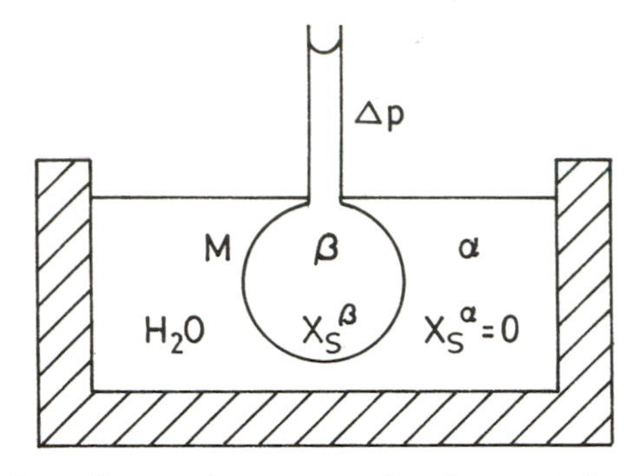

Fig. A1.3 Illustration of osmotic pressure development due to a non-dialyzable solute *S* in aqueous solution *β*

we obtain

$$\Delta p = \frac{RTX_S}{v^\beta} \quad \text{(A1.54)}$$

which is the osmotic pressure which would be measured by the manometer in Fig. A1.3. Notice the close relationship between this and the ideal gas law. This treatment and its experimental verification has been attributed to van't Hoff.[3]

Now a situation involving exchange between dilute aqueous solutions of strong electrolytes such as NaCl and NaR is considered, where Na^+ and Cl^- are dialyzable but anion R^- is not. The equilibration follows schematically from the initial membrane configuration

$$\begin{array}{lcl} \beta & \mathrm{M} & \alpha \\ Na^+, R^- & | & Na^+ \, Cl^- \end{array} \quad \text{(A1.55)}$$

to the final membrane configuration

$$\begin{array}{lcl} \beta & \mathrm{M} & \alpha \\ Na^+, Cl^-, R^- & | & Na^+ \, Cl^- \end{array} \quad \text{(A1.56)}$$

Here one must introduce the electrochemical potentials η_i as discussed in §3.6 and Chapter 8 as a replacement for the chemical potentials μ_i, i.e.

$$\eta_i = \mu_i + z_i F\psi \quad \text{(A1.57)}$$

where z_i is the ionic charge, F is the Faraday, and ψ is the electric potential. The zero of potential in phase α is defined so that $\Delta\psi = \psi^\beta$.

Now the equality of electrochemical potentials η_{Na^+} and η_{Cl^-} yields the relation

$$\mu^\beta_{Na^+} + F\phi^\beta = \mu^\alpha_{Na^+} \quad \text{(A1.58)}$$

or for a Henrian solution

$$F\psi^\beta = RT \ln \frac{X^\alpha_{Na^+}}{X^\beta_{Na^+}} \quad \text{(A1.59)}$$

Similarly

$$F\psi^{\beta} = -RT\ln\frac{X^{\alpha}_{Cl^-}}{X^{\beta}_{Cl^-}} \tag{A1.60}$$

whence

$$X^{\alpha}_{Na^+}X^{\alpha}_{Cl^-} = X^{\beta}_{Na^+}X^{\beta}_{Cl^-} \tag{A1.61}$$

Assuming that $X^{\beta}_{R^-}$, $X^{\alpha}_{Na^+}$ and $X^{\alpha}_{Cl^-}$ are given, and recognizing from solution electroneutrality that

$$X^{\beta}_{Na^+} = X^{\beta}_{R^-} + X^{\beta}_{Cl^-} \tag{A1.62}$$

the generalization of Gibbs–Duhem equations (A1.41) or (A1.52) is

$$X^{\beta}_{Na^+}\,d\eta^{\beta}_{Na^+} + X^{\beta}_{Cl^-}\,d\eta^{\beta}_{Cl^-} + X^{\beta}_{R^-}\,d\eta^{\beta}_{R^-} = v^{\beta}\,dp^{\beta} \tag{A1.63}$$

This integrates exactly via equations (A1.57) and (A1.61) in the Henrian limit to the remarkably simple form

$$X^{\beta}_{Na^+} = \tfrac{1}{2}v^{\beta}p^{\beta}/RT \tag{A1.64}$$

Thus via equations (A1.60), (A1.61), and (A1.62) the osmotic pressure and the membrane or Donnan potential, which are intimately related, can be expressed in terms of $X^{\beta}_{R^-}$, $X^{\alpha}_{Na^+}$ and $X^{\alpha}_{Cl^-}$.

Significant contributions to this development are attributable to van't Hoff.[3] Nernst[4] and Bernstein[5] considered elements of the kinetic case where the anions (say) have different but non-zero mobilities (cf Chapter 8).

APPENDIX 2

General methods for partial differential equations

A2.1 THE VECTOR CALCULUS

In deriving Fick's second equation in one dimension (§ 1.8) we introduced two mathematical concepts of general interest. Fick's first equation

$$J_x = -D\frac{\partial C}{\partial x} \tag{A2.1}$$

contains the idea of the 'gradient' of a scalar (C) while the mass balance

$$\frac{\partial J_x}{\partial x} + \frac{\partial C}{\partial t} = 0 \tag{A2.2}$$

contains the idea of a 'divergence' of a vector (J_x). Generally, in three dimensions

$$\text{grad}\, C = \nabla C = \left(i\frac{\partial}{\partial x} + j\frac{\partial}{\partial y} + k\frac{\partial}{\partial z}\right)C \tag{A2.3}$$

where i, j, and k are the unit vectors in cartesian coordinates. The operator ∇ = del is thus a vector. In a different coordinate system the unit vectors will be different (e.g. in spherical coordinates) and the derivatives will change accordingly (see below).

Now in equation (A2.2) the divergence of the vector J_x equals a scalar $(-\partial C/\partial t)$. Thus the general divergence of a vector should also equal a scalar. Such a result can be achieved through the scalar (or dot product) of two vectors. For example, in the dot product of unit vectors

$$i.i = j.j = k.k = 1 \quad \text{and} \quad i.j = j.k = k.i = 0 \tag{A2.4}$$

Thus the del operator acts upon a vector

$$\boldsymbol{J} = iJ_x + jJ_y + kJ_z \tag{A2.5}$$

in the form

$$\nabla \cdot \boldsymbol{J} = \frac{\partial J_x}{\partial x} + \frac{\partial J_y}{\partial y} + \frac{\partial J_z}{\partial z} \tag{A2.6}$$

which is clearly inclusive of the one-dimensional case (A2.2). Alternatively, by extension of equation (A2.1) the vector $\boldsymbol{J}$ can be expressed

$$J = -D\nabla C \tag{A2.7}$$

to obtain

$$\nabla \cdot J = -D\nabla \cdot \nabla C = -D\left(\frac{\partial^2}{\partial x^2} + \frac{\partial^2}{\partial y^2} + \frac{\partial^2}{\partial z^2}\right)C \tag{A2.8}$$

The combined operation 'div.grad', or $\nabla.\nabla = \nabla^2$ (or 'del squared') is called the Laplacian and now appears in the Fick second equation as

$$\frac{\partial C}{\partial t} = D\left(\frac{\partial^2 C}{\partial x^2} + \frac{\partial^2 C}{\partial y^2} + \frac{\partial^2 C}{\partial z^2}\right) = D\nabla^2 C \tag{A2.9}$$

For the steady state ($\partial C/\partial t = 0$) one must solve the differential equation

$$\frac{\partial^2 C}{\partial x^2} + \frac{\partial^2 C}{\partial y^2} + \frac{\partial^2 C}{\partial z^2} = 0 \tag{A2.10}$$

which is one form of Laplace's equation. For spherical-polar coordinates this can be transformed by well-known methods to

$$\frac{\partial^2 C}{\partial r^2} + \frac{2}{r}\frac{\partial C}{\partial r} + \frac{1}{r^2}\left[\frac{1}{\sin\theta}\frac{\partial}{\partial\theta}\left(\sin\theta\frac{\partial C}{\partial\theta}\right) + \frac{1}{\sin^2\theta}\frac{\partial^2 C}{\partial\psi^2}\right] = 0 \tag{A2.11}$$

where θ is the polar angle and ψ the azimuthal angle (Fig. A2.1). Some solutions in spherical-polar coordinates are presented in Appendix 3.

Fuller accounts of this subject matter are to be found in Bird *et al.*,[6] Carslaw and Jaeger,[7] and Adda and Philibert.[8]

A2.2 LAPLACE TRANSFORMS

The Laplace transform method of solving linear partial differential equations (no powers of derivatives) with constant coefficients is widely applicable to problems in electromagnetism, heat conduction and diffusion. Its role is to identify an operator L which transforms the unknown time-dependent solution function to another one which satisfies a time-independent differential equation and to

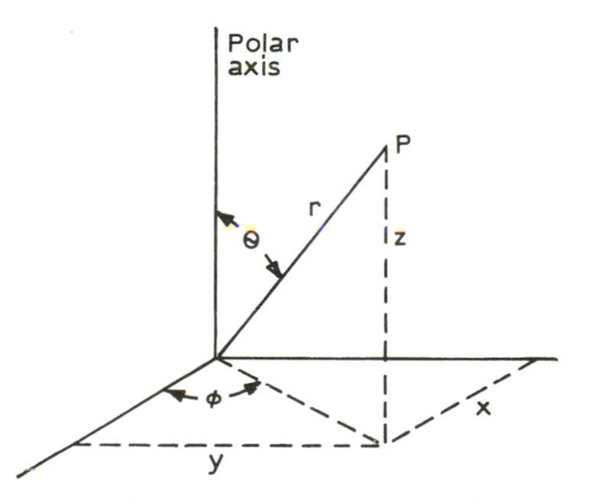

Fig. A2.1 Relation between rectangular and spherical polar coordinates of a point *P*

specify the method of application of the boundary and initial condition to this transformed function. In many problems, the inverse transformation L^{-1} can be obtained from extensive tables compiled for that purpose (see below). The identification of the transformation, developed mainly by Heaviside, yields the integral operator relations in terms of the concentration $C(x, y, z, t)$,

$$L\{C(x, y, z, t)\}_{t \to p} = \overline{C}(p) = \int_0^\infty \exp(-pt)\, C(x, y, z, t)\, \mathrm{d}t \tag{A2.12}$$

and identically for the inverse operation $L^{-1}\{\overline{C}(p)\}_{p \to t}$. A table of transforms contains results such that if, for example,

$$C(t) = C_0 \qquad \overline{C}(p) = \int_0^\infty C_0 \exp(-pt)\, \mathrm{d}t = \frac{C_0}{p} \tag{A2.13}$$

or

$$C(t) = \sin \omega t \qquad \overline{C}(p) = \int_0^\infty \exp(-pt) \sin \omega t\, \mathrm{d}t = \frac{\omega}{p^2 + \omega^2} \tag{A2.14}$$

The general theorems which allow one to easily effect the transformation of the differential equations are

$$L\{C_1 + C_2\} = L\{C_1\} + L\{C_2\} \tag{A2.15}$$

$$L\left\{\frac{\partial C}{\partial t}\right\} = pL\{C\} - C_0 \tag{A2.16}$$

where $C_0(x, y, z)$ is the initial condition for C and

$$L\left\{\frac{\partial^n C}{\partial X^n}\right\} = \frac{\partial^n \overline{C}}{\partial x^n} \tag{A2.17}$$

Other theorems of occasional utility are given by Carslaw and Jaeger.[7]

As an example of the methodology consider the solution of Fick's second equation in one dimension, for a semi-infinite couple,

$$\frac{\partial C}{\partial t} = D \frac{\partial^2 C}{\partial x^2} \tag{A2.18}$$

subject to

$$t = 0 \qquad x > 0 \qquad C = 0 \tag{A2.19}$$

and

$$t > 0 \qquad x = 0 \qquad C = C_0 \tag{A2.20}$$

Applying the above theorems we obtain

$$D \frac{\partial^2 \overline{C}}{\partial x^2} = p\overline{C} + 0 \tag{A2.21}$$

The decaying solution of this equation is

$$\overline{C} = A \exp[-(p/D)^{1/2} x] \tag{A2.22}$$

Table A2.1 Some Laplace transforms applicable to diffusion studies*

$\bar{C}(p)$	$C(t)$
C_0/p	C_0
$1/(p+\alpha)$	$\exp(-\alpha t)$
$\omega/(p^2+\omega^2)$	$\sin \omega t$
$p/(p^2+\omega^2)$	$\cos \omega t$
$\exp(-(p/D)^{1/2}x) \quad x>0$	$\dfrac{x}{2(\pi D t^3)^{1/2}} \exp(-x^2/4Dt)$
$\exp[-(p/D)^{1/2}x]/(p/D)^{1/2} \quad x>0$	$\left(\dfrac{D}{\pi\tau}\right)^{1/2} \exp(-x^2/4Dt)$
$\exp[-(p/D)^{1/2}x]/p \quad x>0$	$\operatorname{erfc}[x/2(Dt)^{1/2}]$

*Expanded tables are given by Carslaw and Jaeger[7] and Oberhettinger and Badii.[9]

and from boundary condition (A2.20) and its transform (A2.13)

$$A = C_0/p \tag{A2.23}$$

Finally referring to the table of transforms (Table A2.1) gives (cf §1.9)

$$L^{-1}\{\bar{C}\} = C = C_0 \operatorname{erfc}[x/2(Dt)^{1/2}] \tag{A2.24}$$

A2.3 FOURIER SERIES AND INTEGRAL TRANSFORMS

In §1.13 the ideas of Fourier series and Fourier integrals, which arise naturally as particular solutions of Fick's second equation, were introduced. There are, however, theorems of a more general nature which pertain to classes of periodic functions and functions whose global integral (for example in one dimension)

$$\int_{-\infty}^{\infty} |f(x)|\,dx \tag{A2.25}$$

exists. For periodic functions in one dimension one assumes that there exists a series representation in harmonics of the form

$$f(x) = A_0 + a_1 \cos x + a_2 \cos 2x + \ldots + b_1 \sin x + b_2 \sin 2x \tag{A2.26}$$

Integrating term by term from $x = 0$ to 2π yields

$$A_0 = \frac{1}{2\pi}\int_0^{2\pi} f(x)\,dx \tag{A2.27}$$

If both sides are multiplied by $\cos nx$ and integrated over the same range, then

$$a_n = \frac{1}{\pi}\int_0^{2\pi} f(x)\cos nx\,dx \tag{A2.28}$$

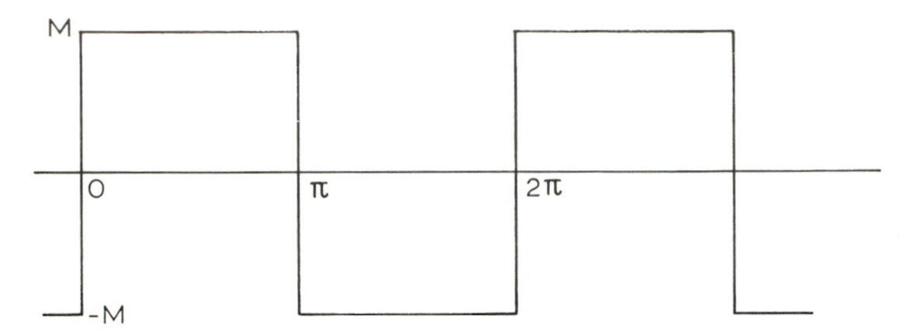

Fig. A2.2 Schematic periodic step function which is subject to Fourier series analysis

and

$$b_n = \frac{1}{\pi}\int_0^{2\pi} f(x)\sin nx\,\mathrm{d}x \tag{A2.29}$$

are proved. Note that if $f(x)$ is even all b_n are zero and if $f(x)$ is odd all a_n are zero. As an example consider the step function of Fig. A2.2. Here the origin has been chosen to define an odd function and the baseline such that $A_0 = 0$. Thus focussing on the sine series and the b_n gives

$$b_n = \frac{1}{\pi}\int_0^{\pi} C_0 \sin(nx)\,\mathrm{d}x - \frac{1}{\pi}\int_0^{2\pi} C_0\sin(nx)\,\mathrm{d}x = \frac{2C_0}{\pi n}(1-\cos n\pi) \tag{A2.30}$$

This implies that $b_n = 0$ for n even and $b_n = 4C_0/\pi n$ for n odd, so f can be written as the series

$$f = \frac{4C_0}{\pi}\sum_{n=0}^{\infty}\frac{1}{2n+1}\sin(2n+1)x \tag{A2.31}$$

It is to be noted that the configuration to be analysed need not actually be periodic, only that the initial and boundary conditions match all or part of a period of the model function. For example, relation (1.70) was derived to represent diffusion out of a plate of thickness h which has an initial condition corresponding to $x = 0$ to π in Fig. A2.2 and a fixed concentration of zero at 0 and π for all time. Thus if $x \rightarrow x/h$ in equation (A2.31), the initial condition corresponding to equation (1.70) is obtained (substitute $t = 0$). This solution was obtained by separation of variables to yield equation (1.69) and the coefficients were obtained by setting $t = 0$ so that all exponents go to 1, and carrying out a Fourier analysis exactly as described here. A collection of formulas for Fourier expansions has been compiled by Oberhettinger.[10]

The Fourier integral theorem is a generalization of the Fourier series theorem presented above and is applicable to initial conditions which are non-periodic.[7] These must, however, be finite as specified in relation (A2.25). The Fourier integral transforms for space variables are the analogy of the Laplace transform for the time variable and can be stated in one-dimensional Fourier transform operator form as

$$F\{f(x)\}_{x \to \xi} = \bar{f}(\xi) = (2\pi)^{-1/2} \int_{-\infty}^{\infty} \exp(i\xi x)\, f(x)\, dx \tag{A2.32}$$

and

$$F^{-1}\{\bar{f}(\xi) = f(x) = (2\pi)^{-1/2} \int_{-\infty}^{\infty} \exp(-i\xi x)\, \bar{f}(\xi)\, d\xi \tag{A2.33}$$

where $\bar{f}(\xi)$ and $f(x)$ are the Fourier transforms of each other. In comparing the real exponents of the Laplace transform (A2.12) with the imaginary exponents in equations (A2.32) and (A2.33), one notes that in the diffusion equation the space variables are differentiated twice while the time variable is differentiated once, so real solutions can be assured in all cases. For even functions f we can let $\exp(i\xi x) \to \cos \xi x$ in the foregoing, as specified by $F \to F_c$, which is the Fourier

Table A2.2 Fourier transforms of relevance to diffusion studies*

$f(x) \quad (x > 0)$	$g(y) = \int_0^\infty f(x) \cos(xy)\, dx$
$1 \quad 0 < x < a$ $0 \quad x > a$	$y^{-1} \sin(ay)$
$(a^2 + x^2)^{-1/2}[(a^2 + x^2)^{1/2} + a]^{-1/2}$	$\pi(2a)^{-1/2} \operatorname{erfc}(ay)^{1/2}$
$x^{-1/2}[a + x + (2ax)^{1/2}]^{-1}$	$\pi(2a)^{-1/2} \exp(ay) \operatorname{erfc}(ay)^{1/2}$
$\exp(-ax)$	$a(a^2 + y^2)^{-1}$
$\exp(-ax^2)$	$\frac{1}{2}\pi^{1/2}a^{-1/2} \exp[-(y^2/4a)]$
$x^{-1} \sin(ax)$	$\begin{cases} \frac{1}{2}\pi & y < a \\ \frac{1}{4}\pi & y = a \\ 0 & y > a \end{cases}$
$\operatorname{sech}(ax)$	$\frac{1}{2}\pi a^{-1} \operatorname{sech}\left(\frac{\pi y}{2a}\right)$
$f(x) \quad (x > 0)$	$g(y) = \int_0^\infty f(x) \sin(xy)\, dx$
$x^{-1} \exp(-ax^2)$	$\frac{1}{2}\pi \operatorname{erf}(\frac{1}{2}ya^{-1/2})$
$x^{-1} \sin(ax)$	$\frac{1}{2} \ln \left\| \frac{y + a}{y - a} \right\|$
$x^{-1} \cos(ax)$	$\begin{cases} 0 & 0 < y < a \\ \frac{1}{4}\pi & y = a \\ \frac{1}{2}\pi & y > a \end{cases}$

*A comprehensive set of tables is given by Oberhettinger.[11]

cosine transformation, and for odd functions we can let $\exp(i\xi x) \rightarrow \sin \xi x$, as specified by $F \rightarrow F_s$, which is the Fourier sine transformation. Some Fourier transforms are given in Table A2.2. Equation (1.74) represents an elementary application.

Note that in complex problems Fourier and Laplace transforms can be used successively (cf §1.15).

APPENDIX 3

Special mathematical functions

A3.1 THE DIRAC δ-FUNCTION

This special limiting or singular function is not to be confused with the Krönecker δ-function, the unit diagonal matrix (cf §A4.1). Consider relation (2.10) representing the single particle distribution function for a tracer in random walk diffusion, i.e.

$$P = \frac{1}{2(\pi Dt)^{1/2}} \exp\left(-\frac{x^2}{4Dt}\right) \tag{A3.1}$$

which satisfies

$$\int_{-\infty}^{\infty} P(x)\, \mathrm{d}x = 1 \tag{A3.2}$$

Now in the limit as $t \to 0$, $P(x = 0) \to \infty$, while the half-width of the distribution goes to zero, the area under the curve remaining equals to unity. This limit of P is one representation of the Dirac δ, a function of infinite height, zero width and unit contained area, i.e.

$$\int_{-\infty}^{\infty} \delta(x)\, \mathrm{d}x = 1 \tag{A3.3}$$

Another representation is

$$\delta(x) = \lim_{g \to \infty} \frac{\sin gx}{\pi x} \tag{A3.4}$$

Since it is singular, δ cannot be the end result of a calculation, i.e., integration over its argument must always be carried out. According to Schiff[12] δ can be transformed under an integral according to

$$-\delta(x) = \delta(-x) \tag{A3.5}$$

$$\delta' = \frac{\mathrm{d}\delta}{\mathrm{d}x} = -\delta'(-x) \tag{A3.6}$$

$$x\delta(x) = 0 \tag{A3.7}$$

$$x\delta'(x) = -\delta(x) \tag{A3.8}$$

$$\delta(ax) = a^{-1}\delta(x) \tag{A3.9}$$

$$f(x)\,\delta(x-a) = f(a)\,\delta(x-a) \tag{A3.10}$$

The latter expression is the most important for it implies that

$$\int_{-\infty}^{\infty} f(x)\,\delta(x-a)\,\mathrm{d}x = \int_{-\infty}^{\infty} f(a)\,\delta(x-a) = f(a)\int_{-\infty}^{\infty} \delta(x-a) = f(a) \tag{A3.11}$$

A3.2 THE ERROR FUNCTION

The error function is of the greatest importance in diffusion theory. It takes the integral form

$$\theta(x) = \operatorname{erf} x = \frac{2}{\sqrt{\pi}}\int_0^x \exp(-\xi^2)\,\mathrm{d}\xi \tag{A3.12}$$

with the properties that

$$\operatorname{erf}\infty = 1 \quad \text{and} \quad \operatorname{erf}(-x) = -\operatorname{erf}(x) \tag{A3.13}$$

The error function complement is defined as

$$\operatorname{erfc} x = 1 - \operatorname{erf} x = 1 - \frac{2}{\sqrt{\pi}}\int_x^{\infty} \exp(-\xi^2)\,\mathrm{d}\xi \tag{A3.14}$$

For $x < 1$ the exponential can be expanded as a series and integrated so

$$\operatorname{erf} x = \frac{2}{\sqrt{\pi}}\sum_{r=0}^{\infty}\frac{(-1)^n x^{2n+1}}{(2n+1)n!} \tag{A3.15}$$

The first order term is

$$\operatorname{erf} x = \frac{2}{\sqrt{\pi}}x \tag{A3.16}$$

which is equivalent to the approximation used in deriving equation (1.63) with

$$x \rightarrow x/2(Dt)^{1/2}$$

For $x > 1$ one can develop the series

$$\operatorname{erfc} x = \frac{1}{\sqrt{\pi}}\frac{\exp(-x^2)}{x}\left(\frac{1}{2^0x^0} - \frac{1}{2^1x^2} + \frac{1.3}{2^2x^4} - \frac{1.3.5}{2^3x^6} + \ldots\right) \tag{A3.17}$$

so for very large x

$$\operatorname{erfc} x = \frac{1}{\sqrt{\pi}}\frac{\exp(-x^2)}{x} \tag{A3.18}$$

Expressions and tables for the derivatives $d^n\,\mathrm{erf}/\mathrm{d}x^n$ and integrals (i^n erf) are given in Carslaw and Jaeger.[7] However, approximate formulas such as equation (1.61) are adequate for most purposes. Note that when in an infinite diffusion couple solution, $x \rightarrow x/2(Dt)^{1/2}$, (cf equation (A3.1)) that the derivative of the initial condition step function is proportional to the δ-function, i.e.,

Table A3.1 The error function complement

x	$1 - \operatorname{erf} x$	x	$1 - \operatorname{erf} x$
0	1·0	1·0	0·157 299
0·05	0·943 628	1·1	0·119 795
0·1	0·887 537	1·2	0·089 686
0·15	0·832 004	1·3	0·065 992
0·2	0·777 297	1·4	0·047 715
0·25	0·723 674	1·5	0·033 895
0·3	0·671 373	1·6	0·023 652
0·35	0·620 618	1·7	0·016 210
0·4	0·571 608	1·8	0·010 909
0·45	0·524 518	1·9	0·007 210
0·5	0·479 500	2·0	0·004 678
0·55	0·436 677	2·1	0·002 979
0·6	0·396 144	2·2	0·001 863
0·65	0·357 971	2·3	0·001 143
0·7	0·322 199	2·4	0·000 689
0·75	0·288 844	2·5	0·000 407
0·8	0·257 899	2·6	0·000 236
0·85	0·229 332	2·7	0·000 134
0·9	0·203 092	2·8	0·000 075
0·95	0·179 109	2·9	0·000 041
		3·0	0·000 022

$$\frac{\mathrm{d}}{\mathrm{d}x}\left[\operatorname{erf}\frac{x}{2(Dt)^{1/2}}\right]_{t\to 0} = \frac{2}{2\sqrt{\pi D t}}\exp\left[\frac{x^2}{2(Dt)^{1/2}}\right]_{t\to 0} = 2\delta(x) \tag{A3.19}$$

Table A3.1 gives $\operatorname{erfc} x$ in numerical form.

The error function $\theta(x)$ is closely related to the error integral $\phi(x)$ from probability theory

$$\phi(x) = \frac{1}{(2\pi)^{1/2}}\int_0^x \exp(-\eta_2/2)\,\mathrm{d}\eta \tag{A3.20}$$

Evidently

$$\theta\left(\frac{x}{\sqrt{2}}\right) = 2\phi(x) \tag{A3.21}$$

Probability paper, which linearizes $\phi(x)$, is widely available (for example Keuffel and Esser, stationery suppliers; Fig. 2.27). Normalized diffusion data for an infinite couple with a constant diffusion coefficient appears as a straight line on such a plot. In using probability paper quantitatively to extract average diffusion constants from the slope one must take care to recognize the transformation (A3.21). Seith[13] presents a nomograph for extracting diffusion coefficients from experimental data.

A3.3 STIRLING'S FORMULA

This formula is used to simplify factorials (!) of large numbers. It states the approximation

$$\ln n! = \left(n + \frac{1}{2}\right) \ln n - n + \frac{1}{2} \ln 2\pi + \frac{1}{12n} - O(1/n^3) \qquad (A3.22)$$

which for n of molecular magnitudes is accurately approximated by

$$\ln n! = n(\ln n - 1) \qquad (A3.23)$$

A3.4 SPHERICAL HARMONIC SOLUTIONS OF LAPLACE'S EQUATION

It may be easily verified that equation (A2.11) in spherical-polar coordinates has a class of solutions of the form

$$C = C_0 + \sum_l A_l r^l Y_{lm}(0, \psi) + \sum_l B_l r^{-l-1} Y_{lm}(0, \psi) \qquad (A3.24)$$

where l is an integer, and Y satisfies the operator equation (equation (12.4) defines Λ)

$$\Lambda Y_{lm} = -l(l+1) Y_{lm} \qquad (A3.25)$$

The so-called 'spherical harmonic' Y has the form

$$Y_{lm}(\theta, \psi) = \frac{1}{N_{lm}} \exp(im\psi) \sin^{|m|}\theta \frac{d^{|m|} P_l(\cos\theta)}{d(\cos\theta)^{|m|}} \qquad |m| \leqslant 1 \qquad (A3.26)$$

where

$$\frac{1}{N_{lm}} = \pm \frac{1}{(2\pi)^{1/2}} \left[\frac{2l+1}{2} \frac{(l-|m|)!}{(l+|m|)!}\right]^{1/2} \qquad (A3.27)$$

(+ sign for $m \leqslant 0$ and $(-1)^m$ sign for $m > 0$) and the Legendre polynomials are

$$P_l(x) = \frac{1}{2^l l!} \frac{d^l}{dx^l} (x^2 - 1)^l \qquad (A3.28)$$

The following are the first few spherical harmonics

$$Y_{00} = \frac{1}{(4\pi)^{1/2}} \qquad Y_{10} = \left(\frac{3}{4\pi}\right)^{1/2} \cos\theta \qquad (A3.29)$$

$$Y_{1,\pm1} = \mp \left(\frac{3}{8\pi}\right)^{1/2} \sin\theta \exp(\pm i\psi) \qquad (A3.30)$$

$$Y_{20} = \left(\frac{5}{4\pi}\right)^{1/2} \left(\frac{3}{2}\cos^2\theta - \frac{1}{2}\right) \qquad (A3.31)$$

$$Y_{2\pm1} = \mp \left(\frac{15}{8\pi}\right)^{1/2} \sin\theta \cos\theta \exp(\pm i\psi) \qquad (A3.32)$$

$$Y_{2,\pm2} = \frac{1}{4}\left(\frac{15}{2\pi}\right)^{1/2} \sin^2\theta \exp(\pm 2i\psi) \qquad (A3.33)$$

$$Y_{30} = \left(\frac{7}{4\pi}\right)\left(\frac{5}{2}\cos^3\theta - \frac{3}{2}\cos\theta\right) \qquad (A3.34)$$

A full account is to be found in Hobson.[14]

APPENDIX 4

Matrices and determinants

A4.1 MATRICES

Matrices are ordered arrays of numbers which possess something like the rules of combination of ordinary numbers. In multicomponent diffusion theory they are usually of the square ($n \times n$) form

$$\begin{bmatrix} a_{11} & a_{12} & \dots & a_{1n} \\ a_{21} & a_{22} & & \\ \vdots & & & \\ a_{n1} & & & a_{nn} \end{bmatrix} \qquad \text{(A4.1)}$$

or of the column form, which represents a vector

$$\begin{bmatrix} b_1 \\ b_2 \\ \vdots \\ b_n \end{bmatrix} \qquad \text{(A4.2)}$$

Matrices can be multiplied if the number of columns in one is equal to the number of rows in the others. For example, equations (4.1) and (4.2) can be multiplied. The general rule of multiplication used in diffusion theory is

$$[a_{ik}] \cdot [b_{kj}] = [c_{ij}] = \left[\sum_{k=1}^{n} a_{ik} b_{kj}\right] \qquad \text{(A4.3)}$$

where the symbols represent individual ordered elements in the initial and product matrices, respectively. For example, in reference to multicomponent diffusion theory let $a_{ik} = D_{ik}$, the diffusion matrix, $b_{kj} = b_k$, the column gradient matrix (or vector $-\nabla C_k$) and $c_{ij} = J_i$ the column flux matrix, then we see that each of the Fickian flux matrix elements

$$J_i = -\sum_{k=1}^{n} D_{ik} \nabla C_k \qquad \text{(A4.4)}$$

is properly represented by the matrix equation

$$\begin{bmatrix} J_1 \\ J_2 \\ \vdots \\ J_n \end{bmatrix} = \begin{bmatrix} D_{11} D_{12} \ldots D_{1n} \\ \vdots \\ \\ D_{n1} \ldots\ldots D_{nn} \end{bmatrix} \begin{bmatrix} -\nabla C_1 \\ -\nabla C_2 \\ \\ -\nabla C_n \end{bmatrix} \tag{A4.5}$$

in accord with the multiplication rule (A4.3). This can be written in short notation as

$$J = D(-\nabla C) \tag{A4.6}$$

Now it is clear in the theory of equations that if the fluxes and gradients are independent the set (4.4) can be solved simultaneously for the $(-\Delta C)$ as linear functions J, or in matrix notation

$$(-\nabla C) = D^{-1} J \tag{A4.7}$$

An examination of the conventional solution of the inversion problem suggests a simple formal approach to the problem. Software for inverting matrices is available for most micro, and of course, mainframe computers. These involve the evaluation of matrix determinants and determinants of submatrices (or minors). The minor of an element of a matrix is the submatrix formed by deleting the column and row in which the given element appears. The evaluation of determinants is presented in § A4.3.

There are a number of square matrices of general interest. A *diagonal* matrix is a square matrix which has elements only on the main diagonal. If all the elements are the same then the matrix is designated as *scalar*. The particular diagonal matrix which has unity as all elements is called the *identity* or *unit* matrix I. For every square matrix

$$AI = IA = A \tag{A4.8}$$

This is easy to verify via rule (A4.3). The identity matrix is often denoted by the Krönecker delta symbol δ_{ij} which is defined by

$$\delta_{ij} \begin{cases} = 0 & \text{when } i \neq j \\ \\ = 1 & \text{when } i = j \end{cases} \tag{A4.9}$$

A symmetric matrix is one where $a_{ij} = a_{ji}$ for all i,j. A *skew-symmetric* or *anti-symmetric* matrix satisfies $a_{ij} = -a_{ji}$ for all i,j. This definition insists that all diagonal elements $a_{ii} = 0$. The *transpose* of a matrix A is one obtained by interchanging the rows and the columns and is designated by A^{T}. For symmetric matrices

$$A = A^{\mathrm{T}} \tag{A4.10}$$

and for skew-symmetric matrices

$$A^{\mathrm{T}} = -A \tag{A4.11}$$

A4.2 DETERMINANTS

The value of the determinant of a 2×2 square matrix

$$\begin{vmatrix} a_{11} a_{12} \\ a_{21} a_{22} \end{vmatrix} \tag{A4.12}$$

is

$$\det A = a_{11}a_{22} - a_{12}a_{21} \quad \text{(A4.13)}$$

The value of the 3×3 determinant

$$\begin{vmatrix} a_{11}a_{12}a_{13} \\ a_{21}a_{22}a_{23} \\ a_{31}a_{32}a_{33} \end{vmatrix} \quad \text{(A4.14)}$$

is defined as

$$\det A = a_{11}a_{22}a_{33} + a_{12}a_{23}a_{31} + a_{13}a_{21}a_{32} - a_{31}a_{22}a_{13} - a_{32}a_{23}a_{11} - a_{33}a_{21}a_{12} \quad \text{(A4.16)}$$

which appears in the solution of linear equation sets like equation (A4.7). Now $\det A$ can be factored to

$$\det A = a_{11}(a_{22}a_{33} - a_{32}a_{23}) - a_{12}(a_{21}a_{33} - a_{31}a_{23}) + a_{13}(a_{21}a_{32} - a_{31}a_{22}) \quad \text{(A4.17)}$$

$$= a_{11}\begin{vmatrix} a_{22}a_{23} \\ a_{32}a_{33} \end{vmatrix} - a_{12}\begin{vmatrix} a_{21}a_{23} \\ a_{31}a_{33} \end{vmatrix} + a_{13}\begin{vmatrix} a_{21}a_{22} \\ a_{31}a_{32} \end{vmatrix} \quad \text{(A4.18)}$$

This suggests how the determinant of a square matrix of any order can be decomposed into 2×2 determinants. If we let M_{ij} be any minor and A_{ij} its cofactor such that

$$A_{ij} = (-1)^{i+j}M_{ij} \quad \text{(A4.19)}$$

to account for the cyclic sign change which occurs in relation (A4.18), then by generalization of relation (A4.18) it can be stated that the value of a determinant is equal to the sum of the products of each element of any row and its cofactor (Kramer's rule). Obviously this process can be repeated so that a determinant of arbitrary order can ultimately be reduced to a sum of 2×2 determinants. Returning now to the evaluation of det D^{-1} in relation (A4.7), for each element can be written

$$D_{ij}^{-1} = |D_{ij}|/|D| \quad \text{(A4.20)}$$

where $|D|$ is the determinant of the matrix of coefficients $[D_{ij}]$ and $|D_{ij}|$ is the determinant of the minor corresponding to the element D_{ij}. A pair of $n \times n$ determinants satisfy an important product theorem which is stated

$$|AB| = |A| \cdot |B| \quad \text{(A4.21)}$$

That is, the determinant of a product matrix is equal to the product of the determinants of the constituent matrices.

A4.3 EIGENVALUES AND EIGENVECTORS

With each square matrix A of order n can be associated a polynomial function

$$f(\lambda) = |A - \lambda I| = \begin{vmatrix} a_{11} - \lambda & a_{12} & a_{1n} \\ & a_{22} - \lambda & \\ a_{n1} & \cdots & a_{nn} - \lambda \end{vmatrix} \quad \text{(A4.22)}$$

which by evaluation of the determinant yields the form

$$C_0\lambda^n + C_1\lambda^{n-1} + \ldots C_{n-1}\lambda + C_n = 0 \qquad \text{(A4.23)}$$

This is called the characteristic equation of matrix A. The n roots of this are called the eigenvalues of the matrix. If we define the trace of the matrix as the sum of the diagonal elements, i.e.

$$\operatorname{tr} A = \sum_{i=1}^{n} a_{ii} \qquad \text{(A4.24)}$$

then a number of expressions are simplified. Consider the characteristic equation

$$\begin{vmatrix} a_{11} - \lambda & a_{12} \\ a_{21} & a_{22} - \lambda \end{vmatrix} = 0 \qquad \text{(A4.25)}$$

or

$$\lambda^2 - \operatorname{tr} A\lambda + \det A = 0 \qquad \text{(A4.26)}$$

This possesses the roots (see §4.6.1)

$$\lambda = \tfrac{1}{2}\{\operatorname{tr} A \pm [(\operatorname{tr} A)^2 - 4\det A]^{1/2}\} \qquad \text{(A4.27)}$$

It can easily be proved in general that

$$\sum_{i=1}^{n} \lambda_i = \operatorname{tr} A \quad \text{and} \quad \prod_{i} \lambda_i = \det A \qquad \text{(A4.28)}$$

The eigenvectors are column vectors X_i such that the matrix product

$$[A - \lambda_i I]X_i = 0 \qquad \text{(A4.29)}$$

This corresponds to a set of n linear homogeneous equations in X_i (cf equations (A4.4) and (A4.5)). Their geometric interpretation appears in specific examples within the text. A number of important theorems follow:

1. if the eigenvalues of a matrix are distinct, the eigenvectors are linearly independent
2. if A is a real symmetric matrix, then the eigenvalues of A are real
3. if A is a real symmetric matrix, then the eigenvectors of A associated with distinct eigenvalues are mutually orthogonal vectors.

If we write equation (A4.29) in the form

$$AX_i = \lambda_i X_i \qquad i = 1, 2, \ldots n \qquad \text{(A4.30)}$$

and recognize that λ_i can be written as the diagonal matrix

$$\Lambda = \begin{bmatrix} \lambda_1 & 0 & 0 \\ 0 & \lambda_1 & \\ 0 & & \ddots \; \lambda_n \end{bmatrix} \qquad \text{(A4.31)}$$

then equation (A4.30) can be written as

$$AX = X\Lambda \qquad \text{(A4.32)}$$

If the eigenvalues are distinct so that the X_i are independent, equation (A4.32) can be multiplied from the left by X^{-1}, i.e.

$$X^{-1}AX = \Lambda \tag{A4.33}$$

This operation on a matrix by its eigenvectors is called diagonalization.

A4.4 QUADRATIC FORMS

In Chapter 3 we introduced the entropy production rate in the bilinear form (equation (3.58))

$$T\sigma = \sum_i J_i X_i \tag{A4.34}$$

Together with the flux equation (3.62), i.e.

$$J_i = \sum_k L_{ik} X_k \tag{A4.35}$$

Substitution yields the quadratic form

$$T\sigma = \sum_i \sum_k L_{ik} X_k X_i \tag{A4.36}$$

It is assumed here that the square matrix L is symmetric. If it is not it can always be symmetrized by redefining the off-diagonal elements as L_{ik} and L_{ki} with the symmetric values

$$\tfrac{1}{2}(L_{ik} + L_{ki}) \tag{A4.37}$$

If this is positive for all possible values of the X_i one refers to the quadratic form or the matrix of coefficients as being *positive definite*. By completing squares it can be proven that:

1. a necessary and sufficient condition for a real symmetric matrix A to be positive definite is that all the leading principal minors of A be positive. For example, with

$$L = \begin{bmatrix} L_{11} L_{12} \\ L_{22} L_{22} \end{bmatrix} \tag{A4.38}$$

$$L_{11} > 0 \quad L_{22} > 0 \quad \det L > 0 \tag{A4.39}$$

   These are not, however, independent
2. if A is a real symmetric *semidefinite* (i.e. non-negative) matrix, then $|A| = 0$
3. Every real *indefinite* form assumes positive as well as negative values.

Full accounts of the subject matter are to be found in Pettofrezzo[15] and Eves.[16]

APPENDIX 5

Finite difference algorithm for diffusion equations

The coupled set for ternary systems

$$\frac{\partial C_1}{\partial t} = \frac{\partial}{\partial x}\left(D_{11}\frac{\partial C_1}{\partial x}\right) + \frac{\partial}{\partial x}D_{12}\frac{\partial C_2}{\partial x} \tag{A5.1}$$

and

$$\frac{\partial C_2}{\partial t} = \frac{\partial}{\partial x}\left(D_{21}\frac{\partial C_1}{\partial x}\right) + \frac{\partial}{\partial x}D_{22}\frac{\partial C_2}{\partial x} \tag{A5.2}$$

with variable coefficients $D(C_1, C_2)$ are considered for finite and infinite systems. For the former the initial and boundary conditions can be written as

$$C_i(x_+, 0) = C_{i0} \qquad C_i(x_-, 0) = C_{i1} \tag{A5.3}$$

and

$$C_i(x_0, t) = C_{i0} \qquad C_i(-x_0, t) = C_{i1} \tag{A5.4}$$

together with

$$\partial C_i/\partial x = 0 \quad \text{at} \quad x = \pm x_0 \tag{A5.5}$$

For an infinite system we let $x_0 \rightarrow \infty$ in equation (A5.5).

To solve equations (A5.1) and (A5.2) by the finite difference method the range in x and t is divided into intervals Δx and Δt, respectively, the concentrations at time $I\Delta t$ at the points $(J-1)\Delta x$, $J\Delta x$, $(J+1)\Delta x$, are denoted by $C(I, J-1)$, $C(I, J)$, $C(I, J+1)$ respectively, and the concentration at time $(I+1)\Delta t$ and point $J\Delta x$, is denoted by $C(I+1, J)$. Then all the the derivatives in equations (A5.1) and (A5.2) are replaced by finite difference approximations using a Taylor series expansion so that, for example, equation (A5.1) becomes after differentiation

$$\begin{aligned} C_1(I+1, J) = {} & C_1(I, J) + [\Delta t/(\Delta x)^2] \\ & \times \{[D_{11}(J+1) - D_{11}(J)][C_1(I, J+1) - C_1(I, J)] \\ & + D_{11}(J)[C_1(I, J+1) - 2C_1(I, J) + C_1(I, J-1)] \\ & + [D_{12}(J+1) - D_{12}(J)] \times [C_2(I, J+1) - C_2(I, J)] + D_{12}(J) \\ & \times [C_2(I, J+1) - 2C_2(I, J) + C_2(I, J-1)]\} \end{aligned} \tag{A5.6}$$

The equation for C_2 is obtained by exchanging subscripts. This will be designated as equation (A5.7).

By Taylor expansion at the boundaries about the intervals Δx, $2\Delta x$, conditions (A5.3)–(A5.5) become

$$\begin{aligned}
C_1(I+1,1) &= 4C_1(I+1,2) - C_1(I+1,3)/3,\\
C_2(I+1,1) &= 4C_2(I+1,2) - C_2(I+1,3)/3,\\
C_1(I+1,M) &= 4C_1(I+1,M-1) - C_1(I+1,M-2)/3,\\
C_2(I+1,M) &= 4C_2(I+1,M-1) - C_2(I+1,M-2)/3
\end{aligned} \tag{A5.8}$$

For infinite systems, M is chosen large enough that diffusion does not reach points $\pm M\Delta x$ during time t. For finite couples, points $\pm M\Delta x = x_0$ are the actual ends of the couple.

Using equations (A5.6)–(A5.7) the computations are carried out at intervals of Δx over the range of given x and this is repeated in steps of Δt to the given t. Before each step at Δt, D_{11}, D_{12}, D_{22}, and D_{21} are first calculated from previous concentrations obtained at time I. To minimize errors entering into the approximations above and to obtain accurate results, proper choices of magnitudes for Δx and Δt are required, particularly at the early stages of the diffusion when a discontinuity exists at the interface. All test solutions for infinite couples rapidly converge to a parabolic penetration behaviour, thus establishing the stability of the finite difference methodology. Their stability can be further established via the employment of successively finer subdivisions and by checking the convergence using the h^2 extrapolation.[17]

APPENDIX 6

Data sources

In the following we present a short list of references which contain useful compendia of diffusion data or formulae for estimating such data when direct measurements are not available.

1 SOLIDS

Diffusion in and through Solids, by R. M. Barrer (Cambridge University Press; 1951).

Diffusion in Metallen, by W. Seith (Springer, Berlin; 1955).

La Diffusion dans les Solides, Vol. II, by Y. Adda and J. Philibert (Presses Universitaire de France; 1966).

Metals Reference Book, by C. J. Smithells (Butterworths, London; 1967), p. 654.

Diffusion in Multicomponent Systems, by J. S. Kirkaldy, in *Advances in Materials Research*, Vol. 4, edited by H. Herman (John Wiley, New York; 1970).

Diffusion in Semiconductors, by B. L. Sharma (Clausthal-Zellerfeld, Germany; 1970).

Diffusion and Point Defects in Semiconductors, by B. I. Boltaks (Nauka, Leningrad; 1972).

Diffusion Mechanisms in Ferrous Alloys, by M. A. Krishtol (Metallurgiya, Moscow; 1972).

Hydrogen Diffusion in Metals, in *Diffusion in Solids, Recent Developments*, by J. Volkel and G. Alefeld, edited by A. S. Nowick and J. J. Burton (Academic Press, New York; 1975), p. 232.

Diffusion in Reactor Materials, by G. B. Fedorov and E. A. Smirnov (Atomizat, Moscow; 1978), translation by NBS.

Correlations for Diffusion Coefficients, by A. M. Brown and M. F. Ashby, in *Acta Met.*, **28**, 1085 (1980).

Smithell's Metals Reference Book (sixth edition), edited by E. A. Brandes (Butterworths, London; 1983), pp. 13-1–13-97.

Tracer Diffusion in Concentrated Alloys, in Diffusion in Crystalline Solids, by H. Bakker, edited by G. E. Murch and A. S. Nowick (Academic Press, Orlando; 1984), p. 330.

2 LIQUIDS

Transport Phenomena, by R. B. Bird, W. E. Stewart and E. N. Lightfoot (John Wiley, New York; 1960).

Diffusion and Heat Conduction in Liquids, by H. J. V. Tyrrell (Butterworths, London; 1961).

Multicomponent Diffusion, by E. L. Cussler (Elsevier, Amsterdam; 1976).
Viscosity and Diffusivity, A Predictive Treatment, by J. H. Hildebrand (John Wiley, New York; 1977).
Diffusion in Liquids, by H. J. V. Tyrrell and K. R. Harris (Butterworths, London; 1984).
Diffusion: Mass Transfer in Fluid Systems, by E. L. Cussler (Cambridge University Press, 1984).

3 GENERAL

Diffusion in Solids, Liquids, Gases, by W. Jost (Academic Press, New York; 1952).
Diffusion and Defect Data (journal), Volumes 1–50.
Diffusion in Polymers, edited by J. Crank and G. S. Park (Academic Press, London, 1968).
Zeolites and Clay Minerals as Sorbents and Molecular Sieves, by R. M. Barrer (Academic Press, London; 1978), Chapter 6.

REFERENCES TO APPENDICES

1 L. S. DARKEN AND R. N. GURRY: 'The physical chemistry of metals', 1953, New York, McGraw-Hill
2 I. PRIGOGINE AND R. DEFAY: 'Chemical thermodynamics', 1954, New York, Longmans
3 J. H. VAN'T HOFF: *Z. Phys. Chem.*, 1887, **1**, 481. Translation: in 'Cell membrane permeability and transport', (ed. G. R. Kepner), 1979, Stroudsburg PA, Dowden, Hutchinson and Ross
4 W. NERNST: *ibid.*, 1888, **2**, 613. Translation: *ibid.*
5 J. BERNSTEIN: *Pflügers Arch.*, 1902, **92**, 521. Translation: *ibid.*
6 R. B. BIRD, W. E. STEWART, AND E. N. LIGHTFOOT: 'Transport phenomena', 715 *et seq.*, 1960, New York, John Wiley
7 H. S. CARSLAW AND J. C. JAEGER: 'Conduction of heat in solids', 2nd edn, 56, 297, 455 *et seq.*, 482, 1959, Oxford, Clarendon Press
8 Y. ADDA AND J. PHILIBERT: 'La diffusion dans les solides, Vol. 1', 1966, Paris, Presse Universitaires de France
9 F. OBERHETTINGER AND L. BADII: 'Tables of Laplace transforms', 1973, New York, Springer-Verlag
10 F. OBERHETTINGER: 'Fourier expansions: a collection of formulas', 1973, New York, Academic Press
11 F. OBERHETTINGER: 'Tabellen zur Fourier transformation', 1957, Berlin, Springer-Verlag
12 L. I. SCHIFF: 'Quantum mechanics', 51–2, 1949, New York, McGraw-Hill
13 W. SEITH: 'Diffusion in Metallen', 1955, Berlin, Springer
14 E. W. HOBSON: 'The theory of spherical and ellipsoidal harmonics', 1931, Cambridge University Press
15 A. J. PETTOFREZZO: 'Matrices and transformations', 1966, N.J., Prentice-Hall
16 H. EVES: 'Elementary matrix theory', 1966, Boston, Allyn and Bacon
17 J. CRANK: 'The mathematics of diffusion', 201, 1970, Oxford University Press

Classified bibliography

1 CHEMICAL AND STATISTICAL THERMODYNAMICS

Mixtures, by E. A. Guggenheim (Oxford University Press; 1952).

Thermodynamics of Alloys, by C. Wagner (Addison-Wesley, Reading, Mass; 1952).

Chemical Thermodynamics, by I. Prigogine and R. Defay (Longmans Green, London; 1954).

The Statistical Mechanics of Simple Liquids, by S. A. Rice and P. Gray (Interscience, New York; 1965).

Selected Topics in Statistical Mechanics, by J. G. Kirkwood (Gordon and Breach, New York; 1967).

Thermodynamics: An Advanced Treatise for Chemists and Physicists, (fifth-edition) by E. A. Guggenheim (North-Holland, Amsterdam; 1967).

Statistical Thermodynamics: Vol. 1, by A. Munster (Springer-Verlag, Berlin; 1969).

Thermodynamics of Solids, by R. A. Swalin (John Wiley, New York; 1972).

Physical Chemistry of Metals, by L. S. Darken and R. W. Gurry (McGraw-Hill, New York; 1953).

Statistical Physics, by L. Landau and E. M. Lifshitz (Pergamon Press, New York; 1980).

2 THERMODYNAMICS OF IRREVERSIBLE PROCESSES

Thermodynamics of the Steady State, by K. G. Denbigh (Methuen, London 1951).

Introduction to the Thermodynamics of Irreversible Processes, by I. Prigogine (Charles C. Thomas, Springfield Ill.; 1955).

Nonequilibrium Thermodynamics, by D. D. Fitts (McGraw-Hill, New York; 1962).

Non-Equilibrium Thermodynamics, by S. R. de Groot and P. Mazur (North-Holland; Amsterdam, 1962).

Nonequilibrium Thermodynamics in Biophysics, by A. Katchalsky and P. F. Curran (Harvard University Press, Cambridge, Mass.; 1965).

Thermodynamics of Steady States, by R. J. Tykodi (Macmillan, New York; 1967).

Thermodynamics of Irreversible Processes, by R. Haase (Addison-Wesley, Reading, Mass.; 1969).

Non-equilibrium Thermodynamics and its Statistical Foundations, by H. J. Kreuzer, (Clarendon Press, Oxford; 1981).

3 DIFFUSION IN SOLIDS

Electronic Processes in Ionic Crystals, by N. F. Mott and R. W. Gurney (Oxford University Press; 1940).

Elasticity and Anelasticity in Solids, by C. Zener (Chicago University Press; 1948).
Atom Movements, edited by J. H. Holloman (ASM, Cleveland; 1951).
Diffusion in and through Solids, by R. M. Barrer (Cambridge University Press; 1951).
Imperfections in Nearly Perfect Crystals, edited by W. Shockley (John Wiley, New York; 1952).
Diffusion in Solids, Liquids, Gases, by W. Jost (Academic Press, New York; 1952).
Diffusion in Metallen, by W. Seith (Springer-Verlag, Berlin; 1955).
Diffusion in Metals and Alloys in the Solid Phase, by S. D. Gertsriken and I. Ya Dekhtyar (Fizmatgiz, Moscow; 1960).
Diffusion in Solids, by P. G. Shewmon (McGraw-Hill, New York; 1963).
The Chemistry of Imperfect Crystals, by F. A. Kroger (North-Holland, Amsterdam; 1964).
R. E. Howard and A. B. Lidiard, 1964, *Rep. Prog. Phys.*, **27**, 161.
Atomic Migration in Crystals, by L. A. Girifalco (Blaisdell, New York; 1964).
Diffusion in Body-Centred Cubic Metals, (ASM, Metals Park, OH; 1965).
La Diffusion dans Les Solides, Vols I, II, by Y. Adda and J. Philibert (Presses Universitaires de France, Paris; 1966).
Metals Reference Book, by C. J. Smithells (Butterworths, London; 1967).
Lattice Defects and their Interactions, edited by R. R. Hasiguti (Gordon and Breach, New York, 1968).
Diffusion Kinetics for Atoms in Crystals, by J. R. Manning (D. Van Nostrand, Princeton, NJ; 1968).
Correlation Effects in Diffusion in Solids, in *Physical Chemistry — an Advanced Treatise*, Vol. 10, by A. D. Le Claire, edited by H. Eyring, D. Henderson and W. Jost (Academic Press, New York; 1970).
Vacancies and Interstitials in Metals, edited by A. Seeger (North-Holland, Amsterdam; 1970).
Diffusion in Semiconductors, by B. L. Sharma (Clausthal-Zellerfeld, Germany; 1970).
Tracer Diffusion Data for Metals, Alloys and Simple Oxides, by J. Askell (Plenum Press, New York; 1970).
Point Defects in Metals, by A. C. Damask (Gordon and Breach, New York; 1971).
Atomic Transport in Solids and Liquids, edited by A. Lodding and T. Lagerwall (Verlag der Zeitschrift fur Naturforschung, Tubingen; 1971).
Anelastic Relaxation in Crystalline Solids, by A. S. Nowick and B. S. Berry (Academic Press, New York; 1972).
Defects in Crystalline Solids, by B. Henderson (Edward Arnold, London; 1972).
Points Defects and Diffusion, by C. P. Flynn (Clarendon Press, Oxford; 1972).
Interatomic Potentials and Simulation of Lattice Defects, edited by P. C. Gehlen, J. R. Beeler Jr. and R. I. Jaffee (Plenum Press, New York; 1972).
Diffusion and Point Defects in Semiconductors, by B. I. Boltaks (Nauka, Leningrad; 1972).
Diffusion Mechanisms in Ferrous Alloys, by M. A. Krishtal (Metallurgiya, Moscow; 1972).
Surface Self-Diffusion of Metals, by G. Neumann (Diffusion Information Center, Bay Village, OH; 1972).

Interdiffusion Processes in Alloys, by I. B. Borovskii, K. P. Gurov, I. D. Marchukova and Yu. E. Ugaste (Nauka, Moscow; 1073).
Thermodynamics and Kinetics of Diffusion in Solids, by B. S. Bokshstein, S. Z. Bokshstein and A. A. Zhukhovitskii (Metallurgiya, Moscow; 1973). (NBS translation).
Atomic Diffusion in Semiconductors, by D. Shaw (Plenum Press, London; 1973).
Geochemical Transport and Kinetics, edited by A. W. Hofmann, B. J. Giletti, H. S. Yoder Jr. and R. A. Yung (Carnegie Institution of Washington; 1974).
Introduction to Diffusion in Semiconductors, by B. Tuck (P. Peregrinus, Stevenage; 1974).
Diffusion Processes in Solid-Phase Welding of Materials, by V. M. Fal'chenko, L. N. Larikov and V. R. Ryabov (Mashinostroenie, Moscow; 1975) (NBS translation).
Diffusion in Solids: Recent Development, edited by A. S. Nowick and J. J. Burton (Academic Press, New York; 1975).
Point Defects in Solids, edited by J. H. Crawford and L. M. Slifkin (Plenum Press, New York; 1975).
The Mathematical Theory of Diffusion and Reaction in Permeable Catalysis: The Theory of the Steady State, VI, by R. Aris (Clarendon Press, Oxford; 1975).
Low Temperature Diffusion and Application to Thin Films, A. Gangulec, P. S. Ho, and K. N. Tu (Elsevier–Sequoia, New York; 1975).
Diffusion Processes in Ordered Alloys, by L. N. Larikov, V. V. Geichenko, and V. M. Fal'chenko (Naukova Dumka, Kiev (NBS translation); 1975).
Solid State Diffusion, by J. P. Stark (John Wiley, New York; 1976).
Defects and Diffusion in Solids, by S. Mrowec (Elsevier, Amsterdam; 1980).
Impurity Diffusion Processes in Silicon, by F. F. Y. Wang (North-Holland, Amsterdam; 1981).
Mass Transport in Solids, edited by F. Bénière and C. R. A. Catlow (Plenum Press, New York; 1983).
Smithell's Metals Reference Book (sixth edition), edited by E. A. Brandes (Butterworths, London; 1983), Chapter 13.
Mass Transport in Solids, NATO Advanced Study Institute (Plenum Press, New York, 1983).
Transport in Nonstoichiometric Compounds, edited by G. Petot-Ervas, Hj. Matske and C. Monty, (North Holland, Amsterdam; 1984).
Diffusion in Crystalline Solids, edited by G. E. Murch and A. S. Nowick (Academic Press, Orlando, FA; 1984).
Defects and Defect Processes in Non-metallic Solids, by H. Hayes and A. M. Stoneham (John Wiley, New York, 1985).
Atomic Transport in Concentrated Alloys and Intermetallic Compounds (AIME, Warrendale, PA; 1985).
International Seminar on Solute-Defect Interaction: Theory and Experiment, edited by S. Saimoto, G. R. Purdy, and G. V. Kidson (Pergamon Press, Toronto, 1986).
A. R. Allnatt and A. B. Lidiard, 1986, *Rep. Prog. Phys.*, **50**, 373

4 DIFFUSION IN LIQUIDS, GASES AND POLYMERS

The Kinetic Theory of Gases, by O. E. Meyer (translated by R. E. Baynes) (Longmans Green, London; 1899).
Dissertation, D. Enskog, Uppsala University, 1917.

The Theory of Rate Processes, by S. Glasstone, K. J. Laidler and H. Eyring (McGraw-Hill, New York; 1941).
Kinetic Theory of Liquids, by Ya. I. Frenkel (Izvestiya ANSSSR; 1945).
Diffusion in Solids, Liquids, Gases, by W. Jost (Academic Press, New York; 1952).
Transport Phenomena, by R. B. Bird, W. E. Stewart and E. N. Lightfoot (John Wiley, New York; 1960).
Diffusion and Heat Flow in Liquids, by H. J. V. Tyrrell, (Butterworths, London; 1961).
The Structure and Properties of Ionic Melts, by R. W. Laity (Aberdeen University Press, Aberdeen; 1962).
Molecular Theory of Gases and Liquids (revised edition), by J. O. Hirschfelder, C. F. Curtiss and R. B. Bird (John Wiley, New York; 1964).
Diffusion Processes, by M. H. Jacobs (Springer-Verlag, Berlin; 1967).
Diffusion in Polymers, edited by J. Crank and G. S. Park (Academic Press, London; 1968).
The Mathematical Theory of Non-Uniform Gases (third edition), by S. Chapman and T. G. Gowling (Cambridge University Press., London; 1970).
Transport Phenomena in Liquid Metals and Semiconductors, by D. K. Belashchenko (Atomizdat, Moscow; 1970).
Atomic Transport in Solids and Liquids, edited by A. Lodding and T. Lagerwall (Verlag der Zeitschrift fur Naturforschung, Tubingen; 1971).
Modern Theory of Polymer Solutions, by H. Yamakawa (Harper and Row, New York; 1971).
Diffusion Processes, edited by J. N. Sherwood, A. V. Chadwick, W. M. Muir and F. L. Swinton (Gordon and Breach, London; 1971).
Glass Science, by R. H. Doremus (Wiley–Interscience, New York; 1973).
Geochemical Transport and Kinetics, edited by A. W. Hofmann, B. J. Giletti, H. S. Yoder Jr. and R. A. Yund (Carnegie Institution of Washington, 1974).
Liquid State Chemical Physics, by R. O. Watts and L. J. McGee (John Wiley, New York; 1976).
Viscosity and Diffusivity, A Predictive Treatment, by J. H. Hildebrand (John Wiley, New York; 1977).
The Properties of Liquids and Gases (third edition), by R. C. Reid, J. M. Prauznitz and T. K. Sherwood (McGraw-Hill, New York; 1977).
Ionic Liquids, by D. Inman and D. G. Lovering (Plenum Press, New York; 1981).
Diffusion and Conductance in Ionic Liquids, A Linear Response Treatment, by H. G. Hertz (Akademische Verlagsgesellschaft, Wiesbaden; 1982).
Diffusion in Liquids, by H. J. V. Tyrrell and K. R. Harris (Butterworths, London; 1984).
Diffusion: Mass Transfer in Fluid Systems, by E. L. Cussler (Cambridge University Press; 1984).

5 TRANSPORT IN MEMBRANES AND ZEOLITES

Diffusion and Membrane Technology, by S. B. Tuwiner, L. P. Miller and W. E. Brown (Reinhold, New York; 1962).
Diffusion and Membrane Technology, ACS Monograph Series, by S. B. Tuywiner, L. P. Miller and W. E. Brown (Reinhold, New York; 1962).
Diffusion and Membrane Technology, by S. B. Tuwiner (Reinhold, New York; 1962).

Membranes and Ion Transport, Vol. 2, edited by E. E. Better (Wiley–Interscience, London; 1970).
Membrane Transport, A. Kleingeller and A. Kotyk (Czech Academy of Sciences, Prague; 1971).
Membranes — A Series of Advances, edited by G. Eisenman (Marcel Dekker, New York; 1975).
Current Topics in Membranes and Transport, by F. Bronner and A. Kleingeller (Academic Press, New York; 1975).
Zeolites and Clay Minerals as Sorbents and Molecular Sieves, by R. M. Barrer (Academic Press, London; 1978), Chapter 6: Diffusion in Zeolites.
Biological Membranes, Vols. 3 and 5, edited by D. Chapman and D. F. H. Wallach (Academic Press, London; 1976, 1984).
Membrane Transport in Biology, Vol. 1, by G. Giebisch, D. C. Tosteson and H. H. Ussing (Springer-Verlag, Berlin; 1978).
Cell Membrane Permeability and Transport, edited by G. R. Kepner (Dowden, Hutchinson and Ross, Stroudsburg, PA; 1979).
The Properties and Applications of Zeolites, Session 1, Diffusion Processes, edited by R. P. Townsend (The Chemical Society, London; 1980).
Basic Principles in Membrane Transport, by S. G. Schulz (Cambridge University Press; 1980).

6 DIFFUSION IN MULTICOMPONENT SYSTEMS

Isothermal Diffusion in Multicomponent Systems by J. S. Kirkaldy in *Advances in Materials Research*, Vol. 4, edited by H. Herman (John Wiley, New York; 1970).
Multicomponent Diffusion, by E. L. Cussler (Elsevier, Amsterdam; 1976).
D. G. Miller, V. Vitagliano, and R. Sartorio, 1986, *J. Phys. Chem.*, 1986, **90**, 1509.

7 TECHNIQUES OF DIFFUSION MEASUREMENT

Quantitative Electron Microprobe Analysis, by R. Theisen (Springer-Verlag, Berlin; 1965).
Experimental Methods for Studying Diffusion in Liquids, Solids and Gases, in *Physical Methods of Chemistry, Vol. I*, by P. J. Dunlop, B. J. Steel and J. E. Lane, edited by A. Weissberger and B. W. Rossiter (John Wiley, New York; 1972).
Quasielastic Neutron Scattering for the Investigation of Diffusive Motions in Solids and Liquids, by T. Springer (Springer Tracts in Modern Physics, Springer, Berlin; 1972).
Practical Scanning Microscopy, edited by J. I. Goldstein and H. Yakowitz (Plenum Press, New York; 1975).
Electron Microprobe Analysis, by S. J. B. Reed (Cambridge University Press, 1975).
Physical Aspects of Electron Microscopy and Microbeam Analysis, edited by B. M. Siegel and D. R. Bearman (John Wiley, New York; 1975).
Non-traditional Methods in Diffusion, edited by G. E. Murch, H. K. Birnbaum and J. R. Cost (AIME, Warrendale PA; 1984).
Microbeam Analysis, edited by J. Armstrong (San Francisco Press; 1985).

8 APPLICATIONS TO PHASE TRANSFORMATIONS

Liquid Metals and Solidification (ASM, Cleveland; 1958).

Decomposition of Austenite by Diffusional Processes, edited by V. F. Zackay and H. I. Aaronson (Interscience, New York; 1962).
Principles of Solidification, by B. Chalmers (John Wiley, New York; 1964).
Oxidation of Metals, by K. Hauffe (Plenum Press, New York; 1965).
Oxidation of Metals and Alloys (ASM, Metals Park, OH; 1971).
Theory of Metal Oxidation, Vol. 1, by A. T. Fromhold (North-Holland, Amsterdam; 1976).
Introduction to High Temperature Oxidation of Metals, by N. Birks and G. H. Meier (Edward Arnold, London; 1983).

9 DIFFUSION-CONTROLLED PATTERN FORMATION

Leisegang Rings and Other Periodic Structures, by E. S. Hedges (Chapman and Hall, London; 1932).
Stabilité Structurelle et Morphogénèse, by R. Thom (Benjamin, New York; 1972).
Thermodynamics of Self-Organizing Systems, by D. B. Chambers, Ph.D. Thesis, McMaster University, 1973.
Instability and Dissipative Structures in Hydrodynamics: Advances in Chemical Physics, Vol. 32, edited by I. Prigogine and S. A. Rice (John Wiley, New York; 1975).
The Mathematical Theory of Diffusion and Reaction in Permeable Catalysis: Questions of Uniqueness, Stability and Transient Behaviour, Vol. 2, by R. Aris (Clarendon Press, Oxford; 1975).
Self-Organization in Non-Equilibrium Systems, by G. Nicolis and I. Prigogine (John Wiley, New York; 1977).
Life, Logic and Bootstrap Physics, by J. S. Kirkaldy (JASAK, Ancaster, Canada; 1978).
Pattern Formation and Pattern Recognition, edited by H. Haken (Springer-Verlag, Berlin; 1979).
Dynamics and Instability of Fluid Interfaces, edited by T. S. Sorenson (Springer-Verlag, Berlin; 1979).
Chemical Instabilities, edited by G. Nicolis and F. Baras (D. Reidel, Dordrecht, Holland; 1984).

10 MATHEMATICS AND SOLUTION METHODOLOGY

The Theory of Spherical and Ellipsoidal Harmonics, by E. W. Hobson (Cambridge University Press; 1931).
Tables of Functions with Formulas and Curves, by E. Jahnke and F. Emde (Dover Publications, New York; 1945).
Bessel Functions for Engineers, by N. W. McLachlan (Oxford University Press, London; 1955).
Tabellen zur Fourier Transformation, by F. Oberhettinger (Springer-Verlag, Berlin; 1957).
Conduction of Heat in Solids, by H. S. Carslaw and J. C. Jaeger (Clarendon Press, Oxford; 1959).
Conduction Heat Transfer, by V. S. Arpaci (Addison-Wesley, Reading, Mass., 1966).
Elementary Matrix Theory, by H. Eves (Allyn and Bacon, Boston; 1966).
The Mathematics of Diffusion, by J. Crank (Oxford University Press; 1970).

Fourier Expansions: A Collection of Formulas, by F. Oberhettinger (Academic Press, New York; 1973).
Tables of Laplace Transforms, by F. Oberhettinger and L. Badii (Springer-Verlag, New York; 1973).
Applications of the Monte Carlo Method in Statistical Physics, edited by K. Binder, (Springer-Verlag, Berlin; 1984).

Index

B

C

D

F

G

H

M

P